THE SOCIETY OF NAVAL ARCHITECTS
AND MARINE ENGINEERS

TRANSACTIONS

Volume 108

2000

Published in 2001 by

The Society of Naval Architects and Marine Engineers
601 Pavonia Avenue, Jersey City, New Jersey 07306

The volumes of *Transactions* contain professional information of permanent value. Information on the various functions and activities of the Society and a list of members may be found in the 2001-2002 SNAME Membership Directory and Information Book.

Printed by Port City Press, Baltimore, MD.

ISSN 0081 1661

ISBN 0-939773-29-5 (Hard Cover)
0-939773-30-9 (Soft Cover)
0-939773-31-7 (CD)

Table of Contents

2000 Annual Meeting Activities

Reports

Business Session

Refereed Papers

The following papers were selected for presentation at the

2000 Annual Meeting by the Society's Papers Committee

Refereed Papers (cont'd)

Commemoratives

2000 MEMBERS OF COUNCIL

President: Jose Femenia
Treasurer: David H. Klinges

Executive Director: Philip B. Kimball
Assistant Treasurer: Alfred A. Bozzuffi

Robert G. Allan- *Vice President (Pacific 2000)*
Robert Axelrod- *Honorary Vice President*
Robert F. Beck- *Chair, Journal of Ship Research* Committee
Donald M. Birney- *Chair, Pension Committee*
Alfred A. Bozzuffi- *Assistant Treasurer*
Alan J. Brown- *Vice president (Northeast 2000)*
Edward J. Campbell- *Past President*
Connie J. Carruthers- *Section Chair, Arctic*
Edward W. H. Clendenning- *Chair, Sections Committee*
Donald P. Courtsal - *Chair, Audit Committee, Honorary Vice President*
Joseph J. Cuneo- *President-elect*
Charles R. Cushing- *Chair, Fellows Committee*
John C. Daidola- *Chair, Technical and Research Steering Committee*
William G. Day, Jr.- *Vice President (Mid-Atlantic 2000)*
John M. Dempsey Jr.- *Chair Application Committee*
Walter B. Devine- *Vice President (Pacific 2001)*
Michael E. Dockter- *Section Chair, San Diego*
John A. Fowlis- *Section Chair, Pacific Northwest*
C. L. French- *Past President*
Edward T. Gaines- *Vice President (Central 2002)*
William H. Garzke, Jr.- *Chair, Education Committee*
Robert B. Geary- *Chair, Papers Committee*
Robert Gunter- *Chair, Hampton Roads*
A. Dudley Haff- *Honorary Vice President*
Roy L. Harrington- *Honorary Vice President*
Arthur J. Haskell- *Past President; Chair, Development Committee*
John H. Higginbotham- *Chair, Membership Committee*
Richard B. Houseman- *Section Chair, Eastern Canadian*
Dwight G. Hutchinson- *Section Chair, Southeast*
Paul C. Jackson- *Section Chair, Northern California*
Bruce Johnson- *Chair, Electronic Media Committee*
Robert G. Keane, Jr.- *Vice President (Mid-Atlantic 2002)*
Samuel T. Kelly- *Section Chair, Los Angeles Metropolitan*
Philip B. Kimball- *Executive Director*
J. William Kime- *Past President*
Ronald K. Kiss- *Past President; Chair, Awards Committee*
David H. Klinges- *Honorary Vice President and SNAME Treasurer*
Jennifer E. Kollmer- *Section Chair, New England*
John T. Lehman- *Section Chair, Philadelphia*
Edward M. Lewandowski- *Chair, Marine Technology* Committee
Thomas P. Mackey- *Vice President (Central 2000); Chair, Nominating Committee*
Monroe D. Macpherson-*Honorary Vice President*
Robert G. Mende- *Honorary Vice President*
Richards T. Miller- *Honorary Vice President*
William B. Morgan- *Honorary Vice President; Chair, Publications Committee*
Edward K. Moll- *Vice President (Northeast 2002)*
Richard A. Mueller- *Section Chair, Great Lakes/Great Rivers*
Perry W. Nelson- *Past President*
Peter G. Noble-*Chair, Planning Committee*
David A. O'Neil- *Past President*
Michael G. Parsons- *Vice President (Central 2001)*
Ellsworth L. Peterson- *Honorary Vice President*
W. R. Phillips, Jr.- *Past President*
Donald E. Ridley- *Honorary Vice President*
Robert J. Riviere- *Section Chair, Gulf*
Bruce S. Rosenblatt- *Vice President (Northeast 2001); Chair, Annual Meeting Committee*
Lester Rosenblatt- *Past President*
Robert J. Scott- *Chair, Scholarship Committee*
William J. Sembler- *Section Chair, New York Metropolitan*
Nathan Sonenshein- *Honorary Vice President*
Mark H. Spicknall- *Chair, Journal of Ship Production* Committee
Robert C. Staiman- *Section Chair (Chesapeake 2001)*
Arnold M. Stein- *Chair, Member Insurance Committee*
Barbara Ann Stone- *Section Chair, Texas*
Richard W. Thorpe- *Chair, Advisory Public Service Committee*
Daniel L. Walker- *Section Chair, Canadian Atlantic*
Thomas S. Winslow- *Vice President (Pacific 2002)*
Curtis B. Woolard, Jr.- *Vice President (Mid-Atlantic 2001)*
David R. Zoller- *Chair, Finance Committee*

2000 EXECUTIVE COMMITTEE

Jose Femenia, President
Robert G. Allan
Alan J. Brown
Edward W. H. Clendenning
Joseph J. Cuneo
John Daidola
William G. Day, Jr.
Walter B. Devine
C. L. French
Edward T. Gaines
John H. Higginbotham
Bruce Johnson
Robert G. Keane, Jr.
Philip B. Kimball
J. William Kime
David H. Klinges
Thomas P. Mackey
Edward K. Moll
William B. Morgan
Perry W. Nelson
Peter G. Noble
David A. O'Neil
Michael G. Parsons
W. R. Phillips Jr.
Bruce S. Rosenblatt
Lester Rosenblatt
Richard W. Thorpe
Thomas S. Winslow
Curtis B. Woolard, Jr.
David R. Zoller

Section Officers

September 1, 1999 to August 31, 2000

ARCTIC
Chair: Archibald C. Churcher
Vice Chair: Frank G. Bercha
Secretary/ Treasurer: Colin Revill

CANADIAN ATLANTIC
Chair: James L. Colwell
Vice Chair: Daniel Walker
Secretary/ Treasurer: Mark Cusack

CHESAPEAKE
Chair: William H. Cowardin, Jr.
Vice Chair: Robert C. Staiman
Secretary/ Treasurer: Thomas A. LaFone

EASTERN CANADIAN
Chair: John C. Keast
Vice Chair: CDR Eric Guy Bramwell CAF
Secretary: Robert L. Wolfe
Treasurer: Stephen W. Leake

GREAT LAKES/GREAT RIVERS
Chair: Richard A. Mueller
Vice Chair: Michael Hawkins
Secretary/ Treasurer: Dale Karr

GULF
Chair: Darren L. Savoye
Vice Chair: Robert J. Riviere
Secretary: John Richards
Treasurer: Joe Picciola

HAMPTON ROADS
Chair: Claude Thompson
Vice Chair: Robert Gunter
Secretary/ Treasurer: John L. Harrington

LOS ANGELES MET
Chair: Salvatore Di Francia
Vice Chair: Samuel T. Kelly
Secretary /Treasurer: Dick McCabe

NEW ENGLAND
Chair: AllenW. Pac
Vice Chair: Jennifer Kollmer
Secretary/ Treasurer: Christine C. Tate

NEW YORK MET
Chair: Michael T. Martin
Vice Chair: William J. Sembler
Secretary: Martin Toyen
Treasurer: Kent Merrill

NORTHERN CALIFORNIA
Chair: George J. Buffleben
Vice Chair: Partick J. Goss
Secretary/ Treasurer: Paul Jackson

PACIFIC NORTHWEST
Chair: Robert Van Slyke
Vice Chairs:
John Fowlis- British Columbia
Paul Smith- Puget Sound
Gregory S. Lind- Columbia River
Secretary/ Treasurers:
Chris Mulder-British Columbia
Terrence K. Dai – Puget Sound

PHILADELPHIA
Chair: Marjorie M. Dunn
Vice Chair: John T. Lehman
Secretary/ Treasurer: Mitchell D. Walker

SAN DIEGO
Chair: Dominic Burns
Vice Chair: Lee Dunclift
Secretary: David German
Treasurer: Michael Dockter

SOUTHEAST
Chair: Paul Lesky
Vice Chair: Brian Chinn-North
Dwight Hutchinson –South
Secretary/ Treasurer: Randall R. Fiebrandt

TEXAS
Chair: Steve Scanio
Vice Chair: Barbara Stone
Secretary/ Treasurer: Richard Ervine

SNAME Administrative Staff

Philip B. Kimball, Executive Director

Associate Executive Director (Finance and Administration)........Allan Gray
Information Systems Manager........Joseph Caggiano
Technical Papers/Web Site Administrator........Connie Cali-Poutre
Exposition Assistant........Brenda Elmore
Publications Manager........Susan Evans
SNAME Sevice Center Clerk........Tommie-Anne Faix
Administrative Asst. (Scholarships, T&R support, PE exam)........Erlinda Faustino
Marketing and Advertising Administrator........Mary Ellen Flattes
Exposition Manager........Susan Giver
Accounting Supervisor........Carl Honnegar
Annual Meeting Manager........Carolyn Marus
Accounts Payable/ Receivable........Anne Mikaelian
Service Center Rep (Publication Sales)........Claudio Pujols
SNAME Mailroom Coordinator........Raymond Rivera
SNAME Service Center Manager........Eileen Romanelli
Technical Director........Alan Rowen
Service Center Rep.(Students)........Linda Miller
Service Center Rep (Membership)........Debra Sciancalepore
Member Services/Section Liaison Manager........Deborah Sirotiak
Assistant to the Executive Director........Barbara Trentham
Technical Advisor........Charles Wilson

2000 Annual Meeting

The Society of Naval Architects and Marine Engineers' Annual Meeting was held in beautiful Vancouver, Canada at the Vancouver Convention and Exhibition Centre. Chosen not only for it's proximity to the Pan Pacific Hotel and the Waterfront Hotel but also for the beautiful vistas of Burrard Inlet. Best of all, the inlet is home to a 70 million ton a year cargo shipping industry as well as being home to 25 cruise ships. A comprehensive technical program, elegant social events, wonderful tours provided by members of the Pacific Northwest Section, and unseasonably gorgeous weather made for a hugely successful 108th Annual Meeting.

The meeting initiated with the President's Reception at the scenic Plaza of Nations in downtown Vancouver where a delicious and diverse buffet was served. The Pacific Northwest Section arranged for delightful entertainment by members of the Tsleil-Waututh Nation dance troupe "Children of Takaya" and a special display of historic and local small craft, as well as a display of photographic art of the Haida native culture throughout the room. Earlier that day many attendees took advantage of courses offered both in Naval Architecture and Human Factors in Safety Engineering. Many others attended committee meetings.

Thursday began with a keynote address by Mike Kelsey on yacht building with a specific emphasis on aluminum yacht building, followed by a bagpipe led procession to the exhibit hall for the ribbon cutting of the 2000 International Maritime Exposition. From there many began attending the technical programs while the students enthusiastically participated in their student congress. There the students used some of Mr. Kelsey's techniques when they constructed ships of aluminum foil, a coat hanger and a paper grocery bag. A record breaking 165 soda cans were carried in one ship.

The technical program ran Thursday and Friday offering 21 papers reproduced in this volume. The highlight of this meeting was a series of papers that focused on small boats indigenous to the Pacific Northwest. On Friday the T&R sessions included an update on the work being done by Ad Hoc panels on Ballast Water Management, Passenger Vessel Safety, Load Lines, Collisions and Groundings as well as the Environmental Concerns of Ferry Systems in the San Francisco Bay.

As is the tradition, many awards were presented at the President's Luncheon—thirty-five Fifty-Year Golden members were announced, two distinguished service awards, a top recruiter award, one honorary member certificate, three certificates of recognition for T&R, a graduate paper honor prize, a graduate paper award, an undergraduate paper award, an undergraduate honor prize and the Dr. James A. Lisnyk Ship Design Competition Award. In addition, the Vice Admiral E. L. Cochrane Award, for the best paper published in Marine Technology, and the ABS-Capt. Joseph H. Linnard Prize, for the best paper contributed to the annual meeting, were also presented. The winners of these awards can be found in "Marine Technology/SNAME News, January 2001, pages 10-21. Thursday evening saw many Alumni Receptions given in various locations.

After another successful day of technical sessions and expo visitations, the Annual Banquet took place Friday evening in the Ballroom of the Waterfront Hotel. Attendees were treated to an entertaining and informative speech by Captain Graham Westgarth followed by the presentation of the Davidson, Taylor and Land Medals.

The Annual Meeting concluded with a Dinner Party held Saturday evening at the Pan Pacific Hotel where the beautiful backdrop of the Convention Centre sails and the Burrard Inlet added to the evening's enjoyment. President Jose Femenia, President-elect Joseph Cuneo, and four past Presidents—Edward J. Campbell, Arthur J. Haskell, Ronald K. Kiss, and David A. O'Neil were in attendance.

EXECUTIVE DIRECTOR'S REPORT TO COUNCIL, 2000

Necrology Report
The deaths of the following members were reported during FY 00

Christopher Adams	Joseph L. Kersey
Ioannis Agouras	William F. Kirkpatrick
John D. Atkin	Thomas F. Larkin
Anthony C. Brown	Leonard J. Lucas
Donald T. Burkhardt	Ernest A. Maier
William J. Burns	Alfred G. Mitchell
T. I. Catherwood	Philip W. Monck
Robert F. Craig	John N. Naegle
Robert D. De Robertis	Kenneth A. Payne
Robert J. Derusha	Mel Pestana
Donald R. Dilley	Wilfred H.E. Peters
Wayne D. Eberhardt	James B. Robertson Jr.
John Fenton	David R. Saveker
Adolph Gemmet	Frank H. Sellars
Marvin H. Gluntz	James L. Simmons
Warrren A. Goodwin	Carl A. Sipe
E.W. Hazard	Pasquale Spinosa
Geerd N. Hendel	James A. Sweeney
Robert N. Herbert	Gabriel F. Villela
Bryant A. Hilliard	William Watson
L.V. Honsinger	William E. Weisert
William H. Jory	Pel J.White
Karl W. Kahlenberg	Dennis M. Yawarsky

1. HEADQUARTERS

a. Headquarters was in attendance and exhibited at many conferences during the year including the WorkBoat Show, IBEX, CMA, OTC, ASNE, ICMES, SPS, MarCas and the Annual Meeting. Each staff member has now had a chance to attend and have the experience of working at the booth and meeting members. We average 12 new memberships per conference.

b. The practice of communicating via e-mail is being vigorously promoted in order to save time and cost. All standing committee chairs and T&R committee/panel chairs are encouraged to comply by sending their reports electronically.

c. On July 17, the July issue of Marine Technology and SNAME News was mailed to the membership. This issue was the first in a long time to be on time and under budget. It was prepared in house while the search for a Publications Manager was under way. Also included in the poly bag were the Annual Meeting brochure, the Propeller/Shafting Symposium 2000 brochure and the Legislative brochure, all of which were also produced in house.

d. We bartered with DBC again this year for a booth and room at the December WorkBoat Show in New Orleans. The introductory course will be given for members interested in preparing for the PE exam in April 2001. The course will be conducted entirely by SNAME with the registration income coming directly to us.

e. Plans for the International Maritime Exposition are complete for Vancouver, and we not only met our goal of 150 exhibitors, but exceeded it! Look forward tomorrow to honing your skills at the model boat maneuvering basin with a number of radio controlled boats from which to chose. The attendees will also have a treasure hunt with many prizes to be awarded on Friday afternoon.

f. The new web site was up and running on the promised date of September 1, 2000. We will continue to update and improve the information, and add new features to make our site 'one-stop shopping' for all. It will be a portal for technical and research information as well as many other useful benefits and services.

g. Instead of revising and reprinting the old Catalog of Publications, we have now produced the catalog ourselves in electronic format, which has been posted on the web site and can easily be modified and updated. We have been printing copies in the office as required for use at trade shows, and have handed out copies at OTC, ICMES, ASNE, and SPS.

h. On June 24 and 25, we again conducted a successful Chairman-Elect Seminar at Saddlebrook Resort in Tampa. There were 26 attendees discussing current and new issues of importance to the future of the Society.

2. STAFF

a. New employees who have joined the HQ staff over the past year include:

(1) Alan Rowen, Technical Director (June 26, 2000)
(2) Susan Evans, Publications Manager (July 5, 2000)
(3) Carl Honegger, Accountant (July 5, 2000)
(4) Raymond Rivera, Mail Room Coordinator (July 17, 2000)

b. Changes in staff positions which occurred this past year include:

(1) Claudio Pujols, in addition to his duties as a Service Center Representative, has been promoted to Publications Administrator
(2) Linda Miller, in addition to her duties as Service Center Representative, has added the responsibilities of Student Membership
(3) Brenda Elmore has been promoted to Exposition Administrator
(4) Connie Cali-Poutre has been promoted to Web Site Administrator and will continue to handle Annual Meeting papers.

c. This past summer Christine Gill, a junior at Webb Institute, worked under the Director's Assistantship Program. Christine assisted with web site development, T&R bulletin production and supervised the scanning of five years of Section papers and presentations.

d. We continue to hire students from local high schools to assist us on a part time basis. This program continues to be successful, and these students spread the word to their fellow students about SNAME and the marine industry.

3. SNAME PROGRAMS AND ACTIVITIES

a. On January 29, 2000, a mini-conference was held at HQ by President Femenia to discuss nuclear propulsion for high-speed ships. The conference concluded with the formation of an ad hoc Panel 10 on "Alternative Propulsion Plants for High-Speed Ships". The panel's findings are expected to result is in a symposium in 2001.

b. The ferry report entitled, "Ferry Transit Systems for the Twenty-first Century", dated January 10, 2000, was prepared by SNAME to refute the allegations of the Bluewater Network, an environmental group which alleged high emissions from ferries in San Francisco Bay. This report was quickly posted on the SNAME web site and distributed to 216 periodicals and newspapers, ports of authority, government agencies, and legislative representatives for all to better understand the facts.

c. SNAME and ASNE are continuing the joint student membership program, which has one year to run.

d. The 'Professional Delegate' Program was established early in 1999 to provide for SNAME involvement on many worthwhile national and international technical symposia and standards meetings by paying a portion of the travel costs for experts to attend and participate. Eight delegates have participated in meetings this year. This program will continue this year to the extent of the funding available.

e. As part of the internationalization of the Society, Reciprocal Agreements or strategic alliances are in place with RINA, IMarE, and The American Boat and Yacht Council (ABYC) with interest from STG in Berlin, CIMarE in Canada and The Chinese Society of Naval Architects and Marine Engineers.

f. In December 1999, SNAME and ASNE representatives met in Washington to discuss continuing cooperation between the two societies. Highlights of the meeting include the following:

(1) Set up a joint committee to investigate simultaneous annual meetings and combined exposition in the future,
(2) Continue the cooperation an promotion of the dual student membership program and Lisnyk competition,
(3) Conduct a study to improve the retention of students,
(4) Consider a joint 'Dual Use' conference in 2002 and
(5) Conduct a joint environmental conference in 2001.

g. A new Membership Recruitment Drive will be conducted this fall similar to the program three years ago. We have been very careful to remove members' names from the mailing lists based on the experience last time, and apologize in advance should you receive a recruitment letter. We are confident that with many new benefits and services in place we can retain a greater percentage of new members recruited this time by publicizing and promoting these benefits in direct mail campaigns.

h. The Offshore Technology Conference was a success in May and our share of the revenue as a sponsor was greater than last year.

4. SNAME PROPERTIES

a. The continuation of forced entries and burglaries in the first half of the year required the installation of magnetic locks, updating of the card access system, a CCTV system, and motion sensors. This work was successfully completed by Nelson Construction for approximately $20,000. No further incidences have been reported since May 2000.

b. A small space on the 2nd floor had been vacated on January 1, but a new tenant was found on April 1, 2000.

c. We created another office utilizing the entry foyer by the elevator on the 4th floor. A new entry door to HQ was installed opposite the elevator, and a new SNAME name sign was prepared and installed by Nelson Construction.

d. The portable partitioning in the office was rearranged to create five spaces where there had been only three. This has now provided an office or work station for everyone.

e. At the recommendation of Don Birney, we have hired Nelson Construction to manage the building. This arrangement, which started this summer, is cost effective and working well.

f. At the June 22 meeting, the Executive Committee approved a motion to put the building on the market. We have now entered into an exclusive six-month agreement with Dan Brown of TRS Associates, Inc. to market the building.

g. In February the cooling tower failed, and was replaced by Bailey Refrigeration on March 17, for approximately $30,000.

5. AWARDS

The following awards were presented at the Annual Banquet on Friday, October 1, 1999.

THE WEBB MEDAL

Presented to: Walter Maclean
for "outstanding contributions to education in naval architecture, marine or ocean engineering"
By: Naresh M. Maniar

THE BLAKELY SMITH MEDAL

Presented to: WIlliam Bennett, Jr.
For "outstanding accomplishement in ocean engineering"
By: Roger W. Mowell

THE VICE ADMIRAL 'JERRY' LAND MEDAL

Presented to: William A. O'Neil
for "outstanding accomplishment in the marine field."
By: William O. Gray

THE DAVID W. TAYLOR MEDAL

Presented to: Robert J. Scott
for "notable achievement in marine engineering."
By: Jose Femenia

The following awards and presentations were made at the President's Luncheon on Friday, October 1, 1999:

50-YEAR MEMBERSHIP CERTIFICATES

"representing 50 years of membership."

Presented to:

Hermann A. Allen (in absentia)
Robert J. Anderson (in absentia)
Ralph E. Anselmi (in absentia)
Daniel A. Ball (in absentia)
Gordon H. Boyd (in absentia)
David F. Butler (in absentia)
Nicholas G. Callinicos (in absentia)
Sheng-Lun Chuang (in absentia)
John A. Cochrane (in absentia)
William E. Dennis (in absentia)
Walter B. Devine
Robert T. Dewhirst (in absentia)
Stephen C. Dvorak
F. Leif Eareckson (in absentia)
Edward English (in absentia)
Philip W. Farley (in absentia)
John E. Flipse
William G. Haddock, Jr. (in absentia)
Richard E. Henning (in absentia)
William E. Heronemus (in absentia)
Robert D. Hesley (in absentia)
Moses W. Hirschkowitz (in absentia)
Donald R. Hudson
William E. Jacobsen (deceased)
George F. Jacobson (in absentia)
Ephraim F. Kaufman (deceased)
Albert N. Kreihs (deceased)
Robert W. Long (in absentia)
Daniel M. Mack-Forlist (in absentia)
Lucas B. Mayer (in absentia)
John Z. Mcardle
Roger W. McAleer (in absentia)
Rene P. Meric, Jr. (in absentia)
Marvin O. Miller (in absentia)
Alfonso M. Montilla (in absentia)
William Munch, Jr.
Charles A. Narwitz
William G. Neal, Jr., (in absentia)

William O. Nichols
James E. Nivin
Richard B. Nessley (in absentia)
Morton C. Patten (in absentia)
John H. Redpath, III (in absentia)
James B. Rukin (in absentia)
Kwan Seeto (in absentia)
Charles S. Sinclair (in absentia)
Walter Smedley (in absentia)
Frank R. Stobinski (in absentia)
Robert Strasser (in absentia)
George B. Stunz, Jr. (in absentia)
Arthur H. Sulzer
John R. Taft (in absesntia)
William duBarry Thomas
Jan-Olof B. Traung
Harrison O. Travis, Jr. (in absentia)
Robert F. Vollack (in absentia)
Thomas J. Walker (in absentia)
Lawrence W. Ward
Robert M. Wilkinsonr (in absentia)
Carl P. Wilson (in absentia)

CERTIFICATES OF APPRECIATION

Monroe D. Macpherson
for "outstanding, continuing, dedicated service to the Society as Chairman of the Society's Oversight Committee for Professional Licensing" and to
Charles R. Cushing
Alexander Delli Paoli
Walter M. Maclean
Fred E. Shumaker
for "outstanding, continuing, dedicated service to the Society as a member of the Society's Oversight Committee for Professional Licensing"

DR. JAMES A. LISNYK SHIP DESIGN COMPETITION AWARDS
The 1999 Dr. James A. Lisnyk Ship Design Competition Awards were presented on May 18, 2000, at a luncheon held by the Society of Naval Engineers in Arlington, Virginia, in conjunction with the ASNE Symposium.

DISTINGUISHED SERVICE AWARDS
for "dedicated personal service and/or technical contributions to the Society and its activities at the National and section levels."

Presented to:
Alfred A. Bozzuffi
Ronald F. Briggs
Naresh Maniar
Richard C. Rodi
Havard B. Stewart (in absentia)
By: Ronald K. Kiss

STUDENT PAPER AWARDS
for "outstanding student papers presented at a Section meeting during the 1999-2000 Program Year."

GRADUATE HONOR PRIZE
Presented to: Shih Wei Liao
Dominique Roddier

UNDERGRADUATE PAPER HONOR PRIZE
Presented to: Jennifer S. Grimsley (in absentia)

UNDERGRADUATE PAPER
Presented to: Douglas S. Snediker (in absentia)
Michael J. Telfer (in absentia)

By: Robert J. Scott

VICE ADMIRAL E. L. COCHRANE AWARD
Presented to: Michael G. Dyer

By: Philip B. Kimball

AMERICAN BUREAU OF SHIPPING CAPTAIN JOSEPH H. LINNARD PRIZE
for "the best paper contributed to the proceedings of the Society"
Presented to: Douglas Faulkner

By: Robert B. Geary

HONORARY MEMBER CERTIFICATES

Presented to: Roy L. Harrington
Daniel Savitsky
Charles Wilson

By: Jose Femenia

T&R CERTIFICATE OF RECOGNITION
Presented to: John E. Ancarrow, Jr.(in absentia)

By: John C. Daidola

GEORG P. WEINBLUM LECTURE

The twenty-second Weinblum Memorial Lecture was given by Professor Bernard Mole, on November 17, 1999, in Hamburg, Germany.

APPOINTMENTS

SOCIETY REPRESENTATIVES TO OTHER ORGANIZATIONS AND OTHER APPOINTMENTS

ACCREDITATION BOARD FOR ENGINEERING AND TECHNOLGY

Board of Directors
Prof. Edwin G. Wiggins (to June 31, 2000)
Capt. Robert E. Apple (fill in)

Engineering Accreditation Commission
Dr. Stanley E. Dunn
Prof. Richard Harbach
Engineering Accreditation Alternate
Capt.. Robert E. Apple

TECHNOLOGY ACCREDITATION COMMISSION

No representation as ABET is in a rotating process with societies for representation on the TAC

Program Evaluators:
Prof. Harold C. Alexander
Mr. Robert E. Apple
Mr. Elwood Baumgart
Dr. Rameswar Bhattacharyya
Mr. Norman D. Birrell
Prof. Neil Bose
Mr. Richard Broad
CDR Errol M. Brown, USCG
Prof. Boris S. Butman
Prof. Ralph G. Davis
Mr. Alexander Delli Paoli
Dr. Stanley E. Dunn
Prof. Jose Femenia
Prof. James Harbach
CDR Dwight G. Hutchinson, USCG (Ret.)
Capt. Harry A. Jackson, USN (Ret.)
Dr. Robert G. Latorre
Dr. Alaa E. Mansour
Prof. Charles J. Munsch
Dr. Wayne Neu
Dr. Michael G. Parsons
Dr. Robert E. Randall
Dr. Herman E. Sheets
Prof. James R. Sherrard
Prof. Lawrence W. Ward
Dr. Edwin G. Wiggins

COUNCIL OF ENGINEERING AND SCIENTIFIC SOCIETY EXECUTIVES (CESSE)

Mr. Joseph Caggiano
Ms. Susan Evans
Ms. Mary Ellen Flattes
Ms. Susan Giver
Mr. Allan Gray
Ms. Jaime Horowitz (to April 2000)
Mr. Philip B. Kimball
Mrs. Carolyn Marus
Mrs. Eileen Romanelli
Mrs. Deborah Sirotiak
Mrs. Barbara Trentham

MARINE ENGINEERING COUNCIL UNDERWRITERS LABORATORIES

Open

MARITIME TECHNICAL INFORMATION FACILITY

Mr. Charles R. Cushing

SPE LIAISON SOCIETY EXECUTIVE DIRECTORS

Open

NATIONAL INVENTORS HALL OF FAME FOUNDATION SELECTION

Radm. Nathan Sonenshein, USN (Ret.)

OFFSHORE TECHNOLOGY CONFERENCE

Board of Directors	Mr. Roderick J. Allan
Program Committee 1998-1999	Gregory G. Thory
Program Committee 1999-2000	Gregory G. Thory

PAN AMERICAN INSTITUTE OF NAVAL ENGINEERING

Professor Jose Femenia

SHIP STRUCTURES SUBCOMMITTEE

Dr. Walter M. Maclean

SPERRY BOARD OF AWARD

Mr. Ronald K. Kiss
Mr. James A. Wittmeyer (to December 31, 1999)
Capt. Perry W. Nelson, USN (Ret.) Alternate

U.S. COMMITTEE ON THEORETICAL AND APPLIED MECHANICS

Dr. Michael M. Bernitsas

Respectfully submitted,

Philip B. Kimball

Philip B. Kimball
Executive Director

TREASURER'S REPORT TO COUNCIL, 2000

SNAME
Balance Sheet
As of July 31

ASSETS		
Current Assets	2000	1999
Cash & Cash Equivalents	$325,908	$127,963
Accounts Receivable	228,774	$124,241
Prepaid Expenses :		
Section Advance	279,364	262,511
Deferred Pension Contributions	228,688	182,359
Other Prepaid Expenses	163,383	174,112
Total Prepaid Expenses	671,435	618,982
Total Current Assets	1,226,117	871,186
Investments	20,368,252	14,217,634
Property & Equipment	316,902	312,190
Less Depreciation	(214,417)	(159,794)
Net Property & Equipment	102,485	152,396
Due from SNAME Properties	3,172,414	3,070,164
TOTAL ASSETS	$24,869,268	$18,311,380
LIABILITIES & NET ASSETS		
Accounts Payable	$29,055	$14,129
Deferred Revenue	276,697	185,980
Other Liabilities	116,890	116,890
Capital Lease (Current Portion)	10,000	10,000
Total Current Liabilities	432,642	326,999
Long Term Capital Lease	25,002	49,336
Total Liabilities	457,644	376,335
Net Assets (Net Worth)	24,411,624	17,935,045
TOTAL LIABILITIES & NET ASSETS	$24,869,268	$18,311,380

SNAME
Statement of Activities
for the eleven months ending 7/31/00 and 7/31/99

Revenues	FY '00	FY '99
Member Dues	$892,257	$912,459
Publication Sales	388,665	389,309
Contributions	864,755	144,971
Functions	545,026	460,202
Advertising	30,035	27,200
Other Income	11,124	14,453
Interest & Dividends	76,164	62,336
Total Revenue	$2,808,026	$2,010,930
Expenses		
Program Services		
Publications	$302,481	$258,046
Functions	482,386	311,686
Scholarships, Grants	5,251	12,430
Total Program	790,118	582,162
Management & General		
Salaries & Fees	798,216	721,341
Payroll Taxes, Benefits, Ins.	160,164	154,397
Subtotal	958,380	875,738
Facility Expense	144,578	144,690
Membership Services	124,251	184,799
Professional Fees	52,475	50,208
Headquarter & Section Expense	50,918	36,405
Research & Home Page	101,340	145,013
Depreciation	57,438	26,626
Sundries	42,375	39,895
Other Expenses	4,320	4,015
Advertising	27,882	30,255
Subtotal	605,577	661,906
Total Management & General	1,563,957	1,537,644
Total Expenses	$2,354,075	$2,119,806
Increase (Decrease) Net Assets	453,951	(108,876)

SNAME Properties, Inc.
Balance Sheet at July 31

ASSETS	2000	1999
Cash & Cash Equivalents	$2,335	$725
Accounts Receivable	3,588	0
Prepaid Expenses	48,503	40,256
Total Current Assets	54,426	40,981
Property & Equipment at Cost	3,444,184	3,393,392
Less Depreciation	(1,199,763)	(1,105,692)
Net Property & Equipment	2,244,421	2,287,700
Total Assets	$2,298,847	$2,328,681
LIABILITIES & NET ASSETS		
Accounts Payable	$0	$2,259
Mortgage (Current Portion)	52,000	52,000
Total Current Liabilities	52,000	54,259
Mortgage payable	1,195,315	1,255,165
Due to SNAME	3,172,640	3,070,390
Other Liabilities	24,472	21,482
Total Liabilities	4,444,427	4,401,296
Net Assets (Net Worth)	(2,145,580)	(2,072,615)
Total Liabilities & Net Assets	$2,298,847	$2,328,681

SNAME Properties, Inc.
Statement of Activities
at July 31, 2000 and July 31, 1999

Revenue	7/31/00	7/31/99
Rental Income	$365,430	$355,948
Parking and Other Income	20,460	19,706
Total Revenue	385,890	375,654
Expenses		
Cleaning and Rubbish Removal	$30,053	$30,762
Utilities	53,779	62,363
Maintenance & Repairs	34,638	45,671
Real Estate Taxes	56,365	65,682
Property Management Fees	7,693	12,024
Interest Expense Mortgage	90,009	93,625
Professional Fees	255	7,270
Depreciation	85,497	94,322
Insurance	0	2,640
Leasing Commisssion Expense	14,192	3,437
Security	7,866	4,701
Administrative Expenses	$39,950	$39,990
Total Expenses	420,297	462,487
Increase (Decrease) in Net Assets	(34,407)	(86,833)

Respectfully submitted,
David H. Klinges
Treasurer

Awards and Citations Presented at the 2000 Annual Meeting

FIFTY-YEAR GOLDEN MEMBERSHIP CERTIFICATES WERE AWARDED TO THE RECIPIENTS LISTED BELOW:

Clifford F. Anderson (in absentia)
John L. Horton (in absentia)
F. Everett Reed (in absentia)
Norman H. Brubaker (in absentia)
Jan-Erik Jansson (in absentia)
James B. Rettig (in absentia)
Robert X. Caldwell (in absentia)
John E. Kelly
Walter G. Riebling (in absentia)
Saverio J. Cina (in absentia)
Robert L. Kelly (in absentia)
James H. Robinson (in absentia)
Charles Davies (in absentia)
Douglas G. Lamb (in absentia)
Peter G. Schmidt
Robert F. Doran (in absentia)
David P. Levy (in absentia)
David Seymour
Rodney M. Elden
Samuel C. Loveland, Jr. (in absentia)
Howard D. Springer (in absentia)
Silvano Gandusio (in absentia)
Fendall Marbury (in absentia)
Richard H. Suehrstedt (in absentia)
Roger L. Gintling (in absentia)
Robert H. Murray (in absentia)
Robert L. Wilchar
Jan Hart (in absentia)
Norman A. Nicolson (in absentia)
Lloyd W. Wood (in absentia)
Charles E. Heil
Peter Joseph F. O'Reilly
Charles Zeien
Ran Hettena (in absentia)
Robert F. Rae (in absentia)

DISTINGUISHED SERVICE AWARDS

Presented by: Ronald K. Kiss

CHRISTOPHER J. REYLING (in absentia)

Christopher J. Reyling, a Life Fellow, is a dedicated and active participant in the affairs of the Society on both the National and Section levels. He has served as a Vice President and Member of the Council, and is active in the Society's Technical & Research Program serving as a member of the Ship Design Committee, and as the Chairman of it's Panel SD-1 on Weights Engineering. He has been a leader in the affairs of the New York Metropolitan Section of the Society where he has held all the Section officer positions including the Chairmanship of the Section. He has been a dedicated leader and participant in a number of the Sections very successful Ship Operations, Management, and Economics Symposia, and served on the Steering Committee of the recent ICMES 2000 symposium. Chris is also the author of technical papers presented to the Society and printed in Marine Technology.

WILLIAM J. SEMBLER

William J. Sembler, a Life Member of the Society, is active in the Society's Technical and Research Program serving as a member of the Ships Machinery Committee and as the Chairman of its Panel M-42 on Exhaust Gas Boilers. He is also active in the New York Metropolitan Section where he has served on numerous section committees, and in all the Section officer positions, and has been an active leader in several of the sections outstanding symposia including the latest section sponsorship of the ICMES 2000 Symposium. Bill has also given guidance to the student chapter of the Society at the United States Merchant Marine Academy at Kings Point where he is a Professor of Marine Engineering. In addition to his National and Section activities he is the author of several papers presented to the Society, and of the chapter on "Pumps, Compressors, Blowers, and Ejectors" in the 1992 edition of the Society's publication "Marine Engineering".

T&R CHAIRS
John Daidola recognized the T&R Chairs in attendance.

Robert F. Beck, Chair, Hydrodynamics Committee
Thomas Lamb, Chair, Ship Design Committee
Ralph E. Johnson, Chair, Ship Technical Operations Committee
Robert S. Behr, Chair, Ships' Machinery Committee
Philip G. Rynn, Chair, Hull Structure Committee
William T. Bennett, Jr., Chair, Offshore Committee
Michael L. Powell, Chair, Ship Production Committee
Lee Dana, Chair, Small Craft Committee (in absentia)
Henk Van Hemmen, Chair, Ship Repair & Conversion Committee

T&R CERTIFICATE OF RECOGNITION
John C. Daidola, T&R Chair, awarded Certificates of Recognition for outstanding service in the society's Technical & Research Program.
to:
Thomas P. Mackey
Alan L. Rowen
Kenneth L. Tuttle (in absentia)

STUDENT PAPER AWARDS

GRADUATE PAPER HONOR PRIZE

Robert J. Scott, Chairman of the Scholarships Committee, presented the 2000 Graduate Paper Honor Prize to Michael Griffin For his paper "Numerical Prediction of the Forces and Moments on Submerged Bodies Operating Near the Free Surface" presented to the New England section on January 26, 2000.

GRADUATE PAPER PRIZE

Robert J. Scott, Chairman of the Scholarships Committee, presented the 2000 Graduate Paper Prize to Peter Gooding for his paper "Collision with a Crushable Bow" presented to the New England section on January 26, 2000.

UNDERGRADUATE HONOR PRIZE

Robert J. Scott, Chairman of the Scholarships committee, presented the 2000 Undergraduate Honor Prize to Katherine Croff and Whitney Cornforth for their paper "Development of an Environment-Sensitive Navigation Sysytem for the *AUV AUTOLYCUS*" presented to the New England section on January 26, 2000.

UNDERGRADUATE PRIZE

Robert J. Scott, Chairman of the Scholarships committee, presented the 2000 Undergraduate Prize to Tyson Scofield for his paper "Green Engines: Methods for Implementing the New MARPOL 73/78 Regulations" presented to the New England section on January 26, 2000.

VICE ADMIRAL E. L. COCHRANE AWARD

The Vice Admiral E. L. Cochrane Award is awarded annually for the best paper presented before a section of the Society during the year. After consideration of the outstanding papers presented during the 1999/2000 program year before the seventeen Sections, the Vice Admiral E. L. Cochrane Award for 2000 was presented by Marine Technology Chair Edward Lewandowski to Stanley Stumbo, Kenneth Fox, Lawrence Elliot and Frank Dvorak for their paper "The Preditction, Measurement, and Analysis of Wake Wash from Marine Vessels".

AMERICAN BUREAU OF SHIPPING/ CAPTAIN JOSEPH H. LINNARD PRIZE

The American Bureau of Shipping/Captain Joseph H. Linnard Prize is awarded annually by the Papers Committee for the best paper presented at the 1999 Annual Meeting. The 2000 ABS-Linnard Prize was presented by Papers Committee Chairman Robert Geary to Ick Hung Choe (in absentia), Jang Young Chung (in absentia), Jeom Kee Paik Preben T. Pederson (in absentia), Anil K. Thayamballi Ge Wang for their paper "On the Rational Design of Double Hull Tanker Structures Against Collision".

HONORARY MEMBER CERTIFICATE

Thomas Mackey, Chairman of the Nominating Committee, had the honor of presenting an Honorary Member Certificate to Jose Femenia in recognition of his leadership and effective administration of the affairs of the Society during the past two years.

DR. JAMES A. LISNYK SHIP DESIGN COMPETITION AWARD.

Ronald K. Kiss Presented the 2000 Dr. James A. Lisnyk Ship Design Competition Award to the following students from the University of Newcastle UponTyne:

Dimitrios A. Kanakis
Nikolaos A. Kazitoris
Nikolaos D. Mitroules
IoannisP. Mitsakis
Periklis A. Salessiotis
Ioannis G. Vannas
Dr. Ian L. Buxton, Faculty Advisor from the University Of Newcastle.

MEMBERSHIP RECRUITMENT AWARDS

TOP RECRUITERS OF THE YEAR

John H. Higginbotham, chair of the Membership Committee, presented a 2000 Top Recruiter of the Year Award to Paul J. Roden from the Chesapeake Section. Lee A. Duneclift, from the San Diego Section, received his award in absentia.

SECTION PERFORMANCE AWARDS

The Secion Performance Award for the section to recruit the most members was presented at the Section/Membership Breakfast to Barbara Stone representing the Texas Section.

VICE ADMIRAL "JERRY" LAND MEDAL

The 2000 Vice Admiral "Jerry" Land Medal for "outstanding accomplishments in the marine field." was presented to Mr. Charles R. Cushing by David O'Neil.

Land Medal Presentation

Thank you and good evening.

Your Dinner Program well highlights a career of nearly 50 years for Charles R. Cushing, so I will refrain from repeating all that is said there. If Charles' professional life were to be further abbreviated, these descriptive words could not be left out:

o Deck Officer in the U.S. Merchant Marine
o Pioneer of material handling and containerization
o Protégé of Malcom McLean
o President of a world class company
o Author
o Teacher
o Fellow in four prestigious professional societies
o Holder of the Merchant Marine Academy's highest honors
o Leader in numerous pursuits

But, beyond his career some other words cannot be left out – devoted husband and father.

For our Society, this life member's proactive approach and dedication has helped to produce the results we are seeing in our Professional Licensure Program. His authorship of two chapters in SNAME's 100th Anniversary Historical Proceedings and his chairing of our Fellows Committee represent a few more ways he has dedicated his time and effort to our professional society. When Charles does something, he always adds value.

Charlie began exemplifying the esteemed career of Vice Adm. Jerry Land as a student, such that upon graduation in 1956, the United States Merchant Marine Academy bestowed its own Land medal upon him. Not many have two Land medals. I'm sure Yogi Berra's repertoire would have something apt for us at this time, but I won't go there tonight! While aboard ship as a Deck Cadet during his sea-year, Charles was frequently spotted in the engine room – so you see there was hope for him from the beginning. But seriously, this same insatiable thirst for knowledge about things maritime goes on today, and I believe he is, perhaps, the only practicing naval architect owning a leading American maritime-related engineering firm to work for and hold a Ph.D. since John J. McMullen was active in our business.

Charles was also one of the youngest naval architects to develop his practice to the level of world recognition, beginning this journey only 12 years after obtaining his Bachelor's Degree. I was fortunate to meet him at this point in his career, and we soon collaborated on a system to ship 35,000 horse power engines across the Atlantic, using over-the-road containers. We were helping to make the point in 1968 that containerization was not only the wave of the future, but it was also practical for more than boxed commodities.

We have been both friends and friendly competitors ever since, so it is with a great deal of personal pleasure that I present the Society's recognition for outstanding accomplishment in the marine field – the Vice Admiral Jerry Land Medal – to Dr. Charles R. Cushing.

Land Medal Acceptance

Mr. President, fellow members, honored guests.

Someone once said that making a speech is a solemn responsibility. A man who makes a bad five minute speech to 1000 people only wastes five minutes of his own time, but he wastes 5000 minutes or more than four days worth of other people's time. This should be a hanging offense. With that in mind, let me be brief.

Any recognition that is bestowed upon me tonight must be shared by me with so many people: my professors and instructors at Kings Point, MIT and elsewhere, my shipmates at USL, Grace, Farrell, Isthmian and elsewhere, my colleagues in early container pioneering days at Pan Atlantic, later called Sea-Land and the consultants we worked with at JJ Henry, Bob Mavy, Designers and Planners, George Sharp, Phil Spaulding, to name but a few, my many clients these last 30 years, and particularly my fellow colleagues at CRC Co.

Haile Salassie said

" It is impossible to accomplish lasting and useful work in the world without helping one another,"

This is so true. I have benefited from such help. I believe I and we as professionals, have to repay this debt to our educational institutions, to make young members just entering our profession and society and who are just embarking on their careers, and to cooperate and assist our fellow members to advance the art and science of our profession. Finally, I thank God for my wife Elke, the wind beneath my wings.

Many of you do not know the person in whose name this award is made. It was my privilege when I was a Cadet-Midshipman to have met Vice Admiral Emory Scott Land, known to his friends as "Jerry".

Born in 1879 in Colorado, he was a cowpuncher and hand before he earned a bachelor and masters degree from the University of Wyoming. He then entered the Naval Academy in 1898, excelled scholastically and is best remembered there by scoring the winning touchdown, running through West Point's line, in the Army-Navy game in his senior year.

After postgraduate studies at MIT and while a lieutenant at the Bureau of Construction in 1918, he became a fast friend of the Assistant Secretary of Navy, Franklin D. Roosevelt. Their lasting friendship led to his appointment as Chairman of the U.S. Maritime Commission in 1938, replacing Joseph P. Kennedy.

As war clouds gathered on the horizon, he was the perfect person to administer the greatest shipbuilding program in the world, building over 6,000 merchant ships, the Liberty ships, Victories, T2's etc. Which were crucial to the winning of World War II.

He was the perfect person because:
-He avoided beaurocracy,
-Worked harder,
-Placed a high value on loyalty,
-Had impeccable technical credentials;
-In his own words, he said " I got my start in life by working from sunrise to sunset with my hands for $10 a month,
-and he declared, "I am no "Brass Hat"

When he was elected President of the Society in 1940, he said, "You know, I don't believe in the commission form of government. I think you ought to have a one-man show, and shoot him at sunrise if he doesn't get it right." He passed away in November of 1974, but he live in our memory. It makes me very proud to accept this award named for "Jerry Land, and I thank you all for it.

Let me share with you this Irish blessing.

"These things I warmly wish for you.
Someone to love, some work to do.
A bit of sun, a bit of cheer,
And a guardian angel always near."

DAVID W. TAYLOR MEDAL

The 2000 David W. Taylor Medal for "notable achievement in naval architecture and/or marine engineering" was presented to Thomas G. Lang by President Jose Femenia.

Taylor Medal Presentation

Thomas G. Lang has been a leader and an innovator in the design and development of unique hydrodynamic hull forms. Most notably he is known and respected for his developmental efforts and his enthusiastic promotion of the SWATH (Small Waterplane Area Twin Hull) concept.

Dr. Lang began his career at the Naval Ordinance Test Station, in Pasadena, California where he served as a Member of the Bureau of Weapons Hydrodynamics Advisory Committee. His early accomplishments were related to hydrodynamic design of torpedoes. Early in his professional career, as he completed his doctoral dissertation in the Aerospace Department at Penn State University, he was laying the groundwork for the engineering procedures that lead to his pioneering development work on SWATH.

Dr. Lang was instrumental in his development of the SWATH. He has been granted over twenty patents, including a number specifically related to the SWATH concept. He took this basic technology work and developed design concepts leading to free running model tests in San Diego Bay and then to a series of towing tank tests. He personally solved many of the vexing technical and administrative problems and ultimately oversaw the construction of the first operational prototype SWATH, the 190-ton Kaimalino. It was this ship that established the first full-scale database for this type of hull form, and furthermore demonstrated the viability of SWATH as a stable platform in high sea states underway at substantial speeds.

His forty-nine years of experience in ship research, innovative marine design, analysis, test and program management has been well shared with the profession by virtue of his patents and over eighty technical papers and publications. These cover the subjects of SWATH, hydrofoils, torpedoes and underwater propulsors. He was a key consultant in the hydrodynamic testing of the finned keel and designed the riblets used on Stars and Stripes, which won the America's Cup back from Australia in 1987.

Dr. Lang was elected to the grade of Fellow in SNAME in 1996, and has received numerous awards over his distinguished career. He is most deserving of this award for his notable achievement in naval architecture.

I am honored to present the David Taylor Medal to Dr. Thomas G. Lang.

Taylor Medal Acceptance

It is a great honor to receive this award. I have enjoyed working in the field of naval architecture because it is so creative. My generation has seen the development of many new concepts in this field.

These new concepts include nuclear submarines, a wide variety of hydrofoil boats, surface effect ships, air cushion vehicles, wing-in-ground effect craft, hydroplane boats with air lift, cathedral and deep vee planing hulls with strakes and steps, offshore racing boats, offshore drilling rigs, offshore crew and supply boats, SWATH ships, semi-SWATH, planing SWATH, single-hulled SWATH, high-speed catamaran ferries, trimaran ships and boats, hydrofoil sailboats, sail boards, wing sails, kite sails, mini-subs, man-powered submarines, nuclear-powered ships, electric drives, gas turbines, high-speed diesels, shrouded propellers, supercavitating and superventilating propellers, water jets, mist jets, ram jets, multidirectional thrusters, drag reduction by polymers, riblets, microbubbles and air cavities, and advances in theoretical hydrodynamics and computational fluid dynamics. The list goes on and on. The future holds advances in all of these areas, plus many new concepts that are unknown today.

SNAME has contributed to these advances by publishing journals and books, forming study committees, and holding meetings like this one to bring people and ideas together.

Thank you for this David W. Taylor SNAME award in the exciting fields of naval architecture and marine engineering.

DAVIDSON MEDAL

The 2000 Davidson Medal for "outstanding scientific accomplishment in ship research" was presented to Robert F. Beck by Michael Parsons.

Davidson Presentation

Ladies and Gentlemen. It is my honor and pleasure tonight to present the year 2000 Kenneth S. M. Davidson Medal to Robert F. Beck, Professor of Naval Architecture and Marine Engineering at the University of Michigan. The Society presents the Davidson Medal annually to recognize "outstanding accomplishments in ship research."

My introductory remarks will parallel the citation included in your program. I would like, however, to comment briefly on Professor Beck's accomplishments in two areas: his personal research with his graduate students and his leadership roles in the marine hydrodynamics community in general.

First on Professor Beck's Personal Research:

Bob Beck is an internationally recognized expert on marine hydrodynamics: particularly ship and platform motions. He received his education from Michigan and MIT. His 34-year research career has been characterized by a dedication to teaching and mentoring of advanced graduate students through his research, technical creativity, and consistently high research standards.

Bob's earliest work involved ship maneuvering and seakeeping in open and confined waters, included ship-ship interaction and bank suction. He has made important contributions to panel methods using free surface Green functions. He and his students have developed time domain ship motions simulation methods. His current work involves the development of fully nonlinear ship motions simulations using desingularized techniques.

He has published over 50 important papers on this work in journals, conferences, and workshops. He has co-authored the "Motions in Waves" chapter of Principles of Naval Architecture 1988 and again in the upcoming new edition. He is a Life Member and Fellow of the Society.

Now on Professor Beck's Overall Research Leadership:

Professor Beck's research career has also been characterized by his leadership contributions to the marine hydrodynamics community. He has since 1972 been a member and later chair of the Technical and Research Panel H-5 Analytical Ship-Wave Relations. He currently chairs the Hydrodynamics Committee.

He served many years on the Journal of Ship Research Committee and in 1998 he became the overall editor of that leading scientific journal of our Society. He currently serves at the United States representative to the Executive Committee of the International Towing Tank Conference (ITTC). At the University of Michigan, Bob served as Director of the Marine Hydrodynamics Laboratory for four years during which he led a major expansion and improvement of its research capabilities. He also led the five year/$5M inter-disciplinary Program on Ship Hydrodynamics. This program focused on the fundamental processes that occur in ship wakes with emphasis on the remote sensing of surface disturbances by synthetic aperture radar.

Bob Beck is an equally dedicated teacher of undergraduates and is currently teaching an introductory College-wide first year course using sailing yacht design as the example of engineering activity. He has served as chair of his department and currently serves as one of the four elected members of our College of Engineering Executive Committee.

As one of Bob's faculty colleagues for 28 years and as one of the SNAME Regional Vice Presidents for his Central Region, it gives me great pleasure to present, on behalf of the Society, the Davidson Medal for the year 2000 to Professor Robert F. Beck.

Davidson Acceptance

I would like to thank the Society for this award. I am honored to be in the company of the distinguished previous recipients.

I could not possibly thank everyone who has helped make this award possible without sounding like an Oscar winner. Thus, I will not even try. However, I would like to thank a few people who had major influences on my research.

First, are my college professors who instilled in me the love of hydrodynamics and the tools to try and understand complex fluid flows. I have been lucky enough to have many great teachers over the years from my undergraduate days at Michigan with Professors such as Harry Benford and Dick Couch, to my thesis supervisors at MIT, Professor Martin Abramowitz and Nick Newman, to my post-doctoral work with Ernie Tuck at the University of Adelaide, Australia. Finally, I would like to thank my colleagues in the Department of Naval Architecture and Marine Engineering, particularly Francis Oglivie who was chairman of the Department during the start of my career.

What really keeps me going though are the students, especially my Ph.D. students. They are always asking "why" and thus forcing me to question my own understanding. If you cannot give a clear explanation of a process, no matter how complex, then you do not understand it yourself.

Probably the biggest driving force in my research has been the designers. They are a very greedy group—they always want more and better hydrodynamic design tools that are beyond the state-of-the-art. For example, my research group at Michigan and other groups around the world have worked for years to improve seakeeping prediction methods. Just when we are getting proficient at predicting seakeeping characteristics of "normal ships" in moderate sea states, designers have come up with weird hull forms with ram bows, tumble home, and /or multi-hull or even worse, offshore structures that do not even look like ship hulls. Moderate sea states are no longer enough, we now have to worry about survival conditions in extreme sea states with breaking waves. It is a never-ending challenge for both the designer who has to have answers, even if it is just a guesstimate, and the researcher who prefers a more rational approach but understands the need for answers.

Finally, I would be remiss if I did not thank my family. This award would not be possible without their patience and understanding. My wife, Barbara and daughters, Jennifer and Alison have asked many straightforward questions only to get long winded explanations about trailing voriticity, divergent waves, or some other fluid phenomenon to which they respond politely with that look that says "here he goes again....."

Thank you.

REPORT ON ELECTIONS For submission at the BUSINESS SESSION on Thursday, October 5, 2000

1. ELECTED BY COUNCIL

a) Honorary Member (for life)
Jose Femenia

b) Honorary Vice President (for life)
Edward W. H. Clendenning

c) Fellows

Robert G. Allan, Bilal M. Ayyub, Amos Baki, Ronald F. Briggs, Alan J. Brown, Edward W. H. Clendenning, William G. Day, Jr., Peter A. Fisher, Ian F. Glen, Bruce L. Hutchison, Arnold P. Moore, Peter G. Noble, Henry A. Olson, William J. Rourke, Rubin Sheinberg, Jennifer J. Zeien

2. ELECTED BY BALLOT OF THE MEMBERS TO THE OFFICE OF VICE PRESIDENT for the term of three years (to December 31, 2003): ballot counted September 18, 2000

Pacific Region:	Henry A. Olson
Central Region:	James D. Sharrow
Mid-Atlantic Region:	William H. Cowardin
Northeast Region:	Robert B. Geary

CHANGES TO THE BYLAWS:

B1-3. Members

(a) The grade of Member may be accorded exclusively to persons having competence and important experience in naval architecture, marine engineering, ocean engineering, or another phase of engineering or science closely allied with the marine industry. A candidate who is a graduate of a recognized technical institution or other institution satisfactory to the Society shall have had not less than five years of increasingly important engineering or scientific experience indicative of growth, competency, and achievement at least two of which shall have been in responsible charge of work. A candidate possessing a Professional Engineer's license or equivalent will be deemed to have given evidence of engineering or scientific experience indicative of growth, competency and achievement with at least two years in responsible charge of work. Alternatively, a candidate who is not a graduate of a recognized technical institution shall have had not less than nine years of engineering or scientific experience indicative of growth, competency, and achievement at least two of which shall have been in responsible charge of work

D3-1. Standing Committees. There shall be the following Standing Committees of the Society in addition to the Executive Committee:

Advisory Public Service Committee
Annual Meeting Committee
Applications Committee
Audit Committee
Awards Committee
Development Committee
Education Committee
Electronic Media Committee
Fellows Committee
Finance Committee
Investments Committee
Journal of Ship Production Committee
Journal of Ship Research Committee
Marine Technology Committee
Members Insurance Committee
Membership Committee
Nominating Committee
Papers Committee
Pension Committee
Planning Committee
Professional Engineering Licensure Committee
Publications Committee
Scholarships Committee
Sections Committee
Technical and Research Steering Committee

D3-24. Professional Engineering Licensure Committee. The Professional Engineering Licensure Committee shall consist of at least ten currently licensed P.E. members. The committee shall oversee and provide support for the development of professional engineering examinations in naval architecture and marine engineering (formerly ship design engineering), and ocean/offshore engineering if and when appropriate, for the National Council of Examiners for Engineering and Surveying (NCEES). The Society shall designate a liaison representative to the NCEES Examination for Professional Engineers (EPE) Committee who shall become an ex-officio member of the committee. The committee shall maintain liaison with the Standing Committees on Education, Publications and Electronic Media, have broad representation of the Society's technical membership and submit reports to the Executive Committee and Council as appropriate.

D3-24. Professional Engineering Licensure Committee
D3-24. D3-25. Publications Committee
D3-25 D3-26 Scholarships Committee
D3-26 D3-27 Sections Committee
D3-27. D3-28 Technical and Research Committee

D3-6. Applications Committee. The Applications Committee shall consist of at least six members. Membership on the Committee shall be for a term of three years and shall be staggered such that one-third of the members are appointed each year. The Committee shall have charge of all matters connected with applications for membership in the grades of Member, Affiliate, Associate Member, and Student Member and shall use such literature and forms as may be approved by the Executive Committee. The Committee shall meet a minimum of six times annually, generally bi-monthly, and shall report the names and grades of all new members to the Executive Committee and/or Council.

President Jose Femenia's Annual Address

Good afternoon ladies and gentlemen. It is a sincere pleasure to be here with you in beautiful Vancouver. It is hard to believe that a year has gone by since last year's annual meeting in Baltimore.

I ask you to join me in a moment of silence for the late Prime Minister Trudeau.---------- Our sympathies to all Canadians. He was a good man and a true leader.

Much has transpired in the past year related to our society. The Sections, for the most part, have had a successful year including the sponsoring, or co-sponsoring, of five conferences and symposia.

I totally appreciate the efforts of the section leaderships and the committees that organized the regular section meetings and the symposia and conferences, and on behalf of the Society and the profession-at-large, I want to thank all who worked so hard to deliver very successful technical meetings. I also want to thank the headquarters staff for their hard work on our behalf and for their loyalty to the membership. In particular I want to thank Phil Kimball for the excellent leadership he has given the headquarters staff during the past two years and Barbara Trentham for lending "corporate memory" to the Society and for her guidance and personal assistance to me. There are many more individuals at headquarters, on the Executive Committee, the Council and serving on the scores of other committees and panels that work tirelessly for the Society that should also be recognized, but time does not permit to recognize them individually. To them all THANK YOU. I am indebted to you.

The best way to predict the future is to create it. Creation is the result of fertile minds and the tools and resources to make dreams come true. Fertile minds are given by God, and the technical tools and resources necessary to design and build boats, ships and off-shore structures are developed and enhanced by our society, SNAME.

As a technical society, SNAME has been, and continues to be, very active in creating and enhancing the technical tools of the naval architect, the marine engineer and the ocean engineer. As I've said before, without the creative genius of our profession we would not have modern day society. We must not forget this fact nor should we let the public-at-large to forget. Nor should we sit back and reminisce about history and expect the world to keep showering us with praises and a comfortable living. In today's society you have to produce or fall by the wayside. As a Society we need to take on a more active role in creating the future. We must insure that our profession has the tools to venture into uncharted technical waters. We must insure that when one of us thinks "out of the box" and come up with an unorthodox idea for marine transportation, resource recovery or recreation we have the technical resources to follow our dream and look into a new or better way of doing something. Our T&R structure with its numerous committees and panels has been effective in defining and developing the industry's design tools for over 100 years, mainly by using the volunteer efforts of our members and the support of their employers. A model that has worked well during the 19th and 20th centuries when we had the luxury of time and the world was much slower paced. As we start the 21st century, we see the SNAME committee/panel model being stretched and over-taxed while the technical demands on our profession continue to grow, if not accelerate. At one time life was simple, at least we now think so. Engineers would use creative juices to address the technical problems at hand. When our forefathers realized that their colleagues were addressing similar problems they decided to share information instead of each one reinventing the wheel. Consequently the concept of technical meeting evolved and individual, and companies, were proud of their accomplishments and wrote papers not only to share information but also to brag about their ideas and accomplishments. When it was realized that it might make life easier if guidelines were available, technical organizations wrote guidelines. Often months or even years, and in some cases decades, would pass from the time a need was first defined and the time that the appropriate paper or guideline was published. In the past this was often an acceptable way of doing business. Unfortunately, in today's fast paced world we normally have to respond much quicker. With faxes and e-mails we don't even have the excuse "the mail is slow." Additionally we in the marine industry are now under the influence of corporate and political leaders and special interest groups who have agenda that often differ from that of building and operating the best ships. The commercial ship of today is not simply the vehicle used to carry cargo in its holds but rather it is an element of a transportation system. The recreational craft of today is simply not the rugged, seaworthy sailboat or cruiser of yesterday, but today's recreational vessel is one that has had much more emphasis placed on appearance and marketability. No longer do we simply have the users of ships and boats to contend with, but we have external groups such as environmentalists and regulatory agencies impacting on our actions. I do not want to give the impression that the influence of the environmentalist, the accountant or the regulating agencies is necessarily bad. In many cases I agree with their positions. What I want to say is that we, the marine industry, need to respond quicker to issues and in conjunction with the Society take a more active role in insuring that the non engineers that influence our actions are well educated as to the full range of the facts.

We must insure, quoting from Rick Thorpe and the ˌdvisory Public Service Committee, that "disinformation" is ot spread thus causing political and corporate leaders to make oor decisions related to maritime issues. As a case-in-point, in ıly 1999, when the Blue Water Network issued a report alleg- ıg that fast ferries are more polluting than automobiles and usses, it was imperative that SNAME address the issue and set ne record straight. We should all be thankful for Jim Sweeney nd Ad Hoc Panel #9 for quickly publishing the report, "Ferry ransit Systems and the Environment", which highlighted the rrors in the Blue Water Network report and showed that fast erries can be more environmentally friendly than automobiles nd busses. Unfortunately, the Ad Hoc Panel report was reac- ive and was published after the damage was done. It was a amage control effort rather than a proactive document ntended to inform the world about the environmental and conomical benefits of fast ferry systems. NAME needs to become very proactive in its efforts to nform the public-at-large as well as the political and corporate eadership as to the economic and social benefits of specific narine devices and systems, especially when used in new appli- ations. We need to take full advantage of the media in getting he information to the public and thereby setting the stage for efining our future. We need to take a more active leadership ole in promoting the virtues of marine systems and new devel- pments in marine transportation, ocean resource recovery nd aquatic recreational vehicles. We need to take a more roactive role in helping individual marine visionaries and cor- orations develop concepts and ideas into reality by support- ng or brokering the research activities necessary. SNAME, hrough its T&R structure should be the vehicle hrough which marine-related technical informa- ion gets developed, refined, published and dissem- nated to the profession and the public at large.

To succeed at this task in a timely manner, NAME will have to expand its traditional volun- ary T&R structure by adding more funded proj- cts in order to accelerate and expand the process. The Society will have to become more adept at eeking outside funding in order to support a more xtensive, fast-paced T&R program. In so doing ve will have more influence in defining the future f the marine industry and thus help creating our wn future. We need to expand our role from the olvers of the technical problems to creative hinkers of better ways to use the world's waters for ransportation, resource recovery and recreation. We will then need to articulate the new concepts nd ideas in a manner that they will allow them to each fruition. SNAME should be viewed as not nly the source of technical information but in ddition, it should be viewed a source of inspira- ion for its member to become the creative thinkers hat will define the marine industry of the future.

In the process of helping "engineer " the future of the marine industry, we must never lose sight of the customer. As Mike Kelsey so graphically showed us this morning, the successful industry must be very responsive to the customer. In our case the vessel owner, ranging from governments, to corporations, to individuals. To succeed we must not only wait to hear from them, but we must anticipate their needs and desires and offer them new ideas for products and systems. In other words, we must become more proactive in generating the "new and better ideas". A key element in improving our products and systems is to listen to the men and women who operate and use our creations; from the marine engineer on the floor plates to the mate responsible for cargo operations to the company's operations manager. As a professional society, we need to encourage feedback from the individuals using our boats, ships and rigs in order to make them even better and safer.

In summation, I firmly believe that we can do much to create our future. SNAME is a great organization and can, through its members, be the catalyst for the new ideas that will define the marine industry of the 21st century. I am confident that under President-elect, Joe Cuneo, our society will continue to prosper and grow and continue to be the world's premier professional society representing the marine industry.

Thank you for your attention and support over the past two years.

Banquet Speaker
Captain Graham Westgarth

Ladies & Gentlemen:

Let me start by thanking you for the opportunity to speak this evening. It's certainly an honour to address members of an organization with such a long and distinguished history.

In a way, I feel a little fraudulent in giving this speech as on first agreeing to the invitation, I misunderstood the context of the occasion and thought I was to address the local membership--it was somewhat humbling, not to say terrifying, to learn that there would be in excess of 500 participants, representing an organization with around 12,000 members, from some 87 nations.

If, therefore, you hear any strange sounds during the presentation there's a good chance it could be my knees knocking together.

I have one other area of concern, which is being from a nautical background, it is somewhat disconcerting to be locked in a room with so many engineers. Nevertheless, over the next twenty minutes or so, I will try and do justice to both yourselves and the occasion.

Whatever the outcome however, I feel you should at least pay due respect to the frugality of the organizers.

In the letter of invitation to speak I received from the Society, it was a great relief to note that "all transportation and lodging costs would be borne by the Society". Given that the Teekay office is some three blocks from this delightful setting and my home a little over 15 minutes away, my bill should be somewhere in the region of $3.00 Canadian. On second thought, I'll probably need to stop for a beer on the way home which could elevate the cost to $10.00.

Now, onto more weighty matters, I was advised that the assembled group would welcome a message on the economics of world shipping, or world shipbuilding. I was very glad to hear such a group of people existed. When you mention economics and shipping to a group of bankers they either start shaking in their boots or dissolve into fits of laughter.

I am certainly not presumptuous enough to speak on behalf of the shipping industry in general, but am certainly willing to share the perspective of an independent shipowner as to the economic trends and challenges within the tanker industry.

The real challenge for a company like Teekay is to provide a safe, cost-effective service, whilst providing enduring value for our shareholders. These are easy words to say, but what do they mean in reality?

Before expounding on the various factors behind such a statement, it's probably worth stating that I will be referring to a number of statistics, some of which you may disagree with, however as a famous playwright once said, "It's more important to be thought provoking than accurate". On the other hand, Winston Churchill said, "It's a good thing for an uneducated person to read books of quotations".

I'll leave the audience to decide which of these statements is the most accurate.

So here goes, let's first take a look at the economics of the industry.

Less than 12 months ago, tanker owners were facing one of the worst freight markets ever. Plunging ship values meant that many ship owners were in breach of minimum hull value covenants on bank loans. Things were looking very grim indeed.

Nine months later we are in the midst of the strongest tanker market since the early 1970's with tanker demand up by 10% and freight rates up by some 200%.

The dynamics which have driven this turnaround are continued global economic growth which in turn has led to increased oil production, and therefore, an enhanced demand for transport facilities.

The supply statistics are also in balance. Newbuilding yards are close to being full for the next 30 months, thus capping the in-flow of new tonnage. At the same time with 30% of the world tanker fleet being over 20 years old, and facing unprecedented discrimination from charterers and the ever-growing regulatory pressure, scrappings will continue to occur under all but the most extreme freight market scenarios.

It is thus fair to say, that from an analytical point of view things look very positive for the next two-three years.

The most remarkable thing about this 1800 market turnaround is that when you scan the horizon for the next thunderstorm, there is none!

So, does this mean the tanker business has reached the promised land of perpetual profitability? Unfortunately, not in our view. I'm not sure how the rest of you view this, but I for one, tend to get nervous when there is nothing but blue sky above. Shipping, after all, has been a business in which one has never been allowed to be happy for very long.

While I would be ecstatic if I am proven wrong, I see no reason to conclude that the tanker business has suddenly ceased to be deeply cyclical. Over ordering of tankers will eventually occur. Newbuildings this year are already 50% higher than last year. This oversupply will eventually result in a market downturn, possibly a severe one. It may be three years or five years away, but it will occur.

In a way it's unfortunate that the shipbuilding industry has totally different drivers to the shipowner with the emphasis being on filling yard capacity at whatever price. In a way, this scenario can be likened to the story of Odysseus and the sirens with the shipbuilders being the sirens and Odysseus the shipowner. Unfortunately, now and then, the shipowner tends to release himself from the mast even though he knows it will fuel the fire that will eventually consume the market he is in.

You may ask "Why are we already worrying about this next downturn?" Simply, because despite today's euphoria, we still cannot earn our cost of capital by relying solely on the tanker cycle. Our fragmented industry has historically been a notorious destroyer of shareholder value for the average owner of tanker assets.

In the Aframax sector alone, there are some 168 owners and whilst the top 20 control some 80% of the total supply, with Teekay as the largest owner of around 60 vessels having 10% of this market, the tail still wags the dog.

Returning to the theme of shareholder value, if you had invested in three-month U.S. Treasury Bills over the last cycle, you would have earned 4.7% on average. If you had been an investor in Teekay Shipping, your return would have been 7.1%. You would have beaten the T-Bill return on a nominal basis, but not on a risk-adjusted basis. And, if you had invested in Teekay's peer group of public bulk shipping companies, you would have managed only a 2.8% return!

If the largest, most professionally run companies in our industry cannot earn their cost of capital, what are the chances that smaller operators will survive?

How much longer should we expect investors to support the tanker industry if we don't deliver adequate returns?

This, in essence, concludes my thoughts on the cyclical nature of our business. You will notice that I have not made any suggestions as to how this high degree of cyclicality, much of which is self-imposed by the industry itself, might be dampened. In essence, it's almost impossible to envisage a solution as long as the business drivers of ship ownership and shipbuilding are so fundamentally opposed.

The only way the shipowner can exercise some control over this part of his business therefore, is by the careful management of his sale and purchase programme.

At Teekay, we often remind each other about the fact that our company is really in two distinctly different businesses; namely, the highly cyclical asset owning business, and the non-cyclical service business.

What I have just described in the first part of my comments today was, of course related to our cyclical business. Because of its capital-intensive nature, our asset business is the biggest single driver of our company in respect of ship-day topline, or revenues. Yet, the trade in the commodity of ship hulls is essentially an undifferentiated part of our business, in which anyone with enough money can participate. Since we do not conclude asset transactions very often, this is not the most time consuming part of our business.

We spend far more of our time thinking about the service and operating side of our business. There is a good reason for this: if you get it right, you can actually differentiate your company from others in the eyes of your customer. This is very significant.

Over the next few minutes, I will discuss how a shipowner might think about his service business as a means to build his franchise and, at the same time, enhance his bottom line.

Risk Management and Controls

Running vessels is a risky business. Therefore, the first set of issues weighing on the mind of the shipowner revolves around the safe and competent running of his ships, including:

Long term availability and competency of staff: This, in our view is a key area for success. We can see on a daily basis, the market tightening for not only good sea staff but also qualified experienced shore personnel. Teekay have introduced a number of initiatives in this area, including the extension of stock options to all shore staff and key vessel personnel. Our global staff turnover is currently in the region of 3%.

Ship quality and features: The focus on structural integrity and reduction in manning over the years means that it is absolutely essential for shipowners to have in-place, not only long-term asset protection policies, but also they need to continually seek technological labour saving devices. However, the real secret is to have a well-built vessel with high quality standards applied.

Safety practices: The advent of the ISM Code has seen an increased focus on safety & environmental management. It is absolutely essential that this is not seen as just another piece of legislation but as an opportunity to continually improve standards. Ultimately this will result in bottom line benefits.

The minimum standards in each of these areas are of course, dictated by regulation and it requires a lot of work just to meet these standards. However, in my view it makes excellent business sense to go the extra mile and strive for the highest possible standard in these areas. In doing so, you help your customer minimize his risk. This should get you preferred supplier status, which in turn builds the value of your franchise as new business opportunities are created.

For example, at Teekay we know that when our ships call at a customer's terminal, the appearance of the ship reflects on our customer. A big part of the reason why we place such a high priority on the fabric maintenance of our ships is to assist our customer in maintaining a positive public image. In return, our satisfied customer rewards us with repeat business.

Customer Service

Next on our mind is the area of customer service. Most owners talk about service, but many merely offer the minimum expected product. Service innovation and creativity are value-added features that help our customer's bottom line. These might include:
- convenience
- flexibility
- consistency/predictability
- information
- responsiveness
- and transparency

Service organizations, which is certainly what we at Teekay consider ourselves to be, should have cross-functional teams organized around the customer processes.

A responsive organizational structure staffed with empowered people is essential to being a true service company. This must be driven from the highest levels of the organization.

The customer service aspect of our business has been under-prioritized for years. In tomorrow's world, you will either add value or get out! This statement applies to everyone within our industry.

Cost Effectiveness

The third big issue on the owner's mind is achieving his goals in risk management and customer service in the most cost efficient way. This issue is tackled by trying to build:
- higher vessel utilization
- an effective organization
- economies of scale
- IT (Information Technology) systems
- and process efficiencies

This year Teekay will spend $10 million on continuous improvement initiatives and new infrastructure projects which will allow us to raise our quality of service further. Yet, due to significant economies of scale, our per ship day operating expenses and overhead costs are still expected to fall.

Scale and cost efficiency lay the foundation for providing a superior product at a competitive price. Once again, we need our service suppliers to understand this.

Consolidation

Today's customers in the tanker business are bigger, more focussed on their core oil business, more quality conscious and more sensitive to reputation risk. They increasingly expect their suppliers to deliver value, and to do so, suppliers must have significant resources, be it people, ships or capital. Consolidation in the tanker industry is beneficial for customers because it creates more capable suppliers. Customers want cost effective and efficient operators - they don't want cheap operators.

The highly fragmented tanker industry is gradually responding to this challenge. There is still a long way to go, but the pace of consolidation is quickening. We should fully expect to see the consolidation trend among shipping companies continue.

One important point to note however, is that owners should not pursue consolidation in the hope that it will lead to an erosion of the competitive nature of the tanker market. Even a substantially more consolidated industry will see continued price competition. Owners will not have pricing power. Just look at the oil companies themselves for proof of this.

In essence, a consolidated market structure would see fewer, but financially healthier tanker companies who can deliver a quality product. Companies that are able to meet its customer's needs for flexible service will be rewarded with better profit margins. In turn, these companies will drive further consolidation through their financial strength.

Finally innovation. I think it's fair to say, that shipping on a whole is conservative in nature. I constantly hear that the industry does not support or is unwilling to pay for experimentation or innovative products.

My message to you is that the industry cannot afford not to change and pursue innovation. I have already stated, doing it the old way simply does not pay. Each one of us has to look at what we do whether that be in manufacturing, supply, construction or operation and drive out waste, as well as seeking additional efficiencies by challenging ourselves on a daily basis, and looking at long-term gains.

To sum up, the good news is that if you get your asset purchases more right than wrong, on average through the shipping cycle, you are laying a solid foundation for profitability in the future. The bad news is that no matter how good you are at timing such purchases, you cannot build a barrier against entry by others.

Similarly, if you get the service business right, you can build a franchise that consistently enhances profit margins. The stronger the franchise, the higher the margins. However, such margins are insufficient to overcome the effect of poorly timed asset acquisitions.

The only way therefore, to create long term shareholder value is to be good at both the asset business and the service business. Companies that succeed in both will drive the consolidation of our industry. In doing this they have a real chance of scaling the considerable barrier to achieve sustainable profitability in the shipping business.

Thank you very much for listening.

2000 Annual Meeting Photographs

Following are some formal photographs from the Annual Meeting.
Some events and awardees are not pictured because formal photographs were not available.

50 Year Members

Peter G. Schmidt with Jose Femenia

Charles Zeien with Jose Femenia

David Seymour with Jose Femenia

John Kelly with Jose Femenia

Distinguished Service Award

William J. Sembler with Ronald Kiss

Honorary Member

Jose Femenia

T&R Certificate

Alan Rowen with John Daidola

T&R Certificate

Thomas Mackey with John Daidola

Top Recruiter

Paul J. Roden *with John Higginbotham*

Graduate Paper

Peter Gooding *with Robert Scott*

Undergraduate Paper

Tyson Scofield *with Robert Scott*

Graduate Honor

Michael Griffin *with Robert Scott*

Undergraduate Honor

Katherine Croff and Whiteny Cornforth with Robert Scott

Lisnyk Award

University of Newcastle Upon Tyne Students with Ronald Kiss

ABS Linnard Prize

Ge Wang, Anil Thayamballi, and Jeom Kee Paik with Robert Geary.

Cochrane Award

Stanley Stumbo, Kenneth Fox, Lawrence Elliot and Frank Dvorak with Edward Lewandowski.

T&R Committee Chairs

Robert Beck, Ralph Johnson, Robert Behr, John Daidola, Thomas Lamb, Michael Powell, Henk van Hemmen, Philip Rynn.

The Davidson Medal

Robert F. Beck (center)

with Jose Femenia and Michael Parsons

The Vice Admiral "Jerry" Land Medal

Charles R. Cushing (center)

with Jose Femenia and David O'Neil

The David W. Taylor Medal

Thomas G. Lang with Jose Femenia

Winners of the Studen
Congress Design Competition

Patrick Nester - Webb Institute, Holly Clarke - Florida Atlant University, Brian MacClaren - University of New Orlean Korin M. Strome - Virginia Tech, Jamie Bardsley - Universi of British Columbia, Sarah Vitous - University of Michiga

Past, Present, and Future Presidents

from left: Edward Campbell, David O'Neil, Jose Femenia, Joseph Cuneo, Arthur Haskell, and Ronald Kiss.

First Female Fellow

Jennifer Zeien

Papers Presented at the 2000 Annual Meeting

SNAME *Transactions*, Vol. 108, 2000, pp. 1–20

Wash Waves—Problems and Solutions

M. Leer-Andersen, Visitor, **P. Clason,** Visitor, **P. Ottosson,** Visitor, **H. Andreasson,** Visitor, and **U. Svensson,** Visitor, SSPA Sweden AB

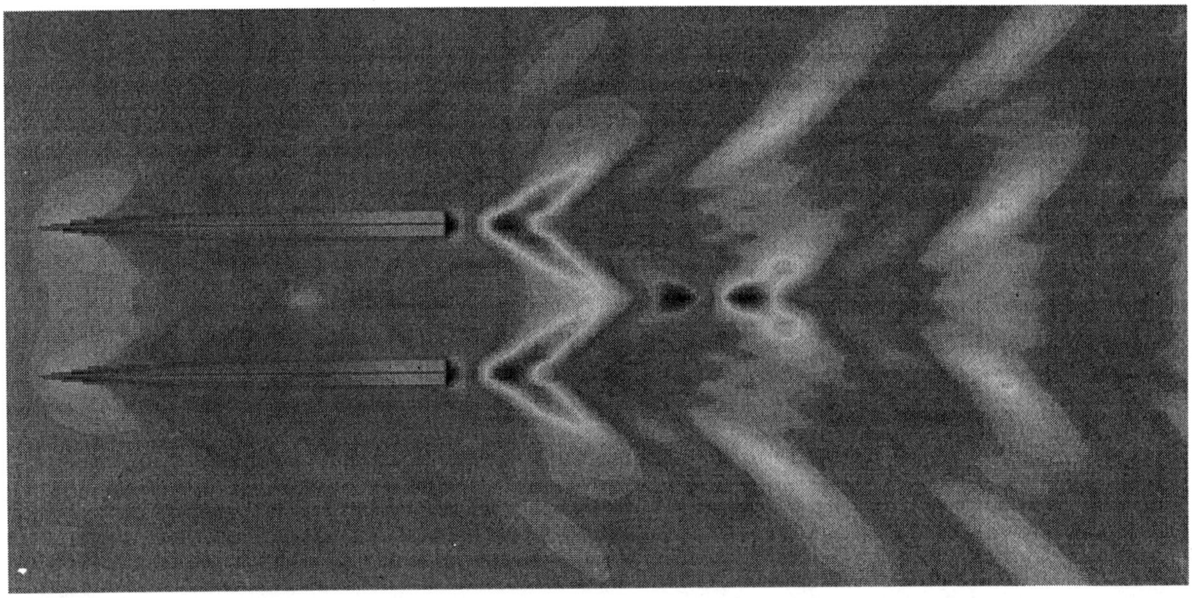

ABSTRACT

This paper will address the increasing problem of ship generated waves in coastal areas. It will include a discussion of the damage and danger that wash waves can present, and what can be done in terms of designing a HSC(High Speed Craft) in such a manner that the wash waves generated are minimised. This will include a comparative evaluation of HSC concepts, what can be done in terms of optimisation, how to evaluate the design and what can be expected in terms of accuracy and limitations when evaluating the design's wash wave. HSC wash wave regulations in Scandinavia will be discussed, and a different approach to regulations will be presented. As a final measure recommendations and guidelines will be given for how to create "route planning" that will be beneficiary for low wash wave impact in coastal and shallow areas.

NOMENCLATURE

a	Wave amplitude
c	Celerity
C	Comparative number
C_w	Wave resistance coef.
DW	Dead weight
H_h	Wave norm
H_s	Significant wash wave height
h	Water depth
L_{wl}	Waterline length
Fn	Froude number
Fn_h	Depth Froude number
P	Effect
R	Resistance
R	Convergence criterion
S	Wetted area
T_h	Average wave period
U	Ship speed
$\mathbf{y}$	Optimisation vector
α	Cusp angle
δ	Optimisation result
λ	Wave length
θ	Propagation angle
$\triangle$	Displacement mass
∇	Displacement volume

1. INTRODUCTION

This paper discusses and offers possible solutions to an increasing problem within the maritime industry. The problem is ship generated wash waves. High speed craft operators in particular experience increasing pressure from both legislation and the public to decrease their wash wave impact on coastal areas.

HSC(High Speed Craft) have existed for several years, but their number, size and operating speed have been increasing continuously, thus adding to the problem. Therefore more and more attention is directed towards this problem, but still most HSC designs do not have a small wash amplitude as a primary design parameter, and few new designs are evaluated on the basis of their wash wave characteristics. Therefore, there is no comparable knowledge of different HSC concepts ability, or lack of, to generate small wash wave amplitude, is unknown. Legislation governing allowable wash wave emission from HSC in Scandinavia does not consider any local tolerance in terms of biological, hazardous or erosional vulnerability. Therefore another methodology is discussed. This paper suggests a new nondimensional number to be used for comparison of different concepts in terms of the radiated wash wave energy. Besides choosing a low wash concept, further gains can be achieved by optimising the chosen concept along the lines of low wash, for which reason an optimisation procedure is presented. The optimisation will lead to a discussion of prediction of wash waves in terms of accuracy and limitations. Finally, if the boat has already been built, and the wash waves generated are at an unacceptable level, "route planning" is the last option.

Although wash waves (Kelvin and Havelock waves) are of primary concern in this article, other ship generated wave systems, such as solitones and the Bernoulli wave will be discussed in areas where they might pose a hazard or have a damaging effect.

2. SHIP GENERATED WAVES

A short introduction to how a moving ship can change the free surface elevation is presented in the following, as many of the terms will be used throughout the paper.

Wash waves This wave system is normally the only free surface elevation that is visible. For deep water, the Kelvin wave theory adequately describes the gravitational wave pattern generated by a ship, see fig. 1. The wave pattern consists of divergent and transverse waves.

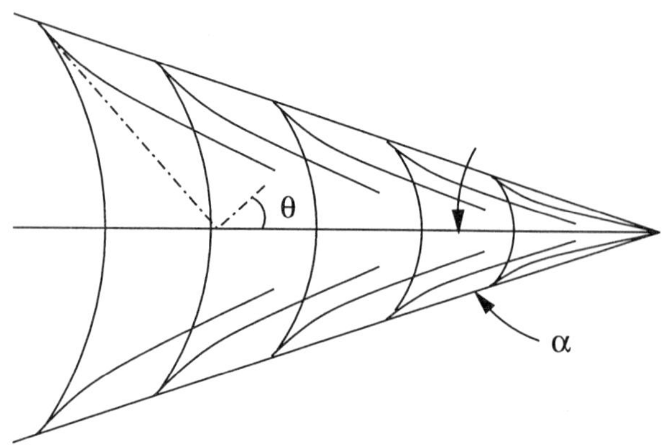

Fig. 1: Kelvin wave pattern (α=cusp ang., θ=propagation ang.)

Strictly, the Kelvin wave only applies if the disturbance is a point source, as a ship clearly is not. In fact, separate Kelvin wave patterns will be created at the bow, shoulders and stern of a ship, but viewed at a distance from the ship these systems will coincide and form the well known Kelvin pattern. The point where the divergent and transverse systems meet will have an angle (cusp angle, α) of 19.28^o (see fig. 1), and will at this point have a propagation angle relative to the ship path of 35.27^o (propagation angle θ), see [5].

When the ship approaches shallow water the wave pattern and amplitude change. For deep water the appropriate Froude number to use is

$$Fn = \frac{U}{\sqrt{gL_{wl}}} \tag{1}$$

and for shallow water

$$Fn_h = \frac{U}{\sqrt{gh}}$$

where h is the water depth. If the water depth is small shallow water theory (Havelock theory) applies for which the wave velocity on the surface is determined by $V = \sqrt{gh}$. As a rule of thumb shallow water theory should be used if $Fn_h > 0.7$. As Fn_h approaches unity, the cusp angle will approach 90° and the propagation angle 0°, and as a consequence the divergent and transverse wave systems will become one system. Above critical speed (super-critical speed), the transverse wave system can no longer follow the ship and will die out, and the divergent wave system will again arch backward with increasing Froude number.

Bernoulli wave The pressure distribution around the hull causes a raising and lowering of the water surface, which results from positive or negative pressure. The surface elevation depends on the speed of the vessel, but is fixed in the horizontal plane. Normally, this surface elevation is difficult to determine, since it disappears in the wash wave pattern of the vessel. At constant vessel speed the Bernoulli wave does not need any maintaining energy. The energy needed to create the Bernoulli wave is recovered at retardation. In shallow and narrow canals, where the presence of the canal walls increases the magnitude of the pressure distribution, the effect of the Bernoulli wave is much amplified. If the vessel speed is high and the canal is narrow and shallow the effect of the Bernoulli wave can be substantial and be a cause of shore wash on channel sides.

Solitones A soliton wave, first discovered by Scott Russel in 1834, has two major differences from the wash wave system. It exhibits only an elevation above the undisturbed free surface and will always be transverse. As it has no through it will, unlike wash waves, produce a net displacement of fluid. The celerity, or wave velocity c of the soliton will exceed that of the a normal transverse wave, as [7]

$$\frac{c}{\sqrt{gh}} = \sqrt{1 + \frac{a_0}{h}}$$

In other words, even at critical speed the soliton will move forward and away from the ship, and if the ship maintains the same Froude number several discrete solitones can be formed periodically. According to [7] solitones were created close to critical speed even in large cross section area tanks, and should therefore not be treated as solely a canal phenomena.

3. ENVIRONMENTAL IMPACTS OF WASH WAVES

Shipping has impacts on the environment in many different ways. This brief description is concentrated on the effects associated with wash waves from ships.

Ship generated waves can, for some operating conditions and local environments, be the main reason for erosion on coasts and bottoms along shipping routes, particularly in shallow areas. Large and/or fast craft that pass bays, beaches or other shallow coastal areas generate wash waves that move towards the coast. These waves can remove and transport sediments, loose bottom material, plants and gravel from beaches and the sea floor. The sediments are transported to other shallow areas or out to deeper waters. This, along with all other reasons for erosion, can lead to reshaping of bays and beaches. Installations in coastal areas, such as buildings, quays, piers, etc. are also affected by eroding wash waves, as well as commercial activities and industrial installations.

Erosion of loose material on beaches creates what is called beach scarps. This is a phenomenon that occurs when the constant, eroding effects of the waves form a scarp. The over-steepened beach scarp finally avalanches into the water and the process is started from the beginning again.

Sea bottoms and shores can be either soft bottom or hard bottom. In archipelagos, shores that are normally exposed to waves are mainly composed of rocks since the sediments have been washed off by eroding waves. Underneath the rock layer there is soil. The plant species on hard bottoms are mainly algae while the diversity of higher plants is smaller. The biological diversity of both plants and animals is generally very high. Seashores and bays not exposed to waves are usually soft bottom areas, as such areas more often that not are net receivers of sediment, and this is where the main part of marine plant life is found.

Erosion of beaches and bottoms can have severe impacts on the marine ecosystems. Strong currents along the bottom affect both plants and sediments. The sediments in soft bottom areas are eroded and removed to other areas. The strong currents remove and affect vegetation, in particular eelgrass, reducing the volume of eelgrass beds, and thus changing the conditions of life for plants and animals. Strong currents can be created by ships, especially if the conditions are such that seiches (eigenfrequency of the free surface elevation in very confined waters) can exist. It is very likely that large deplacement ships under special conditions can create such problems, see [11].

Most shallow soft bottom areas serve as nurseries for younger organisms, and store the main part of the nourishment for the small organisms. If severe erosion of sediments and strong bottom currents changes the conditions of life for marine plants and organisms, the number of species (biological diversity) and the number of specimens can thus be seriously reduced.

3.1. Governmental limits on ship generated waves

The common method to handle critical cases has been to introduce speed restrictions in specified areas and for selected types of ships. A general speed restriction, however, does not give due consideration to the size of the ship nor does it consider the influence of water depth on the waves. An example is the route between Stockholm and Åland/Finland in the Stockholm archipelago. The very narrow route has been operated by conventional ferries in ever increasing sizes. Although there has been a speed limit for the ferries on selected parts of the route there has been considerable erosion and damage.

Danish initiative for HSC In May 1997 the Danish Maritime Authority introduced a regulation for HSCs in Danish waters [2] as defined by the High Speed Craft Code (IMO resolution MSC 36(63)). The regulation states that an HSC operator needs to obtain approval for the intended service before the service starts. In order to obtain approval detailed information about the service must be submitted such as route plan, timetable and relevant documentation showing that the HSC service will not generate waves exceeding a defined wave norm.

The explicit purpose of the regulation is to ensure that the waves generated by the HSC will not cause an unnecessary risk to traffic safety and to leisure activities in coastal areas.

Swedish regulation for HSC In January 1999 Sweden introduced a law [4] which in essential parts copied the Danish regulation. The Swedish law defines high speed craft as ships which have a length of more than 50m and which travel at a speed greater than $U = 3.7 \cdot \Delta^{0.1667}$ (IMO definition) or at least 32 knots.

The operator of HSCs must apply for and obtain an explicit approval for operation before the HSC is entered into service by the Swedish Maritime Administration (SFV). The SFV is to define specific regulations for HSC operation.

Intentions of the regulations are to reduce the negative impacts of HSC traffic, but they are also viewed as a necessary restriction for safety aspects. Deliberate or unintentional violation of the law will result in a fine.

The SFV has further defined the regulations for operation with HSC in SJFS 1999:2 as of January 1999 [1].

The route plan An application for HSC operation must include, in addition to ordinary ship data, a proposed speed and route plan for the intended operation. The route plan must show the entire route in Swedish waters and must explicitly identify four area types along the route. The application will have to be supported by calculations showing that a wave norm is fulfilled for the relevant areas along the planned route. The calculations must be prepared by a ship research institute or similar.

The area types are:

- General speed restricted areas, *i.e.* harbours and areas where separate rules for safety and protection environment apply. The HSC shall operate in these areas in such a manner that the wave norm is not exceeded.
- Archipelagos and other coastal areas where particular safety and environmental concerns must be regarded. The HSC must operate in these areas in such a manner that the wave norm is not exceeded.
- Areas where the HSC will accelerate or decelerate to or from its normal speed in such a manner that the wave norm will be exceeded.
- Areas other than areas defined above where the HSC can operate with out regard to the wave norm.

The wave norm The wave norm defines the maximum acceptable wave height from the long-periodic wash waves in 3.0m water depth. The wave norm is defined as:

$$H_h = 0.5 \cdot \sqrt{\frac{4.5}{T_h}}$$

where H_h is the max wave height (m) and T_h is average wave period (sec).

Effects of regulation In brief terms the regulation states that the wash wave height from a HSC must not exceed 0.35m in 3.0m water depth along a coastline exposed to HSC traffic. The wave norm is based on an evaluation of various wave patterns which shows that subcritical wave height in most cases is less than 0.35 m. The max wave height is of course very dependent on speed and water depth, but with the wave norm the discussion of water depth is avoided. The wave height limit will indirectly induces a speed restriction but it also encourages the operator to evaluate different routes. An alternate route may have a longer sailing distance but the average speed can be higher since it is possible that the HSC can operate further away from sensitivc areas.

It is, however, not yet determined whether the current wave height limitation is an acceptable limit with regard to erosion levels and other damaging effects from the wash waves.

The wave pattern around a ship can be determined using analysis or model tests. A more delicate problem is to determine the wave height in a shallow area when the wave has travelled a considerable distance and with due regard to underwater topography. Models are being developed but they still need further verification.

One of the first operators to be approved under the new law was Destination Gotland who is trafficking the route between Visby and Oskarshamn in the Baltic Sea with the newbuilt "HSC Gotland". Rather than performing a vast range of complicated (and expensive) mathematical analyses, the actual waves were measured on strategic points along the HSC route. The wave measurement report formed the main reference in the final HSC application for the route. The procedure was accepted.

A different solution was used on the route between Gothenburg on the Swedish west coast and Frederikshamn in Denmark where the two competing HSC operators, Stena Line and SeaCat, reached a mutual agreement on maximum speed for their HSCs outside a sensitive area in the Gothenburg archipelagos. Public attention and public opinion were in this case probably more important than maximum speed with regard to current legislation.

Future regulations According to SFV Sweden and the Danish Maritime Authority they have received some criticism for introducing the HSC regulation. The criticism is however not focused on the introduction of the HSC restriction itself but more on the fact that the HSC regulation was not developed and agreed upon in the international community beforehand. It is likely that other countries will follow with similar HSC regulations in the near future or possibly IMO will modify the HSC code to include limitations of the generated wash wave.

An Environmental Impact Assessment (EIA) could help mitigating the impacts by suggesting mitigate and monitoring plans at an early stage of the planning process and to serve as a tool for increased and improved environmental management.

3.2. Environmental Impact Assessment in shipping

Performing an Environmental Impact Assessment (EIA), as a tool for increased and improved environmental management, could help mitigate the adverse environmental impacts from wash waves. An EIA is a process that gives all stakeholders the opportunity to participate in describing, analysing and evaluating the environmental and socio-economic impacts from a certain activity.

Impacts on the coastal and marine environment from shipping and other transports as well as industries are becoming increasingly disturbing, as coastal areas serve as the livelihood for a majority of the world's population. Coastal areas are especially rich in natural resources, such as fish, plants, and coral reefs. The need and the demand for Environmental Impact Assessment (EIA) in coastal and marine areas is increasing internationally as the environmental issues and impacts on society are addressed, as discussed further in [10].

The purpose of an EIA is to form a decision support by clarifying the impacts that a certain activity or project has on the environment, health and the use of natural resources, and to present measures for mitigating these impacts. Both direct and indirect impacts should be described.

The EIA process contains several parts. The planned activity or project must be thoroughly described regarding alternatives, impacts and technical details. A presentation of choices of alternatives, including the zero-alternative must be included in the assessment. This means that the EIA must include a description of the situation if the proposed project or activity was not to be executed, which is used for comparison with the impacts of the proposed project. Alternative locations and measures of mitigation of the adverse effects of the project are also included in the EIA. Ideally the EIA process should influence the choice of alternative. If this is not possible, the EIA should influence on the choice of location and design of the project or activity, or at least attempt to mitigate any adverse impacts [3]. For best results, the EIA should be commenced at an early stage of the project.

Emphasis is put on public participation during the EIA process. The EIA is generally considered as a product produced by experts, a package of information. The ideal is that the EIA is considered a process, where emphasis is put on open information, public participation, and evaluation and monitoring. The development of international EIA guidelines by many of the international development organisations, banks etc. will help the development of the EIA into a long-term instrument for environmental management and monitoring.

When it comes to shipping, performing an EIA prior to e.g. hull design, route planning and physical planning of harbour areas can mitigate environmental impacts. The main impacts from shipping come from operational activities; emissions, operational discharges, wash waves etc. A large part of the impact receiving attention arises from accidental discharges of oil, although the volume of oil from accidents is smaller globally than that of operational discharges. Still, as such environmental impacts are important in the big picture, they must be included in an EIA, as the adverse effects of wash waves would otherwise receive too much attention in the process.

In already studied routes in the port of Gothenburg (see [11]), an EIA would be used for describing and analysing any adverse impacts on the environment from the ships using the route.

The EIA could be applied to an area of operation as described in section 5.2, and some of the results from that section are utilised here. To describe the effects on the environment for a shipping route, an EIA could be performed, describing the following items:

- Detailed descriptions and technical specifications for the suggested activity.
- Alternatives; alternative locations or alternative techniques, including the zero-alternative.
- Short and long term impacts on the environment on all alternatives presented, including erosion, effects on marine flora and fauna, air and water quality.
- Impacts on recreational activities in the area.
- A monitoring and mitigation program to avoid or minimise negative impacts.
- A socio-economic study of the specific ship route.

To give an overview of the EIA method, the following indications on differences in impacts can be given for the two calculated speeds in the example used:

- The calculations (see section 5.2) have shown that at a speed of 40 knots, the height of wash wave that reaches the shore can be up to 1.5m, while at a speed of 20 knots the wash wave height is estimated to 0.22m.
- The result of a higher wash waves can be among others things, disturbed recreational possibilities, and a higher grade of erosion on bottoms and on beaches, which can result in disturbances for marine and coastal flora and fauna.
- The EIA process includes all environmental aspects, which in this example also includes the aspect of higher level of emissions to air.
- Socio-economic aspects included in the EIA are the savings in time for the shipping company as a result of higher speed, while socio-economic costs are connected to increased erosion and costs connected to restoration of eroded areas, as well as decreased incomes from reduced recreational possibilities.

The EIA process always includes a zero-alternative, which means that all analyses of environmental impacts must be compared to the zero-alternative, i.e. no shipping traffic in the route. The wash waves compared to the zero-alternative must be compared to the natural erosion, which is caused by wind, sun, moon and other natural erosion causes.

The result of the EIA will show which alternative is the best for the environment and for society at large. The results can be presented as a final score for each alternative, which can then be compared to each other. A cost-benefit analysis of the socio-economic analysis of the EIA shows which alternative is the best, looking at the total cost for society.

It goes without saying that an EIA for a HSC operator will be quite expensive. Still gains can be large in terms of a much improved operational concept leading to improved public relations and possibly even direct economic benefits.

4. HULL DESIGN

Hull design will set the level of wash wave amplitudes that can be expected from the design. Great diversity between concepts and even within the same type of concept can be found. Therefore, knowledge of which concept to choose and what is possible to further accomplish in terms of optimisation is important.

Furthermore, choice and implementation of propulsion can have a quite significant effect on wash amplitude, as the wave system the propulsion creates will interact with the wash wave system. This problem will not be discussed further in this article, but for reference [12] found differences of up to 40% in wash wave amplitude between propeller and water jet propulsion.

4.1. Concepts

Several HSC concepts are operating world-wide, and several more are conceived or under development. Some of the more established concepts are catamaran, monohull, SWATH, hoover-craft, SES, WIG, foil assisted craft and combinations. Although it is generally accepted that for example a hoover-craft emits less wash wave energy than say a catamaran, no actual means of comparison in terms of wash wave energy exists. Therefore, to assist in choosing the right concept for a new route, the following comparative non-dimensional number, Q, is suggested.

The dimensions that should be included are: U=ship velocity, DW=dead weight, L_{wl}=length of waterline. The need for U is obvious, and the dead weight should be included as different concepts can have very different DW/displacement ratios.

As a first step in the development of Q the fictive "payload effect", $P_{pl} = DW \cdot gU$, is introduced, where g is the gravitational acceleration. The payload effect is a simple measure of the two things that count from an economic perspective: How much can be transported from A to B, DW, and how fast can this DW be transported, U.

The wave height is a good measure of how much wave wash a vessel creates at a certain speed, but to compare vessels at different speeds, wave effect is a better measure. The reason for this is that the energy transported in the waves is not only a function of wave height but also of wave length and wave speed. Since the wave is created by the craft the wave effect is equal to the wave resistance effect, $P_w = R_w U$, where R_w is the wave resistance. This is only true if wave breaking effects are considered as negligible.

The wave resistance is commonly expressed as

$$R_w = C_w \frac{1}{2} \rho S U^2 \quad (2)$$

where ρ is the density and S is the wetted surface area. C_w is the dimensionless wave resistance coefficient, which depends on the form of the craft and the dimensionless Froude number Fn. When vessels are operating in shallow water, C_w also depends on the water depth length ratio, h/L_{wl}. Q can now be defined as

$$Q = \frac{P_w}{P_{pl}} = \frac{C_w(Fn, shipform, h/L_{wl}) \rho S U^2}{2gDW} \quad (3)$$

which is easily seen to be dimensionless.

As an example, C_w has been calculated for different types of HSC with the potential flow solver in the CFD code SHIPFLOW (more detailed information available in section 4.2). SHIPFLOW solves the wave resistance for an inviscid fluid, which means that no viscous effects like spray resistance are included. C_w is obtained by integrating the pressure over the wetted hull surface.

The vessels were selected to obtain a representative spread of typical high-speed vessels. The two first vessels, Mono 1 and Mono 2, are modern high speed mono hulls with similar L_{wl} but slightly different C_b. The third vessel, Cat 1, is a large size catamaran of semi SWATH type. Cat 2 is a smaller modern high speed catamaran. Cat 3 and Cat 4 are two similar modern medium size catamarans with bulb formed bows. Cat 3 is slightly longer than Cat 4 and has a lower C_b. All vessels have a design speed of around 40 knots, and some comparative data are given in table 1.

Vessel	L_{wl}	C_b	DW
Mono 1	88 m	0.39	340 ton
Mono 2	88 m	0.45	400 ton
Cat 1	112 m	0.58	1300 ton
Cat 2	60 m	0.50	200 ton
Cat 3	74 m	0.49	330 ton
Cat 4	70 m	0.60	340 ton

Table 1: Main particulars of ships used in example calculations.

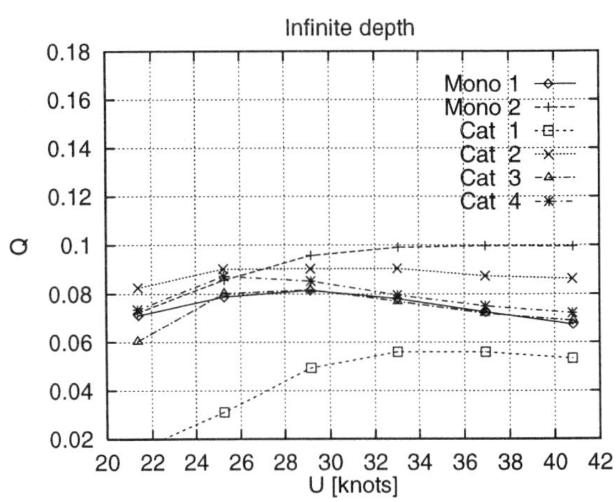

Fig. 2: C as function of velocity for different HSC operating in infinite depth water

Three sets of results are presented with Q given as a function of U. One infinite depth and two shallow water (10m and 15m) depth cases are evaluated.

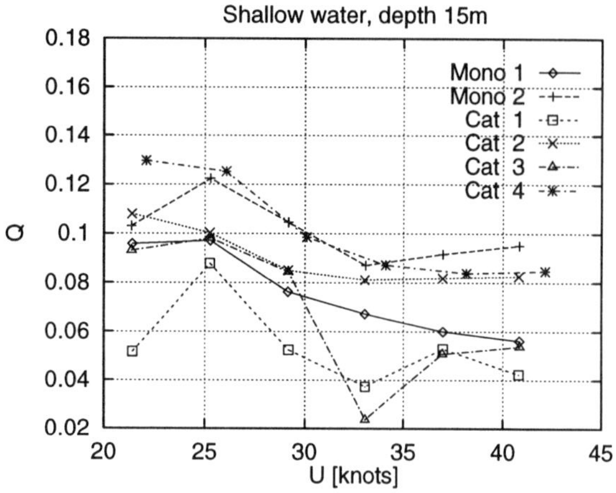

Fig. 3: C as function of velocity for different HSC operating in shallow water, depth 15 m

In figure 2 (infinite depth) Q is similar for both mono hulls, while C differs a lot for the four catamarans. There are probably two reasons for the large spread between the catamarans. The first is the wide difference in L_{wl}, if Q is plotted as function Fn instead of U the difference is less. A second likely reason is that Cat 1 is a semi-SWATH, known for low wave resistance, and cannot be classified as a normal catamaran.

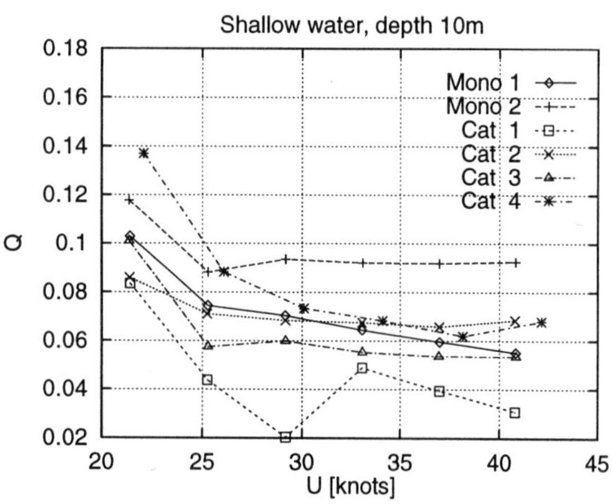

Fig. 4: C as function of velocity for different HSC operating in shallow water, depth 10 m

The ranking of the vessels follows from the infinite depth case to the two shallow water cases, which is a sign, not of Q being h/L_{wl} independent but the ranking seems to be independent. Q for the two mono hulls peaks around 23 knots in figure 3 and 20 knots in figure 4. This corresponds to critical depth Froude number, Fn_h, where it is normal to have an increase in wave resistance. Even though there is a clear ranking between the concepts, it dose not seem to be possible to find a span in Q for each concept since the hull shape and L_{wl} plays an much to important part, and too few vessels are analysed.

Of course Q can only be a tool in the prediction of wash waves. It will always be important to include how much wash a vessel is producing in the calculations since Q will not indicate if maximum wave height or wave energy is exceeded.

4.2. CFD Optimisation

Several methods can be used to optimise a hull for low wash. Most obvious is of course using the well known ship design cycle with emphasis on low wash together with a combination of model trials and CFD calculations. However, in this paper an entirely different approach is presented, as a non-interactive procedure of optimising hull shapes with regard to wash waves using only CFD calculations will be used. A set of frames is given as input to the procedure, after which no additional input is required or possible for the method to optimise the hull. The optimisation is based on results from potential flow solutions of the free surface, and are evaluated in terms of the significant wash wave amplitude from wash wave profiles along the hull(This method was originally presented at [8]).

The optimisation procedure consists of three steps.

1. Definition of hull shape as a function of a finite number of parameters based on a pre-existing hull
2. Calculation of the wash wave height with SHIPFLOW
3. Optimisation of parameters to minimise wash wave height

The above steps to attain a minimised wash wave height are of course iterative.

The first step is accomplished by transforming the original hull to a set of splines. By moving points on each spline (parameters) the hull shape can be changed as a function of these parameters. The result from computation of the modified hull in SHIPFLOW, (description of SHIPFLOW in next section) is then iteratively treated by the optimisation procedure. The optimisation procedure used is Method of Moving Assymptotes(MMA), which has the advantage of an initially very fast convergence, but with slower convergence at a later stage in the iterative procedure. This characteristic is very useful for this kind of optimisation, as the computation time to obtain a fairly optimised hull would otherwise be too costly.

For each iteration the optimisation routine (an internal FORTRAN77 work code from Alfgam Optimering AB, Sweden) needs input of SHIPFLOW results ($f_i(\mathbf{y}),\ i = 0, 1, 2$ for each parameter and one optimised solution) based on hulls defined by the parameters. This procedure is then continued until the parameters have reached a sufficiently constant level. Besides the objective to minimise the significant wash wave height, constraints on the hull shape and the total resistance are incorporated in the optimisation procedure.

Prediction of Wash Waves The prediction of the wash waves based on the hull shape is done with the commercial CFD package SHIPFLOW from FLOWTECH International AB. This package uses a zonal approach to compute the flow, utilising one of three different mathematical models in each zone. At the hull surface a boundary layer (momentum integral type) formulation is used, until the boundary layer expands to a thick boundary layer downstream. This flow domain is computed by solving the Reynolds-Averaged Navier-Stokes equations using a k-ϵ turbulence model, and the remaining part of the flow is treated by a potential flow formulation. Only the potential flow and thin boundary layer solver in SHIPFLOW are used to obtain the wash waves and total resistance. Even though SHIPFLOW also has the capability iteratively to solve for trim and sinkage and to calculate a non-linear solution on the basis of the disturbed free surface, these options are not used in the optimisation because of calculation time.

An example of the panelisation used for both the free surface and the hull is shown in fig. 5, where only half the 'real' flow domain is calculated, as the symmetry of the free surface and the two demi-hulls has been used to decrease the number of panels. The total number of panels used in all runs is approximately 2000 for the free surface and 2000 for the hull.

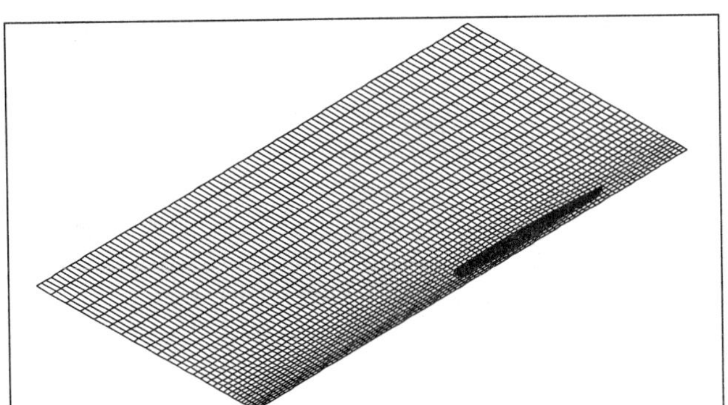

Fig. 5: Panelisation Hull and free surface

Hull Shape Modification To render non-interactive iterative optimisation possible, a program capable of modifying the hull shape as a function of a finite number of parameters is necessary. Several program packages are already commercially available that can modify the hull shape, but unfortunately such programs are based on interactivity from the user, and are as such not suitable for this kind of problem.

In the present work a 'hull shape modification routine' (HSMR) has been written with the following goals in mind:

1. The hull must always be smooth
2. The change of hull form must be a function of only a small number of parameters
3. The frames must never cross the centreline
4. The original hull should not change when all parameters are zero

The first objective is self-evident, and the second is a consequence of keeping the optimisation time at a reasonable level (additional parameters or design variables increase computation time). The third objective is to make calculations in SHIPFLOW possible and finally the last objective is required to actually be able to optimise a given hull.

This program is of course only part of the optimisation procedure, but without question a very important one, as it does not matter if the calculation of the significant wash wave height is accurate, or if the optimisation procedure is very effective, if the hulls generated by the FORTRAN77 code cannot produce a 'better' ship than the original for any set of parameters.

Without describing the HSMR routine in any detail (reference is made to the original paper), the idea is to transform the 3D surface of the hull to two 2D problems, and to alter these by use of cubic splines. An example of this can be seen in fig. 6. A parameter point is where the thick lines cross each other, and moving such a point will alter the entire shape of the hull smoothly.

The object function being optimised is the significant wash wave height for the wave profiles describing a curve on the free surface starting in the hull centreline some distance upstream, following the intersection between hull and free surface from bow to transom and continuing some distance downstream (depending on the Froude number). Additional output from SHIPFLOW used in the optimisation procedure is the total resistance (not only wave resistance, but also frictional resistance) and the displacement.

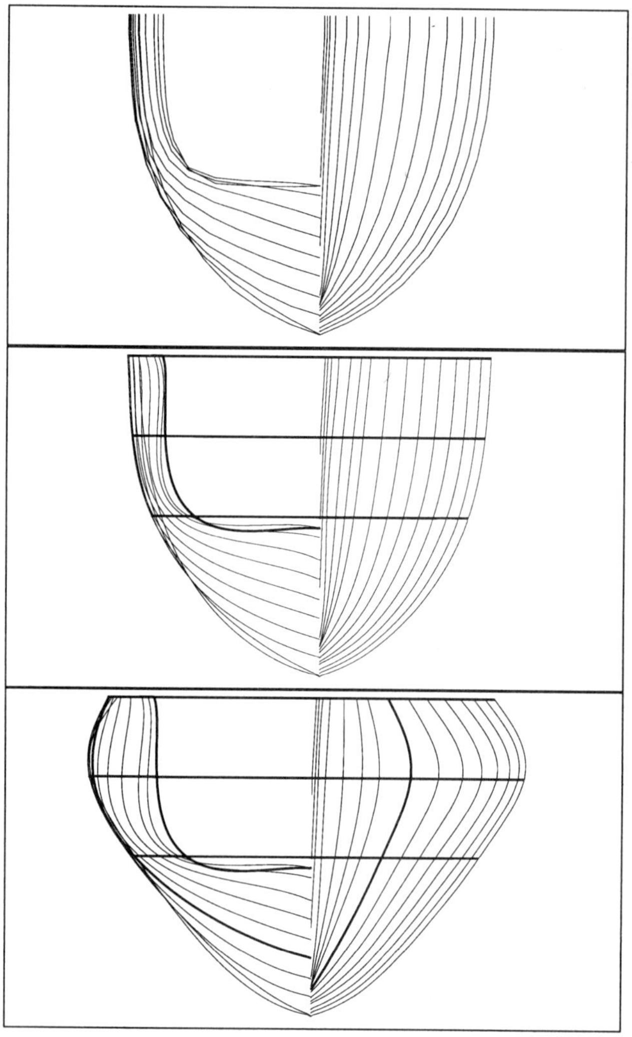

Fig. 6: Top: The original Hull I. Middle: Defined by splines with $\mathbf{y} = \mathbf{0}$. Bottom: Modified hull.

Two constraint functions are implemented. The displacement must be constant

$$f_1(\mathbf{y}^{(k)}) \geq f_1(\mathbf{0}) = \Delta(\mathbf{0})$$

and the total resistance must only increase some fractions of the original total resistance

$$f_2(\mathbf{y}^{(k)}) \leq Cf_2(\mathbf{0}) = CR_T(\mathbf{0})$$

Also, obvious geometrical bounds are applied to ensure that the optimised result can still be considered a ship. All principal dimensions except hull beam are held constant.

Graphically, the optimisation procedure can be sketched as in fig. 7, where the convergence criterion is set to

$$R = \frac{\sum_{i=1}^{N} \left| y^{(k)}(i) - y^{(k-1)}(i) \right|}{\sum_{i=1}^{N} \left| y^{(k)}(i) \right|} \leq 0.01$$

where N refers to total number of parameters.

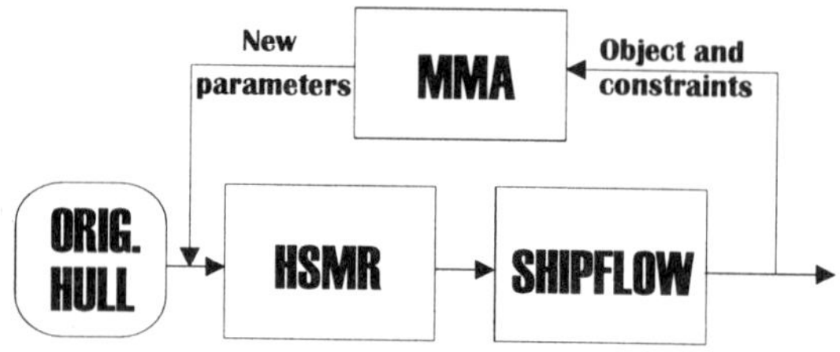

Fig. 7: Optimisation procedure

Example of optimisation As an example a typical catamaran ferry has been optimised (the original can be seen in fig. 6). The results are presented in tabular form, where the parameters movements in the y direction are shown at full scale[mm]. The parameters in the tables are ordered in such a way that Spl (Horizontal spline) 1 is the spline closest to the keel, and the columns are the parameter points on each spline counted from stern to bow. In the table row '1/2' then shows the parameters for the horizontal spline closest and next closest to the keel at vertical spline 1,2 and 3. The object function is presented in non-dimensionalised form as

$$\delta = \frac{H_s}{H_{so}}$$

where H_s is the significant wash wave height for the optimised hull and H_{s0} for the original hull at the same speed. All optimised results have the same or slightly higher displacement than the original.

The results are shown in table 2 with the optimised parameters at (from top to bottom) water-depth 10m for 10, 19.25(crit) and 30knots, and finally for deep water at 35knots. The optimised body plans are shown in fig. 8 for all conditions (same order as table), with the horizontal line showing the design waterline. As an example of the new wave pro-

Spl	1	2	3	1	2	3	δ
Subcritical speed							
1/2	156	165	624	-447	-576	-1000	0.829
3	483	484	511				
Critical speed							
1/2	564	1000	952	-1000	-32	266	0.513
3	840	-990	599				
Supercritical speed							
1/2	130	-603	1000	1000	547	1000	0.719
3	-77	-436	-500				
Service speed, deep water							
1/2	242	-890	1000	1000	1000	1000	0.704
3	-223	-820	-563				

Table 2: Optimised parameters for catamaran hull

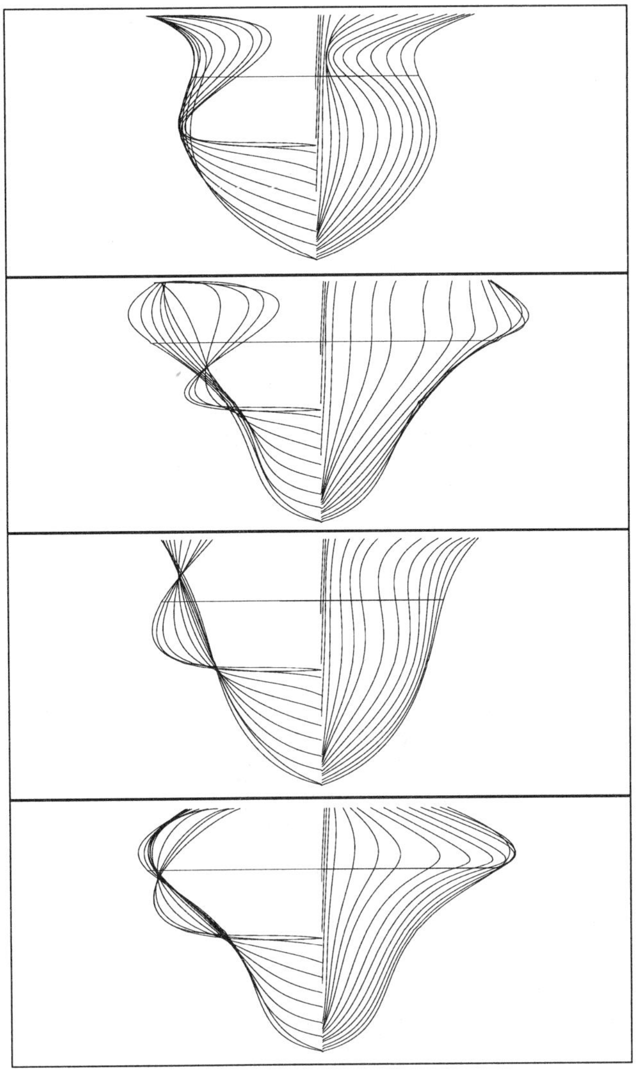

Fig. 8: Optimised hulls at different speeds and Froude numbers

file, the profile for the optimised and original hull (profile between hulls) have been plotted in fig. $9(x/L_{pp} = 0$ is at the FP and $x/L_{pp} = 1$ is at AP).

As can be seen from table 2 the optimised hulls show a large decrease in significant wash wave height between 20-50%. The constraint of total resistance increase was set to 10% in the optimisation, but all of the optimised hulls have smaller total resistance than the original (frictional resistance did not increase sufficiently compared to the decrease in wave resistance, which was approximately proportional to the decrease in H_s). What should be noted from fig. 8 or table 2 is the large diversity between hull shapes under different conditions.

This might not necessarily be the 'true' picture as different initial step lengths (of the parameters) were

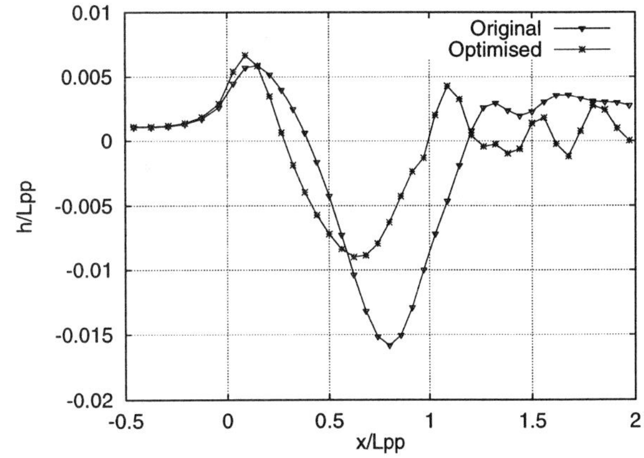

Fig. 9: Wave profiles for optimised and original hull at critical speed

used in order to be able to get a non-trivial result. The lines above the waterline in fig. 8 might seem a little strange, especially for critical and service speed at deep water. However, they have no bearing whatsoever on the resulting wave profiles as no trim and sinkage or non-linear calculations have been used.

The optimised hulls have shown a very large decrease in the significant wash wave height, and have at all speeds transformed the hull to some kind of a semi-swath type hull and/or a hull with a more V-shaped foreship. This type of hull is exactly what should be expected for optimising with the goal of decreasing wash waves. However, the very large decrease for the catamaran is to some extent connected to the change of the transom (the optimisation decreased the breadth of the transom in the waterline). But the transom stern resistance only accounted for at most 30% of the total decrease in wave resistance. Therefore this part of the optimised hull should not be dominant.

The method presented can easily be modified to optimise other object functions, as for instance total resistance, and optimising other hulls can be accomplished with a fair amount of work. As can be seen from the different body plan plots above, the expresiveness of the HSMR routine is very high, but allowing only small changes to the hull shape or local changes can very easily be accomplished. Even higher expresiveness can be accomplished by adding more parameter points, but this will increase computation time, which for the example shown was 48-72 hours on a HP-715 workstation.

4.3. Evaluation

Ability to accurately predict a hull's performance in terms of wash waves is important. For optimisation qualitative accuracy is important, and for the effect on the environment quantitative accuracy is necessary. The wave pattern can be derived quite easily and accurately by use of Kelvin and Havelock theory, but the amplitude of the wash waves is more difficult. Two options are available.

CFD potential flow solvers are the dominant solution, but in the near future it is expected that RANS (Reynolds Averaged Navier Stokes) solvers with VOF (Volume Of Fluid) free surface calculation will emerge (and to some extent already have). The potential flow solution is generally quite good and fast, but close to critical speed the amplitude is often overestimated, and it cannot take breaking waves into account as the previously mentioned VOF method is capable of. The main reason for this problem, especially at critical speed is due to reflection at the far edge of the free surface. However, ongoing development at SSPA is partly aimed at diminishing this problem.

The figures below show examples of comparisons between SHIPFLOW calculations and model trials. In addition to shallow water the comparisons include vertical and inclined banks on the port side of ship to simulate restricted water effects. For the wave profiles presented, the non-dimensional bottom clearance is presented as $H_T = (h - T)/T$, and the non-dimensional distance to bank as $Y_B = (y - B/2)/B$. The solid line represents the SHIPFLOW calculation and the dotted line is the model experiments. Figures 10 to 12 show examples of the comparison for a typical HSC catamaran at sub-, super and critical speed along a longitudinal wave profile. As can be seen from the wave profile comparison the calculation matches quite well with the model experiments except for critical speed. At critical speed, where only transverse waves exists, the solution is often not satisfactory. However, it should be remembered that the presented comparisons are with a bank at shallow water, which will complicate the problem.

5. ROUTE PLANNING

Once the ship is built it is very difficult or at least very expensive to make any changes to the hull. Still, careful route planning in or close to restricted waters can reduce the wash wave problem significantly. If

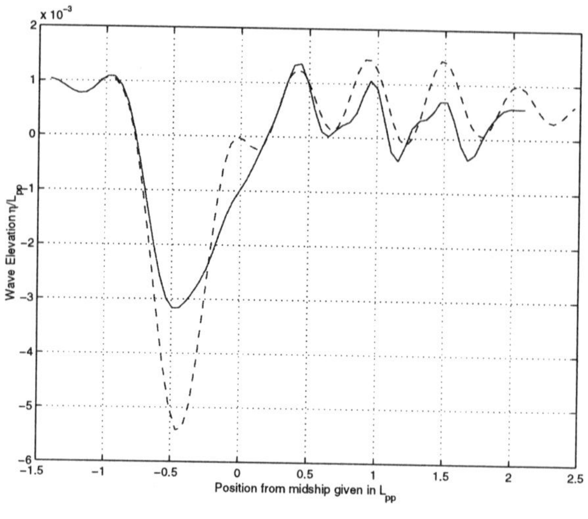

Fig. 10: $Fn_h = 0.74$, $H_T = 0.5$, $Y_B = 0.7$.

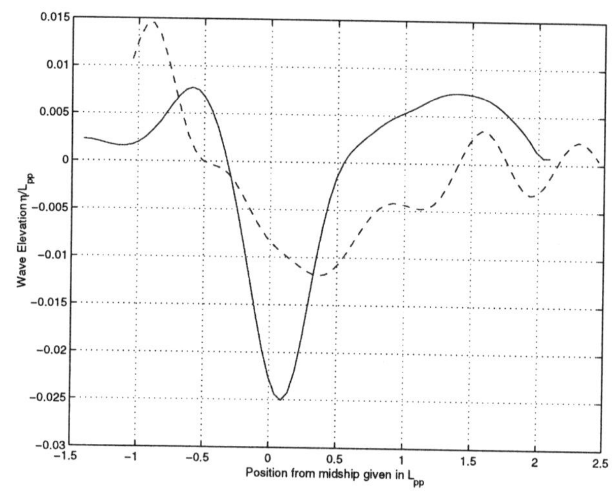

Fig. 11: $Fn_h = 1.03$, $H_T = 0.1$, $Y_B = 0.7$.

done properly, the result might not even prolong the sailing time.

For a ship which is expected to travel in waters which are sensitive to wash waves, a method has been developed for providing the ship officer with a ready-reckoner for how to operate the ship in order to minimise these problems. This method not only provides information/suggestions on how to reduce speed in time but also suggests different directions for the ship before it enters the harbour. It also provides information about expected wave heights in sensitive areas around the approach fairway.

The procedure for developing such a route is as follows:

- Performing model tests, or alternatively, CFD calculations.

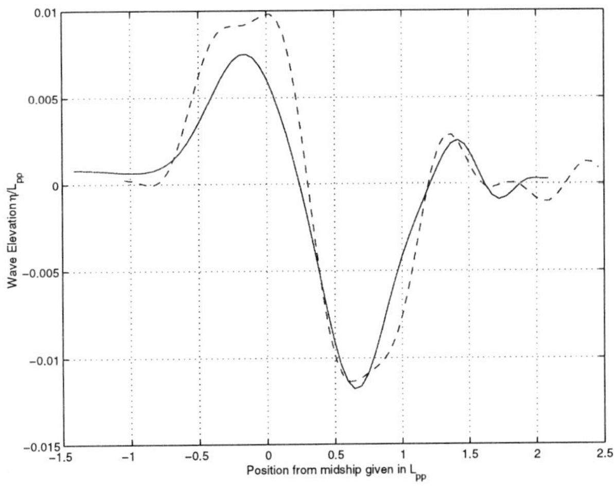

Fig. 12: $Fn_h = 1.47$, $H_T = 0.5$, $Y_B = 0.7$.

- On the basis of the tests/calculations provide tables showing wave heights for different lateral distances to the ship lane and for different speeds.
- Calculate wave height in specific areas.
- On a simplified sea chart plot recommended routes and max speeds for different areas.

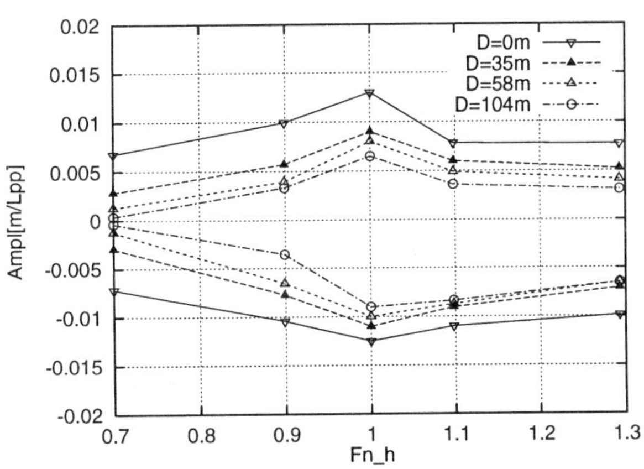

Fig. 13: Modern Semi-swath catamaran

5.1. Estimation of wave heights at different lateral distances

Both for model tests and CFD calculations, a limited area around the ship can be covered. In the former case, the waves may be measured up to a lateral distance of, say, 400 m for a normal size model and test basin. In the case of CFD calculations, the max lateral distance practically covered is normally less than that.

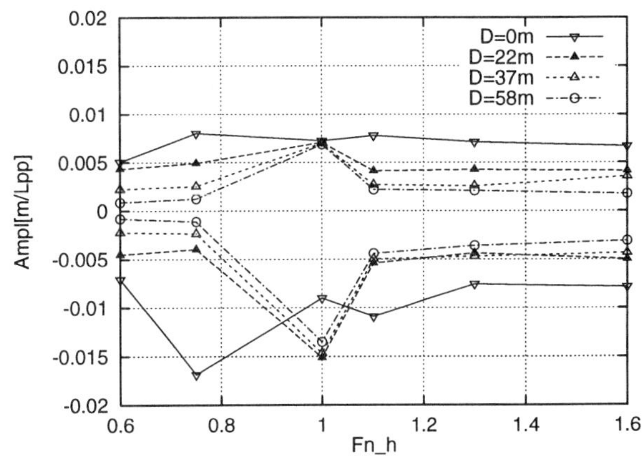

Fig. 14: Fast traditional catamaran

Model tests and to some extent also CFD calculations are time consuming and thus expensive to carry out in each individual case. A systematic study, covering CFD calculations for a number of different types of ships, has therefore been carried out, see [?]. Some examples are provided in fig. ?? and ??, showing the dimensionless max and min wave height on the basis of the Depth Froude Number. Thus, for a specific type of ship it is possible to select a suitable ship from the systematic series and to recalculate the waves using the principal dimensions for the actual ship.

When estimating the wave height at greater distances from the ship and close to a shoreline some semi-empirical data has to be used, as the calculation of the wash waves under such circumstances will be both very expensive and not very accurate. Fig. ?? shows an estimate of how the cusp height varies with the lateral distance from the ship track. This diagram is derived from model tests carried out in connection with different projects carried out at SSPA. No shallow water effects are taken into account in this figure.

When the waves approach shallow water their wave length decreases and their amplitude decreases slightly after which it will increase. To estimate this effect fig. ?? can be used (extracted from [?]), by using the wave length at the origin λ_0. In connection with SHIPFLOW or any other code capable of calculating the near field wash waves, it is thus possible to obtain a rather good estimate of the wash wave amplitude close to shore as a function of decreasing water depth. However it must be emphasized that none of the above methods considers wave breaking. As a rule of thumb the waves will start break-

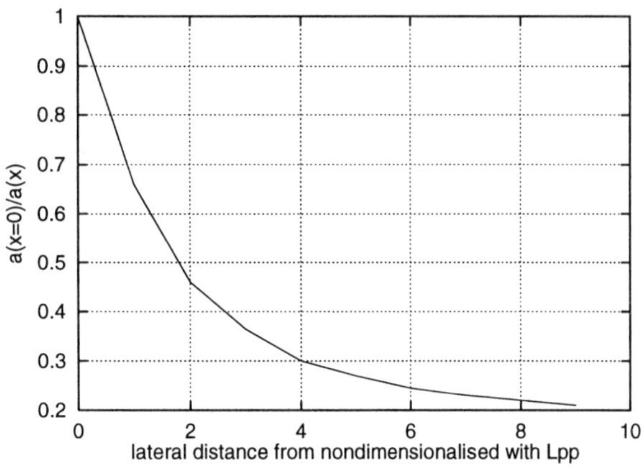

Fig. 15: Decrease of wash wave amplitude as function of lateral distance

ing when $a > \frac{1}{7}\lambda$.

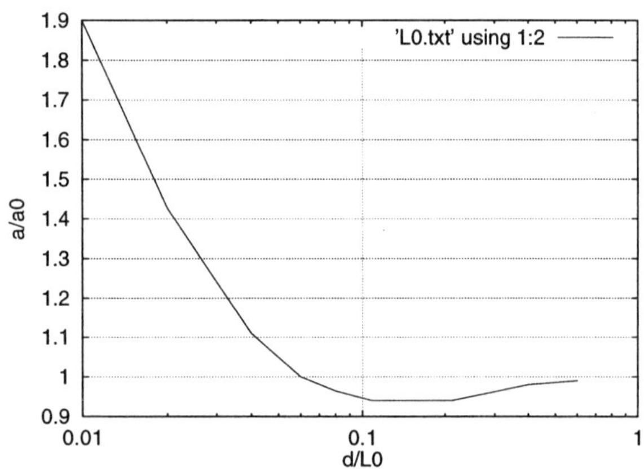

Fig. 16: Wash wave amplitude change a/a_0 as a function of water depth change d/λ_0

If the ship and speed specific wash wave near field is at hand it is then possible to use one or both of the above figures to estimate the wash waves closer to shore or at greater distance from the ship track.

5.2. Recommended headings and speeds

When proposing speed and track the following items are considered:

- Always avoid speeds corresponding to Depth Froude Numbers between 0.7 and 1.3
- When reducing speed or accelerating close to a port or fairway, go through the critical speed as fast as possible
- When passing the critical speed avoid a heading that will "send" the transverse waves into a vulnerable geographical area.

Using the above simple rules guidelines can be given for what speed and heading to use in specific areas along the fairway. However, it must be noted that using generalized curves for wash amplitude as above is very approximative. For better agreement with reality, either model trials or CFD calculations are necessary, but for initial estimates such curves can be used.

Example When approaching Gothenburg from the southern route, see fig. ??, some sensitive areas on the south side of the fairway must be passed. The part of the fairway that might come in question for a route plan starts at island A in fig. ??. The fairway turns slightly to stb'd at island D after which it follows a straight heading ending with a turn to port at islands H and G. The most sensitive area here, with respect to erosion problems, is the area around F. The fairway is also, at least from Island D and inwards, used by a lot of pleasure boats, meaning safety problems when exposed to high wash waves. The fairway has a water depth profile as shown in fig.??, starting at Island A. This bottom profile provides critical speed along the fairway as shown in fig. ??.

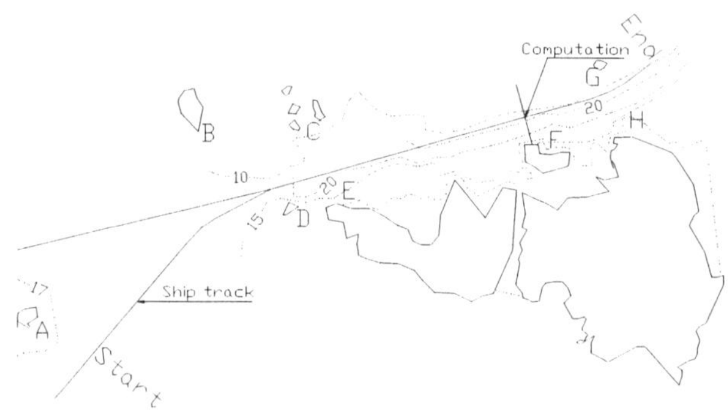

Fig. 17: Approach fairway to Gothenburg harbour.

Also shown in fig. ?? is the lower limit for subcritical speed (or upper limit for deep water), here taken to be 70% of the critical speed. This is a simple but quite suitable Froude number limitation.

The critical speed can preferably be passed, when reducing speed, just before going into the stb'd turn at Island D, thus sending the high transverse waves against the archipelago at C, as this area is not sensitive for high waves. After the speed has been reduced, the test ship is recommended to keep below, say, 20 knots. A detailed speed recommendation might be to follow the lower curve in fig. ??.

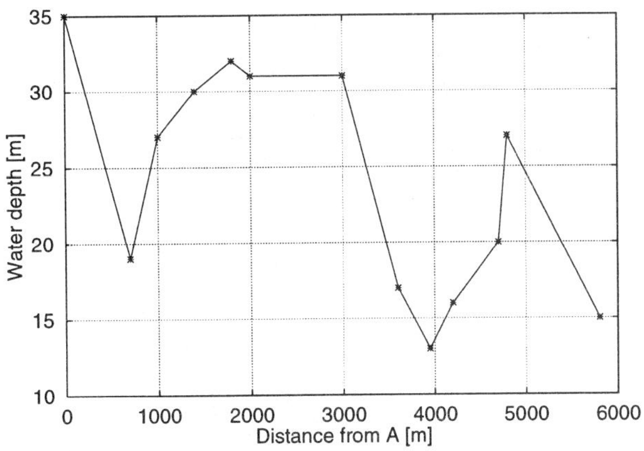

Fig. 18: Variation of depth along approach

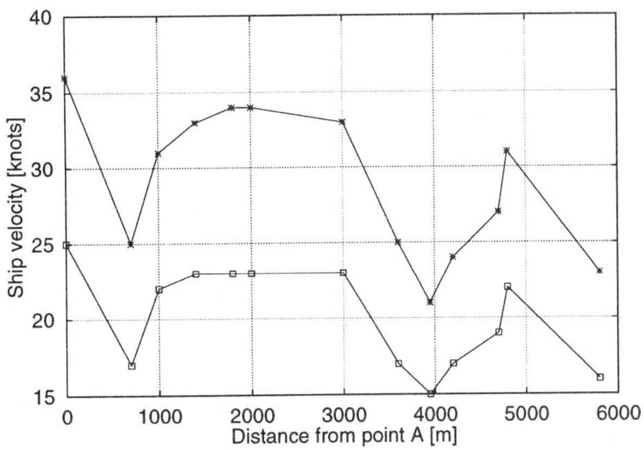

Fig. 19: Critical and lower limit deep water velocity for fairway

The maximum allowable wave height in Swedish waters has been defined previously in this paper. A SHIPFLOW calculation of the fairway at point F (see fig. ??) has been performed at 20 and 40knots. The bottom topography has been modelled by using panels on the bottom with no variation in topography parallel to the fairway (see figs. ??-??). Fig. ?? shows the bottom topography, figs. ?? and ?? show contour plots in the wave elevation with the shore to starboard, and finally fig. ?? longitudinal wave profiles along close to the shore (along line shown in contour plots).

Assuming a period of 6sec, which has been estimated for the 20knots waves in fig. ??, the maximum wave height is $0.5\sqrt{\frac{4.5}{6}} = 0.43$m. The same diagram provides a wave height of 0.21m ($0.002L_{pp}$) at a distance of 25 m away from the shore line. The 3m curve is estimated to be 5m away from the shore line, thus 20m from the calculated wave profile. Fig. ?? provides a decrease of the wave height of about 10% ,

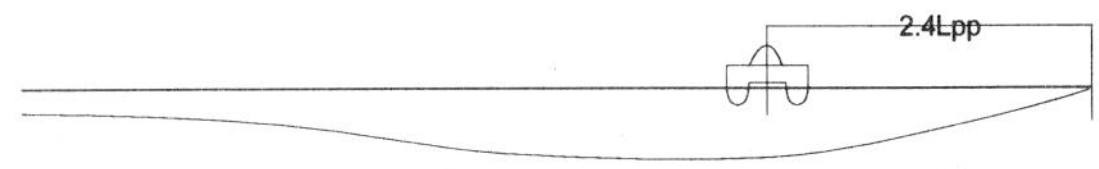

Fig. 20: Bottom topography at point F

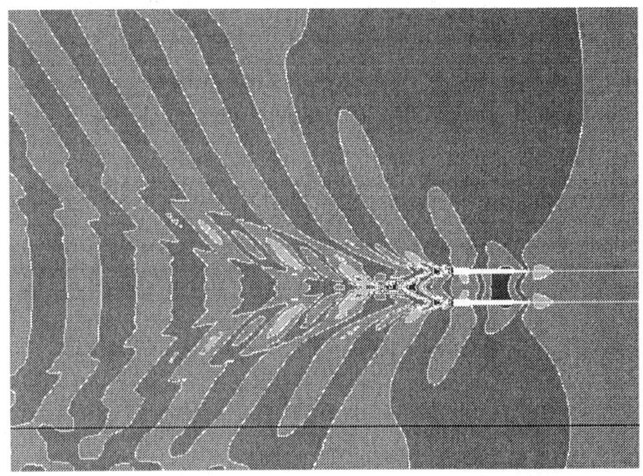

Fig. 21: Wash wave elevation at point F at 20knots

i.e. from 0.21 to 0.19m. The water depth at the calculated wave profile is 15m with a wave length of approximately $0.6L_{pp}$, which indicates an increase in wave amplitude of approximately 15%, i.e. an estimated wave height of $1.15 \cdot 0.19m = 0.22m$, which is well below the maximum allowable from the Swedish HSC rules.

However, performing the same calculation for 40knots, the estimated wave height at a 3m water depth will be 1.5m (ignoring possible wave breaking) and the maximum allowable wave height is 0.39m. Therefore 40knots does not comply with the Swedish HSC rules.

6. CONCLUSION

With the increasing legislation and restriction of HSC operating near coastal and in restricted waters, ship generated waves will become a more important factor both in operating and designing HSC. The legislation in Scandinavia is mainly based on the wave norm. The wave norm leaves little room for increase or decrease in allowable wash amplitudes in areas that are more or less ecologically or leisurely sensitive. Using methods such as an EIA in connection with route planning, would render more effective use of HSC possible.

The most obvious method to reduce wash wave impact is optimisation and choice of the most suitable

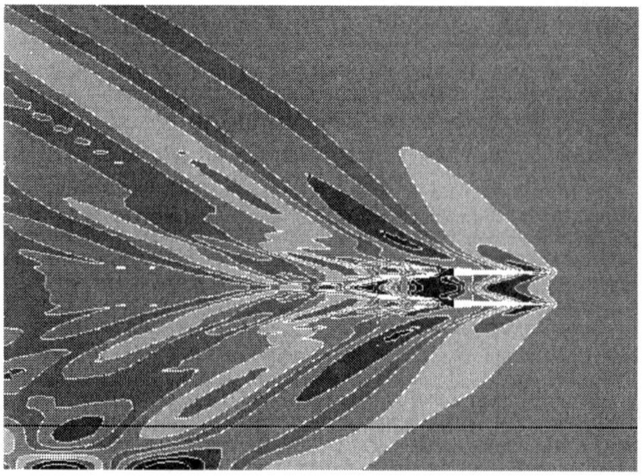

Fig. 22: Wash wave elevation at point F at 20knots

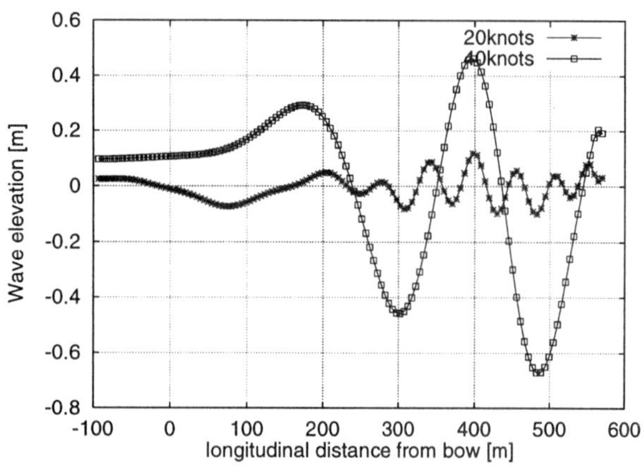

Fig. 23: Wave profiles at cuts shown in fig. ?? and ??

concept. The wash impact can be quite different for different concepts as can be seen in section 4.1. If reducing deplacement is not an option in a low wash design, optimisation procedures such as the one presented in section 4.2 could be used with good result, at least to guide the designer in the right direction. Quite large decreases in wash wave amplitude can be achieved, but of course at a price, which in this case is a hull with more curvature and thus increased building cost.

CFD calculations (SHIPFLOW) of wash waves is generally quite accurate, although critical speed at shallow water does present problems, which in SHIPFLOW 's case should be solved shortly. Further, using CFD (potential flow in this case), with the aid of the suggested emperical methods, can provide important guidelines to route planning

---

[1]Sjöfartsverkets föreskrifter om anmälan av trafik med höghastighetsfartyg" ('SFV regulations for announcing operations with high speed ships'. Technical report, 12 1998.

[2]Bekendtgørelse om sejladssikkerhedsmæssig godkendelse af hurtigfærger" ('notification of operational safety approval for fast ferries'. Technical report, Danish Maritime Authority, 05 1998.

[3]Riktlinjer för miljökon-sekvens bedömningar i utvecklingssamarbetet" ('guidelines for environmental impact assessments in the development cooperation'. Technical report, 1998.

[4]Förordning om trafik med höghastighetsfartyg" ('notification of operation with high speed ships'. Technical Report Swedish Statutes 1998:1355, 10 1998.

[5]John P. Comstock, editor. *Principles of Naval Architecture.* 1978.

[6]A. Defant. *Physical Oceanography*, volume 2. 1961.

[7]T.A. Dinham-Peren I.W. Dand. Hydrodynamic aspects of a fast catamaran operating in shallow water. *Hydrodynamics of High Speed Craft*, 1999.

[8]M. Leer-Andersen. Non-interactive optimisation of hull shapes with regard to minimising wash wave height. *HIPER'99*, pages 146–155, 1999.

[9]Michael Leer-Andersen. Manual for estimating wash waves. Technical report, SSPA Sweden AB, 1998.

[10]Jens Sorensen and Niels West. A guide to impact assessment in coastal areas. The University of Rhode Island, 1992.

[11]Ulrika Svensson. *Miljöeffekter Av Stora Och Snabba Fartyg" ('Environmental Effects from Large and Fast Craft')*, volume Report 984798-01. 1999.

[12]Soeren Taatoe. Wave generation from high speed craft's propulsion systems. Master's thesis, Department of Naval Architecture and Ocean Engineering, DTU, 1997.

Discussion

Chris Cain, Visitor

Your paper is very interesting, and I particularly noted paragraph 5, "Route Planning." First a few general comments concerning route planning methodology:

1. The route planning you suggest should be done when a vessel is being considered for a route.
2. After a vessel has been assigned to a route then more detailed "theoretical" route planning and CFD calculations should be carried out before start of service.
3. At start of service confirmation of the CFD results should be determined by actual on site measurements. Any deviation from predicted results should be highlighted and route adjusted as necessary to arrive within acceptable parameters.

In your conclusions you identify "the increasing legislation and restriction of HSC operating near coastal and in restricted waters." I believe that there is a danger of unrealistic restrictions being applied to HSC which will possibly lead to making them uneconomic. This I believe starts with the question "What is an acceptable ship generated wave, and does this only apply to HSC?"

Since the start of our service with the *Stena Voyager*, we have seen a marked reduction in complaints and incidents following on from an increase in knowledge, and although I do not think we know everything about ship generated waves, we have learnt how to reduce the effects. We now have a situation where the complaints we receive fall into categories which are

1. Resulting from deviations from operating procedures. Can be caused by outside influences i.e. altering course for weather or other vessels.
2. False complaints. We have examples where damage was claimed to have been by the *Stena Voyager* in Belfast Harbour, when the vessel was actually in Scotland.
3. Members of the public ignoring warning signs and putting themselves in dangerous positions along the shoreline. Not only dangerous because of HSC wave effects but also wave effects from traditional vessels.

Undoubtedly, some of the reduction in number of complaints is related to an increase in public awareness. However, most of the reduction is the result of alterations to the operating procedures made since start of service.

Your paper should form the basis of a serious discussion on how to design a HSC for reduced or acceptable impact. There are some who want no development and would rather that fast craft are operating somewhere else, not near them. However, the number of people who choose to travel on fast craft clearly indicate what the public want. The balance has to be made between acceptable effects from a fast craft and the required operating parameters for commercial success.

Henrik Kofoed-Hansen, Member, and
Jens Kirkegaard, Member

Your paper is very interesting and the topic is highly relevant as an increasing number of authorities worldwide require operators of high-speed craft (HSC) to include a risk assessment of the effect of wake wash in their Permit to Operate.

Question 1 The Danish Maritime Authority (DMA) was probably one of the first maritime authorities to introduce legislation for HSC routes. Since May 1997, operators of HSC in Denmark had to document that the ship-generated waves do not exceed a prescribed wave height criterion to ensure safe navigation of small craft in shallow water and leisure activities.

According to your paper (section entitled *Future regulations*, p. 5) you state that the DMA (as well as the Swedish Maritime Authority, Sjöfartsverket) have received some criticism for the introduction of the regulations. According to your paper the criticism has focused on the fact that the regulation was not developed and agreed upon in the international community before the regulations were introduced. Here we have to add that already at the Sub-Committee meeting on Safety of Navigation 15–19 July 1996, DMA contacted a number of other countries' authorities in order to get information about potential wake wash problems and regulations. At that time the wash problems in other countries were of a different character than the problems reported in Denmark, where regulations were strongly required by a number of authorities and institutions. Therefore DMA decided to implement regulations based upon the recommendations made by an Interministerial Committee chaired by the Maritime Administration. During the process DMA (and DHI) has spread out the Danish experiences all over the world and distributed reports to maritime organizations, including IMO's Maritime Safety Committee.

Our company (which assisted DMA in the process) and DMA have only received positive manifestations from a number of authorities and institutions all over the world. Therefore, the raised criticism is new for us and we would kindly ask you to elaborate upon the criticism.

Question 2 Our second question is related to your method for prediction of wake wash in a shallow water environment. From your Section 5.1 *Estimation of wave heights at different lateral distances* we understand you estimate the near-field waves generated by a specific vessel by using the Froude scaling law for ships of the same type. For the far-field assessment you suggest to use a semi-empirical decay function taking into account the diffraction effect (Figure 15) and a linear shoaling relationship (Figure 16, extracted from your reference [6]).

Our questions are:

1. Has this approach been verified by comparison with field and/or laboratory data?
2. We understand the shoaling is calculated on the basis of laboratory data from Sverdrup and Munk (1946)—your reference [6] p. 121. Why do you not use more generally accepted theories (e.g., conoidal wave theory)?
3. Do you expect this method to give accurate results for real coastal bathymetries including e.g., submerged bars and shoals?

Based on our and others' experience it is of crucial importance to resolve the shallow water bathymetry for the prediction of the propagation and transformation of the near-field waves, see examples in Kofoed-Hanse & Mikkelsen (1997), Kofoed-Hansen et al. (1999) and Whittaker et al. (1999).

Additional references

Kofoed-Hansen, H. and Mikkelsen, A. C. "Wake wash from fast ferries in Denmark," Proc 4th International Conference on Fast Sea Transportation (FAST '97), Sydney, Australia, pp 471–478, 1997.

Kofoed-Hansen, H., Jensen, T., Kirkegaard, J., and Fuchs, J. "Prediction of wake wash from high-speed craft in coastal waters." In Proc International Conference of Hydrodynamics of High-Speed Craft, 24–25 November 1999, RINA Hq, London, UK, 10 pp, 1999.

Whittaker, T, Bell, A, Shaw, M, and Patterson, K. "An investigation of fast ferry wash in confined waters." In Proc International Conference of Hydrodynamics of High-Speed Craft, 24–25 November 1999, RINA Hq, London, UK, 13 pp. 1999.

Alexander H. Day, Member

First of all, I would like to start by congratulating the authors on a very interesting study, covering many aspects of a highly topical problem.

I was particularly interested in the work presented in Section 4 on hull design optimization for low wash. I was not surprised that using an objective function based on wave resistance leads to reduced wash waves, as our own recent work using linear (thin-ship) theory has shown similar results (Day & Doctors (2000a)). However, wave resistance is not really suitable as a criterion for characterizing wash waves in general terms since it cannot be directly measured, and is not easily understandable by a skeptical public.

The hull forms generated are really quite radical in shape, and I would be curious to know what improvement in performance the authors would obtain over conventional forms if they constrained the hulls to geometrically simple shapes. This could easily be achieved, for example, by using a simplified "wire-frame" representation of the ship form (see, for example, Day & Doctors (1997)).

The use of a sophisticated CFD package such as Shipflow in conjunction with an optimization approach is clearly highly demanding on computational resource, especially compared with simplified potential-flow methods such as thin-ship theory. This is particularly true in cases in which it is wished to estimate the wave field at substantial distances from the vessel. Calculation of wave heights may well be relevant in optimization studies; our work has shown that one way in which optimization studies can be misleading is that large waves may be moved away from the area under consideration rather than reduced in magnitude, as would be the hope of the designer. Additionally, our results suggest that optimizing the hull form to minimize the waves in the very near-field will not generally lead to the hull form generating the smallest waves in the far-field, and vice-versa (Day & Doctors 2001). I would thus be interested to know if the authors have carried out any comparisons between their approach and simpler and more rapid potential-flow methods; I would also be interested to hear if they have observed these differences in relative performance between the near and the far field.

I would like to ask the authors for a slight clarification regarding Figure 15. It is not clear to me what is meant by "cusp height." I would normally expect that a plot of maximal wave height against lateral distance would show an oscillatory component superimposed upon an exponential-type decay similar to that shown. Is the diagram showing the upper envelope of this curve? I would also like to know if the authors have tried fitting an exponential decay to curves of this form; if so, did they find the exponent to be uniform across different ship designs, and what would be the typical value(s)?

Additional references

Day, A. H. and Doctors, L. J. "Wave-wake criteria and low-wash hull-form design." Submitted to Trans. Royal Institution of Naval Architects (September 2000a).

Day, A. H. and Doctors, L. J. "Rapid Estimation of Near- and Far-field Wave Wake from Ships and Application to Hull-form Design and Optimization." Submitted to J. Ship Research (March 2001).

Day, A. H. and Doctors, L. J. "Resistance Optimization of Displacement Vessels on the Basis of Principal Parameters." J. Ship Research, Vol. 41, No. 4, pp 249–259 (December 1997).

Authors' Closure

First we would like to thank the discussers for their insightful comments and their astute queries. The comments and questions have not only been helpful to close any gaps in the paper itself, but have also provided the authors with new ideas. Furthermore, we feel that the discussers contributions make it possible to provide further information to the issues discussed in the article. We will answer and comment on the discussions in the order presented above.

Captain Cain (CC) notes that 3 different steps are necessary when making a route planning, and we completely agree with that observation. Only, the article is primarily concerned with items 1 and 2, *i.e.* calculation and estimation of wash wave impact prior to the ship starting operation. Even for the state-of-the-art CFD calculations and other methods, especially for restricted and/or shallow water effects in the far field, this is at best a difficult, and at worst, an impossible problem. The article presents methods, which may in some cases be state of the art and in others not, of estimating and improving the wash wave situation. Still, these methods are not sufficiently accurate in many cases for which reason item 3 of CC's methodology is necessary today. The goal must be to be able to eliminate this last item from the route planning for two reasons. First of all, accurate prediction can save expensive full scale measurements, and maybe even more important, will eliminate the possibility that the ship operates for a long duration with a poorly designed route plan (and possibly even a ship that is not suited to its environment).

Another very important point CC makes, which has been stressed in the article, is unrealistic and unfair regulations being imposed on HSC. When blame is to be assigned to beach erosion where HSC operate, rest assured that all attention will be directed towards the HSC craft. They do create the most visible, and probably also the most dangerous waves (concerning safety). However, two very important points need to be emphasized in this respect. Even though erosion occurs along a HSC route, this might as well be created by natural waves or currents. Secondly, large slow deplacement ships can have a much more devastating effect on erosion than HSC craft, not because of their Kelvin wave, but because of their Bernoulli wave. If such a ship moves fairly close to shore, the long periodic Bernoulli wave can create waves with much more energy than is the case for the Kelvin wave pattern for a HSC. As the HSC has relatively little deplacement, this is not a concern for

HSCs. Furthermore, the Bernoulli wave for large deplacement ships has been a wave length that can excite seiches creating eroding flow long after the ship has passed. The article has directed emphasis towards HSC, for no other reason than this is the type of craft currently receiving the pressure from legislation and public. The article is therefore not in any way directed toward isolating the wash wave problem to HSC.

DHI (Danish Hydraulic Institute) has been very much involved in creating the HSC-code currently adopted by DMA and the corresponding agency in Sweden. Henrik Kofoed-Hansen and Jens Kirkegaard from DHI, therefore, had some questions concerning criticism of these rules.

HSC were developed to enhance ship speed, to reduce time for the voyages, to increase cargo capability on a route etc. It would be very surprising if there were no comments on the disadvantages of a regulation that directly or indirectly forces an operator to act in the opposite direction of the intent with the concept. In the case of HSC in Sweden, and we would believe also in Denmark, the public agony towards HSC traffic was of the magnitude that the operators preferred to keep a low profile with their objections. Nevertheless it is our understanding that ship operators would have appreciated a more open round-the-table discussion on reasonable regulations on how to control the disadvantages with HSC traffic prior to the introduction. It is also our understanding that an internationally wider discussion on possible alternatives would have been appreciated.

The efforts performed by DMA and DMI in developing and promoting the Danish legislation for HSC routes should be recognised and appreciated. The positive effect on beach erosion and pleasure craft safety is inevitable. We also believe that most operators realise that the situation in Denmark and in Sweden was such that it called for fairly rapid actions, hence the criticism has not been directed towards the rules themselves, but rather how they have been implemented.

The next comment from DHI concerns wash wave estimation. The method described, with examples in figure 14-15, would not normally be a method SSPA would use for calculating the near field wave pattern and amplitudes. We would instead use SHIPFLOW directly. However, not everyone interested in estimating wash wave amplitude has the possibility

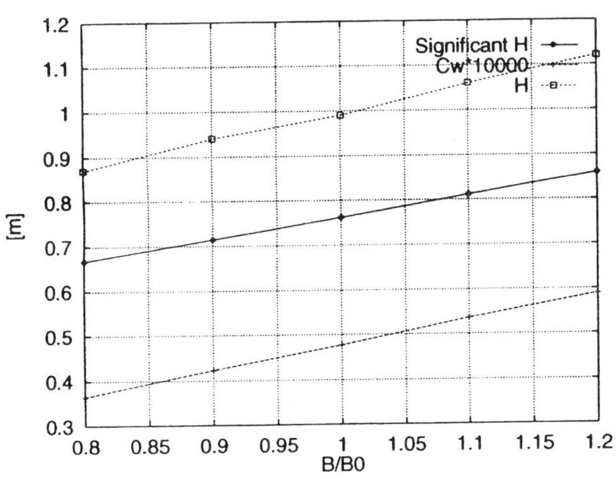

Fig. 24 Different object functions related to wash wave height as a function of demi-hull breadth

to perform SHIPFLOW or other similar calculations, and therefore this method has been suggested. Generally, the answer to questions 1-3 from DHI, will be that the method is intended to be very simple and easy to use, and will as such of course lack accuracy that could be obtained by other methods.

The answer to Question 1 from DHI will have to be both a yes and no. We have not checked how accurate the method is with the use of full scale or model scale data. However, SHIPFLOW calculations of the near-field waves have been extensively validated with good results, and the far field extensions are based solely on model scale experiments. The reason for using Sverdrup and Munk's shoaling effect curve (question 2-3), again relates to the intention of presenting an easy method for estimating wash wave heights originating from a ship, and as such the method can of course not take real coastal bathymetries into account. To take into account bars and shoals, other CFD tools would be necessary, as for example a link from SHIPFLOW near-field calculation with a Boussi-nesq wave model.

Dr. Alexander H. Day (AD) comments on the optimisation criteria, and states that wave resistance (although a good measure of wave energy radiated from the ship) is not a suitable wave optimisation criterion. Before we comment on that issue we would just like to state that the optimisation criterion is in fact the significant wave height.

The reason for using the significant wave height $H_{1/3}$ is that the significant wave height expresses a measure of wave height and is also linked to the energy of the wave system. The most damaging waves (assuming equal wave length and period) are of course the ones with the largest amplitude, and $H_{1/3}$ is the average of the highest one third of all waves.[1] Beside this energy per square meter along a wave profile using $H_{1/3}$

$$E = \rho g \left(\frac{H_{1/3}}{4} \right)^2$$

Beside the above reasons to use $H_{1/3}$, the choice of object function can be chosen quite arbitrarily. In any event, the goal of the optimisation is to derive a hull shape that produces low wash. How low wash is defined mathematically is not important in this respect, as long as the chosen object function does express a function that has minimum at a design with low wash. To illustrate this 5 different hulls have been created, simply by multiplying the hulls y-coordinate with a factor Y/Y_0. Computing the wave field, C_w, H and $H_{1/3}$ can be plotted for the different hulls, as shown in Figure 24. For the above simplified example, the optimised hull would have been the same regardless of the choice of any of the object functions shown in the figure.

A question concerning difference between using a profile in the near field or the far field to calculate the object function was raised by AD. Reference was made to a paper by Day and Doctors stating that large waves can arise farther from the hull than would be then seen very close to the hull. We have unfortunately not had the chance to read this article, as it is not yet published. However, we assume that the phenomenon referred to here is due to interference. The different wave sys-

1. $H_{1/3}$ is usually used for irregular, but can be adopted in this context also.

tems (from bow, shoulders, stern, hull-hull interaction) will interfere and it is possible that this interference can cause higher amplitude further from the hull, and thus create a far-field wave with larger amplitude. However, the optimisation in question will not modify the position of the bow or stern wave system, as all principal dimensions (except breadth) are kept constant. Furthermore, hull-hull wave system interference should also remain fairly constant as the hull C.L. to vessel C.L. is kept constant. For optimisation where principal dimensions (especially length and hull spacing) are changed, we believe that using a wave cut in the far-field could be of importance. However, for this case, we believe that the resulting hulls will remain much the same for the above reasons.

Concerning the fact that less computational methods—such as thin-ship models for the optimisation—have not been performed, this is partly because we feel it is not necessary for the above reasons, and because we at SSPA do not have these models at hand.

The final remark from AD concerns figure (15). It is true that "cusp height" in this respect is a poorly chosen expression. Even though the cusp height usually entails the highest wave, what the figure refers to is of course maximum wave height. The "missing" oscillatory behaviour has simply been removed from figure (15), by using the points between the peaks of the oscillations, as these oscillations would be useless for a figure like figure (15) if used in a general manner. The data shown in the figure have been extracted from several model trials with different models and model speeds.

In closing, we would again like to extend our thanks to the colleagues who have kindly contributed with their insight, and also extend our thanks to the presiding officer for leading us through the presentation.

SNAME *Transactions*, Vol. 108, 2000, pp. 21–43

Predicting the Performance of a Tug and Tanker During Escort Operations Using Computer Simulations and Model Tests

Piotr Waclawek, Visitor, and **David Molyneux,** Member, Institute for Marine Dynamics, National Research Council, Canada

ABSTRACT

Optimizing the performance of an escort tug is an important part of the design process, but the impact of tug performance on total system performance is the critical issue. Whilst tug performance can be predicted with model experiments, this is an impractical method for evaluating the combined tug-towrope-tanker system. Numerical simulation of the total system is a much more practical option, since numerical maneuvering models for ships are well established and can easily include the predicted trajectory of the tanker under the action of a force generated by the tug. The challenging part of the simulation is providing a realistic estimate of the tug performance. The Institute for Marine Dynamics has developed a hybrid modeling process that combines captive model experiments for obtaining hydrodynamic coefficients of the tug with numerical simulation of the complete system. This paper describes the development of the method and presents some results to demonstrate trends in system performance with design variables, such as location of the towing point. It also discusses some operating practices and the associated safety considerations.

NOMENCLATURE

A_l	Underwater lateral area m^2
C_l	Surge force coefficient
C_q	Sway force coefficient
C_{lr}	Centre of lateral resistance
C_{do}	Yoshimura's cross-flow drag coefficient
C_{rN}	Yoshimura's compensation coefficient for yaw rate, yaw moment
C_{rY}	Yoshimura's compensation coefficient for yaw rate, sway force
d	Ship draft, m
dl	Towline extension, m
F	Towline force, N
g	Acceleration due to gravity, m/s^2
GM	Metacentric height, roll, m
I	Moment of inertia, $kg\text{-}m^2$
J	Added mass moment of inertia, $kg\text{-}m^2$
K	Roll moment, N-m
L	Length, waterline, m
l	Length of towline, m
m	Mass (suffix indicates added mass direction), kg
N	Yaw moment, N-m
p	Local angular velocity, roll, deg/s
r	Local angular velocity, yaw, deg/s
R(u)	Resistance, N
U	Forward speed, tug-tanker system, m/s
u	Local velocity, surge, m/s
v	Local velocity, sway, m/s
X	Forward force (surge), N
x_g	Distance from origin in x axis, local coordinates, m
Y	Lateral force (sway), N
z_g	Distance from origin in z axis, local coordinates, m
δ	Direction of thrust vector, deg.
ϕ	Roll angle, deg.
κ	Towline stiffness, N/m
θ	Angle between tanker centreline to towline, deg.
ρ	Water density, kg/m^3
$\omega = \delta - \psi$	Inflow angle, deg.
ψ	Yaw angle, deg.

Suffixes

E	External (wind, current, etc)
G	Global coordinates
H	Hull
P	Propulsion system
R	Rudder
xx	roll
zz	yaw
x	surge
y	sway

INTRODUCTION

The use of escort tugs to control a tanker in the event of its propulsion or steering system failing has grown in the last few years. The reason for this has been the ability to use tugs at relatively high tanker speeds (up to 12 knots). This has been the result of the development of specialized tugs, which generate large hydrodynamic forces from the combination of a hull with a large skeg and an azimuthing propulsion system. The tug can generate towline forces up to two and one half times the bollard pull by using a combination of towrope force, hull drag and propulsion system thrust. The advantage of this system is that when the tanker has a long approach to the terminal, through a sensitive environment, the tug has the capability to control the tanker at speeds that are cost-effective for the shipping operation. These systems have been in use on the approach to the Valdez oil terminal (Alaska) for many years, and have recently been introduced as part of the management system for the Hibernia oil development on the East Coast of Canada.

A classification system for this type of tug has been developed based on the maximum steering force. This is the force that can be generated by the tug, normal to the track of the tanker. Det Norske Veritas has a procedure for classifying the performance of escort tugs in the form [steering force (tonnes), speed (knots)]. For example, a tug with the rating [150, 10] can generate 150 tonnes of steering force at 10 knots. This is combined with a limiting heel angle, to ensure the safety of the tug. However, the determination of the required magnitude of the force is more subjective. This can be based on operator experience, or it can be based on a simulation of the performance of the combined tug, towrope and tanker, including the limits of navigation and environmental forces, such as wind, waves and current.

The size of the combined system of tug, tanker and towrope makes physical modeling of the complete system difficult since a large amount of room is needed for completing maneuvers. A hybrid method, which provides physical data for the most important parts of the system and computer simulation of the total system, is the most reasonable approach. The Institute for Marine Dynamics (IMD) has developed such a process, based on the combination of model experiments on the tug and computer simulation to predict the performance of the combined tug-tanker-tow rope system. The system functions at two main levels. The first is to allow the design team to maximize the steering and braking forces generated by the tug and its propulsion system. The second is to simulate the track of the tanker and tug in an active simulation that can be used to develop operational strategies. The features of the program allow such things as towrope tension, thruster position and roll angle to be predicted and recorded as functions of time. This paper presents some simulations of tanker performance under escort and some results that can be used to discuss the optimization of escort tug design.

DESCRIPTION OF ESCORT TUG OPERATIONS

The operation of large tankers in crowded and confined waterways, with the associated risk of pollution, has increased the demand for escort tugs. The role of the tug is to bring the tanker safely to a halt in the event of a power or steering gear failure. Escort tugs operate in different modes, depending on the circumstances and the design of the tug. In direct mode, the forces acting on the tanker are generated principally from the propulsion system of the tug, and as a result this mode is limited to low speed operation. For indirect mode, the forces generated by the tug are a result of the combined action of the hull at a large yaw angle and the vectored thrust of the propulsion system.

The resulting combination of steering and braking force is dependent on the relative position of the tug to the tanker. To brake the tanker without steering it, the tug is stationed behind the tanker, on the centreline. To brake the tanker and steer it, the tug is positioned off the stern quarter, and to steer the tanker without braking it the tug is positioned off the beam.

The maximum force that can be generated depends on the type of tug being used. For example, a conventional tug, with screw propellers and rudders, will generate maximum thruster force in a direction close to the centreline of the tug. This limitation restricts its application for escort work since it can only generate towline forces in any direction at low speeds. At higher speeds it can only operate close to the centreline of the tanker and as a result braking forces can be generated, but not steering forces.

A tug with fully azimuthing thrusters can apply maximum thrust in almost any angle relative to the tug's centreline, and this results in a lot more flexibility. The propulsion system can be used to steer the tug into a position where significant sway forces can be generated by the tug's hull. The hydrodynamic drag of the hull in this orientation is much greater than that of the conventional tug.

There are competing concepts of propulsion system for escort tugs. One option is twin Voith-Schneider Propellers (VSP) produced by Voith-Hydro. The second option is twin steerable Z-drive units, with or without nozzles. The Z-drive units have conventional propellers, but by steering the propeller, the total thrust can be used to maneuver the tug. The location of the propulsion system also has an effect on escort performance. Traditional tugs have the propulsion system at the stern. As a result, the hull can protect the propellers and the relatively shallow draft of the tug may be an advantage. A tractor tug has the propulsion system at the bow, extending below the bottom of the hull. In this case, the hull has little effect on the hydrodynamic flow into the propulsion system, which means that the resulting force is more uniform in magnitude throughout the range of directions. The penalty, however, is a large increase in draft, and increased resistance caused by the protective structure around the propulsion system. Tractor tugs can use either VSP units or steerable Z-drives. The complication with tractor tugs for escort duty is that they generate most braking force when the propulsion units are furthest away from the tanker. This means that in conventional notation, they are running astern. Speeds up to 12 knots are required for escort tugs. This means that the tug must be directionally stable in either direction.

There are some variations in escort procedure, depending on the location and the operators. For example, in some cases the tug will pick up a towline, as the tanker approaches the escort zone, but then follow the tanker with a slack cable. In others, the tug will only pick up the cable when an emergency arises involving the tanker. The advantage of this system is that there is less wear and tear on the towline, but the disadvantage is that some time is lost while the tug makes the connection to the tanker.

An important safety parameter for the tug is the maximum angle of roll. Escort tugs are required to operate in conditions that most tugs will try to avoid. Escort tugs typically operate with the towline normal to

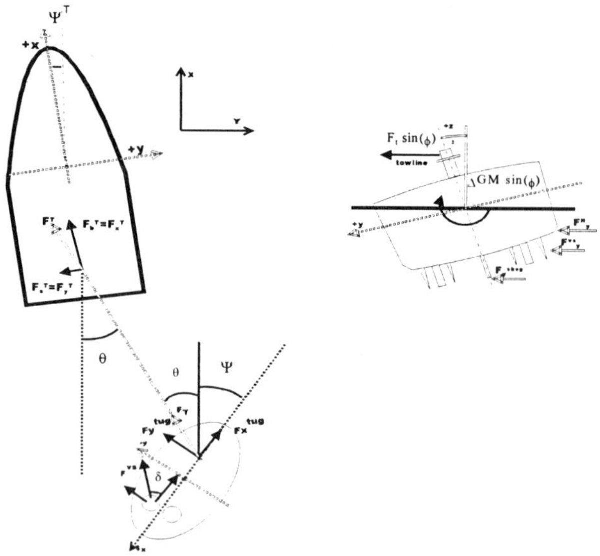

Figure 1, Ship and escort tug configuration

the centreline, when the maximum heeling moment is applied. Roll stability of this type of tug is an important issue. Current stability guidelines are based on the mean heel angle at a steady loading condition, for example when the tug is generating maximum steering force. However, the elasticity of the towrope can affect the heel angle, and maximum angle may be a more reasonable limit. This limit should also make a reasonable allowance for waves and water becoming trapped on the deck. Another aspect of the tug stability during indirect maneuvers is that the propulsion system is often acting to reduce the heel angle of the tug. The dynamic response of the tug if one propulsion system fails during extreme load operations should also be investigated as a possible limit of operation.

The methods described in this paper have been presented for a tug with twin VSP units. This was due to the nature of the project work carried out at IMD at the time the technique was developed. We are in the process of extending the technique to cover tugs with azimuthing Z-drives, and hope to present these results at a later date. The prime extension to the method presented here will be the development of performance data for Z-drives in oblique flow and opposing flow required for transverse arrest mode.

SIMULATION SCENARIO

For the simulations described in this paper, it has been assumed that at the start of the simulation the tug has picked up the towline and is proceeding skeg first, at the same speed as the tanker, but the towline is slack. This set of assumptions has been made to focus on the hydrodynamic performance of the tug and tanker. It is possible to include other options within the simulation, but these are more related to specific operational scenarios.

When an emergency situation arises, the tug is steered into a position to control the tanker. The actual position will depend on the specific situation, but Figure 1 shows a typical position, where both steering and braking forces are generated. It also defines the variables considered.

MATHEMATICAL MODEL OF TUG-TANKER-TOWROPE SYSTEM

The first step in the simulation process was to develop a mathematical model of the system. The three major components that form the dynamic system are tanker, towline and tug. All three components can be modeled as one dynamic system with four degrees of freedom (surge, sway, yaw and roll). In the analysis, it is convenient to treat each of these components in their individual reference systems, and subsequently convert the output into the global coordinates.

IMD has adopted the MMG model (Waclawek, 1996, 1997) for predicting a ship's maneuvering performance. The governing equations for this method are well known and it has the advantage that realistic predictions can be derived when only the general particulars of the ship are known. The equations are given below.

$$
\begin{aligned}
m(\dot u_G - v_G r - x_g r^2 + z_g p r) &= X_H + X_P + X_R + X_E \\
m(\dot v_G + u_G r + x_g \dot r_G - z_g \dot p) &= Y_H + Y_P + Y_R + Y_E \\
I_{zz}\dot r_G + m x_g(\dot v_G + u_G r) &= N_H + N_P + N_R + N_E \\
I_{zz}\dot p - m z_g(\dot v_G + u_G r) &= K_H + K_P + K_R + K_E
\end{aligned} \tag{1}
$$

In all equations, the subscripts are defined as follows. H refers to the hull, P to propulsion, R to rudder, and E to external disturbances. A subscript G denotes a value defined in the global coordinate system.

The MMG model has been programmed into IMD's NavSim™ maneuvering simulation program (Waclawek, 1997). The equations are structured within an open-ended computer architecture, in which the elements on the right hand side are linearly added. The hydrodynamic interaction between the hull, propellers and rudders are recognized in the related right-hand side terms. The major advantage of this model is that it can efficiently compute time domain simulations of the ship's motions with active control of the propeller thrust and rudder angle. Environmental forces, such as wind, waves and current can also be included in the simulation, as well as water depth and bank effects. NavSim has also been adapted to include coefficients derived from experiments with a ship model on a Planar Motion Mechanism (PMM). This gives an extra level of accuracy beyond the coefficients given with the MMG model itself.

Hull Forces

The hull hydrodynamic forces are expressed in three basic terms, in which the ideal fluid effects (lift, drag, and the cross-flow effects) are computed. NavSim uses Yoshimura's model [1988], in which the high speed model is combined with large drift forces in the low speed regime.

$$
\begin{aligned}
X_H &= -m_x \dot u + (m_y + X_{vr}) v r + R(u) + X_H(r=0) \\
Y_H &= -m_y \dot v + Y_r r - m_x u r + Y_H(r=0) \\
&\quad -\frac{\rho}{2} d C_{D0}\left[\int_{-L/2}^{L/2} |v + C_{rY} r x|(v + C_{rY} r x)\,dx - L|v|v\right] \\
N_H &= -J_{zz}\dot r + N_r r + N_H(r=0) \\
&\quad -\frac{\rho}{2} d C_{D0}\int_{-L/2}^{L/2} |v + C_{rN} r x|(v + C_{rN} r x)\,x\,dx \\
K_H &= K_\varphi \sin(\varphi) + K_p p + m_\varphi \dot p - z_H Y_H
\end{aligned} \tag{2}
$$

NavSim has been successfully used for maneuvering research on many hull types (including tankers). Its formulae, however, were derived for large merchant ships, and do not apply to small tugs, especially at the very large yaw angles, typically used in escort operations. The mathematical model for the tug needed to be modified to include the hydrodynamics associated with high speed, high drift angle kinematics. The values of basic hydrodynamic derivatives for the tug can be derived using a Planar Motion Mechanism. The process becomes simpler in the escort configuration because it can be treated as a quasi-static condition, with lift as the dominant component of the tug hull forces.

This quasi-static situation (yaw rate of zero) is prevailing during the indirect escort operations and for most part, u and v related lateral forces are the only components of the hull forces. Thus, from a set of simple static yaw tests the hydrodynamic coefficients of the model are found:

$$
\begin{aligned}
X_H(r=0) &= X_{vv} v^2 \frac{u}{U} \\
Y_H(r=0) &= Y_v v + Y_{v|v|} v|v| \\
N_H(r=0) &= N_{uv} u v
\end{aligned} \tag{3}
$$

One additional aspect of the study was to estimate the influence of the position of the skeg on the behavior of the tug. To this end drift-related lift and drag forces on the tug's skeg are estimated by mean of the known formulae for wing profiles.

Propulsion Forces

Cycloidal propulsion and azimuthing thrusters are considered to be the most effective types of propulsion for escort tugs. A comparative study of tugs with cycloidal propellers and conventional propellers showed the advantage of the cycloidal propeller against the conventional propulsion (Hutchison et al, 1993).

There is, however, very little published data about forces induced by the cycloidal propellers working in all escort regimes. Thus, in order to provide the necessary characteristics of the propulsion system for NavSim's numerical model, data from physical model tests had to be used. The operation of the VSP units in the ship is such that the thrust vectors from each propeller are always parallel. Given this limitation, we can assume that the resultant thrust vector is a function of the speed of the tug, the inflow angle (between the thrust direction and the direction of motion) and the pitch of the propeller blade. From escort tug

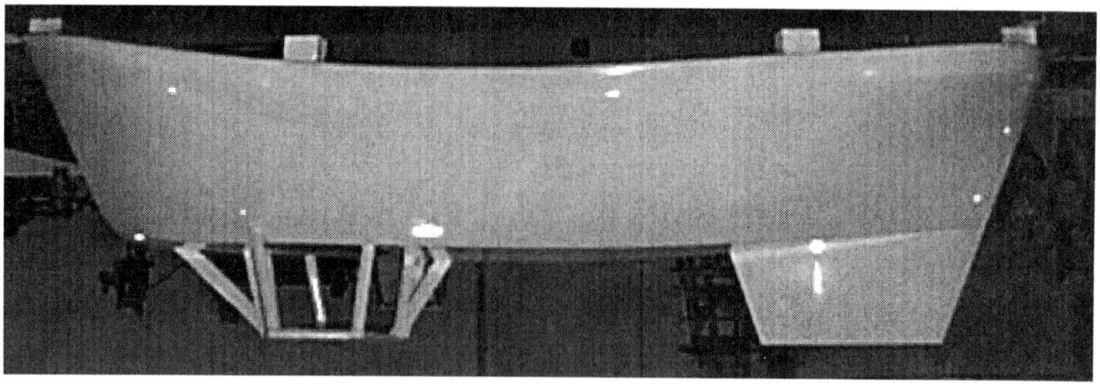

Figure 2, Profile of Model Used for Experimental Determination of Hydrodynamic Forces and Moments

operations, it has been observed that the inflow angle is approximately the same as the yaw angle, for thruster angles between zero and approximately 40 degrees. In this case, the thruster is directed into the direction of motion, regardless of the yaw angle of the tug. This removes inflow angle as a variable within the propulsion system model, and the assumption is reasonable for the range of tug operation from straight-ahead operation to beyond the condition to generate maximum steering force. The other requirement is for the propulsion system in a braking condition, i.e. thrusting astern relative to the direction of the tanker. In this case the simplest condition is a thruster angle set to thrust along the centreline of the tug.

These data will be required for the tug from zero speed (bollard) to the maximum speed of operation when assisting the tanker.

Basic resistance and propulsion data are also required input data for NavSim. These can also be determined from model experiments, if no data can be found for similar hulls. Wake characteristics were not considered in the numerical model, as it has been widely accepted that there is no hull-propeller interaction. Further research is necessary to define the limits of this assumption, since the experiment results obtained for this paper do not entirely support this observation.

Towline Forces

The dynamics of the towline were modeled through its stiffness, which was assumed to be proportional to its extension (dl). The catenary effect of the towline was ignored.

$$F_{towline} = \kappa * dl \qquad (4)$$

Simple kinematic relations in 4 degrees of freedom were derived to define the entire towline connection with respect to the tanker and tug position. The slacking of the rope often happened at the beginning of the simulation or during the transient time. Since no catenary effects are modeled, it causes an instantaneous, and sometimes dramatic on-off rope connection (as if a step-function was used), with obvious reaction of the entire system.

EXPERIMENT AND ANALYSIS METHODS

Description of Tug

The tug used for deriving the experiment data is described in another paper (Allan et al, 2000). The tug is a twin VSP unit tug, in tractor configuration. The general particulars of the tug are given in Table 1. The model was fitted with a skeg and a protective cage for the VSP units. A photograph of the model is shown in Figure 2.

Length, w. l.	38.2m
Beam, w.l.	14.2m
Draft,	3.8m

Table 1, Summary of Tug Dimensions

Data to Support Numerical Simulation

Model experiments on the tug hull were carried out to determine the derivatives required for Equation (3). The objective was to design a test program that would ensure that the Parameter Estimation Method based on neural networks (Waclawek, 1996) could be used.

No propellers were fitted for these experiments and so plugs were used to keep the hull watertight. An oversize bulwark was fitted to prevent the model from flooding at large heel angles. The test methodology was adapted from the method described by Hutchison.

(1993). Experiments were carried out on IMD's Planar Motion Mechanism (PMM). The model was fixed at a given yaw angle and measurements were made of surge and sway forces for the range of operating speeds and yaw angles. The load measurement system was connected to the tug on an axis along the centreline, at the height of the towing staple on the tug. The model was free to roll about the axis through the towing staple, and free to pitch and heave. The most important variables measured were sway force (forward and aft) and surge force, F_x. Yaw moment was calculated from the difference between the forward and aft sway force measurements, and total sway force, F_y was the sum of the two measured sway loads. Pitch angle, roll angle, heave amplitude and carriage speed were also measured. A tug-based coordinate system was used for measured forces and moments. The model was ballasted to obtain the correct value of GM and roll decay experiments were carried out to estimate the roll damping coefficients.

The surge and sway forces were non-dimensionalized, using the coefficients given below, and plotted against yaw angle.

$$C_l = \frac{F_x}{\frac{1}{2} \rho A_l v^2} \qquad C_q = \frac{F_y}{\frac{1}{2} \rho A_l v^2} \tag{5}$$

In addition, the centre of lateral resistance, C_{lr} as a fraction of the waterline length was plotted against yaw angle. Centre of lateral resistance is the point of zero yaw moment for the hull and appendages.

Data required for the numerical model of the propulsion system was also obtained from experiments on the PMM. The set-up for the model is described below. For the tug with zero yaw angle, thruster angles of zero and 180 degrees were set and data was obtained for tug speeds ranging from zero to 12 knots, over a range of power settings, up to full power for the ship. Thruster performance was estimated by subtracting the hull only data from the combined hull and propeller data. These data points were developed into two-dimensional look-up tables within NavSim, based on speed of the tug and power absorbed by the propellers. There was one table for forward propeller directions and one for astern directions.

Data to Predict Tug Performance

The numerical simulation presented in this paper was developed in parallel with a research project to determine the steering and braking force of a specific tug design (Allan et al, 2000). This project required the experimental derivation of the maximum steering and braking forces over the range of operating speeds, for a fully functional model of the tug and the propulsion system. The same partially captive system used for measuring the hull only forces was used for measuring the steering and braking forces. In addition to the parameters measured for the experiments on the hull only, the propulsion performance of the model was measured. For each speed and yaw angle a range of thruster angles and shaft rotation rates were tested, to bracket zero yaw moment about the towing staple. The two thrusters were always set to the same angle, relative to the centreline of the tug. For deriving predictions from the model experiments, thruster angles were set to cover the range required for direct and indirect modes of operation.

The torque absorbed by each propeller was calculated from the motor control voltage and the shaft rate of rotation. Prior to the experiments, the relationship between control voltage, torque and rate of rotation for each motor was derived. A test rig was built for this purpose, with a drive motor, a brake motor and a torque dynamometer connecting the two together.

To obtain predictions of tug performance from the model experiments, the measured data was plotted against thruster angle, for constant speed, yaw angle and rate of propeller rotation. The measured values of surge were corrected for viscous scale effects. Propeller torque was corrected using a method developed by Voith-Hydro. Data for constant speed and yaw angle was plotted and interpolated at the point of zero yaw moment. These results were then cross-plotted against delivered power and predictions were interpolated at the point where the limiting power was reached on either shaft. The results for each speed, yaw angle and mode of operation were then combined into a plot of steering force against braking force.

Figure 3 shows the model being tested on the PMM. Figure 4 shows a plot of the experimentally derived steering and braking force for 10 knots. It also shows the results of the simulation for the same condition.

Figure 3, Model being tested on PMM for experimental derivation of steering and braking forces

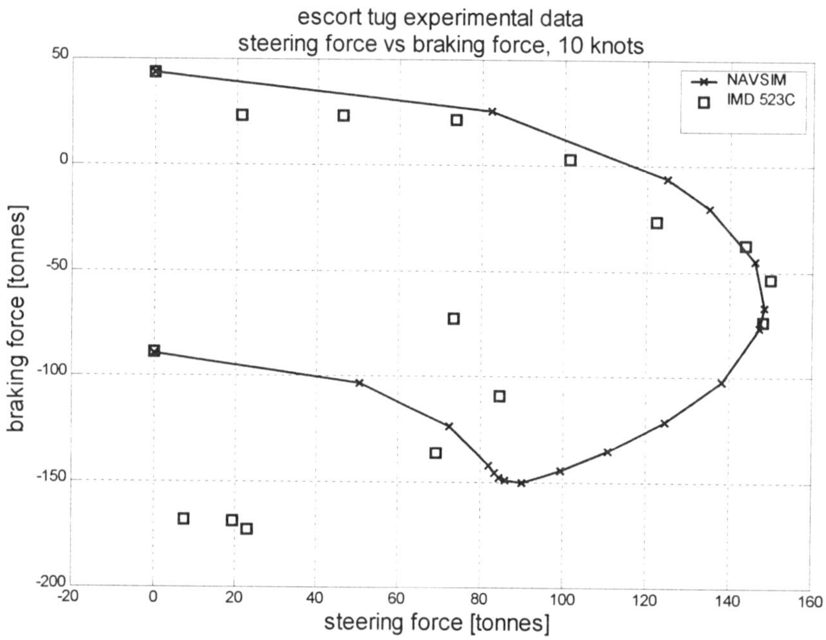

Figure 4, Comparison of simulated results with captive experiments, Braking and Steering Force, 10 knots

DISCUSSION OF RESULTS

The results of the simulation compare very well with the experimental data, when the thruster angle is close to the direction of the flow. This covers the range of performance from maximum force ahead to the maximum steering force. The ability to predict the maximum steering force is very important, because this is considered to be the most import performance parameter of the tug. The maximum steering force limits the tug's ability to control the tanker and it is the most important variable limiting the roll safety of the tug.

The other region where the simulation gives good agreement with the model results is for the tug in a direct braking mode. In this case, the tug is lined along the centreline of the tanker applying astern thrust. This is a pure stopping maneuver and does not require azimuthing thrusters to be angled off the centreline. This however, is not the maximum braking condition that can be obtained in practice. Adding a yaw angle to the tug, by the use of the thrusters, inclines the hull to the flow, and the braking force is increased considerably. This is likely to induce a small steering component, but careful setting of the thruster angles can minimize this. It can be seen from Figure 4 that the maximum braking force obtained (with a small steering force) was approximately 170 tonnes, compared to 90 tonnes in the direct mode.

The simulation results for the maximum braking condition are not as well predicted as the maximum steering force. Further research to improve this prediction is continuing in the following areas. Additional experiments were carried out for the hull for three different skegs as well as the hull with no skeg. From this research, we suspect that the simulation is over-predicting the forces generated by the hull and under-predicting the forces generated by the skeg. This affects the balance of the hull about the towline, which in turn prevents the hull from reaching a stable equilibrium at a sufficiently large yaw angle to generate the maximum braking force.

The other area of further research is defining the forces generated by the propulsion system in more detail. Additional experiments were carried out to determine the influence of inflow angle on thruster performance. The nature of the experiment procedure was such that we could not test the propulsion system without the hull, but the hull forces could be subtracted from the combined hull and propulsion forces in order to estimate the force generated by the propeller alone. From these calculations, for a range of yaw angles and thruster angles, it appears the assumption that the resultant force is independent of inflow angle is reasonable for inflow angles up to 40 degrees. For large inflow angles, from 90 degrees to 180 degrees, the resultant force increases as inflow angle decreases. Maximum force occurs at inflow angles close to 90 degrees and can be up to three times the value of the force at 180 degrees, although it varies with tug speed and power setting. The use of the value for the propulsion force at 180 degrees is underestimating the amount of force that can be generated by the propulsion system. This too is contributing to the inability of the simulation to predict the maximum braking force, with a large yaw angle. Figure 4 shows that the magnitude of the maximum braking force from the simulation is within 25 tonnes of the experimentally derived values, but the direction of the force occurs at a larger angle to the track of the tanker. Despite the limitations of the simulation to predict maximum braking force, there are some important tug design features and tanker-tug system performance characteristics that can be derived from the simulation.

SIMULATION

The NavSim Escort model is designed to carry out simulation in two modes:

- Unrestricted 'real' tanker-tug escort mode, and
- PMM mode, where the tanker is replaced by a towing point moving on a straight line.

The first mode allows us to simulate a real interaction between the two vessels in 4 degrees of freedom The second mode is a reduced version, to enable direct comparison of experiments with the results of the simulations. In this case, the towing point is following a rectilinear path with a constant speed along the centreline of the tank.

In the first mode, the user also defines the rope length, the horizontal position of the staple and its height on both ships. The initial speeds and position for each vessel are also defined. During the simulation NavSim provides real-time control of the tug's propulsion system, where the user can set the thrust magnitude by setting propeller's pitch within a range between 0 and 1. The thrust direction is set between 180 and −180 degrees. The simulation will give predictions of surge and sway forces for each ship and yaw and roll moments. Towrope tension is calculated from the resulting forces. In addition, the kinematic data (displacements, velocities and accelerations) are available for both ships. These results can be plotted for further study.

Steady State Validation

The hydrodynamic derivatives of the simulation model were estimated using the results of the captive experiments on the PMM. To validate this model, a series of simulations were performed, for which the speed of the system was constant and matched the PMM tests. In the numerical model, however, the tug was free to find its steady equilibrium position. The resulting forces, heel, and other kinematic parameters were obtained after the tug's yaw rate had stabilized at zero. The tuning method and some results showing good predictions of forces, yaw and heel angle have already been published (Waclawek, 1999). This earlier work concentrated on the maximum steering force. In the present paper, the work is expanded to include the full range of steering and braking force combinations. The results of the simulations are plotted as steering force against braking force and compared with the model experiments, in Figure 4.

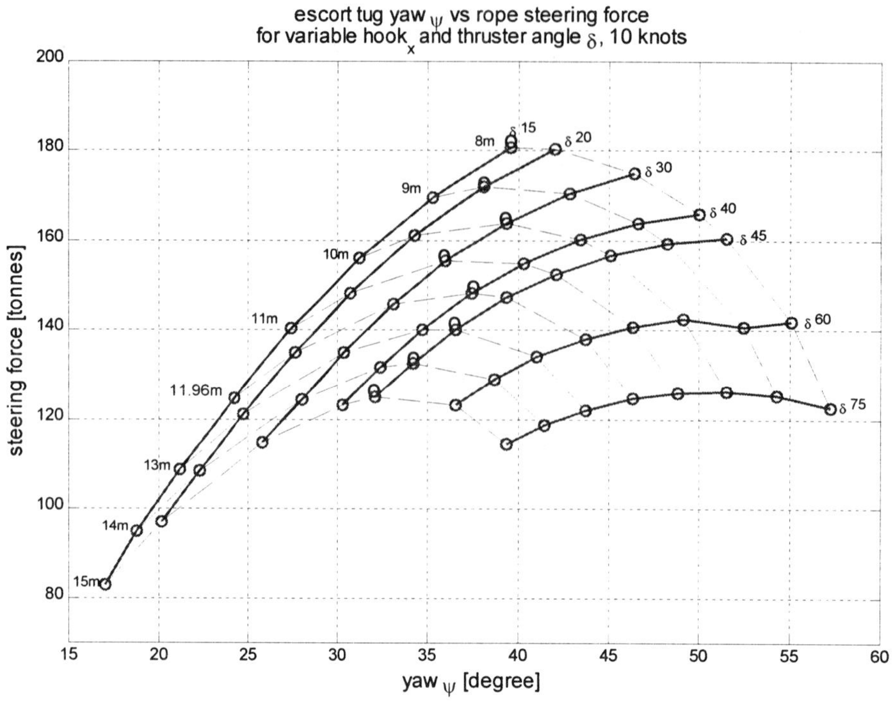

Figure 5, Maxima of steering forces for different staple-x position and thrust angle δ.maximum steering, 10 knots

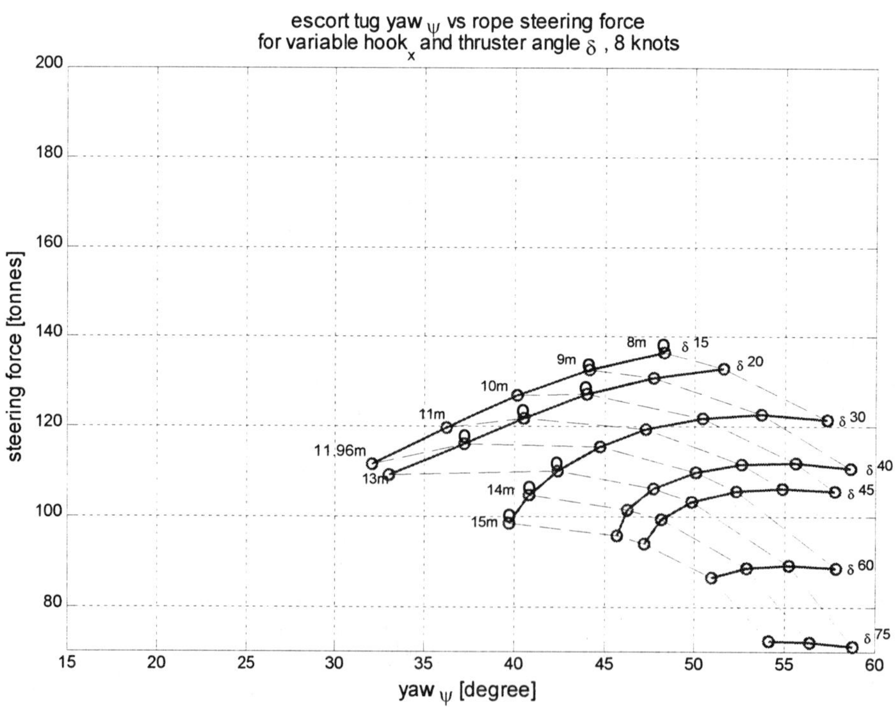

Figure 6, Maxima of steering forces for different staple-x position and thrust angle δ.maximum steering, 8 knots

Location of the Staple on the Tug

Multiple simulations were carried out to determine the maximum steering and braking force obtained for different staple positions and thrust directions. Some results are summarized in Figures 5 and 6, for two speeds (8 and 10 knots). The staple position was varied along the centreline of the tug, from 8 metres forwards of midships (towards fin) to 15 metres forwards of midships. For each staple location the simulations covered a thrust angle range, δ, from 15 to 75 degrees. Bold circles indicate the maximum of steering forces at the corresponding thrust angles for each staple position. For example, from Figure 4, when the thrust is directed at 45 degrees and the staple position is located at 12.0 metres, the maximum steering force obtained was 146 tonnes.

Figures 5 and 6 show the influence of staple location and thruster angle on the maximum steering force. Clearly the maximum value can be changed significantly by varying the location of the towing staple. The thrusters can then be used to tune the magnitude of the steering force, but their effect is much smaller than that due to towing staple position. However, these values of steering force do not take into account such factors as directional stability requirements at other operating conditions and limiting heel angles required to meet classification society regulations. The results do show that there is a wide range of steering force possible for a given tug, so designers should pay careful attention to the location of the towing staple when maximum steering force is required.

Heel angle

The steady state heel angle of the tug depends on the speed of the tug and the vertical and horizontal position of the staple. The maximum acceptable value of heel angle is usually determined by a classification society to prevent conditions that can lead to overturning the tug. Such conditions are always associated with large steering forces and these forces are significantly increased when the tug and tanker are in waves (Birmingham et al, 2000). Figure 7 shows heel angle as a function of the maximum steady steering force at 10 knots, taken from Figure 5. It shows that increasing the steering force by 7 tonnes will 'cost' the tug an additional degree of heel angle.

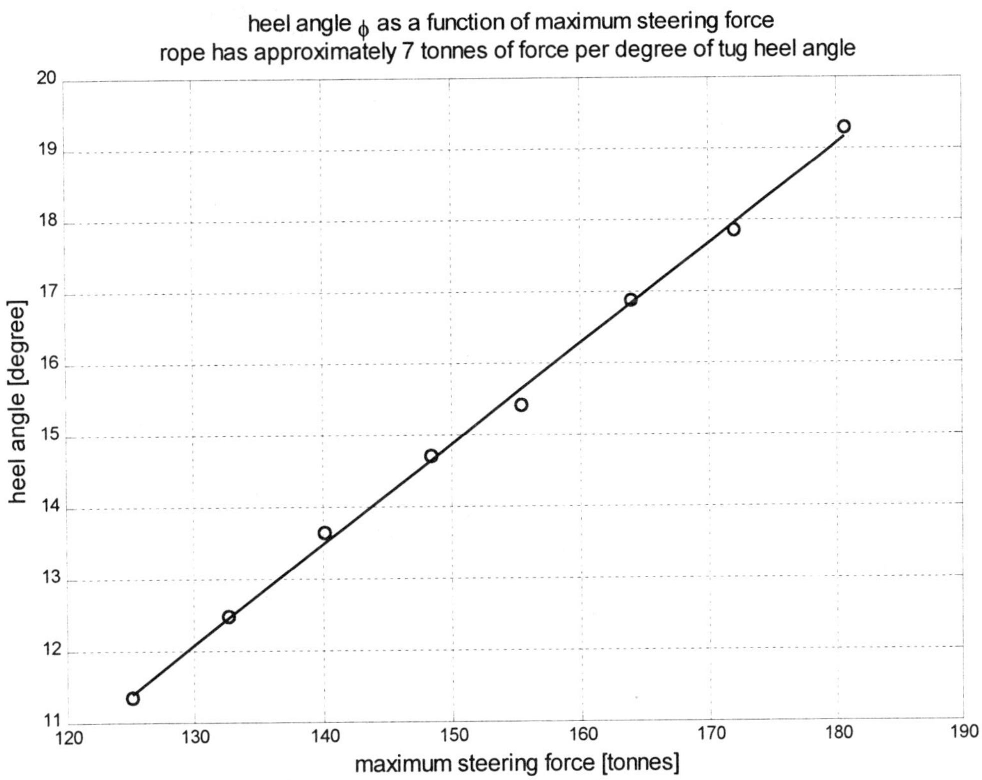

Figure 7, Effect of maximum steering force on heel angle ϕ.

A Simulation of the Complete Tanker-Tug System

Model testing is a relevant source for some of the information needed for predicting ship maneuvers. Numerical simulation also has its place in research because of its virtually unlimited and safe representation of physical settings that can be combined with a very low cost of simulation. In this research the power of numerical simulation becomes the most visible when simulating maneuvers of the complete tanker-tug system. Such a configuration would be very costly and unreliable to carry out if physical modeling was the only option. In the next section the simulation and the results of the indirect mode of escorting tanker are discussed.

The Dynamics of Stopping in Indirect Mode

The simulation model includes recordings of the constantly changing tanker-tug dynamics. In the early phase of the stopping maneuver (indirect mode, calm water) large sway forces are generated on the tug, as a result of yawing the hull about the towrope attachment point. This results in an increase in heel angle as the tension in the towline increases. For simplicity, we assumed that the propeller force remains constant in magnitude and direction. With an ideally performed maneuver the heeling moment reaches its maximum (since the system is at its highest speed) and the tug should soon find itself in the steady state. In reality, resultant sway force hovers, which causes one or more overshoots. As a result, the tug rolls considerably, and the tanker (with its rudder set to zero and no power) receives a substantial force from the towline.

As the tanker starts to slow and turn due to the respective surge and sway components of the tow force, the tug delivers less and less drift-induced steering force and eventually, at some lower speed, the direct mode is a more efficient method of stopping. The example of such a maneuver is presented in Figure 8.

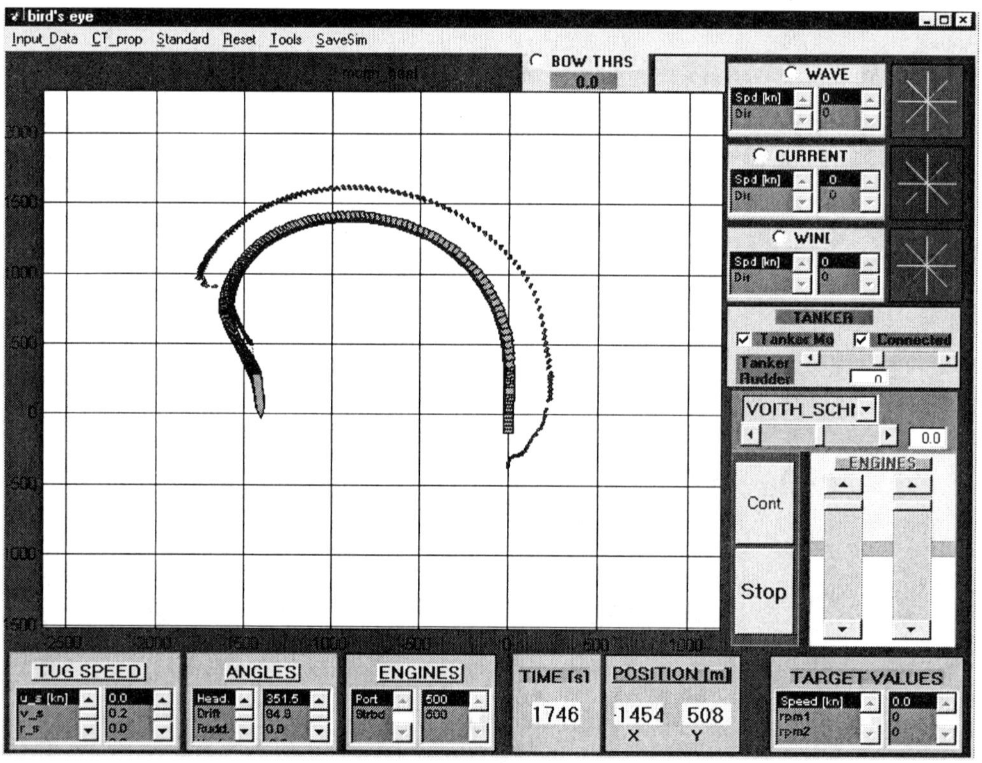

Figure 8, Simulation of tanker-tug stopping maneuver (indirect mode) with no roll angle monitoring.

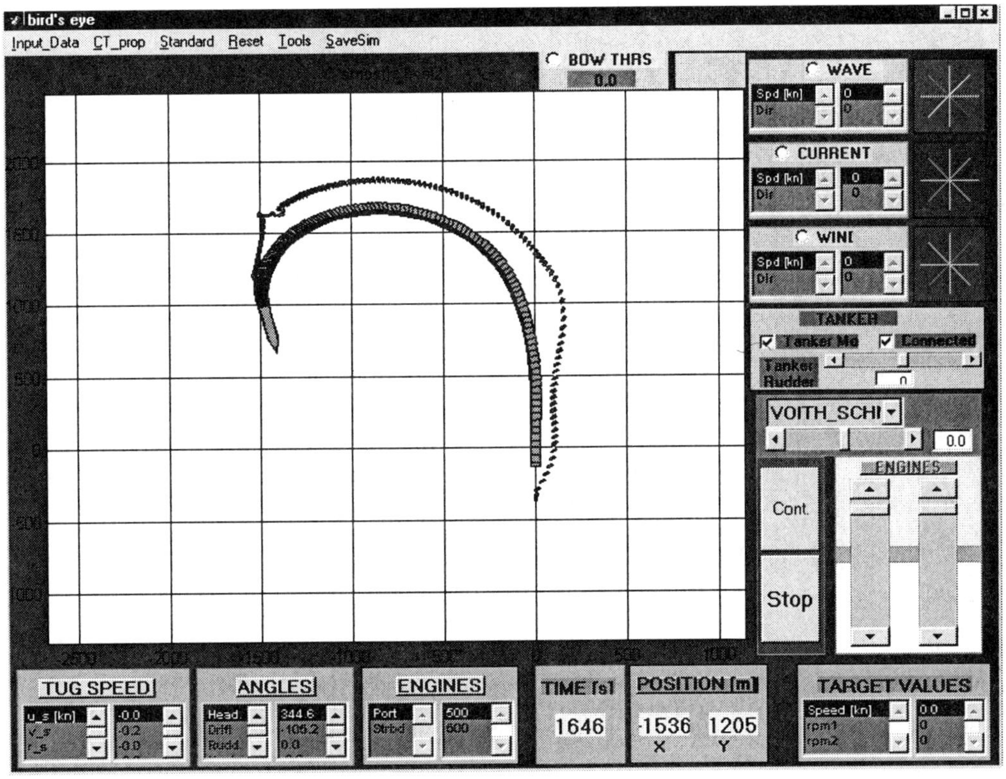

Figure 9, Improved entry into indirect stopping maneuver, with roll monitoring

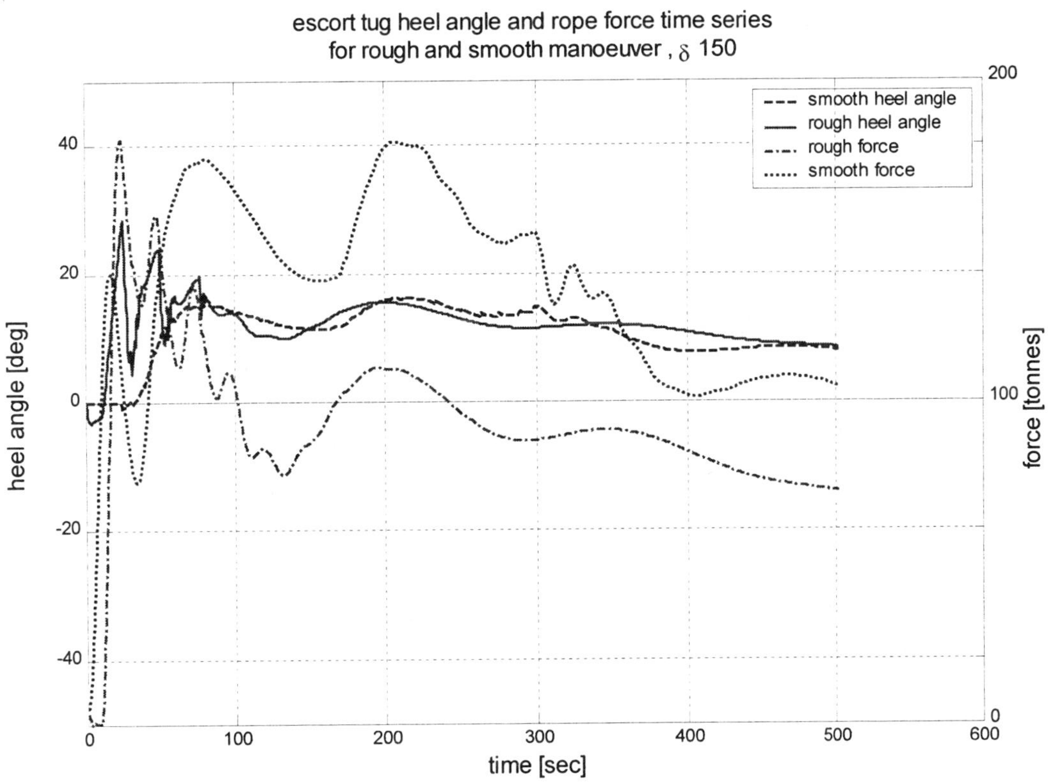

Figure 10, Heel and towline forces for 'rough' and 'smooth' maneuvers

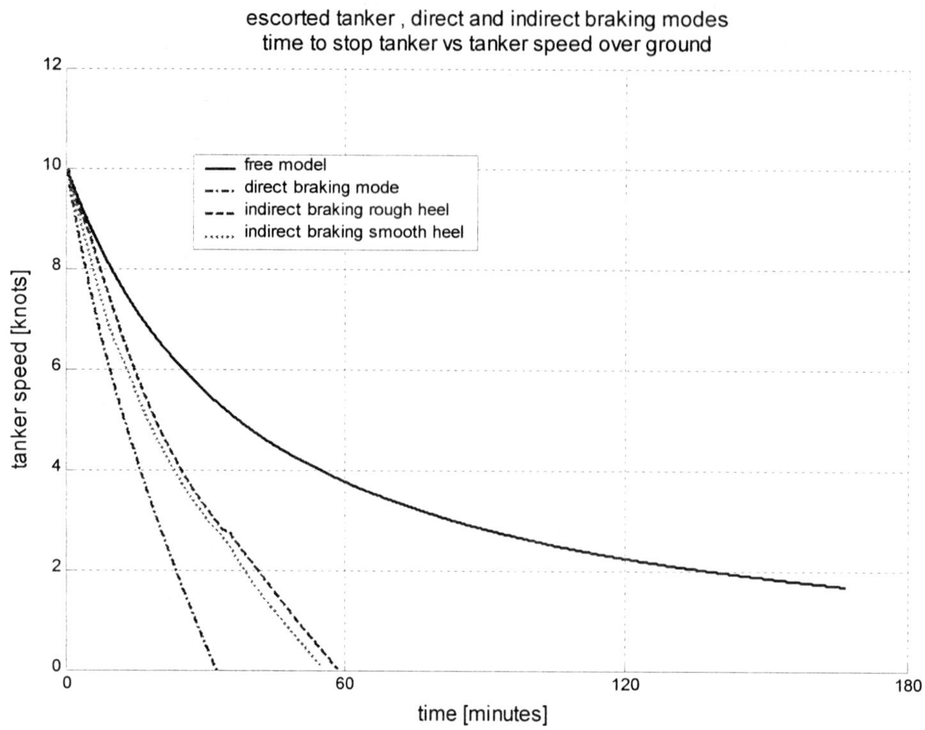

Figure 11, Tanker speed against time, for different stopping methods

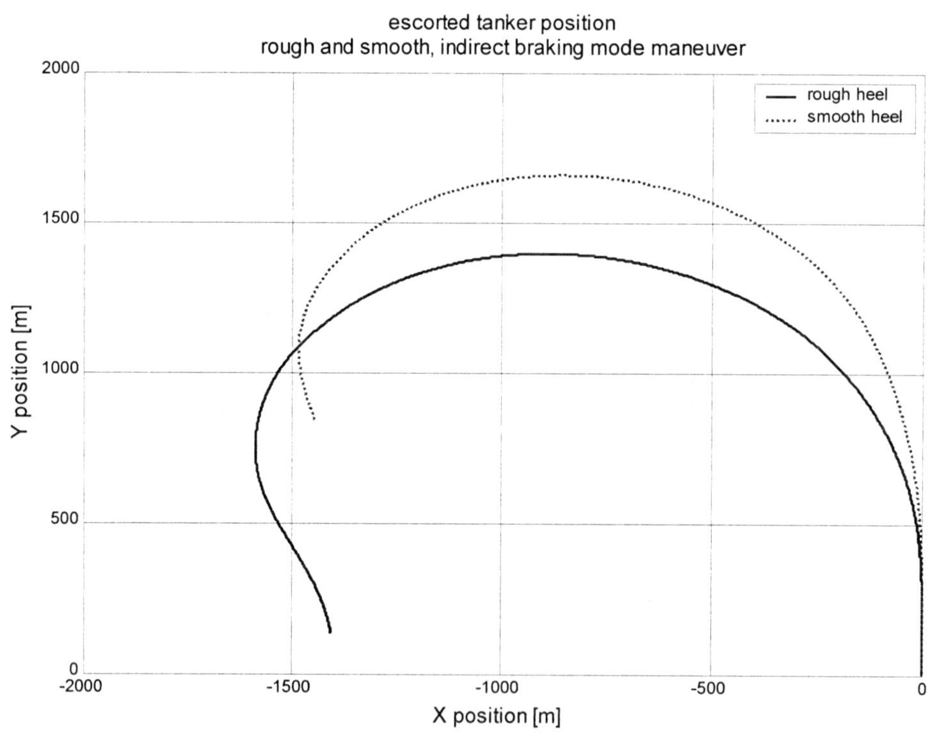

Figure 12, The tanker CG trajectories for 'rough' and 'smooth' escort maneuvers.

This somewhat blind entry to the U-turn maneuver illustrates the dynamics of the system, although it would hardly be seen as a sound and acceptable maneuver performed by the tug skipper. The skipper would rather use the combination of the thrusters' pitch and direction to get a smoother entry to the maneuver and hence avoid the excessive heel. The skipper's skills, desirably supported by a well-designed autopilot, will be crucial in obtaining the maximum steering force in the shortest amount of time, without jeopardizing the safety of the crew. From this perspective, a better example of the entry into the indirect stopping is presented in Figure 9. Heel angle and towline force time series for both maneuvers are shown in Figure 10. The time series of the towline forces show maximum values of dynamic loads well above the static forces. However, further study (including the modeling of catenary effects) to address this issue will be required.

Figure 11 presents the results of the simulation of tanker speed against time for three different tug maneuvers to stop the tanker. In the simplest case, the tug only acts to brake the tanker. For this maneuver, the tug is positioned directly behind the tanker to generate a pure braking force. The other two methods correspond to cases shown in Figures 8 and 9. Also shown is the track of the coasting tanker, with no assistance. All these maneuvers commence at the same initial speed of 10 knots. Figure 11 shows clearly how the different tug strategies impact the tanker's kinematics. The direct braking mode takes the least time to stop the tanker (35 minutes). The two indirect modes show little difference in time to stop. The 'rough' method does give a slightly shorter stopping time, but when other factors such as maximum heel angle are considered, the advantages are minimal. However, the direct mode has stopped the tanker in a straight line, whereas the indirect mode has turned the tanker around. The track of the two indirect modes is shown in Figure 12.

Figure 12 shows how the overall advance and transfer of the tanker is affected by the way the tug carried out the maneuver. In the 'rough' case, when neither the maximum heel angle nor the maximum of towline force were restricted, the tanker needed a distance of 1400 meters ahead and 1600 meters in width to complete the U-turn. The same values for the 'smooth' run show 1700 meters of advance and 1500 meters in width. Note that these are the values of the tanker's CG position and in practice, more space is required for the tanker beyond the centre of gravity, and the tug.

Figure 13 shows the advance of the tanker for the different stopping strategies. The indirect mode can stop the tanker in a much shorter distance, but requires more width, since the tanker is turned around.

In a narrow channel, the tanker has to stop in a straight line, and so the braking mode is the only method that can be applied. A cautiously executed combined indirect/direct stopping maneuvers can prove to be a more efficient option to shorten the

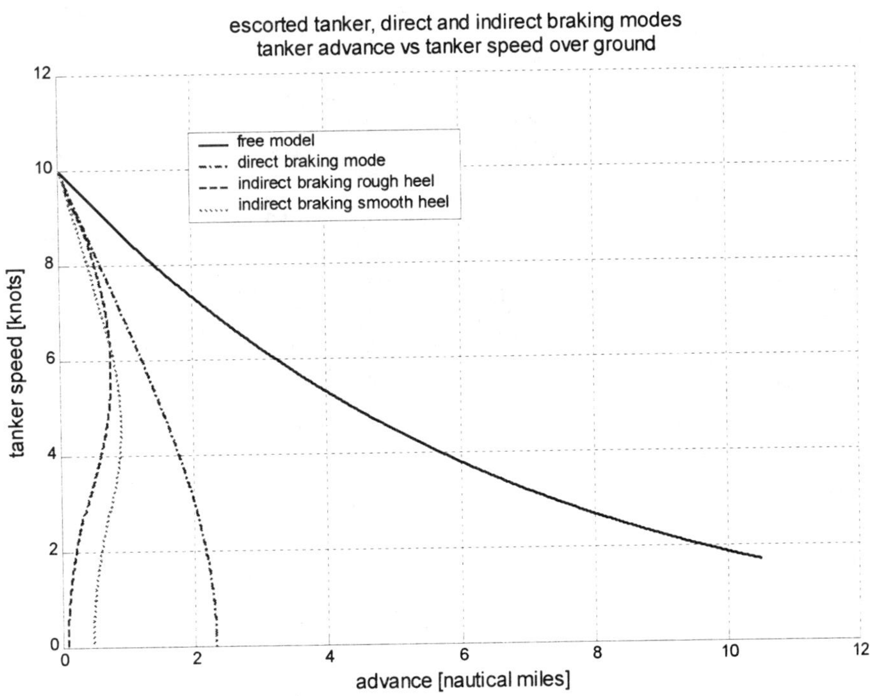

Figure 13, Advance of tanker during different stopping modes

tanker's advance without altering its heading. These trade-offs clearly illustrate the need to evaluate the local environment when determining the best assist method for a given combination of tug and tanker. The results of the simulation can be used to set maximum speed for a given size of tanker and escort tug.

Tanker-Tug Disconnection

A sudden disconnection of the towrope during escort (in the indirect mode) was simulated to investigate the tug's behavior in this situation. To simplify this situation no action is taken by the skipper after the disconnection occurred. In Figure 14 the grey color of the tug's icons represent the tug connected to the tanker (the speed is constant) and the white icons show the tug position a few moments after the disconnection occurs. Time histories of the calculated tug responses are given in Figures 15 and 16. The change of the tug heading is quite high, indicating high value of yaw acceleration. Indeed, the yaw acceleration reaches 4 degrees/s^2 as can be seen from Figure 15. Values of sway and surge acceleration are rather moderate and do not rise above 0.1g. Figure 15 shows how well roll motion is attenuated after the tug has been disconnected. The peak amplitude of the roll acceleration is quite important and reaches 8 degrees/s^2. The discussion of these results is beyond the scope of this paper because no experimental data was obtained on the maneuvering characteristics of the tug. It would be a logical extension of the work described in this paper to obtain experimental values of yaw moment against yaw rate, for the tug. This would help in understanding the dynamics of the tug during transition maneuvers, where maximum roll angles and loads occur.

VALIDATION OF SIMULATION RESULTS

Free-running Model Experiments

An earlier section of the paper compares the captive model data for the combined tug and propulsion system with the results of the NavSim simulation.
Whilst this agreement was good, the simulation is based on setting the thruster angle, relative to the centreline of the tug, and allowing the tug to obtain an equilibrium position on the end of the towline. This

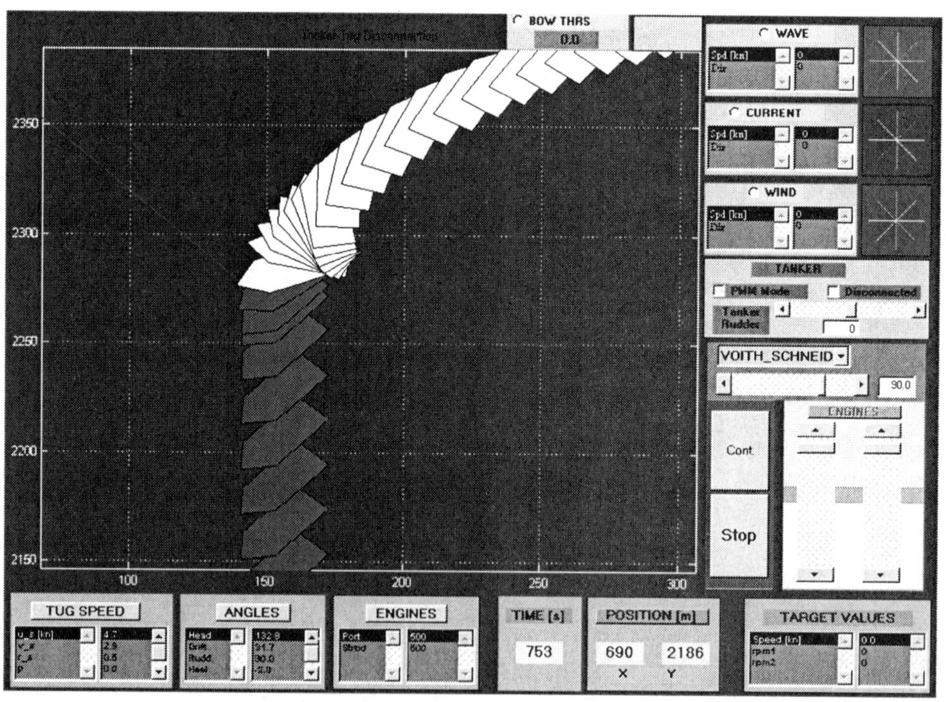

Figure 14, Bird's eye view of tug during sudden disconnect of tow rope was fundamentally different from the captive model test technique used for the PMM. In order to ensure that there were no obvious discrepancies in the two techniques, we wanted to compare the results of the simulation with data that allowed for any potential dynamic conditions experienced by the tug.

The ideal data for validating the numerical simulations of the effect of the escort tug on the tanker would have been full-scale data for the combined tug-tanker system. Unfortunately this was not available to us and so we had to use an alternative method. The next best alternative was model data for the tug operating in a more realistic condition than the captive experiments on the PMM. As discussed above, space constraints prohibited a model of the complete system, and so a simpler model of the tug alone had to be used. The tug model was modified to allow it to be towed by a towline, in the same way that the real tug would operate. The modifications included fitting a towing staple, with adjustable attachment points, a watertight deck and realistic bulwarks. A superstructure was also fitted for some of the experiments. The model in this condition is shown in the frontispiece.

This test represented the quasi-static simplification of the escort tug situation. The carriage was used to tow the tug, but the effect of the tug on slowing down the carriage was ignored. For these experiments, the thruster angles were set to a nominal angle and when the model had reached the test speed, it was free to take up its natural position, relative to the towpoint on the carriage (tanker). Shaft rotation was set to the same values used for the captive tests. The tension in the towline and the angle of the towline, relative to the tanker and the tug were measured. Roll angle was also measured and yaw angle for the tug was calculated.

The presentation of the data in Figure 17 includes a slightly different analysis from that shown in Figure 4. For Figure 17, the yaw equilibrium position was calculated from the PMM data on the basis of zero moment at the same nominal rate of rotation for the propellers as the values used in the 'free-running' tests. This contrasts with Figure 4, where the limiting power on one shaft is used.

The data for the two model test methods are compared with the simulation results in Figure 17. This figure shows that there is good agreement between the two testing methods for maximum steering force and maximum braking force. The three data points with steering force values at approximately 50 tonnes, are quasi-static equilibrium positions for the tug, but they do not correspond to the maximum force that can be generated by the system. Model experiments for this type of tug can ensure that a certain force is obtainable, but it is difficult to ensure that this is the maximum possible value, since other combinations of thruster force and direction may exist which give greater forces than the values tested.

The discrepancies between the simulation results and the model test results are discussed above. This additional set of experiments, with a free-running model, confirmed the relationships between maximum steering force and maximum braking force, determined from the captive experiments. The additional work needed to improve the results of the simulation has already been discussed.

The tethered experiments demonstrated that there were some conditions in which the tug did not maintain a stable position, relative to the tanker. In some cases, the tug would oscillate about a mean position. This indicated a potential for the tug to be directionally unstable with controls fixed, in some operating conditions. This oscillation was at relatively low frequencies, and was not dangerous in terms of the tug motions. It was not clear how realistic this condition was, since it may be controllable with steerable thrusters, which were not used in the model experiments. Further study is required to see if staple position also has an effect on the dynamic stability of the tug.

From these results, it can be seen that the agreement between the two testing methods is good. The advantage of the captive method is that longitudinal position of the towing staple and installed power can be treated as a design variables, which means the final values can be optimized from one set of experiments.

AREAS FOR FURTHER RESEARCH

There has been a high degree of growth in the use of escort tugs over the last few years. Escort tugs are gaining acceptance as a method of reducing the risk of pollution for disabled tankers, without introducing a high penalty to the operating speed of the tanker. This makes them attractive for operational areas where there are long, restricted approaches through sensitive environments. Most of the work has focused on the best type of tug for escort operations, and currently tractor tugs operating with the fin first are seen as the most effective. However, there may be alternative configurations of propulsion system and fin, which can provide acceptable compromises if the tug has other duties to perform. The type of work described in this paper will allow for an effective evaluation of alternative tug designs, in terms of the difference in their ability to generate steering and braking forces and the effect of these forces on total system performance. The model presented here does not include catenary effects on the cable, which will introduce a certain amount of additional damping into the system.

Other areas of research to be explored are related to the behaviour of the tug during the transition from one mode to another. It is during these maneuvers that the maximum roll angles and towline loads occur, and these are important factors affecting tug safety. Other factors such as tug responses during an emergency disconnect can be studied by using a simulation, but these would benefit from more specific information on the tug hydrodynamics. Other transition factors that should be studied are the tug's ability to recover from a sudden large heel angle, where large amounts of water may come onto the deck.

It is also important that the results of the simulation be validated against full-scale data. At present, the method has been checked against published data, where possible, but it is essential that actual operating data be used before it can be fully accepted as either a training tool or the basis for setting standard operating procedures.

The use of simulation may also reduce the need for a complete experimental matrix to define steering and braking force diagrams, such as that shown in Figure 4. Experiments can be used to define the hull forces over the range of speeds and yaw angles, and a few key experiments can be used to check the propulsive performance. This will make the cost of the experiment programs smaller, and encourage more structured research programs in escort tug design and performance. To make this approach truly effective however, more information is required on fin and propulsive system performance at oblique flow angles, in open water and when fitted to a tug.

CONCLUSIONS

The combination of physical model experiments and numerical simulation is a very practical approach to predicting the performance of disabled tankers supported by escort tugs. The physical model experiments provide accurate performance data on the tug, since the model can be made to a suitable scale for propulsion experiments. This ensures that there are no unreliable scale effects influencing the prediction of propulsion system performance. Tanker data can be obtained from maneuvering experiments, or stock data can be used for describing a 'generic' tanker. Modeling the complete system at a reduced scale would not be practical, due to the amount of room needed.

An accurate simulation model should have many benefits to designers, operators and regulators of escort tugs. From the designers' perspective, the simulation can be used to optimize variables such as delivered power, staple location, and hull form features. From the operators' perspective, the simulation can be used to assess the differences in tug performance. For example, what are the practical considerations to be considered when increasing the maximum steering force from 120 to 150 tonnes. It can also be used as a training aid, in developing effective strategies for stopping a tanker in a given geographic area. From the regulators' perspective, it allows regulations for classifying the performance of tugs to be based on realistic variables. Factors that are not considered in current regulations are dynamic effects during transition from one position to another. This raises particular concern since, according to the simulations, these are the conditions where the worst loads and heel angles occur.

ACKNOWLEDGEMENTS

The authors would like to thank many people who have contributed to the work of this paper. Firstly, Mr. Robert Allan, of Robert Allan Ltd. of Vancouver, who supported us during the development of the model test methods and allowed us to use one of his company's designs for the test case. Dr. Jens-Erk Bartels and Mr. Harald Gross of Voith Hydro lent us the models of the VSP units, and provided support during the development of the model test method. Dr. Richard Birmingham and Mr. James Smith, from the University of Newcastle-Upon-Tyne assisted with the planning and execution of the free-running model experiments. We would also like to thank IMD's management for sponsoring this research project. Finally we would like to thank the staff of IMD who carried out the experiments and in particular Mr. Gavin Earle, who ran the simulations presented in the paper.

REFERENCES

Allan, R, Bartells, Dr. J-E and Molyneux, D. 'The Development of a New Standard of High Performance Escort Tug', International Towage and Salvage Conf., Jersey, U.K. May 15-19, 2000.

Birmingham, R., Molyneux, D and Smith, J. 'Modelling the Stability Behaviour of Escort Tugs in Transient Conditions', STAB 2000, Launceston, Tasmania, Australia, February 7 to 11, 2000.

Gale C., Skogman A. et al:, 'Perceived Advantages of Z-Drive Escort Tugs'. The International Towage and Salvage Convention, Southampton, U.K. 1994.

Hendy N. et al. 'Quasi-static and Dynamic Bahaviour of Escort Tugs – A Designer's Viewpoint', BLDSC conference, London, U. K. 1993

Hutchison B., Gray D., Jagannathan S., 'New Insights into Voith Schneider Tractor Tug Capability'. Marine Technology, Vol 30, No 4, 1993

Waclawek, P., 'A Neural Network to Identify Ship Hydrodynamic Coefficients', MARSIM'96 conference proceedings, Copenhagen, Denmark, 9-13 Sept. 1996.

Waclawek, P., 'NavSim™ - an Overview of the Interactive Ship Maneuvering Simulation ', 12 International Conference on Hydrodynamics in Ship Design, Szklarska Poreba, Poland, 17-19 Sept. 1997.

Wacalwek, P., 'Escort Tug-Tanker: Modeling of Escort Mission in Calm Water', HydroNav'99-Maneuvring'99, Gdansk-Ostroda, Poland, 22-24 September 1999.

Yoshimura, Y., 'Mathematical Model for the Maneuvering Ship Motion in Shallow Water', J. Kansai SNAJ, Vol. 210, 1988.

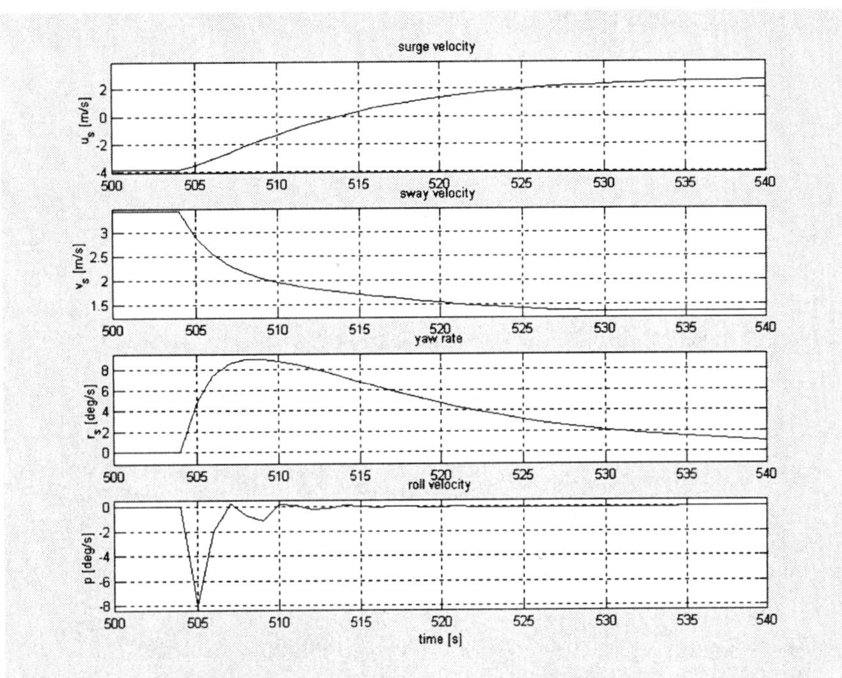

Figure 15, Tug velocity time series for a towline disconnection

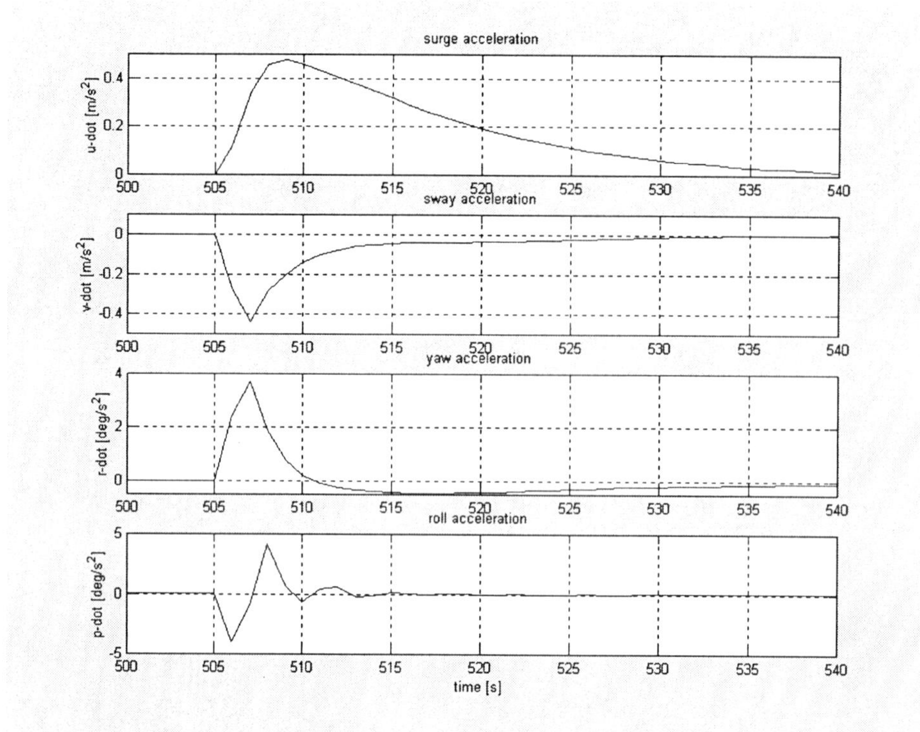

Figure 16, Tug accelerations time series for a towline disconnection

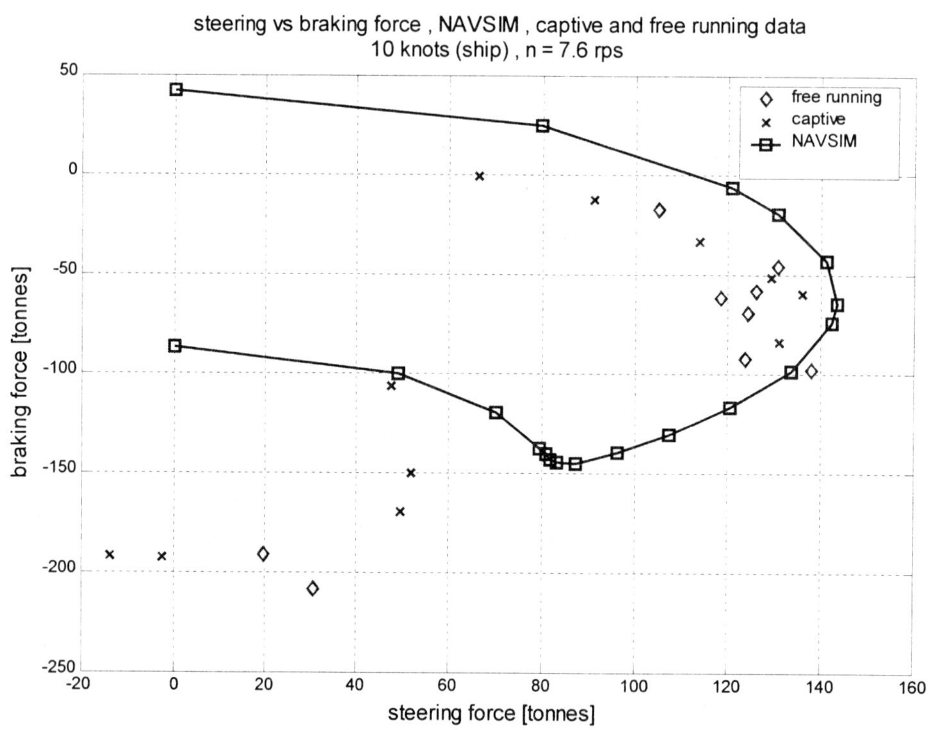

Figure 17, Comparison of Tethered model results with simulation, Braking and Steering Force, 10 Knots

Discussion

Andrew Kendrick, Member

This is a very interesting paper dealing with a relatively new technology, and I look forward to following future work by the authors in this area.

I have a number of questions on the paper, and one mild complaint. While the figures are easy to follow in electronic version, in black and white printout the curves are often difficult to distinguish. It would be nice if the final version could rely on discrimination using something more politically correct than pure color!

The question of heel, both static and dynamic is obviously crucial to safety. I found both the form and the scope of the curve in Figure 7 somewhat surprising. First, the relationship between force and heel appears remarkably linear over a large range of angles. Secondly, the fact that it extends up to 20 degrees of heel I presume is due to the additional bulwarks installed for the tests. I would appreciate the authors' comments on these points.

I would also appreciate clarification on what data was used in undertaking the performance simulations—the NAVSIM values or one of the sets of model test results. Given the substantial differences in the direct braking forces this will have had significant influence on the results presented in Figures 11 and 13.

More generally, can the authors offer any physical insight into how the forces on the tug are being generated by the combination of hull and skeg? Were there, for example, any obvious stall points, and if so what effects do they have on the dynamic behavior of the system? The results in Figure 5 suggest a very smooth performance degradation as yaw angle increases, which I would not have expected.

Richard Birmingham, Member

First, I congratulate the authors for their contribution to this complex area of research. This paper is a most valuable addition to the literature in this field, detailing the comprehensive work of the authors in a most useful way, and demonstrating the current state of the art in simulating these towing operations. I fully support the authors' contention that the combination of computer simulation and model tests is a sound method of exploring the complex behavior of tugs when undertaking escort operations, although this paper does clearly indicate that this is no easy option, with many subtle elements to be taken into consideration.

In considering the way forward for future work the authors acknowledge that the behavior of the tug when in transition from one mode of operation to another requires further work. Figure 10 in the paper indicates clearly that the most extreme responses are at times of transition, in this case at the beginning of the maneuver. This area is already of interest to both the authors and researchers at Newcastle University, and has been the subject of a joint paper (referenced by the authors: Birmingham et al. 2000). However, this area may be of greater importance than we have acknowledged. A tug master recently commented that when on active escort duty the tug is *never in a state of equilibrium*. While this may not be technically correct, the statement does reflect the fact that in these operations the tug master continually has to adjust the controls in order to maintain the required performance. Equilibrium is not a stable state, and any perturbation (from wind, current, the tanker's wake, etc.) is not self-correcting, but has to be adjusted for by the tug master. The tug is therefore in a continual state of transition.

There are two important implications resulting from this observation. Firstly, if a simulation tool is to be used to assess the safety of these operations, then it is the extreme states that are of interest, so an accurate simulation of the tug behavior when in transition conditions is necessary. This requires a detailed definition of the hydrodynamic coefficients for all speeds, angles of yaw and angles of heel—a multi-dimensional matrix that may require more extensive model tests than those already undertaken. Secondly, if the simulation tool is to be used to predict the performance of the tug and tanker system for operational planning purposes, then the input of the tug operator has to be considered. The simulation tool can provide the optimum performance, and hence minimum stopping distances; however, the realization of this performance for actual vessels in service is dependent on the skill of the tug operator. An interesting study would be to evaluate the degree to which operators can achieve optimum performance for extended periods. It may be that the stopping distances derived from the simulation tool should be extended by ten or twenty percent to allow for the difficulties encountered by the tug master.

The comments of the authors on these observations would be most welcome.

David L. Gray, Member

The authors are to be commended for their paper and presentation of this exciting material. Those of us who have spent many years learning about designing, model testing, computer analysis and full-scale trials of escort tugs are pleased to have another source of information on these boats. An escort tug is a relatively new tool and there is still a lot to be learned about their performance.

I would like to ask the authors more about their results from their model of the dynamics of the escort towline. The Glosten Associates has collected an extensive set of data from full-scale trials with VSP-type escort tugs. The data include time series measurements of the forces in the towline. Much of the data are very similar in form to that presented in Figure 10 of the paper. However, what we have not done in trials is evaluate the towline dynamics as a function of length paid-out, line type (i.e., weight and elasticity) or sea state. The dynamic forces in the towline have implications for the rope manufacturers, the winch and deck hardware manufacturers and for the users of the tug.

The computer model that the authors have developed would be a good tool to study these issues. I would like to know if the authors have any comments from their studies of the towline dynamics, specifically as to how the results are related to the line properties mentioned above. Can you specify towline properties in the program and do repeated trials of a particular maneuver to evaluate the influence of the rope on the overall process and on the line itself? Can the nonlinear characteristics of the towline be modeled, specifically the shock loads that result after the line has gone slack? Do you find or do you think you would find that the towline properties affect the braking and steering force capability diagrams, especially in a seaway?

Robert G. Allan, Member

The authors have presented an excellent summary of their work done to date in developing a sophisticated, predictive simulation model for escort operations. As tug escort operations become more prevalent around the world, it is imperative that such models be available for proper analysis of the forces required to control a disabled tanker, and most critically, that these models be as accurate as possible.

There is no question that the accurate determination of the steering and braking forces of an escort tug can at this time only be accurately determined through scale model testing. Ultimately we would hope to see Classification Societies accepting model test results in order to rate escort capabilities rather than requiring highly expensive and inconvenient full-scale testing. As the body of knowledge of tug escort performance grows, it may be possible in the future to be able to simulate that as well, but there are many variables in the design, and just as many, if not more, in the manner in which the tugs are operated. Therefore it is critical that these simulation tools be developed not only to analyze a particular escort situation, but also to be a highly useful training tool for the tug operators and also for the regulators of such operations.

There are some issues raised in this paper on which I am pleased to have the opportunity to comment and some which I would like the authors to address.

In the description of escort tug operations, the authors leave the impression that there is still a potential role for conventional twin-screw tugs in escorting, albeit only in the braking mode. I believe that it is very unwise to promote even this limited suggestion of applicability, as the forces involved are simply too great for a twin-screw tug to be able to operate safely. In particular, the position and type of typical towing gear on a T/S tug is absolutely inappropriate for escort operations. Do the authors have any potential plans for testing conventional twin-screw tugs to assess their (in)appropriateness in escort operations?

At present the only specific escort tug "regulations" are published by DNV. These principally cover towing stability, winch design and some fundamental design issues. However, the stability standards are based on achieving a critical ratio between righting energy and heeling energy, not an actual maximum heel angle. This makes it a reasonable and sensible criteria to analyze, but impossible to apply in practice. Hence, a maximum heel angle is at least a measurable limit.

The tug model used in this study incorporated a lot of very unique hull form features intended to enhance escort performance. These ultimately proved to be very beneficial, but are by no means typical of tugs in this (or any) service. Therefore, I question the validity of the use of this hull as the basis of future predictions in simulation studies, as it would tend to overpredict the performance available from a more standard hull. In particular, the roll stability of this particular hull form, as represented by heel angle vs towline pull is dramatically better than that of a more conventional Voith tug hull form. Would the authors comment on the parameters of tug design that are input into the simulation model, and how these might affect the outcome of the simulation?

Authors' Closure

First, we would like to thank the discussers for taking an interest in our paper and providing their thoughtful comments and suggestions. The prediction of escort tug performance is becoming increasingly more important, and open discussion of ideas is a key factor in developing the necessary techniques.

Dr. Birmingham has identified two important issues around the development of simulation tools for escort tugs. Based on observations on model tests, we have found that the tug tends to be directionally stable in the maximum steering condition, but less so at the maximum braking condition. We were unable to determine what factors would affect or eliminate this instability since we were unable to steer the VSP units during the experiments. The fixed angle of the propulsion vector was one factor that determined the selection of the captive methodology for the experiments. Without this capability, it is not feasible to investigate how control settings, or possibly more importantly, how changes in control settings can influence the magnitude of the towline loads and the resulting motion of the tug. We hope to be able to carry out some experiments in the future which will determine these factors at least at model scale.

The question of a suitable test matrix to enable accurate dynamic prediction of the tug behavior is difficult to answer. The most obvious approach would be to extend the test matrix to include yaw variation, in much the same way that you would include it when determining hydrodynamic coefficients for normal ship maneuvering predictions. The only difference being that the initial yaw angles will be much higher and yaw rates should be appropriate for a tug under the action of a towrope, rather than its normal steering system. We did not include these extra experiments, since our primary objective was the simulation of the total system. The time scale for changes on the tanker is much lower than the time scale on the tug. We hoped that the mean force generated by the tug over a long period of time was sufficient to predict the performance of the tanker. We have implicitly assumed that turning moment as a function of yaw rate is the same as that predicted by the MMG method. Dr. Birmingham is right in that this should be confirmed by additional experiments.

Mr. Gray brings a practical naval architect's perspective to the issues around escort tugs. When discussing the possibility of using a conventional twin tug for escort duties, we were thinking only of the hydrodynamic potential, rather than the typical position of winches. We do not have any plans at this time to include conventional tugs in our studies, since Hutchison et al. showed the relatively poor hydrodynamic performance of a conventional tug compared to a specialized escort tug design. This work did not include propulsion experiments, and such data may be useful for comparison purposes. The only time it might be of practical significance, however, could be in an emergency, and this was the only tug available.

NavSim starts with hydrodynamic coefficients determined from the MMG method. These are effectively average values based on hull form parameters. This first iteration of the prediction is tuned against key experiment results, and the basis of this tuning was principally against the data for the hull alone, with emphasis on the maximum steering condition. These values are then used for the remaining yaw angles. The design we used here was for illustrative purposes, and if the hull was significantly different, then the initial NavSim predictions should be tuned against data for the new hull, rather than using this hull for every condition.

This leads into Mr. Kendrick's comment related to the use of simulation results or model data. The predictions are based on the simulated tug data rather than the measured values. Even though the relationships between steering force and braking force between model and simulation are different, especially around the maximum braking condition, the magnitude of the maximum braking force is correctly predicted by the simulation. This means that the tug will take up a different position relative to the tanker, but should not substantially change the distances given in Figures 11 and 12.

Mr. Kendrick also raises some questions regarding some of the equations in the simulation method. The current version of NavSim uses essentially a linear method for calculating roll restoring moment. While this was simple to apply in principle, given the flared hull form used, it was not completely unrealistic. The heel angle needed for deck edge immersion was 30 degrees, and so we felt comfortable using this simplification, even at large roll angles. We hope to expand the method to include realistic roll restoring moments, and also include factors such as the extra forces created by water flowing over the top of bulwarks.

The best presentation for seeing the performance of the fin is to look at the side force against yaw angle, and I would refer the author to Allen et al. for the data for this hull. This presentation shows that the side force shows a local peak at approximately 40 degrees of yaw, then drops slightly and increases to a maximum at around 70 degrees, where the side force is entirely drag. The 40-degree yaw angle corresponds to the stall point of the skeg, which is a very low aspect ratio foil. Since escort tugs rarely operate beyond these yaw angles, the data given in Figures 5 and 6 are mostly within the linear relationship between side force and yaw angle.

Finally, we would like to apologize to Mr. Kendrick. The figures were prepared using different line styles, but this difference was lost in the pixel resolution when the documents were converted.

SNAME *Transactions*, Vol. 108, 2000, pp. 45–56

The USCG Alternate Compliance Program

Jaideep Sirkar, Member, and **John Hannon,** Member, United States Coast Guard Headquarters, Washington, DC

ABSTRACT

The United States Coast Guard (USCG) is the governmental agency that acts as the flag administration for U.S. flag vessels. The paper discusses the regulatory standards for ship design and construction in the United States and compares it with internationally recognized standards. The standards, and the processes for development, and their implementation are discussed. The USCG has developed processes to delegate various flag state functions to recognized organizations, including classification societies. A major effort by the USCG in delegating flag state functions is the Alternate Compliance Program. This program, together with recent legislation that provide incentives to change registry ("reflag") to the United States flag, has focused attention on regulatory standards for ship design and construction. This paper summarizes the Alternate Compliance Program and the process for and issues related to reflagging to the United States flag. Finally, the paper looks at the changing face of traditional flag state functions and responsibilities.

[1] United States Coast Guard Headquarters, Washington DC

BACKGROUND

The United States Congress (which is the legislative branch in the government of the United States) delegates the authority and the responsibility for regulating the safety of merchant shipping to the United States Coast Guard (USCG). Based on this authority, the USCG develops regulations that include the areas of design, operation, manning of ships and other vessels, for the purpose of safety and environmental pollution. The Society of Naval Architects and Marine Engineers (SNAME) documented (Cleary, 1993) a brief history of the regulatory responsibilities of the USCG. The USCG develops, updates, and publishes the regulations in the Code of Federal Regulations (CFR). The regulations for ship design and construction (for vessels registered in the United States) are generally located in various "subchapters" in 46 CFR Parts 1-199, as follows:

46 CFR Subchapter

D	Tank Vessels
E	Load Lines
H	Passenger Vessels
I	Cargo and Miscellaneous Vessels
I-A	Mobile Offshore Units
O	Bulk Dangerous Cargoes
R	Nautical Schools
S	Subdivision and Stability
T	Small Passenger Vessels
U	Oceanographic Research Vessels
V	Marine Occupational Safety and Health

The USCG also represents the United States at the International Maritime Organization (IMO), which is a specialized United Nations agency whose mission is to provide a forum for seafaring nations of the world to negotiate uniform international regulations for merchant shipping.

DELEGATION OF USCG FLAG STATE RESPONSIBILITIES TO OTHER ORGANIZATIONS

The United States Code Title 46 U.S.C. 3316 authorizes the Coast Guard to accept the reviews of plans, inspections and examinations performed by a classification society. These functions are performed for the purpose of determining compliance with Coast Guard Rules and regulations of classed and unclassed vessels. Since 1929, the American Bureau of Shipping (ABS) has been recognized as the classification society that may perform many of these functions. Public Law 104-324 (Coast Guard Authorization Act of 1996) amended 46 U.S.C. 3316 to permit the use of foreign classification societies who are interested in participating in performing flag state duties on behalf of the USCG. Consequently, the USCG amended 46 CFR Parts 8, 31, 71, 91 and 107, for the purpose of implementing the law, as well as recognizing and authorizing classification societies to perform these functions.

While the United States just recently expanded its scope of delegations regarding statutory functions that shall include numerous classification societies, this practice is fairly commonplace among many other seafaring nations. A recent document (FSI 8/INF.3, 1999) published by the International Association of Classification Societies (IACS) provides a brief overview on the extent of delegations regarding statutory functions made by over 200 administrations. Magill, 1999, presents a summary of the international role of classification societies in regulating merchant shipping.

USCG'S ALTERNATE COMPLIANCE PROGRAM

What is the Alternate Compliance Program (ACP)?

ACP is a five-year old program developed as an alternative method for owners of U.S. flag vessels to fulfill the regulatory requirements for vessel design, inspection and certification. Under this program, the Coast Guard can issue a certificate of inspection based upon reports issued by a recognized and authorized classification society that confirm a vessel complies with applicable international conventions, classification society rules and other specified requirements. ACP is a viable option for owners of tank vessels, passenger vessels, cargo vessels, miscellaneous vessels, and mobile offshore drilling units that engage in international voyages.

ACP may be best defined as:

CFRs (Code of Federal Regulations) = Class rules + International Requirements + Supplement

What does this definition mean? In the traditional process of regulatory compliance the ship owner complied with the CFRs to obtain a certificate of inspection from the Coast Guard. Under the ACP, a ship owner does not comply directly with the CFRs. Rather, an equal level of safety is met by relying on the rules of the classification society, applicable international conventions (appropriate for the type and service of the vessel), and a *Supplement*.

What is a *Supplement*? It is a document that contains (a) various interpretations made by the Coast Guard of international conventions, (b) navigation safety and pollution prevention standards for all vessels in U.S. waters, required by statute, and (c) Coast Guard requirements that are not included in class rules or international conventions.

Why was ACP initiated?

ACP was initiated as a pilot program under the American Bureau of Shipping (ABS), in February 1995. This pilot program, referred to as the Coast Guard's ABS based ACP, was the result of a regulatory reform initiative begun in 1992 **to enhance the competitive position of the U.S. maritime industry** while maintaining a level of vessel safety and environmental protection equivalent to Coast Guard regulations.

Based on the success of the pilot program the Coast Guard formalized the ACP through an interim rule published in December 1996. The interim rule expanded the program to include other recognized and authorized classification societies. One year later, a final rule was published. The Coast Guard anticipates this rule will result in reduced vessel down time, meeting required standards, and greater flexibility in scheduling inspections. ACP allows both the ship owner and the Coast Guard to concentrate on a systems approach versus a sometimes over-burdensome regulatory approach. Simultaneously, ACP reduces or eliminates duplicative inspection tasks performed by the Coast Guard and the classification society, resulting in savings for the ship owner, as well as allowing the Coast Guard, through its port state control program, to redirect its resources to those vessels that pose the highest safety and environmental risks.

Where is ACP today?

Initially, ACP was developed with ABS due to the Coast Guard's extensive experience in delegating certain certification functions to ABS. The passage of the Coast Guard Authorization Act of 1996 (Public Law 104-324) allowed the Coast Guard to delegate certain functions to foreign-based classification societies as well. Consequently, the availability of the ACP was expanded to other recognized and authorized classification societies.

How does a classification society participate in the ACP? Participation involves a three-step process: recognition, authorization to issue international certificates, authorization to participate in ACP.

Recognition – The first phase of participation requires satisfying a set of performance criteria based on: standards developed by the International Maritime Organization (IMO); a satisfactory port state control performance (based on a rolling, three-year average of detentions of distinct arrivals in U.S. waters); and reciprocity for ABS to perform similar functions in the country where the classification society is based.

Authorization to Issue International Certificates - Upon recognition, a classification society may apply to the Coast Guard to issue certain international certificates on behalf of the Coast Guard. Based on a review of the class rules and procedures, the Coast Guard may enter into an agreement with the society to issue international certificates.

Authorization to participate in ACP - In the third and final step of this process, the Coast Guard may authorize a classification society to participate in ACP. This occurs two years after the society has issued its first safety related international certificate under step two above.

More complete details on these recognition and authorization processes are included under *Focus On "The U.S. Supplement"*.

To date, recognized classification societies are ABS, Lloyd's Register (LR), and Det Norske Veritas (DNV). The table below shows the various international certificates these societies have been authorized to issue on behalf of the Coast Guard. Currently, only ABS is authorized to participate in ACP.

	ABS	LR	DNV
Int'l Load Line Certificate	✓	✓	✓
Int'l Tonnage Certificate	✓	✓	✓
SOLAS Cargo Ship SAFCON	✓	✓	
SOLAS Cargo Ship SAFEQUIP	✓		
MARPOL 73/78 IOPP Certificate	✓		✓
MARPOL 73/78 IOPP Certificate (NLS)	✓		
Verification of Compliance with MARPOL 73/78 Annex III	✓+		
Verification of Compliance with MARPOL 73/78 Annex V	✓+		
IMO MODU Code Safety Certificate (1989 Code)	✓		
ISM Code	✓		

Table 1 – International Certificates authorized for issue on behalf of the United States Coast Guard

The number of vessels enrolled under ACP is growing and currently stands at approximately 160, including vessels currently under construction. Clearly, ACP is proving to be a successful and popular program.

What is the future of ACP?

Compared to other major Coast Guard programs, ACP is in its relative infancy. However, all early indicators point toward continued success as the program matures. The Coast Guard fully expects that in time more classification societies will be eligible to participate in the ACP. Therefore, in addition to the benefits of ACP, a choice among classification societies may also be available to ship owners in the not-too-distant future.

Further, the Coast Guard has completely revised its policy on the development of the *Supplement*. Instead of the approach that used a line-by-line comparison of class rules and CFRs, a systems approach has been developed that will provide the ship owner even further flexibility in regulatory compliance within ACP.

RECOGNITION AND AUTHORIZATION PROCESS

To be recognized, classification societies must satisfy two necessary and 23 sufficient conditions. The necessary conditions are reciprocity and port state control performance.

In accordance with the statute, the USCG can only delegate to a classification society those functions that have been delegated to the American Bureau of Shipping in the country where the classification society is headquartered. This principle of reciprocity can be a major hurdle for many classification societies seeking to enter into the ACP.

While the basic principle of reciprocity is simple, the details of determining whether the statutory condition of reciprocity is satisfied can be fairly complex, requiring detailed legal and technical analyses of the facts. These issues could include situations where partial delegations (for the functions related to issuing a particular certificate) have been made to ABS in the home country of the classification society, split registry situations with different levels of delegation for ABS in the different registries within the same country, issues related to existing conditions that allow other classification societies to participate in the home country, without the actual existence of specific authorizations. This situation can be further complicated by the fact that the classification society seeking entry into the ACP will need to work closely with the administration of the home country to provide the necessary information to the U.S. Coast Guard.

The second necessary condition is port state control performance, which requires that a classification society have an acceptable record of vessel detentions in U.S. waters (only those detentions associated with classification society performance). While this can be measured in different ways, currently the USCG measures this performance as the average detention rate (total detentions per distinct vessel arrival in the United States) for the most recent 3 years. Port state control performance is deemed acceptable when the average detention rate of a particular classification society is better than the average rate for all classification societies combined.

Once these two necessary conditions are satisfied, the USCG evaluates the classification societies against 23 minimum standards in order to gain recognition. As stated earlier, these standards are based on guidance developed at IMO. The main elements of these standards include the following: experience, size of classification society, history of rule development and inclusion of casualty experience into the rule development, existence of rules for construction and survey processes, adequate certification under recognized quality schemes, and financial and business independence from shipowners and shipbuilders. It should be emphasized that the recognition of a classification society only demonstrates compliance with a minimum set of standards.

Once the classification society is recognized, they may seek authorization to issue specific international certificates to U.S flag vessels, and perform all the statutory functions (in addition to the classification duties) associated with the issuance of the particular certificate. At this stage, the USCG carefully reviews the rules and the processes of the classification society before authorization. Additionally, the classification society must develop an U.S. Supplement to the rules of the classification society.

FOCUS ON THE "U.S. SUPPLEMENT"

A significant benefit of the ACP is that it allows the USCG and the industry to focus on the "delta". This is the difference between the international ship design and construction standards and the standards required by the USCG for U.S. vessels. This allows the USCG to continuously review those standards that are not entirely harmonized with international standards.

The USCG can thus determine which of those standards should be reviewed for possible elimination; and the *Supplement* can be used to develop strategies for its work program at IMO.

In 1997, the USCG undertook a significant project to identify such standards. The result of this project was documented in a Federal Register Notice of Policy (FR Notice) (Federal Register, 1998). Based on both subjective and objective risk assessments, the policy listed all critical ship safety systems related to ship design and construction. These systems contain only those that are not adequately covered by international standards. The reader is referred to the FR Notice for a complete list of such systems for various ship types.

A few major examples of such systems are listed below:

Lifesaving appliances and arrangements
Boilers and other Pressure Vessels
Fire extinguishing systems
Various alarm systems

OVERSIGHT OF CLASSIFICATION SOCIETIES IN THE ALTERNATE COMPLIANCE PROGRAM (ACP)

Initial Review

As previously described the process of determining whether a class society should be recognized and subsequently authorized, to perform work on behalf of the U.S. Coast Guard is a multi-tiered process.

The first tier essentially reviews the content of a class society's published rules as measured against the comparable U.S. regulatory standards. The attendant work procedures of the society are reviewed for critical functions such as plan review and survey. After this review is complete, a U.S. *Supplement* is developed to address areas where class rules or work procedures omit unique U.S. regulatory requirements or interpretations of international convention.

In addition to this rules review, other functional analysis of the organization itself also occurs in this first tier. Three examples of a functional analysis are: (a) meet certain minimum standards defined in Title 46 CFR Part 8 (includes minimum tonnage classed, minimum number of vessels classed and mimimum number of exclusive surveyors); (b) a record of satisfactory previous performance in assuring vessels classed by the society are maintained to class and convention standards as measured by a three year rolling average of detentions and interventions under the U.S. Port State Control Program; (c) maintains an office in the United States; (d) maintains a Quality Management System (QMS) within their organization to the requirements of the ASQC 9002 (ISO 9002) series or equivalent. Most of these requirements focus on the past and present performance as indicators of future performance, but the QMS system requirement is focused on maintenance of an organizational structure that will continue to foster a culture of quality and safety compliance within the organization. The Coast Guard places considerable reliance on the QMS to promote the continuous and uniform update and application of class rules across the full range of a class society organization. The QMS promotes the use of standard procedures by class society personnel at all levels. Standard procedures ensure such things as the training and qualification of class employees who are called upon to apply class rules and statutory regulations. The QMS gives a level of consistency to the very subjective business of classing vessels. The first tier of class society review sets the stage for what can reasonably be expected in terms of future performance, the second tier sets out to measure and monitor actual performance through a system of independent verifications categorized under the broad heading entitled oversight.

Inspection versus Audit

The process of oversight is to achieve the desired goal of possessing ships that meet the standards of class, international convention and the U.S. Supplement. This is accomplished by examining a representative sample of the larger unseen process described in the class society rules and the QMS. The Coast Guard uses a combination of process audits and material inspections, both scheduled and random. Coast Guard personnel who conduct independent oversight examinations must focus on material condition. It is the best clear indicator of class society performance.

Rather than impose the Coast Guard's way of doing inspection on the class societies, the Coast Guard accepted that each of the class societies' survey process is equivalent. The ultimate goal is to achieve the desired results of safety and environmental compliance envisioned in the minimum standards. How class societies achieve desired outcomes is only constrained by statutory time limits on intervals between periodic examinations or certificates renewals. Oversight is conducted by the Coast Guard's inspectors with an awareness that the process being used by the class society may not be the process preferred by the individual Coast Guard inspector. Class surveyors and Coast Guard inspectors have each been inculcated with an unwavering belief in their respective work processes as the preferred method of ship inspection. Added to this, it is difficult for the Coast Guard inspector to be knowledgeable about the work processes of multiple class societies and because of this, the focus of oversight most often shifts to the tangible outcome of survey, a vessel's material condition. Regrettably, what almost always ensues is a debate over how much the survey process can control material condition outcomes, with the primary blame for failures placed at the feet of the ship owner, the crew or sometimes the personal failings of the individual surveyors. All too often in the past the class society response to a substandard vessel report was to focus on the immediate correction of the vessel's more glaring deficiencies and then perhaps a thought about sacking the surveyor who certified the vessel as a form of corrective action. But there is always a nagging feeling that this victory is short lived. Long-term, sustainable gains in safety cannot come without a joint commitment to safety management by the ship operator (more on the ISM Code later) and to QMS by the class society. In the absence of these commitments there can be no process improvements that will ensure future work is being properly performed, even in the absence of direct inspection and oversight. The real message of oversight should be that we want to fix the systems that allowed problems to occur, more than we want to fix the individual problem that brought the system problem to our attention in the first place.

Quality Management Systems

As previously stated the Coast Guard requires all recognized Class Societies to maintain a QMS to ISO 9000 standards or equivalent. All IACS member class societies are required to maintain a QMS by IACS who uses its own quality standard similar to the ISO 9000 series standard. Member class societies conduct internal audits and submit to external audits by IACS. In addition, some class societies choose to maintain QMS certification from an independent quality registrar and submit to external audits by them. In addition to all of this, some flag administrations conduct independent audits of the class society QMS in conjunction with the work done for their flag's registry. Although the Coast Guard considered independent audits, it chose instead to observe the internal and external audits of the class society scheduled by IACs or a quality registrar. This system of audit observation has proved satisfactory for the U.S. since it minimizes the burden placed on the class society while allowing the Coast Guard to benefit from the knowledge and experience of other auditors. The only negative experience with this approach has been that the Coast Guard observer does not set the scope of the audit observed or select the dates on which the audit will be conducted. Another minor problem has been if the audit includes a vessel visit, and the vessel is not bearing a U.S. flag, then the class society must seek permission from the vessel owner for the Coast Guard observer to attend. However, these problems have not proven to be significant impediments. A key benefit has been the opportunity for the exchange of ideas and perspectives between Coast Guard observers and auditors, be they from IACS, the quality registrar or the class societies themselves. Attending audits has given Coast Guard observers a chance to see the human side of the class society and the people who struggle as individuals to help their organization succeed.

Crossing the Cultural Divide

The idea that Class Surveyors and Coast Guard inspectors cover a lot of the same ground on U.S. ships was one of the reasons that ACP was initiated. What was not well understood or accepted was that they look at those items for fundamentally different organizational reasons. The initial and elementary observation that there was apparent duplication of effort failed to acknowledge that the motivation for conducting a class survey is very different than the motivation for conducting a government mandated safety and environmental inspection. Class Surveyors are not, and will never be, government inspectors who enforce laws. Class societies exist only to provide a valuable service for their clients and in the current formula that client is the ship owner and as such maintaining that business relationship is essential to the survival of every class society. Conversely, the Coast Guard's funding is provided by appropriation through Congress. Working for government means conducting inspections that have been defined in law for the purpose of serving the broader societal goals of maritime safety and a clean environment. Although one might reasonably argue that the class society also serves these goals, they do so more from the pragmatic economic view that safe and seaworthy ships are more reliable and therefore more profitable, which serves the needs of the client. Since governments have now created an international regulatory atmosphere that punishes non-compliance with safety and environmental regulations, class societies must address these realities since they too have an economic impact on vessels and clients. What ACP and all similar flag administration authorizations set out to do is to bend the client based service provided by the class society to also serve the needs of society in general, even if the underlying motivations for that service are different than governments. It is a useful alliance, but one that increasingly fails to acknowledge that class societies themselves now face growing competition in the market place. The pressure to serve the client first and foremost can sometimes place the class societies between the clients' goals and the goals of the Coast Guard. In entering into ACP, both the Coast Guard and the class society tacitly accept that the class society will continue to perform their function as a service to paying clients. The class society will advise, but never dictate, to their business clients when and if they should yield to the mandates of regulations or laws enacted by governments, especially when such an action is clearly detrimental to the business interests of the client. For this role we must continue to rely upon governments as the ultimate enforcers of the law. As a consequence, Port State Control programs have increasingly emphasized the responsibilities of flag administrations as being fully culpable in the decisions of class societies that certify the vessels of their registry.

The ISM Code and ACP

Throughout the years Coast Guard inspections and class surveys have failed to drive substandard vessels from the seas. Often blame for this is laid at the feet of lax, corrupt or incompetent class surveyors and/or flag administration officials. Although this has been regrettably true in some cases, it is really more about what is missing from this equation of responsibility, i.e. the ship owner. It should not be assumed that safe ships can be maintained if the ship owner's only commitment to safety and the environment is simply to do what they are told at the most recent inspection or survey. Luckily, this "catch me if you can" philosophy ends up being poor business for the long term, but it does yield short term gains for those who lack long term commitments. Imagine for a moment that we could graph the status of safety compliance of a substandard ship on a Cartesian plane with the X-axis being the level of safety and the Y-axis being time. For the ships that only meet the minimum standards on the date of survey, the peaks of the line showing the level of safety would rise to minimum standard only on the dates of the dates of surveys and fall to lower levels immediately thereafter. Over time, the graph would come to resemble the waves of the ocean with the peaks representing compliance with the minimum standards, and everything else falling below the minimum. Survey and inspection are really just ways to gauge the commitment of the owner to the safety standards, they are not meant to be a substitute for an effective preventive maintenance schedule carried out by the company with adequate resources. If anything has been learned in the many years of inspecting and classing ships, it is that surveys can supplement, but never replace, the need to have an effective safety management system implemented by an owner committed to safe and environmentally friendly ships.

To attempt to rectify this lack of commitment on the part of some owners, the IMO and the U.S. Coast Guard have adopted the International Convention for the Safe Management of Ships (ISM Code). The code took effect on July 1, 1998 for Passenger Vessels, Oil Tankers, Chemical Tankers, Bulk Carriers and High Speed Craft with all other vessels having to be in compliance by July 1, 2002. Under the ISM Code companies must implement safety management systems (SMSs) that ensure compliance with safety and environmental requirements not just on the dates of surveys and inspections but at all times. The ISM Code's mandate for an effective SMS is now an integral part of the ACP formula

Although some flag administrations perform ISM Code certification directly, in most cases, the class societies that have developed the expertise to perform the audits needed to verify compliance with the ISM Code. The class society performs ISM Code certification as a function wholly separate from class survey and often provides ISM Code certification to vessels that it does not class, or has vessels in class with ISM Code certification from another society. This makes ISM Code different, because in almost all other cases where statutory certificates are issued by class on behalf of governments the certificate is connected in some substantial way to the requirements of class (Remember this was the original issue with duplication of effort). Issuing statutory certificates is often just a matter of performing a few small additional items during a class survey. But not so with the ISM Code, the audit ventures into the territory beyond the mere physical condition of the vessel to judge the human component of safety. Class societies and others contended that because class survey and ISM Code are wholly separate, one could be performed without knowledge of the other. But as time has passed it has become increasingly clear that the results of management decisions under the SMS have had an impact on vessel condition and therefore impact class. The SMS and the physical condition of the vessels are linked together in a continuous cycle of action and reaction. Shared information between surveyor and ISM auditor is emerging as one of the most beneficial aspects of ISM Code when looking toward effective implementation of ACP. Unfortunately, class societies have an ingrained aversion to sharing information with other class societies. Numerous instances have arisen in the last two years where a vessel was known to have serious material problems related to class or had major non-conformities for ISM Code, but this information was not communicated between surveyors and ISM auditors, even when they are within the same class society. There has been much debate over the link between material condition and the SMS but clearly it would be beneficial for the surveyor and auditor to judge this link for themselves instead of blindly adhering to functional or organizational separations. The Coast Guard is working vigorously to mandate communication between class societies for U.S. vessels that share class and ISM Code certification between two different class societies. If this effort proves unsuccessful the next move would be to mandate that ISM Code certification and class certification must be from the same class society for U.S. vessels.

REFLAGGING TO THE UNITED STATES FLAG

One of the beneficial outcomes of ACP has been the streamlining of the reflag process from foreign to U.S.. The process of obtaining a U.S. certificate of inspection for vessels initially constructed and certified by a foreign administration is described in Navigation and Vessel Inspection Circular (NVIC) 10-81 (Change 2). NVIC 10-81 lays some basic ground rules regarding the reflag process that focuses on the need to perform a comparative analysis between the rules under which the vessel was constructed and the U.S. regulatory requirements. Since ACP accepts as its basic formula the use of Class, Convention and Supplement, there is no need to perform the comparison with U.S. regulations, aside from satisfaction of the items in the supplement. The Coast Guard, as basic policy, will continue to not accept for reflag vessels over 10 years of age, so that the vessels reflect the more recent amendments to international convention requirements. Also, streamlining assumes that the class society of the reflagging vessel is also authorized by the U.S. Coast Guard, since that class society would have developed a U.S. supplement for its rules. If this is not the case, then the reflag may be done under ACP using an authorized class society, but they must perform a gap analysis for the individual vessel for compliance with their rules versus the previous class society, so that the approved U.S. supplement can be used. Plan review and initial inspection of ACP reflags would be performed by the authorized class society with oversight by the Coast Guard. In all likelihood, the "hand over inspection" (from Coast Guard inspection to ACP) would take place during the reflag. Drydocking is not a proforma requirement of reflagging, rather it often depends upon the scope of the work needed to align the vessel with rules of the authorized class society or the ACP supplemental items. In any case, ACP has greatly streamlined the previously complex process of reflagging to U.S. flag.

CONCLUSIONS

FULFILLING THE PROMISE OF THE ALTERNATE COMPLIANCE PROGRAM

ACP burst into being among the myriad of programs born of the Clinton Administration's first term verve to reinvent government. Nothing in Washington escapes politics and there are those who still contend that the oversight program was crafted under the pressure of political mandate. Those opposed to ACP will long lament that it was and is an idea that bears the fatal flaws of politics. Yet despite the politics of the moment, ACP was an inevitable consequence of the convergence of international standards and U.S. regulatory requirements. The U.S. and others had worked for decades to raise the international convention standards, and in doing so negated the need to establish domestic standards that adhered to a higher standard. If politics are a proper context for ACP then the most applicable quote comes not from President Clinton but from President Ronald Reagan. During his historic efforts to convince the former Soviet Union to sign agreements of disarmament, President Reagan was often quoted as saying in public that he pursued a policy of "trust, but verify." At the risk of comparing class societies to Soviets, ACP is also a program of trust, but it too must be verified. Perhaps the cost of failure is not so grave as the price of nuclear war, but it is a very real price nonetheless for those who must sail the oceans in the ships certified under the ACP program and for the marine environment.

The Coast Guard and ABS thought that we knew each other before ACP and yet we have learned more about each other in bringing ACP into being than in all the years of our previous relationship. There is still a long way to go, but the feeling of partnership between class society and Coast Guard is stronger than ever before. As new class societies are added to the list of ACP authorizations, new challenges will present themselves. Questions about how many class societies the Coast Guard can effectively manage will be asked and must be answered. But the question will never again be "Is ACP is here to stay?" The only question that will remain is, "How effective can we make ACP?"

The Changing face of Flag State Functions and Responsibilities

The vast majority of flag states have delegated statutory functions to classification societies and other inspection and certifying authorities. While precise figures are impossible to determine, based on a review of FSI, a reasonable estimate is that about 80% of all such statutory functions have been delegated. However, many nations including the United States, United Kingdom, Norway, and others, endowed with strong technical staffs, and a proud maritime tradition, have conducted significant statutory survey and certification functions themselves.

Programs such as ACP, where classification societies assume traditional flag state functions (Thorpe, et al 1997) allow flag state administrations to concentrate scarce resources on port state activities. This shift in the compliance regime from Flag State to port state – is one that is creates new issues that flag administrations have to learn to cope with. These include new working relationships with classification societies.

ACKNOWLEDGEMENTS

The authors gratefully acknowledge the contributions of Molly Moore, of PMG, Inc. for her assistance with editing and proofreading the paper.

REFERENCES

Cleary, W. A., Jr., "Regulation", *A Half Century of Maritime Technology 1943-1993, Benford, Harry , Ed. SNAME,* May 1993.

Federal Register, Vol. 63, No. 30, Page 7495, *Critical Ship Safety Systems Table and Components of a Supplement Under the Alternate Compliance Program,* February 1998.

FSI 8/INF.3, "Delegation of authority to IACS Member Societies on Statutory Surveys and Issuance of Certificates," Paper submitted to the Sub-Committee for Flag State Implementation, International Maritime Organization, by *International Association of Classification Societies*, October 1999

Magill, M. C., "Merchant Ship Regulation and the Role of Classification Societies", *Lloyd's Register formal presentation notes,* November 1999.

Sirkar, J., "The Coast Guard's Alternate Compliance Program (ACP): Yesterday, Today, and Tomorrow", *Proceedings of the Marine Safety Council,* U.S. Coast Guard, May 1998.

Sirkar, J., "U.S. Coast Guard's Alternate Compliance Program," *Panel Presentation at ASTM Seminar,* December 1999.

Thorpe, R., McAlear, R., Leback, W., "The Changing Role of Ship Classification Societies," *SNAME Transactions, Vol. 105,* 1997.

U.S. Coast Guard Navigation and Vessel Inspection Circular (NVIC) 2-95, "U.S. Coast Guard's Alternate Compliance Program," August 1997.

U.S. Coast Guard Navigation and Vessel Inspection Circular (NVIC) 10-81, "Coast Guard's Certification And Inspection of Certain Categories of Existing Vessels," October 1981.

Discussion

Peter J. Williams, Member

There is an internationally agreed upon trend toward ensuring better enforcement of existing rules rather than the development of more rules. At the same time the international campaign to eliminate substandard shipping is gaining momentum. In line with both these important moves is the USCG's ACP which has many parallels with what is happening in the UK, i.e., rationalization and alignment of regulatory requirements with international standards. Commercial cost advantage, which is gained by unscrupulous ship operators registering ships with flags of convenience, must be eroded. This can be achieved by making responsible flags commercially attractive.

There are two necessary conditions to be satisfied for a classification society to be recognized under the ACP. One concerns port state control performance and the other concerns reciprocity. The former has clear links with safety and should therefore be a key factor. The latter, however, has political/free trade overtones which perhaps do not sit comfortably with maritime safety. When a flag State considers what functions it will delegate, it does so in the context of its own circumstances and expertise. There are significant variations amongst the world's flag States. If the functions delegated by the country in which the classification society is headquartered do not align with what the USCG delegates to ABS, true reciprocity will be distorted.

The ISM Code, for example, is very different from other Convention requirements. The paper does not make this distinction but does touch on the importance of the philosophic principle of the ISM Code not being lost during audit. The paper also rightly highlights the link between the physical condition of the vessel and the Safety Management System. The paper then goes on to describe the invidious position the classification society finds itself in when deficiencies in the physical condition are found and linking this to the SMS. In such circumstances, classification societies are, to a degree, auditing themselves (or their competitor society) in public. The problem does not go away if the same society that verifies the ISM Code also classifies the vessel. Indeed, this may even exacerbate the situation.

There is no doubt that the relationship between flag State, classification society, shipping company and the wider shipping industry is changing. The ACP agrees with these changes and supports the trend toward better enforcement of existing regulations rather than developing more regulations. It does not, however, remove the requirement for a ship to be issued with a certificate of inspection from the USCG, which remains over and above international requirements for ships trading to the USA.

With respect to the changing relationships in the shipping industry there is another quote from President Reagan (repeated recently by Al Gore) that may, in the fullness of time, prove appropriate—"You ain't seen nothing yet!"

Blaine Collins, Member, and **Terje Staalstrøm,** Member

General

We commend the authors for preparing a comprehensive summary of the USCG Alternate Compliance Program (ACP).

We are pleased to offer the following comments to clarify certain points and to stimulate a debate regarding the future of the ACP and possible improvements.

Where is ACP today?

The authors have correctly indicated that there are significant differences between recognition and authorization. However, the distinction between the two-year period during which a class society may issue certificates on behalf of the USCG and authorization to participate in the ACP is not clear to industry and results in confusion and uncertainty. Essentially, for shipowners and shipyards, there is no difference between the two. Therefore, we would recommend that the two-year period be referred to as *provisional, initial, primary* or similar terms to indicate that this period is identical to ACP for the shipyard or shipowner. The sole difference can then be communicated and understood to be USCG specified oversight functions during this period that is entirely between the class society and the USCG.

Authorization to participate in ACP

Subsequent to submittal of the paper, DNV has earned USCG authorizations to issue SOLAS Cargo Ship Safety Construction, SOLAS Cargo Ship Safety Equipment and the ISM Code certificates on behalf of the USCG. We appreciate the USCG's attention to detail and updating Table 1 for the presentation and the final paper for the *Proceedings*.

Recognition and authorization process

It is noted that one of the two necessary conditions for recognition is port state control performance. We agree that the port state control performance is a good indicator of a class society's performance.

We note, however, that this is presently based only upon port state control performance in U.S. waters. In our view, this is an anomaly and the basis of evaluation should be further expanded.

First, U.S. ships, the very ships that will participate in the USCG ACP are not included in the USCG port state control regime. Therefore, consideration should be given for comparable assessment of U.S. ships either as part of the USCG annual classification society port state control performance statistics or via a separate assessment program.

Secondly, the USCG port state regime, while rigorous, represents only 10% of the worldwide port state control statistics. Additionally, it is believed that true substandard owners and operators will not trade their vessels to the U.S. Therefore, consideration should be given to evaluating the classification society performance based upon worldwide port state control statistics.

The ISM code and ACP

Unfortunately, it is correct that substandard ships still exist; however, a number of statistics, such as port state control detentions, indicate that the number of substandard ships are on the decline. Naturally, continued focus on substandard ships is required of all parties who have a role in operating and maintaining safe ships and verification of compliance with standards, rules and regulations.

It is incorrect, however, to assume that the Cartesian plane of safety over time reflects only the minimum standards at the peaks, which occur in conjunction with surveys and inspections. Indeed, the main purpose of a classification society renewal survey is to ensure that the ship is at such a level that it may trade until the next major survey, at which time the vessel is expected to meet at least minimum requirements. Therefore, a significant point is that class society rules and survey requirements are based upon ensuring that the minimum level occurs at the end of the period, not at the beginning of the period.

Crossing the cultural divide

We would argue, rather strongly, that the cultural divide does not exist and, as the USCG gains experience with the ACP, that it will clearly recognize that class societies and the USCG work toward the same goals, which, for DNV are safeguarding life, property and the environment. Indeed, there is a long-standing relationship between Flag Administrations and class societies based upon reliance upon class rules and inspections for structural and machinery items. Over time, with the increase in the world international fleet, growth in international trade and the extreme demands this placed on Flag Administrations to carry out statutory safety and load line inspections, Flag Administrations increasingly delegated these statutory functions to class societies. Similarly, class societies continued to demonstrate the objectivity, and effectiveness, of their third party role. Today, there are a number of debates about the role of class, generally related to how class societies may improve their performance or new expectations of owners, flag administrations and other stakeholders. These discussions are welcome and provide essential input for improvement. It is interesting to note, however, that now, more than ever before, the role of classification societies is an important component of maritime safety.

Shipowners do pay the classification society fees for services under the current model. This has been determined most efficient as, ultimately, the shipowner has full responsibility for the safe operation of the ship, including expenses in connection with operating the ship. So, while there may be a natural relationship based upon mutual goals for safety at sea, there is clearly no concession on the part of any major classification society nor is there any conflict arising from economic pragmatism and class societies. Leading class societies will continue to enforce their rules, the international regulations and the U.S. Supplement to their rules for U.S. ships.

Conclusion

The ACP follows the practice of other leading Flag Administrations to delegate statutory functions to classification societies. This places additional responsibilities upon class societies, which they readily accept, for maritime safety. ACP also provides a number of benefits for all maritime stakeholders. DNV is excited about the ACP's success, its future and support of the U.S. marine industry. Finally, DNV looks forward to demonstrating the nonexistence of the "cultural divide" to the USCG.

C. M. Magill, Member

Among all the very technical papers read to learned bodies, such as SNAME, from time to time there is one of a more esoteric nature which gives an insight into regulatory administration—and some of the politics behind it. Technical innovation and development often enables new standards to be established but, invariably, it is politics that are the driving force behind their adoption, implementation, and enforcement. This paper is one that addresses both the technical and political aspects of the subject.

While the background to the initiation of the Alternate Compliance Program is described by the authors, there are some additional points which may assist in understanding the whole picture.

Until the advent of the ACP it was theoretically possible for the owner of a U.S. flag vessel to class the ship with any society of his choice. However, practically—meaning financially—this choice was not actually available. The regulatory standard for the construction of U.S. ships is specified in the Code of Federal Regulations as being that of the American Bureau of Shipping. If a ship were to be built to another classification society's rules then, after that society has completed its appraisal of the design, a further review of the design would have to be carried out by the USCG to verify compliance with the mandatory standard of ABS's rules. The additional costs involved in such duplicate approvals could not be justified by any owner or builder and, to my knowledge, never were.

The political initiative that enabled an element of choice of classification society to be exercised by U.S. shipowners was the recognition that the rules of certain classification societies, together with their plan approval and survey procedures, are based on similar engineering principles to those of the American Bureau of Shipping and their application in total could be considered to provide an equivalent classification standard to that prescribed in the CFR's. This principle had already been recognized for some time in the Transfer of Class Agreement (TOCA) between the IACS member societies but required, by USCG officers, further detailed examination and comparison of critical safety systems requirements of class societies rules against those of the CFR's. The rest is history and has been well described in the paper.

The authors indicate that ACP is a relatively young program in the U.S. and that it is also a live program that can be expected to be further developed and expanded in the future, under the right conditions. An inevitable consequence of this is that as soon as a paper is written it becomes out of date with respect to certain published information. Table 1 is such an example. The U.K. Administration has just recently indicated that additional delegation for the performance of statutory services on British ships will be given to ABS and LR. This delegation will fulfill the established reciprocity requirements of 46 CFR Part 1, et al., and should lead to LR being permitted to deal with MARPOL Annex 1 IOPP and NLS Certification, IBC Code, IGC Code and MODU Code Certification in due course and the table may be expected to be updated accordingly when all the due processes of the CFR's have been completed. Similar recent developments in Norway will no doubt be commented upon by a contributor from DNV.

My final comment is one on the subject of reciprocity. There are valid reasons for having such a provision in the U.S. regulations and no objection could be made to its inclusion as one of the two necessary conditions for recognition and authorization of "foreign" classification societies. However, the understanding of what reciprocity means can be different in countries outside the U.S. For example, in the U.K. it is considered that equal treatment and authorization to the same degree of each of the class societies recognized by the Maritime and Coastguard Agency (MCA) constitutes full reciprocity. But the MCA still retains responsibility for a significant portion of the possible statutory certification work that could be delegated and has stated an intent to continue this position. Thus reciprocity, as seen from the U.S. perspective, is not recognized as being satisfied in the U.K. for aspects such as ISM Code and Safety Equipment Certification. This is, of course, a political matter that can only be resolved at the two government's departmental levels and all that the classification societies that are affected by it can do is plea for consultation between the governments and a review of their respective understanding of the term "reciprocity." Without this, the element of choice which was intended to be introduced into the U.S. regulations may yet continue to be constrained.

SNAME *Transactions*, Vol. 108, 2000, pp. 57–70

Development of a Shallow Draft Ro-Ro Transport Vessel, the M/V *Delta Mariner*

John L. R. Edgar III, Associate Member, The Glosten Associates, Inc., **Horst F. Nehrenheim,** Visitor, The Boeing Company, **Justin W. Slater,** Associate Member, Halter Marine, Inc., **Warren D. Snider,** Life Member, Foss Maritime Company, and **R. W. "Bob" Triplett, Jr.,** Visitor, Friede Goldman Halter

ABSTRACT

The M/V Delta Mariner *was developed by Foss Maritime Company to provide dedicated transportation services for the Boeing Company's Delta IV satellite launch vehicle. This paper examines the design process that evolved to support the fast-paced vessel development cycle, and the product of this development cycle – a 312 foot, shallow-water-capable, stern-loading Ro-Ro vessel. This unique design process involved a team approach, wherein the United States Air Force, the Boeing Company, Foss Maritime Company, The Glosten Associates, Halter Marine, equipment vendors and regulatory agencies interacted cooperatively in order to make possible the timely completion of this project.*

INTRODUCTION

The *M/V Delta Mariner* is a 312 foot, shallow-water-capable, stern-loading Ro-Ro vessel developed by Foss Maritime Company to provide transportation services to the Boeing Company. The vessel's primary mission is to carry Boeing Delta IV rocket components from a new fabrication facility in Decatur, Alabama, to launch facilities at Cape Canaveral Air Station, Florida, and Vandenberg Air Force Base, California.

To meet Boeing's vessel availability schedule, the product development cycle was completed in nine months, from initial concept sketches to first steel erected. An efficient design communication process evolved to support this aggressive schedule, a process that came to be known as "design lead," referring to the formal transfer of decision responsibility from group to group. The process involved mutual contribution to a common design product rather than focusing effort on "airtight" contract documents and rigid design solutions.

Product – Vessel Development Program

The product followed a typical "design spiral" (Taggart 1980), beginning with creation of design requirements, stepping through the concept, contract and detail design stages, then moving into production design and finally, production of the vessel.

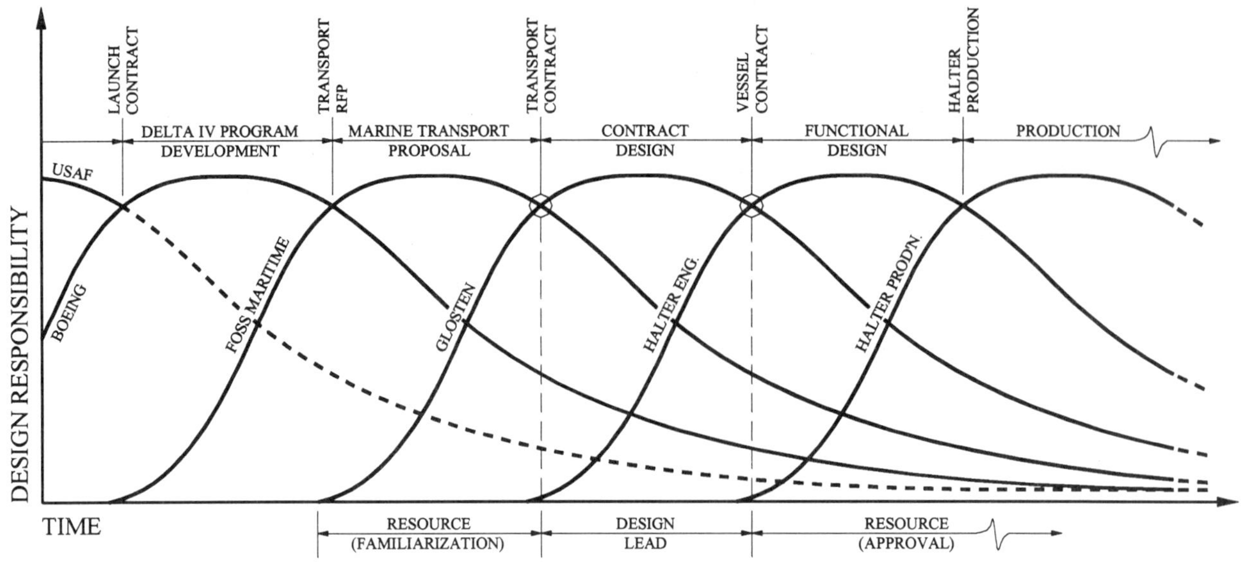

Figure 1 Design Lead Plot

Process – Team Focused Design

A conscious effort was made to create a cohesive team, in which each member would make contributions throughout the development cycle, not just when leading the design. To facilitate the input of expertise and to minimize delays and miscommunication, "design lead" was discussed and formally assigned at each stage of the development cycle.

Figure 1 illustrates the design lead concept. The analogy is essentially an "unrolled" design spiral showing the product transitions between team members. Two important – and often overlooked – design stages are brought out by looking at the design spiral in this way, the "ramp-up" effort in a familiarization role before assuming design lead and the on-going "taper-down" effort in an approval role after relinquishing design lead.

To achieve this team-focused design process, four basic principles were applied throughout the product development cycle:

- Identify project partners early to allow them to be effective members of the project team.
- Recognize and acknowledge the needs of individual team members as partners in a cooperative effort, each with a common goal: successful completion of the project.
- Establish priorities and acceptable tradeoffs between overall project performance, schedule and cost objectives.
- Maintain focus on clear, unambiguous communication between and among all teams.

This paper follows the design lead process as it was applied by the various members of the team involved with developing the *Delta Mariner*.

UNITED STATES AIR FORCE EELV PROGRAM SUPPORT

Product – Satellite Launch Services

In 1995 the U.S. Air Force began the Evolved Expendable Launch Vehicle (EELV) program development and procurement cycle to provide low-cost satellite launch service for national defense and for commercial markets. The EELV program set an ambitious goal of having the launch system in place for a first commercial launch in 2001, with additional launches scheduled through 2020.

Process – Partnership Development

During the first phase of the EELV program, four competitors – Alliant Tech Systems, Boeing, Lockheed Martin, and McDonnell Douglas – completed a 15-month contract to validate the EELV concept. Based on the concept validation effort, the Air Force selected two contractors, Lockheed Martin and McDonnell Douglas (now Boeing), to participate in a second phase, the Pre-Engineering, Manufacturing, and Development phase. In November 1997 the Air Force announced that it intended to introduce and maintain competition in the EELV program by naming both Boeing and Lockheed Martin as launch service providers (Kuhn 1999). Based on an ongoing performance evaluation, Boeing was awarded 19 of the expected 28 government payloads, worth a total of $1.38 billion.

By allowing the EELV launch system to serve a combined defense and commercial satellite market, this program has fostered a unique relationship between the Air Force and the launch vehicle contractors, a *partnership* in which all parties have a vested interest in the overall program success.

THE BOEING COMPANY
SATELLITE LAUNCH SERVICES

Product – Boeing Delta IV Launch Vehicle

The Boeing EELV launch vehicle, the Delta IV, is a new liquid-hydrogen- and liquid-oxygen-burning engine and common booster core (CBC) first stage assembly. Coupled with the already successful Delta II and Delta III upper stages, it creates a family of launch vehicles with payload ranges between 9,285 pounds and 28,950 pounds. The EELV program requires launch services at two launch sites, Cape Canaveral Air Station (CCAS), Florida, and Vandenberg Air Force Base (VAFB), California. Thus the launch providers had to address transportation of launch vehicle components across the United States.

Manufacturing Plant Location

In September 1997 Boeing announced the decision to locate the new manufacturing factory for the Delta IV CBC assemblies in Decatur, Alabama. Decatur – located on the Tennessee River, roughly 450 miles north of Mobile, Alabama – is a prime location for finding skilled labor and the space necessary for constructing the new 1.5 million square foot manufacturing, administrative and support facility. Under any conditions, transport of the CBC assembly is complicated. Although each CBC assembly is relatively light, it is very large – approximately 165 feet long, 19 feet wide and 21 feet high, much larger than all "normal" commercial cargo. Adding to the complication of transporting the CBC assemblies was the "just-in-time" manufacturing scheme requiring a reliable, scheduled transportation system capable of delivering three CBC assemblies, in a single load, to the launch sites for final assembly.

Boeing initiated a number of transportation feasibility studies to review a wide range of shipping options. The shipping options studied include modifying truck or rail equipment to use on specially-prepared overland routes; building a super-sized, custom aircraft to allow transport by air; or developing a marine transportation system using one of several inland water routes connecting the Decatur facility to the Gulf coast.

Marine Transportation System

Based on the studies, Boeing decided to develop a marine system as the primary method for CBC transportation between Decatur and the launch sites, using the Tennessee-Tombigbee (Tenn-Tom) waterway (see Fig. 2). The Tenn-Tom waterway, an elaborate system of canals, locks and rivers leading 550 miles north from Mobile to Decatur, is the most direct shipping route; however, it imposes the two most significant vessel design restrictions – the limited water depth (9 feet in low water conditions) and the restricted clearance heights beneath bridges and overpasses (54 feet in high water conditions).

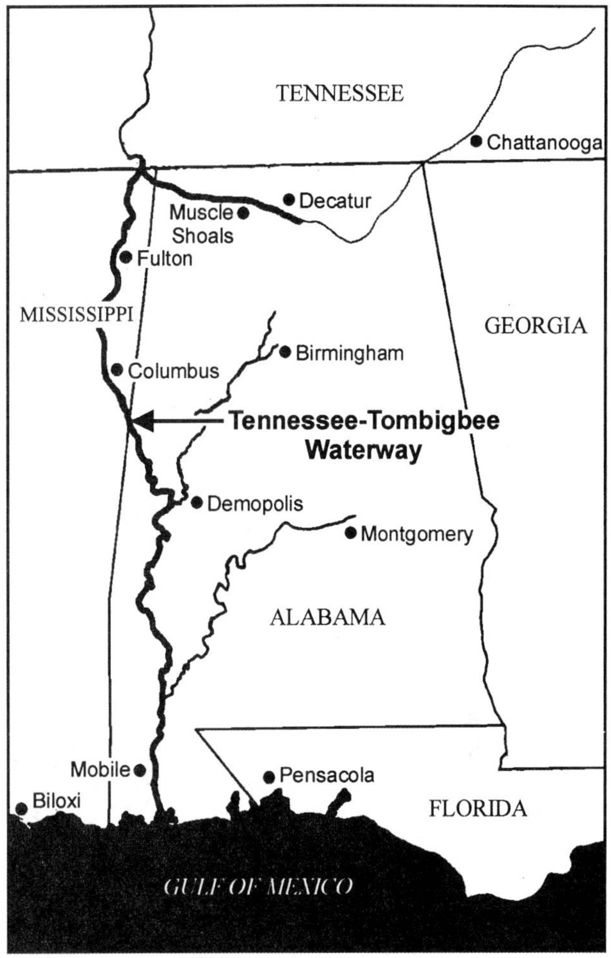

Figure 2 Tenn-Tom Waterway

Process – Marine Transportation Partnership

Having identified a need to develop a marine system, and implementing the partnering strategies fostered in the EELV program, Boeing opted to add a partner to develop marine transportation services rather than obtaining and operating a vessel themselves. Boeing developed a two-part RFP: a general statement of work outlining the program requirements for the 20 year marine transportation services contract (essentially a long term vessel charter agreement), and a vessel performance specification presenting cargo requirements and minimum standards for the vessel. The intent of these documents was to present the requirements to bidders without dictating a specific vessel or system configuration.

The requirements of the marine transportation RFP are outlined below:

- Program Schedule – The vessel must be complete and operational by 30 June 2000 to support transportation system and Delta IV component testing prior to the first commercial launch in 2001.

- Cargo Bay Dimensions – To accommodate three CBC assemblies, stowed side by side with their associated handling systems, the main cargo bay must have clear, stanchion free dimensions of 200 feet long, by 70 feet wide, by 30.5 feet high. In addition to the main cargo bay, 2,200 square feet of cargo area, with a clear height of 18.5 feet, is required to accommodate other Delta IV components.
- Cargo Ramp and Stern Door – The configuration of the cargo ramp and stern door systems are dictated by two requirements, the transport vehicles Boeing uses to move the CBC assemblies and the loading procedures that the transport vehicles impose.
  - *Transport Vehicle* – Boeing manufacturing and assembly facilities use a specially designed Elevating Platform Transporter (EPT), a precision transport vehicle with 18 independently controlled wheels, to move and position the CBC assemblies over relatively short distances. This dual-powered (electric or diesel) hydraulic-actuated vehicle will also be used to move the CBC assemblies on and off the vessel, requiring that the vessel and the stern ramp system comply with restrictions imposed by EPT operations. The primary restriction is brought about by the EPT length (140 feet) and its limited ground clearance which limit the ramp inclination to approximately 6% up or down.
  - *Loading Procedures* – The EPT loading operation requires that each of the three CBC assemblies be loaded straight in and straight out, parallel to centerline. This straight in/straight-out operation requires that the ramp and stern door have a clear width of 64 feet.
- Dock Facilities – The manufacturing facility in Decatur and the launch facilities at Cape Canaveral and Vandenberg all have docks with unique configurations. In particular, each facility has a different dock surface height and vastly differing tide/water level characteristics. The vessel ramp and ballast systems must accommodate these differences with a goal of allowing uninterrupted CBC loading operations.
- Vessel Routing – To serve facilities in Decatur, Cape Canaveral and Vandenberg, the vessel must regularly transit the Tenn-Tom waterway, and the Panama Canal, as well as operate in the Gulf of Mexico and both the Atlantic and Pacific Oceans. The Tenn-Tom Waterway requires extended inland operations at a draft of 8 feet, while the offshore portions require a fully capable offshore vessel – a dual mode transportation system.
- Vessel Service Speed – The just-in-time approach to delivery of CBC components to the launch sites requires that the vessel be capable of achieving an average speed of 14 knots over the entire ocean portion of the route, regardless of sea condition and time of year.
- Reliability – With the marine transportation portion of the CBC assembly process forming an integral part of the manufacturing schedule, reliability of the marine transportation system is critical. A number of requirements are focused on providing increased system reliability and a 20 year service life:
  - *Main & Auxiliary Machinery* – The vessel is required to have two independent main propulsion systems located in separate enginerooms to provide full control in the event of an engineroom fire. Three generators are required for power generation, again to provide redundancy in emergency situations.
  - *Systems Design* – Copper-nickel piping is required on all critical systems. Ballast tanks must have cathodic protection and a high performance coating system. All machinery components must be supported by U.S. based parts and service organizations.
- Regulatory Approvals – The vessel must be U.S. flagged, inspected by the United States Coast Guard, classed by the American Bureau of Shipping, and comply with International Maritime Organization regulations.
- Flight Hardware Transport – Transporting, maintaining and handling the sensitive flight hardware (the CBC assemblies and associated cargo) is a delicate operation, one not typically undertaken by the maritime community. The flight hardware transportation system must comply with cargo acceleration limits, interface with custom securing hardware, and provide monitoring systems, environmental control and security in the cargo bay.
  - *Cargo Acceleration Limits* – Acceleration limits in the main cargo area, based on design criteria used by Boeing in the development of the associated CBC hardware transportation systems, are:

Longitudinal		+/- 1.00 g (collision)
Transverse		+/- 0.40 g
Vertical	1.0 g	+/- 0.62 g

  - *Securing Hardware* – The CBC assembly is transported on two “pallets,” specially designed cradles that are bolted to heavy insert plates in the vessel main deck. Other hardware is transported in custom containers that use standard marine lashings and flush fittings in the vessel deck.
  - *Cargo Monitoring* – Provisions for electronic cargo monitoring equipment, two staterooms for Boeing personnel and a cargo monitoring office are required.
  - *Environmental Control* – The cargo bay must have a salt-water-excluding, forced-air ventila-

tion system capable of providing a 15-minute air change with the stern ramp/door closed.
  - *Security* – The cargo bay must have a continuous video surveillance system and have locked, monitored and alarmed access doors.
- Program Requirements – Because the marine transportation program forms an integral part of the overall Delta IV development program, concise scheduling, risk and issues management planning, as well as regular design and program management reviews are required.

In the fourth quarter of 1998 Boeing released the marine transportation RFP to potential transportation providers, requesting a long term charter rate and conceptual design of the transportation system. The contract award process Boeing put in place follows the partnering strategy fostered by the Air Force. Boeing would award an "advance agreement" – a development contract – to bring a marine transportation partner onto the Delta IV team. This strategy allowed Boeing to transfer design lead onto an experienced marine transportation provider while still maintaining review and oversight control of the transportation system.

FOSS MARITIME MARINE TRANSPORT SUPPLIER

Process – Assemble Design Team

Having identified the Boeing marine transportation system as a worthwhile objective, Foss Maritime Company assembled a design team to prepare the proposal. Foss selected The Glosten Associates to provide design and technical advice to the proposal development effort. Project partners at this stage included Boeing (relinquishing design lead and moving into an oversight and review role), Foss (assuming design lead on the proposal effort) and Glosten (preparing for contract design by providing technical support to the proposal team).

During the concept design and proposal development stage, it became clear that early participation by the building yard would be necessary to meet Boeing's delivery schedule. A series of informal discussions was initiated with a number of shipyards regarding the feasibility of delivering a vessel of this size by June 2000. These discussions led to development of a compressed "mini-bid" process early in the contract design phase. Bids would be obtained from shipyards based on a fairly simple set of specifications and drawings (lines, midship section, deck arrangements, propulsion schematic, weight estimate and major equipment list). Using the mini-bid process, a shipyard would be brought onto the project team (by awarding a notice of intent) early enough to incorporate their design guidance before contract package was completed and a final construction contract signed.

Product – Foss Concept Design Solution

The concept design solution consists of two primary components: a concept design to identify the major vessel features that comply with the vessel performance specification, and a transportation system cost model that allows development of a charter rate based on the concept design and the vessel life cycle costs.

Concept Design

Initially, two vessel concepts were considered:

- A small dual-mode ship, capable of being operated on both the inland waterway portion of the route and the ocean portion of the route.
- A barge (and tug) that could serve the inland portion of the route and could be loaded aboard a submersible ship for the ocean portion of the voyage.

Early in the concept design effort, it became clear that if it were feasible to build a small dual-mode ship capable of operating at the 8-foot draft inland and of achieving the sustained 14-knot speed offshore, that this small vessel would be the simplest, most cost-effective alternative. The barge-to-ship option did not seem to match the vessel implied by the vessel performance specification and Foss's system cost model indicated that the barge-on-ship option was more expensive. Furthermore, the lift operation added additional project risk and increased the delivery voyage duration.

Having made a decision to pursue the small dual-mode ship, the concept effort then focused on refining the life cycle cost model and confirming the feasibility of a small dual-mode vessel.

Vessel Configuration

The goal of the concept vessel configuration was to keep it "simple, simple, simple." Initial designs reflected this thought and the early designs were based on a "powered barge" concept. As soon as the vessel began to take shape around this powered barge, the significant design challenges became evident:

The Tenn-Tom waterway would limit the overall vessel height to 58 feet and require that the vessel be able to operate for extended periods at an 8-foot draft on inland waters. Fortunately, the entire "design cargo load" of three CBC assemblies and associated flight hardware is less than 200 ST, and the voyage profile allows the trip to Decatur with only enough fuel to complete the round trip (approximately 1100 miles) and refuel in Mobile. The 8-foot draft required for inland operations would not be suitable for offshore operations where seakeeping and the propeller immersion would become serious issues. By maximizing the ballast capacity in the hull, a target ocean draft of 12 feet could be achieved. It was clear while the CBC and flight hardware were the primary cargo, ballast water would make up most of the displacement.

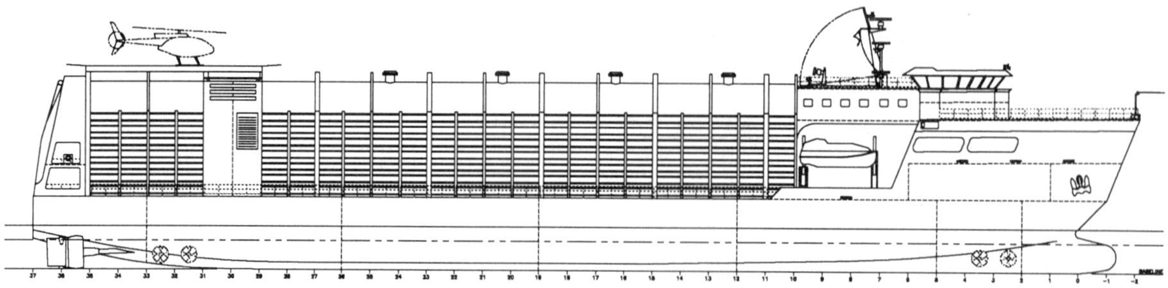

Figure 3 Concept Outboard Profile

To meet the height restrictions, early arrangement concepts located the cargo bays directly behind a low accommodations block. This linear arrangement allowed the entire vessel height to be dictated by the cargo bay height of 30.5 feet above the main deck, but led to a vessel approximately 350 feet long. As costs were generated and reviewed, the low, linear arrangement, though simple, was proving to be too expensive due to its length.

The total height requirements in conjunction with the height of the cargo bay would limit the available hull depth to 20 feet. This was at the upper limit of the ABS allowable length-to-depth ratio, making it likely that significant hull deflections would occur in rough seas. Buckling of the main cargo bay, which spans nearly the entire length of the vessel, was a real concern because of the long transverse spans and the necessarily lightweight structure. To limit the buckling loads, a series of ring frame expansion joints was incorporated in the vessel.

The lower limit on length for the vessel is dictated by the 200 foot long main cargo bay with another 80 feet required for the associated flight hardware. The flight hardware requires only an 18.5 foot clear height (and less than 70 feet in width). Thus by carefully arranging a single deck accommodations block on top of the lower "forward" cargo bay, a vessel was drawn with an overall length just over 300 feet.

Beam, also a cost driver, was to be minimized with an obvious lower limit at 70 feet (the minimum clear dimension of the cargo bay). The final beam was selected to be 82 feet, driven primarily by an expressed desire to meet the clearance limits in approaches to NASA's Stennis static test facility in Southern Mississippi, as well as several other potential cargo facilities. At this stage in the design, the 82 foot beam appeared workable – leaving 6 feet of deck on each side of the cargo bay for uptakes and house structure. As the design progressed beyond the concept stage, this beam selection proved to be the single greatest design and construction challenge, complicating the house and uptake structure, the lifeboat arrangements, the cargo bay ventilation and access routes.

The primary propulsion system concept, conceived to meet the divided engineroom requirements and be as simple as possible, was a twin diesel arrangement, geared to straight shafts and fixed pitch propellers in nozzles. It was clear that getting enough power through the propellers would be a challenge – at an 8-foot inland draft, the propellers would be out of the water and at a 12-foot ocean draft, there would be very little immersion. To comply with draft and speed requirements, it became clear that resistance would have to be reduced, eliminating from consideration the initial simple bluff bow, barge-like hull. Therefore, the concept hull design evolved to include a simple, hard-chined hull with a prismatic bulb and a flat raked stern – see Figure 3.

Cost Model

The cost model consists of two basic parts, the vessel capital costs and the life cycle operating costs. Vessel capital costs were estimated using historical parametric cost data, the driving parameters being principal dimensions, installed power and steel weight. As discussed earlier, this parametric cost model had a significant impact on the subsequent design effort as reductions in vessel dimensions were reflected in capital cost reductions.

The life cycle operation costs relied on Foss's vessel operating and maintenance cost data, as well as specially developed voyage duration models, crewing cost studies and CBC delivery schedules. The life cycle cost model became the primary tool used during the proposal discussions with Boeing, and was used extensively to guide the decisions made for the contract design.

Based on the marine transportation proposals and subsequent extensive interviews with bidders, Foss Maritime Company was awarded the advance agreement to develop the Delta IV marine transportation system. Almost immediately after the award of the advance agreement, Foss Maritime and The Glosten Associates began work on the contract design for the transport vessel.

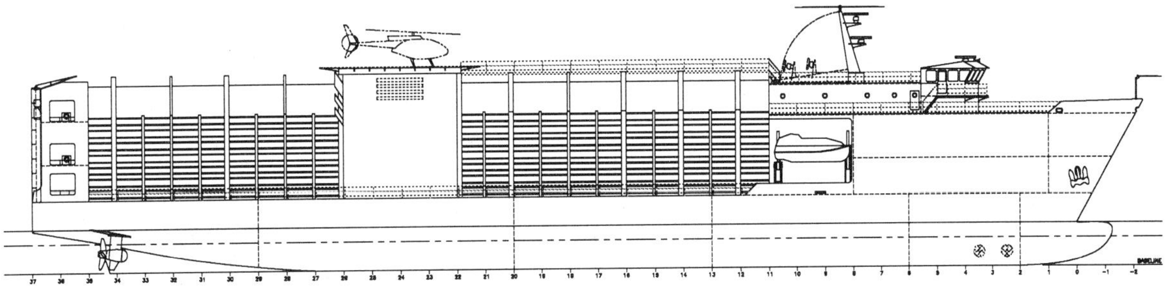

Figure 4 "Mini-Bid" Package Outboard Profile

GLOSTEN ASSOCIATES CONTRACT DESIGN PROGRAM

Process – Develop Design Partnership

The need to bring a shipyard onto the team as early as possible led to two distinct phases of the contract design: development of the mini-bid package and completion of a full contract design package. The mini-bid package was issued in April 1998, and Halter Marine Group (now Friede Goldman Halter) was awarded a notice-of-intent the following month, thus adding the vessel builder to the project team. As the contract design progressed, having Halter's input was valuable and allowed the contract effort to be "tuned" to address Halter's specific building preferences.

Next a concerted effort was made to identify areas in the design that would not be completely resolved at the end of the contract design cycle. These areas included the final crew size, details of the thrusters, final selection of main engine, generator, thruster, ramp and lifeboat vendors, and details of the electronic command control and communication systems. The reasons for delaying resolution of these issues was primarily that time was short and that the solutions could be "bracketed" so that all remaining options were similar and would not affect the contract design.

While some issues were left unresolved, a number of "detail" issues were found to be critical to the feasibility of the overall design. Based on discussions with Halter, the design effort on these items was carried to a level beyond that typically undertaken in a contract design. The primary areas where extra contract design effort was applied were the uptake arrangements, the cargo bay ventilation systems and the vessel stability documentation (intact and damaged).

Halter's efforts were valuable in assessing the cost impact of various options. Two items in particular required Halter's input – the cargo bay expansion joint configuration and the hull form. The decision to delete the cargo bay expansion joints traded added steel weight for a simpler structure. The Marine Research Institute Netherlands (MARIN) indicated that further revision to the hull form would be necessary to achieve the desired propulsion power. The changes were an up-swept bulb (to improved the overall resistance) and a slightly tunneled stern to accommodate the largest possible propellers. These changes in hull form were reviewed with Halter and although slightly more expensive, could be accommodated within the schedule constraints.

Product – Contract Design Documents

By focusing on aspects of the design that are unique to this vessel – hull form, arrangements, tank locations and primary structural systems (including the cargo house) – the mini-bid package presented a broad overview of the vessel without taking time to define the more standard systems (for example, the mechanical systems) or to refine the weight estimate beyond simple parametric estimates – see Figure 4. The contract design package reflected the basic design presented in the mini-bid package, but included significantly more detail and included input from the shipyard on a number of points – see Figure 5.

Hull form refinement, both in terms of an accurate resistance model and to ensure the vessel could operate effectively at the 8-foot river draft and the 12-foot ocean draft, was a critical early design task. Because model testing would be an important part in assessing both these issues, MARIN was contracted to develop and refine a candidate hull form.

The first iteration of the hull form showed clearly that the simple, hard-chine hull that formed the basis for the concept design was not acceptable; the resistance was too high. MARIN recommended a hull very similar to the shallow draft European ferries with a fair form bulb, slack bilges and a slightly vee'd stern run.

Arrangement decisions made in the concept design were refined for the mini-bid package to include underdeck spaces and tanks. While not specifically required by the Boeing RFP, Foss elected not to carry fuel in skin tanks and to provide flooding protection in way of all critical machinery spaces (the main machinery and thruster rooms). Access to spaces below the main deck, (in particular to the engineer's control center and the main machinery rooms) is complicated by the main cargo bay,

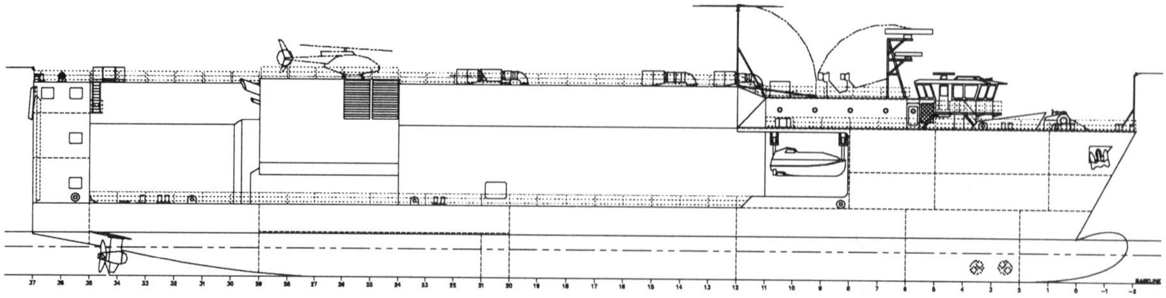

Figure 5 Contract Outboard Profile

which fills the main deck and is not available for crew access due to Boeing security concerns. Access below deck is provided by incorporating a longitudinal access tunnel running on centerline from the forward machinery space to the main machinery rooms.

Based on preliminary conversations with shipyards, the direct shaft propulsion scheme used in the concept design was dropped in favor of azimuthing Z-drive units. Although the initial cost of a Z-drive system was slightly higher, the construction scheduling flexibility afforded by the unitized system was very attractive in this tightly scheduled effort. In addition, the enhanced maneuvering afforded by the Z-drive units is desirable in the restricted inland waterways.

Two very different proposals were received from ramp vendors: a double door system that separated the main watertight boundary from the ramp surface itself and a single ramp/door system. Both options had merit and potential drawbacks, so vendor selection was delayed until the shipyard could be brought in on the decision.

HALTER MARINE DETAIL DESIGN PROGRAM

Process – Define and Build Supplier Team

The timing of this project allowed Halter to incorporate the new strategies identified in an ongoing internal review. Known as the "To-Be" process, this is an in-depth review of ship design and production processes and procedures at Halter. Changes made as part of the To-Be process pushed the project team toward more effective project management, design reviews, improved scheduling and separation of the detail design process into a functional design and a production design.

The functional design phase allows a soft start on the detail design effort, giving the shipyard engineering group time to review the contract package for compliance with yard procedures and preferences, to identify and order major equipment, to understand the overall contract design and identify outstanding design issues, and to develop the vessel build strategy using production as a resource. Foss and Glosten continued to refine the contract design in some areas as Halter proceeded with functional design in others. Particularly in some of the difficult structural areas such as the transitions fore and aft of the stacks, design effort by Glosten continued even after production design of the early innerbottom modules had begun and was nearing completion. Indeed, Glosten's knowledge of the history behind these areas and Glosten's computer finite element model were utilized to satisfy ABS questions even after construction of the cargo bay top was well underway.

With Halter acting as a resource during the Glosten contract design, the early efforts enhanced the design understanding, fostered confidence in the design team and allowed a direct transition to Halter's leading design decisions. A good example of this transition is the lifeboat pocket where efforts to minimize the beam of the vessel created an extremely tight area. Early detail sketches prepared by Halter allowed selection of a lifeboat/davit combination that met all requirements but still fit in the very limited space available. Many of Glosten's comments could be given almost instantaneously because of their knowledge of the design evolution and history. The time saved by this process, compared with Halter's engineers having to "reinvent the wheel" by gaining all the project knowledge themselves, was invaluable to the project.

Foss's lean construction management approach did not focus on detailed drawing reviews but instead worked to provide quick, upper-level reviews of the large volumes of drawings received during peak periods. Furthermore, due to the tight construction schedule, many of the drawings sent for owner review were also issued to the yard for construction the same day; therefore, fast reviews by people familiar with the design history were critical to maintaining schedule. Thus reviews were often completed by the engineer who developed the original design and were guided by asking "will this work?" rather than by checking to make sure an exact copy of every contract design detail had been incorporated. Glosten's willingness to sacrifice some pride of ownership of the design and Foss's willingness to accept Halter's design decisions reinforced Halter's design leadership. With

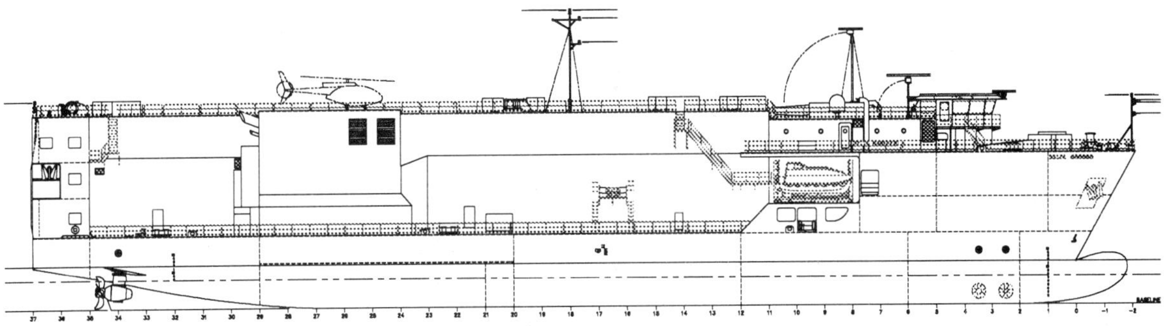

Figure 6 Final Outboard Profile

design leadership, the creative, problem-solving spirit is encouraged – not just to satisfy minimum requirements of the specifications, but to build a vessel that will function properly and that everyone can be proud of in the end.

Vendors on Team

Suppliers have important relationships to both the shipyard and the vessel owner. They must support the shipyard by providing project information, engineering assistance during functional and production design phases, schedule support and warranty support. It is also desirable for suppliers to have a longer-term relationship with the vessel owner, providing parts, repair and technical support for equipment operation over the life of the ship. In some cases, it may make sense for the vessel owner (rather than the shipyard) to procure key equipment items in order to establish this relationship. Either way, the ideal case is to have key suppliers maintain an on-going relationship with both the shipyard and the owner.

To maintain the schedule of the project, the vendors and subcontractors needed to be selected early and made part of the design team. Performance based procurement packages issued to vendors and subcontractors encouraged – and in several cases, required – participation and review of different design alternatives for their systems. Perhaps the best examples of vendor participation in the design process are the stern ramp and the electronics/ vessel management system.

The stern ramp RFP included an option to supply a complete door frame assembly with the ramp. This frame assembly allows the vendor to pre-mount, align and test the door and the hydraulic locking systems in a structure that fits directly into the vessel's structure, forming the aft end of the cargo bay and the upper part of the transom. While the design of the frame interfaces required close cooperation and careful communication between the ramp vendor and Halter's engineers, the resulting door and frame system allowed complete installation, fully welded-out and ready for connection to the pre-packaged hydraulic power unit, in less than two weeks without significant overtime or the need for multiple shifts.

The electronics/vessel management system is an area where the project team did not have all the expertise needed to produce a complete design. Vendor selection relied on a procurement package that outlined the ship's mission requirements, identified areas where equipment or design was firmly fixed, highlighted areas where vendor design input was necessary and presented a "wish list" of equipment that was desirable but not essential. All the potential vendors were asked to provide a two-part proposal. First, a proposal for a baseline electronics / vessel management system meeting all applicable river and open-ocean requirements and including (as much as possible) all firm owner preferences. Secondly, vendors were asked to develop estimates for any vendor recommended upgrades and improvements as well as the optional "wish list" items. The most competitive proposals were reviewed in detailed meetings with each of the short-listed vendors. The final selection was based not only on price but also on evaluations of vendor expertise in solving the problems presented and an internal assessment of the long-term system performance.

Because the procurement packages were sent out early, the team was able to evaluate the vendors' efforts and alternatives. This allowed the critical systems to be procured as early as possible and allowed the vendors' suggestions to be incorporated into the developing design. In addition, early vendor selection allowed the vendors' schedule constraints to be integrated into the overall vessel assembly schedule decisions, preventing the situation where systems and equipment are compromised because of unrealistic overall scheduling.

Regulatory Involvement

A unique (and, at the time, potentially risky) aspect of the *Delta Mariner* construction program was the decision by the design team to participate in the American Bureau of Shipping (ABS) and United States Coast Guard (USCG) Alternate Compliance Program (ACP) (ABS 1999). This was the first vessel under 90 meters to be reviewed completely under this program. In fact, the vessel was actually accepted for review under the ACP prior to all of the rules being finalized. For this new application of the ACP to be successful, participation on the design team by both ABS and the USCG was required. During the early stages of the project, meetings were held with both ABS and USCG to acquaint them with the vessel's unique requirements and to identify issues that were thought to be "gray areas" requiring interpretation. These meetings not only allowed the members of the regulatory agencies to became familiar with the vessel, but also allowed them to become acquainted with the other members of the team. These relationships allowed the design to proceed with a confidence that the chosen path was reasonably assured of later approval, and also improved the ability of the team to solve future design and construction problems.

Further meetings with ABS and the USCG were held approximately twice a month during the functional design, in order to discuss alternatives for compliance with various regulations. This was extremely important since the "dual mode" river and open-ocean operations often involved regulations that, if not directly contradictory, make compliance with both very difficult.

Navigation lights are a good example of this sort of regulatory contradiction. Compliance with the ocean regulations involves lights that are not required for river operation, and it requires vertical separation of lights which is not attainable due the air draft restrictions in the river. Up-front planning and close cooperation between the designers, ABS and USCG allowed a compromise arrangement that meets the intent of the regulations, yet provides sufficient bridge clearance when running in the river (the primary ocean masts fold and stow flat on deck).

Another important aspect of the teaming relationship was USCG Inspection. Although the vessel was reviewed under the ACP program, the USCG wanted to be active in several key areas of the *Delta Mariner* design and construction to help further define guidelines for ACP implementation. The most important of these items were safety and lifesaving issues. To stay involved, the USCG inspector made weekly visits to the vessel to coordinate with the ABS surveyor. Not only were the meetings a chance for the inspector to tour the vessel and review progress, but they also allowed time for the other team members to get information and input into the design and construction of the vessel.

Product – Detail Design Documents

The two-part functional and detail design process incorporated by Halter produced two important products: first the detail design documents themselves (scantling plans, piping diagrams, outfit details), and, second, an unparalleled understanding of the entire vessel and its design constraints.

This design understanding allowed the fast, accurate design decisions necessary to support an aggressive production schedule. The confidence that Halter's design development would be accepted by both the owner and regulatory bodies allowed production drawing development (in particular the three-dimensional structural modeling and pipe routing) and production itself to proceed well before the entire vessel detail engineering package was completed.

HALTER MARINE VESSEL PRODUCTION PROGRAM

Process – Engineering and Production Teaming

During functional design, the input of the production team was needed. The shipyard decided how they were going to build the vessel and from this developed an assembly breakdown and established the size of the assemblies. In addition, they developed the construction schedule. As a consequence, the functional and production design schedules for engineering could be developed.

Team formation continued with ongoing formal communication at weekly meetings involving Halter engineering, production and program management, along with various subcontractors and with the owner included whenever possible. Design and production issues, as well as scheduling and general production concerns, were resolved efficiently in these meetings.

Glosten continued to assist Foss with review of the production design product; therefore, knowledge of the project was maintained and kept up to date among members of the design team. The relationships established during the functional design allowed open discussion of production design issues and resulted in development of quick, mutually acceptable solutions.

This communication also allowed the schedules of the various groups to be adjusted when "work-arounds" were required as a result of difficulties in design or production. For example, the bow section was delayed in the schedule and the stern assemblies expedited when design difficulties with the bulbous bow structure made it apparent that the drawings for the bow could not be completed on time. Other efforts included "staged" issuance of piping drawings for some assemblies where the drawings for the larger diameter pipe were issued and construction begun while design of the smaller diameter systems continued.

The ongoing teamwork and close communication between engineering and production allowed each assembly to include a high level of outfitting. Many systems, particularly involving large diameter pipe, handrails, ladders and structural foundations were incorporated at the assembly level. This early outfitting eased outfitting schedule pressures, providing significant production flexibility with only minimal engineering cost increases.

Product – The *M/V Delta Mariner*

The vessel production program followed a typical work breakdown system as implemented in many modern shipyards. Installation access and delivery schedule of major machinery dictated the primary assembly erection sequencing. The midship hull assemblies of the *Delta Mariner* were erected on the ways first, followed by the forward and aft hull assemblies, the uptake and focsle assemblies, and the accommodations block. The aft main deck and cargo housetop assemblies were last to be erected, providing clear crane access to the main engine rooms and stern thruster rooms until very late in the construction sequence.

SUMMARY AND CONCLUSIONS

The focus on developing a design team, as opposed to unilaterally developing a design, allowed the project to benefit from the skills and talents of all team members throughout the design process. By actively identifying and formally assigning design lead as the project evolved, efficient transitions took place without slowing the overall design process or losing the important design history and continuity.

Finally, by keeping all team members involved in the ongoing design process, the final product was measurably improved (see Fig. 6). The *Delta Mariner* was delivered on 17 May 2000, nearly 2 weeks ahead of schedule, and is presently being prepared for operational testing with Boeing equipment.

At delivery, the final design lead transition takes place, with the entire design team shifting to a support role as the vessel and her crew prepare for operations. We will all look forward supporting the *Delta Mariner* as she delivers the first of many Delta IV launch vehicles in the fall of 2000.

REFERENCES

1. ABS, *U.S. Supplement to ABS Rules for Steel Vessels under 90 Meters (295 Feet) in Length for Vessels on International Voyages,* American Bureau of Shipping, 29 March 1999.
2. KUHN, Chief Master Sgt. T., USAFR. "EELV: A new Rocket for the Millennium." *Airman*, March 1999.
3. TAGGART, R. *Ship Design and Construction.* New York. Society of Naval Architects and Marine Engineers, 1980.

APPENDICES

1. *M/V Delta Mariner* Principal Characteristics
2. *M/V Delta Mariner* General Arrangements.

APPENDIX 1

M/V DELTA MARINER
PRINCIPAL CHARACTERISTICS

Length Overall	312'-7"	River Mode Draft	8'-0"
Length Between Perpendiculars	296'-0"	River Mode Displacement	3,300 LT
Frame Spacing	8'-0"	Ocean Mode (Design) Draft	12'-0"
Installed Power	8,000 BHP	Ocean Mode Displacement	5,600 LT
Design Cargo Deadweight	180 LT	Maximum Cargo Deadweight	3,780 LT

<u>Tank Capacities:</u>

Fuel Oil	163,800 gallons	Potable Water	24,300 gallons
Ballast	4,050 LTSW	Fresh (flushing) Water	39,750 gallons
Lube Oils	6,300 gallons	Waste Oil & Water	6,300 gallons

APPENDIX 2

M/V DELTA MARINER
GENERAL ARRANGEMENTS

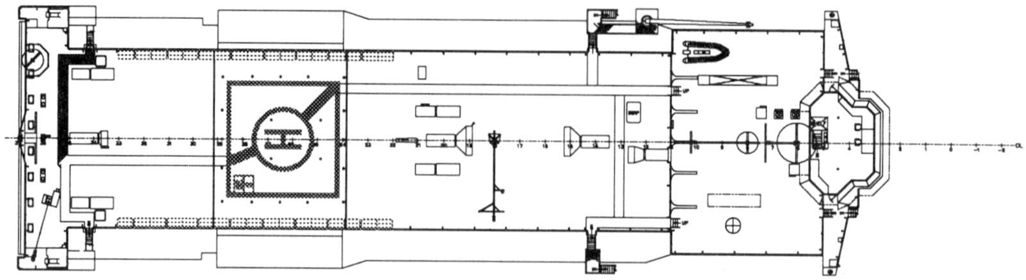

PILOT HOUSE & HOUSE TOP

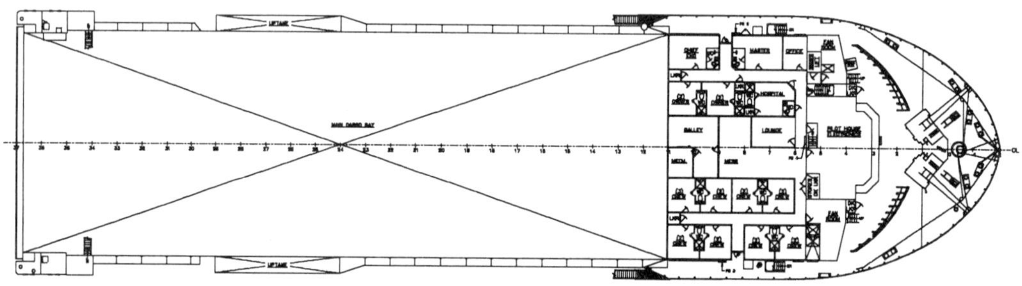

02 LEVEL

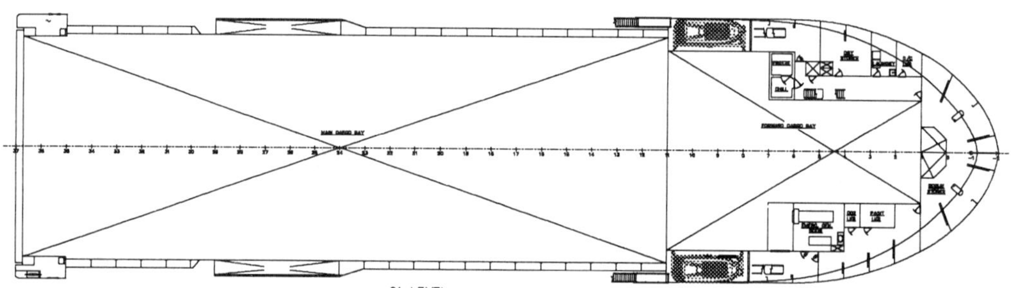

01 LEVEL

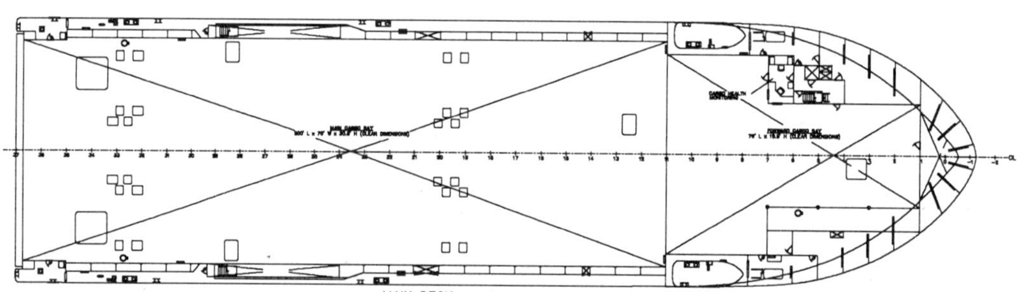

MAIN DECK

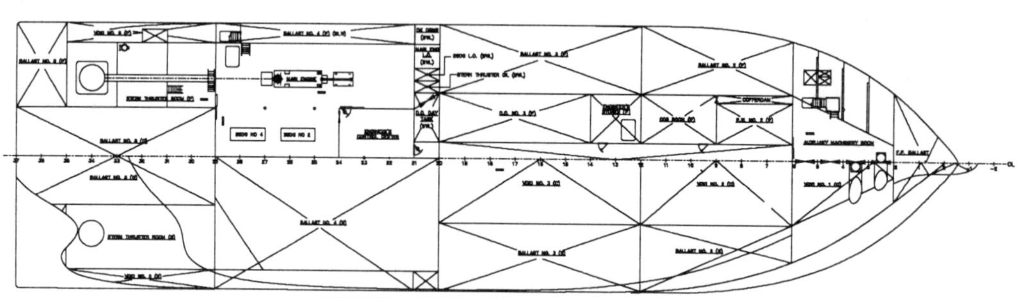

INNERBOTTOM AND HOLD PLAN

Discussion

Jay P. Carson, Member

This is an excellent case study of a collaborative approach to ship design, involving a buyer of transportation services, the ship operator, a design agent and the builder. By clearly assigning roles during each stage of design, and focusing on good communications, it appears that the participants developed a unique vessel to an aggressive schedule. Several techniques are particularly noteworthy, and I would appreciate hearing more from the authors about the following aspects of the project:

1. *Use of a cost model for design decision-making*—It is often difficult to gain consensus among owners, operators, and builders on the costs of design, construction and operation; it appears that all participants contributed data to a cost model that was accepted as sufficiently realistic to guide major design decisions.

2. *Focus on the significant few rather than the trivial many*—Rather than attempt to develop a contract design package with all systems defined, it appears that the participants focused on aspects that were unique to the design or drove mission performance; conventional systems appear to have been defined during functional design.

3. *Design for producibility*—Build strategy and shipyard input appear to have been incorporated into the design through relatively early shipyard involvement.

4. *Concurrent design and review*—By having the buyers, sellers, design agent, and regulatory bodies involved throughout the design, it appears that the conventional heel-to-toe process of design-review-revise was shortened, and that reviewers focused on key features rather than details.

Forming a team from different companies can be a significant challenge, often requiring extensive planning, formal and informal guidelines, coaching, and support from information technology. What activities, if any, were conducted up front to make this a high-functioning group?

Collaboration sometimes requires participants to suspend traditional protections and self-interest. What type of legal agreement(s) were used to provide incentives for collaboration while ensuring that the performance of each participant was complete, accurate and timely?

Finally, what are the top three things that the participants would do differently in managing this project?

Brian Kraut, Visitor, Seacoast Electronics, Inc.

I would like to congratulate the authors on both a well-written and informative paper and on the on-time completion of a very complicated vessel. The success of this project in spite of the technical hurdles and tight time constraints speaks well for the team approach to designing and building this vessel.

The authors have made an excellent point that vendors and subcontractors need to be selected early and made part of the team. Their selection of vendors on the basis of performance-based packages rather than a rigid set of specifications gives vendors much needed flexibility to provide the best solutions and not just the lowest cost equipment to meet the specifications. The contract also allowed for the proposed packages to be fine-tuned as the vendors worked together on the design issues with the other team members. It would be helpful if the authors could describe how they evaluated these bid packages to select their vendor team members.

Seacoast Electronics is pleased to have been chosen as the supplier of the bridge electronics, consoles, and vessel management system for the *Delta Mariner.* We successfully teamed with Nautronix for vessel management and Fleetway for integration and regulatory compliance in the spirit of teamwork described in the paper. The process described by the authors is undoubtedly a model that will be followed many times in the future. It's success depends not only on effective electronic communication but also on personal contact among all parties, which in our case was fostered by events such as our pilothouse mockup review and by Halter providing facilities in their engineering office for our integrator to work alongside their ship designers. With regard to electronic communication of design data, what standards were used and do the authors consider that they were satisfactory?

There were several design challenges due to the unique mission requirements of this ship. One challenge due to the dual ocean and river modes required was the selection of the radars and location of antennas to be effective in both modes. Original plans called for two radars on tall masts for ocean use that could be folded during river transits. Separate low-mounted small radars fore and aft would be used in the river. By discussions with Foss and Halter, Seacoast was able to gain an understanding of the problem and offer an acceptable solution using four scanners, only two displays, and special inter-switching.

Another unique design challenge was in the vessel automation system. The Nautronix vessel management system supplied for this ship is much more extensive and capable than would be found in a comparable cargo ship, largely due to the unique additional requirements for ramp angle monitoring and ballast control during loading and unloading cargo, and for numerous cargo condition and security sensor and display functions, as described in the Boeing Company section of the paper. Seacoast and Nautronix developed these features after contract award, with the active participation and cooperation of Boeing, Foss, Glosten, and Halter. Furthermore, timely development of the traditional machinery control and monitoring features of the VMS would not have been possible without the freely given support of Glosten and Foss, especially with regard to information for owner furnished equipment.

Another design challenge was in the bridge consoles. It was the responsibility of the Seacoast team to design the consoles to meet Foss's requirements. This was an area that required a great deal of teamwork. Three-dimentional CAD console drawings were done by Seacoast and commented on by the other team members. The design evolved over several weeks. The input of all of the team members ensured that the final design would meet all ergonomic, visibility, manufacturability, and regulatory requirements. A meeting between all parties involved in the console design using a full-scale mockup and a 3-D CAD photorealistic rendering was done to finalize exact equipment placement and console shape.

I would like to thank the authors for presenting their approach to the design and construction of this vessel. They have done an excellent job at both describing the team approach and implementing it.

Authors' Closure

Mr. Carson: We appreciate Mr. Carson's thoughtful and thought-provoking discussion.

Mr. Carson commented briefly on four aspects of the project. Our response to these comments is a resounding "Yes, absolutely correct!"

- Participants must contribute to all aspects of the decision-making process, including development of cost models.
- Maintaining focus on participant's unique expertise and design understanding is critical to allow contribution to the design process without getting caught up in trivial details.
- Early yard involvement is the only way to effectively incorporate build strategy into the design process.
- Concurrent design and review effort is essential to identify and resolve key issues efficiently.

We agree that forming a cohesive, high-functioning team can be difficult. For this effort, open and honest communication, facilitated by frequent, widely distributed e-mail, and periodic design review meetings formed the foundation for developing a cohesive team.

Regarding legal agreements, it was our goal, and we achieved that goal, to rely on a motivated design team, the trust gained through prior working relationships, and the incentive offered by the potential for future work to ensure project performance While typical contracts and related legal agreements were in place throughout the project, they were not used as performance "incentives."

Finally, Mr. Carson asks what three things the participants would have done differently in managing this project. This question led to a lot of discussion between the authors and we must acknowledge our good fortune in having had innumerable things go so well in spite of our best efforts. That said, we couldn't agree on just three items—we offer the following four:

1. **Begin design review meetings earlier in the design process**—possibly as early as concept design.
2. **Transition smaller pieces of the design sooner**—to create a gradual taper on the "ramp-up" effort.
3. **Bring the command, communication, and control "integrator" on board earlier**—so that this critical person has time to understand the design before working towards implementation.
4. **Work to eliminate paper**—project communication relied heavily on e-mail and other electronic files. Paper became a communication bottleneck, being both difficult to distribute and to locate.

Mr. Kraut: We are grateful for Mr. Kraut's valuable perspective on this project. As one of the many outstanding vendors and subcontractors forming an integral part of the design team, his comments are both interesting and reinforce the importance of a team approach in a fast-paced design effort.

Regarding bid package evaluation, performance based specifications allowed and encouraged flexibility to develop the "best" solutions. Therefore, selections had to be based on a mix of objective and subjective measures. The evaluation process relied heavily on close cooperation between the owner and the yard to balance near-term yard concerns (first cost, installation support) with longer-term owner concerns (operating costs, maintenance and support).

Electronic communication was a critical element in the success of this project. Fortunately, standardization was not a significant issue for this project—a personal computer and an e-mail account provided full access to any team member. In future projects, we will certainly address standardization (particularly of CAD data) more carefully. Initially, we found CAD data difficult to distribute widely because many team members did not have access to CAD software. To alleviate this bottleneck, the team grew to rely heavily on "neutral" format graphics files (such as Adobe Acrobat) to provide simple, inexpensive viewing access to CAD data.

SNAME *Transactions*, Vol. 108, 2000, pp. 71–98

A Tug & Barge System for Sea and River Service

Henk H. Valkhof, Visitor, Maritime Research Institute Netherlands, **Teun Hoogeveen,** Visitor, Marine Heavy Lift Partners bv., **Reint P. Dallinga,** Visitor, **Serge L. Toxopeus,** Visitor, and **Timo F. Verwoest,** Visitor, Maritime Research Institute Netherlands

ABSTRACT

In 1997, Marine Heavy Lift Partners (MHLP) and the consultancy firm MARVECO discussed with the Maritime Research Institute Netherlands (MARIN) the design and development of a new tug & barge concept. The combination has to travel between the United Kingdom and Germany, where it will be an important part of a highly reliable logistic chain, comprised of production facilities, road transport and waterborne transport. The combination has to cross the North Sea and will make use of the river Waal / Rhine through Holland up to at least Emmerich in Germany. To become this important link it is obvious that the probability of delays has to be minimised as much as possible. The production process cannot be stopped, while at the same time the storage facility on the assembly location has only limited capacity.

Another aim of the research programme was to show that a seagoing tug-barge combination could perform equally well or even better than conventional means of transport. To achieve the targets set, it was necessary to have a combination with the possibility to easily exchange the river- and sea-tug by keeping the barge unchanged and thus avoiding transhipment and related time loss. However, such a concept required extensive studies, not only to guarantee good propulsive properties both in shallow and deep water, but also good seakeeping and manoeuvring properties.

Moreover, the operability of the ship in all conditions (on sea, on the river, in shallow water, etc) was studied and the chance of delays significantly reduced to less than 1 per cent.
This paper will describe the design process from the determination of the main particulars up to the hull form development with the aid of potential flow codes and the series of model tests carried out to determine the calm water, the seagoing and the manoeuvring properties. Furthermore, the logistic aspects, the environment of the sea and the rivers to cross and their implications on the concept will be presented.
Given the very promising results and moreover in view of the increasing congestion of the European roads, particularly in Holland, two of the combinations are under construction now and will more or less start their services in the course of the year 2001.

NOMENCLATURE

WD	- Water depth	[m]
$\mathbf{L_{PP}}$	- Length between perpendiculars	[m]
$\mathbf{L_{WL}}$	- Length on waterline	[m]
B	- Breadth moulded	[m]
$\mathbf{T_F}$	- Draught moulded at fore perpendicular	[m]
$\mathbf{T_A}$	- Draught moulded at aft perpendicular	[m]
T	- Mean draught	[m]
$\mathbf{R_{WD}}$	- Resistance at corresponding water depth	[N]
$\mathbf{R_\infty}$	- Resistance at unrestricted water depth	[N]
$\mathbf{R_s}$	- Ship resistance	[N]
$\mathbf{P_s}$	- Required shaft power	[kW]
$\mathbf{V_s}$	- Ship speed	[kn]
$\mathbf{C_b}$	- Block coefficient	[-]
$\mathbf{F_x}$	- Longitudinal force	[kN]
$\mathbf{F_y}$	- Transverse force	[kN]
$\mathbf{F_z}$	- Vertical force	[kN]
$\mathbf{H_s}$	- Significant wave height	[m]
$\mathbf{T_p}$	- Peak period	[s]
$\mathbf{T_2}$	- Mean zero-upcrossing period	[s]
$\mathbf{U_{10}}$	- wind speed at 10 m height	[m/s]
$\mathbf{R_{AW}}$	- Added resistance in waves	[kN]

INTRODUCTION

The increasing volumes of cargo in the European Union, in particular between Central Europe and the United Kingdom have resulted in congestion on the roads from Germany to North Sea ports like Rotterdam, Antwerp, Zeebrugge. At the same time it can be seen that existing waterways like the river Rhine have much more cargo carrying capacity than is used presently. One of the drawbacks of river transport has been that cargo always needs to be transferred from an inland vessel into a seagoing vessel to cross the sea. Low air draught seagoing ships, capable to sail both on the river and at sea are rather limited in capacity. Up till now the two environments were difficult to match for large ships.

The challenge was to design a system of waterborne transport, which would combine two environments with apparently conflicting operating conditions. At sea, a deep sea propulsion system would be required to cope with wave conditions normal for the southern part of the North Sea, and on the river the propulsion system would have to be able to operate in very shallow draught conditions.

Apart from the design and the development of this concept, it was even more important to attract sufficient volume of cargo of the desired type. To define the type of cargo, a study was done into cargo flows distinguishing between various types of cargo and various types of transportation like trucks, trains, and ships. A different study was initiated to define most likely ports of loading and discharging along the river Rhine and on the British East Coast. Leading logistic operators controlling large flows of cargo in both directions between Germany and the United Kingdom were consulted. It appeared that it was necessary to operate the push-barge system in an existing logistic flow. Therefore a very tight sailing schedule is required, allowing almost zero delay for each shipment independent of seasons, waves and weather.

Environmental conditions are also playing an important role, like the reduction of CO_2 when individual trucks need no longer drive between their destinations in Germany and in the United Kingdom. Other requirements appear to play a role as well. The push-barge should not generate waves of unacceptable significance while sailing on the river, to protect the shores. Of course the height of the bridges and differences in water levels of the river have also been of influence on the design of the combination.

All these considerations have lead to the following system of push-barges:

- a standard river-going pusher tug on the river
- a sea-going pusher tug at sea.

At the stern of the push barge, arrangements are available for both vessels to connect with existing and well-proven coupling systems. On the river, coupling will be effected in a conventional way using winches and wires, while at sea the ARTICOUPLE system has been selected.

With the ARTICOUPLE system the tug and barge can be coupled simply at various combinations of draughts. Two basic types of couplers have been developed, friction-engaged and teeth-engaged, each with their specific characteristics. For this project the K-coupler (teeth-engaged) is used which is heavier in construction and capable to withstand practically any type of weather and sea and are accepted by the international classification societies for normal class notations. A schematic drawing is included in Figure 1.

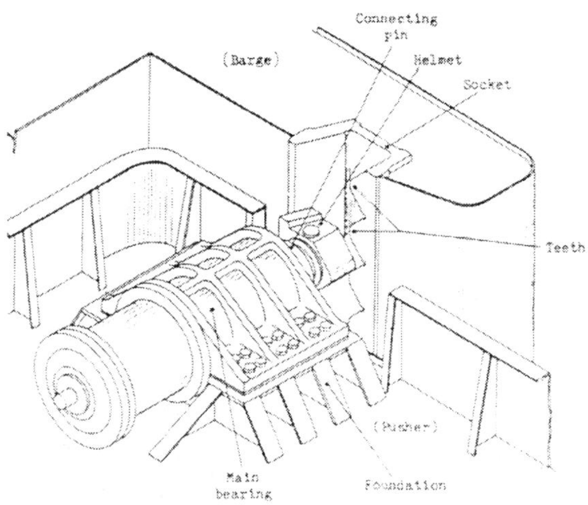

Figure 1: The ARTICOUPLE K-couple version

It might be clear that this special concept requires a thorough study to ensure the high performance needed to successfully operate the combination. Hence, beside the design considerations, CFD-tools (Computational Fluid Dynamics) have been used to optimise the hull form and the connection between tug and barge. After this, a series of calm water model tests were carried out both in deep and restricted water depth conditions. Given the demand for good and safe sea-going properties a series of seakeeping and manoeuvring tests have been carried out. This paper will describe all aspects in more detail on the following pages.

DESIGN CONSIDERATIONS

As mentioned in the introduction the design of the tug & barge concept and its possible success greatly depends on its place in a logistic chain. Below the design considerations are presented in more detail.

Creation of the Tug & Barge Concept

The overall dimensions of the barge are dictated by the sailing restrictions on the river Rhine. The requirements for the upper river Rhine (110 m) and the lower river Rhine (135 m) limited the length of the barge. The height was limited by the ability to pass the numerous bridges and the seasonal variation of the water levels. Under extreme conditions, the push-barge even has to take in additional ballast to pass the bridges in case of very high water levels, while at extreme low water, the push-barge can only partially be loaded due to fairway restrictions. It is possible that additional cargo can be taken on board prior to going to sea. Since no locks have to be passed, there are no width restrictions. Therefore, the width has been selected to achieve good seakeeping properties.

The type of cargo available to the push-barge system determines the interior layout. In this case the cargo available is limited to Ro-Ro cargo (trailers) with the additional possibility to carry containers on the top deck. Therefore, the barge has been fitted with three Ro-Ro decks and elevators. Trailer entry is effected over a ramp via the stern of the barge.

To be able to carry as much cargo as possible, a relatively full barge was designed and the shape of the barge was developed in close co-operation with MARIN. Ro-Ro cargo is relatively light with a relatively large volume. Therefore, the barge was designed as spaciously as possible. The sides were even flared out to create more deck-capacity. Since a lightly loaded barge has a short natural period of roll, wing tanks were installed in the flared sides to lower the GM, in combination with a carefully designed bilge radius and bilge keels. These measures have been taken to keep the roll accelerations within an acceptable range.

The bow has been designed as high as possible without endangering the air draught requirements for the river, to provide as much protection against waves running over the bow and hitting the cargo. During model testing much attention has been given to this element, leading to an improvement of the bow section. For a high vessel of this type, considering also the sailing route, a vertically adjustable flying bridge is an absolute need. The location of such a bridge on the seagoing pusher tug, could result in uncomfortable motions particularly for the crew. It was, therefore, decided to position this bridge on the barge on one of the aft wings, adjacent to the seagoing pusher tug. An additional advantage of this arrangement is that the seagoing pusher tug remains also a tug while sailing alone, and the opportunity is created for a back up tug, which can operate as a tug until it is needed. The crew lives on the tug and the bridge is the slave steering position. The same bridge is also used on the river, allowing any type of available standard driver pusher tug to be used.

Operational Profile

The basic system consists of two push barges, of one seagoing pusher tug and of one river pusher tug. The latter two being standard units adapted only in case of the seagoing pusher by the coupling system and by modifying the bow. This is the smallest operational unit and it will be referred to as one spread. More spreads can be added to the line, or can be made to operate between other ports. The operation of the line is straightforward, see Figure 2. One barge is located in the United Kingdom together with the sea pusher tug. The other barge is located in Germany together with the river pusher tug. At a certain time both units start their respective journeys, one across the North Sea to the continent, and the other one down the river Rhine going with the current. Their departure times have been carefully matched so that they meet at the mouth of the river (port of Rotterdam area). When they arrive at the meeting point, the pusher tugs will change places. This operation will take very little time. The seagoing pusher returns to the United Kingdom with the push-barge from Germany, while the river pusher tug takes the barge arriving from the United Kingdom up river to Germany. Thus the barges follow a route like the number eight, while the individual tugs sail in circles. The pusher tugs always stay with the barges unlike most pusher tug combinations at present. This idea of changing the tug / barge combinations allows each specific "propulsion unit" (read tug) to stay in the type of environment where it operates best.

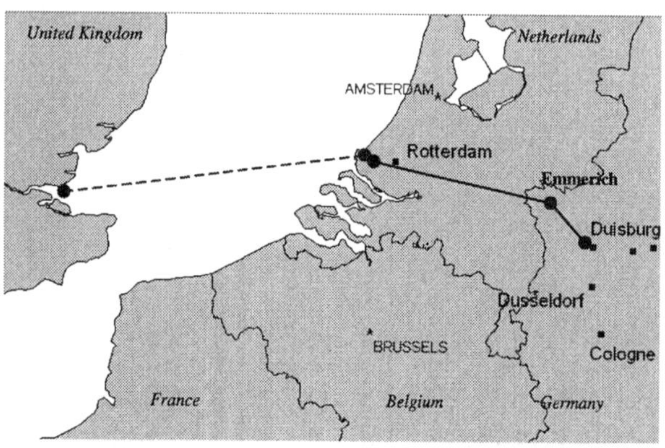

Figure 2: Route on North Sea and rivers

Safety and Reliability

Manoeuvring

To ensure safe and reliable passage, the manoeuvrability of the tug-barge combination has been studied. During its service, the combination has to travel through busy water ways and therefore the ship should be controllable and easily manoeuvrable. Furthermore because of the unconventional concept compared to the current ships on the route between Great Britain and the continent, the tug-barge concept had to be compared to ships of similar dimensions, to prove that it could perform as well or even better than conventional ships. Another topic of interest regarding tug-barge combinations is the level of the connector loads during the passage and during manoeuvring.

The study concerning the manoeuvring of the concept was divided in a preliminary desk study phase and a model test phase.

Seakeeping

Important performance issues in seakeeping relate to the mission of the system: the sustained speed and the risk of sea-fastening problems. Regarding the sustained speed attention focussed on the involuntary speed loss due to the added resistance from wind and waves and a "voluntary" power reduction or change in course motivated by excessive behaviour of the barge or the tug. In this respect green water loading on the fore deck and slamming impacts below the bow of the barge were considered in addition to crew performance problems due to vertical and horizontal accelerations.

To be regarded as a safe and seaworthy system the tug-barge hinges should be sufficiently strong to withstand the connection loads. In addition the crew should be in a position to avoid major discomfort due to excessive tug behaviour by changing course or reducing speed.

IMO Criteria for Manoeuvring

By their draft resolution the IMO proposed standards for overshoot angles, initial turning ability and turning ability, which ships, with a length of 100 m or more or carrying dangerous goods, should not exceed (IMO Resolution A.751(18), see (IMO 1993)). These draft standards are applicable to the tug-barge combination under consideration.

The resolution provides criteria for the turning ability, the yaw checking and course keeping ability, the initial turning ability and the stopping ability of the ship. Explanatory notes are provided to give background information for the uniform interpretation, application and consistent evaluation of the standards, see (IMO 1994).

The turning ability is a measure for the ability of the ship to turn. Two characteristics are used in the criteria: the advance and the tactical diameter.

The advance is the distance travelled in the original direction, from the moment the rudder is laid, to the moment the heading of the ship has changed 90°.

The tactical diameter is the distance travelled perpendicularly to the original course, from the moment the rudder is laid, to the moment the heading of the ship has changed 180°.

To comply with the criteria concerning the turning ability, the advance should not exceed 4.5 times the ship length and the tactical diameter should not exceed 5 times the ship length.

The initial turning ability is a measure for the distance travelled before a certain heading deviation is realised.

To comply with the criteria, the distance travelled should be less than 2.5 times the ship length by the time the heading has changed 10° from the original heading with the application of 10° of helm angle.

The yaw checking and course keeping abilities of the ship are measures for the response of the ship to counter rudder and for the ability of the ship to maintain a straight path in a predetermined direction.

To comply with the criteria, the overshoot angles during zig-zag manoeuvres are considered. The overshoot angle is defined as the additional heading deviation experienced after counter rudder is applied.

Two types of zig-zag manoeuvres are used in the IMO Standards: the so-called 10°/10° zig-zag and the 20°/20° zig-zag. These tests are performed by turning the rudder alternatively 10° and 20° respectively to either side following a heading deviation of 10° and 20° from the original heading respectively.

Limiting values are given for the first overshoot angle values for the 10°/10° and 20°/20° zig-zag manoeuvres and for the second overshoot angle value for the 10°/10° zig-zag manoeuvre.

The first overshoot angle during the 10°/10° zig-zag should not exceed:

- 10°, if L_{pp}/V_s (length-approach speed ratio, with L_{pp} in meters and V_s in meters per second) is less than 10 seconds;
- 20°, if L_{pp}/V_s is 30 seconds or more; and
- (5 + 1/2 L_{pp}/V_s) degrees, if L_{pp}/V_s is 10 seconds or more, but less than 30 seconds.

The second overshoot angle during the 10°/10° zig-zag should not exceed the criterion for the first overshoot angle during the 10°/10° zig-zag by more than 15°.

The first overshoot angle during the 20°/20° zig-zag should not exceed 25°.

The stopping ability is measured by the distance travelled during a full astern manoeuvre after a steady approach at full test speed.

To comply with the standards for stopping, the track reach should not exceed 15 times the ship length.

Because the criteria are not designed specifically for tug-barge combinations some assumptions were made regarding the method of application. One topic of attention is the length of the ship, which is defined as the length measured between the aft and the forward perpendiculars. It was assumed for this tug-barge combination that the length to be used is the distance measured between the aft perpendicular of the tug and the forward perpendicular of the barge. The length between perpendiculars L_{pp} used during the analysis of the manoeuvring characteristics was, therefore, 129.78 m.

Further Design Development through Calculations and CFD

In the early stage of the project MARIN was requested to support MHLP during their concept study by carrying out speed-power, manoeuvring and seakeeping calculations.

Speed-power predictions for the tug & barge concept were carried out to give an indication of the performance to be expected. This was followed by computational fluid dynamics (CFD) calculations using MARIN's non-linear potential theory code RAPID. In the final stage additional speed-power calculations combined with an investigation into the fairway restrictions were carried out to give an indication of the performance on the river in relation to the operational profile.

Besides the speed-power predictions, manoeuvring predictions were carried out to give and indication of the manoeuvrability. These calculations were conducted to determine whether the concept was viable from a manoeuvring point of view and to determine whether additional measures should be taken in order to ensure sufficient manoeuvring characteristics of the design, before extensive hull optimisation and model testing was conducted.

Seakeeping calculations were carried out to give an indication of the performance on the southern part of the North Sea.

Calm Water Speed-Power Calculations

During the early stage of the concept study the ability to attain a certain speed was crucial to ensure the feasibility to operate the tug-barge combination according to the above-described operational profile. Therefore, MARIN was asked by MHLP to start their study with a speed-power prediction for both the river- and sea-going tug & barge combination together with advice on the first set of hull lines of the barge. For the predictions use was made of the Holtrop/Mennen method (Holtrop/Mennen 1982) together with statistical material on tugs and barges available at MARIN. This statistical material mainly exists of the results of model tests, while only limited full-scale information is present in the database of the institute. To determine the effect of the restricted water depth during river-going conditions the method of Schlichting (Schlichting 1934) and Lackenby (Lackenby 1963) was used.

The main particulars of the tug/barge combination used for these first speed/power estimates were:

	River-going	Sea-going	
Length b.p.p.	145.50	141.90	m
Breadth moulded	19.00	19.00	m
Draught moulded	1.70	3.60	m
Displacement	3,850	7,600	m^3

It has to be noted that it was assumed that a present day representative river-going or sea-going tug would push the designed barge.

In the table below and in Figure 3 the results of these calculations are presented, valid for trial conditions in water of 15 degrees Celsius, a clean hull and propeller blades and no effects of wind and waves.

	River WD 2.70 m	Sea Unres. WD
V_s [kn]	P_s [kW]	P_s [kW]
4	275	-
5	520	-
6	880	-
7	1395	-
8	2055	-
9	2935	-
10	4135	1425
11	-	1910
12	-	2525
13	-	3295
14	-	4215

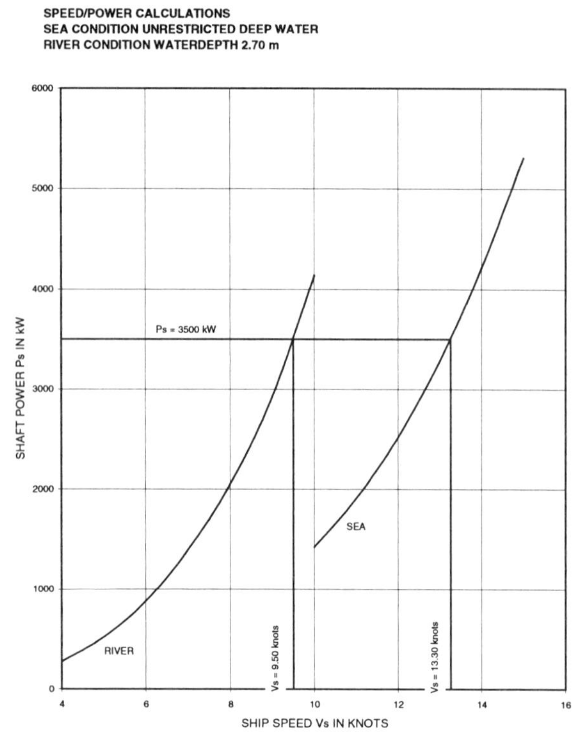

Figure 3: Speed power relation

To compare the results of the sea-going tug and barge combination with those of a self propelled ship a calculation was made for a ship with the following main particulars:

Length b.p.p.	122.00	m
Breadth moulded	19.00	m
Draught moulded	3.60	m
Displacement	7,000	m^3

	Sea-going Unres. WD
V_s [kn]	P_s [kW]
10	1130
11	1588
12	2225
13	3196
14	4726

Assuming the same available shaft power of about 3,500 kW, about the same speed is found, 13.3 knots. However, the payload of the self-propelled ship is definitely lower compared to the payload of the tug and barge combination (estimated to be about 400 tonnes).

CFD Wave Pattern and Flow Calculations

Non-linear potential flow calculations were made to judge the initial hull lines design of the barge and to try to improve the lines from a hydrodynamic point of view. These calculations have been made with MARIN's CFD-code RAPID (Raven 1995, 1993), which computes the wave pattern and the inviscid flow along the hull to exact fully non-linear boundary conditions on the water surface. Primary purpose of the use of RAPID is the minimisation of the wave resistance. An expert judgement of the predicted wave pattern and flow field indicates which modifications of the hull form will reduce the wave resistance. Running RAPID for the modified design can subsequently check this. Additionally, the predicted pressure distribution may indicate possible improvements from the viscous resistance point of view (e.g. reduction of separation). The flow direction on the hull can be used for aligning the bilge keels or knuckle lines with the local flow.

The solution of the RAPID calculation is found in an iterative procedure starting from any initial approximation. Each iteration solves a Dawson like problem, using source panel distributions on the hull at a certain distance above the wave surface. The flow field and wave surface are repeatedly updated until all boundary conditions are met. Simultaneously the dynamic trim and sinkage of the hull are adapted in accordance with the equilibrium between hydrodynamic, hydrostatic and gravity forces. Though wave resistance predictions are not yet sufficiently accurate for use in performance estimates, differences between design variations are well indicated.

After studying the results of the first RAPID calculation, carried out for a speed of 12 knots at the 3.60 m draught, the following conclusions were drawn:

- The bow wave, though rather pronounced is quite normal for this type of ship and block coefficient.
- The wave along the hull was considered moderate.
- It was suggested to slightly reduce the entrance angle of the waterlines in the forebody by a small change of the stem profile, resulting in a small lengthening of the waterlines below the surface.
- It was suggested in view of the steepness of the pressure gradient in the afterbody to soften the transition between the bottom and the afterbody to reduce the buttock angles.

In accordance with the above suggestions the hull lines were modified and a new potential flow calculation carried out.

The results of this calculation showed only a marginal improvement of the forebody wave system, while in the afterbody the pressure gradient shows a small improvement. Along station 1 in the afterbody the local steep wave has been significantly reduced. These effects are shown in Figures 4 and 5 respectively.

Figure 4 shows a comparison of the wave profiles along the hull for both hull forms calculated.

Figure 5 shows the pressure distribution (afterbody view) for both hull forms

Given the strict required payload capacity, it was decided to keep the lines of the second configuration and to use these lines for the seakeeping and manoeuvring calculations.

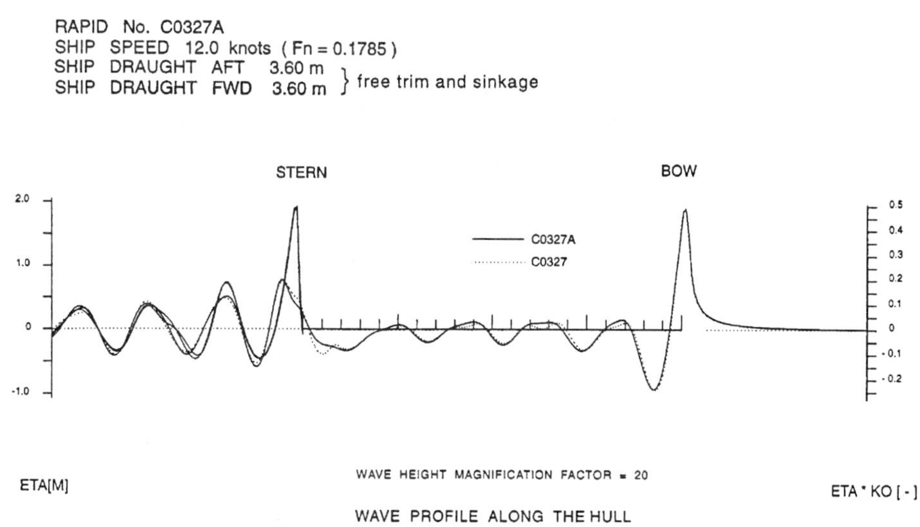

Figure 4: The wave profile along the hull for both hull forms

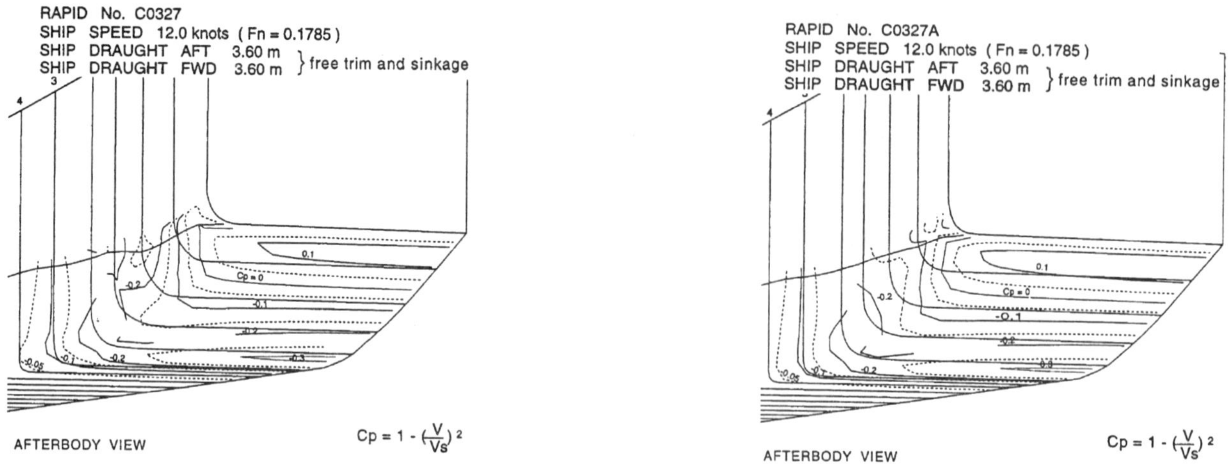

Figure 5: The pressure distribution for both hull forms

Manoeuvring Calculations

During the desk study, time domain manoeuvring simulations were conducted to analyse the manoeuvring behaviour and course keeping ability of the barge in combination with the tug, for deep and shallow water and at two different draughts. These calculations were conducted in order to obtain an early estimate of the manoeuvrability of the tug & barge combination

At this stage, only the design of the barge was available, so that the simulations were conducted with an arbitrary tug boat, representative for barges of the size of the barge under consideration.

The simulations were calibrated for this system using in-house data available at MARIN for tug-barge combinations, to verify and improve the estimations derived from the computer program. Inherent directional stability calculations, zig-zag simulations and turning circle simulations were conducted.

The results from the desk study clearly showed that the tug-barge concept had potentially good manoeuvring characteristics. Therefore, it was decided to continue the project.

Seakeeping Calculations

Calculations were performed to obtain an early impression of the behaviour and performance of the system and to focuss the experimental work on the most relevant issues.

Wave Climate

Regarding the wave climate two sources of information were used. The first is the European database in Global Wave Statistics (Area 20). Annual and seasonal statistics are available for 45 deg sectors on the wind speed and the joint statistics of significant wave height and mean zero-up-crossing period.

The second source is based on hindcast information over a period of 5 years as obtained from the ECMWF in Reading, UK. The information covers the wind speed and direction, the height, peak period and direction of wind sea and swell and the height and mean period of the total wave system over the period between January 1994 and January 1999.

The following table summarises the wind statistics from both sources. Although the mean wind speed is very similar, the GWS data suggest a higher rms value. Consequently the storm frequency is higher than obtained from the ECMWF time histories.

The statistics indicate that BF 8 is exceeded on the (annual) average once per month.

Key wind speed statistics

		Wind	
Weibull Parameters		GWS	ECMWF
A		16.2 knots	16.8 knots
B		1.77	2.09
Mean		14.4 knots	14.9 knots
RMS		8.4 knots	7.5 knots
Freq. of Exceedance			
BF 9	43.8 knots	0.003	0.0001
BF 8	37.5 knots	0.012	0.004
BF 7	32.0 knots	0.036	0.022
Storm Frequency			
> BF 9	1/year	2	1
> BF 8	1/year	12	6
> BF 7	1/year	43	21
> BF 6	1/year	108	45
> BF 5	1/year	196	77

The following table summarises key wave statistics from both sources. The GWS climatology yields substantially higher waves than observed in the ECMWF record. The annual number of storms with waves higher than 5 m seems to be about 10. A wave height of 2.5 m is exceeded almost once a week.

Key wave height statistics

		Significant Wave Height	
Weibull Parameters		GWS	ECMWF
A	m	1.88 m	1.50 m
B	-	1.44 -	1.47
Mean	m	1.70 m	1.31 m
RMS	m	1.20 m	0.91 m
Frequency of Exceedance			
Hs > 5.0 m	-	0.017	0.0055
Hs > 2.5 m	-	0.221	0.106
Storm Frequency			
Hs > 5.0 m	1/year	13	3
Hs > 2.5 m	1/year	87	37

Criteria

Performance and "seaworthiness" criteria were adopted for several aspects of the behaviour of the system.

Regarding the sustained speed the available thrust was regarded as a "criterion" for the involuntary speed loss. It was estimated from the propeller characteristics by assuming that the available torque from the diesel engines is constant at reduced speed.

The seafastening loads of trailers on the trailer deck and the accelerations in all directions at the bridge of the tug were anticipated as reasons for a "voluntary" power reduction or change in course.

The safe working load of the seafastening chains was adopted as a criterion for seafastening problems. The actual loads were calculated from the local accelerations by means of a quasi-static approach, see (Dallinga 1982). Recent work by TNO Automotive (Verhoeff 1999), based on time domain simulations with a detailed multi-body model accounting for vehicle tyre and suspension characteristics, chassis stiffness and the chain snap loads after losing the pre-tension, suggests that the quasi-static approach is reasonably accurate in those cases where snap loads are avoided. Where this is not the case, for instance, when the air-suspension is not de-flated before securing, the lashing load contains a high-frequency component due to snap loads. Figure 6 shows an sample time history. During several of the simulations the high-frequency component is of the same magnitude as the quasi-static load. This implies that the quasi-static approach is not conservative. The present situation considers de-coupled trailers resting on a rigid support. For this reason the quasi-static approach seems acceptable.

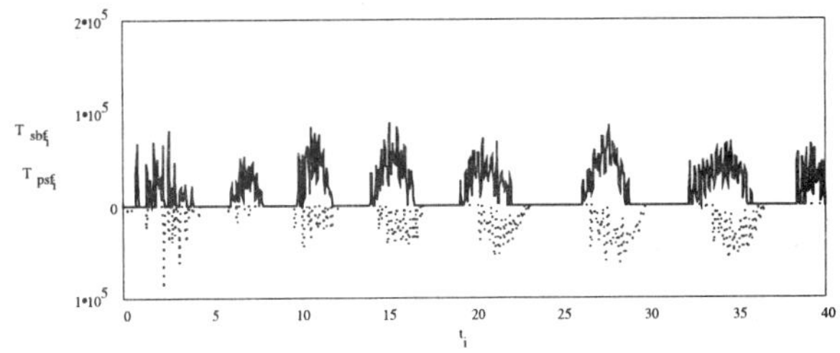

Figure 6: Character vehicle lashing loads

Behaviour

To obtain an early impression calculations were performed on the involuntary speed loss, seafastening loads and tug-barge hinge loads. The results were regarded as sufficiently accurate to select the loading condition of the barge, a first estimate of the maximum sustained speed and the most unfavourable headings for the model tests.

In the early assessment it was assumed that the tug would have only little influence on the behaviour of the barge. It was also assumed that the contribution of the barge would dominate the added resistance due to wind and waves. The calculations for the isolated barge were performed with a strip-theory code.

Experience from earlier investigations suggested that the effects of forward speed on the highest hinge loads is relatively small, allowing the use of a zero-speed 3D diffraction code to obtain a first estimate of the hinge loads.

The effect of the presence of the tug on the global wave loads of the barge were not considered in the present investigation.

Additional Speed-Power Calculations for the River-going Tug & Barge Combination

During the first speed-power predictions for the river-going tug & barge combination it became clear that the initially chosen restricted water depth of 2.7 m together with the minimum draught of 1.70 m could lead to a necessary reduction of the attainable speed due to the risk of grounding. Thus the water clearance between the bottom of the tug & barge and the bottom of the river could determine the maximum speed of the total tug & barge system on parts of the river.

For this reason MARIN offered to carry out additional speed – power predictions for two draughts,

1.70 m and 2.50 m, and four water depths on the route from Duisburg to Rotterdam.

For these calculations the following main particulars were used:

Length b.p.p.	145.90	145.90	m
Breadth moulded	19.00	19.00	m
Draught moulded	1.70	2.50	m
Displacement	3568	5097	m^3

To determine the minimum water depth on the route from Duisburg to Rotterdam and two other relevant water depths, use was made of MARIN report 14755-1-CP (Intervaart 1998) containing a study in the technical concepts of sea-river transport.

In relation to this report an important distinction has to be made between water depth of the river Rhine and the depth of the fairway illustrated in Figure 7.

The fairway cross-section is calculated on the basis of a "minimum fairway cross-section" the size of which is dependent on the narrowest place of a particular waterway section in case of low water. The fairway depth, which forms the basis for the fairway cross-section, is called the "Overeengekomen Laagste Rivierafvoer" (OLR) for the Dutch section of the Rhine and "Gleichwertigter Wasserstand" (Gl.W.) for the German section. The water level is lower than this reference water level only 20 days per year.

According to (Intervaart 1998) the fairway depth from the estuary up to the Dutch/German border and from there up to Cologne is 2.50 m. This is the minimum water depth on the route for the whole year minus 20 days. This water depth can be recognised as the worst case scenario for the barge to operate in. Apart from this water depth, speed-power calculations were carried out for 3.50 m and 4.50 m and for unrestricted water depth. In addition to these water depths it was decided that a minimum keel clearance of 0.25 m for the barge should be guaranteed. Combining the chosen fairway depths with the keel clearance the maximum draught for each fairway depth was calculated (2.25 m, 3.25 m and 4.25 m).

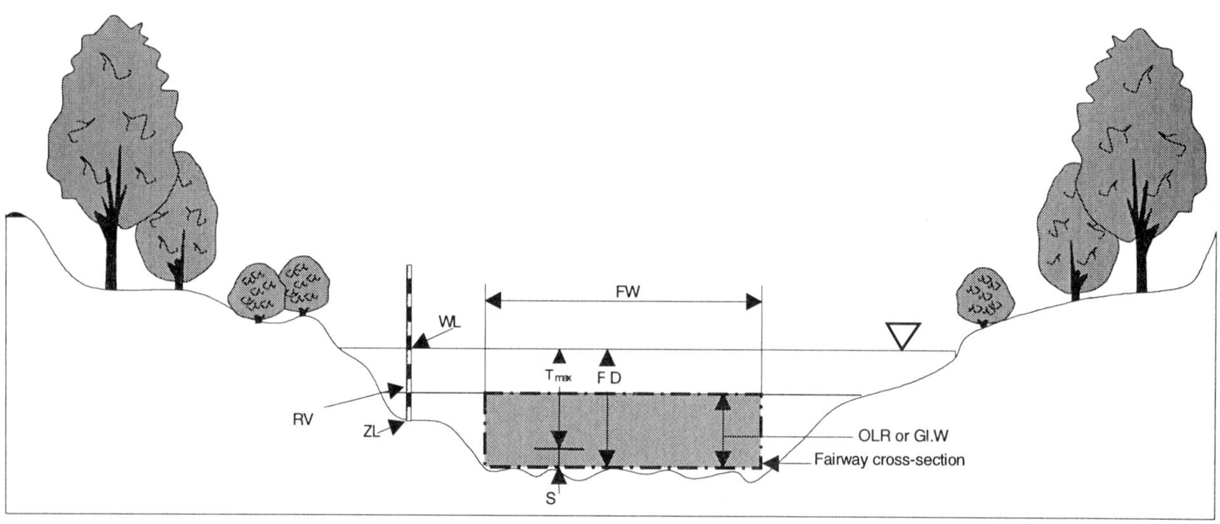

Key:

FD	-	fairway depth
FW	-	fairway width
OLR or Gl.W.	-	"Overeengekomen Laagste Rivierafvoer" or "Gleichwertiger Wasserstand": minimum fairway depth
WL	-	water level
RV	-	water level gauge reference value for the OLR or Gl.W.
ZL	-	"0" level on water gauge
T_{max}	-	Maximum vessel draught
S	-	Mutual distance between the keel and the bed of the fairway cross-section, which was assumed to be 0.25 m

Figure 7: Relationship between OLR or Gl.W., measured water level, fairway depth and vessel draughts

Using these maximum draughts together with Figure 8 the number of days could be determined at which a certain water depth was guaranteed. The draught at the water-level gauge location Emmerich was used in the figure, because this was the lowest point that had to be passed on the route from Duisburg to Rotterdam.

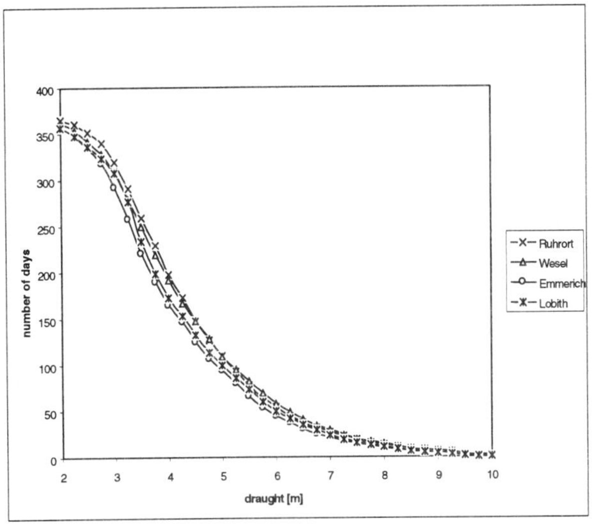

Figure 8: Draught of a vessel and the number of days per year on which this draught was available - various water-level gauge locations, arithmetic mean calculated for the observation period 1982 - 1994 (source: (Intervaart 1998))

Number of days to attain the minimum mean water depths

Chosen water depth	T_{max}	Number of days
Min. fairway depth	2.25 m	Approx. 347 days
3.50 m	3.25 m	Approx. 257 days
4.50 m	4.25 m	Approx. 146 days
Unrestricted depth	-	-

Combining the speed-power prediction for a certain keel clearance with the number of days at which the water depth (corresponding with this keel clearance) is guaranteed, gives a prediction of the number of days at which a minimum speed is guaranteed. This being a major tool to match the performance of the river-going tug & barge combination with the requirement of a very tight sailing schedule.

The results of these calculations are presented below, valid for trial conditions. As already mentioned above, for the predictions use was made of the Holtrop/Mennen method (Holtrop/Mennen 1982) together with statistical material on tugs and barges available at MARIN. This statistical material mainly exists of the results of model tests, while only limited full-scale information is present in the data-base of the institute. To determine the effect of the restricted water depth during river-going conditions the method of Schlichting (Schlichting 1934) and Lackenby (Lackenby 1963) was used. It should be noted that the exact slope of the prediction around the speed limit against grounding and the critical speed is not exactly known.

For a draught of 1.70 m:

	P_s for T = 1.70 m			
V_s [kn]	WD: 2.50 m [kW]	WD: 3.50 m [kW]	WD: 4.50 m [kW]	WD: Unrestr. [kW]
1	6	4	3	2
2	39	27	22	13
3	125	88	70	44
4	284	199	160	99
5	537	376	302	188
6	911	639	512	320
7	1427	1001	803	502
8	2115	1485	1193	744
9	-	2123	1705	1063
10	-	2959	2371	1478
11	-	4233	3271	2017

	T = 1.70 m		
	WD: 2.50 m	WD: 3.50 m	WD: 4.50 m
Speed limit grounding	≈ 7.1 kn.	≈ 10.6 kn.	≈ 13.2 kn.
Critical speed	≈ 8.7 kn.	≈ 10.3 kn.	≈ 11.6 kn.

For a draught of 2.50 m:

	P_s for T = 2.50 m	
V_s [kn]	WD: 3.50 m [kW]	WD: Unrestr. [kW]
1	6	2
2	44	14
3	141	44
4	321	101
5	605	190
6	1025	323
7	1614	509
8	2390	755
9	3436	1075
10	5050	1495
11	7370	2028

	T = 2.50 m WD 3.50 m
Speed limit grounding	≈ 7.9 kn
Critical speed	≈ 10.3 kn

An important conclusion that can be drawn from these predictions is that for a keel clearance of 1.0 m the ship speed is mainly limited by the maximum speed related to risk of grounding of approximately 7.9 kn. with a required shaft power between approximately 2100 to 2400 kW.

Apart from the restrictions in relation to low water levels, high water levels can also cause restrictions in maximum speed or even a stop of operation on parts of the chosen route. To give an indication: when on the route from Duisburg to Rotterdam, water levels at Emmerich are considered again. From 1982 to 1994 a total of 14 unnavigable days due to high water were registered, i.e. an average of 1.2 days per year. The effect of high water on shipping decreases in the Dutch part of the Rhine further downstream.

EXPERIMENTAL PROGRAM

To verify the calculations and to quantify aspects that are not accessible by numerical means, an experimental program was carried out containing resistance and propulsion, manoeuvring and seakeeping tests.

Calm Water Program

The experimental program started with a calm water test program. In the ideal situation resistance and propulsion tests for both the sea-going tug & barge and river-going tug & barge combination would be preferable on both restricted and unrestricted water depth. However, such a total calm water program was not attainable within the available budget.

Using one ship model of the sea-going tug & barge combination for the calm water, manoeuvring and seakeeping tests made a reduction of the costs for model production possible. Therefore, it was decided to carry out self-propulsion tests with the sea-going tug & barge combination in unrestricted water depth. In addition, a resistance test for this configuration was carried out. These two tests would provide information concerning the performance of the sea-going tug & barge combination for the barge with a representative sea-going tug.

To provide the minimum necessary information concerning the performance for the river-going tug & barge combination, it was decided to carry out resistance tests in both restricted and unrestricted water depth for the tug & barge combination with the general shape of a representative river-going tug. The obtained results would be the basis for a prediction of the performance of the river-going tug & barge combination.

Selection of Tug Boats

Based on the size of the barge together with the requirements concerning the performance of the tug & barge combinations, two typical tug boats were designed.

For the river-going tug a generalisation was made of the lines of comparable representative river-going tugs available at MARIN.

The main particulars of the selected river-going tug:

Length b.p.p.	34.310	m
Breadth moulded	13.000	m
Draught moulded	1.700	m
Displacement	581.7	m^3

For the sea-going tug use was made of data from tugs presented in open literature combined with in-house data. The hull form and appendages were designed to ensure good representative powering and manoeuvring characteristics.

The main particulars of the selected sea-going tug:

Length b.p.p.	28.776	m
Breadth moulded	10.030	m
Draught moulded	2.950	m
Displacement	457.0	m^3

The Tug & Barge Combinations

Combining the barge design with the river and sea-going tugs resulted in two configurations used for the calm water tests.

The river-going tug & barge combination:

	Barge	River-going tug	
Ship model Nos	7948	7950	-
Draught T_F	2.50	1.70	m
Draught T_A	2.50	1.70	m

In Figure 9 an overview of the river-going tug & barge combination is given. In this configuration the representative river-going tug is placed behind the barge. To guide the flow around the afterbody of the barge, the notch in the barge is filled with a sponson.

The sea-going tug & barge combination:

	Barge	Sea-going tug	
Ship model Nos	7948	7949	-
Draught T_F	3.40	2.95	m
Draught T_A	3.40	2.95	m
Propeller models No.		5319 R+L, $P_{0.7}/D$=1.066	-
Nozzle models No.		1359 (19B)	-

In Figures 10 and 11 an overview of the sea-going tug & barge combination is given.

The sea-going tug is placed inside the notch of the barge and connected with two hinges to the barge.

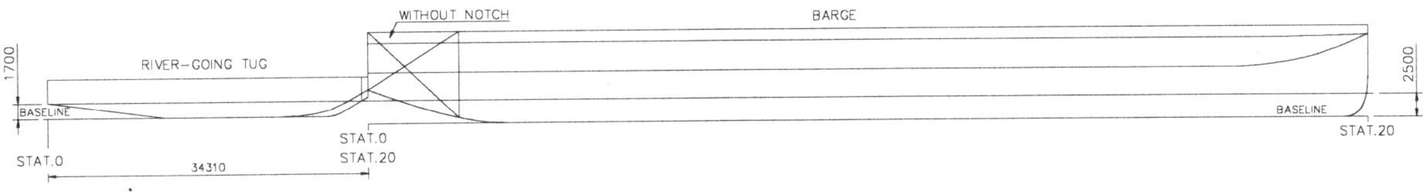

Figure 9: River-going tug & barge combination

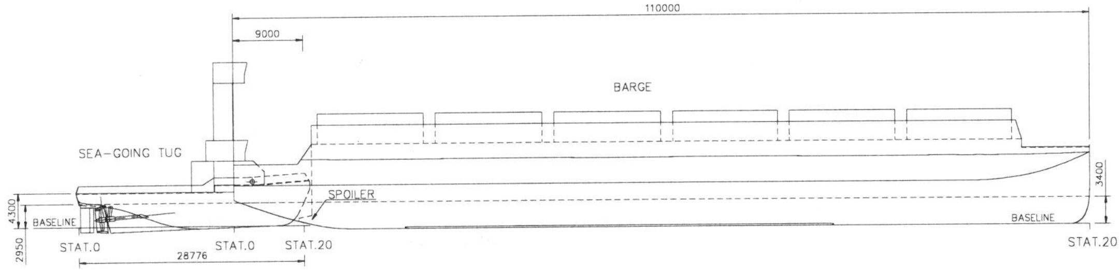

Figure 10: Sea-going tug & barge combination

Figure 11: Sea-going tug & barge model prepared for seakeeping and manoeuvring tests

River-going Tug & Barge Calm Water Tests

With the river-going tug & barge combination resistance tests were carried out in the Shallow Water Basin of MARIN. This basin has a length of 220 m, a width of 15.8 m and an adjustable depth of up to 1.15 m.

The shallow water resistance tests with the river-going tug were carried out at a full-scale water depth of 3.50 m over a speed range of 5.5 tot 8.5 knots. During the resistance tests the carriage velocity, the trim fore and aft of the barge and the resistance of the barge in the tug & barge combinations were measured. By assuming the resistance of the barge independent of the shape of the general representative river tug, the relation between the speed and the required effective power for the barge was obtained. Furthermore, because the trim fore and aft was monitored, the speed at which the barge will ground could be determined. This speed limit was determined at approximately 8.5 knots.

Based on the measured resistance of the barge for the river-going tug & barge combination, an estimate was made of the propulsive power required for the condition on full scale. For this purpose the additional resistance of the river-going tug had to be assumed at approximately 10% of the total resistance. Furthermore, three propellers of 1.50 m were assumed with 19A nozzles representative for this type of tugs. The resulting estimation is valid for the river-going tug

& barge combination sailing in water of 15 degrees Celsius and a mass density of 1000 kg/m^3. (See the table below and Figure 12).

In addition to the resistance test in shallow water, a resistance test with the same configuration was carried out in the Deep Water Basin of MARIN. This basin has a length of 250 m, a width of 10.5 m and a depth of 5.5 m.

During these tests the total resistance of the tug & barge combination was measured, together with the carriage speed and the trim fore and aft, and used as a basis for an additional estimate of the required propulsive power for the river-going tug & barge combination in unrestricted water depth.

The results of both estimations are presented in the following table:

V_s	P_s (Keel clearance 1.0 m on trials)	P_s (Unrestricted water depth on trials)
[kn]	[kW]	[kW]
5	281	-
6	466	-
7	877	-
8	1989	596
9	-	888
10	-	1303
11	-	1883
12	-	2657
13	-	3941
14	-	6132
15	-	9234

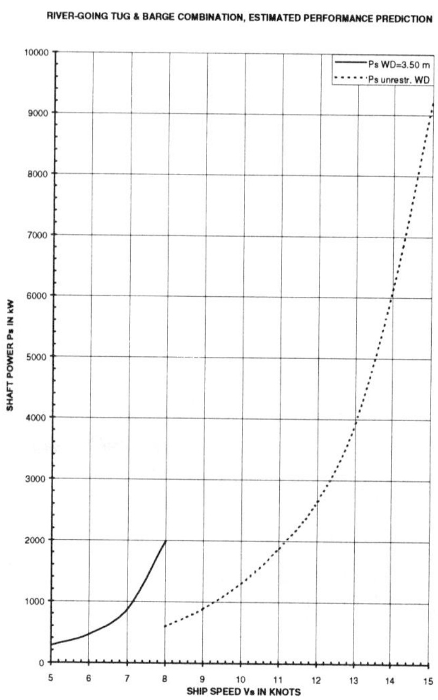

Figure 12: Estimated performance prediction

The results show a speed reduction of approximately 3 knots due to the influence of the restricted water depth with a keel clearance of 1.0 m. Furthermore, when the test results are compared to the previous carried out speed-power prediction it seems that for the speed-power prediction the effect of shallow water on the performance was over-predicted.

During the resistance tests in shallow and unrestricted water depth video recordings and photographs were made of the wave pattern along the hull. Figures 13 and 14 show the wave pattern around service speed for the river-going tug & barge combination in shallow and deep water.

Sea-going Tug & Barge Calm Water Tests

With the sea-going tug & barge combination resistance tests were carried out in our deep water tank. The resistance test results of the river-going and sea-going tug & barge combination in unrestricted water depth can be compared.

	River-going (deep water, bare hull)		Sea-going (deep water, appended hull)	
V_s	R_s		R_s	
[kn]	[kN]	%	[kN]	%
6	-	-	-	-
8	73.8	100	80.6	109.2
10	128	100	134	104.7
12	213	100	227	106.6
14	363	100	393	108.3

When the two configurations are compared it should be noted that for the river-going tug & barge combination only bare hull resistance has been measured where the resistance for the sea-going tug & barge combination the resistance of the appendages is included. However, in addition to this the waterline length of the river-going tug & barge combination is larger and the displacement less. When the appendage resistance is assumed to be approximately 5 to 8 percent the resistance of both configurations remains more or less the same.

In addition to the resistance test a self-propulsion test with the sea-going tug & barge combination has been carried out. During these tests the sea-going tug propelled the tug & barge combination over a speed range of 8 to 15 knots. During the tests torque, thrust and RPM of the propellers was measured, as well as the nozzle trust, the forces on the hinges between the tug and the barge, the trim and sinkage of the barge and the speed of the carriage.

Based on the propulsion test results a performance prediction has been derived. The values for the trial speed of 13.5 knots and the service speed of 12.5 knots are:

Condition	Ps in kW	Speed in knots
Trials	4116	13.50
Service	3373	12.50

The results are valid for the sea-going tug & barge combination as tested, fitted with bilge keels which will be present on the barge, propellers and nozzles similar to the tested propellers and nozzles of the sea-going tug operating in deep unrestricted water of 15 degrees Celsius and a mass density of 1025 kg/m^3, a freshly painted and clean hull and propeller blades and no effects of wind and waves. For the service condition a margin is adopted of 20 per cent, relative to the trial condition.

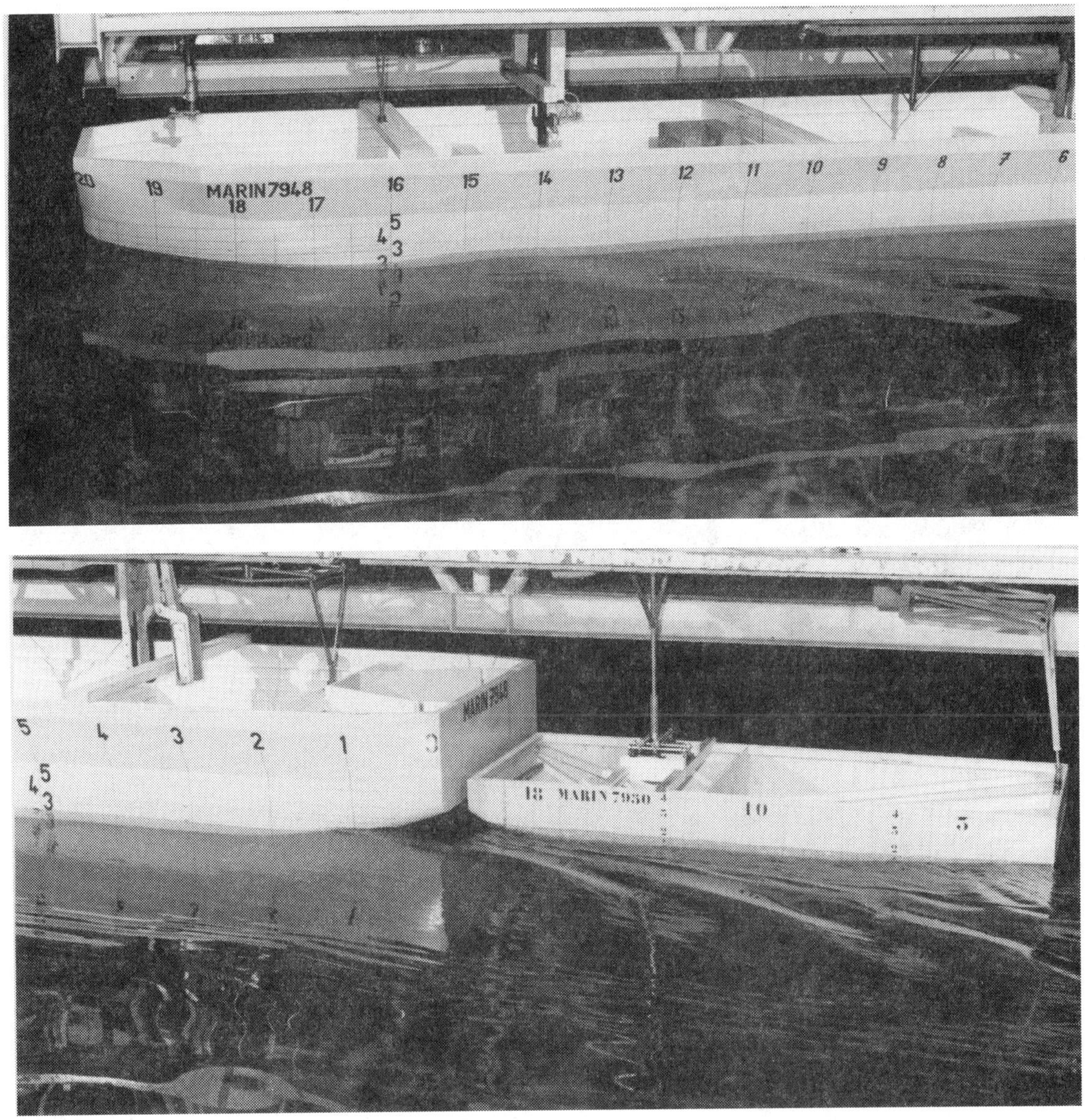

Figure 13: Wave profile of the river-going tug & barge combination at 8.0 knots in a water depth of 3.50 m

Figure 14: Wave profile of the river-going tug & barge combination at 8.0 knots in unrestricted water depth

During the resistance and self-propulsion tests of the sea-going tug & barge combination video recordings and photographs of the wave pattern along the hull were also taken. Figure 15 shows the wave pattern during the self-propulsion tests around service speed for the sea-going tug & barge combination in unrestricted water depth.

Manoeuvring Program

In the second phase of the project, manoeuvring tests were conducted to verify the design of the tug-barge combination. Standard free sailing zig-zag and turning circle manoeuvres were carried out, to determine the yaw checking ability, the course keeping ability and the turning ability of the concept.

During the tests, the motions of the ship model were measured, as well as the connection loads during manoeuvring between the tug and the barge. In Figure 11 the ship model of the sea tug & barge combination as used during the seakeeping and manoeuvring tests, is presented.

Figure 15: Wave profile of the sea-going tug & barge combination at 12.5 knots in unrestricted water depth

All manoeuvring test results were compared to the relevant IMO manoeuvring criteria. This resulted in the following summarising table:

MHLP tug & barge, 12.5 knots (L_{pp}/V_S = 20.2 s)

	IMO limit	Test results
1st overshoot 10°/10°	15°	3°
2nd overshoot 10°/10°	30°	3°
Initial turning ability	$2.5 \cdot L_{pp}$	$1.7 \cdot L_{pp}$
1st overshoot 20°/20°	25°	6°
Advance	$4.5 \cdot L_{pp}$	$3.2 \cdot L_{pp}$
Tactical diameter	$5.0 \cdot L_{pp}$	$3.2 \cdot L_{pp}$

It appeared that the manoeuvrability of the concept complied easily with the recommended values. Additionally, it was found that although the design of the barge changed slightly during the course of the project, the manoeuvrability of the concept predicted by the computer simulations conducted during the desk study agreed very well with the manoeuvring test results. Some discrepancy was found between the predicted turning ability and the actual model test results. This shows that computer simulations can be used efficiently and reliably in the first stage of the design in order to determine the magnitude of the manoeuvring characteristics, but in later stages, model tests are necessary to obtain the actual values.

Because tests were conducted at two different approach speeds, the influence of the speed on the manoeuvrability could be investigated, resulting in the following table:

MHLP tug & barge

Approach speed	12.5 knots	7.5 knots
1st overshoot 10°/10°	3°	2°
2nd overshoot 10°/10°	3°	1°
Initial turning ability	$1.7 \cdot L_{pp}$	$1.7 \cdot L_{pp}$
1st overshoot 20°/20°	6°	3°
Advance	$3.2 \cdot L_{pp}$	$3.1 \cdot L_{pp}$
Tactical diameter	$3.2 \cdot L_{pp}$	$3.5 \cdot L_{pp}$

From this table, it is seen that for the lower approach speed, the yaw checking and course keeping ability increases slightly and the turning ability decreases. For both approach speeds, however, the results easily comply with the IMO recommended values.

Connection loads during manoeuvring

The loads that were measured in the two connection points were measured during each test. In the following table, the maximum of the forces measured during each test are presented:

MHLP tug & barge, connector loads

Manoeuvre	V_s [kn]	F_x [kN]	F_y [kN]	F_z [kN]
Zig-zag 10°/10°	7.5	140	70	30
Zig-zag 20°/20°	7.5	220	60	50
TC 35°	7.5	180	60	30
TC 45°	7.5	200	60	30
Zig-zag 10°/10°	12.5	440	180	140
Zig-zag 20°/20°	12.5	620	260	180
TC 45°	12.5	440	150	80
TC 60°	12.5	440	150	80

From these results, it is clearly seen that the forces for the higher speed are much larger than for the lower speed. Roughly, it is found that the forces are proportional to the velocity squared, as should be expected based on theory.

The connection loads measured during the tests were found to be highest for the zig-zag manoeuvres. The largest forces were found in the longitudinal direction. However, these values were smaller than the forces measured during the seakeeping tests.

Comparison of the manoeuvring characteristics with other ships

A comparison can be made between the manoeuvring characteristics of the tug & barge concept and those of cargo ships of similar dimensions. Results of the model tests with the tug & barge combination as well as results for cargo ships of similar dimensions are presented in Figure 16.

From these graphs, it can be concluded that the yaw checking and course keeping ability of the tug & barge is excellent, while the turning ability of the concept is just above average, but well below the IMO limits.

Seakeeping

To obtain a credible demonstration of the performance and seaworthiness, the model tests aimed at an unbiased impression of the behaviour of the system in service and extreme conditions. This was realised by performing tests at a realistic speed in realistic irregular waves from various directions with a free-running system.

Facility

The model tests were performed in the new Seakeeping and Manoeuvring Basin of MARIN. The 170*40*5 m basin, operational since June 1, 1999, facilitates tests in regular and irregular waves from arbitrary directions. Waves are realised with a flap type wave maker along two adjacent sides of the basin. The 331 60cm-wide flaps are driven by very accurate individual electric servos.

During the tests the free-running model reacted with the tug rudders on the heading, the rate of turn and the transverse position in the basin; the propeller rpm of the tug was constant. During a run the carriage is used as a platform for power supply, relay of measurements and observation. It is designed to follow the model automatically on its trajectory through the basin.

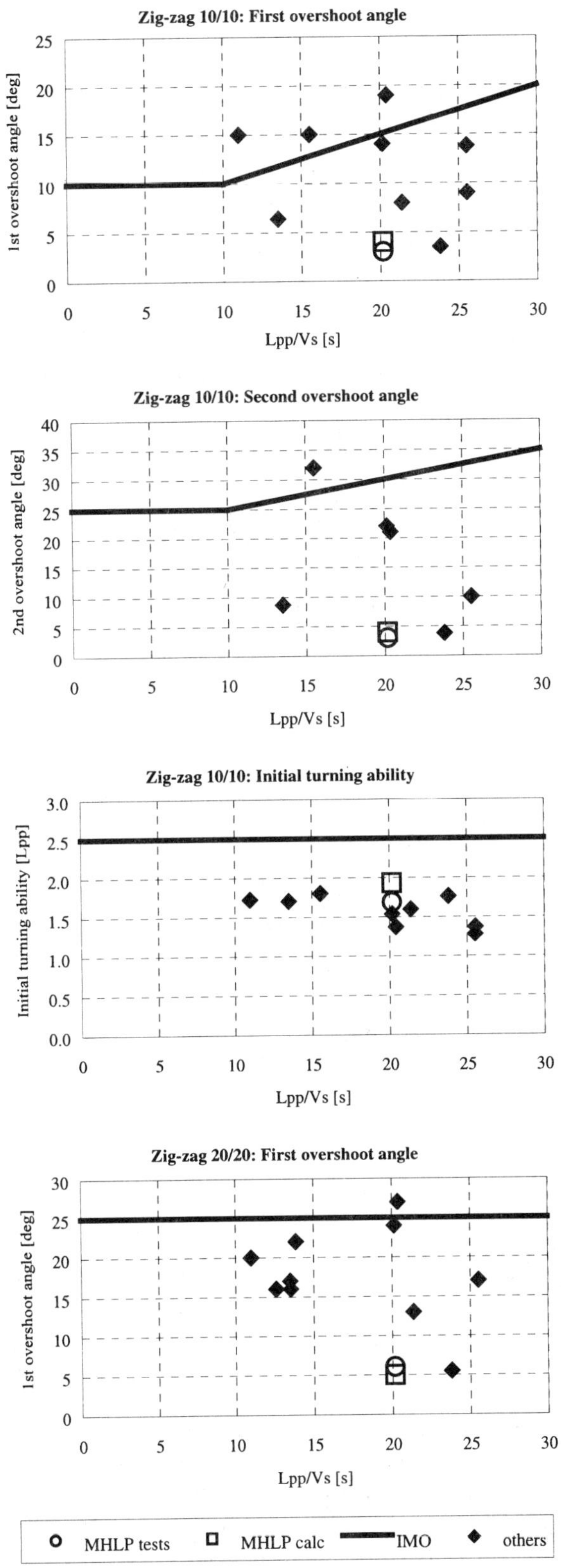

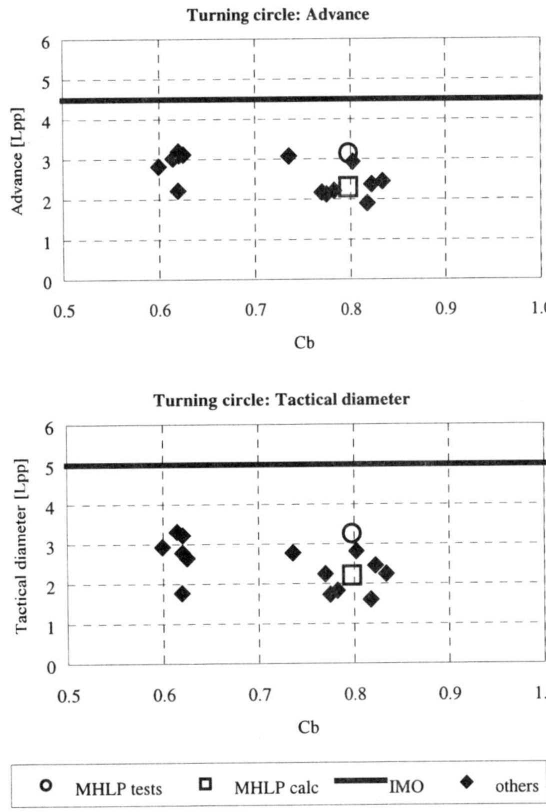

Figure 16: Comparison of manoeuvring characteristics with other ships

The selected scale of 1:15.635 was a compromise on the minimum weight of the tug (that carries the electrical motor to propel the unit) and problems with handling the model in the basin. The maximum speed (12.5 knots or 1.63 m/s on model scale) and maximum significant wave height (5.5 m or 0.35 m on model scale) are well below the basin capabilities (6 m/s maximum speed, 0.45 m maximum significant wave height).

The instrumentation covered:

- the 6 degrees of freedom of the barge;
- the relative pitch angle of the tug;
- the horizontal and vertical accelerations at the tug bridge;
- the 3 force components of the loads in the PS and SB hinge;
- the relative wave elevation at the bow of the barge,
- keel impact pressures below the keel of the barge forward, bow flare impact pressures on board the barge;
- the thrust and torque of the tug propellers;
- the tug rudder angle.

The recorded motions of the barge were used to calculate the accelerations at several reference points to evaluate the risk of lashing problems. One other reference point was adopted to investigate crew exposure to accelerations at an alternative bridge on board the barge.

The experimental program aimed at verification and demonstration of the performance of the system. A limited number of tests focussing on the operational limits was used for this purpose. A typical and an extreme wave condition were used.

The tests in the extreme head seas focussed on shipping of water, slamming in the bow and below the keel and the accelerations at the tug bridge. In the shorter operational wave conditions the 120 and 60 degrees heading represented the most unfavourable wave directions regarding the hinge loads. The tests in stern quartering waves also provided a check on the course keeping in waves, an issue hardly accessible by means of calculations.

The following table summarises the test program.

Sign.Wave Height	Target Speed	Focus
Head seas - 180 deg		
5.0 m	6 knots	Survivability
2.5 m	12 knots	Sustained speed
5 regular waves		Added resistance
Bow-quartering seas - 135 deg		
5.0 m	6 knots	Survivability
2.5 m	12 knots	Sustained speed
5 regular waves		Added resistance
Bow-quartering seas - 120 deg		
2.5 m	12 knots	Hinge loads
Beam seas - 90 deg		
2.5 m	12.5 knots	Normal lashing loads
5.0 m	9 knots	Extreme rolling
Stern-quartering seas - 300 deg		
2.5 m	12 knots	Hinge loads
3 regular waves		RAO Hinge loads
Stern-quartering seas - 315 deg		
2.5 m	12 knots	Course keeping
5.0 m	9 knots	Course keeping
3 regular waves		RAO Hinge loads
Following seas - 0 deg		
5.0 m	9 knots	Tug swamping
Roll decay tests at various speeds		

Performance

The hydrodynamic characteristics of the system were used to predict operability limits in a wave scatter diagram representing the adopted wave climate. In the early stages of the project the calculated transfer functions were used; after the tests these were corrected for the test results.

In the evaluation of the lashing loads a quasi-static approach is adopted. The resulting load can be expressed as a linear combination of the (complex) transfer functions of the local transverse, vertical and roll acceleration, see (Dallinga 1994).

The evaluation of the sustained speed for a particular wave condition in a wave scatter diagram (a combination of significant wave height H_s and mean zero-upcrossing period T_2) requires an estimate of the associated wind speed.

In older wind-wave models the peak period T_p of the growing wave is estimated from the wind speed U_{10} and the significant wave height H_s by means of empirical relations. Janssen et al. (Janssen 1984) suggest for deep water and an unstable atmospheric boundary layer:

$$T_p = 9.25\, H_s^{\,0.688} / U_{10}^{\,0.376}$$

The ECMWF hindcast data provide a check on the applicability of the above estimate in the prevailing typical coastal conditions. Figure 17 shows a cross-plot of the actual wind speed and an estimate from the total significant wave height and the mean zero-upcrossing period. The agreement is reasonable.

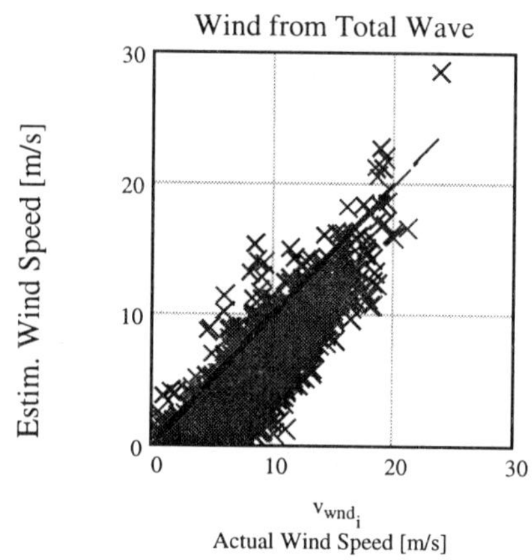

Figure 17: Cross plot actual and estimated wind speed

Results

The initial calculations focussed on the added resistance, the lashing loads and the hinge loads. The results were used to select the most relevant headings and realistic speeds for the model tests.

An interesting characteristic of calculations is the insight that is obtained for a wide range of conditions. Figure 18 illustrates this point with a contour plot of the transfer function of the longitudinal hinge loads as a function of wave direction and wave frequency. The results immediately show that the highest loads are obtained in relatively short period quartering waves.

Based on these results wave directions of 60 and 120 deg were selected for the tests.

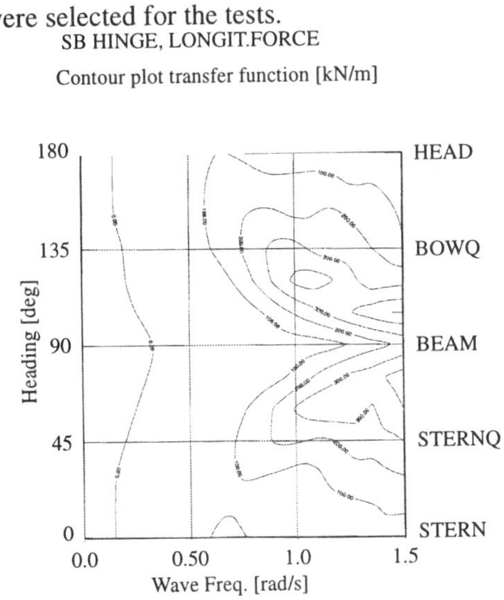

Figure 18: Contour plot

Figure 19 compares the predicted (zero-speed) transfer functions of the longitudinal and vertical hinge loads with the test results (at transit speed). Although the character of the predicted loads is quite good the accuracy of the prediction is disappointing. The relatively high forward speed may explain the discrepancies.

Observations during the tests made clear that the very full bow leads to a considerable dynamic swell-up in waves, see Figure 20. Apart from magnifying the relative wave elevation this leads to a relatively high added resistance in waves. As the phenomenon is not accounted for by the theory, the initial prediction under-estimates the added resistance considerably, see Figure 21.

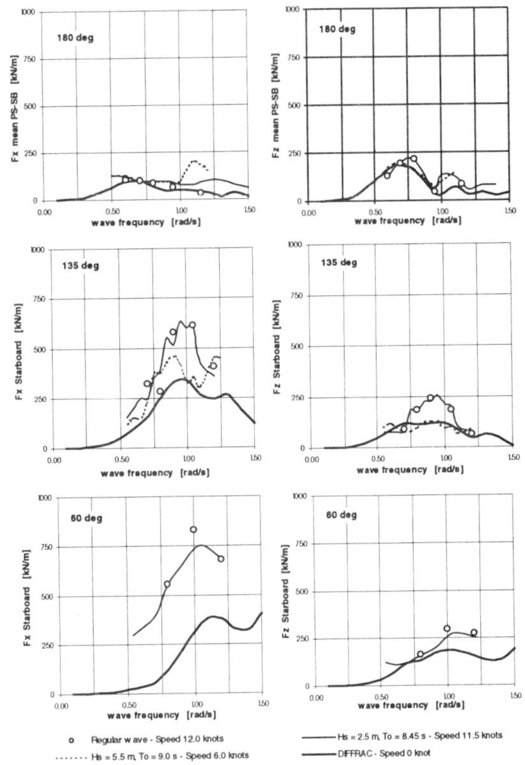

Figure 19: Predicted and measured hinge loads

Although the prediction of the roll damping is not a very strong point of numerical models, the early estimate on the roll response was used to evaluate the need for bilge keels. According to the calculations a pair of 40cm-high bilge keels nearly doubled the roll damping. For this reason they were incorporated in the design. The results from the tests agreed reasonably well with the initial estimate on the roll response.

Figure 20: Dynamic swell-up in head seas

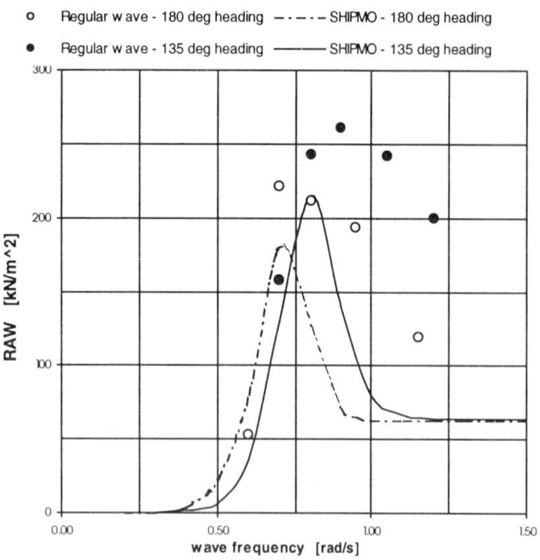

Figure 21: Added resistance

The rather high relative pitch motions of the tug in combination with the high location of the tug bridge lead to relatively high longitudinal accelerations. The following table indicates the acceleration levels at 12 knots forward speed. It also shows the same accelerations at an alternative bridge location above the stern of the barge.

Acceleration levels at 2 bridge locations SDA in m/s^2 Hs = 2.5 m, 12 knots					
Longitudinal		Transverse		Vertical	
tug	barge	tug	barge	tug	barge
Head seas – 180 deg					
2.99	1.06	-	-	1.81	1.60

Bow-quartering seas – 135 deg					
2.90	1.03	1.02	0.87	2.23	1.95
Beam seas – 90 deg					
0.99	0.14	3.95	3.61	1.88	2.51

A characteristic of tests with a scale model is that answers are given on questions that were not asked explicitly. One of the unexpected observations was that the free-board of the selected "typical" tug was insufficient to keep the deck dry. See Figure 22.

Figure 22: Test in stern quartering waves

SUMMARY AND CONCLUSIONS

From the model tests the following conclusions can be drawn:

- When the principle is adopted that the "cargo hold" and the "engine room" can be separated and exchanged, the possibility emerges to navigate seagoing ships on the river with almost double capacity as present low air draught integral ships.
- The operation in an existing logistic chain results in very high requirements to reliability of the vessels, back-up systems and strictly keeping the time schedule.
- The calm water model test program shows that on the river at a draught of 2.50 m and a keel clearance of 1.0 m with a representative river-going tug an average speed of 7.7 knots can be maintained for approximately 257 days a year.
- In addition, the calm water model test program shows that, using a representative tug with similar propellers, the sea-going tug & barge combination can maintain a speed of 12.5 knots in average service conditions with a shaft power of 3373 kW.

Based on the manoeuvring test programme, the following conclusions are drawn:

- Compared to conventional ships of similar dimensions, it could be concluded that the manoeuvrability of the concept was very good. The model tests clearly demonstrated that from a manoeuvring point of view, a tug-barge combination is a safe concept to be used for sea transport.
- The connector loads during manoeuvring are of a smaller magnitude than the forces due to the motions in waves.
- All results found during the manoeuvring model tests depend largely on the selected pusher tug. The pusher tug used during the tests was especially designed to provide adequate steering forces when connected to the barge. If during the service of the combination a different pusher tug is used, the manoeuvrability of the combination will be unknown.

Regarding the performance in waves it may be concluded that:

- The coherence between the initial numerical predictions and the test results seems sufficient to conclude that the limited number of tests covers the most relevant conditions.
- Transverse accelerations in the cargo resulting from roll appeared to be higher than on board a conventional monohull. However, in the prevailing wave climate they remain well below commonly accepted criteria for the present transport mode. Other accelerations appeared to be on the same level than on board of a conventional monohull,
- The initial free-board requirements for the best bow design were under-estimated because of the dynamic swell-up;
- The same swell-up yields a relatively high added resistance in waves,
- The acceleration levels at the bridge of the tug are relatively high; much higher than at a typical location on the barge.

In the final design the above was resolved by heightening and sharpening the bow, with a further increase of the bilge keel height and by moving the control bridge from the tug to the stern of the barge. The free-board and power requirements will be reflected in the selection of a suitable pusher tug.

ACKNOWLEDGEMENTS

The authors like to thank Mr Jan van Velthoven and Mr Piet van Bruggen, who initiated the idea and who developed the system together with Mr Teun Hoogeveen; Thanks also to Mr Leon Plompen for his suggestion to include the system into a logistic chain, providing the basis for a cargo commitment of sufficient magnitude by a major European logistics operator. The enthusiastic and dedicated support from the MARIN team during the model test programs, as well as the support over a period of time during the development by Mr Philip den Ouden of Smit International is highly appreciated.

REFERENCES

INTERVAART, Technical concepts of sea- river transport, *VBD report*, order no. 1034, September 1998 (internal MARIN report 14755-1-CP)
Available via Internet: http://www.waterland.net/nim/intervaart.htm

DALLINGA, R.P., Safe securing of trailers and deck cargo, *RoRo 94*, Göteborg, Sweden, 1994.
HOLTROP, J. and MENNEN, G.G.J., "An approximate power prediction method", *Int. Shipbuilding Progress*, Vol. 29, No. 335, July 1982

IMO Draft MSC Circular 644, *Explanatory Notes to the Interim Standards for Ship Manoeuvrability*, June 1994.

IMO Resolution A.751 (18), *Interim Standards for Ship Manoeuvrability*, November 1993.

JANSSEN, P.A.E.M, et al., An operational coupled hybrid wave prediction model, *J. Geoph. Research*, Vol. 89, pp 3625-3684, 1984.

LACKENBY, H., "The effect of shallow water on ship speed", *The Shipbuilder and Marine Engine-builder*, September 1963

RAVEN, H.C. and Valkhof, H.H., "Application of nonlinear ship wave calculations in design", *6th PRADS symposium*, Seoul 1995

RAVEN, H.C., "Nonlinear ship wave calculations using RAPID method", *6th Int. Conf. Numerical Ship Hydrodynamics*, Iowa City, 1993

SCHLICHTING, O., "Schiffswiderstand auf beschränkter wassertiefe: widerstand von seeschiffen auf flachem wasser", *Jahrbuch STG*, Vol. 35, 1934

VERHOEFF, L., Truck securing on Ro/Ro ships-Dynamic Simulation and Analysis, *TNO report 99.OR/VD.026.1/LV*, 1999.

Discussion

David L. Gray, Member

I wish to thank the authors of this paper for providing us with an excellent presentation of the analysis process that went into the development of this new dual mode tug and barge concept. It bodes well for the future of our industry that Marine Heavy Lift Partners (MHLP) is supporting this high quality analysis work. As always, MARIN has demonstrated that their teams produce top quality results.

With respect to my questions:

The authors state that low air draft seagoing ships capable of sailing both on the river and at sea are rather limited in capacity, and that the two environments are difficult to match for large ships. The paper gives an extensive discussion of the water depth versus number of days versus speed and power considerations; however, the authors do not present any data about the height restriction for the waterways. We note that the problem of sailing draft and air draft are related. At high water when there is plenty of underkeel clearance, these is a loss of air draft, and when there is low water, there is plenty of air draft. So the question must also be asked, is there a minimum clearance from river bottom to the underside of a particular bridge that defines the height from the baseline to the top of the highest fixed structure on this vessel? Presumably the vessel can ballast up or down to best accommodate other conditions on the river. Without knowing the overall height restriction, it is difficult to understand why a low air draft seagoing ship is not a viable alternative. An example of a dual mode river/oceangoing ship is presented in paper #4 by Jay Edgar of The Glosten Associates. It is also a RO-RO vessel, however, designed for a very specialized cargo.

Would the authors please explain the acceleration numbers presented for the two potential locations of the bridge? Are they rms values? If so, the longitudinal accelerations in head and bow quartering seas are about 0.1 g rms for the barge mounted pilothouse and 0.3 g rms for the tug pilothouse. The NORDFORSK criteria for the bridge of a fishing vessel gives a criteria value of 0.05 g for lateral (perhaps also appropriate for longitudinal) acceleration. With the bridge in either location the accelerations will be large, and judging from the wave environment 2.5 m significant wave height is exceeded between 78–90% of the time.

Do the authors have any further comment about the pros and cons of fitting the bridge on the barge? Can they explain how the control umbilical will work, how the watches will be changed, and will the crew be required to cross from the tug to the barge? They also state that since the same bridge will be used on the river, this allows any type of available standard driver pusher tug to be used. How will the control umbilical be made up in this case? Do standard river tugs have this capability or will control on the river be via a telegraph or voice commands?

John P. Hackett, Member

I would like to thank the authors for a fine paper. The project certainly represents a unique undertaking that had numerous engineering and business challenges, and must have been very enjoyable to work.

My comments fall into two general categories: those related to the vessel/system design, and those related to the hydrodynamic verification.

How is the notch in the seagoing barge filled in to become a river barge with a flush transom? I assume the "sponson" (notch fill-in piece) extends to the bottom of the barge and is flush with the transom to produce a smooth buttock line. The paper states the conversion from the seagoing configuration to the river configuration takes "very little time." Does this mean a couple of hours or a day? Where is the sponson stowed when not in use? Is a pier side crane required to install/remove it from the barge? The table on page 6 gives the impression that the down river-bound barge, when it reaches the mouth of the river is ballasted down, as part of its preparation to transit the North Sea. Is this always done, or only if North Sea conditions warrant?

Can the authors provide a body plan of the barge?

Litton Ingalls Shipbuilding has been using CFD for hull form evaluation and optimization since 1995. Based on our experience, I concur with the authors' statement regarding the need for "expert judgment" when setting up the model, interpreting the results, and changing the lines to improve vessel performance. How many different hull design projects does MARIN evaluate with RAPID in a year? On average, how many iterations of the lines and the resulting CFD recalculations take place for a given project? Was either of the tugs mentioned in the paper run through a RAPID analysis? Can RAPID handle the combined tug and barge? Does RAPID have a problem with the stern notch in the seagoing barge? Clearly, as Figures 4 and 5 show, the results in the stern region without the tug are different for the two barge designs. However, since the barge is not going to operate without a push tug, what effect does the presence of the tug have on the barge RAPID results? Could a different optimum barge stern result if the tug was included in the analysis?

As a design naval architect, I can relate to the statement of not having enough money to perform all of the model tests the ship designer would like to see.

Figure 10 indicates a spoiler located at the bottom of the notch. Could the authors discuss the design of the spoiler? Was model testing used to optimize the spoiler design? How tug dependent is the spoiler design?

I am impressed with the size of the seakeeping and maneuvering model of the seagoing tug/barge variant, about 8.3 meters long. Hopefully, the new MARIN seakeeping and maneuvering basin used for part of this test program will mean an end of the need to build two different scale models, a large one for resistance and propulsion and a smaller one for seakeeping and maneuvering.

I am curious about the photos of the tug/barge model in a seaway, i.e., the photo on the opening page and the photo that is Figure 22. The deck of the tug is awash. How did MARIN keep the tug model afloat and maintain proper trim?

I find the figures comparing this project's maneuverability with other vessels', and the IMO criteria very beneficial.

Again, a well done and very informative paper.

Raimo Hämäläinen, Member

I would like to thank the authors for an interesting paper. They have done exacting work to check operability of tug and barge systems. We have built and looked at similar systems in the Baltic area a long time ago in the Kvaerner-Masa Yards. The name of the shipyard was at that time Wärtsilä Marine.

I have the following observations and questions:

1. When she is operating alone, the heeling angles of the tug are normally big. Did you look at that kind of subject and if so, what did you make of it?
2. There have been some accidents in the Baltic area where the character of the cargo has changed from fixed to liquid

due to hard rain or spray water. Did you look at that kind of occurrence, or were the barges covered?
3. You used Schlichting and Lackenby methods in restricted water depth calculations. Why didn't you use large model test data? This should give more reliable results.
4. It is difficult to believe that the power level of a self-propelled ship is worse than a tug-barge case.
5. You have optimized the barge lines by CFD to deep water. Do you think that the best lines in deep water are also best (or close to the best) in shallow water? Can you also compare calculated wave profile results with the test results?
6. What kind of mathematical model have you used in your maneuvering calculations, and what kind of simplifications did you need to do with your model?
7. The maneuvering test results (course stability) are much better than we can expect from that kind of project. Could you give more information or better pictures from steering devices (rudder + nozzle)? Is the tug some kind of big fixed fin behind the full barge?
8. You have used older Holtrop/Mennen methods (1982) in power prediction. Why weren't newer programs used?

Dirk H. Kristensen, Member

The authors are to be commended for providing the Society with an excellent paper describing a comprehensive design program for an Articulated Tug-Barge (ATB) system. These types of vessels are becoming increasingly popular in the United States, making this paper a valuable reference. The paper describes the effective use of analytical tools early in the design process combined with model testing to assess key performance characteristics.

The authors state that early speed-power predictions were estimated using the Holtrop/Mennen method together with "statistical material" on tugs and barges available at MARIN. Could the authors comment on how the results of the H-M method compared with the "statistical material"? Also, does the statistical model test data available include results from other ATB tests?

We note that the sea-going tug is fitted with nozzles. Can the authors comment on whether the nozzled tug was selected because it was thought to be representative of the tugs that may be available for this service? We believe nozzles would be beneficial in this application, particularly in light of the possibility of separated flow and relatively large wake fraction. Can the authors comment on what, if anything, they may have learned regarding the flow regime in the vicinity of the propellers and the suitability of propeller nozzles for this application?

It was interesting to note some of the novel features being considered for this system. Specifically, the use of a sponson to fill in the stern notch while on the river; and the installation of a control bridge on the barge to overcome the problem of high lateral acceleration in the wheelhouse of the sea-going tug. Can the authors comment on whether there has been any further development of these concepts? It would seem that there might be some practical obstacles to overcome. For example, the method of ballasting and connecting the sponson to the barge, and the cost, in time and mooring fees, for the sponson.

How would controls be arranged between the auxiliary wheelhouse located on the barge and the tug? How would personnel be transferred to and from the wheelhouse on the barge and the tug, particularly during heavy weather?

Gabor Karafiath, Member

This paper deals with an unusual tug and barge concept that is designed to operate both on rivers and on the ocean with a common barge component pushed by two different tug types. The river environment presents water depth constraints which necessitate a shallow draft conventional river-going tug which is not suitable for ocean transit. Thus the requirement for an oceangoing tug is generated; however, the seakeeping tests indicate that the typical ocean-going tug has marginal freeboard for this application. Therefore, would it not be simpler and more effective to design and build a special tug that could handle the powering and seakeeping aspects of both the river and the ocean transit?

For the river operation, the barge draft is between 1.7 and 2.5 meters depending on the river water level. At sea the draft is 3.4 meters. How is this draft change accomplished? The possibilities are through additional cargo or through water ballast. The additional cargo option implies the potential for loading delays and the water ballast implies a transport inefficiency.

The paper highlights the large effect that shallow water has on the powering characteristics. The shallow water also affects the wave elevation along the barge as shown by a comparison of the photographs in Figs. 13 and 14. At station 16 there is a significant wave hollow caused by the shallow water. This wave hollow is not present for the deep water operation.

Initial estimates of the performance were prepared using MARIN's historical model test database and the Schlichting/Lackenby method for estimating shallow water effects. These are shown below for comparison to the final test result in Table 1.

It is interesting to note that the initial powering estimate in both shallow and deep water was too high, but the Schlichting/Lackenby method correctly estimated the ratio of delivered power in shallow water to delivered power in deep water. The Lackenby1963 reference summarizes the earlier Schlichting work and presents a speed loss diagram for an expanded range of speeds and water depths.

The final powering estimate is based on shallow water resistance tests. Was there any estimate made of the effect of shallow water on the propulsive efficiency, and do the authors feel that there is a need for self-propulsion tests in shallow water?

Many of the detailed technical issues relating to seakeeping, powering and maneuverability were explored. It would be helpful to have additional information that would compare the performance to a more conventional means of transport. The proposed concept is unique, and model testing has served to highlight and solve some of the problems such as the development of a better seakeeping bow and the need to install the vertically adjustable flying bridge on the barge rather than on the tugboat. The authors are to be congratulated for bringing the concept to this advanced level of design maturity.

Authors' Closure

The authors would like to thank all of the discussers for their time in studying the paper and their useful and positive contributions. In particular, the good experiences of some of the discussers with MARIN's potential flow code RAPID was of signi-

Table 1 Conditions: Draft 2.5 m, Water Depth 3.5 m, Speed 8 Knots

Prediction	Ps Shallow Water	Ps Deep Water	Ps Ratio Shallow/Deep
Initial Estimate	2390 kW	755 kW	3.2
Final Test Estimate	1989 kW	596 kW	3.3

ficant influence on the motivation of the authors to continue with the further development of these and other modern design tools.

Regarding *Mr. Gray*'s discussion, a restriction on the minimum clearance between river bottom and underside bridge existed, but this did not restrict the air draft too much. As an average, only 2 to 3 days a year the water level is too high.

The acceleration figures are given in significant double amplitudes (SDA = 4*RMS). The NORDFORSK criterion in SDA is 1.96 $m \cdot s^{-2}$. This means that for H_s = 2.5 m, the criterion is exceeded for the tug for all conditions. For the barge, the criterion is only exceeded for beam seas.

Please note that a significant wave height of 2.5 m along the route is only exceeded between 11% and 22% of the time.

The steering signals from the bridge on the barge are transmitted to the tug by electrical cables. Use is made of a standard tug fitted with a dedicated control interface.

The crew crosses from the tug to the barge and vice versa using a gangway. This gangway is also suitable for rough weather.

Regarding *Mr. Hackett*'s discussion, the sponson fills the notch in the barge completely, making the bottom of the barge flush with the transom. The sponson stays connected to the river tug during transition. The conversion takes about one hour, with most of the time taken up by attaching the river tug to the barge by steel wires.

The barge is always ballasted down to sea loading condition, even in calm seas. However, the barge has its own ballasting system and can be (de)ballasted during transit if desired.

Unfortunately, a lines plan cannot be provided.

With RAPID, MARIN assesses about 80 different hull forms each year. On average, 3 iterations are applied to each hull form. This means that including intern research projects, about 300 calculations are conducted per year.

RAPID was not used to optimize the tug hull form. Although RAPID can in general handle the combined Tug&Barge hull form, the transition between the barge and the tug has to be smooth to ensure good panel (grid) generation. Because the transition region is mainly dominated by viscous effects, which are not modeled in RAPID, it was decided to only optimize the barge hull.

Since the flow around the notch will be influenced by the presence of the tug, it was decided to calculate the barge without the notch also.

The presence of the tug could, however, have led to a different "optimum" stern of the barge. Given the limited budget, this was not studied during the project.

During the project, a different design of the sea tug became available. To reduce the gap between the notch and the bow of the sea tug, the spoiler was applied. No spoiler optimization was conducted during the project.

Indeed, the same large model was used for the calm water, seakeeping, and maneuvering tests. However, in general, smaller models are preferred for free sailing, due to problems with handling and testing large models.

Basin dimensions restrict the size of the model: when conducting turning circles, the turning diameter should fit within the width of the basin.

Handling by personnel: a large model requires more ballast and more manpower to handle in emergency situations.

The model scale is restricted by the desired wave heights.

Large models need longer test durations and therefore more runs in order to produce reliable statistics.

Applying seals wherever necessary ensured the watertightness of the model. Holes in the bulwark were used to drain excess water from the deck. Additionally, pumps were used to drain any water from the model.

Regarding *Mr. Kristensen*'s discussion, it was found that the Holtrop/Mennen method did not compare too well with the available statistical material. A correlation coefficient of ΔC_A between 0.0002 and 0.0010 was found, which means a variation between 8 and 35 per cent between the tested results was found and those were determined by means of the Holtrop/Mennen method.

Indeed, the available statistical material also includes other ATB test results.

A tug fitted with nozzles was used because all sea-going tugs available and suitable did have nozzles.

Although not separately studied in the test program, the inflow into the propellers is of major importance and during the design phase the following aspects were taken into account:

Connection between the tug and barge to be optimized as much as possible.

Flow disturbance must be avoided as much as possible.

In case of poor inflow, a nozzle/headbox concept leads to local stiffening of the construction and can therefore help to avoid vibration hindrance.

Depending on the operational conditions, particularly those where the propeller will operate under severe loading conditions, a nozzle certainly can help. Still, this has to be studied from case to case.

For more information about the sponson and the bridge on the barge, the authors refer to the answers to the questions of Dr. Hackett and Mr. Gray.

Regarding *Mr. Hämäläinen*'s discussion, heeling angles of the tug alone where not studied within the scope of this project. However, standard, well-proven and reliable tug designs will be selected for the service.

Since the cargo consists of trailers, this aspect was not studied.

It has to be realized that only very limited information is available in the MARIN's statistics on tug and barge concepts operating on restricted water depth conditions. Moreover, also expressed by Mr. Gabor Karafiath (see his comments on the paper), the trends predicted by means of the Schlichting/Lackenby method were in this case very well in line with the model test experiences.

The comparison between the self-propelled ship and the tug and barge combination is made based on a few probably arbitrary assumptions. The major point in the selection of the main particulars was the requirement for the ship to also operate on the inland waterway route to Germany. Moreover, the block coefficient should not be of an unrealistic value.

During the design process it was tried to take into account both deep water and restricted water depth conditions. This was mainly based on the experience available at the institute.

Though it is of course possible to compare the calculated and observed wave profile, this was not allowed by the owner given special reasons.

Use was made of the MARIN computer program SurSim which is capable of simulating arbitrary maneuvers with merchant ships. The calculation of the forces on the hull are conducted using the so-called cross flow drag theory, in which details of the hull form can be incorporated, see reference [1] of the paper. The simulations for this project were conducted by using the hull form and characteristics of the Tug & Barge combination. Interaction effects between the flow from the barge and the tug were not incorporated. To improve the estimated values, the results were correlated to results of maneuvering tests with other ATB-combinations available at MARIN.

From experience it is known that barges when towed can be unstable. This can be improved by applying skegs. However, due to the integral design process, special attention was already paid to the directional stability of the combination in the very early design stage. A suitable standard tug was selected that was applicable to this particular barge.

A mistake was made in the reference of this paper to the older Holtrop/Mennen method. During the calculations the latest code of the H/M method available at MARIN was used. The correct reference is: Hooft, J. P., "The cross flow drag on a maneuvering ship." *Ocean Engineering*, Vol. 21, No. 3, 1994, pp. 329–342.

Regarding *Mr. Karafiath*'s discussion, it would probably be more efficient to design a special tug for both the river and sea condition. However, the operational conditions between the river and sea service are very different. Another aspect is the shallow water depth during the river service which influences the flow around the river tug and therefore the optimum design of the river tug considerably. Additionally, it was decided that use should be made of existing designs of sea or river tugs.

The draft change during the transition between the river and sea condition is accomplished by using water ballast. Additional cargo is not used in order not to lose time during the transition.

Because of budget constraints, the powering performance on shallow water was estimated using the resistance test results combined with an estimation of the propulsive efficiency η_D on shallow water.

A comparison between the performance of the Tug & Barge under consideration with more conventional modes of transport has been made on page 6 of the paper.

SNAME *Transactions*, Vol. 108, 2000, pp. 99–122

The Evolution of Escort Tug Technology . . . Fulfilling a Promise

Robert G. Allan, Member, Robert Allan Ltd.

ABSTRACT

The legislation of the use of tugs to render assistance to tankers in sensitive waterways was initiated without a thorough understanding of the forces and actions involved, particularly at higher speeds. The use of indirect mode towing to generate very high line tensions demands a tug with hull characteristics which did not previously exist, indeed the very act of rendering indirect escort assistance requires a manouevre which is the antithesis of what was previously considered safe towing practice. However, the towing and tug design industries have responded in the past decade with the development of a series of very large and powerful tugs dedicated to escort service, which have now proven that tugs can indeed be an effective tool in preventing a marine oil spill if properly deployed, and if properly matched to the demands of the route and the service.

This paper traces the development and testing by the author's company of a series of unique hull forms and associated appendages which can be adapted to escort tugs with either cycloidal or azimuthing stern-drive propulsion to provide a higher standard of escort performance than was previously available, reviews research currently underway to provide yet further refinements in this exciting new aspect of tug design, and concludes with an examination of the need for a uniform international standard by which the performance of all escort tugs can be measured.

NOMENCLATURE

ABS	- American Bureau of Shipping
ASD	- azimuthing stern-drive tug
BP	- bollard pull
C_ℓ	- surge coefficient
CLR	- centre of lateral resistance
CPP	- controllable pitch propeller
C_{wq}	- sway coefficient
DNV	- Det Norske Veritas
F_B	- braking force
F_S	- steering force
H_S	- significant wave height
IMD	- Institute of Marine Dynamics
L_{wl}	- length waterline
MCR	- maximum continuous rating
NMD	- Norwegian Maritime Directorate
PMM	- planar motion mechanism
RFP	- request for proposal
VSP	- Voith-Schneider propelled tug

INTRODUCTION

The west coast of North America can fairly be considered as the birthplace of tanker escort towing. Initiatives taken in the 1970's by Washington State, and more recently by the State of Alaska in the wake of the ***Exxon Valdez*** oil spill in 1989 mandated the use of large, powerful tugs to escort tankers in restricted waterways. Unfortunately at the time the requirement was initiated the towing industry was not equipped with tugs well-suited to this purpose, and even more importantly large sectors of the industry did not fully realize that fact.

Many in the towing industry quite rightly stated that to expect tugs to perform the escort tasks being demanded of them would put lives and boats at risk. This concern was absolutely legitimate if one were limited to tugs of conventional form and powering. However there have been some exciting and innovative responses to the escort towing challenge, and over the past decade a new generation of large, very powerful, and most importantly very safe tugs have been created to undertake the task of protecting our coastlines through effective escort towing, and more exciting developments are in the works.

This paper traces the development of this class of tug, with a focus on the program of model-testing and computer simulation studies undertaken by the author's company which has led to the design and construction of one of the most powerful and effective escort tugs in the world, due for delivery from a Spanish shipyard in May, 2000. Also discussed are other developments which have enabled more conventional tugs to safely provide a much improved escort towing capability.

The paper concludes with a look at the need for a uniform International Standard by which the performance of escort tugs can be accurately evaluated and rated.

SHIP-HANDLING VERSUS ESCORTING: AN IMPORTANT DISTINCTION

There remains much confusion within the towing industry and certainly more within the lay community about what really is meant by the "escorting" of tankers. The definition which is becoming generally accepted is: *the deployment of a tug in a position from which it can rapidly and safely effect steering or braking control over a ship which has lost its propulsion and/or steering system in a confined waterway, and (most critically), <u>at speeds in excess of 6 knots</u>*. Typical harbour towage and ship-handling nearly all takes place at speeds of 6 knots or lower.

INDIRECT TOWING

The seminal paper on the subject of escort towing [1] was written by Hutchison and Gray et al in 1993. This work, describing the development of the world's first true "escort tug", the ***Lindsay Foss*** (Figure 1), presented for the first time the now standard "butterfly diagram" of forces which can be developed by a tug at all points of the compass and over the full range of escort speeds.

At the speed range of concern in escort towing, maximum forces can only be generated in the "indirect mode" where hydrodynamic forces are used to effect control over the ship, and propulsive forces are used principally to maintain the tug in an attitude of maximum lift, and then position the tug for either steering or braking the attended ship. It is important to note that it still requires the full power of the tug to reach the position of maximum steering force.

Figure 1 MV *Lindsay Foss*—First of a New Generation of Escort Tugs

Figure 2 illustrates the force conventions used to define the steering and braking force components.

The speed issue is critical, because at lower speeds a disabled tanker can be brought under control by a fairly conventional tug, and many tug owners rightly claim that they have been doing that kind of "escorting" for generations. At the higher speeds however a conventional tug's propulsion system cannot provide sufficient thrust to effect control, the hull form lacks the ability to generate large hydrodynamic forces (in the right direction!), and the stability characteristics are simply inadequate to resist the overturning forces generated.

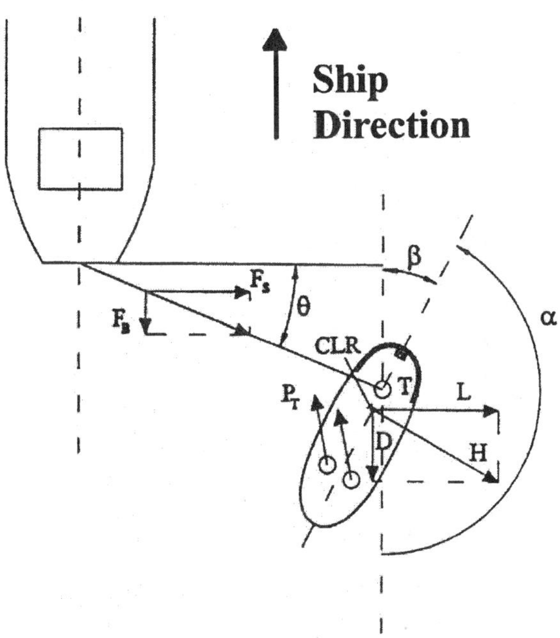

Figure 2 Force Conventions for Indirect Towing

where:

α	=	oblique angle
β	=	yaw angle
θ	=	tow line angle
CLR	=	centre of lateral resistance
D	=	drag force
F_B	=	braking pull
F_S	=	steering pull
H	=	resultant hydrodynamic force
L	=	lift force
P_T	=	propulsion force (thrust)
T	=	position of towing staple

INITIAL HULL FORM RESEARCH

Beginning in 1997, the author's company was involved in a series of design proposals for major escort tugs. As part of these projects, a series of model tests were performed to evaluate a series of hull forms and appendages with the objective of providing significant improvement over the more conventional hull forms in use at that time. The focus of these proposals initially was to verify if an azimuthing stern-drive (ASD) tug could adequately perform the escort functions. The tests undertaken are described in detail in [2], and can be summarized as follows.

Model Testing Program

Preliminary Tests

In order to ensure that the escort towing criteria set for the various projects could be met, a preliminary model test program was undertaken using an existing scale model of a 25 metre, 3,000 hp Z-drive tug. Although this vessel was considerably smaller than the various projects required, the hull form had proven to be very efficient and was considered as a good basis for further development. The objective of the model test program was to evaluate the effects on lift, drag, and heel of a variety of hull appendages and determine a) which of the hull form modifications showed the most promise, and b) make preliminary predictions concerning the escort capability of the final design.

The test program included the measurement of lift, drag, heel, and yawing moment for the model at fixed yaw angles from 0 to 90°. The hull configuration tested included:

a. Existing hull and skeg (as the benchmark for all

further comparisons)
b. Hull without any appendages
c. Box keel; depth = 0.1 x hull draft
d. Deep box keel; depth = 0.2 x hull draft
e. Double deep box keel; depth = 0.4 x hull draft
f. Hull side "wedges"; flare angle = 10°
g. Hull side "sponsons"; flare angle = 45°
h. Bilge keels
i. End plate attachment to box keel

Prototype Design - Model Tests

Based upon the success of these initial tests, yet recognizing the limitations of their accuracy due to the non-conforming parent hull, a further series of tests were commissioned to verify the performance of the first of the design proposals. A 1:25 scale, self-propelled model of the proposed tug (Figure 3) was built and tested at the Ocean Engineering Centre of BC Research.

Figure 3 Self-Propelled Model for Initial Testing of ASD Escort Tug

The model was outfitted with scale sized, fully operational Z-drive units, and towed from a position corresponding to the forward towing staple. Measurements of lift, drag, and heel were made at various yaw angles using varying angles and amounts of thrust to negate the yawing moments about the tow-point.

Further tests were made to confirm directional stability both ahead and astern, measure the resistance of both the bare and appended hulls, and to measure drag in the direct and transverse arrest modes.

Proposed Hull Form

The general hull form for the proposed ASD tugs which evolved as a consequence of the model test program is shown in Figure 4.

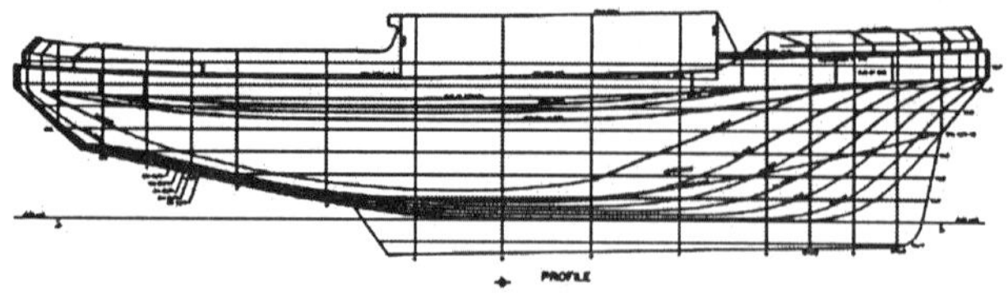

Figure 4 Proposed Hull Form for ASD Escort Tug

This hull form included the following features:
- Hull side sponsons:
  - provides large reserve buoyancy at higher heel angles
  - provides significant righting moment in indirect towing mode
  - provides a natural vertical strake to which to attach fendering
- Deep box keel:
  - provides very high lateral resistance and lift for escort steering
  - very useful deep ballast tank to increase GM in escort mode

The same general hull form in varying sizes was offered for all the escort tug proposals.

CASE STUDIES: COMPARISON OF PROPULSION SYSTEMS

Based on the encouraging success of the model test program described above, a detailed simulation study was undertaken to support a proposal for the design and construction of major escort tugs for Alaska. The details of these studies are also covered in depth in [2], and are summarized as follows:

Escort Tug Comparisons—A Case Study

The requirement was for two escort tugs, with the following minimum performance criteria defined:
- Minimum 14 knot free running speed at 90% MCR
- Minimum bollard pull = 86.2 tonnes

· Steering force at 10 knots ≥ 115.6 T
· Braking force at 10 knots ≥ 154.2 T

Robert Allan Ltd. were retained to prepare two design concepts in parallel in order to respond to this enquiry, for:

a. a Voith-Schneider propelled (VSP) tug, and
b. an ASD tug.

VSP Tug

The design for the VSP tug was developed in conjunction with Voith Schiffstechnik GmbH. Voith prescribed the overall dimensions, the underwater hull form and skeg geometry required to achieve the specified steering and braking forces. Above the waterline the hull configuration was refined to address the operational requirements, including incorporating the sponsons proven in the ASD model tests to enhance escort performance, and widening the stern to ensure maximum fender contact area.

Figure 5 illustrates the final configuration of this tug as proposed.

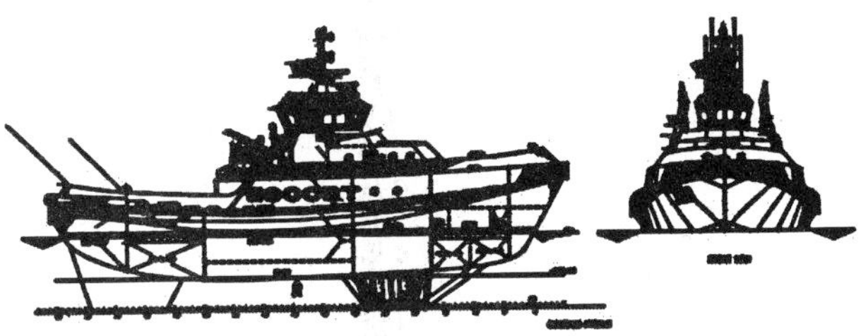

Figure 5 Proposed VSP Tug Configuration

ASD Tug

The design of the ASD tug was developed based on the results of the model test work described earlier. Figure 6 illustrates the configuration of this tug as proposed.

In this instance the hull dimensions were selected as the minimums necessary to suit the performance requirements. Waterline length was selected to meet the speed requirement, and the other dimensions evolved as appropriate to satisfy the buoyancy, trim, and stability performance requirements. Thus this tug was somewhat smaller than the VSP tug.

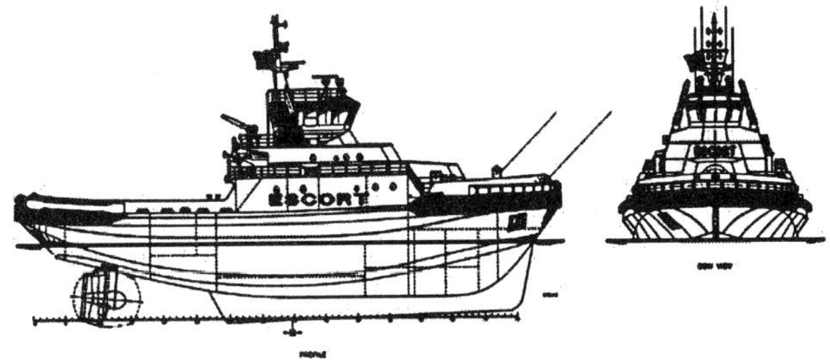

Figure 6 Proposed ASD Tug Configuration

Escort Tug Performance Predictions

Predictions of maximum steering, braking, and combined forces achievable with both designs were made.

Both designs achieved the required minimum steering and braking forces. In both cases heel angles were limited to those which demonstrated compliance with the DNV stability criteria for escort tugs [3], although this was not stipulated as a design criteria.

Escort Towing Simulations

A computer simulation of the ability of the two tugs to arrest a disabled tanker within predefined geographic limits and under defined environmental conditions was conducted. Two specific locations were described, each with different criteria and different transit speeds.

All analyses were performed using a typical 265,000 T. DWT tanker as the attended ship.

Escort Towing Performance Comparison: Summary

a. <u>Valdez Arm</u>:

In this location where allowable transfer was 2,285 metres, it is evident that <u>in some instances</u> the best tug response to a rudder failure is not to oppose the induced turn but either to apply maximum retarding forces (i.e. stop the tanker as quickly as possible) or to assist the turn. Simulations of the failures using an "oppose-the-turn" response were done for comparative purposes, in accordance with the RFP.

At 10° rudder failure the VSP tug performed better than the ASD tug, however both types controlled the tanker in less than half the allowable transfer distance.

At both 20° and 35° rudder failures, the ASD tug outperformed the VSP tug, and was able to return the tanker to course just within the maximum allowable transfer. In three out or four simulated scenarios, the VSP tug did not control the tanker within the limiting transfer distance in spite of having marginally higher indirect steering forces.

This result strongly underlines the importance of overall tug performance, showing clearly that high bollard pull is an essential ingredient of effective escort tug performance, equally important to high indirect steering and braking forces.

b. Valdez Narrows:

In this highly constricted area, with a transit speed of only 5 knots, the higher bollard pull of the ASD tug was a significant advantage. The ASD tug controlled the tanker within 60 metres, whereas the VSP tug was unable to bring the tanker under control within the 274 metre limit for a starboard rudder failure.

Summary of Performance Comparison

Table 1 overleaf summarizes the selected dimensions and hull data for the two tugs, the resulting comparative maximum performance data, and the results of the simulations performed for the Alyeska proposals. The data highlights the impact that all performance criteria have on escort performance, not only high F_s and F_b values, but also high bollard pull and high medium towing speed towing performance.

Figures 7 and 8 following show the comparison of tanker transfer and tanker speed as a function of time under the influence of the two different tugs. These plots show clearly that in spite of a slight advantage in indirect steering forces to the VSP tug, the effect of this difference on the tanker at high speeds is not discernible. Where the overall difference in maximum transfer and in total time elapsed in the escort rescue is apparent is at the latter stages of the scenario where velocity is below 3 knots and the significantly higher bollard pull of the ASD tug shows a distinct advantage.

COMPARISON of VESSEL CHARACTERISTICS & PERFORMANCE
- ASD & VSP ESCORT TUG PROPOSALS

Item No.	*Description*	*ASD Tug Design* (to DNV heel limits)	(to max. thrust limit) *	*VSP Tug Design*
1	Length Overall	44.63 metres		45.50 metres
2	Length Waterline	40.50 metres		42.00 metres
3	Beam, Maximum (ex fenders)	15.00 metres		15.90 metres
4	Beam, Waterline	13.50 metres		14.60 metres
5	Depth, Moulded	6.80 metres		6.20 metres
6	Hull draft	4.50 metres		3.81 metres
7	Maximum draft	5.75 metres		6.65 metres
8	Full Load Displacement	1438 Tonnes		1484 Tonnes
10	Total Lateral area	183.2 m^2		172.4 m^2
11	Power	7460 Kw		7460 Kw
12	Free Running Speed; full load	15.3 knots		**15.5** knots
13	Bollard Pull	**127.0** Tonnes		86.1 Tonnes
14	Fs max @ 5 knots	**101.3** Tonnes	***120.9*** Tonnes	83.9 Tonnes
15	Fs max @ 8 knots	**103.4** Tonnes	***131.8*** Tonnes	98.0 Tonnes
16	Fs max @ 10 knots	115.7 Tonnes	***147.5*** Tonnes	**132.9** Tonnes
17	Fb max @ 5 knots	132.0 Tonnes	***160.1*** Tonnes	**139.7** Tonnes
18	Fb max @ 8 knots	**140.7** Tonnes	***179.4*** Tonnes	127.9 Tonnes
19	Fb max @ 10 knots	146.7 Tonnes	***186.9*** Tonnes	**156.0** Tonnes
	Simulation Results:			
	(a) - Valdez Narrows - 5 knots			
	- max. allowable transfer	274 metres		274 metres
20	- max transfer - port rudder	**64** metres		167 metres
21	- max transfer - stbd. rudder	**57** metres		285 metres
	(b) - Valdez Arm - 10 knots			
	- max. allowable transfer	2285 metres		2285 metres
22	- max transfer - 10 port rudder	1035 metres		**840** metres
23	- max transfer - 10 stbd rudder	1200 metres		**620** metres
24	- max transfer - 20 port rudder	**2165** metres		2750 metres
25	- max transfer - 20 stbd rudder	**2250** metres		3000 metres
26	- max transfer - 35 port rudder	**1850** metres		2170 metres
27	- max transfer - 35 stbd rudder	**1930** metres		2350 metres

N.B.

1 Items in bold text are the highest (best) rated performance between the two vessels

2 Bollard Pull predictions for the VSP tug were obtained directly from Voith Hydro , based on model 36GII-260 drive units.

Table 1

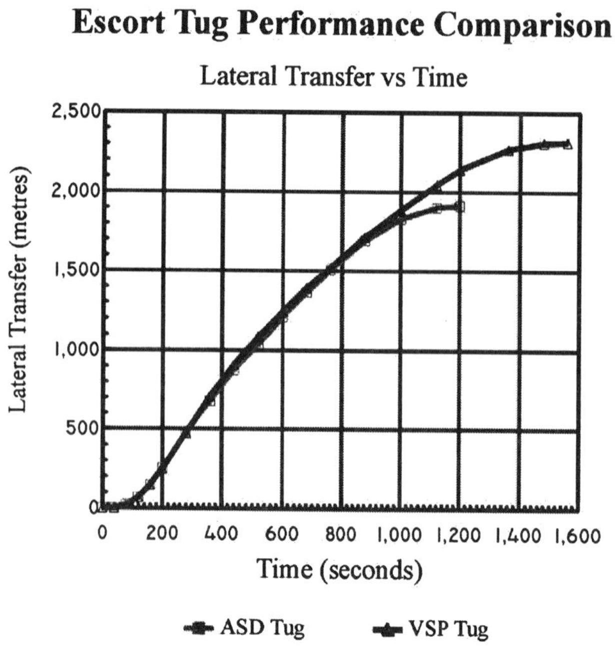

Figure 7 Simulation of Tanker Transfer Versus Time

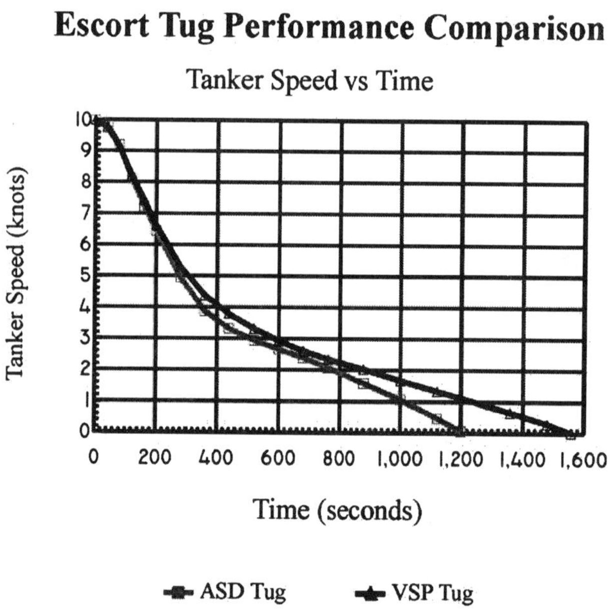

Figure 8 Simulation of Tanker Speed Versus Time

THE EUROPEAN EXPERIENCE

Although the largest and most dramatic escort tugs at the time of writing have been built for service in the Pacific Northwest, in Europe there has been a continuing parallel development of a series of some large and powerful escort tugs of fairly conventional type, and some more innovative smaller escort tugs.

The major European tugs, principally used to escort tankers in the vicinity of refineries in the north of Scotland or in Norway, can be categorized in two groups, principally according to their propulsion configuration, as either ASD or VSP propelled.

ASD Tugs

A number of large and powerful ASD tugs (Figures 9 and 10) have been built recently as escort tugs. These however are simply enlarged versions of conventional terminal or anchor-handling tugs, and do not really address the demands of escort towing in their hull form or layout. Hampered by low freeboard aft, and by high towing positions forward, these tugs are able to effect escort towing simply by brute force and are reliant on the intact buoyancy of their extended forecastles to withstand large heeling moments. Their towering construction results in stability characteristics not well-suited to escort service, particularly in rough weather.

Figure 9 MV *Thorax*—90 Tonne BP ASD Tug

Figure 10 MV *Hopetoun*—125 Tonne BP ASD Tug

Figure 12 MV *Protector*—54 Tonne BP VSP Tug

VSP Tugs

In Norway a series of innovative tugs have been built by Bukser og Bjergning A/S commencing with the VSP tug ***Bess*** (Figure 11) built in 1996. This vessel uses a quite unconventional hull form, designed to operate principally in the "skeg-first" direction, with a fairly high speed. Thus it is able to run alongside tankers at better than average tug speeds, however its low freeboard limits its seakeeping capability. The more recent deliveries of this style of tug have reverted to a more bidirectional hull form, at some loss of free-running speed.

Figure 11 MV *Bess*—55 Tonne BP VSP Tug

Figure 13 MV *Master*—47 Tonne BP VSP Tug

The skeg-first operating principles of these hulls have been adopted by Crowley in their most recent ***Protector*** and ***Master*** series of harbour and coastal VSP tugs (Figures 12 and 13).

A NEW STANDARD OF PERFORMANCE IN ESCORT TUGS

Background

In 1997 Robert Allan Ltd. were contracted by Johannes Østensjø dy AS of Norway to prepare a design for a new Voith Water Tractor escort tug for service on the west coast of Norway.

Operational and Performance Requirements

The basic design criteria and performance specifications for the new escort tug, operating in a "skeg-first" mode were as follows:

- Maximum length:
  - about 40 metres to ensure suitably of operation in harbour
- Escort steering force at 10 knots:
  - 150 tonnes, to DNV Escort Class
- Bollard pull:
  - not less than 90 tonnes
- Free-running speed:
  - 15 knots minimum
- Seakeeping to suit unlimited operation on west coast of Norway and in North Sea:
  - able to operate safely in a skeg-first direction in up to 6 metre significant seas
  - able to run free and tow conventionally in a "drives-first" mode in more extreme seas
- Be equipped and capable of superior performance in fire-fighting (Fire-Fighter 1) and Oil Spill Response and Recovery (to OIL-REC Class)
- Be classed to full ocean-going requirements of DNV and the Norwegian Maritime Directorate (NMD)

Hull Form Concepts

The preliminary hull lines for this concept were developed by Voith Schiffstechnik GmbH, and were very similar to those used in the latest designs of Voith Water Tractors for American and European customers. However the hull form was modified by Robert Allan Ltd. to incorporate the performance enhancing features proven in earlier studies, namely:

- Sponsoned hull sides
- Refined skeg geometry

Initial performance predictions provided by Voith indicated that a vessel of at least 40 metres length was required to generate the 150 tonnes F_s, with an installed power of 6,920 kW. However, no previous vessel ever built had achieved 150 tonnes escort towing force in compliance with the DNV escort stability criteria.

Preliminary Model Tests

Due to the novel nature of the vessel proposed, and the need to ensure good seakeeping as well as the demanding escort towing performance criteria, a program of extensive model testing was commissioned. Initial tests were conducted on a 1:24 scale model without propulsion at the Ocean Engineering facilities of BC Research in Vancouver. The tests program comprised:

- Naked and appended hull resistance measurements
- Flow lines
- Seakeeping and basic motions and accelerations in waves
- Added resistance in head seas

All tests were performed with the model proceeding in both directions.

As a consequence of these tests, the lines were refined and more flare was introduced, with a view to improving the speed and reducing deck wetness. It was also determined as a consequence of the concurrent ongoing development of the design, including weight and stability models, that the vessel needed to be somewhat longer and a little beamier in order to meet the DNV escort stability requirement at 150 tonnes F_S with a reasonable margin.

Model Tests at IMD, Phase I

A further phase of tests was initiated at the Institute of Marine Dynamics (IMD), St. John's, Newfoundland to enable a detailed analysis of the steering and braking force measurements, and to quantify the changes initiated as a result of the preliminary model test program, including the effects of self-propulsion.

The test methodology used by IMD was adapted from the method described by Hutchison et al [1]. Figure 14 shows the model being tested in calm water, utilizing the planar motion mechanism (PMM) for simultaneous measurement of forces and moments on the model.

The full dynamic handling tests showed that there were some potential problems preventing the initial tug design from meeting its target steering force at 10 knots.

Figure 15 shows maximum steering force as a function of towing staple position for a speed of 10 knots. Voith's recommended starting point for the staple location, based on past experience, was approximately 15% aft of the forward end of the waterline, corresponding to slightly forward of the expected position of the centre of lateral resistance (CLR).

At this position however, the maximum steering force obtainable was only 122.6 tonnes. When the CLR location was studied further, as per Figure 16, the actual CLR location was found to be approximately 23% aft of the forward end of the waterline. Towing from this location, the maximum steering force increased to 143.4 tonnes. The reason for the discrepancy with the original estimate was that the new hull form generated significantly more lift force than a conventional Voith hull, thus moving the centre of lateral resistance of the combined hull and skeg aft. However, the total force was still slightly below the required value of 150 tonnes.

Figure 14 Testing of 41 Metre VSP Escort Tug Model at IMD

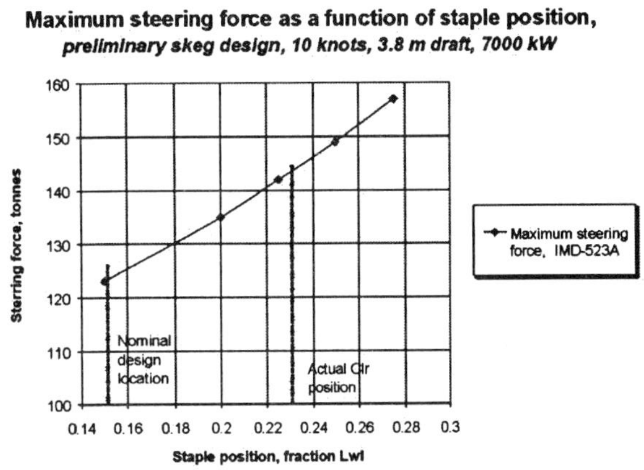

Figure 15 Maximum Steering Forces Versus Tow Staple Position—Initial Data

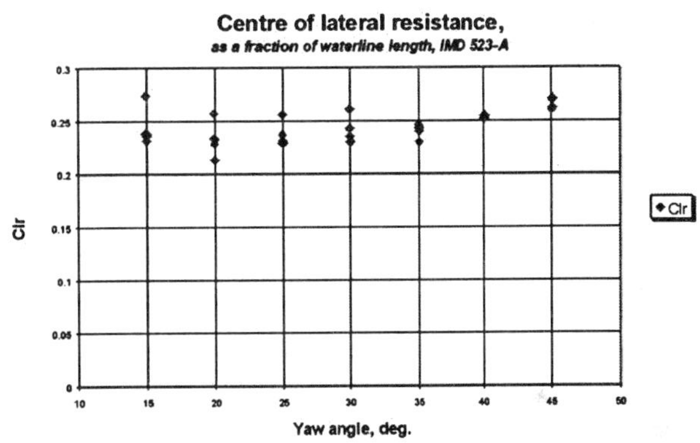

Figure 16 Location of CLR Versus Yaw Angle

Voith Model Testing Program

The initial model test program described above showed the predicted forces could not quite be achieved with the first design configuration. The in-depth analysis of this unexpected shortfall in steering forces showed that despite an improvement in the lift coefficient (C_{wq}) for these special ship lines, the centre of lateral resistance (CLR) moved aft considerably through the significant angles of attack.

To verify this important basic data and to exclude any possibility of scale effects, Voith performed repeat measurements with the same configuration at a smaller scale of 1:40 in its own tank. These repeat measurements proved the validity of the data measured by IMD and furthermore excluded any scale effects between both models.

To meet the design target steering force, the design of the skeg was modified by increasing its lateral area by about 1,92 m^2 at the forward part. With this modification, the total lateral area was increased and the CLR shifted forward by about 2%, improving the balance of forces. With the addition of this area, a higher overall lift coefficient (C_{wq}) was achieved. Examination of the flow lines at the intersection of hull and skeg showed that with the new skeg form, turbulence at the hull/skeg interface had nearly disappeared, which it is believed would account for much of the improved performance. Based on these final measurements the position of the towing-staple was optimized.

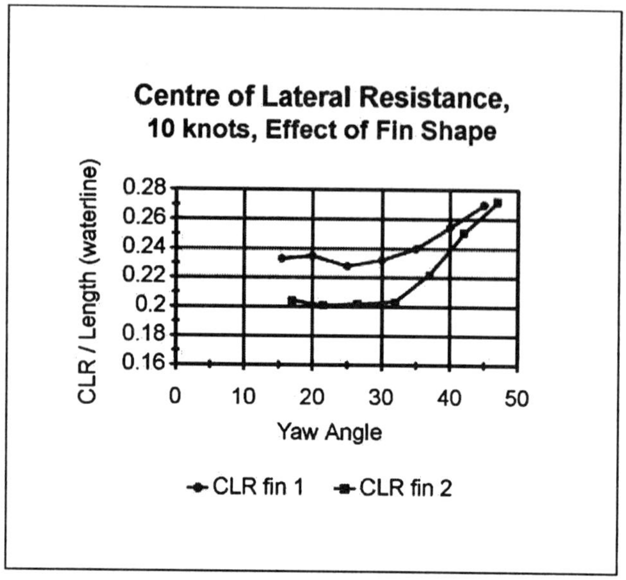

Figure 17 Effect on CLR of Modified Skeg Design

Model Tests at IMD, Phase II

Steering Forces

After much discourse between Voith, IMD, the Owners, and Naval Architects, agreement was reached on the final skeg configuration and the corresponding towing staple position. A second series of tests were then performed at IMD on the modified skeg design.

Results of tests with the second skeg are shown in Figures 17 and 18. Figure 17 gave the location of the CLR at 20.5% L_{wl} aft of the forward end of the waterline. Figure 19 shows that at 10 knots the hull can obtain a maximum steering force of 150 tonnes at 3.8 metre draft, and at 4.05 metre draft, the maximum steering force increased to approximately 160 tonnes.

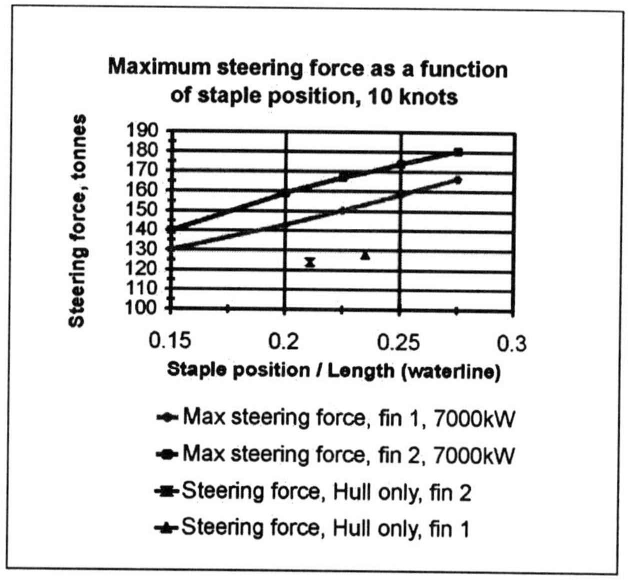

Figure 18 Maximum Steering Force Versus Staple Position

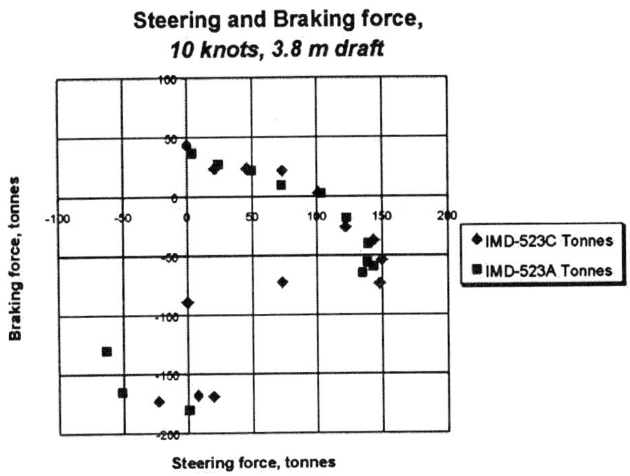

Figure 19 Maximum Predicted F_s

and towline angle in waves did not change relative to the calm water values. However, the action of the waves increased the tension in the cable quite significantly. In waves with a significant height of 3 metres, the peak load in the cable was 120 tonnes above the mean level in calm water, as illustrated in Figure 20.

Roll angle was also affected by waves. Water flowed onto the deck, and as a result, the roll angle of the model in waves was greater than the roll angle in calm water. In waves with a significant height of 1.5 metres, the water on deck was just coming through the scuppers. The results showed that for significant wave heights of 1.5 metres, the mean roll angle did not change much, relative to the calm water value of 12.7°. For 3.0 metre waves however, the roll angle increased to 16.5°. In this case the water would occasionally come over the top of the bulwarks. The tug was however very stable in roll, despite the additional water on the deck.

Seakeeping Tests

The model was also tested for a range of speeds from 4 to 12 knots and in significant wave heights from 1.5 metres to 5.0 metres. JONSWAP spectra were used to represent coastal sea conditions.

The performance of the tug in waves over the range of conditions tested was very good. With the drives forward, it rode well in all sea conditions and only at the 8 knot condition in 5 metre waves did any water break over the bulwarks. When the model was going with the skeg forward, there was a slight increase in the number of waves coming onto the deck, equivalent to 8—10 waves in a 10 minute period. The amount of water that accumulated on the deck was not enough to affect the operation of the ship. The capability of the tug to operate at 10 knots, skeg-first in seas up to 6 metres H_S was ably demonstrated.

Maximum Steering Force in Waves

In addition to the conventional seakeeping tests, the model was tested in waves for escort towing mode. In this test the model was towed on a line, with the VSP units set at the angle determined from the captive experiments with the tug taking a position to apply maximum steering force. The purpose of this test was to see how waves changed the performance of the tug in escort mode.

The model was tested in waves with significant heights of 1.5 metres and 3.0 metres. The mean force

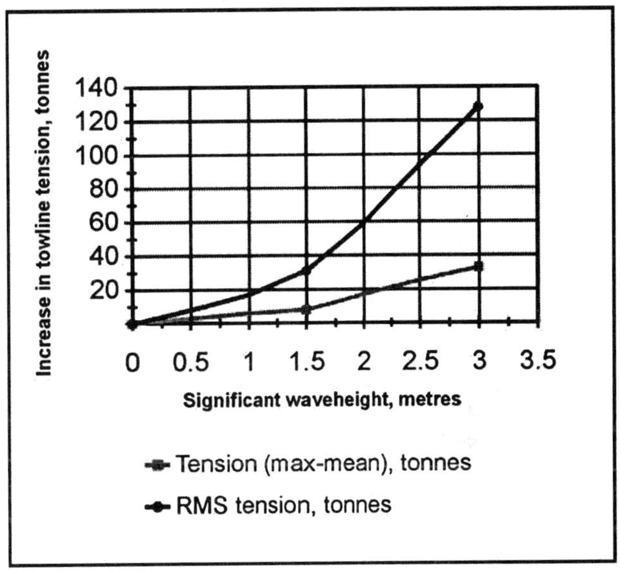

Figure 20 Increase in Tow-Line Tension with Sea State

Drop-Out Tests

The dynamics of the model were observed during a drop-out test. In this case, the model was at the maximum steering force condition, when one motor was cut, to simulate a power failure. Both port and starboard engine failures were simulated.

The model behaved extremely well during the drop-out test. It moved steadily from one equilibrium position to the other, and did not show any dangerous dynamics during the experiments. As the motor was cut, the drift of the model was such that the towline tension was reduced, and the roll angle decreased. As the model reached its new equilibrium position, the tension increased without inducing large roll angles in the model.

Final Performance Predictions

Based on the multitude of tests performed, the vessel's predicted main performance parameters were as follows:

- Speed, skeg first = 15.6 knots
- DNV rated steering force, 10 knots = 150 tonnes
- Maximum steering force, 10 knots = 160 tonnes
- Maximum braking force, 10 knots = 183 tonnes

Final Design Configuration

The final arrangement of the Østensjø Escort Tug is as shown in Figures 21 and 22. Principal features of the design include:

- A unique hull form for demonstrable improvements in escort tug safety, stability, and seakeeping capability
- A wheelhouse providing excellent all-round visibility to support operation in both directions
- A "split-level" accommodation arrangement to maintain a low overall profile
- A fully-enclosed winch house
- A storage "garage" for oil-skimming equipment
- Spacious and comfortable crew accommodation

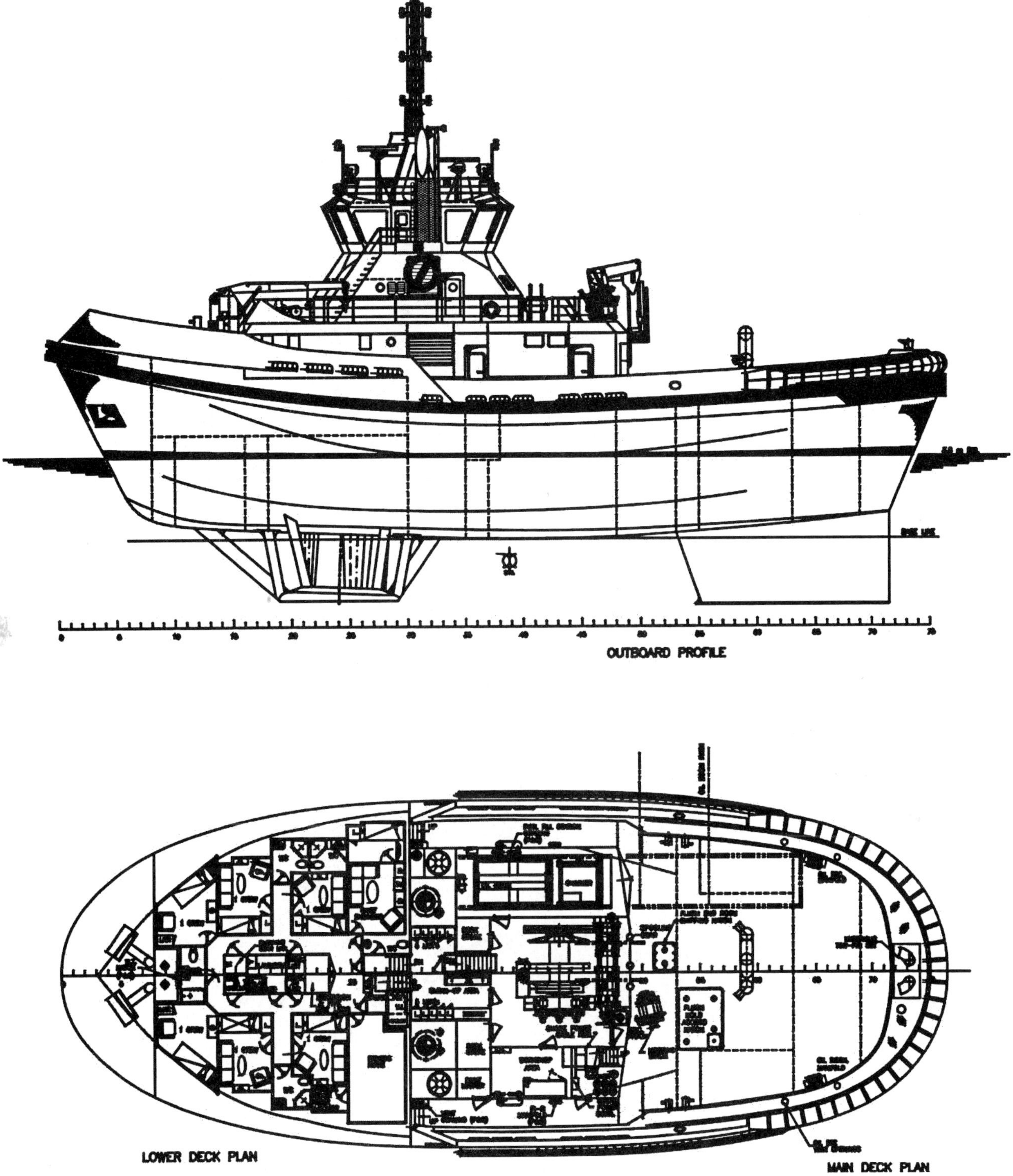

Figure 21 Final General Arrangement for Østensjø Escort Tug

Figure 22 3D Graphics of Escort Tug Concept

Significant Results from Model Testing Program

It is unusual today to conduct model testing on a new tug. There is so much data available for conventional tugs that the expense of a model test is not justifiable. However in the field of escort towing where the forces acting on the hull and the interactions of hull, skeg, and propulsors are just starting to be properly understood, it would be foolhardy not to test for such a major investment, especially where performance guarantees may be required. In this circumstance where novel hull form features were also being proposed it was doubly necessary to test, and ultimately the success of this project has depended heavily on the model test results.

The hull form developed generated significantly more steering force than conventional designs without the sponsons. As a result, the centre of lateral resistance was further aft than initially expected, which affected the equilibrium position for the tug. Therefore in order to achieve the required steering force, modifications to the skeg design and towing staple design were essential.

The major advances in the science of escort towing which arose from this project are:

a. proof of new hull form concepts to provide higher steering and braking forces in conjunction with excellent stability and seakeeping;

b. determination of the critical nature of the relative position of the CLR and tow-point, and more importantly how the CLR shifts position between bare hull and self-propelled measurements;

c. the ability to calculate accurately the roll-centre of the tug under high steering force, critical to establishing maximum F_S values permissible under the DNV escort rating.

IMPROVING THE ESCORT PERFORMANCE OF EXISTING VESSELS

The vessel described in the previous section, and its precedents in both the USA and in Europe have all been custom-designed and built to provide the very highest standard possible in escort towing capability. It might well be argued that no less is suitable for this high-risk operation. However there are many areas where well-found and capable tugs exist, and where the escort towing forces required are not in the same order of magnitude. In such applications it is possible to equip some tugs with appendages which will significantly enhance their escort performance. The following are some examples of the types of modifications possible.

Escort Skeg

Some more conventional ASD tugs can be fitted with an enlarged skeg with a forward bias in order to provide an increased lateral area, and a centroid of area closer to the forward tow-point, as illustrated in Figure 23. This type of modification however can only be contemplated if the tug has ample freeboard and a very high standard of stability. The vessel illustrated is one of a highly successful series of 4,000 bhp, 50 tonne BP ship-assist tugs being built in Canada for the worldwide market. The ninth and tenth tugs in this series were built to serve an oil refinery in Newfoundland, and an escort capability was required. Using a computer model based on the results of the various physical model tests undertaken on tugs of very similar hull form, the following were the changes in F_s-F_b capability predicted for the tug fitted with a 750 mm deep "escort skeg".

Description	F_{s10} (tonnes)	F_{b10} (tonnes)	% Thrust	Heel Angle (degrees)
Original Hull / Skeg	~ 27.0	32.5	100%	~13
With "escort" skeg	36.8	43.8	60	11.6 (DNV max)
With "escort" skeg	57.0	67.9	93	15 (deck edge)

Table 2 F_s-F_b Predictions With/Without Skeg

The results of the trials and delivery voyage of these tugs in the late winter of 1999 brought high praise from the Builder, the Owner, and the Trials Master for the effects of the large skeg on the behaviour of the tugs, including:

- No noticeable loss of directional stability going forward
- No noticeable loss of manoeuvrability
- A dramatic improvement in directional stability going astern
- A significant augment in roll-damping
- No noticeable loss of free-running speed

Figure 23 ASD Assist Tug Fitted with Escort Skeg

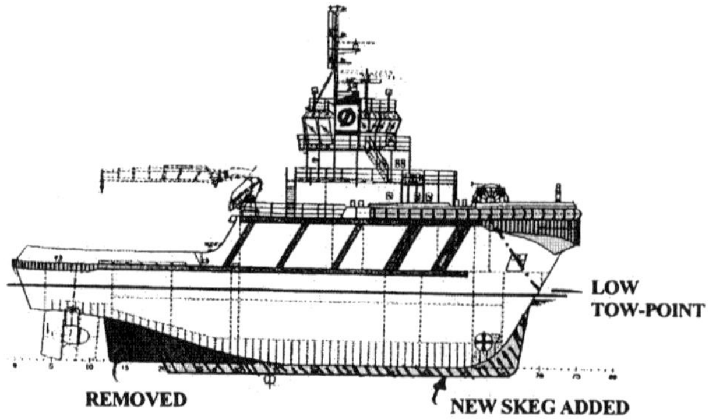

Figure 24 MV *Thorax*—Modified Skeg Configuration

Low Tow-Point

The heel of any tug under the influence of an escort towing hawser is directly proportional to the heeling lever, namely the height of the tow-point above the centre of rotation. Obviously every effort must be made in designing an escort tug to keep the tow-point as low as possible, but this is often diametrically opposed to the desire to provide a safe and dry working deck for the crew. The tug ***Thorax*** (Figure 9) is a perfect case in point. The high bow height provides a very safe working environment for the crew in the worst of North Sea conditions, but as a consequence the heeling arm is very large and the tug heeled dramatically under high transverse load.

The Owners retained Robert Allan Ltd. to recommend changes that could be made in order to improve its escort capability. The principal change was the removal of the existing skeg aft, and the construction of a new skeg to both increase lateral area and to shift CLR forward, as shown in Figure 24.

The Owners also fitted a "low tow-point" to the tug, such that the towing hawser was led from the winch down a type of a hawse-pipe on the stem, emerging just above the waterline through a heavily reinforced bolster.

The combination of these two changes dramatically improved not only the escort towing capability of the tug, but also the significant roll-damping provided by the deep skeg made the tug much more comfortable in all conditions for the crew.

Figure 25 shows the predicted steering and braking forces for the original tug, and for various skeg and tow point variations.

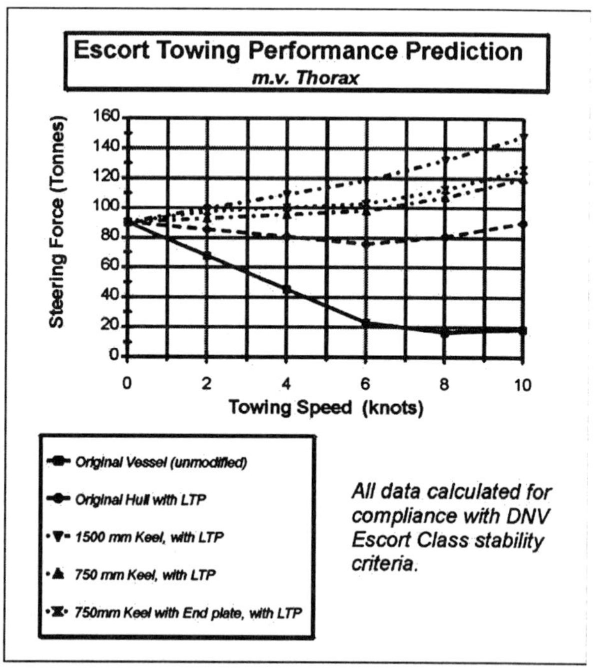

Figure 25 Predicted F_s-F_b Values for Modified MV *Thorax*

To date only preliminary trials have been performed without the benefit of line tension monitoring, however all reports are extremely positive concerning the vessel behaviour and steering/braking performance.

Sponsons

A more radical change would be to fit sponsons to an existing vessel, particularly one where the freeboard and stability characteristics simply were not adequate for the job. However this is a major change to the hull, and the global effects on tug behaviour are less easy to quantify without benefit of model tests.

A STANDARD OF PERFORMANCE FOR ESCORT TUGS

The performance of escort tugs needs to be certified in the same manner as does a tug's bollard pull capability, in order that the purchaser of the services has an assurance of performance. The performance being "sold" is (or should be) all of the following:

- The maximum F_s-F_b capability at a range of specified speeds
- The bollard pull
- The free-running speed, and
- An assurance of an adequate standard of stability and overall safety of the tug in the defined escort conditions

Measurement and certification of speed and bollard pull are routine. However the full-scale testing of maximum steering and braking force capability is a major technical and financial challenge. It requires the availability and the utter cooperation of a suitable ship, with suitably-rated deck gear and fittings. The loads imposed on deck fittings during escort are significant, and many cleats and bitts have already been parted from their roosts during tests of escort tugs. At present only DNV and ABS have specific class notations for Escort Tugs. DNV require (Part 5, Chapter 7, Section 16) that full-scale testing be performed before a final escort certificate can be issued. It can be argued that if extensive model-testing has been performed and documented to demonstrate the performance capability of the tug, then full-scale testing could be waived, or perhaps at most some confirmatory measurements be taken at less than full power to enable a check on the accuracy of the test data, without having to test all systems to the maximum in order to gain a certificate.

With respect to tug stability during escort operations, only DNV at present has a definite criteria. The requirement (defined in Part 5, Chapter 7, Section 16, D200), is in two parts as follows:

a. $R(abs)s \geq 1.25$:

where R(abs) = ratio between righting and heeling areas between the point of equilibrium and a 20 degree heeling angle (Figure 26)

b. $(A+B)_n \geq 1.4\ (B+C)$

where A+B = area under the GZ curve, and B+C = area under the heeling moment curve, taken to the angle of downflooding or 40 degrees, whichever is less (Figure 27).

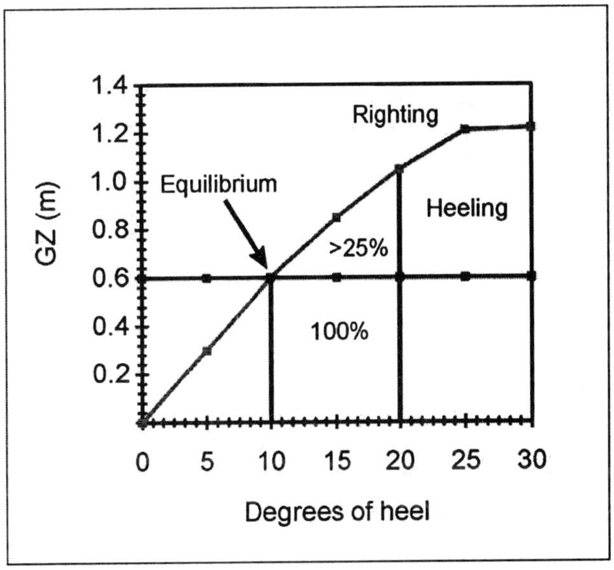

Figure 26 DNV Righting Energy Criteria—Part I

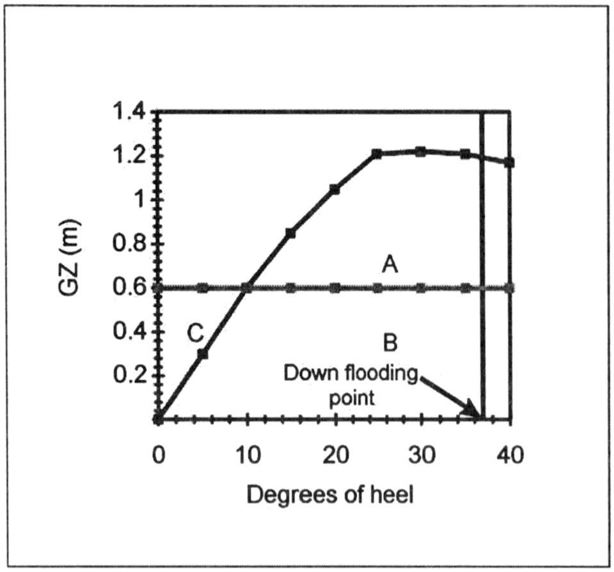

Figure 27 DNV Righting Energy Criteria—Part II

Many discussions on the use of escort tugs have advocated that deck edge immersion be used as the upper limit of induced heel (and hence the angle at which maximum steering and braking forces be calculated), that being a readily observable limit for the operators on the tug. However, it is also a point beyond which the righting energy of the vessel usually experiences an abrupt change in character, and is often a point of maximum righting energy. Given the high transient forces which can be developed during escort in a seaway as described previously, some margin of righting energy is required to ensure fully safe operations. Criteria such as the DNV heel limits, once defined and approved, can be easily marked as a limit on the wheelhouse clinometer, and thereafter can be easily observed by the tug master. If necessary an audible alarm could be added. The master can concentrate on the towing task in front of him instead of checking over his shoulder to see if the deck edge is awash. Using an energy-based criteria provides both a sensible approach to the problem, and a margin of safety for the dynamics of the operation. It remains to be determined whether the DNV criteria is the right one or not, as in the opinion of some it is too conservative. This is an area in which considerable further research is required.

NEW DIRECTIONS

To date the majority of dedicated escort tugs built have been VSP tugs. The use of ASD tugs in this role has suffered somewhat because the tugs that have been built for this role were large but fairly conventional in hull form, and not well-suited to the development of high indirect steering forces.

However the research described herein on the effectiveness of skegs and sponsons on both types of hull forms provides a sound basis for the development of a true ASD escort tug. Currently Robert Allan Ltd. are engaged in a program of joint research with the Institute of Marine Dynamics on such a vessel, to have a performance specification comparable to that of the Alyeska or Østensjø tugs. This tug, designated the ***Rastar*** (Figure 28) will represent a further major advance in escort tug technology, for the first time combining very high BP (125+ Tonnes) with high indirect hydrodynamic force generation capability.

Figure 28 Proposed Escort Tug—*Rastar*

SUMMARY AND CONCLUSIONS

In spite of considerable scepticism concerning the capability of tugs to provide meaningful control over a disabled tanker, the industry has responded with some extremely capable designs over the past five years. It has been amply demonstrated that a purpose-designed tug can be deployed to apply significant forces to a tanker. It has also been demonstrated that each potential scenario for escort towage needs to be carefully analyzed to determine precisely what magnitude of forces are required for each unique set of environmental, geographical, and ship conditions. Once these forces are defined, then the tug characteristics required to support them can be defined.

Tugs equipped with either VSP or ASD/CPP

propulsion can be highly effective in escort towing. Conventional screw-propelled tugs should not be used in escort towing where speeds are in excess of 6 knots, which is the generally accepted lower speed limit of escort service and so defined for purposes of this paper.

An extensive model-testing program is essential to ensure that a new design of escort tug can really fulfill its design objectives, unless that design is closely modelled on a previous vessel (of which there are few of this class!). The relative position of towing point to the CLR is critical to both developing maximum forces and to ensuring a failsafe operation in the event of a propulsion failure on the tug.

Some more conventional ASD tugs can be modified for escort service by fitting appropriately defined appendages, but it is important that the tug's overall design, particularly its freeboard and stability characteristics be suitable for the intended service.

There is a definite need for a unified, international standard for defining and measuring the real escort performance of tugs. Full scale testing is viable, but is also very expensive and potentially dangerous, taking place at a time when the crew is likely not fully familiar with the vessel and its behaviour. Model test results with some confirmatory tests should be a viable alternative, as could computer simulations where the hydrodynamic coefficients used in the simulation can be demonstrated to have been developed through model scale or full scale trials of a suitably similar vessel.

* * *

REFERENCES

[1] Hutchison, Bruce L., Gray, David L., Jagannathan, Sridhar, *New Insights into Voith Schneider Tractor Tug Capability*, The Society of Naval Architects and Marine Engineers, Pacific Northwest Section, Seattle, Washington, March 5, 1993.

[2] Allan, Robert G. *Developments in Escort Tug Technology*, ITS '98, Proceedings of the 15th International Tug & Salvage Convention, November 1998.

[3] *DNV Rules for Ships*, 1999; Part 5, Chapter7, Section 16, D200.

Discussion

Gary C. Faber, Member

I would like to thank the Society and Rob Allan for giving me the opportunity to discuss the captivating and always controversial subject of high-speed escort operations.

In my role as Vice President—Marine Operations at Crowley Marine Services, I am responsible for, among other things, the design, construction and operation of our fleet of Voith tractor tugs and ASD tugs. Our fleet consists of 1 each 50 ton BP ASD, 6 each 50 ton BP VSP tugs, 2 each 60 ton VSP tugs, 2 each 105 ton BP VSP tugs, and most recently, 3 each 150 ton BP ASD tugs. This diversity of vessels with their respective capabilities allows Crowley to be involved in a variety of escort operations.

In his paper, Rob Allan has not only shown us the advances in technology and design for escort tugs, he has accurately and succinctly identified several of the key components of a successful escort program. Development of escort scenarios with their respective preferred maneuvers, model testing and a unified definition of performance are vital to the development of a proper program.

The fundamental component in a successful program is that the stakeholders carefully analyze and agree on potential escort scenarios before the tug is designed.

These scenarios present the potential casualties and the methods or preferred maneuvers for making a save. The save methodology will determine the forces that the tugs will be required to generate and thus the design of the appropriate tug.

Critical to this process is the involvement of all stakeholders, not just the ship and tug operators. It is essential that the regulatory agencies, environmental agencies and interested citizen groups be involved in the escort scenario development. There must be a consensus regarding the mission objectives for the escort program and the tugs; otherwise, the design of the tug will not be appropriate for the job at hand. Too many times, the tug is asked to perform maneuvers outside the design parameters because after the fact, mission objectives were changed. The result is generally less than expected performance in the save scenarios and increased risk to the safety of the crew and the vessels involved.

When this occurs, everyone is disappointed. Rob Allan has shown us how to do it right the first time. I sincerely hope that we all heed his advice.

David L. Gray, Member

As always, Robert Allan has written and presented an excellent paper on tugboats and now on escort tugs and escort tug technology. His emphasis on model testing and, most importantly, on designing a safe and mission-specific escort tug is good to hear. An escort tug is a new tool, and even though Robert Allan has laid claim to the next generation of escort tugs (and hopefully not the final generation), there is still a lot to be learned about their use in an emergency. All of us who have been involved in the specification, design, testing, operation, and training of escort tug users are anxious to learn that these vessels can be used safely and effectively in real life emergencies. Robert Allan's emphasis on safety should be reinforced. The design and use of these high performance boats should not be undertaken cavalierly.

I would also like to mention that the escort tug itself is only one part of the safety enhancement for the transport of oil in confined waterways. There is also the heightened awareness and preparedness that is brought about by defining the emergency scenarios, defining the tug's mission, setting up the lines of communication, defining the responsibilities of the participants, developing escort plans, checking the deck equipment, cooperating in practice exercises and participating in training programs and classroom activities. The OPA 90 rules for the escort of tankers and the process contained in the ASTM Guideline for Escort Vessel Selection are good examples of the full range of efforts that need to be made to go along with the addition of a well-designed escort tug. Other examples include the development of the tanker escort system in place in Prince William Sound, Alaska, the tanker escort plans that have been developed for Puget Sound and the process that has resulted in the escort vessel selection matrices and alternative compliance procedures for San Francisco Bay and Los Angeles/Long Beach.

With respect to the safety of the escort tug, I would like to ask Robert to comment on the performance of the escort tug model in waves. Presumably the DNV energy-based criteria for defining escort tug capability is related to the safe use of the tug in a seaway. In light of the fact that there was a small increase in roll angle in waves and your comment in that, "The tug was, however, very stable in roll, despite the additional water on deck," do you believe these effects are the result of meeting the DNV criteria? Do you believe that there is still a margin with respect to stability, for higher waves, breaking waves or towline failure? Is the DNV criteria doing what you think needs to be done with respect to stability? Do you have any other observations and comments regarding the behavior of the model in waves up to 3 meters and the safety of the actual vessel in these conditions?

David Molyneux, Member, and Piotr Waclawek, Visitor

First, we would like to compliment the author on the paper he has presented. The paper outlines the key developments needed for high-speed escort work, and this is of course the key feature of this type of tug. The requirement to develop forces to control the tanker at 10 or 12 knots requires design features in the tug that are radically different from those required for the more modest speeds of 4 or 6 knots. The author has also presented the propulsion system options available for escort tug operation, and their impact on hull shape and, more importantly, on the ability of the tug to control the tanker. On the basis of the data given in the paper, both the VSP and ASD configurations are practical options for escort work. One thing that will change between the two options is the optimum location of the towing point on the tug. Since the ASD design will have a tow point much nearer the end of the design waterline than the VSP tug, this may be another factor in selecting one option over another. The balance between the hydrodynamic forces and the towrope forces are key to the performance of an escort tug.

It is our observation, based on model test data, that the hull forms for the two options can generate very similar forces, and the magnitude of the force depends on the speed, the yaw angle and the total area of the fin (in the lateral plane of the hull). It would not be unreasonable to assume that the high aspect ratio fin (VSP) would create more force than the low aspect ratio one (ASD). However, by breaking down the force components generated by the hull and the fins, we have found that the fins generate more force than would be predicted by simple engineering approaches. We believe that "classical" assumptions used for predicting the forces due to the foil are not valid, because of the presence of the hull. We suspect that acceleration of the flow, as it passes around the hull, has a large influence on the flow at the fin, and we intend to investigate this further using a combination of numerical and experimental approaches. If this is the case, then the low aspect ratio fin can be as effective as a high aspect ratio, since more of its area is in the region of high-speed flow, close to the hull. We feel that without a full understanding of the flow, in what would normally be considered an extreme off-design condition, we will never be able to fully optimize the hydrodynamics of escort tug design.

A similar hydrodynamic argument can be applied to the choice of propulsion systems. A big advantage of a VSP system is that it operates in a "clean" flow in any direction, by extending beyond the hull. However, when operating in escort mode, the required thruster angles are sufficiently large, relative to the centerline that again the flow into the drive unit is relatively unaffected by the hull. The other major factor to consider is that at high speeds, the propulsion system is only generating 10 to 20% of the total force required. So, provided the propulsion system has enough power to allow the tug to be maneuvered with tension in the towrope, the overall efficiency of the propulsion system is not a key factor in the tug's performance. The major advantage of the ASD design is the reduction in overall draft, which will make this type of tug more attractive in shallow harbors. We would appreciate the author's comments on these observations, based on his practical experience.

We would also like to make some comments on the options available for developing standards for escort tugs. At the present time, the tugs are regulated on the basis of the speed and maximum steering force. These parameters are fine from a point of view of ensuring the safety of the tug but they do not consider the safety of the overall system, which includes the tanker and its cargo. We contend that a realistic system for escort tugs should include some estimate of the effect the tug would have on the ship. For example, it should be possible to define standard parameters to quantify system performance (e.g., advance with minimum transfer to stop the ship, advance and transfer for minimum advance) based on "typical" tanker performance. Such a rating system would then include size of ship and speed for which the tug could act effectively. The advantage of such a system would be to have a realistic assessment of tug requirements for a given situation. Wind and sea conditions could be built into the system, so that practical factors would also be considered. From the point of tug safety, the current regulations would be a layer under such a proposal. Another factor that should be considered is the dynamic loads acting on the tug, rather than the mean steady loads. Also, the tug should be controllable with the towline at full load. Under such a system, it becomes clear that maximum steering force at one speed is not the only tug performance parameter to be considered, but as the tug changes orientation with speed, factors such as bollard pull, become significant.

We see two major disadvantages to this option, both of which could be resolved by some rigorous full-scale measurements. First, there is the requirement to come up with reasonable performance criteria for the tanker-tug system. This requires analysis of full-scale performance data on the combined tug and tanker system. It is our experience that there is not much of this data around, and what there is, is carefully guarded. The other disadvantage is that the only practical method of developing the required information at the design stage would be through the use of simulation. While the addition of model data could improve the level of prediction, any simulation tool would have to be rigorously validated against full-scale data. The same data set could be used for both purposes. So, we ask the author to plead with his clients for full-scale data on the performance of tankers with escort tugs to be collected.

Finally, we are pleased that the author appreciates the benefits of research when making changes to a design, beyond the current range of experience. We appreciate the cost of this research and are continually looking for ways to improve our ability to make accurate and cost-effective performance predictions.

Author's Closure

I thank *Mr. Faber* for his comments, and agree completely with his statement that the escort tug is but one critical component in a complete escort operating system. I also endorse his comment that there must be a thorough analysis of all potential escort scenarios before the escort tug is designed, otherwise the tug designer may well have an inappropriate frame of reference for the vessel requirements.

Mr. Faber's company is to be congratulated for their vision and thoroughness in defining and fulfilling a complete escort towing system for their operations in Alaska in particular, and for their emphasis on crew readiness and training in the critical performance aspects of their equipment.

Mr. Gray's comments are much appreciated, founded as they are in a significant familiarity and expertise in the subject of escort towing. I completely agree with his comments regarding the need for critical attention to the safety of all aspects of the escort operation. At the end of the day, no matter how well designed the tug, it is still at the mercy of very significant forces induced by its attended tanker and the environment, and the crew must be extremely well aware of the limits of the capabilities of their tug.

Mr. Gray asks about the performance of the subject tug model in waves, and the role that meeting the DNV stability criteria might have on that behavior. As discussed in the paper, the model was tested on a towline in 1.5 meter and 3.0 meter significant waves. (It was also tested free-running in seastates up to 6.0 meters.) The significant observation from these tests was the effect of sea state on towline tension, which increased dramatically between 1.5 and 3.0 meter seas, and the impact of the increase in line tension and sea state on roll. The tests indicated that the tug could safely perform escort functions in 3 meter seas (or more), but that the consequent heel angles, if treated as a static value would not satisfy DNV's criteria. However, as Mr. Gray surmizes, it is for just this circumstance that meeting a righting energy criteria such as that of DNV is valuable. If the vessel were certified for maximum Fs at a heel angle equivalent to deck edge immersion, then the reserve buoyancy and reserve righting energy for operation in severe seas would be greatly diminished, as would the effectiveness of the tug in general.

Based on the observed behavior of the model there was definitely still ample reserve stability, even in severe sea states. This observation has been borne out by the experience of the tug *Ajax* on its delivery voyage from Spain in late October 2000. The tug encountered hurricane force winds in the Bay of Biscay and beam seas in excess of 15 meters, and behaved extremely well, although roll angles were severe.

The DNV stability criteria is a step in the right direction with regard to establishing a sensible stability criteria for escort tugs; however, it is not the final answer. It is still a quasi-static, two-dimensional analysis. What is lacking is a criteria which evaluates the reaction of the tug in yaw under influence of the towline, and most critically the instantaneous release of the towline. At present, model testing is the only way to properly evaluate the behavior of the tug under this most critical of actions.

I thank *Mr. Molyneux* and *Mr. Waclawek* for their thoughts and for their commitment to expanding the knowledge base of this critical subject.

The comment that an ASD tug will have a tow-point much nearer to the end of the waterline is not necessarily true. Unfortunately some previous papers on this subject have used ASD tugs; tugs neither designed for nor well suited to escort service as a basis of comparing the effectiveness of propulsion systems. A properly designed ASD escort tug would have its tow-point located just forward of the CLR, as would a VSP tug. This is essential for a "fail-safe" operation.

I agree that the design and efficiency of the skeg (fin) requires further research. However, investigations into skeg -design must also take into account the other more mundane functions required of these vessels as well, and the impact on general maneuverability and control of the tug.

While it is true that only 10–20% of the total towline force generated is contributed by the drive system, the full power of the tug is still required in order to obtain the position of maximum steering force, otherwise the tug would remain at about a 45 degree angle to the axis of the ship.

I disagree with the comment that the "major advantage" of an ASD tug is the reduction in overall draft. In fact, the major advantage of the ASD design is the much higher bollard pull, which can be generated for the same power, enabling the tow to be brought under control more quickly as speed diminishes.

Finally, I would not favor the proposal to develop an escort rating system which includes the effect of the tug on particular sizes and speeds of ships. There are too many variables at play in an escort operation, not the least of which is the geography within which it takes place. Therefore I rather favor the type of notation currently used by DNV and others, using simply the maximum Fs and the associated speed at which the tug meets particular stability criteria. These forces and speeds can then be input into any escort model.

SNAME *Transactions*, Vol. 108, 2000, pp. 123–146

Probabilistic Approach for Intact Stability Standards

Vadim L. Belenky, Member, Consultant, Mentor-on-the-Lake, Ohio

This paper proposes an approach for probabilistic intact ship stability standards and reviews assessment methods that have been developed during the last 40 years. This approach treats capsizing as a random event occurring in stochastic wind and wave environments as well as in an random operating environment, including loading conditions, unique to the employment of each type of vessel. Probabilistic stability standards could be treated as a general reliability problem, considering capsizing as a system fault; however, modifications to the general approach must be made because the fault is fatal and extremely rare. A stochastic approach to determining the risk of capsizing requires a tool that can simulate all of the modes of capsizing in irregular seas: pure loss of stability as a result of extreme external forces and/or decreasing stability in quartering seas, broaching, breaking waves action, green water shipping, etc. This paper examines available solutions to be applied to calculate probability of capsizing. The paper pays special attention to practical calculation of nonlinear ship rolling in high irregular seas. The paper summarizes the author's work as well as other relevant research, including Russian and Japanese papers that are not available in English.

NOMENCLATURE

A	Added mass
A_B	Work of breaking wave
$f(..)$	Probability distribution density
h_s	Significant height of waves
h_w	Instantaneous height of the wave
I_x	Transverse moment of inertia
K	Kinetic energy
k_w	Wave number
L	Length of a ship
L_W	Length of a wave
$\vec{L}$	Vector of loading conditions
M	Mass of a ship
P	Probability, potential energy, weight of cargo
$P(X)$	Probability of capsizing
$P(\overline{X})$	Probability of non-capsizing
X	Event of capsizing
R_x	Transverse inertia radius
$\vec{S}$	Vector of assumed situations
W	Displacement weight
α_m	Effective (reduced) instantaneous angle of wave slope
λ	Risk function
λ_a	Averaged risk function
λ_U	Intensity of upcrossing
ϕ	Roll angle
ϕ_{max}	Angle of maximum of the GZ curve
ϕ_V	Angle of vanishing stability
$\dot{\phi}_{cr}$	Critical crossing roll velocity (the roll velocity of crossing the level that is enough to capsize a ship)
ζ	Instantaneous wave elevation
ρ	Water density
χ	Heading angle

1. INTRODUCTION: PROBABILISTIC VS. DETERMINISTIC STABILITY STANDARDS

Contemporary intact stability standards combine physical and weather criteria with statistics. Rahola [1] was the first to develop this statistical approach. The statistical stability standards are based on previous experience - both positive and negative. The principal disadvantage of the statistical approach is its relationship with this previous experience. So, changes in the fleet would result in a large time lag, until the necessary experience could be collected. Negative experience in stability is quite expensive.

The weather criterion is a deterministic physical one that is based on balance of heeling and restoring moments. Since it uses physical relationships between external action and capsizing, the weather criterion does not depend on previous experience. However, another problem appears. How can we choose a particular value of external heeling moment? Irregular seas and gusty wind produce it, but we have only probabilistic descriptions of these figures. If we design our ship so that it can stand any possible heeling moment, the stability will be excessive - she may never encounter such severe conditions during all her service. At the same time, this surplus stability significantly affects other qualities of a ship including profit [2].

There have been numerous efforts to find a "rational" value for heeling moment. Nonetheless, this value has to be chosen more or less arbitrarily. Practically, this means that different ships have quite different levels of safety. This difference might be very large, see [3] for some comparative calculations. These calculations were carried out for 16 cargo ships; length varied from 79.6 to 189.1 m. Their stability was in a "critical" state according to IMO criteria: one criterion was met without any surplus; other criteria were satisfied with some surplus. Large differences in capsizing probability were observed; however, these ships are meant to be equally safe from a deterministic point of view.

2. FRAMEWORK OF PROBABILISTIC STABILITY STANDARD

Attempts to develop probabilistic standards for intact ship stability have been made since the early sixties, see papers by Krappinger [4], Sevastianov [5,6,7], Caldwell and Jang [8], Kobylinski [9,10], McTaggart [11], Alman et al [12]

2.1 Dependence on Time

Probability assumes infinite time unless time is explicitly specified. This is a result of the application of the stochastic process concept to waves and roll. Probability theory assumes that a stochastic process lasts for an infinite period of time. This means that if no special measures are taken, the probability of capsizing is associated with an infinite time of exposure. However, real world risks are strongly dependent on the time of exposure. An example follows.

Consider two different ships, A and B, that have different time histories of λ with the same maximum and minimum capsizing risk levels, see Figure 1 that is taken from [7]

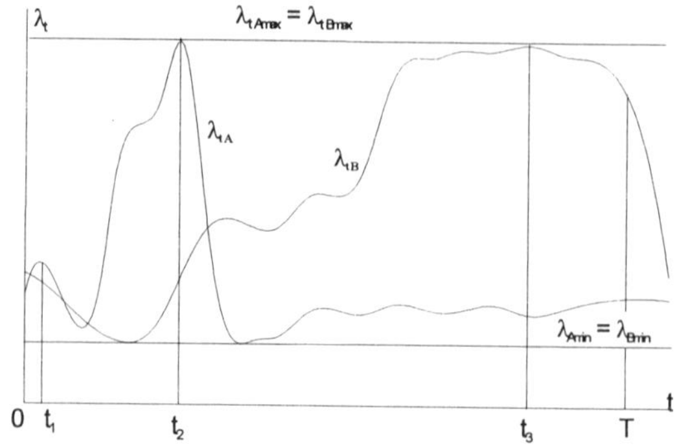

Figure 1. Time history of risk measure

If we compare capsizing risk at times t_1 and t_2, we may conclude that ship A is more dangerous than ship B. If the moment t_3 is chosen, our conclusion will be different. If we look at the entire time history, it is clear that ship B is worse, because she experiences high risk for a longer time within the range from 0 to T. So, in order to compare the stability safety of two ships, we introduce an integral measure.

$$\lambda = \frac{\int_0^T \lambda_{Ins}(t)dt}{T} \qquad (1)$$

This value expresses averaged risk measure per unit of time. It is usually called a "risk function". The probability of non-capsizing during time T can be expressed as follows (see [5] or [7]):

$$P_T(\overline{X}) = \exp[-\lambda T] \qquad (2)$$

Also, these publications show that λ is a more convenient criterion for stability standards than the probability of capsizing or non-capsizing.

2.2 Problem of Sensitivity

One of the most serious obstacles in developing probabilistic methods is the sensitivity to capsizing probability as a stability criterion. It is especially clear when trying to establish a numerical value for a probabilistic standard. For example, a standard value of probability of non-capsizing for 20 years of service is set up as:

$$P_{T=20}(\overline{X}) = 0.99\,.$$

It looks safe enough, but if we consider it as a stability standard, we have to apply it to all the vessels covered by the standard. If there are 100 ships covered by the standard, the probability of the event that there will be no capsizing for 20 years is:

$$P_{T=20}^{N=100}(\overline{X}) = \left[P_{T=20}^{N=1}(\overline{X})\right]^{100} = 0.99^{100} = 0.36.$$

That is, the probability of capsizing of at least one ship from the fleet is more than 60%! Such a probability does not look safe enough. Probability of noncapsizing of one ship $P_{T=20}^{N=1}(\overline{X})=1$ never can be achieved. So, a reasonable value for non-capsizing probability is somewhere in between 0.99 and 1. It is next to impossible to develop a method for non-capsizing probability calculation that would be sufficiently accurate!

Consequently, the probability of non-capsizing cannot serve as a practical stability criterion. The practical choice would be the risk function λ itself. The risk function does not have the above sensitivity disadvantage, since it is related exponentially with probability of non-capsizing, (equation (2)), as shown in figure 2.

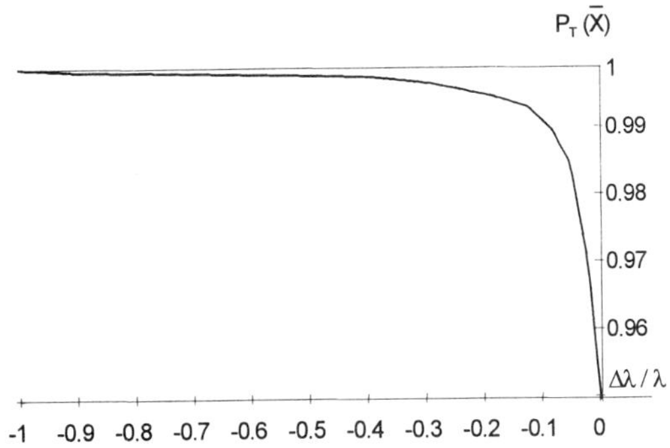

Figure 2. Sensitivity of probability of non-capsizing vs. risk function

2.3 Loading and Operational Peculiarities

To complete the general scheme of the standard, it is necessary to take into account changes in weather and loading conditions during a voyage or a ship's service life, so the risk function (1) has to be averaged through all possible weather and operational states (Sevastianov [5] or [7]).

Let's consider a significant amount of time of ship operation (voyage, mission or whole lifetime) and divide it into intervals when weather and operational conditions (such as speed and course) can be considered to be stationary. This interval we shall call the "assumed situation" and denote S_i. The risk function within this interval is also constant and is denoted by $\lambda^*(X|S_i)$. To get the true average over the time, it is necessary to account for the portion of time each "assumed situation" takes. This share equals $P(S_i)$, the probability of the occurrence of the i-th "assumed situation":

$$\lambda_a = \sum_{i=1}^{n} P(S_i)\cdot\lambda^*(X|S_i)$$

The risk of capsizing, however, depends also on loading conditions such as displacement, *KG*, etc. Following the previous pattern we form groups of loading conditions and denote them as L_j. Generally speaking, loading conditions are also random within some boundaries. The capsizing risk function now also depends on loading conditions: $\lambda(X|\ S_i\ ,L_j)$. Since there is no evident relationship between the assumed situation and the loading conditions (value of *KG* does not depend on weather or speed), these two factors are considered to be independent. Then we can treat them separately and consider the risk function $\lambda^*(X|S_i)$ as already averaged over all possible loading conditions:

$$\lambda^*(X|S_i) = \sum_{j=1}^{m} P(L_j|S_i)\cdot\lambda(X|S_i,L_j)$$

where $P(L_j|S_i)$ is the probability of encountering loading conditions L_j during the situation S_i. The final formula for the averaged risk function then is:

$$\lambda_a = \sum_{i=1}^{n} P(S_i)\sum_{j=1}^{m} P(L_j|S_i)\cdot\lambda(X|S_i,L_j) \quad (3)$$

Sevastianov [6] showed that a complete analogy exists between the framework considered here and general reliability theory that was originally developed for electronics and radio engineering. The main difference is that the failure of at least one element always means fatal failure of the system. Another important detail is that this element sometimes can fail in several different ways depending on the input. Then, the framework presented here can be considered as a particular application of general reliability theory for stability accidents.

The remainder of the review will consider these three components: the probability of the assumed situation, the probability of the loading condition, and the capsizing risk function at constant operational, weather and loading factors. Taken together, these three components give the probability of capsizing per unit of time.

3. PROBABILITY OF CAPSIZING OF A SHIP IN BEAM SEAS AND WIND

Implementation of probabilistic stability standards requires well-developed methods for calculating the probability of capsizing per unit of time. It is impractical to use simulation as the only tool to calculate capsizing probability, because a large number of cases must be considered. The probability of a given capsize might be too small (up to 10^{-9} [3]), so that results are difficult to obtain in reasonable computational time.

For example, if the risk function equals 10^{-9} per hour, this means that the average time before capsizing is 10^9 hours or 1.1 million years. Even if the computer is as fast as 1 second per 1 year (that is 63 million steps of simulation and this is beyond of capabilities of contemporary workstations), it would request 32 hours of calculation time to complete just one realization. To get a more or less reliable statistical estimate of the risk function we might need more than 100 realizations and this is for just one sea state and loading!

However we still can use simulation to calculate averaged values like roll variance or even density of distribution. So we have to combine simulation with theoretical methods.

There are several methods for calculating capsizing probability. It is convenient to classify these methods based on the type of analysis used to identify transition from upright stability. These are:

- Phase Plane Analysis
- Dynamic Heel Limit
- Motion Stability
- Non-linear Dynamics
- Piecewise linear Method

3.1 Phase Plane Analysis

Phase plane (roll vs. roll velocity) is a convenient way to present peculiarities of ship motion at a glance. Since capsizing involves transition to up-side down equilibrium, both equilibria have to be considered. The line dividing areas of attraction towards these equilibria is called the separatrix. Figure 3 shows the phase plane of rolling in calm water. The separatrix is shown in bold.

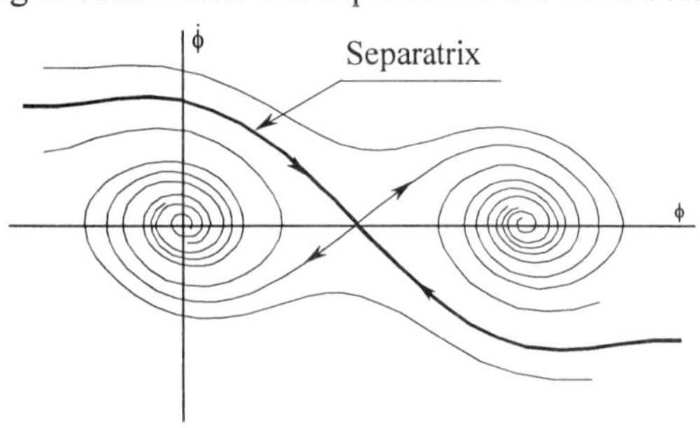

Figure 3. Autonomous roll phase plane

Capsizing is considered to occur if the phase trajectory of forced motion crosses the separatrix; this is the classical definition of stability in terms of phase plane mapping. (A ship is considered to be stable if after being inclined, she returns to her initial position after the cause for the inclination ceases.) Sevastianov [13], Pham [14], and Umeda [15] used this approach.

The most difficult part of the problem is the relationship to time. Capsizing is associated with crossing the separatrix by phase trajectories in the phase plane $(x=\phi, y=\dot{\phi})$. The phase trajectory can be considered to be a two-dimensional random vector and a separatrix as a boundary. Let's imagine that we know the combined distribution of the roll angle and roll velocity. Then, following Sevastianov [25] we can consider an upcrossing problem of a 2D stochastic process, where the theoretical mathematical solution is known.

Upcrossing is a random event where a stochastic process reaches a given boundary and its first derivative is greater than zero (the last condition guarantees that the process really does cross the border). If the process has just one dimension, the border would be a constant, $x=b$. The most important thing about upcrossings is that they can be considered as rare events in the majority of practical cases. This means that the number of upcrossings during a given time obeys Poisson's law and that the time between two neighboring upcrossings is exponential. For stationary processes we can write:

$$f_{U=0}(t) = 1 - \exp(-\lambda_U t) \qquad (4)$$

This gives us the necessary connection to time. Applying upcrossing theory looks quite natural, since capsizing itself is a rare random event and has the same probabilistic qualities.

Intensity of upcrossings can be easily calculated if we now combine the probability density of the process and its first derivative:

$$\lambda_U = \int_0^{\infty} \dot{x} f(x = x_b, \dot{x}) d\dot{x} \qquad (5)$$

If we consider capsizing as crossing the separatrix, we deal with a two dimensional stochastic process and with the border, which is a line $y_b=y(x_b)$ or $x_b=x(y_b)$. The intensity of upcrossing through this line can be expressed as:

$$\lambda_U = \int_{-\infty}^{\infty}\int_{-\infty}^{\infty}\int_{y_B}^{\infty} \dot{x} f(x = x_b(y), y, \dot{x}, \dot{y}) dy d\dot{x} d\dot{y} \qquad (6)$$

Note, that formula (6) gives intensity of upcrossings only, and corresponds to capsizing to starboard only. A complete value of capsizing risk function must include capsizing to port. It can be defined analogously by consideration of downcrossings through another portion of the separatrix. Taking into account the relationship of coordinates at the phase plane $(x=\phi, y=\dot{\phi})$, equation (6) can be rewritten as:

$$\lambda_U = \int_{-\infty}^{\infty}\int_{\dot{\phi}_b}^{\infty} \dot{\phi} \cdot f(\phi = \phi_b(\dot{\phi}), \dot{\phi}, \ddot{\phi}) d\dot{\phi} d\ddot{\phi} \qquad (7)$$

To calculate the integral in equation (7), we need to know the joint density distribution of roll angle, velocity and acceleration. This is a separate problem and we will return to it later in the section 5.3.

Here it has to be mentioned that accuracy of the joint density distribution might be important to the accuracy of the upcrossing rate calculation.

This method which is based on the classical definition of stability might look obsolete since we do not consider wave forces when defining the boundary between capsizing and safety. However, this error cannot be large, because a ship is not that sensitive to external forces causing the ship to be inclined at a large angle [29]. At the same time, numerical integration in equation (7) might not be that precise because of the reduction in accuracy of the joint distribution probability density for large values of roll angle and velocity. (The probability of reaching these values is very small and it is very difficult to verify them with model tests or even full scale trials [29].)

3.2 Dynamic Heel Limit

Capsizing is considered to occur if the dynamic angle of heel exceeds a certain value. This can be directly related to weather criteria for the case of irregular waves and gusty wind. Dudziak [16] and Bielanski [17] used such approach.

If we use a weather criterion, the capsizing is associated with satisfying the following inequality:

$$K(t)-K(t_0)>P(t)-P(t_0) \qquad (8)$$

where $K(t)$ and $P(t)$ are values of kinetic and potential energy of the rolling ship, $K(t_0)$ and $P(t_0)$ are these values at an initial moment of time, with:

$$K(t)=0.5(I_x+A_{44})\dot{\phi}^2 \qquad (9)$$

$$P(t)=\int_0^{\phi(t)} GZ(\phi(t))d\phi \qquad (10)$$

If we combine the distribution density of roll angles and velocities, $f(\phi,\dot{\phi})$ is known and a new stochastic process can be considered:

$$\xi(t)=[K(t)-K(t_0)]-[P(t)-P(t_0)] \qquad (11)$$

Its distribution density can be derived as a result of the composition of $f(\phi,\dot{\phi})$. Then, we can find the capsizing risk function as the intensity of upcrossing of the process $\xi(t)$ through zero.

An evident advantage of such an approach is the compatibility with an existing stability standard. However, calculations of joint probability density of $\xi(t)$ and its derivative would be complex, since it requires performing numerical probabilistic composition several times.

3.3 Motion Stability

Capsizing is considered to occur if forced stochastic rolling loses motion stability. This approach was used by Price [18], Nekrasov [19,20,21], Zelenin [22].

The motion stability concept is based on the concept of small perturbations. A regime of motion would be considered stable if, after being disturbed by a very small change in roll angle or roll velocity, the system will return to its original mode of motion.

Conventional stability analysis includes building a separate equation for perturbed motion. If its solution has a tendency to rise, it usually means that the studied regime is not stable (i.e., the difference between the original motion and the perturbed one grows). If the solution decays, it usually means that the motion is stable. Numerically this is usually expressed in terms of stability indexes: if they are inside the unit circle it usually means the response is stable.

There is a theoretical possibility of using upcrossings to calculate capsizing probability based on motion stability theory. To achieve this we might consider stability indexes as function of time. Virgin [26] used such an approach for regular waves. Then, we can consider stability indexes as a two-dimensional stochastic process and study upcrossings through the unit circle. The problem of how to define stability indexes in irregular seas is part of the general theory of nonlinear random dynamical systems, see the fundamental monograph of Arnold [27] where a systematic consideration of the subject can be found.

So far, there is no method known to the author that would allow calculation of capsizing probability during given amount of time.

3.4 Non-linear Dynamics:

Capsizing is considered as a transition to motion in the vicinity of an upside down position of equilibrium. This approach was realized with extensive use of methods of contemporary nonlinear dynamics, see Hsieh *et al* [23], Lin and Solomon [24].

3.5 Piecewise Linear Approach

The author [3, 28, 29] studied capsizing as a random rare transition to another equilibrium. A piecewise linear representation was used to calculate capsizing probability. The method was developed for utilization in formula (3), so it explicitly takes time into account. A piecewise linear representation was used for decreasing branches of the GZ curve. The increasing branch was left nonlinear, see Figure 4.

It was proven [29] that external excitation does not play a significant role when the maximum level of the GZ curve is crossed. This allows for a relatively simple solution for capsizing. It was associated with crossing of the maximum level with the rolling velocity exceeding a certain critical value that can be calculated using a piecewise linear representation of the GZ curve. Then:

$$\lambda(X|S_i,L_j) = P_{T=1}(\phi > \phi_{max})P(\dot{\phi} > \dot{\phi}_{cr}) \quad (12)$$

where $P_{T=1}(\phi > \phi_{max})$ is the probability of exceeding the maximum level of the GZ curve and $P(\dot{\phi} > \dot{\phi}_{cr})$ is the probability of exceeding the critical roll velocity. The probability of the upcrossing can be calculated using upcrossing theory, if probabilistic characteristics of the nonlinear irregular rolling are known. This creates a strong motive for studying nonlinear stochastic rolling.

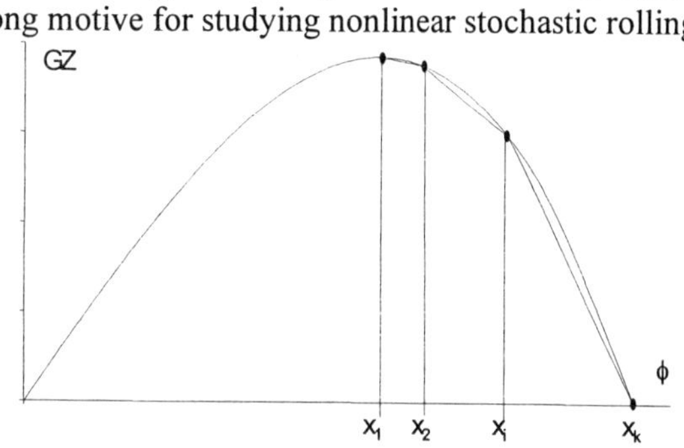

Figure 4. Piecewise presentation of the decreasing brunch of the GZ curve

The method was generalized for multiple degrees of freedom, see [3] where wind drift was taken into account using the following equation for sway motion:

$$(M + A_{22})\ddot{y} + F_{YH}(\dot{y}) = F_{YE}(t) + F_{YA}(t) \quad (13)$$

Here F_{YH} is a force of lateral resistance, F_{YE} is the sway wave excitation force and F_{YA} is the horizontal component of the aerodynamic force. The sway equation (14) was linearized and solved analytically, then the sway component of the heeling moment was taken into account in the rolling equation.

The above method has an evident advantage in that upcrossing theory is used against the level of the maximum of the GZ curve. This value is not that large and, hence, probabilistic characteristics of upcrossings can be calculated more accurately. The disadvantage of it is that the GZ curve has to be constant with respect to time. This make it impossible to use this method for any case involving changes of the GZ curve such as taking into account heaving or following and quartering seas. However, it is possible to modify the methods for these cases, see section 4.

Summary

There are several different ways to calculate probability of capsizing in beam seas per unit of time. The majority of them require that upcrossing theory be used to get a connection with time. The best accuracy is achieved when the level for upcrossings is low, because of the accuracy of the joint distribution of probability density of roll angles and velocities.

4. PROBABILITY OF CAPSIZING IN QUARTERING SEAS AND DUE TO BREAKING WAVES ACTION

4.1 Modes of Capsizing

Capsizing is a complicated phenomenon. Changing heading, speed and course angle relative to the wave direction changes the physics of capsizing significantly. There are at least three different way to capsize in quartering seas: pure loss of stability, capsizing due to parametric excitation and surf-riding followed by broaching. We also have to take into account that waves in open seas can experience breaking. Breaking wave generate an impact that is sometimes enough to capsize a small vessel. If we try to take into account green water shipping, things become even more complicated. If we call these modes of capsizing "scenarios", formula (3) has to be amended:

$$\lambda(X|S_i,L_j) = \sum_{k=1}^{n} P(Sc_k|S_i,L_j) \times \times \lambda_S(X|S_i,L_j,Sc_k) \quad (14)$$

Here $\lambda_S(X|S_i,L_j,Sc_k)$ is the risk function of capsizing caused by scenario Sc_k in assumed situation S_i with loading condition L_j. $P(Sc_k|S_i,L_j)$ is the probability of realization of such a scenario in the above conditions.

For example, surf-riding can be a prerequisite to broaching. So, we first need x to calculate the probability of surf-riding using the method developed by Umeda [30,31] and then we can calculate the probability of developing broaching and capsizing due to broaching (methods are not available yet). Following this approach, the following simple formula is derived:

$$\lambda(X|S_i,L_j) = P_{BW}\lambda_{CB} + + (P_{BW} - 1)[P_{SF}P_{BR}\lambda_{BR} + (1 - P_{SF})\lambda_{PL}] \quad (15)$$

Here:

P_{BW} – probability to encounter breaking waves

λ_{CB}- risk function of capsizing caused by breaking waves

P_{SF} – probability of surf-riding

P_{BR} – probability of broaching

λ_{BR} – risk function of capsizing caused by broaching;

λ_{PL} – risk function of capsizing caused by pure loss of stability (including decreasing stability in quartering seas, actions of waves and wind).

All probabilities are meant to be time dependent and calculated for T=1.

Formula (15) can be considered to be a general one. It can be used for any course, not only for quartering and following seas. For example, for beam seas, probabilities of surf-riding and broaching equal zero. At the same time there is always a non-zero probability of encountering breaking waves.

4.2 Probability of Surf-Riding

Unfortunately, there are very few methods to calculate these probabilities. Time dependence introduces mathematical difficulties. The following papers contain research results concerning these probabilities. We already mentioned Umeda [30,31] for calculating the probability of surf-riding. He considered three types of longitudinal motions: surging, surf-riding under any initial conditions and surf-riding under certain initial conditions.

These regimes can be easily identified for regular waves using the phase plane method. When the heading is significantly less than the phase velocity of waves, we observe only periodic surge. Then, as speed increases and becomes comparable to the wave phase velocity, some of the initial conditions (position of the ship relative to the wave and instantaneous speed) may lead to surf-riding. Further increases in the speed lead to surf–riding under any initial conditions.

To translate the surf-riding problem into a probabilistic frame Umeda [30,31] uses an envelope representation of irregular waves:

$$\zeta_W(t) = 0.5h(t)\cdot\cos(\omega_c t + \varphi(t)) \qquad (16)$$

Here $h(t)$ is the current wave height which is time-dependent, in order to present variations, ω_c is the carrier frequency and $\varphi(t)$ is the uniformly distributed random phase. This is the conventional scheme of an envelope representation of narrow banded stochastic process.

The probability of surf-riding is calculated as:

$$P_{SFR} = P^A_{SFR} + P^B_{SFR}\cdot P\{(x,\dot{x})\in A_{SFR}\} \qquad (17)$$

where P^A_{SFR} is the probability of surf-riding under any initial condition, P^B_{SFR} is the probability of surf-riding under certain initial conditions, $P^A_{SFR}P\{(x,\dot{x})\in A_{SFR}\}$ is the probability that the current position and speed of a ship belong to the domain of surf-riding initial conditions A_{SFR}.

So, there is a way to treat surf-riding in a probabilistic way. However, there are some questions remaining. It is not clear how accurate these calculations are. Another question is whether we have to account for time and, if so, how it should be done, but at least, there is a method that can be used for some estimates.

4.3 Pure Loss Stability: Effective Wave Concept

In a series of works, Boroday [32,33], Boroday and Nikolaev [34], Boroday and Netsvetaev [35] derived formulae for the probability of pure loss of stability by solving the general hydrodynamic problem of a ship heading in quartering seas.

In another series of work Umeda *et al* [36], Umeda and Yamakoshi [37,38,39] used the phase space approach to find out the probability of pure loss of stability in quartering seas.

These works used the effective wave concepts developed by Grim [40] for describing changes of the GZ curve in quartering and following irregular seas.

The idea is simple – the profile of irregular seas along the waterplane is substituted by a sinusoidal wave with length equal to the ship length. This effective wave is immovable, its crest is always amidships, but its random amplitude is set to have the same effect on stability as the original irregular profile.

$$\zeta_{eff}(t) = h_{eff}(t)\cos\left[\frac{2\pi}{L}x\right] \qquad (18)$$

Here L length of a ship. The random amplitude of the effective wave $h_{eff}(t)$ is represented by a Fourier series

$$h_{eff}(t,\xi_G) = \\ = \sum_{i=1}^{n} h_i\cos(k_{W_i}\xi_G(t)\cos\chi - \omega_{ei}t + \varphi_i) \qquad (19)$$

Amplitudes h_i are to be found from transfer functions that were derived to minimize the difference between the effective wave and the original irregular wave profile [40] along the ship waterplane:

$$h_i = \frac{k_{W_i}L\cos\chi\sin(0.5k_{W_i}\cos\chi)}{\pi^2 - (0.5k_{W_i}L\cos\chi)^2}\cdot r_i \qquad (20)$$

Now we can express the changing GZ curve in quartering seas in a very simple way:

$$GZ_W(\phi,t,\xi_G) = GZ(\phi) + \\ + \Delta GZ_W(\phi, h_{eff}(t,\xi_G), L_w = L, v_s) \qquad (21)$$

The effective wave concept is very important, because it offers a clear physics-based transition from regular to irregular waves for problems of stability changes in quartering seas.

4.4 Pure Loss Stability: Phase Plane Analysis

Now, the probability of capsizing can be found by the integration of the safe basin in the phase space $(\phi, \dot{\phi}, \xi_G)$ for all possible values of h_{eff}:

$$P(X) = 1 - p_e \times \times \int_{-\infty}^{\infty}\int_{-\infty}^{\infty} \iint_{S(\phi,\dot{\phi},x,h_{eff})} f(\phi,\dot{\phi},\dot{x},h_{eff}) d\phi d\dot{\phi} d\dot{x} dh_{eff} \quad (22)$$

Here $S(\phi,\dot{\phi},\dot{x},h_{eff})$ is the safe basin and p_e is the probability of a minimum of the effective wave when the ship encounters a wave crest. For details see Umeda et al [36]. To calculate the risk function, we have to consider the problem of upcrossing through the border of the safe basin.

4.5 Pure Loss Stability: Piecewise Linear Approach

The author also used the effective wave concept for development of a risk function calculation method. The method is based on piecewise linear presentation of the GZ curve changing in time, see fig. 5 [41]

An analytical solution for capsizing was developed using a two-dimensional piecewise linear representation of the decreasing branch of the GZ curve. This solution, in combination with a numerical simulation of nonlinear rolling in the increasing range of the *GZ* curve, provides the desired solution. Time is taken into account explicitly in the same way it has been done for beam seas - using upcrossing theory. The rolling equation beyond the level of maximum of the *GZ* is:

$$(I_x + A_{44})\ddot{\phi} + 2N_\phi\dot{\phi} + + W \cdot (a_i\phi + b_i t + c_i) = W \cdot GM \cdot \alpha_m(t) \quad (23)$$

Here the coefficients a_i, b_i and c_i are random values, where their probabilistic characteristics can be found by the Monte-Carlo method. For the particular values of these coefficients equation (23) is a piecewise linear differential equation and can be solved analytically within ranges of linearity:

$$\phi(t) = A_i \exp(\lambda_{1i}t) + B_i \exp(\lambda_{2i}t) + + p(t) + q_{1i}t + q_{2i} \quad (24)$$

where A and B are arbitrary constants, $\lambda_{1,2i}$ are eigenvalues, and p and $q_{1,2i}$ are elements of the partial solution. Solution of equation (24) allows finding the dependence of the capsizing event on a_i, b_i, c_i and roll velocity at the moment of upcrossing. The formulae are quite long and are not included here. See [41,42] for details.

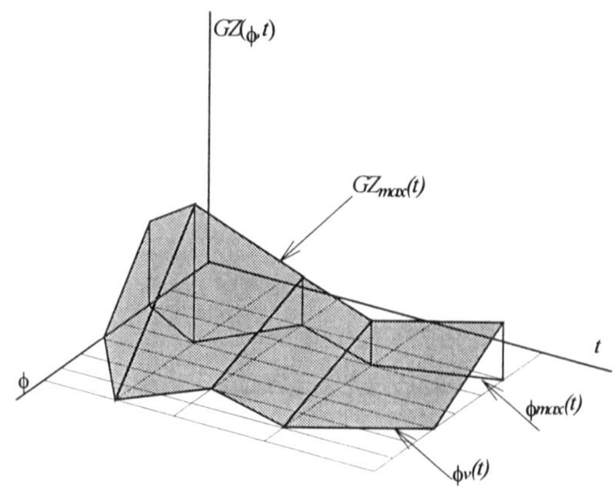

Figure 5. Piecewise linear presentation of changing GZ curve.

A two-dimensional piecewise presentation of the GZ curve makes it possible to account for stability changes in quartering seas. This is also a way to overcome limitations of the one-dimensional piecewise linear method noted in section 3.5

4.6 Capsizing in Breaking Waves: Risk Analysis

Estimation of risk of capsizing caused by breaking waves can be found in Dahle *et al* [43], Dahle and Myrhaug [44], Dahle *et al* [45]. This method has three major components:

- Systematic model experiments where the model is subject to breaking wave action. The purpose of these tests is to identify potentially dangerous waves.
- Calculation of the statistical frequency for the ship to encounter potentially dangerous waves, based on wave statistics related to geographical region and season of the year.
- Calculation of the probability of capsizing if a ship is hit by a potentially dangerous wave. It considers wave direction, time of exposure and results of the above systematic model tests.

The final result is given in the form of statistical frequency of capsizing per year:

$$F(X) = F(Wv) \cdot P(X \mid Wv) \quad (25)$$

where $F(Wv)$ is the frequency of occurrence of a potentially dangerous waves, $P(X|W)$ is the probability of capsizing.

$$F(Wv) = \sum_j \sum_k w_{jk} P_{3jk} P_{4kj} \quad (26)$$

where

j is number of significant wave heights H_{sj}

k is number of zero-crossing periods T_{zk}

$$w_{jk} = \begin{cases} t_{jk}/\tau_{jk} & if: \ t_{jk}/\tau_{jk} < 1 \\ 1 & if: \ t_{jk}/\tau_{jk} \geq 1 \end{cases} \qquad (27)$$

t_{jk} is the time of exposure to sea state (hrs)

τ_{jk} is the duration of exposure to sea state (hrs)

P_{3jk} is the conditional probability of the appearance of steep and high waves for a given sea state, the actual condition of breaking.

P_{4kj} is the probability of occurrence of H_{sj} and T_{zk}

$$P(X \mid Wv) = P_5 P_6 \sum_j \sum_k P_{1jk} P_{2jk} \qquad (28)$$

P_{1jk} is the probability of a dangerous wave direction in sea state *jk*

P_{2jk} is the annual fraction of exposure to sea state *jk*

P_5 is the probability of a ship being hit when the direction of rolling coincides with the direction of the hit.

P_6 is the conditional probability of capsizing if a ship is hit by a breaking wave. It has to be defined using results of a model test. Only two values are considered: 0 for "safe" conditions and 1 for "unsafe" ones, see fig. 6.

Comparing formulae (3) and (25) we see certain similarities. That is, P_{2jk} is analogous to $P(S_i)$ and P_{1jk} to $\lambda(X|S_i,L_j)$. Equation (3), however, was intended to cover all possible reasons of capsizing, not only breaking waves. This creates a possibility of incorporating Dahle into the Sevastianov scheme of a probabilistic standard as one scenario if the connection to time can be resolved.

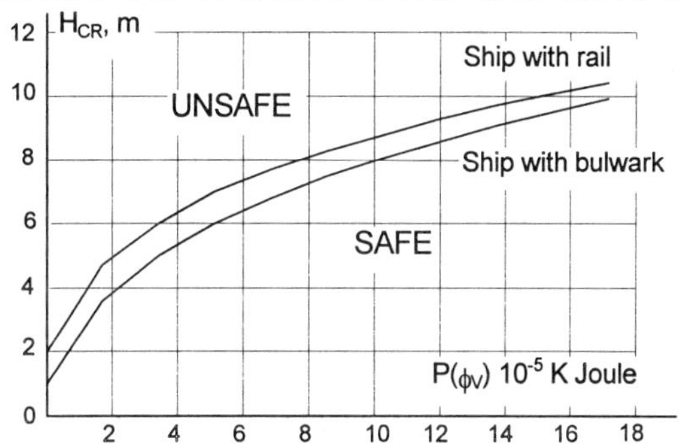

Figure 6. Safe and unsafe regions as found from model tests of M/S Helland Hansen, taken from Dahle *et al* [45]

4.7 Capsizing in Breaking Waves: Dynamic Heel Limit

Belenky and Mordachev [46] tried to estimate the probability of encountering breaking waves and of capsizing due to such an encounter, using an energy balance approach.

The idea was as follows: the wave cannot transfer to the ship more energy than it has itself. Since we do not know mechanics of the interaction between the breaking wave and the ship, we have to use empirical data (from experiment, as above) in order to estimate which portion of the wave energy can be absorbed by a ship during capsizing.

Then, the capsizing condition (9) has to be altered by adding the work of the breaking wave A_B:

$$K(t) - K(t_0) + A_B > P(t) - P(t_0) \qquad (29)$$

The work of the breaking wave absorbed by the ship can be expressed as a portion of the full amount of wave energy (theory of small waves was used in the first expansion):

$$A_B = \kappa_B \frac{1}{8} \rho g h_W^2 L_w L \qquad (30)$$

Here κ_B is portion of wave energy used for capsizing.

Then, based on the inequality (29), the capsizing risk function can be found as the intensity of upcrossing through the zero level of the process

$$\xi(t) = [K(t) - K(t_0)] + A_B - [P(t) - P(t_0)]. \quad (31)$$

A sample of the development based on such an approach can be found in [46]. Note that here we will again need the joint distribution density of the roll angles and velocities.

4.8 Summary

We reviewed several methods to estimate the capsizing risk function in the complicated conditions of quartering seas and breaking waves. Despite some theoretical difficulties, it seems that we are now able to produce some estimates for these figures. Accuracy of these estimates depends on the adequacy of the mathematical models used. The accuracy may depend even more strongly on the accuracy of the probabilistic characteristics of nonlinear rolling in heavy irregular seas. This is the subject for the next section.

5. NONLINEAR ROLLING IN IRREGULAR HIGH SEAS

We reviewed the above methods of calculation of capsizing probability. All require a certain amount of knowledge of the probabilistic qualities of nonlinear rolling in irregular seas. For example, straight forward use of the classical definition requires integration of the joint probability density of roll angle and roll velocity. Limits of integration lie within the separatrix of free roll motion. Using a method by Umeda *et al* [15], we need the distribution of amplitudes of nonlinear rolling that can be obtained from the above mentioned joint distribution of roll angles and velocities. The energy balance method and motion stability approach (see appropriate references given above) also requires the distribution of roll angles and velocities. Finally, knowledge of these figures is necessary to apply a piecewise linear method since upcrossing theory requires them as an input data. So, the problem of an adequate probabilistic description of the nonlinear rolling in stochastic waves should be considered as a problem of separate importance within the framework of our review.

The most difficult part of this problem is the non-linearity of rolling. Probabilistic qualities of linear or linearized ship motions are well known. Unfortunately this knowledge cannot be applied here since capsizing is a strictly nonlinear phenomenon. It means that we know very little about nonlinear rolling and have to start from basic assumptions.

Nonlinear irregular rolling probably can be considered as a stationary stochastic process, if waves and wind characteristics, heading, speed and course do not change. This is a reasonable assumption since the absence of stationary qualities has to be caused by changes in the environment or in the ship herself.

5.1 Ergodic Qualities

Another thing to check is ergodicity. It is a quality of stochastic process that allows finding mean values, variances and so on, using only one long realization of the process, instead of taking into account the whole ensemble, which is formally required by the theory of probability for a general stochastic process. The simplest way to check it is using computer simulation, see Belenky [47,48]. Elevations of irregular waves were simulated using Fourier series:

$$\zeta_w(t) = \sum_{i=0}^{N} a_i \sin(\omega_i t + \varphi_{0i}) \quad (32)$$

The amplitudes of the sine functions, a, were calculated from a given spectrum with the frequency set to ω_i. The initial phase angles φ_{0i} have random values with uniform distribution.

The set of initial phases, φ_{0i} are generated once before the simulation; each set makes one realization of irregular waves. Simulated roll response is used to calculate roll variance after a certain amount of time. The roll variance is presented as a function of the time range it was calculated for. Then, the next realization of waves is generated and the entire procedure is repeated. Figure 7 (a) shows behavior of these functions for a sample linear rolling equation: all curves converge to a certain point in the vicinity of the theoretical value of roll variance. This is exactly characteristic of ergodic stochastic behavior.

Nonlinear systems, however, show completely different behavior – see figure 7 (b), where the real GZ curve is implemented. Variances did not converge even after 20,000 seconds – which is more than 5 hours of full-scale time. Although there is a theoretical possibility that the variances will converge sometime in the distant future, it is beyond engineering interest. This means that we cannot use ergodic assumptions in practical calculations and have to deal with the whole ensemble of realization.

Degtyarev and Boukhanovsky [47,48] confirmed this outcome with independent computer simulations using a different mathematical model for irregular waves and another form of the rolling equation. They also proposed a physical explanation of this phenomenon.

The author also confirmed the existence of this phenomenon using model experiments in the towing tank at the National Research Institute of Fisheries Engineering of Japan in 1999. In these experiments, the model was placed in irregular waves. The model had a low free board and was equipped with a digital gyroscope to measure roll angle and velocity. The model was fixed in sway and yaw during the test, but was free to heave and roll. The duration of the test was up to 40 minutes, corresponding to 6 hours of full-scale time. A longer test time would not be meaningful, since it would be beyond the period of time when the waves could be considered to be a stationary stochastic process.

The test was repeated for 13 different realizations of irregular waves developed from the same spectrum. Despite the significant volume of statistics, observed differences in variances calculated by different realizations of rolling was up to 30%. The same realization of waves repeated twice led to variance deviations less than 1%.

To exclude the influence of sway resistance, the experiments were repeated without lateral restraints and the carriage was moved along with the model so that it could drift freely. The results were essentially the same; however, the limited length of the towing tank did not allow the test time to be more than 10 minutes.

Since the model had a low freeboard, the deck was flooded almost every roll period, which made rolling essentially nonlinear. So, the observed differences in variances could be interpreted as a confirmation of the above idea.

5.2 Distribution Density: Markov Process Approach

The next issue to study is the distribution density of nonlinear roll angles and roll velocities. Again, non-linearity plays a significant role: although irregular waves can be considered as Gaussian process, non-linear rolling is not necessarily Gaussian.

The problem of distribution of the response of a nonlinear system cannot be solved for a general input process. However, this problem can be solved for a particular type of an input process, such as a Markov process.

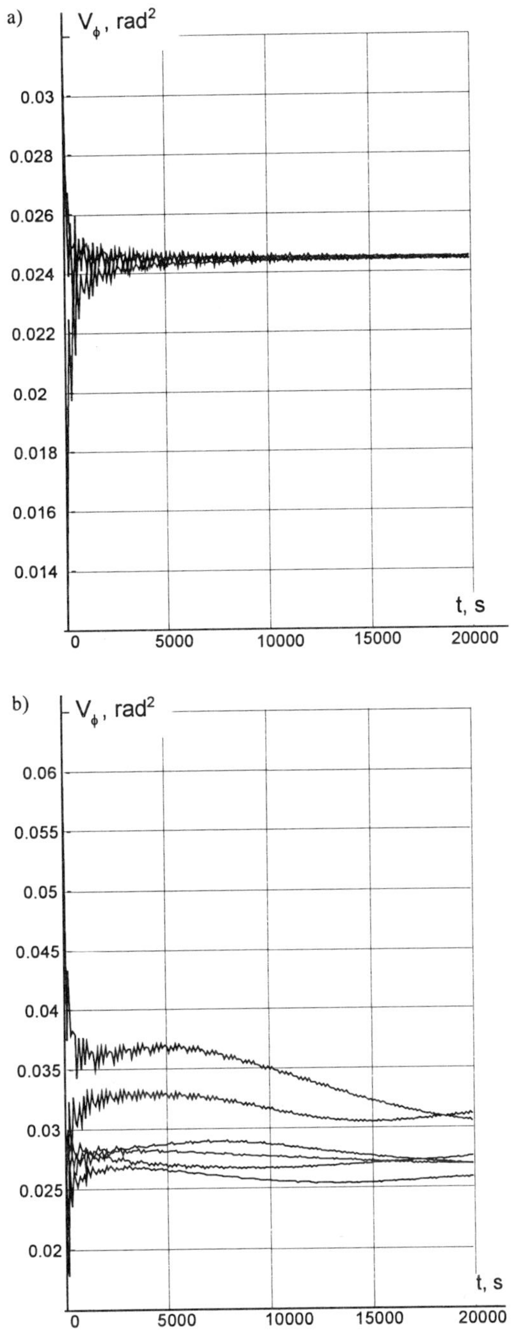

Figure 7. Variance vs. time linear system (a), taken from [47]; (b) nonlinear system, taken from [48]

Current values of a Markov process depend only on the value in the previous moment. The same is true for a multidimensional Markov process. So, if we consider a two-dimensional Markov process, whose components are roll angle and roll velocity, then we can consider the previous moment as the initial moment and previous values as initial conditions.

To find combine distribution density of roll angle and velocity we can use the Fokker–Plank-Kholmogorov equation, which is a partial differential equation.

A brief review of the application of a Markov process to the rolling problem is given by Roberts [49]. The general problem of applying Markov processes analysis to non-linear dynamic system excited by white noise is considered in Caughey [50]. The stationary solution of the Fokker-Plank-Kholmogorov equation of the case of non-linear rolling was reviewed in Haddara [51]. The analytical solution can be derived for the case of non-linear damping only, see Caughey [52]. The Galerkin method was applied to solve the Fokker-Plank-Kholmogorov equation for the rolling equation with both non-linearities by Haddara and Zhang [53] Combination of the averaging method with the Markov process analysis was used by Roberts [49], Haddara and Nassar [54].

Roberts [55] used the Stratonovich-Khasminovskii limit theorem Application of the averaging technique allows for consideration of the slowly-varying amplitude and phase as the components of a two dimensional Markov process. Roberts emphasized that the principle advantage of introducing the averaging is that the Fokker-Plank-Kholmogorov equation can be reduced to one dimension. Such an equation can be solved for any type of non-linearity and it is not necessary to assume the input excitation is white noise. The spectrum of the excitation can be given. Haddara and Nassar [54] combined the averaging technique with the Galerkin method.

Markov process analysis is not free from some limitations: consideration of the system with several degrees of freedom by this way leads to such complex formulae, that the analytical solution loses its sense. Numerical solution of Fokker - Plank - Kholmogorov equation nowadays has no advantages in comparison with the Monte-Carlo method.

5.3 Distribution Density: Simulation

Methods of numerical analysis become more and more popular nowadays because of the development of fast computers.

Degtyarev [56] used an interpolation method for calculation of virtually any probabilistic characteristics of nonlinear rolling. Nechaev [57] used the Monte-Carlo simulation and encountered significant deviations of the rolling distribution from Gaussian law in the case of a biased ship. On the other hand, rolling was Gaussian for a non-biased ship.

The author [58] found that these deviations mainly depend on the shape of the GZ curve in the vicinity of an upright equilibrium. This can be demonstrated using the following samples. Figure 8 shows results of a Monte-Carlo simulation calculated for conventional types of the GZ curve, shown in fig 9, both are taken from [59]. On the other hand, the rolling distribution deviates from Gaussian significantly (see Fig. 10) if GZ curve is "S-shaped" (Fig 11). Both these figures are also taken from [59].

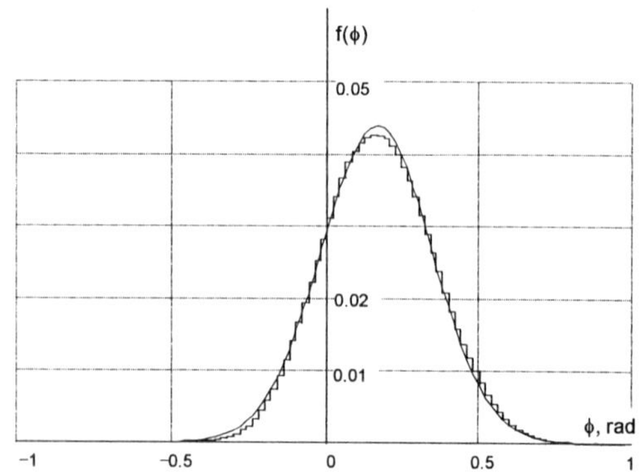

Figure 8. Almost Gaussian distribution – conventional GZ curve

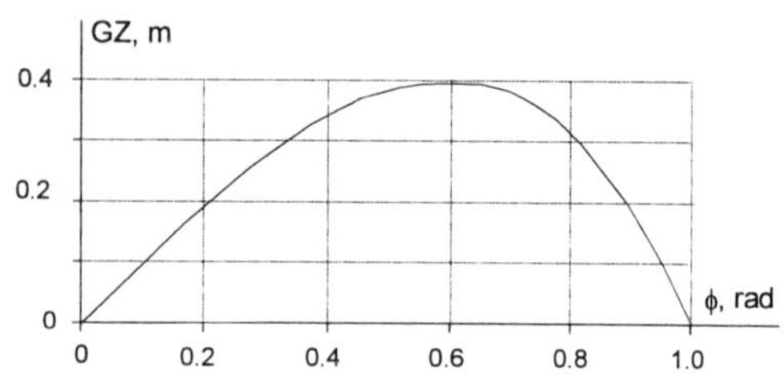

Figure 9. Conventional GZ curve

This observation has a clear physical explanation. Large roll angles are comparatively rare. If the GZ curve is conventional, its non-linearity becomes visible in the vicinity of the maximum. The non-linearity is insignificant in the vicinity of the original equilibrium where the ship spends the majority of her time. Bias increases the chance for non-linearity to influence since the maximum becomes closer in one of the directions. That is why Nechaev [57] has seen dependence of distribution on bias. When we look at the "S-shaped" GZ curve (fig.11), we see that the non-linearity is located quite close to the equilibrium, so that its influence is significant. See [48]

Summary

The author, A. Degtyarev and A. Boukhanovsky [47,48,59] made several attempts to develop a method of calculation of nonlinear rolling distribution based on a technique other than the Markov process. However, the rapid rise in computing speed now makes direct Monte-Carlo simulation quite an ordinary problem, even for desktop personal computers. So, simulation seems to be the most appropriate way to get probabilistic characteristics of nonlinear rolling in irregular seas. The main advantage of the simulation is that we get results free from any additional assumptions. The only limitation is that we have to simulate many realizations to avoid the influence of practical non-ergodicity.

The adequacy of mathematical derivations is very important; however, even the best mathematical methods are meaningless if input data is inaccurate. The next section is a review of input variables and forms of their presentation. Loading conditions, meteorological and operational factors are discussed.

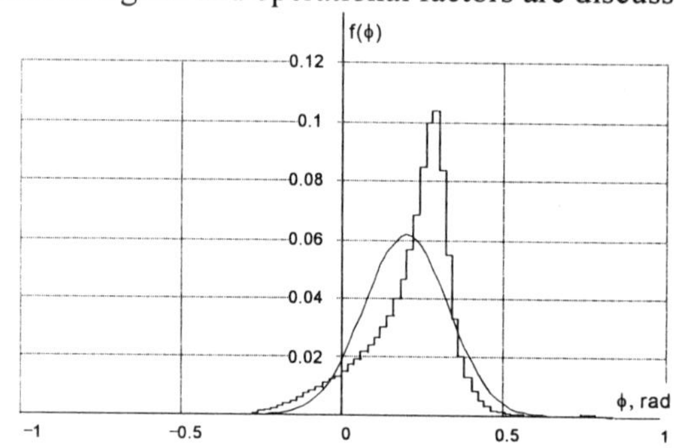

Figure10. Non-Gaussian distribution of nonlinear rolling with "S-shaped" GZ curve

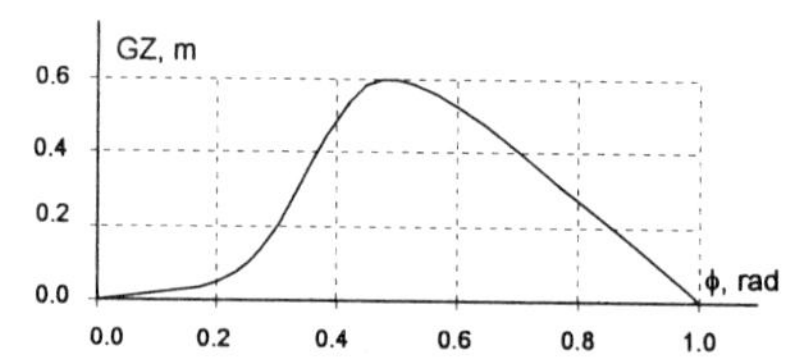

Figure 11. "S-Shaped" form of the GZ curve

6. LOADING CONDITIONS AND ASSUMED SITUATIONS

Sevastianov [5] or [6] proposed two stochastic vectors to describe loading conditions and assumed situation.

The loading condition also can be represented as a stochastic process.

$$\vec{L} = \vec{L}(W, KG, R_x) \qquad (33)$$

where

- weight displacement $L_1 = W$;
- height of the center of gravity $L_2 = KG$;
- radius of gyration around the central longitudinal axis $L_3 = R_x$.

The first mention of time dependence of stability characteristic can be found in Mockel [60], where $GM(t)$ was considered. Rakov and Sevastianov [61] considered two components: $GM(t)$ and $W(t)$ time dependent functions. Probabilistic interpretation of these figures was carried out by Sevastianov [62].

The components of the loading condition vector are not independent. There is a correlation between the displacement, KG, and the mass distribution parameter. The degree of this correlation is different for each vessel and depends on the current service of each ship.

So we shall consider loading as a vector $\overline{OM}$ in the co-ordinate space O, W, KG, R_X The end point M of the vector $\overline{OM}$ is moving in space during the lifetime of the ship; see figure 12 taken from [61].

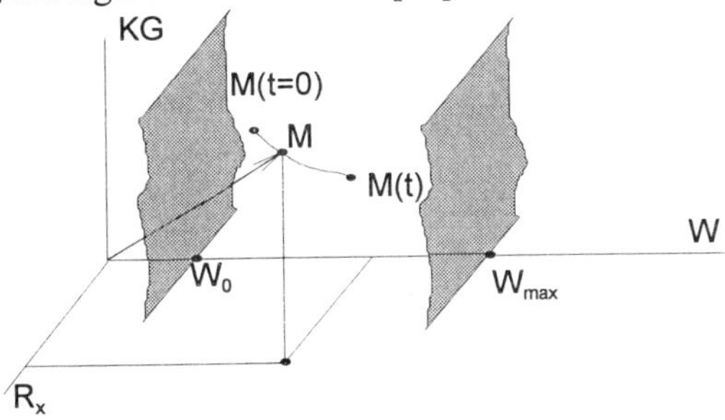

Figure 12. Vector presentation of loading conditions. W_0 - light displacement; W_{Max} - maximum displacement

Each of these parameters can be represented as a sum of three components, which correspond to the empty vessel, deadweight, and special loading (icing, for example).

6.1 Loading Conditions: Light Weight Vessel Component

It is known that corrosion even of the shell plating of a ship below the waterline can reach 0.1-0.2 mm per year. It means, for example, that for a medium size trawler, which has about 300 m^2 wetted surface area, about 5 tonnes of weight is lost during a 10 year period. KG will increase up to 0.3-0.4 m [62]. Another reason for changes in the lightweight loading conditions is an increase in wetness of wooden deck plating and other wooden details. The moisture content of air dried wood, which is normally used in shipbuilding, is about 18 - 20%. After the first several years of service, this increases up to 30 - 40% and then becomes stable. This would add 1-2 tons of light displacement and about 1 cm of KG for medium size vessels. An even more significant increase in light displacement can be caused by an increase in moisture content of the ship's internal insulation. This could add 10 tonnes for a medium size vessel. The affect of such changes on KG depends on the location of such spaces in the ship [62].

Guralnik and Kulagin [63] studied the affect of aging on ship stability. This paper contains formulae that allow numerical estimation of these previously described changes. They found that current stability standards might be violated just because of aging of a ship. It is convenient to illustrate this effect using an ellipse of probability of the loading condition vector $\bar{L}$; see figure 13 taken from [63].

Long term probabilistic stability assessment should take into account that weight characteristics of empty vessel change with time.

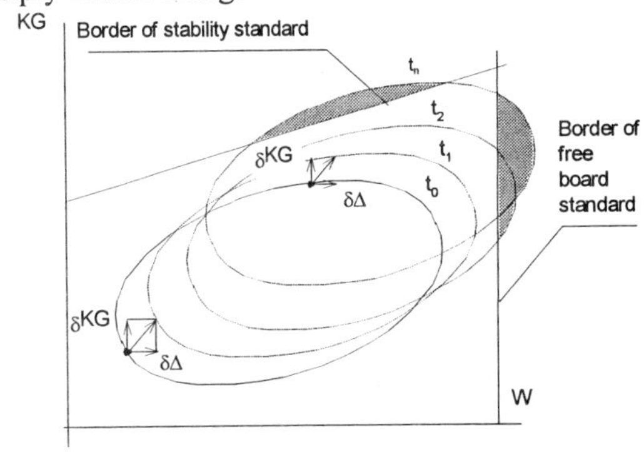

Figure 13. Changing of ellipse of probability of loading condition vector due to ageing

6.2 Loading Conditions: Deadweight Component

Meylunas [64], Meylunas and Braslavskaya [65] studied deadweight components as random variables for cargo ships. These papers contain a general solution for a two-dimensional stochastic vector of loading conditions (W, KG).

$$W(t) = W_0 + P_{Fc} \cdot \eta + P_{St}^0 \cdot \xi \tag{34}$$

$$M_Z(t) = W_0 \cdot KG_0 + P_{Fc} \cdot z_c \cdot \eta + M_{ZSt}^0 \cdot \xi \tag{35}$$

Here $M_Z(t) = KG(t)W(t)$, the moment of the weight. The index zero stands for an unloaded ship, including however, 10% of all stores. Values P_{St}^0 and M_{ZSt}^0 are weight and its moment of 100% of stores, ξ is a random value of percentage of stores left aboard the ship. Meylunas [65] proposed uniform distribution for this value; however, there is nothing in the model that would prevent use of another distribution. The value P_{Fc} is the cargo weight capacity and z_c is the elevation of the center of the cargo mass. Both of these are stochastic values. The value η is a weight capacity coefficient and is a random value between 0 and 1

$$\eta = \frac{P_c}{P_{Fc}} \tag{36}$$

Here P_c is the weight of the current cargo load.

Two types of cargo operation with full use of tonnage and cargo space were considered by Meylunas [65]. The first type describes transportation of mass cargo, the second type concerns general cargo.

The first type of operation does not use all cargo volume;, however, usually all weight capacity is utilized. So, Meylunas [65] assumes that $\eta = 1$.

Different kinds of mass cargo have different specific weights that could be considered as a random number. So both weight and elevation of mass of the cargo could be considered as deterministic functions of random values of specific weight of the cargo. Meylunas [64] found a probabilistic characteristic of this value on the basis of statistics from 960 voyages.

For the second type of operation - transportation of general cargo - it is assumed that almost all the cargo volume is used. In this case, the specific weight of the cargo defines how many tons can be carried. In other words, the specific weight defines the weight capacity coefficient and the type of cargo defines the rest of the values. Probabilistic characteristics of specific weight for general cargo were calculated by Meylunas [64] on the basis of statistics gathered from 348 voyages.

Meylunas [65] developed a completely practical method of calculation with combined distribution density of deadweight and *KG*. One of the main advantages of the method –is that the statistical component can be readily updated using modern data.

The same problem was tackled by Moisseyeva [66] for fishing vessels. She analyzed statistics on a day-to-day basis. Apollinariev [67] developed a mathematical model and simulation software for big fishing trawlers. The model used a generator of random catches that was built on existing catch statistics. Then, the model simulated fish processing and calculated changing in loading conditions. Unfortunately, this work did not produce a general method for calculating loading conditions for fishing vessels as was done by Meylunas for mercantile marine. Also catch statistics have a tendency of fast aging caused by both ecological and economical changing in World fishery.

Probabilistic characteristics of deadweight components strongly depend on the type of vessel. The method for cargo vessels is ready to use now.

6.3 Meteorological Components of Assumed Situation

We assume a situation with a set of random values that describes different weather and operational parameters:

$$\vec{S} = \vec{S}(u_{Am}, \sigma_u, h_s, T_m, v, \psi, \psi_A, \vec{P}_e) \qquad (37)$$

Here:

- averaged wind velocity $S_1 = u_{Am}$;
- standard deviation of pulsation components of the wind velocity $S_2 = \sigma_u$;
- height of significant waves $S_3 = h_s$;
- averaged period of visible waves $S_4 = T_m$;
- speed of the vessel $S_5 = v$;
- heading to waves $S_6 = \psi$;
- heading to wind $S_7 = \psi_A$;
- additional external forces (or resultant vector of such forces $S_8 = \vec{F}_e$).

The parameters from the above list could be divided in two subsets: operational (heading, speed, course, additional forces) and meteorological. Practical application of formula (4) requires knowledge of a joint distribution of all these parameters.

Mordachev [68,69] has shown that it is always possible to decrease the number of parameters dramatically if ports of departure and destination are known.

The location of these two ports as well as meteorological data on wind and waves for the season of interest allows defining the best route for the ship to follow.

Meteorological information with various levels of detail and accuracy is available for various geographical locations. Usually it is presented in the form of significant height of waves, average period of waves, average wind velocity, and direction of winds for the season. This is the minimal set of parameters for very approximate estimates: a fourth meteorological parameter, σ_u, can be estimated using the so-called “gust coefficient”. It relates the average velocity of wind and standard deviation σ_u [70].

Unfortunately, these data cannot be considered as sufficient. It is well known both from oceanographic research and maritime practice that different geographical locations in the ocean have very different types of wave environments.

Of course, geographical factors, including shorelines, currents, wind direction, affect the wave energy distribution. So, the same significant wave height and average period does not necessarily mean the same type of waves.

The spectrum depends not only on geographical location and season. However it changes with time: a developing wave spectrum is different from a fully developed wave spectrum. The duration is also important. As noted earlier, time is an important factor in stability safety. Boukhanovsky, Degtyarev, Lopatukhin and Rozhkov [71,72,73] proposed scenarios for weather, representing it as an impulse stochastic process, see figure 14.

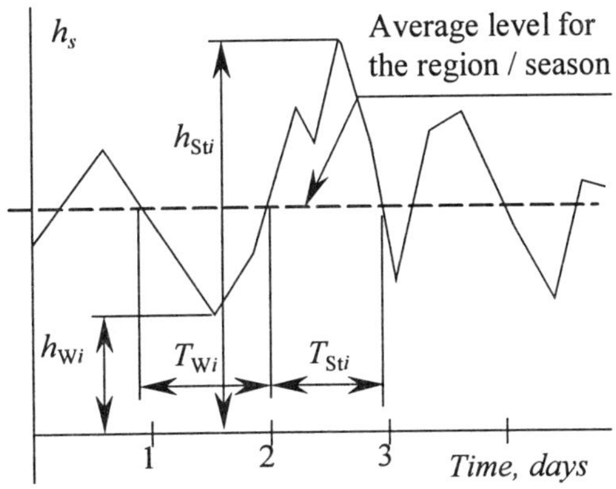

Figure 14. Weather Scenario

This allows for classification of storms and weather windows. An example of such a classification for the Black Sea is given in [72,73]. Every class of storms has its own set of spectra and has typical consequences for its changing. Oceanographic statistics also allow calculation of probabilities of transitions between different classes of storms and development of a transitions matrix. This completes the description of the weather scenario as a two-dimensional Markov chain.

We need to use stochastic processes not only for instantaneous wave elevation but also for the changing spectra to model the sequence of storms and good weather over time. These models are based on meteorological data that are connected to geographical region. These regions are to be defined by the route between two ports and depend on mission of a ship.

6.4 Operational Components of Assumed situation - Human Factors

A route on the map could only define a course and speed in general. The ultimate decision on speed and course is made by a human operator, who takes into account a number of factors that probably never could be formalized.

It was found that influence of human factors on operational parameters can be taken into account using the so-called "boundary model" of behavior of an operator.

The idea behind the "boundary model" is to simulate extreme behavior of an operator, assuming that the real behavior is somewhere within these borders. Three basic motivations were considered:

- Profit: for mercantile marine it is desirable to reach the destination in a minimal time
- Fear: be careful and follow safety regulations and good seamanship
- Human ability to make mistakes: sometimes peoples' behavior does not follow either of the above motives.

Based on these features, three possible strategies could be considered:

- "Mercantile" strategy: minimizing time for cargo vessels
- "Careful" strategy: follow all safety regulations, rules of good seamanship and be completely informed on current risks
- "Mistaken" strategy: follow the most dangerous path, without any profit considerations

Let's imagine that a ship enters a region with known wind and wave characteristics. For the cargo ship at the first expansion, these strategies could be:

- Crossing the region with the shortest possible route with maximum speed
- Tacking in order to avoid dangerous courses in beam and following seas
- Tacking in order to have the most dangerous beam and following courses

The above model is very simple, but it allows for calculation of particular values for speed and course. Practical models could be (and should be) more complicated and based on study of everyday practices for particular type of vessels in particular routes.

Three strategies would lead to three different sets of assumed situations and as a result to three values of the risk function, say λ_C – "careful strategy", λ_M – "mercantile" strategy and λ_F –"mistaken" strategy.

These values allow for estimation of the influence of human factors for a certain ship or design solution.

$\Delta_M = \lambda_M - \lambda_C$ - gives some estimation of a ship's ability to resist human's premeditated mistakes caused by neglect of the danger of capsizing because of mercantile motives.

$\Delta_F = \lambda_F - \lambda_C$ - gives some estimation of a ship's ability to resist human non-premeditated mistakes. This value can be considered as some estimation of the degree of a ship being "foolproof ".

These estimates are very important because stability safety in the real world cannot be considered without human factors. Discussing stability safety margins, we have to take into account how this particular design would act on human errors. It is very possible that a ship with a larger value of λ_C, but with lower values of Δ_M and Δ_F would be preferable in the real world.

Summary

Loading conditions and the assumed situation are very important in stability assessment. They have at least the same level of importance as knowledge of capsizing mechanics. If we do not have reliable input data, any capsizing prediction method would produce wrong results, regardless of its level of sophistication.

Aging of ships, random weights characteristics of cargo, and wave and wind changes with time are among the effects that have to be taken into account. Special attention should be paid to human factors. Since behavior of the human operator is largely unpredictable, we would rather define possible boundaries of behavior and try to analyze what would happen if the operator deviated from prescribed practice.

7. FINAL COMMENTS AND CONCLUSIONS

We have completed a state-of-the-art review of probabilistic study in the field of intact stability. It was presented as an "information chain", where the outcome of one is input data for another. This is the way to look at the practical side of theoretical research. In other words, theory produces a method, an algorithm that has to be fed by input data and included in the system.

A probabilistic approach to an intact stability standard seems to be quite natural. However, it must be developed as part of a larger system of shipbuilding regulations. Generally it requires that:

- Input data is available;
- Accuracy of calculations is adequate for practical purposes and calculations consume a reasonable amount of resources;
- Output can be easily interpreted for regulation purposes.

The paper tried to look at the results from these points of view. Summaries are given at the end of each section of the paper and will not be repeated here. Technical conclusions are given below.

Probability of capsizing must explicitly depend on time. Stability standards are a particular case of the reliability problem. Up-crossing theory is an ultimate tool to relate probability of capsizing to time.

Piecewise linear representation allows combining simulation techniques for moderate roll angles with analytical solution for large roll angles. Capsizing probability due to breaking wave action, broaching and pure loss of stability might be included into probabilistic stability standards. The correct simulation of nonlinear irregular rolling requires averaging over several realizations.

Long term probabilistic stability assessment should take into account change of light displacement characteristics with time. Changes in the wave spectra with time might be taken into account along with variations with geographical region.

It is possible to develop a " boundary" model of human operator behavior in order to introduce a "fool proof" stability criteria.

REFERENCES

1. RAHOLA, J. 1939 *The Judging of the Stability of Ships and the Determination of the Minimum Amount of Stability*. Helsinki,.
2. SEVASTIANOV, N.B. 1979 The Advantages of Probabilistic Approach to Stability Regulations as Compared Deterministic One. *Trans. of Kaliningrad Technical Institute "Seakeeping of Fishing Vessels"*, 81 (1979): 3-9. (in Russian).
3. BELENKY, V.L. 1995 Analysis of Probabilistic Balance of IMO Stability Regulation by Piece-wise Linear Method. *Marine Technology Trans. of Academy of Science*, 6 (1995): 5-55.
4. KRAPPINGER, O. 1962 Uber Kenterkriterien. *Schiftechnik*, (1962): 145.
5. SEVASTIANOV, N.B. 1963 On Probabilistic Approach to Setting Stability Standards. *Trans. of Kaliningrad Institute of Technology*, 18 (1963) (in Russian).
6. SEVASTIANOV, N.B. 1982 Probabilistic Stability Regulation as a Problem of Reliability Theory. *Trans. of Register of the USSR*, 12 (1982): 94-100, (in Russian).
7. SEVASTIANOV, N.B. 1994 An Algorithm of Probabilistic Stability Assessment and Standards. *5th STAB Conference*, 5 (1994).
8. CALDWELL, J.B. and Y.S. JANG. 1986 Risk and Reliability Analysis Applied to Ship Capsize: a Preliminary Study. *The Safeship Project. Ship Stability and Safety. International Conference, RINA.*
9. KOBYLINSKI, L. 1975 Rational Stability Criteria and Probability of Capsizing. *Proc. 1st STAB Conference*, (1975): Paper 1.
10. KOBYLINSKI, L. 1994 Methodology of the Development of Stability Criteria on the Basis of Risk Evaluation. *5th STAB Conference*, 3 (1994).
11. McTAGGART, K. 1998 Ongoing Work Examining Capsize Risk of Intact Frigates Using Time Domain Simulation. *Proc. of 4th International Stability Workshop, Institute for Marine Dynamics, St. John's, Newfoundland.*
12. ALMAN, P. R., P. V. MINNICK, R. SHEINBERG and W. L. THOMAS. 1999 Dynamic Capsize Vulnerability: Reducing the Hidden Operational Risk. *Preprints of SNAME Annual Meeting*, (1999): Paper 10.
13. SEVASTIANOV, N.B. and N. H. PHAM. 1979 Boundary between the Domains of Stable and

Unstable Free Ship Motion in Drift-Rolling Regime. *Trans. of Kaliningrad Institute of Technology "Seakeeping of Fishing Vessels"*, 81 (1979): 17-25. (in Russian).

14. PHAM, N. H. 1981 Determination of Capsizing Risk Function of Vessels Affected by Irregular Seas and Gusty Wind. *Trans. of Kaliningrad Institute of Technology, "Seakeeping of ships"*, 93 (1981): 45-53. (in Russian).

15. UMEDA, N., Y. IKEDA and I. SUZUKI. 1992 Risk Analysis Applied to the Capsizing of High-Speed Craft in Beam Seas. *Proc. of the Conference on Practical Design and Mobile Units (PRADS-92)*, Elsevier Science Publishers.

16. DUDZIAK, J. and A. BUCZKOWSKI. 1978 Probability of Ship Capsizing under the Action of the Beam Wind and Sea as a Background of Stability Criteria. *Polski Rejestr Statkow, Prace Studialno-Roswojowe*, 13 (1978).

17. BIELANSKI, J. 1994 Ship Rolling Under Action of Beam Wind and Sea as a Background of Comparison IMO Stability Criteria. *Proc. of International Workshop on the Problems of Physical and Mathematical Stability Modeling OTRADNOYE'93 Kaliningrad Institute of Technology*, 2 (1994) Paper 7.

18. PRICE, W.G. 1975 A Stability Analysis of the Roll Motion of a Ship in an Irregular Seaway. *International Shipbuilding Progress*, 22: 247 (1975): 103-112.

19. NEKRASOV, V. A. *Probabilistic Tasks of Seakeeping*. Leningrad: Sudostroenie, 1978. (in Russian).

20. NEKRASOV, V. A. 1994 Stochastic Stability Theory of Ship Motion *Proc. of 5th STAB Conference*, 5 (1994)

21. NEKRASOV, V.A. 1997 Capsizing of Ship in Low Cycle Resonance. *Proc. of 6th STAB Conference*, (1997): 65-73.

22. ZELENIN, L. K. 1989 Probability Characteristic of Ship Capsizing in Irregular Waves. *Proc. of Scientific Methodological Seminar in Ship Hydrodynamics SMSSH'89, Bulgarian Ship Hydrodynamic Centre, Varna, Bulgaria*, 2 (1989): paper 38.

23. HSIEH, S. R., TROESCH A.W. and SHAW S. W. 1994 A Nonlinear Probabilistic Method for Predicting Capsizing in Random Beam Seas. *Proc. R. Soc.*, 446 (1994): 195-211.

24. LIN, H. and Y.C.S SOLOMON. 1995 Chaotic Roll Motion and Capsizing of Ship under Periodic Excitation with Random Noise. *Applied Ocean Research*, 17(1995): 185-200.

25. SEVASTIANOV, N. B. 1977 On Theory of Multidimensional Stochastic Processes (Upcrossing Problem). *Trans. Of Kaliningrad Institute of Technology "Analysis of Seakeeping of Fishing Vessels"*, 73 (1977): 3-14. (in Russian).

26. VIRGIN, L.N. 1987 The Nonlinear Rolling Response of a Vessel Including Chaotic Motions Leading to Capsize in Regular Seas. *Applied Ocean Research*, 9:2(1987): 89-95.

27. ARNOLD L., *Random Dynamical Systems*. Berlin Heidelberg: Springer, 1998.

28. BELENKY, V.L. 1989 A New Method of Statistical Linearization in Severe Rolling and Capsizing Problem. *Proc. of Scientific Methodological Seminar in Ship Hydrodynamics SMSSH'89 Bulgarian Ship Hydrodynamic Center, Varna, Bulgaria*, 2 (1989) paper 39.

29. BELENKY, V.L. 1993 A Capsizing Probability Computation Method. *Journal of Ship Research*, 37:3 (1993): 200-207.

30. UMEDA N. 1990 Surf-Riding of a Ship in Irregular Seas. *Journal of Kansai Society of Naval Architects of Japan*, 213 (1990) (In Japanese).

31. UMEDA, N. 1990 Probabilistic Study on Surf-Riding of a Ship in Irregular Following Seas. *Proc. of 4th STAB Conference (preprints)*, (1990): 336-343.

32. BORODAY, I.K. 1967 Statistical Characteristics of Stability and the Probability of Capsizing of a Ship Running on Any Course in Irregular Seas. *Documents of the USSR Expert of the IMCO, Working Group on Stability of Fishing Vessels.*

33. BORODAY, I.K. 1968 Statistical Characteristics and Probability of Capsizing of Ship Heading with Arbitrary Course Relatively Irregular Seas. *Trans. of the Register of USSR "Theoretical and Practical Problems of Stability and Survivability"*, (1968): 21-47. (in Russian).

34. BORODAY, I. K. and E. P. NIKOLAEV. 1975 Methods of Estimating the Ship's Stability in Irregular Seas. *Proc. of 1st STAB Conference*, (1975) paper 3.5.

35. BORODAY, I.K. and Yu. A. NETSVETAEV. 1982 *Seakeeping of Ships*. Leningrad: Sudostroenie. (In Russian).

36. UMEDA, N., Y. YAMAKOSHI and Ts. TSUCHIYA. 1990 Probabilistic Study on Ship Capsizing due to Pure Loss of Stability in Irregular Quartering Seas. *Proc. of 4th STAB Conference (Preprints)*, (1990): 328-335.

37. UMEDA N. and Y. YAMAKOSHI. 1991 Assessment for Probability of Ship Capsizing due to Pure Loss of Stability in Quartering Seas. *Journal of Society of Naval Architects of Japan*, 169 (1991): 15-25 (in Japanese).

38. UMEDA, N., M. ARIJI and Y. YAMAKOSHI. 1991 Assessment for Probability of Ship Capsizing due to Pure Loss of Stability in Quartering Seas. (2nd Report). *Journal of Kansai Society of Naval Architects of Japan*, 216 (1991): 129-138 (in Japanese).

39. UMEDA, N. and Y. YAMAKOSHI. 1991 Probability of Ship Capsizing due to Pure Loss of Stability in Quartering Seas. *Naval Architecture and Ocean Engineering*, 30(1991): 73-85.

40. GRIM, O. 1961 Beitrag zu dem Problem der Sicherheit des Schiffes im Seegang, *Schiff und Hafen*, 6 (1961): 490-497.

41. BELENKY, V.L. 1999 Probabilistic Assessment of Ship Stability in Quartering Seas. *Proc. 4th Japan – Korea Joint Workshop on Ship and Marine Hydrodynamics, Fukuoka, Japan*, (1991): 291-296.

42. BELENKY, V.L. 2000 Piecewise Linear Approach to Probabilistic Stability in Quartering Seas. *Proc. 7th STAB Conference* A (2000): 503-510.

43. DAHLE, E.A, D. MYRHAUG and S.J. DAHL. 1988 Probability of Capsizing in Steep and High Waves from the Side in Open Sea and Coastal Waves. *Ocean Engineering*, 15(1988): 139-151.

44. DAHLE, E.A. and D. MYRHAUG. 1993 Risk Analysis Applied to Smaller Vessels in Breaking Waves. *Spring 1993 RINA Meeting.*

45. DAHLE, E.A., B. ENERHAUG and D.MYRHAUG. 1995 Risk Analysis Applied to Capsize of Smaller Vessels in Breaking Waves with Examples from Russian Waters. *Proc. of Int. Symp. on Ship Safety in a Seaway: Stability, Maneuverability, Nonlinear Approach (SEVASTIANOV SYMPOSIUM), Kaliningrad*, 1(1995):paper 2.

46. BELENKY, V.L. and S.V. MORDACHEV. 1997 On Probability of Ship Capsizing due to Breaking Wave Action. *Proc of 6th STAB Conference, Varna, Bulgaria*, 1(1997): 323-333.

47. BELENKY, V.L., A.B. DEGTYAREV and A.V. BOUKHANOVSKY. 1997 Probabilistic Qualities of Severe Ship Motions. *Proc. 6th STAB Conference*, 1(1997): 163-172.

48. BELENKY, V.L., A.B. DEGTYAREV and A.V. BOUKHANOVSKY. 1998 Probabilistic Qualities of Nonlinear Stochastic Rolling. *Ocean Engineering*, 25:1(1998):1-25.

49. ROBERTS, J.B. 1980 A Stochastic Theory for Nonlinear Ship Rolling in Irregular Seas. *National Maritime Institute, NMI*, R 99, (1980) Sept.

50. CAUGHEY, T.K. 1963 Derivation and Application of the Fokker-Plank Equation to Discrete Nonlinear Systems Subjected to White Random Excitation. *J. Acoust. Soc. Am.*, 35:11(1963): 1683-1692.

51. HADDARA, M.R. 1974 A Modified Approach for the Application of Fokker-Plank Equation to Nonlinear Ship Motion in Random Waves. *Int. Shipbuilding Progress*, 21:242 (1974): 283-288.

52. CAUGHEY, T.K. 1964 On the Response of a Class on Nonlinear Oscillators to Stochastic Excitation. *Proc. Colloq. Int. Cent. Nat. Rech. Sci.*, 148 (1964).

53. HADDARA, M.R. and Y. ZHANG. 1994 On the Joint Probability Density Function of Nonlinear Rolling Motion. *Journal of Sound and Vibration*, 169:4(1994): 562-569.

54. HADDARA, M.R. and M. A. NASSAR. 1986 A Stochastic Model for the Analysis of Rolling Motion in a Realistic Seaway. *International Shipbuilding Progress*, 33 (1986): 144-150.

55. ROBERTS, J.B. 1980 The Effect of Parametric Excitation on Ship Rolling Motion in Random Waves. *National Maritime Institute, NMI*, R 100, (1980) Oct.

56. DEGTYAREV, A.B. 1992 Mathematical Modeling of the Ship Motion with the Help of Interpolation Method. *Proc. of 1st Int. Conference in Commemoration of the 300th Anniversary of Creating Russian Fleet by Peter the Great (CRF'92), St.Petersburg, Russia*, (1992): 254-261.

57. NECHAEV, Yu. I. 1989 *Modeling of Ship Stability in Waves*. Leningrad: Sudostroenie. (in Russian).

58. BELENKY, V.L. 1994 Piece-Wise Linear Methods for the Probabilistic Stability Assessment for Ship in a Seaway. *Proc. of 5th STAB Conference*, 5 (1994).

59. BELENKY, V.L., A.B. DEGTYAREV and A. V. BOUKHANOVSKY. 1995 Probabilistic Qualities of Severe Rolling. *Proc. of Int. Symp. on Ship Safety in a Seaway: Stability, Manoeuvrability, Nonlinear Approach (SEVASTIANOV SYMPOSIUM), Kaliningrad, Russia* 1(1995): paper 7.

60. MOCKEL, W. 1960 Behaviour of Trawlers at Sea. *Fishing Boats of the World*. 2(1960): 404-417.

61. RAKOV, A. I. and N. B. SEVASTIANOV. 1981 *Fishing Vessels' Design*. Leningrad: Sudostroenie. (in Russian)

62. SEVASTIANOV, N.B. 1970 *Stability of Fishing Vessels*. Leningrad: Sudostroenie. (in Russian).

63. GURALNIK, B. and V. KULAGIN. 1995 Load Changes of the Ship in Light Due to her Aging and their Correction in Planning Stability Trials. *Proc. Int. Symp. on Ship Safety in a Seaway: Stability, Maneuverability, Nonlinear Approach (SEVASTIANOV SYMPOSIUM), Kaliningrad, Russia* 1 (1995): paper 16.

64. MEYLUNAS, V. F. 1971 Some Statistic Data on the Distribution of Weight Cargo Coefficient and Specific Volume of Cargo. *Trans. of Register of the USSR*, 1(1971): 241-250. (in Russian).

65. MEYLUNAS, V. F. and A. I. BRASLAVSKAYA. 1976 Probability Field for Displacement and Height of Center of Gravity of Seagoing Vessels in Operation. *Trans. of Register of the USSR*, 4 (1976): 46-58. (in Russian).

66. MOISSEYEVA, M.E. 1971 Analysis of Loading Conditions and Stability of Big Trawler "Mayakovsky". *Rybnoe khozaistvo*, No 7(1971): 27-30. (in Russian).

67. APOLLINARIEV, V.I, and N.B. SEVASTIANOV 1992 Stability Estimation Based on Simulative Modeling. *Proc. of the Conference on Practical Design and Mobile Units (PRADS-92)*, Elsevier Science Publishers, Ltd., 2 (1992): 1161-1172.

68. MORDACHEV, S.V., N.B. SEVASTIANOV and V.L. BELENKY. 1994 Assumed Situation in Probabilistic Norms of Stability. *Proc. of 5th STAB Conference*, 6 (1994).

69. MORDACHEV, S.V. 1995 Assumed Situations and Human Factor in Probabilistic Approach to Stability Assessment. *Proc. of Int. Symp. on Ship Safety in a Seaway: Stability, Maneuverability, Nonlinear Approach (SEVASTIANOV SYMPOSIUM), Kaliningrad, Russia,* 1(1995) Paper 8.

70. LUGOVSKY, V.V. 1975 *Dynamics of sea*. Sudostroenie: Leningrad. (in Russian).

71. BOUKHANOVSKY, A. V., A. B. DEGTYAREV, L. J. LOPATUKHIN and V. A. ROZHKOV. 1997 Climatic Spectra of Wind-Produced Waves. *Sudostroenie*, 4 (1997): 15-18. (in Russian).

72. BOUKHANOVSKY, A. V., A. B. DEGTYAREV, L. J. LOPATUKHIN and V. A. ROZHKOV. 1998 Climatic Wave Spectra *Proc. of 2nd International Shipbuilding Conference, St.Petersburg, Russia,* B (1998).

73. BOUKHANOVSKY, A. V., A. B. DEGTYAREV, L. J. LOPATUKHIN and V. A. ROZHKOV. 2000 Stable States of Wave Climate: Applications Risk Estimation *Proc of 7th STAB Conference,* B (2000): 831-846.

Discussion

William A. Cleary, Jr., Member

Dr. Belenky has provided this diverse Society with an excellent updating of the progress and possible application of the newer mathematical methodologies as they are currently being examined in the interest of improving Intact Stability Minimum Safety Regulation—one of the 17 safety disciplines currently offered by the International Maritime Safety Organization of the United Nations.

The positive effect of this paper on stability regulation

I accept that probability is necessary to solve the many and varied problems in Intact Stability Review and I am enthusiastic at the skill with which Dr. Belenky has placed each of the new higher-order math methods in its proper location among the whole field of activity. However, I was a bit discouraged by the mathematically dictated decision in para 2.2 that "the probability of non-capsizing cannot serve as a practical stability criterion." No matter what is used to develop the new international stability standard, we still have to place on each ship an easily understood and plainly readable standard for voyage planning and every day stability safety management. Dr. Belenky has also recently produced a Great Lakes Section paper summarizing the most recent STAB conference in Australia, which included information that some modern stability researchers are working on bridge stability information systems.

In paragraph 2.3 there is a statement endorsing the "averaging of all possible weather and operational states" which appears to be in contrast to a later statement in paragraph 4.6 that it will be necessary to identify "potentially dangerous waves. . ." and to identify how they are "statistically related to geographical region and season. . . ." I prefer the latter approach. The new standard must assure safety in moderate seas (e.g., < 8 m) and must permit ships caught in severe seas (> 8 m) to protect themselves (perhaps advance warning will be necessary). If *all* oceans are averaged, no notice will be taken by the shipping industry as was the case when Dr. Hogben published a recent paper noting that there had been a rise in seaway heights. The "rise" itself and the average taken over the whole ocean were both so modest that the shipping industry just yawned.

In Section 3 it was encouraging to note the author's assertion that a combination of simulation and purely theoretical methods will be necessary to find the proper results of the information we seek.

Sections 3 and 4 discuss only 4 potential capsize situations but do show which of 10 different math methods would be most useful in each case. My 1982 *Marine Technology* (July) article included a partial list of 34 ship/seaway capsize situations without including multiple direction wind/seaway which I have since learned seems to be a commonplace occurrence (e.g., a distinct long wavelength swell induced by a steady wind for many days with a fresh new wind-sea on top of the swell which is steeper AND is from a 45–60° angle to the original swell direction—the wind has either "backed" or "filled").

While it remains important to examine the 4 which are of current research interest, what is the author's current confidence that these 4 will be enough to cover all ship/seaway capsize scenarios for displacement ships? High speed ship stability evaluation will, of course, require starting all over again and will have many more steps in the required analysis due to different high-speed hull configurations and hundreds of variations of the hull types.

Section 5 is to me the most important development because the answers in high winds and high irregular seas need to be the basis of the next major step in minimum stability standard-making. When discussions on high speed in high seas stability came to IMCO/IMO 20 years ago, many delegates agreed that the maximum seaway to be considered would have to be about 7 meters which was considered the practical limit for linear approximation. It is good to see that mathematicians are now able to lead us into this most important area.

In Section 5.3 a question arises in the discussion of the "normal" GZ curve vs. the "S" shaped GZ curve. My experience would cause me to view the "S" shaped GZ curve as more normal because of the worldwide practical agreement that initial GM should be only a little above the minimum because it promotes an easier roll (a more comfortable ride). I do not think it wise to assume that *most* ships prefer the higher GM. Thus, combating this industry preference will remain "a hard sell."

Section 6 mentions 4 of the many ship maintenance and operation variations which cause the GZ curve to change—both slowly over the lifetime of the ship and to vary on each voyage, each day due to fuel usage, each wave group due to hydrostatic change caused by the seaway, each course change due to the different hydrostatic and dynamic reactions, etc., etc.

Section 6.2 needs to include another shiploading variable, which is directly related to the "natural" rolling period in effect at any moment during a voyage. The *Radius of Gyration* (vertical, longitudinal, transverse) is now able to be constantly monitored using any ordinary spreadsheet program (i.e., Excel or equal). It can be set up for any ship in less than a day and—once in use—can constantly tell the operator exactly what the current period of roll is. The reason this is important is that all the major ship types have drastically altered the "spread of weights (masses)" from the "normal" designs of the 1970's and early 80's. Some of the more obvious ship design changes which have occurred have been:

- All aft machinery and accommodation spaces.
- Cargo (containers) stacked high above the weather deck (instead of inside, protected by hatch covers, tarpaulins, battens and wedges.
- Presence or absence of cargo handling gear, etc.

Also the new cargo arrangements have drastically changed the *aerodynamics* of ships with regard to wind heel, gust reaction, leeway in harbors, etc.

Can probability be useful in all the 17 safety disciplines?

The short answer is *Yes* for those affected by seaway because as Dr. Belenky points out—the oceans are probabilistic—and *No* for most of the rest unless we agree on a new probabilistic basis...

In making decisions on this point we need to keep in mind that all the regulations have 3 levels of specificity which are 1) General; 2) Specific (Deterministic); and 3) Fully statistical (Probabilistic).

Some of the initial regulations of the 1929 SOLAS and 1930 Load Line Conventions are still general (nonspecific) ideas which are not felt to need change. Others, hull strength, for instance, were highly specific—even rigidly so, right from their first adoption and remained so for over 35 years. Hull strength was amended in the 1966 Load Line Conference and in my opinion we went from too strict to no regulation at all since there is no active overview at IMO with regard to the variations of hull strength as decided by the classification societies. Stability has enjoyed all 3 levels; in 1930 when stability was included in the Load Line Convention it was very general; after the RAHOLA thesis (1939) Europe and then the rest of the world began to adopt more specific stability regulations. Now, we are examining fully statistical probability standards. Even so, there are other stability casualties (e.g., the many possibilities for internal flooding) which will remain deterministically regulated.

Equivalent safety

The Rahola thesis is Dr. Belenky's reference [1] and it begins by stating that the most correct way to develop a stability standard is to analyze all possible capsize mechanisms—which would require knowledge of all seaways as well as detailed knowledge of all ships. Rahola then states that the 1939 knowledge of all seaways is not developed enough to permit the best method to be used. He then developed his thesis by examining the results of approx 38–40 European coaster ships up to about 130 m in length which were known to have capsized. Although this *prescriptive* standard has been very successfully used for over half a century, as Dr. Belenky notes, it did not provide equal safety for all types and sizes of ships. It was increased by about 10% for fishing vessels but was never decreased for large ships.

Deterministic stability safety

In spite of the obvious need to utilize probability (and other advanced math methods) to understand seaway actions on ships for the purpose of developing better regulatory stability standards, there exists an *equally dominant need* to understand capsize and capsize preventability *on a single ship to single wave basis*; also on a single ship to a group of waves basis and a ship to multiple seaways basis. Although one end result of safety review is minimum safety regulations, it is equally important to provide each ship with its own dynamic capsize limitations in a wide variety of different seaways. This activity will always remain a specific information, deterministic activity.

High-speed capsize safety

It needs to be emphasized that the discovery of capsize methodologies is still unknown for all types of high-speed ships. As stability researchers begin to examine capsize safety in this field, each ship will have to be fully examined in *both* high-speed hydrodynamics and high-speed aerodynamics as it bounces from wave to wave.

Additionally, there is a special need for examining the seaway directly along the intended track of the high-speed ship—several hours of trackline. This means a multi-nation effort to measure and broadcast continuously wave height, direction, steepness, probable fetch development before moderation, etc.

Until these actions are widely accepted, high-speed voyages, other than those of short limited duration, will remain an unsafe adventure and not a reasonably safe ocean voyage.

John F. Dalzell, Member

The author is to be congratulated on what appears to this non-expert reviewer to be an able summary of the status of work in support of the goal of developing probability-based intact stability standards.

The author makes it clear that, while much progress has been made, considerable development work remains until methods suitable for practical design application become available. It also seems likely that the application of such methods will require considerably more effort than the current ones do.

It is clear that the time dependence of the problem must be considered. The simple expression due to Sevastianov for the probability of non-capsizing as a function of time, (Eq 2), appears to be the result of generalizing the classical statistical approach to rare events. The result is a variant of reliability theory. The Sevastianov development is not accessible to the writer, but it is suspected that there may be a number of rational ways to arrive at this model. The model has two parameters, time duration, T, and a "risk function," λ. The risk function is the overall mean rate of capsizing from all operational and weather states. It seems clear that the application of an averaged risk function within the essentially classical formula is rigorous so long as the weather and operational conditions considered (Eqs 14 or 15) are statistically independent. In short, this reviewer finds it easy to accept the Sevastianov model as a statistically sound and practical framework. Of course, as the author points out, the estimation of the averaged risk function, λ, is likely to be rather difficult, or at least time consuming.

After introducing the Sevastianov model for the probability of non-capsizing in Section 2.1, the author goes on to say that the probability of non-capsizing cannot serve as a practical criterion! Accordingly, this reader has some difficulty with the "big picture" implied by the paper.

If the probability of non-capsizing cannot serve as a practical criterion, how are we supposed to understand the averaged risk function that seems to be defined by the theory for this probability? Additionally, if some specified maximum value of the risk function is the "standard," how are the regulators and industry to arrive at reasonable design values for it without considering the associated probabilities of capsizing and non-capsizing?

It is undoubtedly a naïve point of view, but the writer would start from a socially and economically acceptable value for the odds of capsizing during a design ship life. The probability of non-capsizing is trivially related to these odds, and a simple inversion of the Sevastianov model would then yield a target value of the design risk function.

William L. Thomas III, Member

Dr. Belenky is to be congratulated for his comprehensive review of the various approaches toward intact stability assessments. His comparative discussions behind the respective approaches are very useful as guidance toward the implementation of future methodologies.

I would like to briefly discuss the comments made in the paper regarding the merits of using physics-based simulation tools in stability assessments. A review of the many methods cited in your paper reinforces my thoughts leads me to conclude that a simulation-based approach is necessary because it is essential, and there is really no alternative way to model the complex capsize physics for each ship that is under evaluation. Recent capsize experiments at the David Taylor Maneuvering and Seakeeping Basin have illustrated that capsize behavior can be substantially different for novel new hull forms. The standard righting moment energy-wind moment energy relationships developed by Sarchin and Goldberg, for example, are not adequate for tumblehome designs because the new novel hull forms are likely to fall beyond the database of experience used to formulate the existing criteria. Physics-based simulation tools are the only way to make a comprehensive and economical stability assessment of a new hull design in a timely manner.

Just as differences can be expected in the seakeeping qualities of dissimilar ships, capsizing, broaching, and surfriding behavior can vary significantly as well. The recognition by several countries that the existing standard stability criteria failed to address dynamic capsize led to the formation of the Cooperative Research Navies Dynamic Stability Working Group which developed the time domain dynamic stability tool FREDYN. An overview of the physics in FREDYN will be provided by another paper in this conference (which incidentally discusses capsize probability) by Drs. Kevin McTaggart and Jan DeKat. For the purposes of this discussion, it will suffice to state that the approach taken by FREDYN is to model the relevant capsize, broaching and surfriding physics using a quasi-nonlinear approach, producing adequate predictions without having to resort to highly detailed methods requiring larger computational efforts.

The heart of any simulation-based assessment approach which takes into account the many modes of capsize dynamics rests in the simulations performed in the individual sea condition. Once capsizing is assessed at this level, the calculation of capsize probability can be performed by applying the results against an operability profile of the ship and then weighing the results according to the probability of occurrence of the individual sea condition. When these results are compiled for the complete range of sea conditions present in the operational area of the ship, the capsize probabilities can be assessed.

The approach taken by the Hydromechanics Department at David Taylor is to investigate the capsize risk in a specific seaway uses a Monte-Carlo approach with FREDYN to conduct multiple simulations. The validity of the FREDYN predictions is first validated in comparisons with capsize model test data in regular waves to ensure that the simulations adequately model the surfriding, broaching, and capsize behavior of the model.

For the capsize assessment, simulations in irregular waves are typically carried out for all headings in 15 degree increments and at 5 knot speed intervals throughout the speed range of the ship. Each simulation is at least 30 minutes in duration and is typically performed for 25 realizations to insure that the seaways do not repeat. For each seaway, a capsize index (Tci) is calculated as a measure of the total capsizes in comparison to the total number of simulations such that:

$$\text{Tci} = \frac{\text{Total Number of Capsizes}}{\text{Total Number of Simulations}} \qquad (1)$$

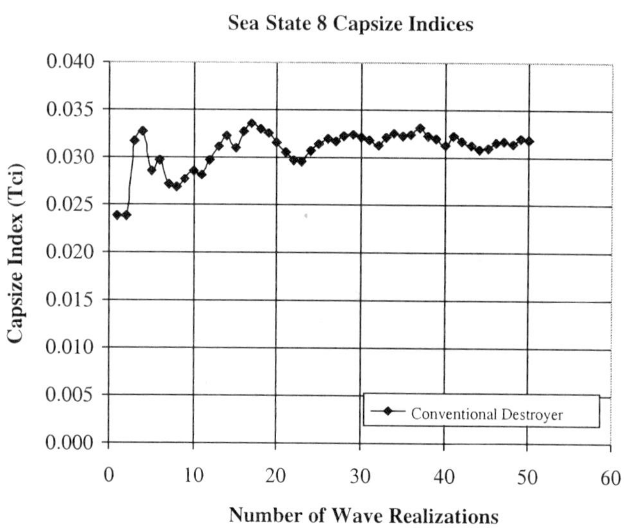

Fig. 1 Capsize indices as a function of wave realizations

Figure 1 indicates how well the value Tci, for one sea condition, converges as the number of realizations increase. In the case of a conventional destroyer, a satisfactory value is obtained with about 25 realizations. This appears to be somewhat less than the author's findings. Furthermore, my experience has been that ships having worse dynamic capsize characteristics tend to require fewer realizations to bound the value of Tci sufficiently for design evaluation purposes.

Armin W. Troesch, Member

The author has made a commendable attempt to bring to light the more recent investigations of stochastic vessel rolling leading to capsize. By necessity, a survey paper such as this can not explain all the implicit assumptions used by the various references listed in the paper. It would be helpful, however, if the author would clarify one statement, which appears several times when discussing beam seas. In Sections 3.1 and 3.5, it is mentioned that the wave force or external excitation does not play a significant role in causing the ship to be inclined at large angles. This suggests that near the critical roll angle the dynamics are influenced more by system inertia and stiffness than by excitation. The implication is that the roll model is one of a lightly damped, lightly excited, resonant system. Physically, this means that the excitation requires several (many?) cycles to build up the necessary roll motion to approach the capsize boundary.

First, does the beam seas model represent a lightly damped, lightly excited system? And second, does this model compare well with practical experience? In other words, does capsize in beam seas (experimental or full scale) appear to be a resonant phenomenon or something that can happen in one or two cycles, implying large forcing?

Jeffrey M. Falzarano, Memb"er

The author is to be congratulated on an excellent paper in this most important subject area. The strength of this paper as stated in the abstract is that it makes available a wealth of Russian and Japanese language material not generally available in North America or Europe.

However, I would be remiss if I did not draw attention to additional non-Russian or Japanese works in this area that have not been mentioned in the text. These include the extensive works of Professor Francescutto of the University of Trieste (e.g., Francescutto 1993). Also the extensive body of work contained in the edited volume by Shlesinger & Swean (1997) which summarizes the results of an Office of Naval Research initiative in this area. Finally, I will also take this opportunity to mention the two investigations in this area that we have been undertaking over the last several years at the University of New Orleans, Marine Dynamics Laboratory. The first is a study of the critical dynamics (boundary between capsizing and no-capsizing) in a realistic (random) seaway (see e.g., Vishnubhotla et al. 2000). The second is an opportunity we had to participate in the ONR Mobile Offshore Base initiative where we recently had an opportunity to apply some of these advanced analysis techniques to the transit draft behavior of this large semi-submersible hull form (Falzarano 1999)

In conclusion, I believe that this paper will provide many of the missing links in order that a physics-based probabilistic dynamic ship and offshore platform stability criteria or at least analysis procedure will eventually be developed and widely used in our profession. Finally, I would like to thank the author for the opportunity to comment on his very excellent paper.

Additional references

Francescutto, A. "Non-linear Ship Rolling Motion in the Presence of Narrow-banded Excitation," *Nonlinear Dynamics of Marine Vehicles*, ASME Winter Annual Meeting, Edited by Falzarano and Papoliuas, 1993.

Shlesinger, M. and Swean, T. *Stochastically Excited Nonlinear Ocean Structures,* World Scientific Press, Singapore, 1997.

Vishnubhotla, S., Falzarano, J., and Vakakis, A., "A New Method to Predict Vessel/Platform Critical Dynamics in a Realistic Seaway," *Philosophical Transactions of the Royal Society,* Theme issue on The Nonlinear Dynamics of Ships, Edited by Spyrou and Thompson, 2000.

Faizarano, J., et al., "MOB SBU Transit Draft Dynamics and Stability Analytic Study," *Proceedings of /the Third International Workshop on Very Large Floating Structures,* Edited by Kim and Ertekin, Honolulu, 1999.

Peter Gooding, Member

[The views expressed here are those of the discussers, and not necessarily those of the U.S. Coast Guard or the Department of Transportation.]

This paper provides an excellent background for understanding the probabilistic approach to ship design. The author's approach is very thorough for looking at the intact stability case. The methodology and different aspects reviewed by the author show the vast amount of information necessary to obtain an acceptable intact stability design.

The author points out that the current standards rely on statistics and deterministic calculation for determining this design. However, the author points out that this method, naturally, can be switched to a probabilistic approach. The author does state that in order to switch, industry and regulators must obtain the following information:

- Proper input data
- Accurate and practical calculations
- Easy interpretation for regulation purposes

In order for the regulations to support a change to a probabilistic approach, designers need to consider ways to provide the data, calculations, and interpretations. In addition, the calculations must always remain straightforward and a closed form result must be available to continue in this direction. The author's input on any missing information (data or calculations) would help improve this change in intact design methodology.

Author's Closure

The author is, indeed, very grateful to all the discussers for their valuable comments and questions. They emphasized most complicated aspects of the problem, giving the author a brilliant opportunity to clarify his views.

Regarding *Mr. Cleary*'s comments, a prospective stability regulator should not be discouraged with the statement that it is difficult to use capsizing probability as stability criterion. It does not mean that probability, as an approach cannot be used for practical regulation. This is a simple matter of computational convenience and accuracy. Probability of capsizing per unit of time is preferable over just capsizing probability without reference on time (the last one implies that the time equals infinity).

Averaging of the above probability should be made accordingly to all weather condition in the *given geographical region* with respect of their lasting time.

Accounting of changing of radius of gyration is both necessary and possible; however, the author is unaware of any practical method to calculate its probability distribution at the design stage.

Regarding *Mr. Dalzell*'s comments, the author agrees with the discusser that acceptable level of risk is a matter of social agreement and economic evaluation rather than engineering subject. In other words, engineering has to provide the society with the risk calculation method and this is up to society to decide what level of safety for which price is acceptable.

However, direct usage of non-capsizing probability as stability criteria might be difficult, since this number is quite close to unit for the vast majority of practical cases. Capsizing risk function is preferable: it is related with non-capsizing probability exponentially which gives the criterion enough sensitivity.

$$P_T(\overline{X}) = 1 - exp(\lambda T)$$

Here λ is capsizing risk function, the averaged probability of capsizing per unit of time.

Regarding *Mr. Thomas*'s comments, I absolutely agree with the discusser that numerical simulation is the most efficient and accurate way to calculate capsizing probability. I also agree that direct counting of number of capsizing is quite adequate for design evaluation purposes, when we are interested in short-term stability performance in extreme seas.

However, if we might need long-term capsizing probability, for example, for stability regulation purposes, the corresponding method should be capable for calculation in moderate seas as well. Despite these situations that yield much smaller capsizing probability, the probability that a ship would encounter such situations might be significantly higher in comparison with high-risk ones. So accumulation of many low-risk situations over a significant amount of time might change the overall probabilistic estimate.

Since probability of capsizing in moderate seas might be small, its direct counting might be quite time consuming. So the best method in this case might be a combination of numerical simulations with analytical solutions. Numerical simulation is necessary here to calculate probabilistic characteristics of nonlinear irregular ship motions.

Regarding *Mr. Troesch*'s comments, decreasing of the influence of the exciting force near critical roll angle might be considered in a statistical sense. Rolling equation is linearized in the vicinity of critical roll angle and yields negative instantaneous GM, so the linearized system has a repelling force instead of attraction. This leads to dramatic decreasing of variance of partial solution and, hence, contribution from the external force becomes small. Comparing:

	Positive GM_0	Negative GM_1
Variance:	$V_\phi = \sum_{i=1}^{N}\left[\frac{a_i^2}{\left(\omega_i^2 - \omega_{\phi 0}^2\right)^2 + 4\delta^2\omega_i^2}\right]$	$V_\phi = \sum_{i=1}^{N}\left[\frac{a_i^2}{\left(\omega_i^2 - \omega_{\phi 0}^2\right)^2 + 4\delta^2\omega_i^2}\right]$
Frequency:	$\omega_{\phi 0}^2 = \frac{GM_0\Delta}{I_x + A_{44}}$	$\omega_{\phi 1}^2 = \frac{GM_1\Delta}{I_x + A_{44}}$

Here Δ is weight displacement, I_x moment of inertia, A_{44} added moment of inertia, δ is roll dumping coefficient, ω_i and a_i are frequency and amplitude of excitation. Such linearization itself does not include any implication on damping value other than the system is still capable for oscillation near initial equilibrium, so

$$4\delta^2 < \omega_{\phi 0}^2$$

In general, there are several scenarios for capsizing in beam seas:

- Resonance. It requires several cycles to build up. This scenario is most probable in swell type spectrum, when waves are close to regular.
- Hit by high-energy wave and wind gust. This is weather criterion scenario and it seems to be quite adequate for high freeboard vessels.
- Hit by high-energy wave and wind gust with progressive deck flooding. The same as above, but for low freeboard vessels.
- Parametric resonance caused by heaving.
- Breaking wave action.

Consideration of each of these scenarios might require a specific math model.

Regarding *Mr. Falzarano*'s comments, the author would like to apologize for not including all the relevant references in the review. Unfortunately the scope of the presentation does not allow including all literature on the subject. So the author tried to put more efforts in a review of the works that are not available in English and hence might be less known for the Society. As a result, this review, of course, could not be considered as perfectly complete. Also, the author had to severely restrict himself on including references from neighbor fields such as deterministic nonlinear dynamics, wave and wind statistics, some aspects of seakeeping etc., as well as works on stability of high-speed crafts, yachts and other non-conventional vessels. However, some of these references should be included in the review, anyway.

The author is grateful to the discusser for supplying additional references that makes the review more complete.

Regarding *Mr. Gooding*'s comments, we might be closer than ever to a probabilistic standard for intact stability. Now, almost all elements we need are available. The main problem left is an integration of all these elements into the system; in other words, the main theoretical work is done, practical application is the next challenge. In particular the following is the subject of work.

Input data—Ship information is available completely, even in electronic format for recently designed vessels. Lines and weight distribution data are necessary. So, what we need are data filters between design and simulation software. Getting proper data on weather would be more difficult, since oceanographic information is not always presented in the form suitable for seakeeping calculations. Extracting such information from oceanographic data sheets is definitely a job to do.

Calculation methods—There are some to choose from; all of them nowadays exist in a form of computer software. Integration and validation work is needed in order to come up with a calculation method suitable for regulation purposes. Most probably, it will be presented in the form of computer software approved by the regulator.

Interpretation for regulation purposes—First of all, it means setting a standard for the risk to a publicly acceptable level, taking into account both the social and economical aspect. As a first step, the currently achieved level can be calculated and accepted as a reference point.

Then, the next generation of intact stability standard can be proposed with better awareness of capsizing risk.

SNAME *Transactions*, Vol. 108, 2000, pp. 147–177

Capsize Risk of Intact Frigates in Irregular Seas

Kevin McTaggart, Member, Defence Research Establishment Atlantic, Dartmouth, Nova Scotia, and
Jan Otto de Kat, Member, Maritime Research Institute Netherlands, Wageningen, The Netherlands

ABSTRACT

Risk analysis provides a rational method for safe design and operation of complex engineering systems. This paper describes the application of capsize risk analysis to the safe design and operation of intact naval frigates. Input probabilistic descriptions of wave conditions and ship operating conditions are combined with time domain simulations of ship motions to determine capsize risk of an intact ship. The maximum roll angle of a ship in an irregular seaway is dependent on the seaway realization, and this effect is included in the risk analysis. Sample calculations for a naval frigate demonstrate application of the risk analysis procedure. Risk analysis results can be used to develop simplified design guidelines for given ship types. In addition, operator guidelines can be developed for minimizing capsize risk.

NOMENCLATURE

A_{jk}^{∞}	infinite frequency added mass
a_i	amplitude of wave component
a_X	Gumbel distribution scale parameter
$\hat{a}_X$	estimate of a_X
$\hat{a}_X^b$	biased estimate of a_X
B	ship beam
B_{jk}^{∞}	frequency independent damping
$B_{jk}(\omega_e)$	frequency dependent damping
b_X	Gumbel distribution location parameter
$\hat{b}_X$	estimate of b_X
C	ship capsize
C_{Dx}	local cross-flow drag coefficient
C_{jk}	frequency independent stiffness
C_{VP}	vertical prismatic coefficient
D	duration
D_p	propeller diameter
D_s	simulation duration
$F_{jk}(t)$	radiation force component
F_{FK}	Froude-Krylov force on ship
F_H	quasi-steady transverse hull force
$F_{H,NL}$	nonlinear component of F_H
Fn	Froude number
F_{PR}	propeller thrust
$F_X(X)$	cumulative distribution function of X
F_x, F_y, F_z	translational forces on ship
$\overline{GM}_{fluid}$	corrected metacentric height
$\overline{GZ}$	righting moment arm
g	gravitational acceleration
H_s	significant wave height
$H_{s,max}$	maximum H_s for given wave period
$\widetilde{H/\lambda}$	nominal wave steepness
I_{ij}	ship rotational inertia component
$J(t)$	propeller advance coefficient
$K_{jk}(t)$	retardation function
K_{xx}	hull moment coefficient
K_T	propeller thrust coefficient
$\overline{KG}$	height of CG above baseline
L	ship life or length
$[M]$	ship generalized (6×6) mass matrix
M_x, M_y, M_z	rotational moments on ship
$M_{x,H}$	roll moment from hull lift and drag
m_0	ship mass
N_C	number of capsizes
N_p	propeller RPM
N_s	number of simulations
N_w	number of wave components
N_X	number of discretized values of X
$\{n\}$	generalized normal vector
$P(C_D)$	probability of capsize during D
$\hat{P}(C_D)$	estimate of $P(C_D)$
$P(C\|X)$	probability of capsize given X
p	pressure
p, q, r	ship rotational velocity components
$p_X(X_i)$	probability of occurrence of X_i
$p_{X\|Y}(X_i\|Y)$	conditional probability of X_i
$Q_X(X)$	exceedence probability of X
$Q_{X\|Y}(X\|Y)$	conditional exceedence probability
$S(t)$	instantaneous wetted surface of ship
$S_\omega(\omega_i)$	wave spectral density
T	local draft
T_{mid}	midships draft
T_p	peak wave period
T_z	zero-crossing period
T_ϕ	natural roll period
t	time
t_p	propeller thrust deduction coefficient
t_s	trim by stern
U, V	ship horizontal velocity components
$\overline{U}_a$	mean wind speed
$u_G, v_G, w_G,$	velocity at ship centre of gravity
V_s	nominal ship speed
v	local transverse velocity
X	random variable
X_i	discretized value of random variable X
x, y, z	ship-based axes
$\{\ddot{x}(t)\}$	ship acceleration vector
x_e, y_e, z_e	earth-based axes
$x_k(t)$	displacement for mode k
α	wave slope angle
α_{max}	maximum absolute value of α
β	incident sea direction relative to ship
γ	Euler's constant (0.5772...)
$\Delta\omega_i$	discretized wave frequency interval
ϵ_i	wave component random phase
$\dot{\eta}_t$	local transverse wave orbital velocity
ζ	water elevation
θ	ship pitch angle
κ_i	wave component wave number
λ	wavelength
μ_X	mean value of variable X
σ_X	standard deviation of variable X
Φ_I	incident wave velocity potential
ϕ	roll angle
$\phi_{max,D}$	max roll angle for duration D
$\phi_{max,L}$	max roll angle for ship life L
$\phi_{max,i}$	max roll angle of rank i
ω	wave frequency
ω_e	wave encounter frequency
ω_i	frequency of wave component i

INTRODUCTION

In recent years, a consensus has emerged among engineers that risk analysis is the most rational approach for safe design and operation of complex systems. Blockley and contributors [1] give a comprehensive overview of risk analysis and its application to engineering systems. For application to ships, Kobylinski [2] gives an example of capsize risk analysis, while Mansour et al. [3] discuss structural design using risk analysis.

For stability of warships, risk analysis offers an alternative to the rules developed approximately 40 years ago by Sarchin and Goldberg [4], which continue to be used by several navies. Although the rules of Sarchin and Goldberg have resulted in very safe operation for naval warships, there are several reasons for moving toward new risk-based approaches. Sarchin and Goldberg's stability criteria were developed based on experience with ships in World War II. Modern ships have considerably different hull forms for which the older criteria are not necessarily valid. For example, modern warships have wide transom sterns, making them more vulnerable to loss of static stability in following seas. For novel hull forms such as trimarans, lack of operational experience makes risk analysis the only viable approach available for developing safe designs. Another important reason for using risk analysis is that it provides information regarding what combinations of speed and heading should be avoided to minimize capsize risk. In the past, lack of accurate simulation tools and large computational requirements have been the major deterrents for application of risk analysis to ship capsize. Fortunately, time domain models of ship capsize now appear to be sufficiently accurate and fast.

This paper describes ongoing work to develop rational risk analysis procedures for the stability of intact naval frigates. This work is an extension of earlier work presented by McTaggart [5, 6]. The present research is being conducted in parallel with Canada's ongoing participation in the Cooperative Research Navies Dynamic Stability Project, which has been described by de Kat [7] and de Kat et al. [8].

NUMERICAL SIMULATION MODEL

The numerical model FREDYN has been developed to simulate the large amplitude motions of a steered ship in severe seas and wind. The model consists of a non-linear strip theory approach, where linear and non-linear potential flow forces are combined with maneuvering and viscous drag forces. The non-potential force contributions are of a nonlinear nature and based on (semi)empirical models.

The derivation of the equations of motions is based on the conservation of linear and angular momentum. These are given in principle in the inertial (earth-fixed) reference system, defined by the system of axes (x_e, y_e, z_e). Euler's method is applied for deriving the equations of motion in terms of a rotating, ship-fixed coordinate system. The equations of motion are given by:

$$[M]\{\ddot{x}(t)\} = \left\{ \begin{array}{c} \sum F_x \\ \sum F_y \\ \sum F_z \\ \sum M_x \\ \sum M_y \\ \sum M_z \end{array} \right\} - \left\{ \begin{array}{c} m_0(w_G q - v_G r) \\ m_0(u_G r - w_G p) \\ m_0(v_G p - u_G q) \\ (I_{zz,0} - I_{yy,0})qr \\ (I_{xx,0} - I_{zz,0})pr \\ (I_{yy,0} - I_{xx,0})pq \end{array} \right\} \quad (1)$$

The matrix $[M]$ is the generalized (6×6) mass matrix of the intact ship and $\{\ddot{x}(t)\}$ is the acceleration vector at the center of gravity; p, q and r represent the rotational velocities for roll, pitch and yaw, respectively. The summation signs in the right hand side represent the sum of all force and moment contributions, which result from:

- Froude-Krylov force (nonlinear)
- Wave radiation (linear)
- Diffraction (linear)
- Viscous and maneuvering forces (nonlinear)
- Propeller thrust and hull resistance (nonlinear)
- Appendages – rudders, skeg, active fins (nonlinear)
- When present: wind or internal fluid (nonlinear)

Large angles are retained in the matrices for transformation between the ship-fixed and the earth-fixed coordinate system. The combination of the integrated hydrostatic and dynamic wave pressures represents the total Froude-Krylov force (or moment), given by the vector:

$$\{F_{FK}\} = \int_{S(t)} p\ \{n\}\ dS \quad (2)$$

where $\{n\}$ is the generalized normal vector and

$$p = \rho\frac{\partial \Phi_I}{\partial t} + \rho g z \quad (3)$$

$S(t)$ represents the instantaneous wetted hull surface in the presence of the undisturbed, incoming wave with velocity potential Φ_I. Linear wave theory is used to describe the sea surface and wave kinematics. In the case of irregular waves, the model makes use of linear superposition of sinusoidal components with random phasing.

Linear, 3D, transfer functions are used in the determination of the diffraction forces, and the wave radiation forces are based on linear retardation functions and convolution integrals (including forward speed terms). The potential, non-lifting hydrodynamic force terms are given by:

$$F_{jk}(t) = -A_{jk}^{\infty}\,\ddot{x}_k(t) - B_{jk}^{\infty}\,\dot{x}_k(t) - C_{jk}\,x_k(t) + \int_0^t K_{jk}(t-\tau)\dot{x}_k(\tau)\,d\tau \qquad (4)$$

where $F_{jk}(t)$ is the contribution to the j mode force from motion in the k-mode, A_{jk}^{∞} is the infinite frequency added mass, B_{jk}^{∞} is the frequency independent damping, C_{jk} is the frequency independent stiffness, $K_{jk}(t)$ is the retardation function, $x_k(t)$ is the time dependent ship displacement for mode k, $\dot{x}_k(t)$ is the time dependent ship velocity, and $\ddot{x}_k(t)$ is the time dependent ship acceleration.

The retardation functions can be obtained as follows in terms of the 3D hydrodynamic coefficients:

$$K_{jk}(t) = \frac{2}{\pi}\int_0^{\infty}\left[B_{jk}(\omega_e) - B_{jk}^{\infty}\right]\cos(\omega t)\,d\omega_e \qquad (5)$$

where ω_e is wave encounter frequency. Viscous effects include roll damping due to hull and bilge keels, wave-induced drag due to wave orbital velocities, and non-linear maneuvering forces with empirically determined coefficients. The quasi-steady hull forces resulting from the motions in the horizontal plane consist of a linear and non-linear part:

$$\begin{aligned} F_H(t) &= F_{H,L}(U,V,r,\theta;t) \\ &\quad + F_{H,NL}\,(v(x,t),T(x,t), \\ &\qquad C_{Dx}(U(t),x),U,V,r,\dot{\eta}_t) \qquad (6) \end{aligned}$$

where U and V are the time-dependent longitudinal and transverse velocity components, θ is the pitch angle, $v(x,t)$ is the local transverse velocity at (sectional) location x, $T(x,t)$ is the local draft, C_{Dx} is the local cross-flow drag coefficient, and $\dot{\eta}_t$ is the local wave orbital velocity in transverse direction. The roll moment resulting from lift and hull drag forces is given by:

$$\begin{aligned} M_{x,H}(t) &= F_{H,L}\,y(t) + F_{H,NL}(t)\,y(x,t) \\ &\quad + K_{ur}\,U(t)\,r + K_{up}\,U(t)\,p \\ &\quad + K_{pp}|p|p \qquad (7) \end{aligned}$$

The propeller thrust is estimated by:

$$F_{PR}(t) = F_{PR}\left[N_p, t_p, K_T(J(t), N_p, D_p)\right] \qquad (8)$$

where N_p is the propeller RPM (assumed constant), t_p is the thrust deduction coefficient at the propeller, and K_T is the propeller thrust coefficient. The propeller thrust coefficient K_T is a function of the instantaneous advance coefficient $J(t)$, propeller RPM N_p, and propeller diameter D_p.

The hull resistance is based on the calm water characteristics as a function of instantaneous speed and sinkage. Appendage forces are estimated by using wing theory in the case of a rudder or active fin, and by using pressure drag in the case of a skeg. Interaction between rudder and hull is considered, and the instantaneous angle of attack depends on the ship motions, rudder angle, propeller race and local wave orbital velocities. The equations of motion are solved in the time domain using a 4th order Runge-Kutta scheme.

Wind forces are modelled using standard semi-empirical methods. McTaggart and Savage [9] provide wind tunnel results for a naval frigate which can be used in simulations.

Simulation of Random Seaway

The numerical model can simulate motions of a ship in a regular or random long-crested seaway. To simulate a long-crested random seaway with a given wave spectrum $S_\omega(\omega)$, a random phase method is used (see Chakrabarti [10]). Using the convention that waves propagate along the earth-fixed x_e axis, the instantaneous water elevation is then given by:

$$\zeta(x_e,t) = \sum_{i=1}^{N_w} a_i\,\sin(\omega_i - \kappa_i x_e + \epsilon_i) \qquad (9)$$

where N_w is the number of discretized wave components, a_i is the amplitude of component i, ω_i is the component frequency, κ_i is the component wave number, and ϵ_i is a randomly generated component phase between 0 and 2π. Assuming small amplitude linear wave theory, the amplitude and wave number of the wave components are as follows:

$$a_i = \sqrt{2\,S_\omega(\omega_i)\,\Delta\omega_i} \qquad (10)$$

$$\kappa_i = \frac{\omega_i^2}{g} \qquad (11)$$

where $\Delta\omega_i$ is the discretized interval of wave component i.

For a realistic representation of a random seaway, at least 20 discretized wave components must be used. The random phases ϵ_i are determined by a random number generator, and depend upon an input integer seed number. It will be shown later that maximum roll angle in a randomly generated seaway can be very dependent on the input seed number for the seaway.

PHYSICS OF CAPSIZE FOR NAVAL FRIGATES

Critical wave and operational conditions may cause a frigate to capsize according to the following mechanisms:

- Static loss of stability
- Dynamic loss of stability
- Broaching

Static Loss of Stability

This refers to the quasi-static loss of transverse stability (associated with an excessive righting arm reduction) in the wave crest. This mode occurs typically in regular or irregular following to stern quartering waves with low encounter frequencies. The ship can capsize when it experiences temporarily a critically reduced (possibly negative) righting arm for a sufficient amount of time, while the wave crest overtakes the ship slowly and the ship is surging or surf-riding periodically. For this mode of capsize to occur in irregular waves, one encountered wave of critical length and steepness is sufficient to cause the sudden catastrophic event. Experimental evidence of this for a frigate hull form has been observed by de Kat and Thomas [11].

Dynamic Loss of Stability

A frigate can lose stability dynamically in conjunction with extreme rolling motions and lack of righting energy under a variety of conditions. This capsize mode may be associated with the phenomena described below.

- *Dynamic Rolling*: This mode of motion occurs at forward speed in stern quartering seas, which can be of regular or irregular nature. Here all six degrees of freedom are coupled, where in addition to roll, surge, sway and yaw can exhibit large amplitude fluctuations. The motion is characterised by asymmetric rolling: the ship rolls heavily to the leeward side in phase with the wave crest (approximately) amidships and rolls back to the windward side in the wave trough, albeit with a shorter half-period and smaller amplitude. Due to the associated surging behaviour, the ship spends more time on the wave crest than in the trough, resulting in a periodic but asymmetric reduction on the crest and restoring (in the trough) of the righting arm. The actual roll period may exceed the natural roll period significantly. In the case of a capsize, the roll motion typically builds up over a number of wave encounters to a critical level, and the ship will usually capsize to leeward. Experimental evidence of this for a frigate can be found in de Kat and Thomas [11, 12].

- *Parametric Excitation*: Parametric excitation results from the time-varying roll restoring characteristics of a ship typically found in longitudinal waves. The periodic changes in static righting arm during the repeated passage of a wave crest followed by the trough can cause large amplitude roll motions, which occur at approximately the natural roll period and simultaneously at twice the wave encounter period. The wavelength must be of the order of the ship length. In such circumstances, parametric rolling - also referred to as low cycle resonance - can result in capsizing. It can occur in regular and irregular waves. It has been observed in head seas, but parametric excitation in astern seas is typically more critical in terms of capsizing. In particular, when a ship travels at the mean group speed in following seas, parametric excitation can occur during the passage (in a regular fashion) of a wave group with a sufficient number of encountered waves of critical height and length (de Kat [13]). Ships with a wide transom stern may be more prone to parametric rolling; typically this is not a critical capsize mode for frigates.

- *Resonant Excitation*: In principle large amplitude roll motions can result when a ship is excited at or close to its natural roll frequency. Roll resonance conditions are determined by the combination of $\overline{GZ}$ curve characteristics, weight distribution, roll damping, heading angle (e.g., beam seas), ship speed, wavelength and height.

- *Impact Excitation*: Steep, breaking waves can cause severe roll motions and may overwhelm a vessel. The impact due to a breaking wave that hits a vessel from the side will affect the ship dynamics and may cause extreme rolling and capsizing. Possible damage to deck structures and subsequent water ingress are not considered here. This capsize mode is relevant especially to smaller vessels in steep seas.

- *Bifurcation*: In laboratory conditions, the roll response may jump from one steady state to another steady-state condition at the same frequency following a sudden disturbance. Bifurcation is usually associated with large amplitude motions. The experimental studies on these and other types of non-linear systems behaviour have so far been restricted to regular waves. 'Double period bifurcation' in long ($\lambda = 2L$), high waves has been observed for a frigate by de Kat and Thomas [11, 12].

Broaching

Broaching is related to the loss of course keeping in waves. A variety of broaching modes exist in regular and irregular waves:

- Successive overtaking waves while the ship is travelling at low speed;
- Low frequency, large amplitude yaw motions;
- Broaching caused by a single wave.

The first mode may occur in steep following seas at low ship speed, where the ship is gradually forced to a beam sea condition during the passage of several steep waves. The other modes occur at higher speed, typically at a Froude number $Fn > 0.3$. The third mode is usually characterised by quasi-steady surf-riding at wave phase speed and steadily increasing yaw angle; this broaching mode has been also been observed for frigates in combination with bow submergence during surf-riding. Experimental evidence of this for a frigate can be found in de Kat and Thomas [11, 12].

Other Factors

For steep wave conditions most likely to cause capsize, the wind direction will typically be approximately collinear with the waves. Consequently, wind will not strongly influence wave-induced capsizing in astern seas. In beam waves, however, it can be more important. Furthermore, capsizing may be associated with a combination of modes, e.g., static loss of stability after a surf-riding and a high speed broach.

PREDICTION OF SHIP CAPSIZE RISK

The complex nature of ship capsize means that a risk analysis must consider all conditions encountered, and that the analysis cannot be restricted to a limited range of headings (e.g., beam seas) or to only very large wave heights. The present approach considers the possibility of ship capsize for all encountered seaways, headings, and ship speeds. For a ship in a seaway of duration D (e.g., one hour), the probability of capsize $P(C_D)$ is:

$$\begin{aligned} P(C_D) &= \sum_{i=1}^{N_{V_s}} \sum_{j=1}^{N_\beta} \sum_{k=1}^{N_{H_s}} \sum_{l=1}^{N_{T_p}} p_{V_s}(V_{s-i})\, p_\beta(\beta_j) \\ &\quad \times p_{H_s,T_p}(H_{s-k}, T_{p-l}) \\ &\quad \times P(C_D|V_s, \beta, H_s, T_p) \end{aligned} \tag{12}$$

where V_s is ship speed, β is the relative wave heading, H_s is significant wave height, and T_p is peak wave period. Each independent variable X in the above equation has been discretized into N_X different values. The last term of Equation (12) denotes a conditional probability given a set of operational and seaway conditions. Similar approaches have been presented by Kobylinski [2] and Dahle and Myrhaug [14, 15]. An important assumption of the above equation is that desired ship speed and heading are independent of wave conditions. This assumption is conservative because ship operators will alter speed and course to reduce capsize risk.

Equation (12) does not explicitly include wind effects, which are considered to be of secondary importance relative to waves for naval frigates; however, wind effects can be included in the risk analysis by considering wind conditions to be dependent on wave conditions. A reasonable approximation is to assume that winds are collinear with waves. Data from Bales et al. [16] suggest that mean wind speed can be approximated as a linear function of significant wave height. The following approximation has been derived for the North Atlantic:

$$\overline{U}_a = H_s \times 1.823/\text{s} + 3.45 \text{ m/s} \tag{13}$$

where $\overline{U}_a$ is mean wind speed at an elevation of 19.5 m.

Once the probability of capsize for duration D has been computed using Equation (12), the associated annual probability of capsize can be computed as follows:

$$P(C_{annual}) = 1 - [1 - P(C_D)]^{1\ year/D} \tag{14}$$

The application of Equation (12) for prediction of capsize risk requires a suitable method to determine capsize risk for given conditions $P(C_D|V_s, \beta, H_s, T_p)$. For a FREDYN simulation of a ship in irregular waves, a random phase approach is used to generate a wave realization. The wave realization is dependent upon an integer seed number provided as input for generation of random wave phases. The occurrence of capsize can be highly dependent on the input seed number. Upon initial consideration, it would seem appropriate to determine $P(C_D|V_s, \beta, H_s, T_p)$ using N_s simulations with different seed numbers as follows:

$$P(C_D|V_s, \beta, H_s, T_p) = \frac{N_C}{N_s} \quad (15)$$

where N_C is the number of simulations for which capsize occurs. The main disadvantage of this approach is that extrapolation cannot be used to estimate capsize risk in cases with no observed capsizes. A more useful approach is to apply a statistical fit to the maximum roll angles $\phi_{max,D}$ from N_s simulations of duration D. In the present discussion, variable X denotes the maximum roll angle in a seaway with given duration and conditions. McTaggart [17, 18] has shown that a Gumbel distribution is appropriate for modelling the maximum roll angle in a one hour seaway. Properties of the Gumbel distribution are described in various texts, including Madsen et al. [19] and Thoft-Christensen and Baker [20]. The cumulative distribution function of the Gumbel distribution is as follows:

$$F_X(X) = \exp\left[-\exp\left(\frac{b_X - X}{a_X}\right)\right] \quad (16)$$

where a_X and b_X are scale and location parameters, with b_X being the 36.8'th percentile of X. The mean μ_X and standard deviation σ_X of the Gumbel distribution are related to the distribution parameters as follows:

$$\mu_X = b_X + \gamma\, a_X \quad (17)$$

$$\sigma_X = a_X \frac{\pi}{\sqrt{6}} \quad (18)$$

where γ is Euler's constant (0.5772...).

Various methods are available for determining Gumbel distribution parameters, including the method of moments and a least squares fit procedure. The method of moments relates the distribution parameters to the computed mean and standard deviation for the variable X. The second method uses a least squares linear fit between the variable X or its transformation (e.g., $\ln X$) and the cumulative distribution function $F_X(X)$ or its transformation (e.g., $\ln[-\ln(F_X(X))]$). A significant advantage of the least squares method is that it can be applied to a limited probability range of greatest interest. When simulating ship motions using the program FREDYN, a simulation will terminate when the ship roll angle exceeds a specified maximum value of 90 degrees; thus, a Gumbel least squares fit provides a suitable approach for handling the limited range of valid maximum roll angles.

For a random variable having a Gumbel distribution, a linear relationship exists between X and $\ln[-\ln(F_X(X))]$. The Gumbel parameters a_X and b_X can be determined by a least squares linear fit minimizing the error in either X or $\ln[-\ln(F_X(X))]$. For predicting ship capsize risk, McTaggart [18] shows that it is preferable to minimize the error in $\ln[-\ln(F_X(X))]$, which can be done using the following equation:

$$\ln[-\ln(F_X(X))] = \frac{(b_X - X)}{a_X} \quad (19)$$

When determining the least squares fit parameters from N_s simulations in a given seaway, the simulated maximum roll angles can be ranked in ascending order to assign cumulative distribution values as follows:

$$F_X(X_i) = \frac{i}{N_s + 1} \quad (20)$$

where X_i is the sorted maximum roll angle for given conditions with rank i in ascending order. The linear regression fit has a slope of $-1/\hat{a}_X^b$ and an intercept of $\hat{b}_X/\hat{a}_X^b$, where $\hat{a}_X^b$ is a biased estimate of a_X, and $\hat{b}_X$ is an estimate of b_X. Monte Carlo simulation indicates that the least squares estimate $\hat{b}_X$ has negligible bias, and that a corrected estimate $\hat{a}_X$ can be obtained as follows:

$$\hat{a}_X = \hat{a}_X^b \left(1 - \frac{0.64}{\sqrt{N_s}}\right) \quad (21)$$

Once the Gumbel parameters have been determined using a least squares fit for a given seaway, the associated probability of exceeding a specified capsize roll angle (e.g., 90 degrees) can be easily estimated using Equation (16).

To reduce required computing time, it is useful to use simulations of a shorter duration D_s to estimate statistics of maximum roll for a longer seaway duration D. The following relationship can then be used:

$$F_X(X(D)) = [F_X(X(D_s))]^{D/D_s} \quad (22)$$

Simulation results indicate that simulation durations of 30 minutes give good estimates of maximum roll statistics for seaway durations of one hour. If a maximum value from simulation duration D_s has a Gumbel distribution, then the maximum value for a seaway duration D will have the following properties:

$$\sigma_X(D) = \sigma_X(D_s) \quad (23)$$

$$\mu_X(D) = \mu_X(D_s) + \frac{\sqrt{6}}{\pi} \ln\left(\frac{D}{D_s}\right) \sigma_X(D_s) \quad (24)$$

Figure 1 shows simulation results and fitted Gumbel distributions for a frigate at a speed of 10 knots, heading of 75 degrees, peak wave period of 12.4 s, and wave height of 9.5 m. The wind is collinear with the waves and has a steady speed of 40 knots, as given by Equation (13). The results indicate that a Gumbel distribution provides a very good fit to the conditional exceedence probability of maximum roll angle. An added benefit of using a fitted distribution for each seaway is that Equation (12) can be revised to give exceedence probabilities for all maximum roll angles in all seaways as follows:

$$\begin{aligned} Q_{\phi_{max,D}}(\phi_{max,D}) &= \\ \sum_{i=1}^{N_{V_s}} \sum_{j=1}^{N_\beta} \sum_{k=1}^{N_{H_s}} \sum_{l=1}^{N_{T_p}} & p_{V_s}(V_{s-i})\, p_\beta(\beta_j) \\ & \times p_{H_s,T_z}(H_{s-k}, T_{p-l}) \\ & \times Q_{\phi_{max,D}|V_s,\beta,H_s,T_p}(\phi_{max,D}|V_s,\beta,H_s,T_p) \quad (25) \end{aligned}$$

Of major practical importance is the required number of simulations for predicting capsize risk in given conditions. McTaggart [18] indicates that 10 simulations of 30 minute duration can provide sufficient data for estimating hourly capsize probability in given conditions. Subsequent work examining a very wide range of conditions has indicated that a larger number of simulations (e.g., 20) should be performed if the hourly probability of capsize is greater than 0.001. In some cases with hourly capsize probabilities greater than 0.001, the actual distribution of maximum roll angle can deviate significantly from a Gumbel distribution. To overcome this problem, the Gumbel fit can be limited to an upper range of maximum roll angles (e.g., the upper 30 degree range) to ensure that the fitted distribution provides a good fit in the range of greatest interest for ship capsize. For consistent results, the limited range of roll angles should include at least 5 values from simulations. Figure 2 gives examples of Gumbel fits to all data points and to the upper 30 degree range of maximum roll angles.

Estimates of capsize risk using fitted distributions have associated errors related to sample size (number of simulations). For estimates of Gumbel parameters a_X and b_X obtained using a least squares fit procedure, their variances have been estimated using Monte Carlo simulation:

$$\sigma^2(\hat{a}_X) \approx \frac{1.45}{N_s} a_X^2 \quad (26)$$

$$\sigma^2(\hat{b}_X) \approx \frac{1.16}{N_s} a_X^2 \quad (27)$$

The following equation can then be used to estimate the sample size error of predicted capsize risk for given conditions:

$$\begin{aligned} \sigma^2(\hat{P}(C_D|V_s,\beta,H_s,T_p) &\approx \\ & \left(\frac{\partial \hat{P}(C_D|V_s,\beta,H_s,T_p)}{\partial \hat{a}_X}\right)^2 \sigma^2(\hat{a}_X) \\ + & \left(\frac{\partial \hat{P}(C_D|V_s,\beta,H_s,T_p)}{\partial \hat{b}_X}\right)^2 \sigma^2(\hat{b}_X) \quad (28) \end{aligned}$$

The following summation then provides an estimate of sample size error of predicted capsize risk for all conditions:

$$\begin{aligned} \sigma^2(\hat{P}(C_D)) &= \sum_{i=1}^{N_{V_s}} \sum_{j=1}^{N_\beta} \sum_{k=1}^{N_{H_s}} \sum_{l=1}^{N_{T_p}} p^2_{V_s}(V_{s-i}) \\ & \times p^2_\beta(\beta_j)\, p^2_{H_s,T_p}(H_{s-k}, T_{p-l}) \\ & \times \sigma^2(\hat{P}(C_D|V_s,\beta,H_s,T_p)) \quad (29) \end{aligned}$$

It should be emphasized that Equations 26 to 29 provide approximate estimates to the sample size error in predicted capsize risk. The actual error in predicted capsize error could be larger because the actual distribution of maximum roll angle in a seaway can deviate from a Gumbel distribution.

To minimize the computational time for capsize risk analysis, time domain simulations are only performed for conditions in which hourly capsize risk is non-negligible ($> 10^{-6}$). For a given combination of ship speed V_s, heading β, and peak wave period T_p, maximum roll statistics are computed for the maximum significant wave height H_s given T_p. Additional time domain simulations are performed at lower significant wave heights until the predicted hourly capsize risk is below 10^{-6}. For small wave heights, maximum roll statistics are assumed to be proportional to H_s using values from the lowest wave height for time domain simulations.

SOURCES OF WAVE STATISTICS

The present risk analysis procedure requires the joint distribution of significant wave height and peak wave period as input. Bales et al. [16, 21] have published wave scattergrams developed using the SOWM wave hindcast model. BMT Global Wave Statistics [22] is another commonly used data source for wave scattergrams.

The present study originally used wave data for Area 15 (western portion of North Atlantic) from

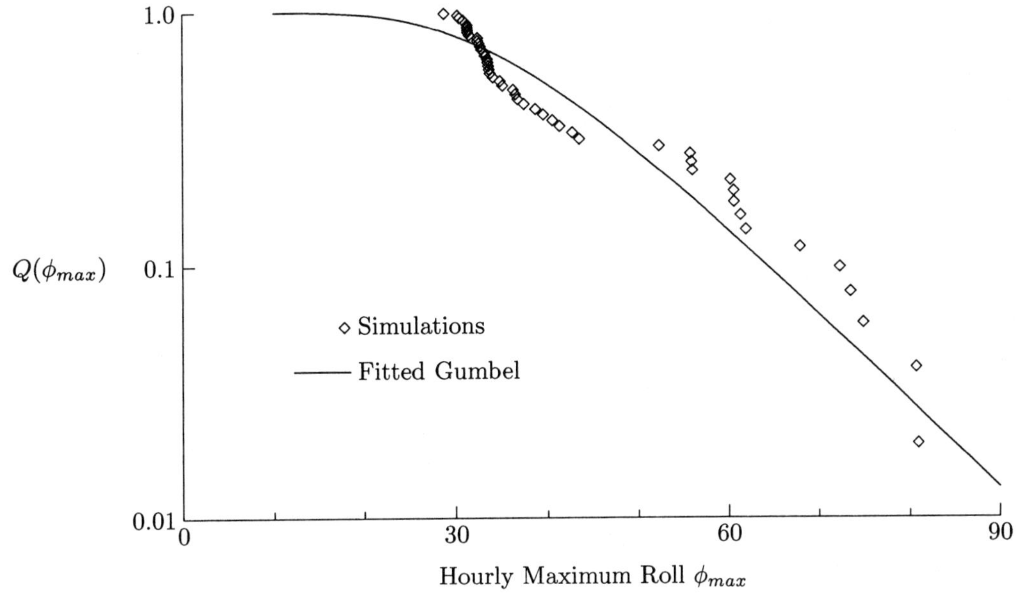

Figure 1: Roll Exceedence Probability for Frigate in One Hour Seaway, V_s = 10 knots, β = 75 degrees, T_p = 16.4 s, H_s = 9.5 m, $\overline{U}_a$ = 40 knots

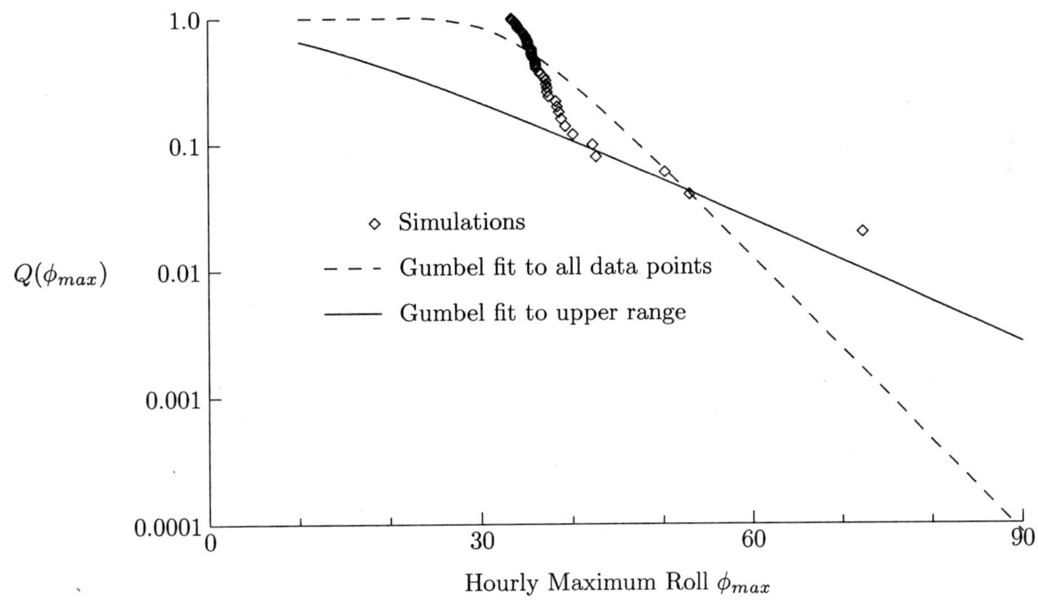

Figure 2: Roll Exceedence Probability for Frigate in One Hour Seaway, V_s = 10 knots, β = 45 degrees, T_p = 13.9 s, H_s = 10.5 m, $\overline{U}_a$ = 44 knots

BMT Global Wave Statistics; however, the resulting FREDYN simulations often experienced numerical instabilities, and predicted capsize risks appeared to be unrealistically high. It was postulated that these problems were caused by unrealistically high wave steepnesses in the BMT data, which are based primarily on visual observations of wave conditions.

To address concerns regarding wave steepnesses arising from BMT Global Wave Statistics, comparisons were made with data from Bales et al. and Buckley [23]. Note that BMT uses zero-crossing wave period when presenting wave data while Bales et al. and Buckley use peak wave period. When comparing the different data sources, the following equation based on a Bretschneider spectrum can be used:

$$T_p = 1.408\, T_z \tag{30}$$

The introduction of nominal wave steepness $\widetilde{H/\lambda}$ below also facilitates comparison of data from the different data sources:

$$\widetilde{H/\lambda} = \frac{2\,\pi\, H_s}{g\, T_p^2} \tag{31}$$

Using numerous wave buoy observations, Buckley has developed limiting envelopes of significant wave height versus peak wave period. Buckley's analysis indicates that nominal wave steepness $\widetilde{H/\lambda}$ has a maximum possible value of 0.049. Figure 3 gives maximum nominal wave steepness versus peak wave period using using data from BMT, Bales et al. and Buckley's limiting envelope for the Northern Hemisphere. The nominal wave steepnesses based on the BMT data are much steeper than Buckley's limiting envelope. In contrast, the North Atlantic data from Bales give steepnesses which are consistent with Buckley's envelope, with the exception of a single observation of $H_s = 8.5$ m and $T_p = 9.7$ s. For the present capsize risk analysis study, the Bales North Atlantic scattergram is used, with the $H_s = 8.5$ m, $T_p = 9.7$ s observation changed to $H_s = 6.5$ m, $T_p = 9.7$ s. Table 1 gives the adapted wave scattergram used for risk analysis in this paper.

EXAMPLE RISK ASSESSMENT FOR HALIFAX CLASS FRIGATE

A risk assessment has been performed to evaluate the hourly capsize probability for a Canadian Halifax class frigate (also know as Canadian Patrol Frigate) operating throughout the year in the North Atlantic. Table 2 gives dimensions for the frigate. The ship is assumed to capsize when the roll angle exceeds 70 degrees, the nominal angle of downflooding.

Table 2: Main Particulars for Halifax Class Frigate

Length, L	124.5 m
Beam, B	14.7 m
Midships draft, T_{mid}	4.64 m
Trim by stern, t_s	0.0 m
Displacement, $\triangle$	4077 tonnes
Vertical centre of gravity, $\overline{KG}$	6.44 m
Metacentric height, $\overline{GM}_{fluid}$ (corrected for free-surface)	1.224 m
Natural roll period, T_ϕ	11.7 s

Input Probability Distributions

The ship has discretized calm water speeds of 10, 17, and 24 knots, with a distribution as shown in Figure 4. Figure 5 shows the distribution of relative wave headings, which are assumed to be evenly distributed with a discrete interval of 15 degrees. The assumption of evenly distributed relative wave headings is likely conservative because a ship will typically alter course in severe conditions to reduce the risk of capsize. For numerical simulations, headings of 3.75 and 176.25 degrees are used rather than the nominal headings of 0 and 180 degrees, which actually represent ranges of 0-7.5 and 172.5-180 degrees respectively.

Annual wave climate statistics for the North Atlantic are taken from Bales [16, 21], as discussed in the previous section. Figure 6 shows the distribution of significant wave heights and Figure 7 shows the distribution of nominal wave steepnesses.

Simulations of Motions for Relevant Conditions

The first stage of the risk analysis procedure involved simulations of ship motions for relevant combinations of ship speed, heading, significant wave height, and peak wave period. The following guidelines determined the number of 30 minute simulations for each condition:

- The maximum wave height given peak wave period for simulations was based on the maximum height of observed waves,

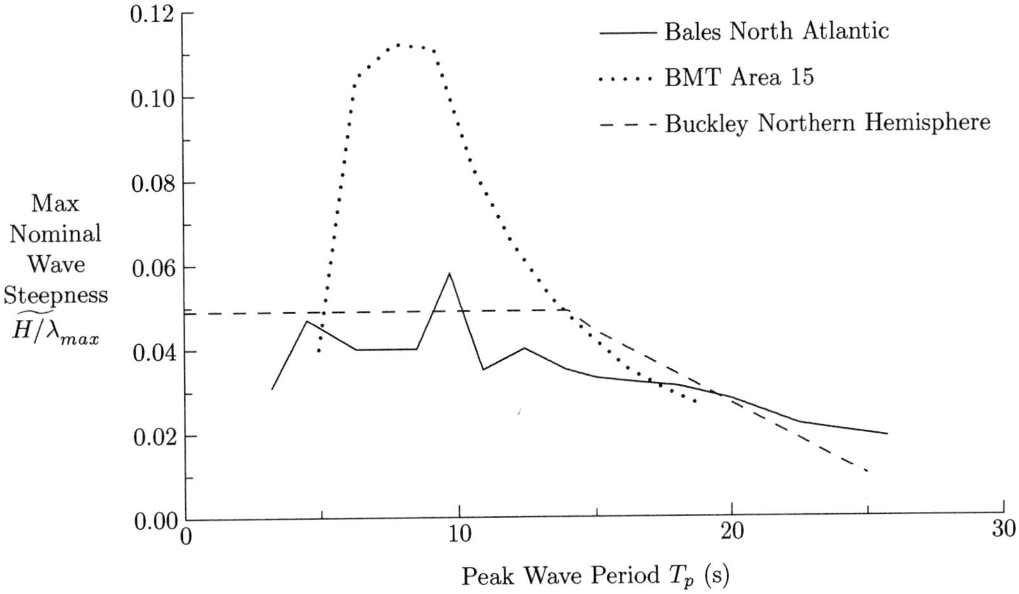

Figure 3: Maximum Nominal Wave Steepness Versus Peak Wave Period

Table 1: North Atlantic Wave Scattergram Adapted from Bales et al. for Capsize Risk Analysis

	Wave Period T_p (s)														
H_s (m)	3.2	4.5	6.3	7.5	8.5	9.7	10.9	12.4	13.9	15.0	16.4	18.0	20.0	22.5	25.7
0.5	541	341	5488	5622	3892	4861	2667	2612	1180	1321	592	213	52	4	
1.5		289	3929	6631	5475	5361	3627	3428	1617	1707	945	569	164	10	
2.5			29	2150	5501	5695	3829	2968	1598	1267	780	421	124	3	
3.5				5	822	4582	3873	3393	1457	1098	623	350	97	6	
4.5					13	613	3115	3381	1293	1008	556	316	75	12	
5.5						22	631	3014	1263	993	467	308	48	5	
6.5						2	27	1336	1103	934	429	287	40		
7.5								229	722	778	347	230	28		6
8.5								16	218	599	310	200	31		7
9.5								2	26	187	316	163	20	2	1
10.5									1	41	202	106	21	1	
11.5										7	67	95	15		1
12.5											13.5	57.5	17.5		
13.5											13.5	57.5	17.5		
14.5												13	13	1.5	0.5
15.5												13	13	1.5	0.5
16.5													2.5	2.5	
17.5													2.5	2.5	
18.5															0.5
19.5															0.5

Total number of observations: 130,339

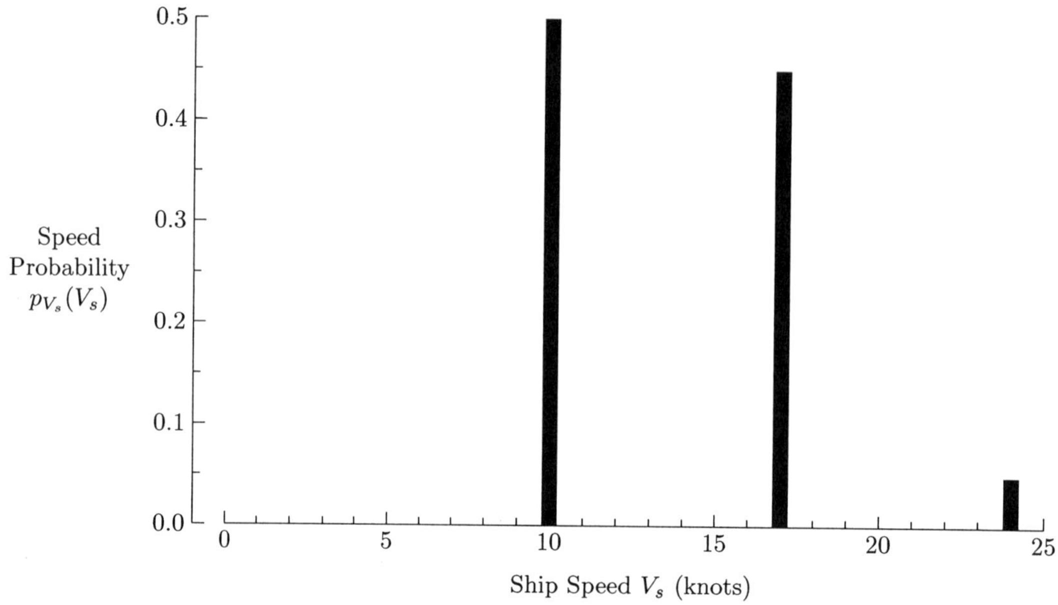

Figure 4: Probability Mass Function for Discretized Ship Speeds

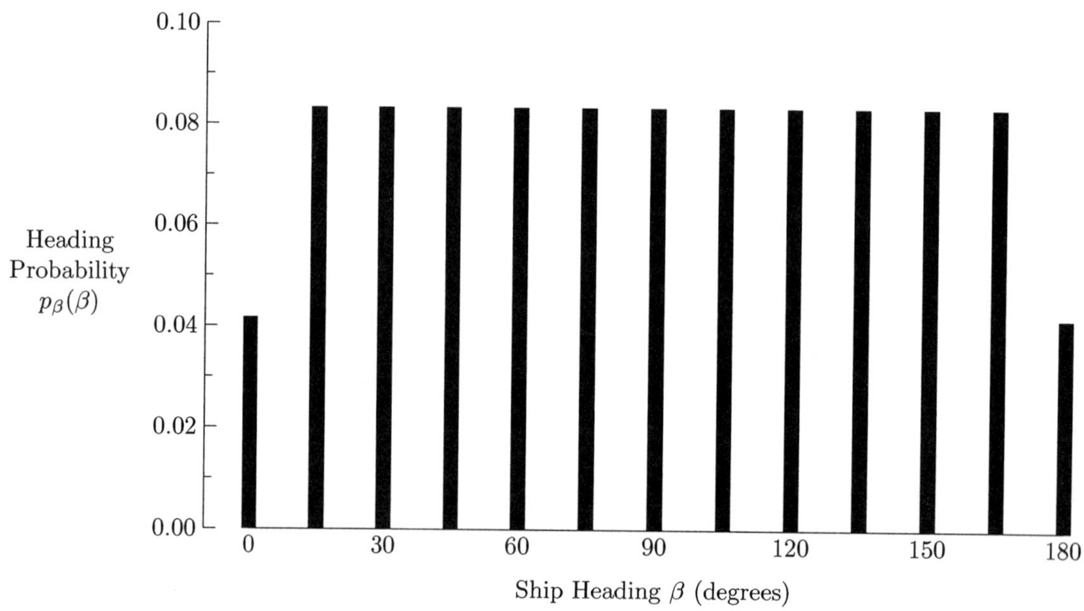

Figure 5: Probability Mass Function for Discretized Ship Headings

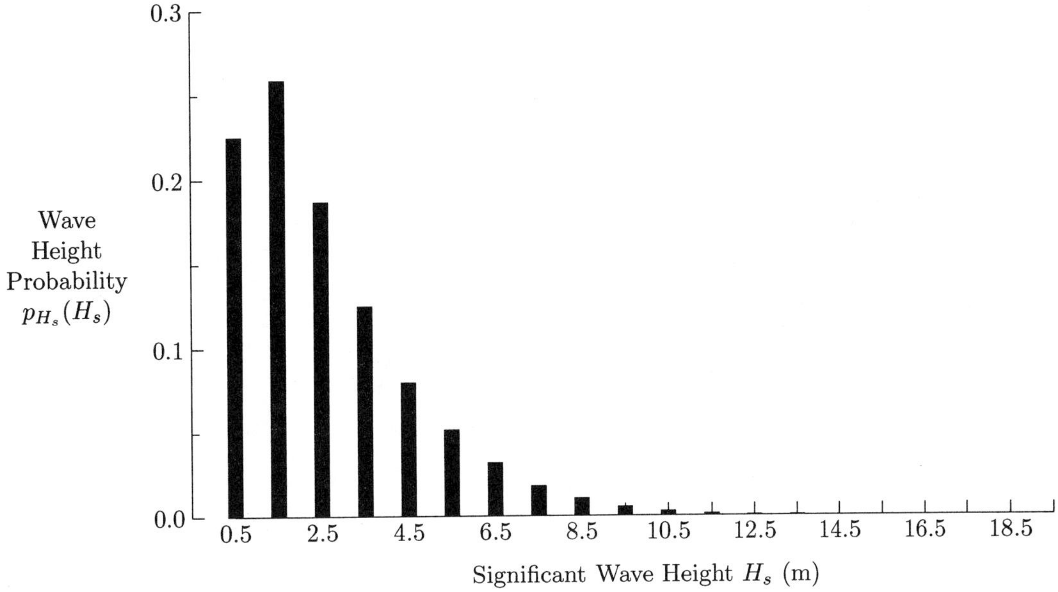

Figure 6: Distribution of Significant Wave Height for Annual North Atlantic

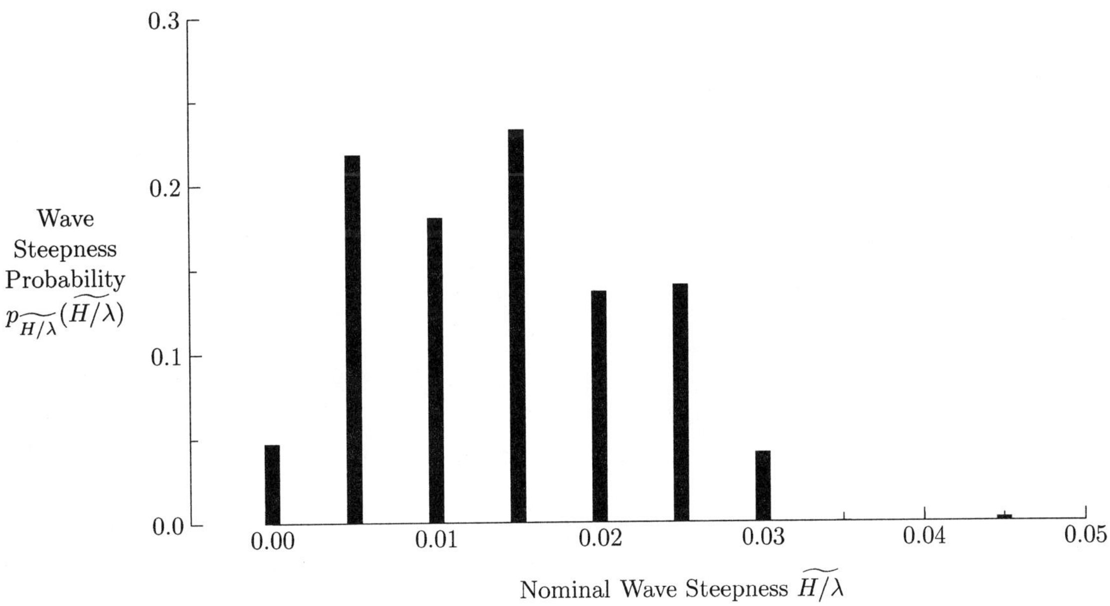

Figure 7: Distribution of Nominal Wave Steepness for Annual North Atlantic

- Ten initial simulations were performed for a given condition. If the predicted hourly capsize probability exceeded 0.001, then the number of simulations was increased to 50.

- For a given combination of ship speed, heading, and peak wave period, computations were performed for the largest wave height, followed by computations for sequentially smaller wave heights. If the hourly capsize risk for a given wave height was below 10^{-6}, then linear response was assumed for smaller wave heights, with no time domain simulations being required.

Each irregular seaway was modelled by a Bretschneider spectrum, with uni-directional waves being assumed. Winds were assumed to be steady and collinear with waves, with wind speed given by Equation (13).

The motion simulations were performed on a 300 MHz Dell Pentium II personal computer, which runs FREDYN approximately 15 times faster than real time. The total required CPU time was approximately 15 days.

Predicted Capsize Risk

Figure 8 shows hourly and annual exceedence probability versus roll angle. The annual exceedence probabilities are based on the hourly values with the assumption that the ship is at sea 33 percent of the time. Assuming uniformly distributed ship headings and that capsize occurs when maximum roll angle exceeds 70 degrees, the hourly capsize probability is 2×10^{-7}, with a corresponding annual capsize probability of 6×10^{-4}. Both the hourly and annual predicted capsize probabilities have estimated coefficients of variation of 0.21 due to sample size errors.

Figures 9 to 11 show conditional probabilities of hourly capsize given speed and heading. When all headings are considered, Figure 9 suggests that the safest ship speed is 17 knots. At the lowest ship speed of 10 knots, the ship is likely prone to loss of steering control in severe seas and subsequent broaching. The highest speed of 24 knots makes the ship vulnerable to loss of static stability stability and surf-riding. Figure 10 indicates that capsize risk is negligible for headings of 135 degrees and greater. This finding is consistent with good seamanship practice, which is to head into waves at moderate speed when conditions are severe. Figure 11 indicates that a speed of 24 knots and heading of 75 degrees is the most dangerous speed-heading combination for the present ship, although other speed-heading combinations also have relatively high risk levels.

Table 3 gives conditional hourly capsize probabilities given significant wave height and peak wave period. Note that the highest capsize probabilities are along the limiting wave heights for each wave period. Figures 12 to 14 give conditional probabilities of significant wave height, peak wave period, and nominal wave steepness given capsize. The conditional probabilities in each of these three figures have a sum of unity. Figure 12 indicates that significant wave heights of 9.5 m and larger present the greatest capsize risk. Similarly, Figure 14 suggests that a nominal wave steepnesses of 0.04 is most likely to be the cause of capsize.

Closer Examination of Most Likely Conditions to Cause

One of the most useful benefits of risk analysis is that it indicates how likely a given combination of ship speed, heading, and wave conditions is to be the cause of capsize. This likelihood is given by the following conditional probability given capsize:

$$\begin{aligned} P(V_s, \beta, T_p, H_s | C_D) \; &= \\ & p_{V_s}(V_s)\, p_\beta(\beta); p_{H_s,T_p}(H_s, T_p) \\ & \times \frac{P(C_D | V_s, \beta, T_p, H_s)}{P(C_D)} \end{aligned} \qquad (32)$$

The most likely conditions given capsize can indicate what conditions should be avoided by a ship operator. From a research perspective, the most likely conditions given capsize indicate where effort should be directed to ensure accuracy of capsize prediction methods. Table 4 gives the ten most likely sets of conditions given capsize for the example frigate. The conditions in this table account for 78 percent of the total predicted capsize risk of the ship. The mean and standard deviations of hourly maximum roll are based on Gumbel distributions fitted to the upper 30 degree range of maximum roll angles from 30 minute simulations. Where necessary, the range of maximum roll angles has been expanded to include at least ten data points from simulations.

Figure 15 gives plots of hourly exceedence probability and maximum wave slope versus maximum roll angle for the four most likely conditions given capsize. The fitted Gumbel distributions give good agreement with maximum roll angle from simulations for the upper range

The wide range of maximum roll angles shown in Figure 15 prompted further investigation into why the range of maximum roll angles can vary so greatly

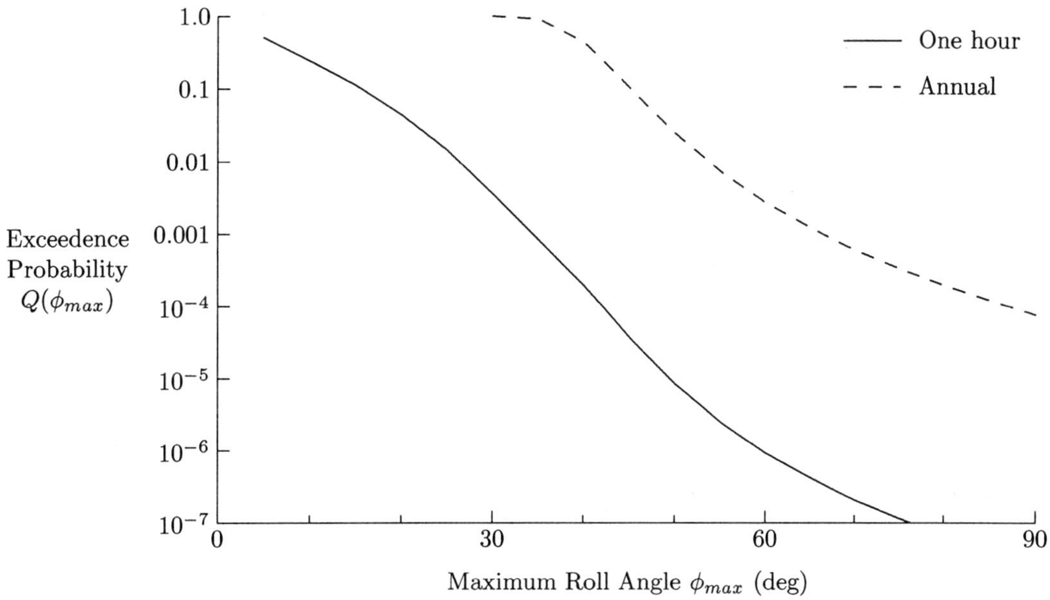

Figure 8: Hourly and Annual Exceedence Probabilities for Maximum Roll

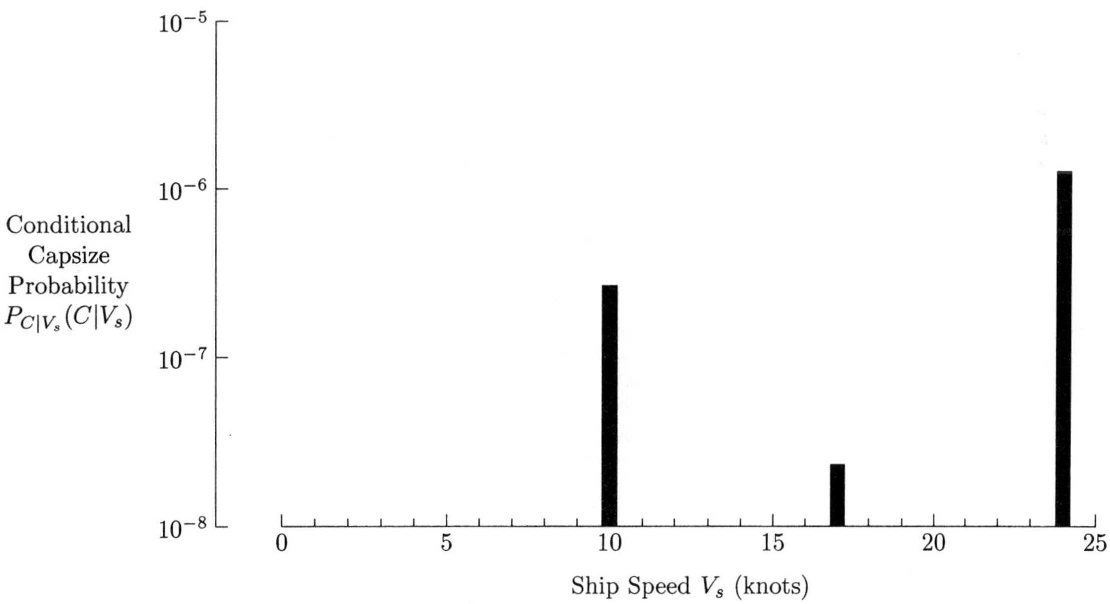

Figure 9: Conditional Hourly Probability of Capsize Given Speed

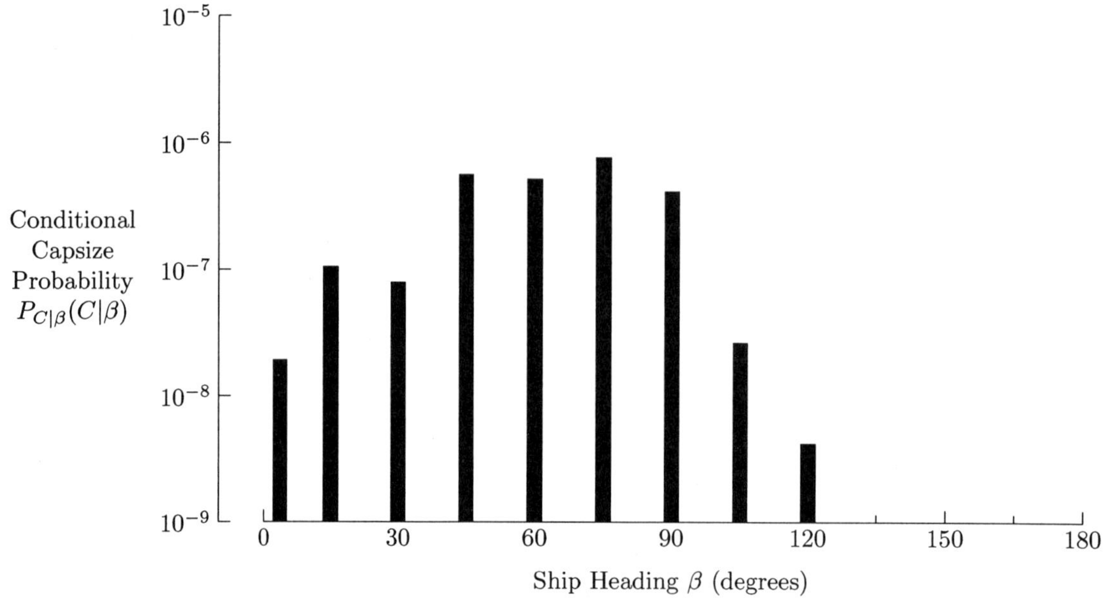

Figure 10: Conditional Hourly Probability of Capsize Given Heading

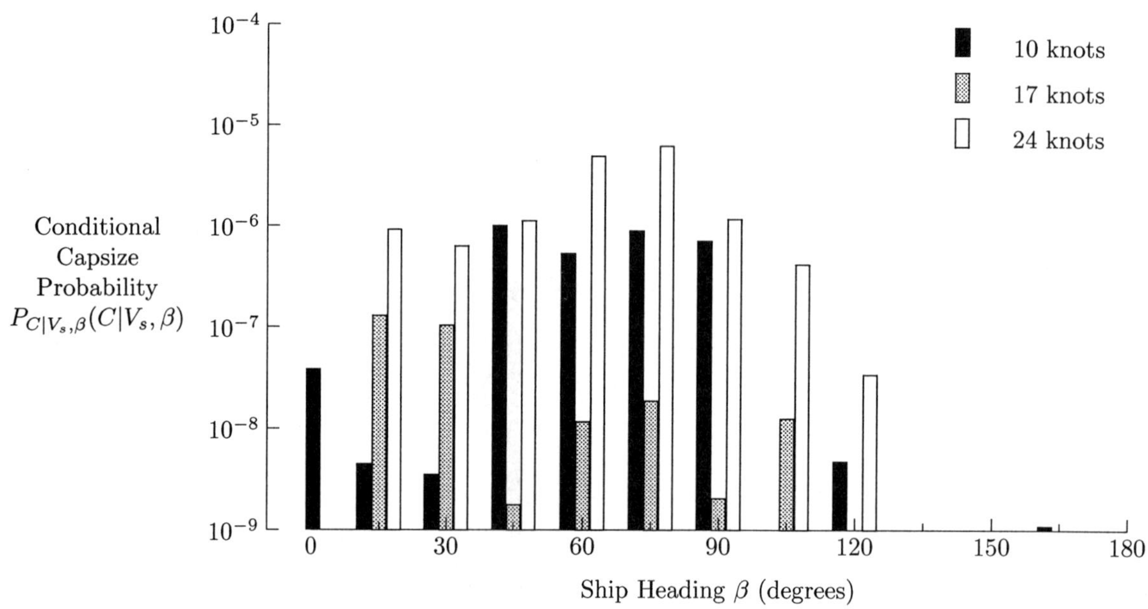

Figure 11: Conditional Hourly Probability of Capsize Given Speed and Heading

Table 3: Conditional Hourly Probability of Capsize Given Significant Wave Height and Peak Wave Period

Conditional probabilities for wave height and period combinations
Cell values are capsize probability in seaway given Hs and Tp
Blank cells indicate wave height-period combination doesn't occur
Cell value 0 indicates < 1E-9

Hs (m)	Wave Period Tp (s) 3.2	4.5	6.3	7.5	8.5	9.7	10.9	12.4	13.9	15.0	16.4	18.0	20.0	22.5	25.7
0.5	0	0	0	0	0	0	0	0	0	0	0	0	0	0	
1.5		0	0	0	0	0	0	0	0	0	0	0	0	0	
2.5			0	0	0	0	0	0	0	0	0	0	0	0	
3.5				0	0	0	0	0	0	0	0	0	0	0	
4.5					1E-9	0	0	0	0	0	0	0	0	0	
5.5						0	0	0	0	0	0	0	0	0	
6.5						3E-4	2E-6	0	0	0	0	0	0		
7.5								2E-9	0	0	0	0	0		0
8.5								8E-7	0	0	0	0	0		0
9.5								0.008	7E-9	5E-9	0	0	0	0	0
10.5									0.001	2E-8	0	0	0	0	
11.5										4E-6	5E-9	4E-9	0		3E-9
12.5											7E-8	2E-6	0		
13.5											4E-6	1E-6	0		
14.5												9E-6	3E-9	6E-7	6E-6
15.5												3E-4	4E-5	4E-7	6E-6
16.5													2E-4	7E-5	
17.5													5E-4	3E-4	
18.5															0.001
19.5															0.005

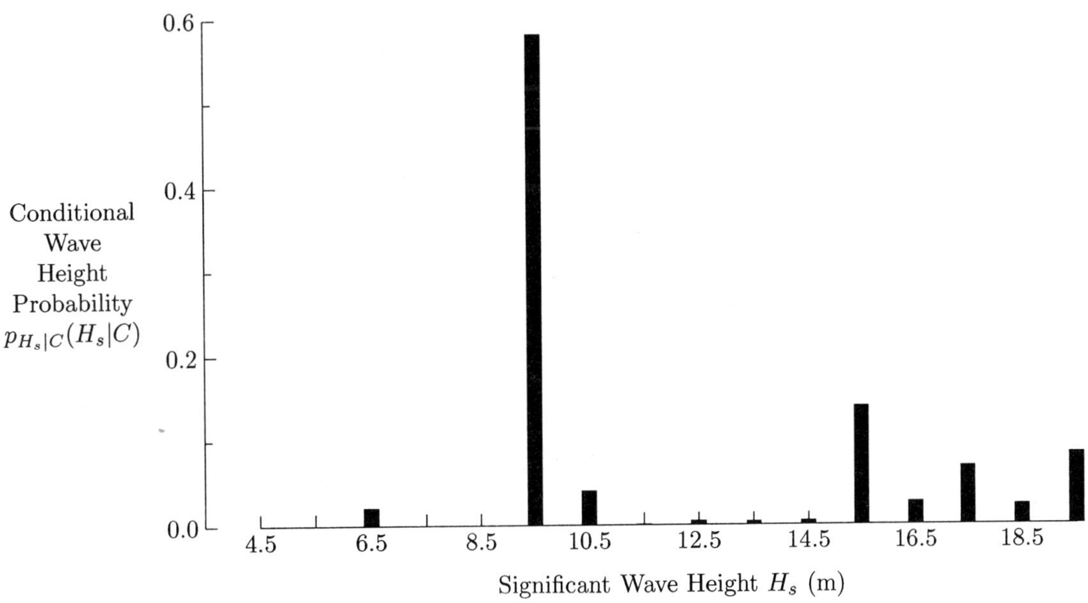

Figure 12: Conditional Probability of Significant Wave Height Given Capsize

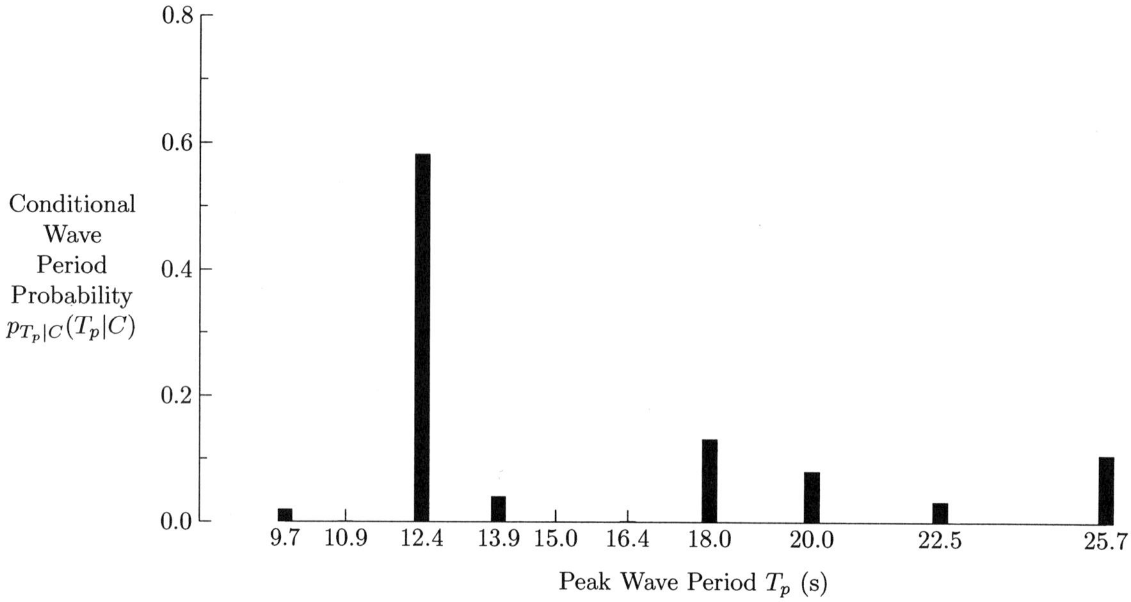

Figure 13: Conditional Probability of Peak Wave Period Given Capsize

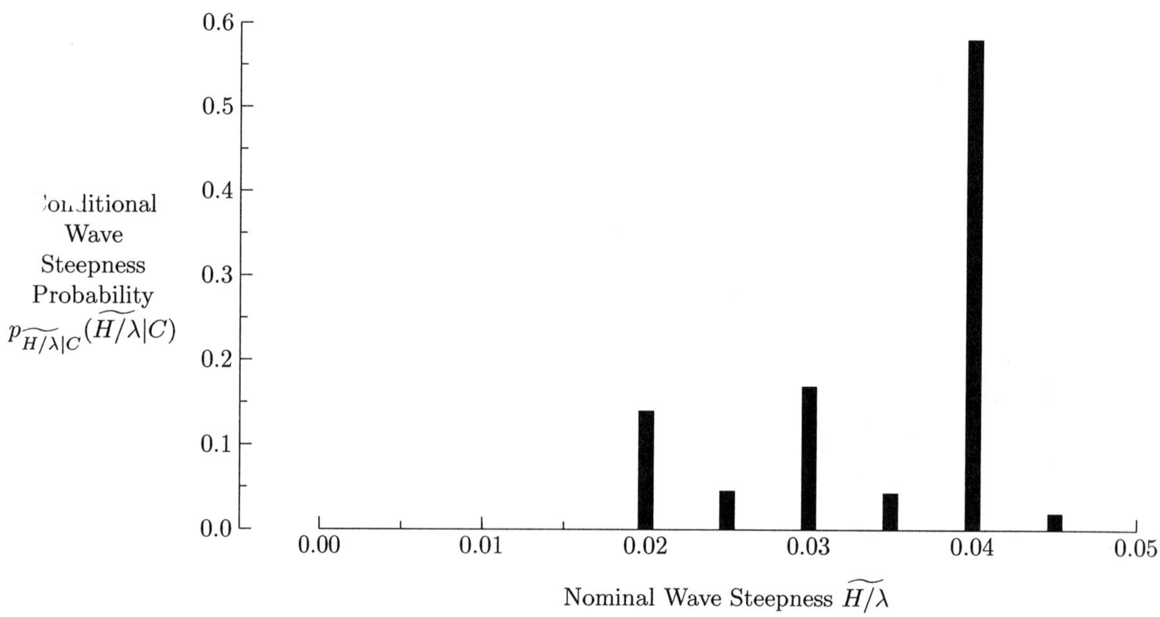

Figure 14: Conditional Probability of Nominal Wave Steepness Given Capsize

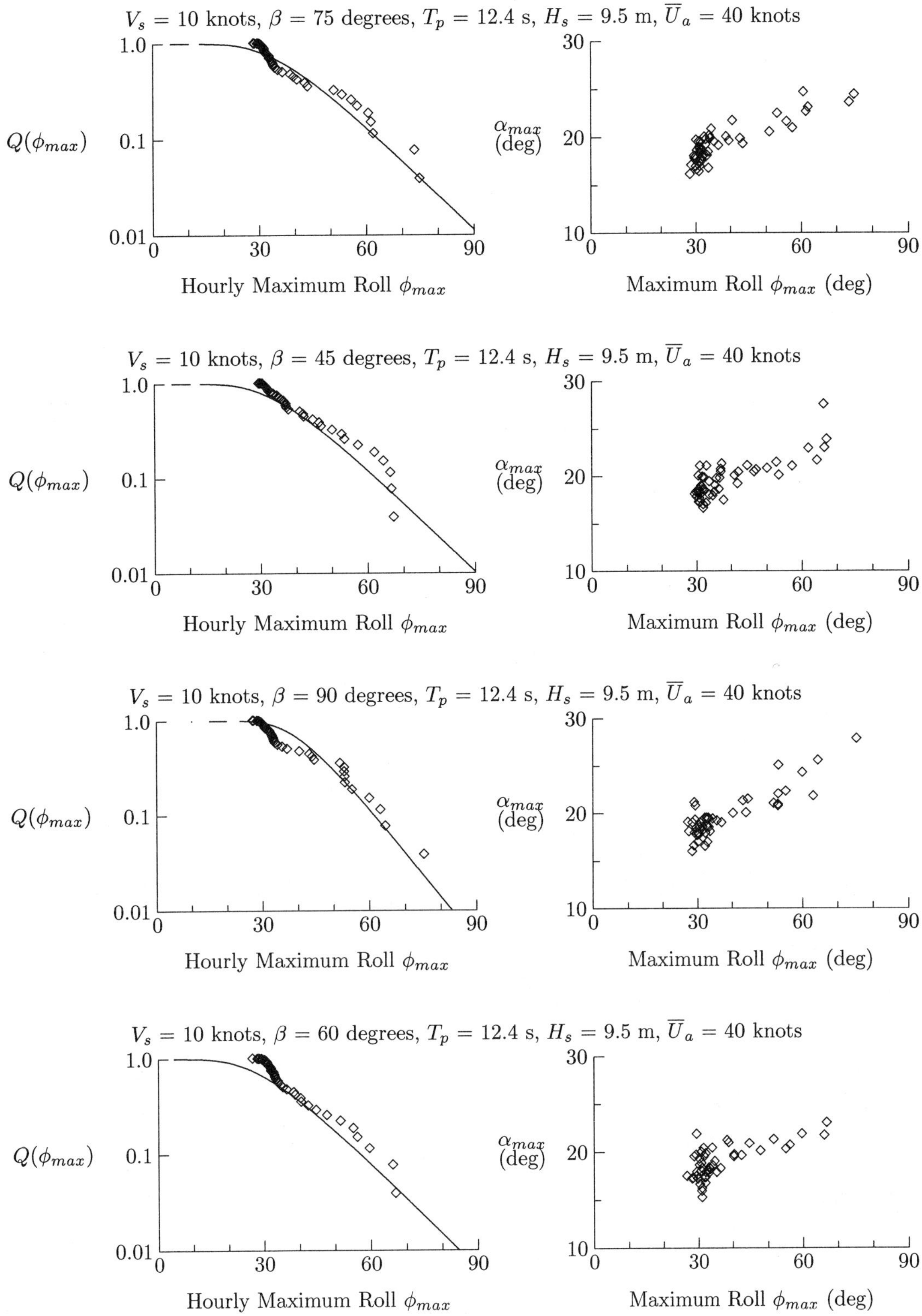

Figure 15: Hourly Exceedence Probability and Maximum Wave Slope Angle Versus Maximum Roll Angle for Most Likely Conditions Given Capsize

Table 4: Most Likely Conditions Given Capsize

$P(V_s, \beta, T_p, H_s \vert C)$	V_s (knots)	β (deg)	T_p (s)	H_s (m)	$P(C \vert V_s, \beta, T_p, H_s)$ Hourly	Hourly max roll (deg) Mean	σ
0.181	10	75	12.4	9.5	0.0587	43.3	15.4
0.166	10	45	12.4	9.5	0.0539	42.3	15.3
0.125	10	90	12.4	9.5	0.0405	45.8	11.9
0.106	10	60	12.4	9.5	0.0343	37.3	15.1
0.076	24	75	18.0	15.5	0.0380	50.1	9.5
0.036	10	45	13.9	10.5	0.0234	23.7	18.8
0.031	24	75	25.7	19.5	0.4040	69.8	3.1
0.023	24	60	18.0	15.5	0.0115	46.8	7.7
0.020	24	90	25.7	19.5	0.2614	68.8	2.5
0.017	24	60	20.0	17.5	0.0443	56.7	6.8

for simulations of given ship speed, heading, peak wave period, and significant wave height. It was found that maximum roll angle can be very dependent upon the maximum absolute wave slope angle encountered during a simulation. Figure 15 indicates that a very large maximum roll angle is usually associated with a large wave slope angle. The most interesting aspect of Figure 15 is the very high sensitivity of maximum roll angle to wave slope at higher values of α_{max}. This high sensitivity indicates that special attention should be given to accurate modelling of steep wave conditions. The present FREDYN numerical model uses simple superposition of linear waves to simulate a random seaway. For real ocean waves, wave deformations will occur at large amplitudes and will likely affect ship motions. An encouraging aspect of the results in Figure 15 is that the largest wave slope angles approach 30 degrees, the breaking limit for a fifth-order Stokes wave (see Sarpkaya and Isaacson [24]).

APPLICATION TO SHIP DLSIGN AND OPERATION

The presented risk analysis procedure is suitable for application to both ship design and operation. The general nature of the procedure means that it can be used if reliable input data and an accurate time domain ship motion model are available.

Direct Application to Design

The presented risk analysis procedure can be used to determine whether a given ship meets a prescribed acceptable level of safety against intact capsize. Table 5, citing data from Reference 25, gives some typical risks which can be used to select an appropriate target risk level. It appears that an acceptable annual risk level for ship capsize would be approximately 10^{-4}. Using Equation (14), the corresponding acceptable hourly capsize risk is 4×10^{-8} for a ship at sea 33 percent of the time.

Table 5: Annual Probability of Early Death (from Reid [25])

Cause	Annual Risk ($\times 10^{-6}$)
Air travel	9
Coal mining	210
Construction work	210
Road accidents	300
All causes, man aged 30	1000

The risk analysis presented in the previous section raises the question of how the dependence of capsize on ship heading should be considered in design. The assumption of uniform heading distribution for a ship in all conditions is overly conservative because a ship captain will go to safer headings under severe conditions; however, there is still a possibility that a ship will be operated at dangerous headings in severe waves. One possible approach is to specify one acceptable risk level that is based on a uniform distribution of headings, and a second acceptable acceptable risk level that is based on the ship maintaining a limited range of headings. For example, it could be prescribed that a ship must meet both of the following requirements:

1. For uniform distribution of headings, hourly capsize risk must be less that 10^{-5},
2. For headings within 30 degrees of head seas, hourly capsize risk must be less than 10^{-8}.

It should be emphasized that the risk levels given in the above criteria are merely initial estimates of suitable values.

A more rigorous approach to treating the influence of operator action on capsize risk is to consider ship speed and heading to be conditional upon wave conditions. For example, the following equation treats distributions of ship speed and heading to be conditional upon significant wave height:

$$\begin{aligned} P(C_D) &= \sum_{i=1}^{N_{V_s}} \sum_{j=1}^{N_\beta} \sum_{k=1}^{N_{H_s}} \sum_{l=1}^{N_{T_p}} p_{V_s|H_s}(V_{s-i}|H_{s-k}) \\ &\quad \times p_{\beta|H_s}(\beta_j|H_{s-k})\; p_{H_s,T_p}(H_{s-k},T_{p-l}) \\ &\quad \times P(C_D|V_s,\beta,H_s,T_p) \qquad (33) \end{aligned}$$

Ongoing work examining ship operational profiles will provide data for determining the conditional probability distributions $p_{V_s|H_s}(V_s|H_s)$ and $p_{\beta|H_s}(\beta|H_s)$.

As indicated previously, approximately 15 days of computer CPU time were required to perform the numerical simulations for the frigate risk example. For practical application, it is very desirable to reduce the elapsed computing time to approximately one day. Faster computer processors will continue to reduce the time required for time domain simulations of ship motions. Processors currently available for desktop computers are approximately three times as fast as the processor used in the present calculations. Parallel processing offers great potential for reducing required time for a risk analysis. Desktop computers with four processors are becoming commonly available, and could likely reduce the time required for a risk analysis by a factor of approximately three. Considering all of the above factors, it is likely that the required elapsed computer time for a capsize risk analysis could be reduced to approximately two days within the near future.

Modifications to Existing Ships

Risk analysis can assist with making decisions regarding modifications to existing ships. For warships, modifications are often proposed which would increase the height of the centre of gravity. The influence of proposed modifications on ship stability could be examined by comparative risk analysis. Results from risk analysis could also be used to identify modifications which could possibly make a ship safer. For example, the roll exceedence probabilities given in Figure 8 for the Halifax class suggest that the capsize risk level could be reduced by a factor of approximately ten by increasing the angle of downflooding from 70 to 90 degrees.

Development of Simple Design Criteria

Due to the complexity of quantitative risk analysis, it is important to have simple design criteria as an alternative. The simple design criteria must ensure that the resulting ship will achieve a target level of safety. For naval frigates, de Kat et al. [8] have proposed the following simple criteria:

1. The righting lever ($\overline{GZ}(\phi)$) in calm water must remain positive up to an angle of at least 90 degrees,
2. The positive area under the righting lever curve must exceed a specified value dependent on vertical prismatic coefficient C_{VP}. For $C_{VP} \leq 0.55$, the required positive area is 1.00 m·rad. For $C_{VP} \geq 0.70$, the required positive area is 0.67 m·rad. For intermediate values of C_{VP}, linear interpolation should be used.

Simple design criteria such as those given above are intended for a specified range of ship parameters. Application of risk analysis to a number of representative ships can ensure that the simple design criteria give adequate levels of safety.

Operator Guidance

Risk analysis results can provide very useful information to ship operators regarding which combinations of speed and heading present the greatest risk of capsize. Alman et al. [26] discuss development of operational guidelines based on time domain simulations. The risk analysis for the Halifax class frigate indicates that an operator can reduce capsize risk to negligible levels by steering a heading within 45 degrees of head seas.

Work is currently underway to provide real-time guidance to ship operators for avoidance of dangerous combinations of speed and heading. Proposed systems would combine current sea state information with a database of predicted ship motions. Figure 16 gives an example of output from a guidance system for a severe seaway with significant wave height of 9.5 m and peak wave period of 12.4 s. The shaded areas in Figure 16 represent capsize risk levels predicted for the Halifax class frigate. Despite the severity of the seaway, the operator still has a wide range of safe heading-speed combinations.

CONCLUSIONS

A procedure has been developed for determining capsize risk of intact ships in irregular seaways. The availability of a fast and accurate program for predicting ship motions makes the procedure viable for practical application. The maximum roll angle of a ship in an irregular seaway is very dependent on wave process realization. This dependence is modelled using a Gumbel fit to maximum roll angles from several different wave process realizations.

Sample calculations for a naval frigate demonstrate the application of the method. The required computational time of approximately 15 days could likely be shortened considerably using faster processors and parallel processing.

Results from risk analysis can be used directly to determine whether a ship meets a target level of safety against capsize. Other applications include development of simplified design guidelines for given ship types and provision of operational guidance for avoiding conditions likely to cause capsize.

REFERENCES

[1] D. Blockley, ed., *Engineering Safety*, McGraw-Hill, London, 1992.

[2] L.K. Kobylinski, "Rational Approach to Ship Safety Requirements," in *Second International Conference on Marine Technology* (Szczecin, Poland, 1997).

[3] A.E. Mansour, P.H. Wirsching, M.D. Lucket, A.M. Plumpton, and Y.H. Lin, "Structural Safety of Ships," *Transactions, Society of Naval Architects and Marine Engineers* **105** (1997).

[4] T.H. Sarchin and L.L. Goldberg, "Stability and Buoyancy Criteria for U.S. Naval Surface Ships," *Transactions, Society of Naval Architects and Marine Engineers* **70**, 418–458 (1962).

[5] K.A. McTaggart, "Risk Analysis of Intact Ship Capsizing," in *United States Coast Guard Vessel Stability Symposium '93* (New London, Connecticut, March 1993).

[6] K.A. McTaggart, "Capsize Risk Prediction Including Wind Effects," in *Proceedings of the Third Canadian Marine Hydrodynamics and Structures Conference* (Dartmouth, Nova Scotia, August 1995), pp. 115–124.

[7] J.O. de Kat, "The Development of Survivability Criteria Using Numerical Simulations," in *United States Coast Guard Vessel Stability Symposium '93* (New London, Connecticut, March 1993).

[8] J.O. de Kat, R. Brouwer, K. McTaggart, and L. Thomas, "Intact Ship Survivability in Extreme Waves: Criteria from a Research and Navy Perspective," in *STAB '94, Fifth International Conference on Stability of Ships and Ocean Vehicles* (Melbourne, Florida, 1994).

[9] K. McTaggart and M. Savage, "Wind Heeling Loads on a Naval Frigate," in *STAB '94, Fifth International Conference on Stability of Ships and Ocean Vehicles* (Melbourne, Florida, 1994).

[10] S.K.Chakrabarti, *Hydrodynamics of Offshore Structures*, Springer-Verlag, 1987.

[11] J.O. de Kat and W.L. Thomas III, "Extreme Rolling, Broaching, and Capsizing – Model Tests and Simulations of a Steered Ship in Waves," in *22nd Symposium on Naval Hydrodynamics* (Washington, August 1998).

[12] J.O. de Kat and W.L. Thomas III, "Broaching and Capsize Model Tests for Validation of Numerical Ship Motion Predictions," in *Fourth International Stability Workshop* (St. John's, Newfoundland, September 1998).

[13] J.O. de Kat, "Irregular Waves and Their Influence on Extreme Ship Motions," in *Twentieth Symposium on Naval Hydrodynamics* (Santa Barbara, 1994).

[14] E.A. Dahle and D. Myrhaug, "Risk Analysis Applied to Capsize of Smaller Vessels in Breaking Waves," *Transactions, Royal Institution of Naval Architects* **136**, 237–252 (1994).

[15] E.A. Dahle and D. Myrhaug, "Capsize Risk of Fishing Vessels," *Schiffstechnik (Ship Technology Research)* **43**, 164–171 (1996).

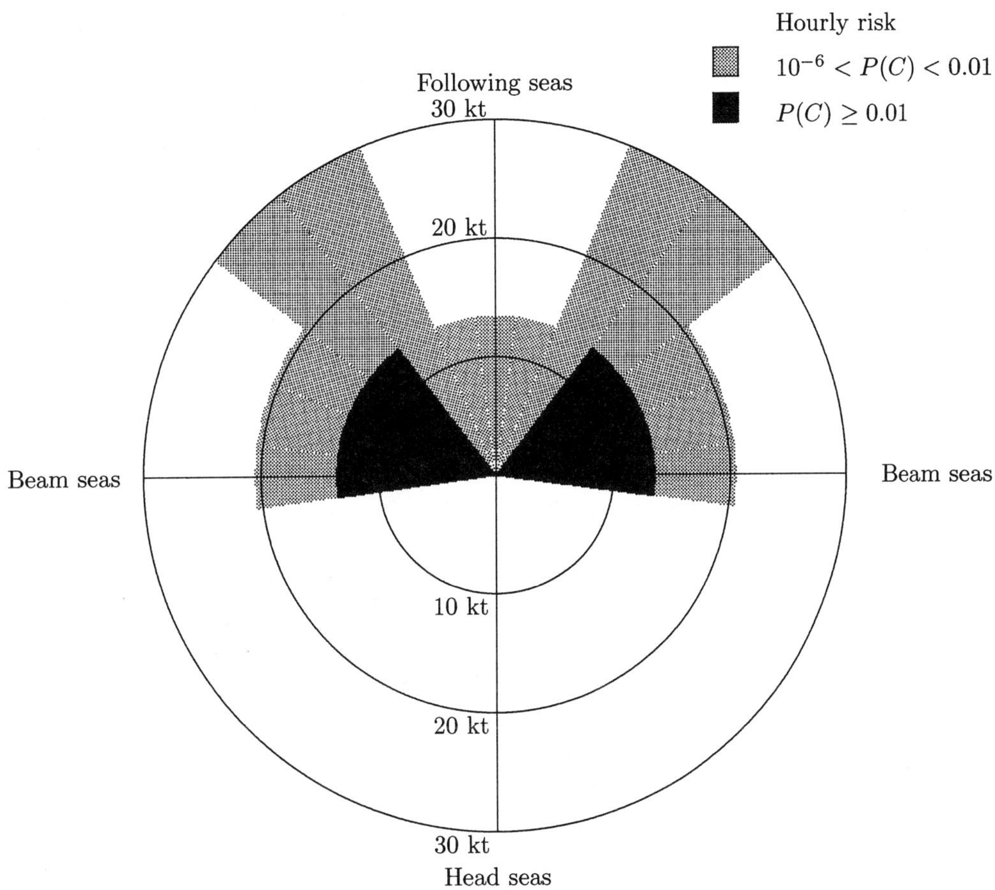

Figure 16: Hourly Capsize Risk in Severe Seaway with $H_s = 9.5$ m and $T_p = 12.4$ s

[16] S.L. Bales, W.T. Lee, and J.M. Voelker, "Standardized Wave and Wind Environments for NATO Operational Areas," Report DTNSRDC/SPD-0919-01, DTNSRDC, July 1981.

[17] K.A. McTaggart, "Ship Capsize Risk in a Seaway Using Time Domain Simulations and Fitted Gumbel Distributions," in *18th International Conference on Offshore Mechanics and Arctic Engineering - OMAE99* (St. John's, Newfoundland, July 1999).

[18] K.A. McTaggart, "Ship Capsize Risk in a Seaway Using Fitted Distributions To Roll Maxima," *Transactions of the ASME, Journal of Offshore Mechanics and Arctic Engineering* (accepted for publication).

[19] H.O. Madsen, S. Krenk, and N.C. Lind, *Methods of Structural Safety*, Prentice-Hall, Englewood Cliffs, New Jersey, 1986.

[20] P. Thoft-Christensen and M. J. Baker, *Structural Reliability and Its Applications*, Springer-Verlag, 1982.

[21] S.L. Bales, "Development and Application of a Deep Water Hindcast Wave and Wind Climatology," in *Royal Institute of Naval Architects Wave and Wind Climate World Wide Symposium* (1984).

[22] British Maritime Technology Limited, *Global Wave Statistics*, Unwin Brothers, London, 1986.

[23] W.H. Buckley, "Stability Criteria: Development of a First Principles Methodology," in *STAB '94, Fifth International Conference on Stability of Ships and Ocean Vehicles* (Melbourne, Florida, 1994), Vol. 3.

[24] T. Sarpkaya and M. Isaacson, *Mechanics of Wave Forces on Offshore Structures*, Van Nostrand Reinhold, 1981.

[25] S.G. Reid, "Acceptable Risk," in *Engineering Safety*, edited by D. Blockley (McGraw-Hill, London, 1992).

[26] P.R. Alman, P.V. Minnick, R. Sheinberg, and W.L. Thomas III, "Dynamic Capsize Vulnerability: Reducing the Hidden Operational Risk," *Transactions, Society of Naval Architects and Marine Engineers* **107** (1999).

Discussion

Neil Hogben, Member

Mr. Rowe has mentioned in his discussion comparisons I have made between BMT and Buckley data for the relation between wave height and period.

Most recently, in seeking to explain the steepness comparison in Figure 3 of the present paper, which I also believe needs clarification, I derived the results shown in Figure 1 contained herein. These show that maximum steepness levels based on cut off values of H_s in the book *Global Wave Statistics,* which correspond to a probability level of about 0.001 agree quite well with the Buckley and Bales data in Figure 3 over most of the period range.

In 1995 from a correspondence with Bill Buckley I discovered close agreement between Hogben (BMT) and Buckley formulae for the relation between wave height and period. The findings have been summarized by Faulkner & Buckley in reference [1], and I draw attention specially to Figure 4 of that reference comparing H_s–T_p relations for various types of sea. Table 1 below shows the close agreement between a formula used in *Global Wave Statistics* as a basis for modeling the H_s–T_z scatter in open ocean areas (including area 15) and a corresponding Buckley formula for "Long Seas."

Additional references

[1] D. Faulkner, W.H. Buckley, *Critical Survival Conditions for Ship Design,* Proceedings of RINA Conference on Design for Abnormal Conditions, Glasgow, 1997.

[2] N. Hogben, *Overview of Global Wave Statistics,* Encyclopedia of Fluid Mechanics, Volume 10, Chapter 11, Gulf Publishing, Houston, 1990.

Table 1 Comparison of BMT and Buckley Formulae for the relation of wave height and period

	T_p(s)	
H_s (m)	BMT (Open Ocean)[1]	Buckley (Long Seas)[2]
1	8.62	8.50
2	10.59	10.49
3	11.95	11.86
4	13.02	12.94
5	13.91	13.85
6	14.68	14.63
7	15.37	15.33
8	15.99	15.97
9	16.56	16.55
10	17.09	17.08

1. Derived from formula for a mean line used in modeling the H_s–T_z scatter in open ocean areas (including area 15) of *Global Wave Statistics* [2], namely:

$$T_p = 1.4\ T_z = 1.4\ e^A\ H_s^B$$

where A = 1.818 and B = 0.297 in open ocean areas.

2. Derived from Buckley formula for "long seas" (ref. 1):

$$T_p = (a\ H_s)^{1/b}$$

where a = 1168 and b = 3.3 for "long seas."

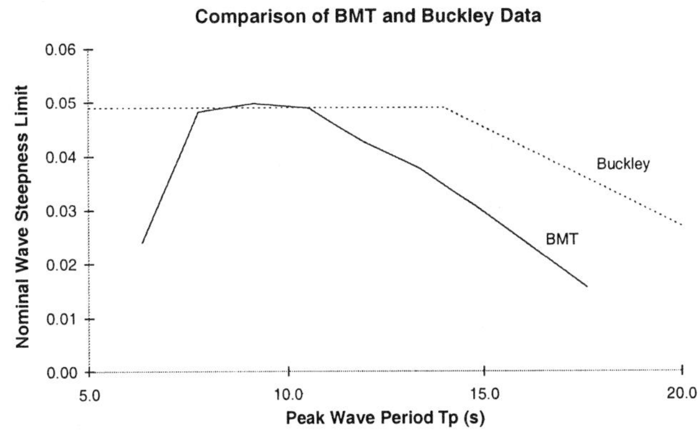

Fig. 1 Steepness Limits for BMT Area 15 (All Seasons—All Directions) (From the book *Global Wave Statistics,* and Buckley Northern Hemisphere)

Stephen J. Rowe, Member

The authors suggest (section "Sources of Wave Statistics," and Figure 3) that *PC Global Wave Statistics* gives unrealistically high wave steepness values. This is at variance with BMT experience, and we believe the data presented in Figure 3 needs further explanation.

Following a discussion with one of the authors of this paper (Kevin McTaggart), it would appear that the curve in Figure 3 labeled BMT was derived from the *PC Global Wave Statistics* data transfer file, taking for each period class the highest wave height with a non-zero value in this file. This transfer file expresses the probability of each height-period class as a real number, with 6 places of decimals. This is in effect taking the maximum wave height to be the highest class with a probability greater than 10^{-6}.

The height-period scatter tables in *Global Wave Statistics* (both the book and PC version), are derived from analytical distribution functions which have been best fitted to the main body of the scatter tables to provide smoothing and reliability enhancement (references [1–3]). These distributions were not optimized to represent the extreme tails of the distribution, and they give non-zero (but exponentially small) probabilities up to infinite wave height. In practice, physical processes would limit the maximum wave height to a finite value. The effective cut off of 10^{-6} used to generate Figure 3 is far into the tail of the distribution, and we consider that the corresponding maximum wave heights are unrealistically high.

Dr. Hogben has shown, as reported in his discussion, that using the scatter tables in the *Global Wave Statistics* book, with a cut off of wave height at a probability of about 10^{-3}, steepness limits for BMT Area 15 agree quite closely with the Buckley and Bales data in Figure 3.

Dr. Hogben also refers to findings from correspondence with Buckley (summarized in ref. [4]) showing close agreement between the Hogben (*Global Wave Statistics*) and Buckley formulae for the relationship between wave height and period for various types of sea. In particular, he cites the formula used as a basis for modeling the height-period scatter in open ocean areas (including area 15) of *Global Wave Statistics*, and shows that it is virtually identical to a corresponding Buckley formula for "Long Seas."

Consequently, we are confident that in general the zero crossing period values given in *Global Wave Statistics* are good, and it would be unfortunate if this paper gave the impression that all *Global Wave Statistics* periods are too short and the waves are too steep.

For the particular problem of vessel capsize described in this paper, it is the very extreme areas of the distribution, with very low probabilities of occurrence, which combine in the simulation with a near certainty of ship capsize in these conditions to give rise to much of the overall risk of capsize. In the case of *Global Wave Statistics* these probabilities are not particularly reliable owing to the fitting process (which was not primarily intended to fit the tail). In the case of the hindcast data from SOWM the probabilities at this level will often be identically zero because the hindcast did not generate any cases of this wave height, but this is also potentially incorrect because the hindcast was not run for long enough to generate a reliable distribution tail at these probability levels. It is a difficult problem!

Additional references

[1] N.M.C. Dacunha, N. Hogben, K.S. Andrews, "Wave Climate Synthesis Worldwide", Proceedings of International Symposium on Wave and Wind Climate Worldwide, RINA, London, April 1984.

[2] N. Hogben, N.M.C Dacunha, G.F. Olliver, "Global Wave Statistics", British Maritime Technology Limited, 1986.

[3] S.J. Rowe, "PC Global Wave Statistics—User Manual", BMT Fluid Mechanics Limited, 1988.

[4] D. Faulkner, W.H. Buckley, "Critical Survival Conditions for Ship Design", Proceedings of RINA Conference on Design for Abnormal Conditions, Glasgow, 1997.

C. Kuo, Member

The authors have presented a method of matching a ship's environmental operational conditions with time domain simulation of its motions in order to identify situations that may lead to capsize in intact conditions. The computing effort is certainly impressive. It is difficult to discuss the details and I shall therefore confine myself to asking two questions and to make a plea from the standpoint of an educator. These would be considered under the following headings:

(A) Comparing risk levels of different activities

There is a tendency to use accident statistics to compare activities which have little or no similarities, e.g., in Table 5, Air Transport and Coal Mining. Indeed, the values vary so much as to make them not very meaningful. It should be remembered that even to compare aircraft failure statistics with shipping failures can be misleading. The airline industry uses "passenger—kilometer" as a measure without duration of exposure being considered, but critical times are landings and takeoffs which are of short duration. The airline industry compares very favorably with shipping because ships travel slower and generally over shorter distances. In the present study, would it not be better to use marine activities as a reference, e.g., fishing, supplying the offshore drilling operations, etc?

(B) Operation guidance

The authors have suggested that useful information can be provided to ship operators, i.e., those on the bridge of the ship, regarding "dangerous" combinations of ship speeds and headings. I would like to know in what form would this information be provided, e.g., by hardware or alarm? In general, the "dangerous" conditions occur in adverse weather conditions and the use of graphs or tables can be counterproductive. An experienced ship captain uses his experience to avoid these conditions and any guidance must then be both easy to use and acts as a "second opinion" to the user.

(C) Plea on the use of risk terminology

My concern is with uses of the terms associated with "risk." Risk is not well understood and special care should be exercised in its usage so as to avoid general confusion. I wish to draw the authors attention to the following aspects:

Capsize is a hazard and a hazard is often defined as "potential to cause harm." Capsize can lead to loss of life and the ship, and thus fits the term "hazard" well. Risk (R) measures the significance of hazard and is made up of two parameters of consequence (C) of a hazard being realized and probability P of the hazard occurring, i.e.,

$$R = f\ (C,P)$$

It is usual to classify the significance of risk in the intolerable, tolerable and negligible risk regions. Risk assessment is the process which locates the hazards in the various regions and this is done by qualitative and quantitative methods. For further details please see my book entitled *Managing Ship Safety,* published by LLP Ltd (ISBN 1-85978-841-6).

In this paper, risk has been used to mean the same as hazard, e.g., "capsize risk" would be a lay person's loose use of the term. "Risk analysis" is used many times when "risk assessment" might be better. I hope the authors will take on board this constructive plea because I will be very pleased to see risk-based methods being applied to many areas of naval architecture.

Andrew Kendrick, Member, and **Aaron Dinovitzer,** Member

The authors' work provides an interesting quantification of the relative risks of capsize under a range of operating conditions. In its current form it has the potential to offer valuable operational guidance and to be used as a training aid for seafarers. In the future, work of this type may offer the opportunity to improve current ship stability criteria. However, there would appear to be a long haul ahead before this ultimate goal can be achieved. In particular, target or acceptable risk levels need to be established prior to the widespread application of risk-based analysis. The comments below are intended to offer some constructive suggestions on how future research efforts might be prioritized.

The intact stability criteria most commonly used in Western navies for frigate and destroyer design are based on those developed by Sarchin and Goldberg, as vividly described in a paper from last year's *Transactions*.[1] No vessel designed on this basis has ever capsized. A very coarse estimate based on the number of ship-years experience gained with these criteria suggests an individual annual risk level of below 10^{-5} (assuming a frigate actually capsized tomorrow). Therefore, in general, any revisiting of design criteria should assume that their overall safety level is more than adequate, at least for conventional ship types. However, a case could be made that they might be overly conservative.

A logical use of the methodology presented in the paper would be to conduct sensitivity studies on the stability parameters now in use. This would reveal which could be relaxed without affecting the outcomes significantly, and which, on the other hand, might be tightened in order to provide significant safety benefits. An example proposed in the paper is to require a positive righting arm at up to 90 degrees of heel. This may well have considerable stability benefits, but would also be a source of major layout difficulties for many current and future designs. Stability criteria must be considered in the context of other drivers of ship design.

The paper makes it clear that capsize probability is influenced strongly by the sea state parameters, a factor that has been highlighted in other studies of capsize.[2] For smaller vessels, which provide most of our recent practical examples, capsize has often been associated with breaking waves in relatively confused sea conditions, which may not fit into the spectral assumptions used in the authors' work. The importance of gathering better statistics on extreme seas has been emphasized by several recent studies (ISSC 97), and would appear to be a necessary component of any attempt to develop true risk-based capsize criteria.

Similarly, more attention needs to be given to the human factors elements of risk. The paper notes that operators will typically adjust heading in severe sea conditions. It is important to know which elements of the situation dominate this response, both to calibrate the risk models and to know which elements of operator guidance to focus on. This may be particularly valuable in avoiding broaching situations, where the operator may not easily perceive the level of risk.

A final comment is that the main stability design driver for most warships is damaged, rather than intact stability. It would be interesting to hear the authors' views on how damage stability can be included in a set of risk-based criteria, or whether they will always have to provide a deterministic overlay to any system.

Additional references

1. Alman, P. et. al., "Dynamic Capsize Vulnerability: Reducing the Hidden Operational Risk" Transactions, SNAME, 1999.
2. Morrall, A. "Capsizing of Small Trawlers" *The Naval Architect*, March 1980.

William H. Buckley, Member

My comments are strongly influenced by the fact that I am a proponent of a much more deterministic approach to capsizing matters than the one employed by the authors. This approach arises from the fact that I find that the highly nonlinear behavior of ships in extreme seas serves to permit the identification of critical (i.e., deterministic) design conditions, whereas the reverse is true of ship motion responses which are essentially linear. Having acknowledged my personal bias on this subject I have the following comments and questions.

Physics of capsize for naval frigates

The authors' description of potential capsize modes is of considerable value because of its comprehensive nature and the extent of the supporting research involved in preparing it. I believe it serves to identify the daunting range of seaway conditions which must be investigated to assure freedom from capsize in extreme seas. In order to help avoid being overwhelmed, however, I have proceeded to define a range of extreme values of H_s versus T_p together with the associated wave spectra (i.e., an envelope of Survivability seaway conditions) for which freedom from capsizing should be demonstrated. (See Faulkner & Buckley [1] section 2.2.) Never having been to sea in such conditions, I have turned to published and other sources of information from commercial ship masters who have encountered them, e.g., see Buckley [2], from which I have drawn the following conclusions as to ship handling in extreme astern seas:

1. The masters proceeded down seas as slowly as they could and while still maintaining a 90° heading to the wave crests. This was done to minimize the amount of time spent with the crest of large, steep waves near midships when lateral stability is at a minimum and to minimize rolling at the same time.
2. Their ability to maintain heading was in direct conflict with their desire to proceed at low speed. Moreover, excessive speed could have led to surfing and subsequent broaching.
3. Because of these conflicting requirements there were occasions when the desired heading could not be closely maintained. Thus some misalignment must be considered for design purposes.

Based upon these considerations a critical design condition for which freedom from capsize is required is proposed in Buckley [2].This condition requires that the most severely elevated, breaking wave in time series wave height data from a hurricane be considered. These observations lead to my first question. Can your time domain simulation (which does involve the needed control derivatives) be employed to analyze frigate or other ship motions resulting from such a wave encounter?

Sources of Wave Statistics

In this section of the paper the authors use a nominal wave steepness parameter ($H/\lambda = 2\pi \ Hs/gT^2$) which I find unsettling. I quite agree that the value 0.049 can be derived from my seaway's of limiting boundary. However, I have avoided using this parameter because it does not identify the critically steep, breaking wave cited above which is likely to result in severe ship motions as a result of its encounter. A method for identifying such potentially critical wave height events is described in Buckley [3], Sections 3.0 and 4.0. The same method is recommended for analyzing frill scale and/or model scale motion data to identify the nonlinearity of the data and to identify any associated extreme motion events.

I agree with the authors' statements that BMT wave steepness statistics tend to be unrealistic, but with one notable exception. Upon comparing BMT and NOAA buoy data for concurrent ocean areas, those areas which correspond to climatic steep seas and those which correspond to climatic long period seas were found to be consistent with one another. In other words, while BMT steepness values did not agree well with buoy data, those ocean areas which were climatically steep and those which were climatically long period in each data set were in agreement. The Bales hindcast climatology data which the authors recommend using for evaluating wave steepness did not agree well with buoy data in regard to their probability distributions of Hs in concurrent ocean areas. On the other hand, with the exception of the Gulf of Mexico, the BMT and NOAA buoy Hs probability distributions were in good agreement. See Buckley [4, 5] for these comparisons.

All of this leads to my second question. Why have the authors apparently not utilized NOAA buoy wave data? The enormous number of measured wave spectra available from a wide variety of wave climates is unparalleled. On the other hand, the utilization of "hindcast" wave data (i.e., Bales data) was undertaken because NOAA buoy wave data were not available at the time this methodology was proposed. See Anon [6]. This situation has since changed markedly.

My last question relates to the identification of critical wave conditions and to the validation of methods for determining ship responses in such wave conditions. First, I note that:

a. Ship motions in extreme seas cannot be investigated full scale in natural environments for a number of reasons, e.g., rarity, safety, unsteady conditions, etc. As a result it must be done in seakeeping tanks at model scale using suitable wave-making equipment.

b. While it is recognized that ultimately computer simulations will offer quicker and less expensive results than tank testing, such simulations must be validated for those *critical wave conditions* likely to cause capsizing in the various modes identified in the paper. *Computer simulations cannot be considered a reliable option in the absence of such validation.* My last question then concerns the wave-making capabilities of the two facilities with which the authors are familiar, i.e., the Offshore Engineering Basin in St. John's, Newfoundland, and the MARIN Seakeeping Tank facility in Wageningen, The Netherlands. Can either or both of these facilities replicate the Survivability envelope wave spectra associated with Figures 6 and 7 and Appendix A of Buckley [7], and if so, at what scale?

References

1. Faulkner, D. and W.H. Buckley, "Critical Survival Conditions for Ship Design", RINA Conf., Design and Operation for Abnormal Conditions, 21–22 October 1997.
2. Buckley, W.H., "Critical Capsizing Conditions in Survivability Seaway's", RiNA ConE, Design and Operation for Abnormal Conditions, Glasgow, 21–22 October 1997.
3. Buckley, W.H., "Analysis of Wave Characteristics in Extreme Seas", Ship Structure Committee Report SSC-353, 1991.
4. Buckley, W.H., "Extreme and Climatic Wave Spectra for Use in the Structural Design of Ships", Naval Engineers Journal, September, 1988.
5. Buckley, W.H., "Matching Vehicle Characteristics to Seaway Environments", Proceedings of ASNE Conference on High Performance Vehicles, 24-27 June 1992. (pp. O/A83–91).
6. Anon, "Seakeeping in the Ship Design Process", Prepared jointly by the Naval Ship Engineering Center and the Naval Ship Research and Development Center, 1975.
7. Ref. (23) of the Paper.

Pierre C. Sames, Visitor, and **Thomas E. Schellin,** Member

We congratulate the authors for their fine work. They tackled a challenging subject of practical significance and showed that considerable progress has been achieved in the area of predicting phenomena that may lead to capsizing. First, we have one question. When determining risk, is it not correct to examine the product of the probability of occurrence and the consequence for each event in terms of monetary units or human life? Only this product yields the risk.

The authors employed a Gumbel fit to extreme roll angles to predict the exceedence probability of a predefined critical roll angle. For vessels that do not have a well defined downflooding angle, the determination of a critical roll angle that causes capsize may be unreliable. In addition, this kind of an analysis removes all aspects of dynamic ship behavior in waves at extreme roll angles, and the results depend on the range of roll amplitudes used for the Gumbel fit. The simple criterion of (15) appears to be more realistic, specially if more simulations can be performed for a given sea state.

The authors outlined additional steps to improve the predicted accuracy, e.g., the introduction of a conditional probability for ship speed on wave height. Indeed, such an approach was selected in the long-term prediction of bow door loads for a RO-RO passenger ferry within a European research project.[1] Speed reduction was estimated based on additional resistance in waves and an assumed minimum speed in storm conditions. The resulting distribution for ship speed depending on significant wave height is presented in our Table 1 below. It should be emphasized that ship speed was assumed to be independent on wave period.

Performing a long-term statistical analysis based on ship speed dependent response functions yielded drastic reductions of long-term values compared to values obtained from the standard approach with constant ship speed equal to 2/3 service

Table 1 Ship speed distributions as a function of wave height

v [m/s]	Zero upcrossing period T_z [s]										
Sig. Wave Height H_s [m]	3.5	4.5	5.5	6.5	7.5	8.5	9.5	10.5	11.5	12.5	13.5
14.5	2.50	2.50	2.50	2.50	2.50	2.50	2.50	2.50	2.50	2.50	2.50
13.5	2.50	2.50	2.50	2.50	2.50	2.50	2.50	2.50	2.50	2.50	2.50
12.5	2.50	2.50	2.50	2.50	2.50	2.50	2.50	2.50	2.50	2.50	2.50
11.5	2.50	2.50	2.50	2.50	2.50	2.50	2.50	2.50	2.50	2.50	2.50
10.5	2.50	2.50	2.50	2.50	2.50	2.50	2.50	2.50	2.50	2.50	2.50
9.5	3.10	3.10	3.10	3.10	3.10	3.10	3.10	3.10	3.10	3.10	3.10
8.5	4.40	4.40	4.40	4.40	4.40	4.40	4.40	4.40	4.40	4.40	4.40
7.5	5.60	5.60	5.60	5.60	5.60	5.60	5.60	5.60	5.60	5.60	5.60
6.5	6.95	6.95	6.95	6.95	6.95	6.95	6.95	6.95	6.95	6.95	6.95
5.5	9.35	9.35	9.35	9.35	9.35	9.35	9.35	9.35	9.35	9.35	9.35
4.5	11.25	11.25	11.25	11.25	11.25	11.25	11.25	11.25	11.25	11.25	11.25
3.5	12.50	12.50	12.50	12.50	12.50	12.50	12.50	12.50	12.50	12.50	12.50
2.5	13.50	13.50	13.50	13.50	13.50	13.50	13.50	13.50	13.50	13.50	13.50
1.5	13.50	13.50	13.50	13.50	13.50	13.50	13.50	13.50	13.50	13.50	13.50
0.5	13.50	13.50	13.50	13.50	13.50	13.50	13.50	13.50	13.50	13.50	13.50

speed. This predicted long-term response is shown in our Figure 1 for the sample RO-RO ferry. It may be of interest to note that the average speed of the vessel based on these conditional ship speed distributions turned out to be 96 percent of the ship's service speed.

Upon closer examination of the probabilities of capsize predicted by the authors, we noticed that certain wave conditions had a dominant effect on the predicted probabilities. For their chosen frigate, results show that a single sea state contributed more than 50 percent to the total probability of capsize, compare Table 4. Is this generally so? Why does this sea state, characterized by H_S = 9.5 m and T_P = 12.4 s, dominate the results? Also, the resulting hourly maximum roll angles do not seem to be too extreme when compared with roll angles in other sea states, e.g., for H_S = 19.5 m and T_P = 25.7 s.

Table 3 shows that a sea state characterized by H_S = 6.5 m and T_P = 9.7 s yielded an hourly probability of capsize of 3.e–4, equal to the hourly probability of capsize for the sea state with H_S = 15.5 m and T_P = 18.0 s. What is the significant difference between these two sea states that justified that only one of these sea states was listed in Table 4, showing the most likely conditions given capsize?

The authors state that seakeeping analyses can provide useful information to ship operators in knowing which combination of speed and heading presents the greatest risk of capsize. Systems based on presenting real-time guidance to ship operators for avoiding dangerous combinations of speed and heading by combining current sea state information with a database of predicted ship motions are promising. Indeed, design concepts for such a system are currently being developed [2]. One of the problems is to supply a sufficiently comprehensive database of predicted ship motions. It would be interesting if the authors have some ideas on how such a database should be developed; specifically, what sea conditions, wave headings, etc. should be analyzed? Will the ship's response in regular waves be significant in this respect? If the ship's response in natural seaways is to be computed, how should such a seaway be defined and what are the important parameters that need to be considered? Can the authors point out principles towards a rational approach in this regard?

Additional references

1. Work program of DEXTREMEL (1997), "Design for Structural Safety under Extreme Loads," Brite Euram III, project BRPR-CT97-0513, http://research.germanlloyd.org/Proiects/DEXTREMEL/index.html
2. Germanischer Lloyd research project GL-SSISS (1998), "Safe Shipping in Heavy Seas," German Ministry of Education, Science, Research, and Technology (BMBF), contract MTK 06129.

Vadim L. Belenky, Member

The authors are to be congratulated for a really excellent paper. The subject of the research is very important: real seaway requires probabilistic methods to be applied for dynamic stability problem.

I would like to comment on some key points in this paper.

The authors connect probability with the time of exposure to adverse conditions. This is very important because if there is no explicit time dependence in the capsizing model, all probabilities correspond to infinite time "by default."

The authors found that amplitude of rolling depends on seed value of random phase generator. In other words, roll response is sensible to initial conditions. This is one of the fundamental nonlinear qualities that might lead to possible absence of ergodicity or cyclic non-stationary. It is not clear yet, if the process simulated by FREDYN is non-ergodic or it just had cyclic nonstationarity. Anyway, one realization, even long enough cannot be used for statistical calculations. So, this could be considered as one more confirmation of practical non-ergodicity of nonlinear rolling.

Another issue is a method for estimation of accuracy of

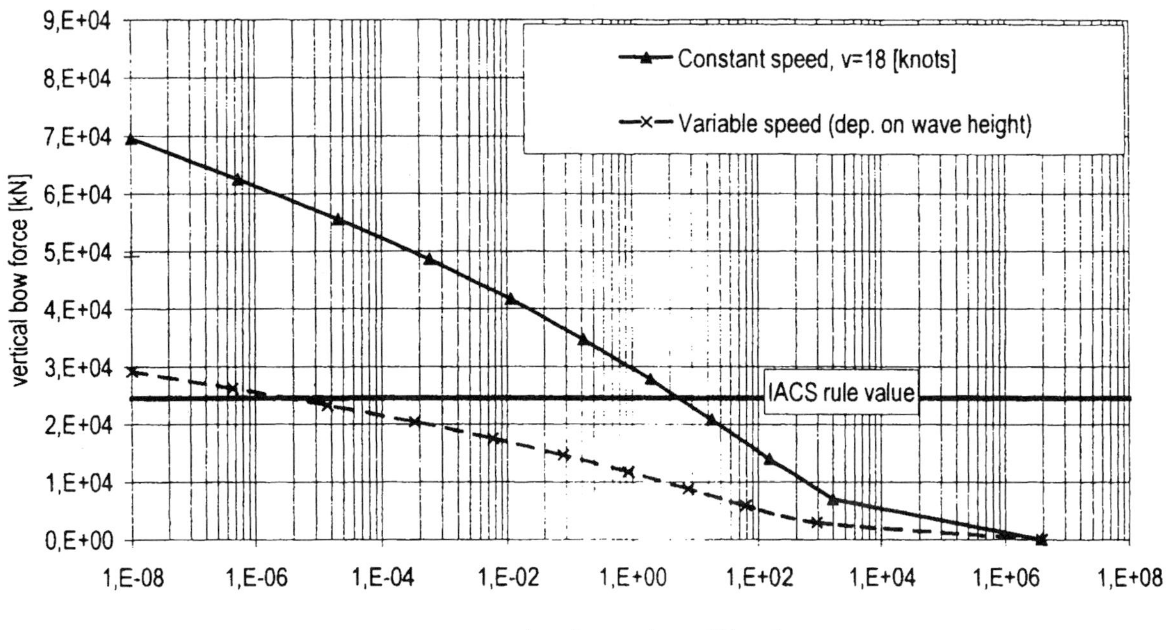

Fig. 1 Ship speed dependent long-term values of predicted bow door loads

Gumbel fitting. It is very important, because a capsizing is a very rare event (up to 10^{-6} per hour); also, statistics of large roll angle are quite small, so without proper error control it is very easy to make a serious mistake, and there is no warranty that it will be done on the safe side.

Authors' Closure

We are very pleased with the level of interest that our paper has generated. The discussers have made many comments that we feel will enhance our future efforts.

Mr. Buckley, who favors a deterministic approach, proposes that a critical design condition be selected for which freedom from capsize is required. Due to the many possible modes of ship capsize, we feel it is hard to know *a priori* what the critical design condition should be for a given ship. A probabilistic approach has the advantage of considering a variety of relevant operating conditions. To assess the extreme motion dynamics of a ship, one could envisage the combination of probabilistic calculations to identify critical conditions, followed by experimental assessment in the most onerous conditions. This could entail deterministic testing in pre-selected (conditioned) wave trains.

The behavior in breaking waves is an important issue that deserves in-depth scrutiny. At the moment no mathematical model is capable of handling the complete nonlinear ship dynamics in breaking waves. We agree with Mr. Buckley that wave buoy data provide the best source of wave climate information. Utilization of such data will require significant effort to synthesize the data into wave climate scattergrams.

Physical model experiments remain essential for validating numerical models of ship capsize in waves. Wave generation facilities at MARIN and the Institute for Marine Dynamics in St. John's, Newfoundland, are able to simulate wave conditions on Mr. Buckley's limiting steepness envelope to maximum significant wave heights of 0.45 m and 0.60 m, respectively. For a maximum significant wave height of 18 m full-scale, the corresponding maximum model scales are 1/40 and 1/30.

The discussions from *Dr. Hogben* and *Mr. Rowe* stem primarily from Figure 3, which suggests that BMT *Global Wave Statistics* gives unrealistically high wave steepnesses for the western portion of the North Atlantic. This figure prompted a very useful dialogue between ourselves and the discussers, who indicate that the unrealistically high steepnesses are caused by a statistical fit procedure with an unbounded upper tail for wave height given wave period. This explanation gives us greater confidence in the underlying wave data for *Global Wave Statistics*. As users of the data, we would be very pleased if the *Global Wave Statistics* software could be improved to model the physical limit on wave height given wave period.

We are enthusiastic about examining the sensitivity of capsize risk to currently used stability parameters, as suggested by *Mr. Kendrick* and *Mr. Dinovitzer*. To date our work suggests that a frigate will have an acceptably low capsize risk if it has a positive righting arm at 90 degrees of heel in calm water. Many frigates in operation essentially meet this requirement, with the qualifier that downflooding commences at heel angles less than 90 degree angles. We agree that the human factor will have a major influence on ship handling and, for instance, broaching risk in adverse weather conditions. The computed probability figures in the papers are not true in an absolute sense—not only because of numerical modeling limitations, but also because the approach ignores active operator involvement. Efforts are underway within the CRNavies group to increase active operator involvement as part of the research activities.

We agree that warship design is generally governed by damage stability rather than intact stability requirements. Our initial risk analysis work considers intact ships for several reasons. Our numerical capsize model FREDYN is relatively mature for intact ships and has been extensively validated. FREDYN can now model damaged ships in waves; however, the program requires further refinement and validation. Risk analysis of intact warships is very important, in part because a warship spends the vast majority of its time in an intact condition, during which it can be exposed to very severe seaways. The present work on intact ships gives us useful background for considering the more complex case of damaged ships.

Professor Kuo's extensive work on ship safety gives him a valuable perspective on our work. He suggests that we consider risk levels associated with marine industries, for which Reid (Ref. [25] of the paper) gives annual fatality rates of approximately 0.002. We feel that target risk levels for warship capsize must be considerably more stringent. The consequence of a warship capsize will likely be the loss of all crew members and the ship itself. The large number of crew and the very high economic value of a warship justify the expense of achieving a low target probability for capsize. Fortunately, several factors contribute favorably to the safety of warships. Naval engineering directorates carefully monitor the stability characteristics of warships. Ship officers and crew are well trained in ship handling. Unlike commercial ships, particularly fishing vessels, warship operators are rarely subjected to economic pressure to increase cargo beyond safe levels.

When a ship is in severe wave conditions, we hope to be able to provide operational guidance in the form of a polar speed-heading plot (e.g., Figure 16) on a video display. A polar plot can indicate which combinations of speed and heading lead to high capsize risk for given wave conditions. Ship mounted radar systems appear promising for determining wave conditions in real time to be used as input for operational guidance. We feel that operational guidance systems could be particularly useful when a warship is tasked to complete a time-sensitive mission, such as search and rescue, in severe wave conditions.

We thank Professor Kuo, *Dr. Sames* and *Dr. Schellin* for correctly stating that the term risk includes not only probability but also consequence. Our paper focuses on capsize probability and largely ignores the consequence, which we nominally consider to be the loss of ship and crew for capsize of an intact warship.

Drs. Sames and Schellin express reservations regarding usage of a Gumbel fit to determine capsize risk in given conditions. We would like to emphasize that we apply the Gumbel fit to observed maxima from time domain simulations, thus retaining the influence of dynamic roll behavior. For capsize risk in given conditions, we have recently done extensive comparisons using fitted Gumbel distributions and distribution-free estimates (Eq 15). The fitted Gumbel distributions give very similar results to the distribution-free estimates; however, the fitted Gumbel distributions offer the advantage of extrapolation beyond observed roll maxima values. This extrapolation can be very useful when deciding whether more simulations should be performed for a potentially dangerous condition.

"Capsize" is a matter of definition; strictly it denotes the heel angle at which a ship does not return to the upright equilibrium condition. It is not so much related to a critical roll angle that causes capsize. For our example case, we have taken a roll angle of 70 degrees, the nominal angle of downflooding, to infer capsize. The Gumbel fit provides a convenient way to predict the probability of occurrence of any extreme roll angle.

The wave condition with $H_S = 9.5$ m and $T_p = 12.4$ s is responsible for the majority of the capsize risk because of its high nominal wave steepness (see Figure 3) and because the mean wavelength (approximately 1.5 times the ship length) is in a range where a frigate can be susceptible to capsize. The discussers correctly note that hourly capsize risk is 0.0003 for two different wave conditions, one with $H_S = 6.5$ m, $T_p = 9.7$ s, and the second with $H_S = 15.5$ m and $T_p = 18.0$ s. Only the second of these wave conditions is listed in Table 4, the most likely conditions given capsize, because it has a higher probability of occurrence, as indicated in the wave scattergram of Table 1.

Regarding the provision of operational guidance, we feel that the ship motion database used for predicting capsize probability is readily applicable. A database from time domain simulations incorporates nonlinearities which are not included in regular wave results. The assumption of a long crested seaway will be acceptable and somewhat conservative under most circumstances. For some operations, such as helicopter operations and replenishment at sea, directional spreading will limit the degree to which motions can be reduced by optimizing speed and heading. Fortunately, seaways are usually moderate when directional spreading is significant; thus, motions can be predicted through linear superposition of frequency domain results. When developing a motion database, the speed and heading intervals should be sufficiently fine, with 5 knots and 15 degrees being proposed values. If operational guidance is required for survivability against capsize and structural failure, then time domain simulations should be performed for relevant combinations of significant wave height and peak wave period, as we have done in our paper. For operational guidance in moderate, directional seaways, frequency domain calculations for approximately 20 wave frequencies can provide an adequate basis for motion predictions using linear superposition. If a ship has widely varying displacement, then the motion database will have to include several different loading conditions.

SNAME *Transactions*, Vol. 108, 2000, pp. 179–195

The Impact of Bioconcretious Structures (Rusticles) for the RMS *Titanic*: Implications for Maritime Steel Structures

D. Roy Cullimore, Member, University of Regina, and **Lori Johnston,** Visitor, Droycon Bioconcepts Inc.

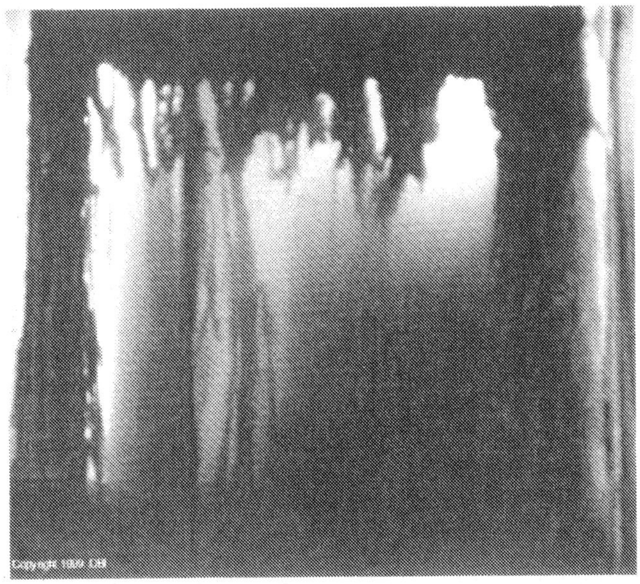

ABSTRACT

Since the sinking of the RMS Titanic in 1912, there has occurred a growing mass of iron-rich bioconcretious structures called rusticles, on the steel surfaces of the ship. Recording the growth rates has only been possible since the ships discovery in 1986, but a cycle has been established in which iron and other elements are being biologically extracted from the steel into these rusticle structures and exported into the oceanic environment as biocolloids. The rate of biological extraction by the rusticles is critical in the determination of ship structural integrity. To ascertain the effects of rusticle activity, three grades of steel (A6, AH36 and EH36) were employed as 6" x 1" x ¼"coupons. These coupons were arranged on IPSCO steel test platforms and placed at sites on and around the RMS Titanic wreck site in August 1998. They remain at the site and are due to be recovered during a subsequent expedition. Laboratory studies indicate that rusticles can be grown on steel, under a variety of conditions, including the application of anodic or cathodic charges. Radiographic analysis has revealed that the iron movement into and throughout a rusticle is focused along narrow pathways within the internal structures. Methodologies have been developed to strip the encrusted rusticle growths from the steel to allow evaluation of the rate of the biological removal of iron from the steel using a patented Blended Chemical Heat Treatment™ process. Laboratory studies indicate that the rusticles are able to extract iron, phosphorus and sulfur from embrittled steel, particularly at anodic sites. A methodology has been developed to manipulate biological growth rates and extraction mechanisms through the use of variable charges. The evaluation of rusticles has direct application to the shipping industry relating to the structural integrity of submerged steel structures including tankers, pipelines, and offshore drilling rigs. This breakdown will negatively impact the environment, particularly oceanic organisms, through toxic releases.

Rusticle Determination

Since the discovery of the *RMS Titanic* in 1985, there have been a number of opportunities to examine and study this deep-sea wreck. The ship sits in under approximately 4000 meters of $1^{0}C$ water, with pressures in excess of 6000psi. One of the first features that stood out in the images of the ship was the mass of rust-like growths, coined rusticles (Pellegrino, 2000). These occurred as a growing mass of iron-rich bioconcretions on the steel surfaces of this once elegant ship. These bioconcretions were first noted growing on the outside and within the ship's structures. The scale of these growths led to the adoption of the term "rusticle" as a derivation of two words, "rust" and "icicle". The term "rust" was elected because the growths had a predominantly rusty color while the texture resembled flakes of rust growing on steel (Cullimore, 1999). The most dramatic, but not necessarily the largest growths, hung over the sides of the hull and parallel the structure of icicles.

In character and form, these growths resemble the speleothems that have been observed in natural limestone caves. These rare speleothems, originally considered to be secondary mineral growths, appear to be of subaqueous origin, and show many similarities to the deep oceanic growths observed at the *RMS Titanic*. In their simplest form, both the rusticles and the speleothems could be described as elongate structures, incorporating organic filaments coated by, or included in, a shell of iron oxide and/or calcite. In the Lechuguilla Cave, New Mexico, the speleothem growths are described as hanging down in ways that closely resemble the rusticles at the *RMS Titanic* (Davis, Palmer and Palmer. 1990). These cave growths are irregular, consisting of iron oxide stalactites and calcite encrusted columns. They are all of ancient origin (greater than 100,000 years) and are now dry and inactive. Microscopic examination reveals that the speleothems are primarily iron oxide deposits covering organic filaments. These encrustations were deposited via oxidative reactions that may have been initiated by bacteria. Other unusual cave features related to rusticles in form, but not based on iron oxides, include pool fingers. These stalactiform subaqueous growths are calcite-encrusted organic strings, interconnected by curved bridge structures. In Wind Cave, South Dakota, hollow subaqueous calcite speleothems known as "helictite bushes" that grow in an upward branching pattern, have been identified. These growths again, closely resemble rusticles and pool fingers. They include fossil bacterial traces, but may be more closely related to the submarine "white smokers" than to either rusticles or the pool fingers (Davis, 1989, 1991; LaRock and Cunningham, 1995).

The bioconcretious rusticles vary in color, texture, size and form. The variations of color, particularly the brilliant orange-brown color of the rusticles, is due to the highly oxidized ferric iron content. Closer examination of the rusticles by Pellegrino and Cullimore (1997), Wells and Mann (1997) and Mann (1997) revealed that the rusticles are complex structures involving water channels, reservoirs, complex iron plate-like structures, thread-like spans, porous matrices and ducts connecting to the outside. Within the rusticle structure, there appears to be a number of different microbial strains occupying specific sites. These were identified using the Biological Activity Reaction Tests (BART™, Droycon Bioconcepts Inc., Canada) to include sulfate reducing bacteria (SRB), iron related bacteria (IRB), heterotrophic aerobic bacteria (HAB), denitrifying bacteria (DN), and archaeobacteria, together with a range of fungi. Six different forms of rusticles were noted, however all bore the common characteristics of diverse and site-focused bacterial consortia.

The supporting structures appeared to be dominated by a mesh-like, heavily mineralized matrix in which goethite was dominant. The presence of goethite in rusticles was confirmed by Garzke *et al,* 1997. In addition, iron oxide sulfate complex, known as green rust ($Fe^{+2}_{3.6}$ $Fe^{+3}_{0.9}(O^{-}, OH^{-}, SO_4^{-})_9$) was found (Garzke *et al,* 1997). A large hanging rusticle, recovered from the ship in 1996, was analyzed by electron diffraction x-ray, which revealed that iron was the dominant atom within the range of atoms tested. The relationship was (dominant atom first): Fe > Na > S > Cl > Ca > Mg > Si > P > Mn. There was a considerable variation in the elemental composition for the various samples analyzed, reflecting the heterogeneous nature of the structures within the rusticles. Where goethites dominate the structure, the iron (Fe) concentrations would be very high, while other components within the rusticle (e.g., the water channels and porous regions) would have lower iron levels. Rusticles also vary widely in size and form. Video imagery reveals rusticle sizes ranging from tiny tubercles or encrustations to massive, braided or rope-like bioconcretions, exceeding 3-4 meters in length. The rusticles form also varied from flat, plate like growths to convoluted and intricate growth patterns.

Relevance of Rusticle Activities to the Maritime Industry

The investigations to-date on the rusticles has revealed that they are capable, under suitable conditions, of extracting iron from steel at significant rates. This biological extraction has the potential to seriously compromise the physical structure of a ship. The time frame for such compromise would appear to be based on the visual evidence. This evidence has been gathered from various sunken vessels (e.g., *Bismarck, Yorktown, Derbyshire and* the *RMS Titanic),* located at various sites around the world. There remains the potential for the covert growth of rusticles within ships during the normal operational life span. Covert growth would mean that the rusticles would thrive at sites within the body of the vessel, particularly those that are not commonly inspected, and where conditions are conducive for growth. In a ship's structure, areas that are most vulnerable include welded areas and areas that have severe stress concentrations. (Mansour, Wirsching, Lucket, Plumpton and Lin. 1997). It is estimated that 2 bulk tankers are lost every month, with 45% of these losses due to heavy weather and structural damage. This category is further described as "strained crack in hull", concluding that structural failure is the major cause for the rising number of bulk carrier losses. This can be seen from the December 12, 1999 sinking of the Erika during a storm off the French coast. The 25-year-old tanker Erika broke in two, spilling about 15,000 tons of fuel oil, polluting 250 miles of beaches, killing or maiming 300,000 sea birds. The classing agency for the tanker, RINA, reported the initial findings into the cause of the accident pointed to a small structural failure in hull structure. This structural failure led to further cracking and finally to the collapse of the hull (Hauley, Barbara. 2000). These structural failures can be a result of corrosion and fatigue cracking. This, in conjunction with biological attachment from within the tanker itself, can result in the loss of a ship's integrity (Ma, Orisamolu, Bea and Huang. 1997).

A range of factors would be important in considering the potential for these rusticles to grow rapidly enough to compromise the normal life-span and sea-worthiness of the ship. These factors could include, but are not necessarily limited to, suitability of steel surfaces on which rusticles can form and function, a high level of humidity or a water saturated environment, oxidative conditions, greater than 1.4% salt concentration in the water, temperature gradient, turbulence, nutrients, electrically charged surfaces and neglect. A typical example of a condition where these rusticles could infest and compromise the integrity of the ship would be between hulls and in compartments where there would be a confined environment in which conditions could become conducive to growth.

The most likely sites for a rusticle infestation to occur would require a number of variables to be achieved. These would include surfaces or areas where the steel is poorly protected with paint, embrittled by stress, electrically charged in any way, involved in rhythmic movement of water over the site, positioned on a temperature gradient, and/or areas where available water contains sufficient nutrients to support growth. Where a site is not subjected to regular inspections for example, on a monthly basis, and/or the rusticle growth is suppressed through the use of biocides or physical removal, the rusticle growths can then begin to extract iron from the steel and weaken the afflicted steel structures. It is a common practice to presume that the appearance of rusty encrustations are merely the result of physical-chemical activity and are an inevitable part of the normal deterioration that may be expected. Traditionally, the appearance of rust within an enclosed chamber has not been viewed as a living mass that is "eating" away at the steel, but rather that the rust is an inevitable chemical event for which solutions may be ineffective over the long term.

In the water well industry, it is now acknowledged that the bulk of the plugging and clogging events that occur down a well are actually biologically derived. Comparable studies have revealed that it is the same groups of bacteria that are involved in these events both down in water wells and deep down at the site of the *RMS Titanic.* Similar rusticle structures are observed at both sites. The question therefore becomes whether steel fabricated ships floating on the surface, or the *RMS Titanic,* a splintered steel structure lying on the ocean's floor, are subject to the same bacterial challenges as water wells, which involve steel structures set into the ground water. The arrival of non-indigenous organisms, such as the Zebra Mussel (*Dreissena polymorpha*), plaguing North American water systems appears to have arrived as covert passengers in or on water going vessels. It is then predicted that there are other unwanted passengers traveling in the bilge areas and other damp dark places within ships that can cause problems to the ships themselves. In water wells, these nuisance bacteria can so substantially reduce water flow into a well that it has to be abandoned. The developing practice in the water well industry to prevent these infestations has evolved into the Sustainable Water Well Initiative in Canada. The objective is to extend the operating life span of wells by routine testing and suitable preventative maintenance and/or radical treatment procedures, depending upon the state of the bacterial infestation. In the Canadian prairies, there are 200,000 operating wells that have been found to have, on average, a life span of 15 years. The capitalization of these wells has been estimated to be minimally CAN$1 Billion and the annual replacement and/or rehabilitation costs are CAN$67 Million.

In the maritime industry it is recognized that inspection of a ship must be done on a routine basis. It is estimated that 20%, 320 ships of the world fleet of 1,744 tankers between 10,000-50,000-dwt are 25 years of age or over, while more than 50 ships still in operation were built in the 1960's. The aging population of tankers makes routine inspection a critical issue. This inspection can be very expensive and time consuming. The sheer size of today's vessels makes this task daunting. Although there are stated guidelines set by IACS (International Association of Classification Societies, 1994), to follow when ship structures are inspected, without the knowledge of the full breadth of the problem these guidelines allow error, and in turn, possible loss of structural integrity. The growth of rusticles may be a major factor affecting the operational life span of ships. Studies to date have found that by using the biological knowledge gained through the rusticle investigation, expanded guidelines could be implemented. The inspection and diagnosis would involve the following four stages:

I. Confirmation that rusticles do grow covertly in ocean going ships in manners that affect the sea worthiness or structural integrity of the ship.
II. Development of thorough monitoring procedures that would include video inspection, bacteriological examinations for aggressivity, chemical analysis, etc.
III. Implementation of preventative maintenance procedures to suppress the rusticle activity, or viable treatment alternatives.
IV. Design strategies in ship construction and operational procedures that would reduce the risk of biological degradation to acceptable levels.

In the maritime industry, a similar scale of impact on the life span of ships would also be causing heavy replacement costs, pollution or unscheduled repairs. These costs could be partially curtailed if the true level of these microbial impacts was found to be significant and controllable. In the shipping industry, there is also an average life span for ships but sudden catastrophic sinkings continue to occur on a regular basis. By improving detection and control procedures to include microbial events that can weaken the structure of the ship, it should be possible to lengthen the life span of ships in much the same manner as water wells are becoming more sustainable (Table One).

Rusticles are a real, yet largely unrecognized, threat to the long-term structural integrity of ships. A strategy could be implemented to augment the existing inspection criteria to include the monitoring and preventative maintenance procedures to control biofouling. The initial focus for such investigations would center on the bilge water and any "pooled" water that may become entrapped within the ship's structures. The signatures for these potentially aggressive bacteria that cause losses in steel strength include corrosive or extraction processes. Table One is a list of the criteria that outlines the detection of biological activity that may have the potential to compromise the integrity of ship's steel structures. The list of criteria is based upon experiences in other industries.

There is a need for dedicated research and development to address the problem of rusticles and other microbial communities having the capacity to cause sudden losses in steel strength and integrity, with subsequent loss of the ship due to structural failure.

Examination of Rusticle Biomass on the *RMS Titanic*

The examination of rusticle growth rates on the *RMS Titanic* has only been possible through annual or biennial expeditions to the wreck site. Through the use of video and high-resolution imagery, these growth rates and patterns of the rusticles can be closely analyzed. In 1996, the first examination of rusticles from the *RMS Titanic* occurred. This examination has shown clear evidence of rusticle growth on the outside of the bow section of the ship, however, the outside of the stern, and the interior of the wreck, cannot be clearly quantified because of the lack of adequate examination. One aspect of this examination was to estimate the percentile coverage of the various parts of the bow section together, and the estimated thickness of the rusticles at these various sites (Table Two). From the 1998 Expedition, comparative assessments through video imagery showed clear evidence that the rusticles continue to grow and there is evidence that the biomass is approximately 30% greater than the mass observed in 1996.

Video surveys of the bow section of the *RMS Titanic*, in August 1996 and 1998, allowed a quantitative estimate of the total volume of rusticles. This estimate is based on the area of the steel covered by rusticles at various points on the ship. Calculations reveal that the bow section of the ship had a total mass of 650 tons of rusticles in 1996 and increasing to 881 tons in 1998 (Table Three). The iron content in the rusticles had presumably been extracted from the ship's steel structures and was now accumulated within the rusticles. The iron content has risen from 178 tons of iron in 1996 to 242 tons in 1998. While there is a significant amount of iron present within this mass of rusticles, there remains a concern as to the rate at which iron is being released from the rusticles into the oceanic environment. Iron concentrations vary in rusticles along with their size, weight and density of the ducts on the surface of the rusticles (Table Three). These estimates are based on examination of rusticle specimens recovered from both the 1996 and 1998 Titanic Expeditions.

Focused Accumulation Sites for the Iron within the Rusticles

One concern arising from the examination of the *RMS Titanic*, was the manner in which the iron is accumulating in the rusticles. It is known that the rusticles have a very large surface area and a highly porous concretious structure, however, little was known of the sites where the iron accumulates. Accumulation occurs as various forms of ferric oxide and hydroxide, dominated by goethites. To examine the sites of iron accumulation in the Rusticles, a Veterinary Grade Radiograph machine operated at 100kV, 1/20 sec, using Kodak high-speed film was used. Five rusticles, collected from the 1996 and 1998 Expeditions, were investigated using radiographic examination (Plate One). It was found that the iron within the rusticle structure was not evenly distributed throughout, but was concentrated into two major regions. The first region where iron was dispersed resembles "cloud-like" structures. The second region had very dense channeling of the iron into localized regions that spread web-like throughout the rusticle. The first impression was that the region-channeled iron resembles a primitive blood system.

The radiographic images confirmed the complex nature of the rusticle structures and the fact that the rusticles had entrapped artifacts such as coal fragments and glass shards that rained over the ship immediately after the collision with the ocean floor. Microbiological examinations found that the microbial consortia within the rusticle were not dispersed but remained concentrated within localized regions of the rusticle. No correlation could be established between the sites of iron concentration and the various bacterial consortia present in the rusticle.

It has been determined, based on extraction results, that a cycle has been established in which iron is being biologically extracted from the steel of the ship into the rusticle structures. The iron is then exported into the oceanic environment as "red dust" and "yellow colloids". The rate of extraction is increasing beyond the predicted 1996 rate of 0.1 tons of iron being mined by the rusticles per day.

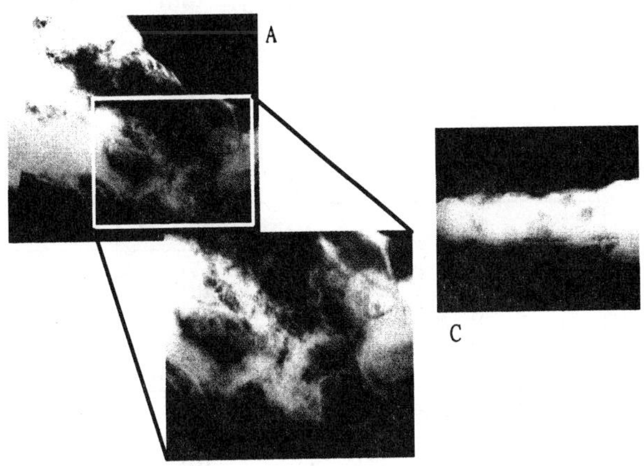

Plate One
Radiographic images of five rusticles recovered from the 1996 and 1998 Expeditions. Three radiographic images are presented for one "shoulder" rusticle recovered in 1996 from the bow section of the ship. The first image A reveals the density of the iron (the lighter the image, the greater density there is of iron. Image B shows a close up that reveals there are complex structures within the rusticle body while a side scan (C) shows that iron is dispersed evenly in this direction.

Table One

Proposed Criteria for the Evaluation of Bilge and "Pooled" Water in Ocean Going Steel Fabricated Ships to Detect Corrosive and Extractive Processes that Could Compromise the Ship's Structural Integrity.

Criterion	Serious Risk	Borderline Risk	Background
Total Suspended Solids, ppm	> 0.5	< 0.5 to > 0.05	< 0.05
Particle Sizes, average, microns	> 6.0	< 6.0 to > 3.5	< 3.5
Total Iron, mg/L	>3.0	< 3.0 to > 0.5	< 0.5
SRB – BART, time lag, days	< 2.0	>2.0 to < 8.0	> 8.0
SRB – BART, reactions*	BT, BB or BA	BB or BT	CG only
IRB – BART, time lag, days	< 4.0	>4.0 to < 8.0	> 8.0
IRB – BART, reactions*	BL, BR, BC	GC, RC, BG, BR	FO, CL
HAB – BART, time lag, hours	< 18	>18 to < 48	> 48
HAB – BART, reactions*	UP	UP	UP, DO
Solids from 100ml, mg	> 15	<15 to > 5	< 5

Note that the * refers to any one of a range of the reactions shown can be considered significant. Total suspended solids are measured using a laser particle counter with the sample freshly stirred. The measurement should be repeated after 30 minutes and if there is a greater than 70% reduction in the ppm then this would indicate dense particles in the sample (greater risk). The BART stands for biological activity reaction test (see site www.dbi.sk.ca for more details). SRB stand for sulfate reducing bacteria, IRB for iron-related bacteria and HAB for heterotrophic aerobic bacteria. All three groups are commonly found in rusticles. Solids are recovered by the membrane filtration of 100ml of sample through a 0.45-micron filter, dried at 60°C for one hour and weighed. The increase in weight reflects the mass of suspended material deposited on the filter.

Table Two

Dimensions, weights, number of ducts, duct density, artifact inclusions and iron concentration on some of the rusticles recovered during the 1996 expedition.

Rusticle #	Height mm	Width mm	Length mm	Mass grams	Duct #	Area/duct duct/mm^2	Artifacts
BS1	94	127	278	1,929	8	1 / 4,413	Rivet caps
BB1	90	131	372	2,100	6	1 / 8,122	
TP1	92	92	232	893	8	1 / 2,668	Coal fragments
NO1	84	122	234	1,485	6	1 / 4,758	Coal fragments, stones
SP1	82	110	272	1,194	6	1 / 4,987	Coal fragments, glass
FP1	16	92	155	115	3	1 / 4,753	Glass shards
FP2	35	50	65	100	1	1 / 3,520	Coal fragments
SL1	60	100	195	634	2	1 / 9,750	Coal fragments
LL1	58	80	102	269	4	1 / 2,040	Glass shards
RT1	45	60	106	93	3	1 / 93	

The dimensions of the rusticles are shown as maximum dimensions and the shapes of the rusticles were more ovoid. This was taken into account in calculating the area per duct. The average opening area of a duct was 7mm^2 and ranged from 3 to 13mm^2. Ducts were determined by immersing the drained rusticle back into water at which time the air would be released through the ducts and could be enumerated. Direct observation, non-intrusive dissection and/or radiography recognized artifacts in the rusticles.

In determining the rusticle mass on the bow section, a comprehensive analysis of the movement of iron as a result of biological activity and growth of rusticles was undertaken in a laboratory setting. One component of the life cycle of the rusticles became obvious when culturing the rusticles in the laboratory generated a red dust (RD) and yellow biocolloidal clouds (YBC) during their growth cycle. These were found to have gravimetric iron contents of 20±5% for the RD and 8±3% for the YBC. It was determined that the iron was moving from the steel into the rusticles, where the common iron content was 30±6%. The extraction process therefore causes the iron to be excreted from the rusticles as RD and YBC, and dispersed into the oceanic environment. There are a number of dynamic factors in this movement of iron from the steel to the oceanic environment. These may be summarized in the following equations:

The movement of iron from the steel into the rusticles may be expressed by the amount of iron accumulated (Fe_R) in the rusticle. This is reflected as a part of the calculated rusticle biomass (R_m) determined by calculated mass volume corrected by the rusticle density (R_d), based upon the mean iron concentration (Fe_c) determined by analysis. Equation One calculates the total iron retained in the rusticles at any given time.

Equation One

$$Fe_R = (R_m * R_d) * Fe_c$$

While the iron concentrated in the rusticles (Fe_R) is calculated for a specific observation event, it does not represent the total amount of iron removed from the steel over the period of time that preceded this event. A daily release of iron from the rusticles can be calculated based upon the factorial daily rates of releases of RD and YBC. The daily releases of iron (Fe_{dr}) may be calculated using Equation Two.

Equation Two

$$Fe_{dr} = (Fe_R * R_{RD}) + (Fe_R * R_{YBC})$$

The daily rates of iron removal are based on the factorial releases of iron in the RD (R_{RD}) and YBC (R_{YBC}). These two factors are calculated from the daily release of biomass of RD (RD_{BM}) and YBC (YBC_{BM}) and the mean factorial concentration of iron in the RD (RD_{FE}) and YBC (YBC_{FE}). These are calculated in Equations Three and Four respectively.

Equation Three

$$R_{RD} = RD_{BM} * RD_{FE}$$

Equation Four

$$RY_{BC} = (YBC_{BM} * YBC_{FE})$$

Clearly, the key component in the determination of the movement of iron out of the steel is the rate at which the rusticles are growing over the steel structures on the ship. There is now a chronological video record of the *RMS Titanic* wreck site from 1986, 1987, 1991, 1994, 1995, 1996 and 1998. This allows a sufficient base line of information to calculate the rate of rusticle growth on the bow of the ship.

Theoretically, the movement of the iron from the steel structures of the *RMS Titanic* may be expected to move out into the oceanic environment in a logical manner. The iron would originate from the steel (Fe_S) and eventually become dispersed throughout the biosphere (Fe_{BS}) with some of the iron terminating in oxidized iron deposits (OFe_D). The critical event is the differentiation of the iron that enters the global biosphere (Fe_{BS}) becoming entrapped locally at the wreck site, or at some other location, as oxidized iron (OFe_D) deposits, such as pig iron deposits. This is depicted in Equation Five.

Equation Five

$$Fe_S \rightarrow Fe_R \rightarrow RD_{FE} + YBC_{FE} \rightarrow Fe_{BS} + OFe_D$$

In the laboratory culture of rusticles, a common occurrence was the massive ejection of red dust into the water. Entrapment of the red dust occurred using a 0.45-micron membrane

Table Three
Estimation of the percentage coverage, thickness and mass of the rusticles covering the outside of the bow section of the *RMS Titanic* based upon observations made during the 1996 and 1998 Expeditions.

Position	% Coverage	Thickness (mm)	Biomass (tons)	Iron (tons)
Plating '96	80%	16	429	107
Bulkhead '96	90%	6	221	71
Plating '98	85%	20	570	142
Bulkhead '98	95%	8	311	100

The biomass is based upon a recorded density of 1.4 and 1.6 for rusticles recovered from the plating and bulkheads with an average iron content of 25% and 32% respectively. This data has been based in part upon video images of the ship taken in 1996 and 1998 and partly upon the analysis of rusticles recovered from the site. Note these data are estimates and are subject to confirmation by further study.

Table Four
Weight Percentage in Coupon

Grade of Steel	Carbon	Manganese	Phosphorus	Sulfur
A6	0.188	0.77	0.013	0.0077
AH36	0.067	1.36	0.016	0.004
EH36	0.133	1.47	0.015	0.0127

filter. Weight determination showed that, for the two rusticle tanks studied, there was a per diem release of red dust at a rate of between 0.018, indicating 0.02% of the total mass of the rusticles. This would mean that over any one year, given a steady release of red dust, the total mass released in this form would be equivalent to 7% of the biomass of rusticles infesting the ship. It can be extrapolated that, in addition to the mass of rusticles infesting the ship and extracting iron, there has been a substantial loss of iron to the oceanic environment. In addition to the 242 tons of iron accumulated by the rusticles on the bow section by 1998, it can be expected that a similar amount has, historically, been removed from the steel and exported to the oceanic environment by the rusticles. This is a critical component planned for future studies.

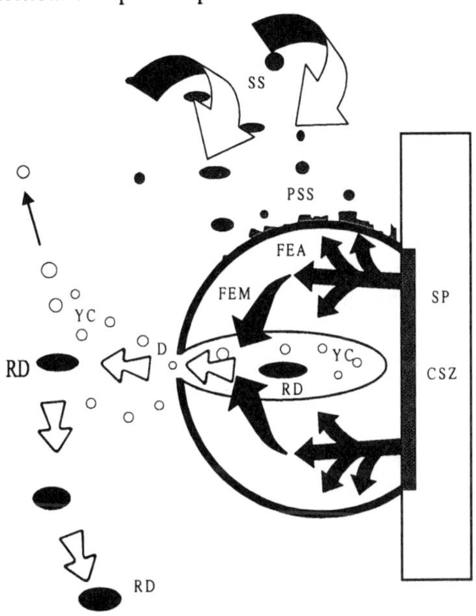

Figure One
Proposed manner in which iron from the *RMS Titanic* is being extracted, accumulated and released into the oceanic environment. Presentation of the movement of iron from the ship's plating (SP) within the corroded steel zone (CSZ) into the rusticle (partial circle). The iron is distributed within the rusticle mainly as ferric accumulates (FEA) that are gradually mobilized again (FEM) and move to the water passageways. Water is expressed from the rusticles through ducts (D) as red dust (RD) and yellow colloidal (YC) materials. Some of the nutrients to support the rusticles comes from the "sea snow" (SS) some of which perches on the rusticle (PSS).

There is clearly a potential for some of the iron to enter into the surface biosphere due to such events as the consumption of fish and other animals from the oceanic environment. Both the iron in the RD and YBC are likely to be consumed by the oceanic biota and, through that means, enter the food chain (Figure One). It is therefore possible that some of the iron from the *RMS Titanic* could actually find its way back into the blood of humankind as a very important part of the blood hemoglobin.

Water Wells and Rusticles

A parallel can be draw between the rusticles at the *RMS Titanic* and growths that occur within the various forms of wells that interface with ground waters (Cullimore, 1993, 2000). The difference being that in water wells, the growth appears on the inside of the stainless steel well screens, while the *RMS Titanic* appears to have the bulk of the growth on the outside of the steel hull. The origin of the iron accumulating in these growths was, for the rusticles growing in the wells, mostly coming from the ground water and surrounding geology. On the ship, the iron appears to be coming from the steel structure of the ship itself.

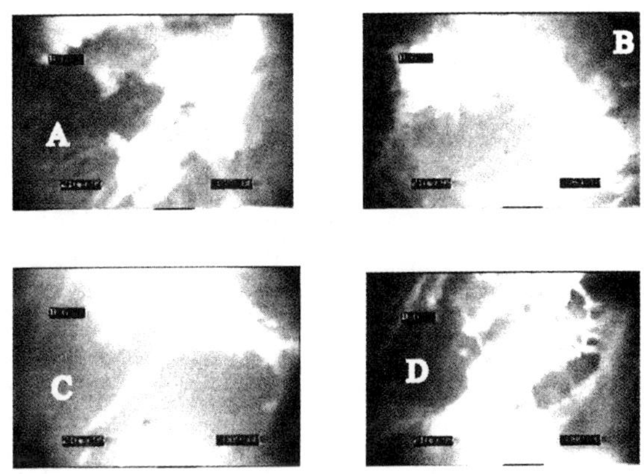

Plate Two
Video imagery of growths occurring within a 4" horizontal extraction well used as a part of a bioremediation process. Four video images are presented showing different forms of growth that can rapidly form in wells. Many of the structures seen in these images resemble those seen in rusticles. For example, these plates show complex woven slime masses (A), slime arches (B), long slime threads (C) and open woven masses with interlacing threads (D). The width of the images is approximately 40mm and these growths developed in eight months.

Down-hole video inspection of water wells that are becoming biologically plugged, reveal similar growths and concretions to those seen in the above-mentioned caves and at the *RMS Titanic*. Injection and extraction wells at bioremediation sites reveal startlingly similar concretious growths. For example, a bioremediation site in Holland utilizes horizontal extraction wells as a part of the bioremediation process, extracting chlorinated hydrocarbons, such as trichloroethylene and polycholoroethylene. The bioconcretious growths in the 4" well screen appear within the first two years of operation, causing a 50% loss in hydraulic production. These growths appear similar in structure and growth patterns to those observed at the *RMS Titanic* (Plate Two).

Laboratory Evaluation of Corrosive Processes Using the Corrosion-BART™ Test

The first stage in the investigation has been to develop a methodology to determine whether or not microbial growths in the form of rusticles can cause iron extraction to occur from a targeted steel plate. To achieve this outcome, a Biological Activity Reaction Test (BART™, Droycon Bioconcepts Inc., Canada; Cullimore and Alford, 1990) was modified (Figure Two). The BART™ test uses a 15ml water sample which creates a reductive/oxidative (redox) front with a rising nutrient gradient selective to the bacteria being investigated (e.g., sulfate reducing bacteria). To contact the mild steel coupon, six-3mm holes were drilled through the basal cone of the test vial and the vial itself is placed under pressure onto a 26mm rubber washer to prevent leakage.

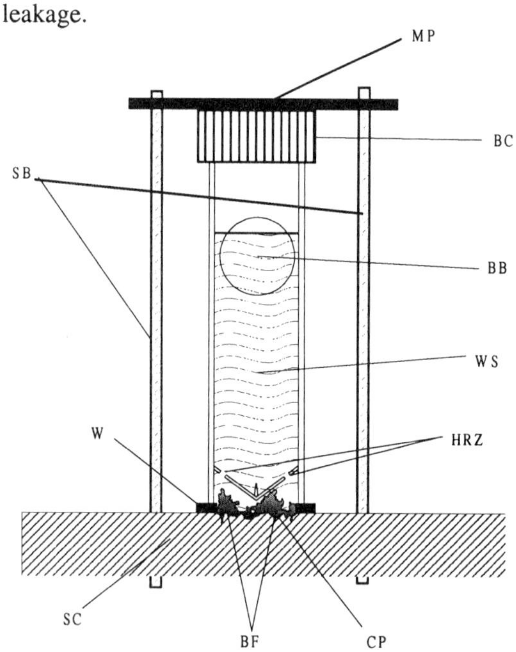

Figure Two
The biological activity reaction test (BART™, Droycon Bioconcepts Inc., Canada) has been modified to determine corrosion of the surfaces of steel coupons.

The BART test vial is modified to encourage the growth of bacteria within the water sample (WS) to grow on the steel coupon (SC). To achieve this, the patented BART test is placed on a circular rubber washer (W) and pressed down using a main plate (MP), which is screwed tightly on by two screw bolts (SB). The water sample is added prior to this procedure by unscrewing the BART cap (BC) and adding 15ml of the water sample, at which time the BART ball (BB) floats up to create a redox gradient. Holes (HRZ) are cut in the base of the test vial to allow the bacteria to grow directly on the steel and form biofilms (BF). Corrosive pitting (CP) forms at the contact sites of the steel coupon (CP).

To conduct the test, a 1% (w/v) suspension of rusticles recovered from the site of the *RMS Titanic* was made using a 4% sterile solution of sea salts. The 15ml suspension was aseptically injected through the cap into the test vial. This set up a redox gradient selective for a specific group of bacteria. The tests were conducted at room temperature ($21^{0}C$) for 60 days. Each steel coupon was examined for biological corrosion and/or extraction by reflectance microscopy (x10 and x40 diameters). It was observed that biological growths appear to be attached to the steel surfaces on which the test vial had been positioned. Examination of the growths by both photomicrography and radiography revealed a range of structures that resembled rusticles in color, texture and form. The observed rusticle-like structures grew from the surface of the mild steel to an average height of 2mm occupying 60% of the volume (after 60 days), having a wet volume of approximately 0.37ml, a dry weight of 0.09g with an iron content of 22%. This resulted in approximately 0.2g of iron being biologically extracted from the steel in the 60-day period. Removal of the rusticle structures attached to the steel and the subsequent cleaning of the steel occurred using a chemical immersion technique. The steel coupon was immersed in a 1.0% solution of hydrochloric acid for sixty minutes followed by a sterile distilled water rinse. The steel was again microscopically examined for evidence of biocorrosion and/or extraction. Pitting was observed covering 20% of the steel's surface with the pits ranging in diameter from 0.1 to 0.8mm with a depth of 0.4 to 1.0mm.

This experiment revealed that, under reductive conditions, the biocorrosive processes could be initiated within sixty days, at room temperature. These processes generated biological growths that resembled rusticles on a micro-scale. The next stage of this study will be to apply this methodology to coupons of steel from the *RMS Titanic.* This will allow somewhat controlled conditions to observe the forms of corrosion and/or extraction that occur on this particular steel. Field experience has revealed that corrosion and/or extraction tends to be dominated by lateral flaking and dissolution of the steel directly under maturing rusticle growth.

Interaction Between Rusticles and Modern Maritime Steels

To ascertain the effects of the steel grades on rusticle activity, three grades of steel, A6, AH36 and EH36, were employed as 6" x 1" x ¼" coupons. The coupons had been hammered, twisted, burnt in bunker oil or tempered to compare with a control coupon. Analysis of these steels was performed by IPSCO (Table Four).

These coupons were arranged on specifically designed IPSCO Steel Test Platforms (Figure Three). The design allowed the coupons to be arranged in a "staircase" pattern (SC). The triangular-shaped platform could also be placed with the coupons close to the rusticles (FL), with the larger surfaces of the coupons parallel or perpendicular (UP) to the flat surface on which the Test Platform is being placed. The third option was elected, allowing the rusticles to grow down over the larger steel surfaces on the coupon.

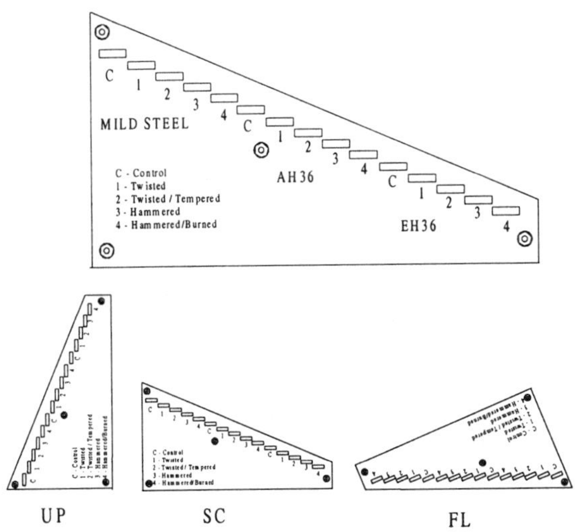

Figure Three
Diagram of the arrangement used to position the steel coupons on the IPSCO Steel Test Platforms. The upper diagram displays the side view of the platform that had a length of 485mm and a height of 265mm. Fifteen steel coupons were inserted in a stepwise arrangement to give an overall width of 160mm. Three grades of steel were identified in five different states. The lower diagram shows the three manners in which the platforms could have been installed at the site. These were upright (UP), in a staircase arrangement (SC) and as a flat layer just above the floor (FL).

Four IPSCO Steel Test Platforms were placed on, and around the wreck site, in August 1998 (Figure Four). The four platforms are located as follows:

(1) At the stern section, approximately three meters in front of the sheered double bottomed hull supporting the starboard exposed reciprocating engine;
(2) At the hull section, on the boat deck, port side just aft of the bridge;
(3) At the hull section, port side of the stem at the very front of the bow on a pile of rusticle hulks on the ocean floor;
(4) At the hull section, immediately beneath the port side well deck on the ocean floor near some intense rusticle growth on steel cables hanging over the side of the ship.

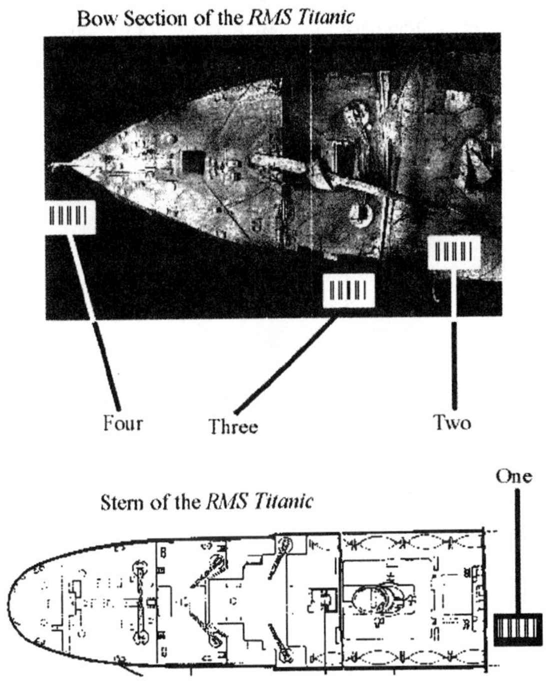

Figure Four
Plan view of the positions of the four IPSCO Steel Test Platforms placed, in August 1998, at sites on, and around, the *RMS Titanic*. The first platform (One) was placed close to the engine room on the ocean floor (lower diagram). The other three were placed on the port boat deck just behind the bridge (Two), at a point on the ocean floor below the port well deck (Three) just to the port of the stem insert point into the ocean floor in a field of rusticle debris (Four).

The four Test Platforms were observed one week after placement, and in each case there was evidence that the rusticles were beginning to infest the steel coupons. When the IPSCO Test Platforms are recovered from the site, each coupon will be subjected to a sequence of activities that will allow the total dry weight and iron content of the rusticles to be determined. Each coupon will be weighed and measured to determine how much of the iron has been extracted from the steel. The rate of growth of the rusticles would therefore be determined by a mass balance analysis based upon two possible scenarios:

1. That the weight of iron lost from the steel coupon would equate to within 5% of the weight of the iron recovered from the rusticle. This would indicate that the growth of the rusticle has not yet reached a state of maturation in which iron is being exported as RD and/or YBC. A linear relationship can therefore be projected between the growth rate of the rusticle and the biomass recovered from the steel coupon.

2. That the weight of iron lost from the steel coupon exceeds greater than 15% of the weight of the iron recovered from the rusticle. This would indicate that iron was already being exported to the oceanic environment from the rusticles through the releases of RD and/or YBC. This could be calculated using the established rate of RD and YBC releases determined in laboratory studies (see Table Four). An extrapolated growth rate for the rusticles could be calculated using Equations One through Four.

In the laboratory, rusticles have been grown using various grades of steel under a variety of conditions including the application and manipulation of electric charges. It has been found that rusticle growth can be stimulated by the application of a charge across a steel plate, when there is greater than 1 .4 % sea water solution used as the conductive "bridge". Radical rusticle growth has been observed on charged steel when a direct current of 12 volts at 1.8 amperage was administered at a constant rate (Ostryzniuk, Alford and Monea, 1999). This can be sufficient to cause the combination of electrical charge and biological activity (cultured rusticles) to biologically bore through a 3mm mild steel plate in five hours (Plate Three). This would mean that, under the favorable circumstances similar to those created in this test, mild steel could be removed at a rate of 0.6mm/hr. When establishing these patterns of charge, it was noted that the opposing charged plate remained pristine and showed no evidence of infestations by rusticles. This leads to the postulation that the rusticles are growing at variable, charged sites (e.g., the very stem of the bow) on the *RMS Titanic* but not at sites that have an opposing charge.

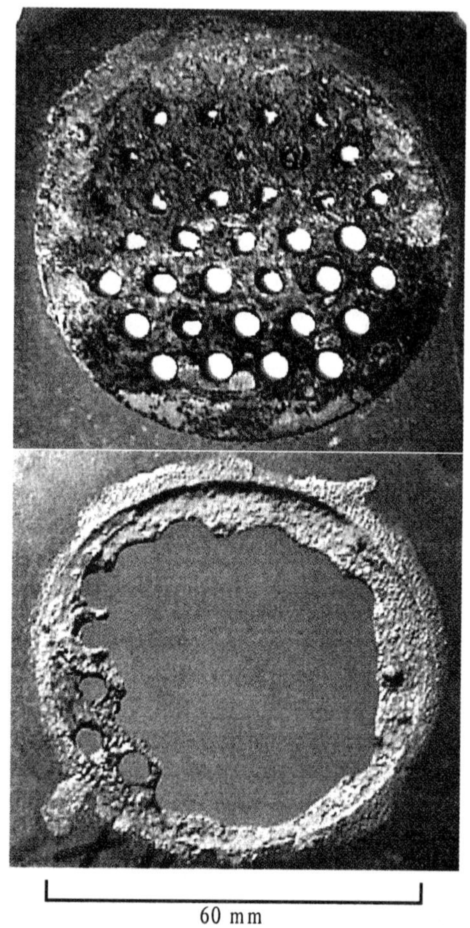

Plate Three
Impact of biological activity (cultured rusticles) to biologically bore through a 3mm mild steel plate in five hours under electrically active conditions. Two 1/8" steel plates were subjected to a 10% suspension of rusticle growths over a 60mm diameter area of the plate perforated over 25% of the area with 3mm holes. The Plates were also electrically charged. The plate on the right also was subjected to a 12% saline produced water and perforated completely in five hours. The plate on the left was not exposed to the produced water but regular deionized water.

Potential Costs and Environmental Damage Caused by Covert Rusticle Growth

There is a clear chain of evidence that suggests the ability for microbial growths such as rusticles, in the form of slimes, tubercles, nodules and encrustations, to extract elements such as iron from the steel into these structures. This type of functioning bioconcretion is evident at the *Titanic* site. These growths are contributing to the deterioration of the *RMS Titanic* through biodegradation; it is a natural extension to consider the implications of these biofunctions at other sites where conditions may be conducive for growth. It becomes clear that the rusticles have become dominant in an environment where there is high salt concentration (i.e., highly conductive), available organic material such as the "sea snow" found at the wreck site, and steel surfaces.

In examining existing engineered steel structures that occur within the oceanic environment, it is evident that current management strategies, over the lifetime of such structures, would be ineffective and restrictive to the determination of observable occurrences of rusticle-like growths. There are two main points at which these growths may become key to the ongoing life of a structure. The first critical point is that these growths may be generated at covert sites, remaining unobserved, yet create catastrophic failure of the structure. This could result in loss of buoyancy, leaking and/or total loss of structural integrity resulting in the sinking of the vessel. This would render forensic proof for this type of failure difficult to determine. The second critical point relates to abandoned steel fabricated structures, such as steel drums, pipelines and oil rigs, where these growths now have the potential to seriously compromise the structural integrity of these structures.

Historically, there has been very little understanding or examination of the impact of microbial activity on abandoned steel structures. The extent and significance of these biologically induced events must rely heavily on the limited knowledge that has been gained with respect to biocorrosion, and on the understanding of the function and aggressivity of microbial communities (consortia) to establish and proliferate in extreme environments. The covert microbial growth and degradation of steel structures has the potential to cause serious environmental problems.

Potential maritime examples range widely but perhaps one that has a long term and serious concern is the disposal of steel drums containing nuclear waste into the marine environment over the last fifty years. The chain of events established by this practice involves the loading of an industrial steel drum with radioactive material and allowing it to free fall to the ocean floor. At that time, popular opinion was that the ocean floor was a sterile environment divorced from the terrestrial environment. In the event of a free fall of the loaded drum, the first concern is generated by the collision between the drum and other objects on the ocean floor, as well as with the ocean floor itself. These impacts may cause seam failures and/or the embrittlement of the mild steel of the drum itself. These sites become focal points for initial attachment and growth of microorganisms. The form and extent of the growth would be dependent on the nutrients in the environment (positive impact) and the level of radiation being generated by the wastes (negative impact). In the case of a negative impact due to radiation, there are two microbial mechanisms that can counter this influence on their growth. The first of these mechanisms is that some microorganisms, such as the bacteria belonging to the genus *Deinococcus*, have a very high resistance to radiation (e.g., 1 to 2 megarads) due to unique cell wall surface structures. Under suitable environmental conditions these bacteria could therefore grow over the surface of the steel containment drum, causing a loss of integrity of the steel surface. An alternative mechanism occurs where the indigenous microbial consortia are protected from the radiation by the nature of the concretious growths that these microbes form. The microbially induced concretious growths contain a high metal content, retarding the impact of the radioactive contents of the drum. Initial growth may occur through the metal rich biocolloids, such as red dust from the growth of rusticles, which attach or come to rest on the outer surfaces of the steel structure. Consequent growth may then generate conditions that cause corrosion of the steel structure, followed by perforation and leakage of the contained radioactive waste into the environment. Given the rate of growth of rusticles on the *RMS Titanic,* which is not radioactive although probably electrically active is very slow, measurable in years and decades rather than days and weeks. The embarrassment of steel structures such as drums is likely to occur over an extended period of time, yet the potential for large-scale environmental contamination to occur continues to be immense.

The potential for corrosion to occur by rusticle-like concretions on drums filled with toxic material becomes more realistic. This is founded on the dual premise that: (1) there is no restricting radiation field to suppress microbial activity, and (2) the contained toxic material may well be selectively toxic to humankind and the biota but not toxic to many species in the microbial community. For example, cyanide is clearly perceived to be a powerful toxic compound, however, there are many microorganisms that are resistant to cyanide and some can even degrade it as part of the process of natural self-purification commonly seen when toxic spills occur. The sequences involved in the event of the degradation of toxic containment vessels would follow a pattern of impact, embrittlement, localized corrosion, and finally, perforation. The perforation would be followed by localized leakage with a probable accumulation of the toxic material within the concretious growths. The bioaccumulation events would occur at the viable cells present and active within the concretious growth. Finally, biodegradation and/or dispersion, depending upon the nature of the toxic material, would occur. Growths such as rusticles would therefore both accelerate, through corrosion and perforation of the steel structure, and reduce, through bioaccumulation with or without biodegradation, the release of toxic materials from the compromised structure.

Underwater pipelines offer a number of problems relating to the potential biological compromise that could involve the activity of rusticles. The contents of the pipeline, whether abandoned or in use, may be critical to the nature of the growth. For example, an underwater gas pipeline may suffer from corrosive physical pitting or perforation, leaking methane into the surrounding environment where it would be utilized by the methane degrading (methanotrophic, MTB) bacteria. These MTB may be incorporated into the concretious consortium-forming rusticles. Consequently early micro-leaks of gas or oil may serve as the primary organic feed stock for the rusticles that then become much more aggressive and, in so doing, structurally weaken the pipe causing sudden and dramatic pipeline failures. Abandoned pipelines would carry only static perched material that may form a smaller feed stock for microbial growth, but the lack of forced flow and pressures may allow the microorganisms to gain entry into the pipeline itself and initiate reductive corrosive processes from directly within the pipeline. Concurrently, the pipeline could also be challenged with oxidative corrosive processes being generated by rusticle-like structures. The form of these attacks on the integrity of the steel pipeline would be very much affected by the nature of the variable electrical charges along the pipeline, as anodically charged sites tend to be the focal point of the microbial activity. Many pipelines are protected, during the active life span, from these types of events through a process of cathodically charging the outer surfaces, which reduces microbial attachment and growth. Once the pipeline is abandoned, corrosion prevention practices are halted, and the gradual process of microbially induced corrosion and deterioration continues unimpeded.

Where there is off shore drilling operations such as oil and gas, the steel structures being used for such operations are known to be subjected to corrosive processes generally thought to have originated from infestations of sulfate reducing bacteria (SRB). The presence of this type of activity is clearly signaled by hydrogen sulfide-driven electrolytic corrosion, the presence of copious black sulfide rich slimes and often, the "rotten" egg odor of the gas. This group of microorganisms forms a major component of the consortia of rusticles and can be found cloistered within the bioconcretions formed by this process. Rusticle growths on the outside of a drilling rig would clearly be recognized, treated accordingly and removed during routine maintenance procedures. The same thing may not happen in the event of rusticles growing within the ballast tanks of these rigs. Where this process is occurring, there is a risk that the growth may compromise the structural integrity of the ballast tanks in a manner that could lead to sudden and catastrophic failures.

Abandoned steel rig structures, particularly when immersed in seawater beneath the light penetration zone, are more likely to become infested with bioconcretious structures resembling rusticles. Residual hydrocarbons would form a major part of the feed stock supporting the growth, and the primary focal regions for growth are likely to be oxidative over steel that is, in some manner, embrittled to facilitate microbial attachment and growth. The rate at which these rusticle-like structures would begin to form and compromise the steel cannot be estimated since the experiential base is limited to a few sunken ships with the *RMS Titanic* being the one most vigorously investigated. Ships sunk in shallower depths such as the *RMS Britannic* show very different forms of biological encrustations with a much greater diversity of life forms. The relative significance of the relationship between these different species and the rate of corrosion challenge to the steels on the vessels remains to be determined.

The potential environmental costs of biological activities on submersed steel structures are a combination of positive and negative impacts. Positive impacts relate to the recycling of the iron from the structures back into the natural oceanic cycles. This relates, in part, to the *RMS Titanic* investigation, forming perhaps a fitting conclusion to the investigation of the ship herself. Of particular interest is the fact that iron has now become recognized as a key limiting factor in the rate of oceanic carbon dioxide fixation through the phytoplankton. The iron released from these submerged steel structures may therefore play a significant role in increasing the ability of the oceans to fixate carbon dioxide and so, therefore, in some small way, reduce the negative impact of greenhouse gases on global warming. From there, the iron would gradually move into the global biosphere to the point that, perhaps some time in the distant future, there will be iron atoms from the *RMS Titanic* now being pulsed around the human body and carrying oxygen to meet our respiratory needs. The same may be said over the fullness of time for the many other steel ships now resting on the ocean's floor and gradually deteriorating. This may be viewed as a positive long-term environmental effect.

Negative impacts are more diverse and difficult to predict. These impacts may relate more to the releases of more toxic, potentially carcinogenic or recalcitrant chemicals of concern, that were previously confined by the steel but now becomes released, at least in part, as a result of biological activity. These effects become very difficult to predict, challenging to contain, and may impact in many different manners over various time scales. In the past, when a ship sank at sea, there was little concern for its impact on the environment. Today, a ship sinking means simply that the contents of that ship have now become subject to a different pattern of release that is not in the domain of humankind. Impacts, therefore, become speculative since the knowledge has yet to be gained and in that regard the *RMS Titanic* is one of the signposts pointing the way to understanding these effects.

There is a continuing need for dedicated research and development to address the problem of rusticles and other microbial communities having the capacity to cause sudden losses in steel strength and integrity. These losses must be recognized and effectively managed throughout the life span of the steel structure, but also post usage. This post usage monitoring will, in effect, require a more complete understanding of biologically induced corrosion events and how these activities are impacting the environment.

Acknowledgments

The authors wish to acknowledge the financial support of the Society of Naval Architects and Marine Engineers that allowed this project to move forward. Also acknowledged are the many people who at different times provided important information or facilitated the collection and analysis of the rusticles. These include: Bill Garzke (Gibbs & Cox Ltd.), George Tulloch, Greg Andorfer (Maryland Science Center), P.H. Nargeolet (Infremer), David Elisco (Stardust Visual), Charlie Haas, Dr. Charlie Pellegrino and the staff of Droycon Bioconcepts Inc.

References

Cullimore, D.R. and Alford, G.A. (1990) *Method and Apparatus Producing Analytic Culture,* U.S. Patent 4,906,566.

Cullimore, D.R. (1993) *Practical Manual of Groundwater Microbiology.* CRC Press Inc.,Boca Raton, FL, U.S. ISBN 0-87371-295-1 p 77 - 116 and 271 B 315

Cullimore, D.R. (2000) *Microbiology of Well Biofouling.* Lewis Publishing and CRC Press Inc.,Boca Raton, FL, U.S. ISBN 1-56670-400-6 p 137 - 280.

Cullimore, D.R. (1999) *IV Titanic, the Connection between Rusticles and Clogging*, published in *The Application of Heat and Chemicals in the Control of Biofouling Events in Wells* (Alford, G and Roy Cullimore, eds), Lewis Publishing and CRC Press Inc.,Boca

Raton, FL, U.S. ISBN 1-56670-385-9 p 162 - 166.

Davis, D.G., Palmer, M.V. and A. N. Palmer (1990) *Extraordinary Subaqueous Speleothems in Lechuguilla Cave, New Mexico.* National Speleological Society Bulletin, 52: 70 B 86.

Davis, D.G. (1989) *Helictite BushesCa Subaqueous Speleothem?* National Speleological Society Bulletin, 51: 120 - 124.

Droycon Bioconcepts Incorporated. ***BART***^{TM}- *Biological Activity Reaction Test User Manual.* 1999.

Garzke, W.H., Brown, D.K., Matthias, P.K., Cullimore, R., Wood, D.,Livingstone, D, Leighley, H.P., Foecke, T., and A. Sandiford (1997) *Titanic, The Anatomy of a Disaster, A Report from the Marine Forensic Panel (SD-7)* Proceedings of the 1997 Annual Meeting of the Society of Naval Architects and Marine Engineers, Ottawa, Canada.

Hauley, Barbara (2000) *Safety Charter Targets Old Tankers, Flags of Convenience.* Maritime Weekly. Maritime Reporter and Engineering News. March 2000. Pp20-21.

IACS (International Association of Classification Societies). *Bulk Carriers-Guidelines for Surveys, Assessment and Repair of Hull Structure.* 1994.

LaRock, E.J. & K.I.Cunningham (1995) *Helictite Bush Formation and Aquifer Cooling in Wind Cave National Park, South Dakota.* National Speleological Society Bulletin, 57: 43 - 51.

Ma, Kai-tung, I.R. Orisamolu, Robert G. Bea, Rong T. Huang. (1997) Towards *Optimal Inspection Strategies for Fatigue and Corrosion Damage. (SD-7)* Proceedings of the 1997 Annual Meeting of the Society of Naval Architects and Marine Engineers, Ottawa, Canada.

Mann, H (1997) *A Close-up Look at Titanic's Rusticles* Voyage 25 47 - 48.

Mansour, A.E., P.H. Wirsching, M.D. Lucket, A.M. Plumpton and Y.H. Lin. *Structural Safety of Ships. (SD-7)* Proceedings of the 1997 Annual Meeting of the Society of Naval Architects and Marine Engineers, Ottawa, Canada.

Ostryzniuk, N, Alford, G. and Monea, M. (1999) *Apparatus using Electro-Magnetic Forces (EMF) to Control Biofouling.* U.S. Patent Pending.

Pellegrino, C. and Cullimore, D.R. (1997) *A Study of the Bioarcheology of a Physically Disrupted Sunken Vessel* Voyage 25 39 - 46.

Pellegrino, C (2000) *Ghosts of the Titanic,* Harper Morrow NY

Wells, W., and H. Mann (1997) *Microbiology and Formation of Rusticles from the R.M.S. Titanic* Resource and Environmental Biotechnology 1: 271-281.

Discussion

David Livingston, Member

I would like to congratulate the authors on this most informative paper.

I was fortunate to participate with Dr. Cullimore on the expedition to *Titanic* in 1996 and 1998 and had the opportunity to dive to the wreck. This, together with the historical association with Harland and Wolff, ensures that I have a considerable interest in the phenomena of Rusticles and have sought to discover any occurrences on vessels or structures within our shipyard.

To date I have been unable to identify any occurrences, but a description of some of the structures where they have not occurred, or have not been recognized, may be of some interest.

The Thompson Dry Dock in Harland, and Wolff Shipyard was completed in 1910 during the build of *Olympic,* and she was the first vessel to occupy this dock closely followed by *Titanic*. It is reasonable to assume that the steel used for the construction of the sliding caisson gate was from the same source and of similar quality and standard to that used on *Titanic*.

This dock is still in daily use and the structure of the gate is unchanged. The gate is in good condition and operates efficiently.

Every 10 years it is un-stepped and dry docked for painting and general maintenance. Between these periods little work is done on the gate. To date there is no evidence of the formation of rusticles; however, the gate is due for its 10-year survey in the near future and the opportunity will be taken to look in the more inaccessible places.

The Flap or "Box" Gate of the Belfast Dry Dock was constructed in 1968. It also has a 10-year overhaul when it is removed and dry docked. This work is currently underway and afforded the opportunity to examine this structure in some detail. Again there is no evidence of rusticles.

Other significant structures recently in the shipyard were two semi-submersible Aker H3 Rigs. These units are around 25 years old and were dry docked for the first time. Interestingly, one had been laid up, under "care and maintenance," for some 12 years; the exterior of the pontoons are unpainted and protected by impressed current cathodic protection. Again, neither of these vessels exhibited any evidence of rusticles.

William H. Garzke, Jr., Member

Dr. Cullimore is a valued member of Panel SD-7, Marine Forensics and I welcome this opportunity to comment on this important work. When Dr. Robert Ballard discovered the *Titanic* wreck in 1985, he found the bow portion of the wreck covered with an icicle-type growth that he termed "rusticles." This has become the common reference word to bioconcretious structures. As I have been involved in the discovery of other deep ocean wrecks, these growths have been found on such wrecks of the MV *Derbyshire* (lost in September 1980), the German battleship *Bismarck* (lost in May 1941), and the USS *Yorktown* (lost in June 1942). However, such growths have not been observed on the wrecks of HMHS *Britannic* (lost in November 1916) and the RMS *Lusitania* (lost in May 1915). Panel SD-7 is working with two exploration teams that will examine the latter two wrecks to determine any traces of such growths. The authors' comments are welcomed on this phenomenon.

In 1997, Drs. Cullimore and Leighly introduced the maritime industry to microbial corrosion in the paper, *Titanic*, The Anatomy of a Disaster." However, continuing research now indicates that such corrosion is not limited to steel structures in the deep oceans. In a paper presented at the Corrosion 97 Conference, authors Rong Huan, Beverly McFarland, and Ronald Hodgman described the effects of microbial-induced corrosion in the cargo oil of crude oil tankers describing the activities of aerobic and anaerobic bacteria in oil tankers.

When the internal structures of cargo oil tanks are exposed to crude oil, gases, and seawater, the result has been the pitting of structure and reduction in material thickness. Hence, the steel structure of the tanker was being gradually reduced. If left unchecked, the result would be much the same as the authors described for the barrels containing radioactive waste—a structural failure of the tanker and a possible oil spillage. The Chevron Shipping Company discovered that the corrosion in its tankers was from the activities of microbial communities (rusticles) and not a single type of microbe. These microorganisms use hydrocarbons for food and in the process produce acids and gases that induce corrosion of the containing vessel.

The microbial ecology in the bottom of cargo oil tanks is subject to successional changes. As aerobic microbes become established, they deplete the concentration of oxygen and produce acids and gaseous by-products such as carbon dioxide. The presence of these acid-producing bacteria then enable anaerobic microbes to begin growing once the oxygen content is above 0.5 mg L^{-1} [1]. Anaerobic bacteria are a strict sulfate-reducing type and cannot grow or produce hydrogen sulfide unless there is low oxygen as well as acids and the proper hydrogen content present. Stagnant conditions are known to promote microbial-induced corrosion in uncoated tank bottoms. Bacteria tests from waters in San Francisco Bay, Los Angeles Harbor, and Singapore Harbor have shown their presence in seawater collected from the harbors of Singapore, San Francisco, Los Angeles and Nagoya.

As the authors have pointed out, inspection of tankers and bulk carriers is becoming a matter of increasing critical importance with the number of losses being sustained in older vessels of these types. Steel structures in the deep oceans are also potential victims of microbial corrosion. More research needs to be done to find ways to combat these microorganisms before catastrophic failures occur.

Additional reference

[1] Huang, Rong. T., McFarland, Beverly L., and Hodgman, Ronald Z., *Microbial Influenced Corrosion in Cargo Oil Tanks of Crude Oil Tankers*, Paper No. 535, NACE International, 1997.

SNAME *Transactions*, Vol. 108, 2000, pp. 197–212

Effect of Stacking Sequence on the Compressive Strength of Composite Laminates

Loern Halverson, Associate Member, MARCO Marine Seattle, and **Mark Tuttle,** Visitor, Professor, University of Washington Mechanical Engineering Department

ABSTRACT

The increasing use of unidirectional and bi-directional plies in composite boatbuilding will require more advanced engineering analysis to predict the strength of these high performance structures. This paper demonstrates analysis methods used in advanced composite research, and specifically problems associated with panels loaded in compression.

In this study experimental and analytical methods were used to investigate the effect of stacking sequence on the compressive strength of laminates. Sixteen-ply laminates made of 0°, 45°, 90°, and –45° unidirectional plies were tested to verify which stacking sequences were strongest. All laminates were quasi-isotropic, theoretically having identical in-plane properties for each stacking sequence. Nevertheless, experimental measurements show that ultimate strength varied by as much as 23% depending on stacking sequence. Various modes of failure were also observed, dependent on stacking sequence. Some specimen types failed by matrix cracking and others failed by fiber cracking. Holographic interferometry was used to observe out of plane displacement.

Analysis using Finite Element Methods was done in an attempt to predict the observed trend in compressive strength. Buckling strength, interlaminar matrix stresses, and fiber stresses were all analyzed. Results showed that although buckling strength, matrix stresses, and ply stresses could explain individual failures, a single failure criterion did not universally explain the experimentally observed trends. Although the materials tested here are commonly used in the aerospace industry, the problems identified also apply to advanced composites increasingly used in marine applications.

BACKGROUND INFORMATION

Traditionally, the use of composites in the marine industry has been relatively conservative. Economics and fabrication methods caused simple fiberglass lay-ups with resin-rich mixtures to be the norm for many years. However, today there is an increased use of highly engineered laminates, and advanced fiber materials are becoming more common. There is an increasing demand for lighter weight, higher strength, and increased stiffness in designs. Many of the materials and processes developed by the aerospace industry have recently found their way into composite boatbuilding.

Laminates made with chopped strand or mat, commonly used in the marine industry, are relatively straightforward to analyze. The material properties are isotropic, or the same in every direction or axis. Woven roving, also commonly used, has differing properties along its warp and fill directions, giving the designer the ability to tailor the properties of the finished laminate to the actual loads of the structure. [1] Recently, the use of unidirectional fiber plies is giving designers even more freedom to create a finished laminate of custom properties with the highest strength and lightest weight.

Production methods are also improving. Automated lay-up and resin infusion methods are continuously being developed. Some manufacturers are beginning to use pre-preg laminates. [2]

The use of advanced laminate schedules, production methods, and materials will create a need for greater engineering understanding of the behavior of the laminates. As materials become more anisotropic with the use of unidirectional fibers and lower resin ratios, predicting material behavior will become increasingly complex.

The purpose of this study is to understand the effects of stacking sequence on the compressive strength of laminates. Calculating the tensile strength of composite laminates is relatively straightforward, but failure due to compressive loading is more complex since bucking and/or delamination phenomena must be considered. The intent here is to show that the increased use of advanced composites in vessel construction will require more engineering knowledge of the laminates used. Compression in multi-ply laminates is particularly complex.

Experimental testing was performed using both graphite/bismaleimide (IM7/5260) and graphite/ polyimide (IM7/K3B) specimens, machined from parent laminates. The parent laminates were cured using an autoclave and vacuum bag techniques, and involved a symmetrical stacking sequence with equal numbers of 0°, +45°, 90°, and –45° plies in all cases. Hence, all laminates are classified as "quasi-isotropic". This type of laminate is commonly used in practice, in part because mechanical properties (e.g., stiffness or strength) are expected to be identical in all directions within the plane of the panel. As will be seen, this is not true in the case of in-plane compressive strengths.

This paper uses the quasi-isotropic laminate analysis and experimental results as an example of methods that can be used with advanced composites engineering. It is not intended to inhibit designers from the use of these composites, as bending and tensile analysis methods are well documented and proven. Unidirectional low resin composites offer the greatest opportunity for the designer to tailor the material for the highest strength, best stiffness, and lowest weight combination. However, this paper will demonstrate that compressive analysis is a complex problem and care must be used when designing for loads near failure.

RESULTS SUMMARY

Tests demonstrated the compressive strength of quasi-isotropic laminates can vary by as much as 23%, depending on the stacking sequence (determined by loading direction). This observed trend showed an increase in compressive strength as the 0° plies were located toward the midplane of the laminate. This was opposite of the expected results. Results were expected to show an increase in compressive strength as the 0° plies were located toward the outside of the laminate, presumably increasing global buckling strength.

Both experimental and analytical/numerical studies were performed during this project [3]. During the experimental study, compressive tests of 16 ply quasi-isotropic laminates were performed. Two different material systems were used: IM7/K3B and IM7/5260. Specimens were tested in a fixture designed to minimize or remove the effects of free edge stresses. Holographic interferometry was used to observe out-of-plane displacements during compressive tests. Observations of deformations under load and indications of failure initiation mode were recorded for each laminate type. Three series of laminates were tested, and specimens of these laminates were loaded to total failure to measure ultimate compressive strength.

The analytical portion of the study consisted of two separate investigations. The first analysis was performed to determine whether the observed trend in compressive strength could be explained on the basis of free edge stresses. The fixture used to hold specimens during testing was designed to minimize free edge stresses, but analysis was still completed to show that free edge effects were removed.

Both a mechanics of materials approach and Finite Element Analysis (FEA) were used to calculate the stresses developed at a free edge. Initial calculations were for laminates in tension and compared favorably to other published results. The method was then applied to the experimentally tested laminates loaded in compression. It was assumed that laminates with high compressive free edge stresses would exhibit high ultimate compressive strength, whereas those with the most-tensile free edge stresses would exhibit the lowest compressive strength. The strength trend predicted by the mechanics of materials approach did not agree with the experimental results, suggesting that free edge effects had no effect on the observed compressive strength.

The second analytical portion of this study was a Finite Element Analysis of the buckling modes, interlaminar stresses, and ply stresses of the actual tested specimens. Observations of the compressive experimental tests showed that global buckling was not the cause of failure in the tested specimens. Hence, it was hypothesized that localized buckling at the ply level is responsible for the observed trend in compressive strength. Finite element analysis allowed calculation of relative buckling loads for each specimen type. These loads increased as the 0° plies were located toward the center of the laminate, as in this study's original experiments. To explain why this behavior changed in later experimental tests, interlaminar stresses were calculated for each specimen. These stresses showed that some lower strength laminates developed high interlaminar shear stresses. However, interlaminar shear stress alone did not predict the experimental results. Individual ply stresses were also calculated and compared to measured trends. The stresses calculated by FEA methods gave much insight into how each specimen bears compressive loads, but did not provide conclusive evidence to predict the range of experimental trends measured.

Quasi-isotropic Laminate Description

All laminates tested in this study were quasi-isotropic, consisting of equal number of 0°, +45°, -45°, and 90° plies. These continuous-fiber laminated composites are composed of multiple plies of unidirectional fibers. All quasi-isotropic laminates tested in this study were also symmetric. Specific stacking sequences used will be described in following sections.

Quasi-isotropic laminates theoretically have similar properties in every direction in the plane of the laminate. Each ply can be considered transversely-isotropic, defined by having identical properties along two axes and different properties along their principal axis (fiber direction). Quasi-isotropic laminates have plies oriented at equal divisions of 180°, and have an equal number of plies at each of the fiber angles. Examples of quasi-isotropic laminates are $[0/60/120]_T$, $[0/45/90/-45]_T$, and $[0/+45/-45/90]_s$ (using standard composite laminate notation described below). Quasi-isotropic laminates theoretically have the same properties at any loading angle along the *xy* plane defined by their surface. This proves true for a laminate in tension (at any point away from any free edges). However, in compression this study shows that quasi-isotropic laminates have different strengths based on the loading direction, or "stacking sequence".

Composite Lamination Notation

Laminates are described in this paper using the following notation. Uni-directional plies are described by the orientation of fiber direction. The *x* direction is commonly the primary loading direction and is labeled 0°. Angles are measured counter-clockwise. Plies with their fibers perpendicular to the *x* axis are labeled 90°. Subscripts are used to describe if the notation describes a symmetric laminate ($_S$) or if the total number of plies are listed ($_T$). Numeric subscripts are used if 2 or more plies are placed together, or if sequences are repeated. For example, the following symmetric laminate, $[0/+45/-45/90]_{2s}$ is equivalent to:

[0/+45/-45/90/0/+45/-45/90/90/-45/+45/0/90/-45/+45/0]

EXPERIMENTAL RESULTS

Experimental Test Description

A total of 85 specimens were loaded in uniform axial compression. In each test a fixture was used to maintain alignment of the specimens during loading. Compressive loads were applied with a Tinius Olsen universal testing machine. During loading, holographic interferometry was used to observe out-of-plane displacements of the specimens.

Compressive Specimens

The compressive specimens were cut from three different composite panels, using two different material systems. All panels were 16 ply quasi-isotropic laminates of graphite fiber. All panels were made with IM7 graphite fibers. Two resins were used, K3B and 5260. K3B is a thermoplastic polyimide manufactured by DuPont, and 5260 is a bismaleimide resin manufactured by BASF. Table 1 lists all the tested laminates.

Table 1: Tested Laminates

Specimen Type	Stacking Sequence	Panel Number	Material
A	$[0/45/90/-45]_{2s}$	#716/717	IM7/K3B
B	$[45/0/-45/90]_{2s}$	#716/717	IM7/K3B
C	$[90/45/0/-45]_{2s}$	#716/717	IM7/K3B
D	$[45/90/-45/0]_{2s}$	#716/717	IM7/K3B
E	$[0_2/45_2/-45_2/90_2]_s$	#806	IM7/5260
F	$[45_2/0_2/90_2/-45_2]_s$	#806	IM7/5260
G	$[45_2/90_2/0_2/-45_2]_s$	#806	IM7/5260
H	$[90_2/45_2/-45_2/0_2]_s$	#806	IM7/5260
I	$[0/45/90/-45]_{2s}$	#Q1-3-6	IM7/5260
J	$[45/0/-45/90]_{2s}$	#Q1-3-6	IM7/5260
K	$[90/45/0/-45]_{2s}$	#Q1-3-6	IM7/5260
L	$[45/90/-45/0]_{2s}$	#Q1-3-6	IM7/5260

Note that specimens of four different stacking sequences could be made from each panel by cutting the specimens at 45° increments from the panel's 0° axis.

All specimens were cut using an abrasive diamond cutting wheel. Specimen dimensions were 1.00" x 4.90", and were 0.090" thick.

Compressive Test Fixture

The compressive test fixture was designed to maintain the alignment of the compressive specimen during loading. The specimens were clamped at the top and bottom, and constrained by simple supports on each side (knife edge supports). These constraints were chosen to remove the potential effect of the free edge stresses on the ultimate compressive strength of the specimen. The fixture is shown in Figure 1.

Figure 1: Compressive Test Fixture

Laser Holographic Interferometry

Out of plane displacements were observed using holographic interferometry. Holographic interferometry is a non-destructive experimental method that uses two collimated light beams and a holograph to observe out-of-plane deformations[4-6].

For the compressive tests in this project, a Newport Corporation HC-1000 Holocamera was used to create a hologram of the specimen in a clear thermoplastic plate with a 7-mW HeNe laser. The loaded specimen is observed through the hologram formed in the thermoplastic plate. Fringes are created from the interference between the light reflected from the specimen and the hologram in the thermoplastic plate. These fringes represent out-of-plane displacement of the specimen. Besides being a non-destructive method and allowing the specimens to be viewed in real-time, holographic interferometry has the advantage of not being critically sensitive to surface finish. Relatively rough surfaces will still reflect enough light to generate fringes. For a summary of modern holographic interferometry techniques, see reference [7].

After a holograph of the specimen is made, the actual specimen is viewed through the thermoplastic plate. Any displacement of the specimen towards the holograph will cause interference between the reflected object beam and the holograph illuminated by the reference beam. This interference appears as light and dark fringes as the interfering beams convert phase relations into differences in amplitude.

Test Procedure

Two types of tests were performed. The majority of the tests were compressive tests to failure. Holographs were taken continuously as a specimen was loaded. The sensitivity of laser holography made it necessary to make a new hologram every 1000 lbs because the out-of-plane displacement created so many fringes that they became blurred together. Each specimen was continuously observed with holography until ultimate compressive failure. The process was videotaped so it could be reviewed.

A second series of tests were performed to confirm repeatability of the observed displacements. A series of specimens were loaded to approximately 90% of ultimate failure while being observed using holography. The specimens were then unloaded, removed, and reclamped into the fixture. A second loading experiment confirmed that the displacements observed in the first loading were repeated.

Compressive Strength Results

A total of 64 specimens were tested to ultimate compressive failure. The experiments were performed to observe how the location of the 0° plies affected the ultimate compressive strength. Holographic images were also inspected to note any characteristic deformation shapes for each specimen type and to note any correlation between observed "bulges" and final failure locations.

Ultimate Compressive Strengths

The first series of tests were of specimens A, B, C, and D. These specimens were all cut from Panels #716 ($[0/45/90/-45]_{2s}$) and #717 ($[0/45/90/-45]_{2s}$). All were IM7/K3B material. The averaged compressive test results are shown in Table 2.

Table 2: Compressive Test Results, Specimens A-D

Specimen	Stacking Sequence	Ultimate Compressive Strength
A	$[0/45/90/-45]_{2s}$	73,900 PSI
B	$[45/0/-45/90]_{2s}$	79,750 PSI
C	$[90/45/0/-45]_{2s}$	87,300 PSI
D	$[45/90/-45/0]_{2s}$	102,000 PSI

The test results show that in specimens A through D, compressive strength increases as the 0° plies move toward the midplane of the laminate. This agrees with trends seen in preliminary tests.

The second series of tests were performed on specimens E, F, G, and H. All were cut from Panel #806($[0_2/45_2/-45_2/90_2]_s$) and were of IM7/5260 material. These specimens have similar plies paired together. This stacking sequence was chosen in an effort to magnify the trends. The averaged compressive test results are shown in Table 3.

Table 3: Compressive Test Results, Specimens E-H

Specimen	Stacking Sequence	Ultimate Compressive Strength
E	$[0_2/45_2/-45_2/90_2]_s$	66,800 PSI
F	$[45_2/0_2/90_2/-45_2]_s$	95,400 PSI
G	$[45_2/90_2/0_2/-45_2]_s$	87,600 PSI
H	$[90_2/45_2/-45_2/0_2]_s$	81,500 PSI

Unlike specimens A through D, specimens E though H do not show an increase in strength as the 0° plies move toward the center. Note that none of the E through H specimens correlate directly to the stacking sequence of specimens A through D. Specimens E and H have ±45° plies adjacent to each other, and specimens F and G have the 0° and 90° plies adjacent to each other. The results do not lead to any conclusions about compressive strength based on ply location.

A third series of tests were performed with identical laminate schedules as specimens A, B, C, and D, but with the IM7/5260 material of specimens E, F, G, and H. These specimens are labeled I, J, K, and L, and were cut from Panel Q1-3-6 ($[45/0/-45/90]_{2s}$). The averaged compressive test results are shown in Table 4.

Table 4: Compressive Test Results, Specimens I-L

Specimen	Stacking Sequence	Ultimate Compressive Strength
I	$[0/45/90/-45]_{2s}$	71,100 PSI
J	$[45/0/-45/90]_{2s}$	86,000 PSI
K	$[90/45/0/-45]_{2s}$	73,500 PSI
L	$[45/90/-45/0]_{2s}$	95,400 PSI

It was expected that specimens I through L would show the same trend as specimens A through D. However, the test results show that specimens I through L have a different compressive strength trend. The tests show specimen type J ($[45/0/-45/90]_{2s}$) is stronger than specimen type K ($[90/45/0/-45]_{2s}$), where in previous tests the corresponding specimen type C ($[45/0/-45/90]_{2s}$) is stronger than specimen type B ($[90/45/0/-45]_{2s}$). Accordingly, this third series of tests does not agree with the theory that compressive strength increases as the 0° plies are located toward the midplane of the laminate. Averaged results from each specimen type can be compared in Figure 2.

Specimen Deformation and Failure Observations

Tests of specimen types A through D and types E through H were observed with holographic interferometry. Each test was recorded through the holocamera by videotape while the load was applied to the compressive test fixture. This allowed tests to be reviewed and compared easily. Six tests of specimens A through D were recorded, and four tests of specimens E through H were recorded.

Approximate measurements of load vs. displacement were taken with a dial indicator on the crosshead of the testing machine. All specimens showed linear deformation under load, and the modulus of elasticity for each specimen was the same for each specimen type (approximately 5.4 Msi). Measurements were not accurate enough to record behavior near ultimate failure.

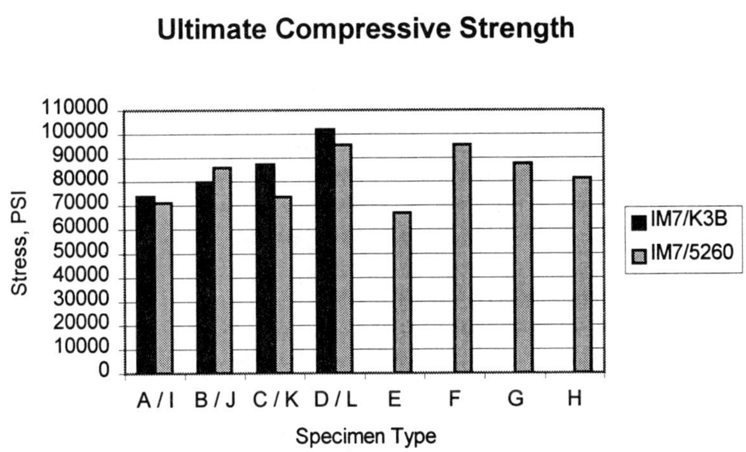

Figure 2: Ultimate Compressive Strength Results

Global buckling could not be seen in any test. The holograms showed random elliptical "bulges" on 90% of the specimens, but did not show any predictable mode shape. It was concluded that the explanation of the trend in compressive strength was not a global buckling problem. Figure 3 shows a typical interferometry image.

Figure 3: Specimen C (#15) 6,300 lbs

The elliptical "bulges" observed were theorized to be local buckling of one or more plies at a void or discontinuity of properties.

Experimental Results Summary

Specimens A, B, C, and D

Tests of these specimens showed that ultimate strength increased as the 0° plies were located toward the midplane of the laminate, as was expected from preliminary tests by the Boeing Company. Over 8 tests of each laminate type were performed. Laminate material was IM7/K3B.

Failures of the outer ply before ultimate failure was observed in all A type specimens. Inspection of the plies under a microscope showed fiber failure in the outer 0° ply. This was accompanied by audible cracking sounds, and a visible crack across the surface of the specimen. This cracking began to occur up to 1,000 psi before final failure. Figure 4 shows a micrograph cross section of a type A specimen. The fiber cracking of the outer ply can be seen.

Figure 4, Specimen A fiber cracking

Type B specimens also exhibited audible cracking, but were not always accompanied by visible cracks on the surface of the laminate. Microscopic inspection of type B specimens with visible cracks showed that fiber failure had occurred in the 0° ply located just under the outer 45° ply. This fiber failure in the single ply occurred up to 1,000 psi before ultimate failure.

Specimens C and D showed no characteristic deformation shape between tests. These specimen types had the highest compressive strength, and tests did not exhibit any visible cracking.

Specimens E, F, G, and H

These specimens paired similar plies together in an effort to magnify the trends seen in the first tests. Ultimate strength tests showed that the trend in strength did not increase as the 0° ply was located toward the midplane of the laminate, as it did in laminates A through D. The actual laminate schedule switched the order of the 0°, 45°,90°, and -45° plies so there were no laminates corresponding to specimens tested in the A through D series. These laminates were made of IM7/5260 material.

All type E specimens showed splintering of the outer 0° plies after failure, but did not show the pre-failure cracking of the outer ply observed in specimen type A. Most specimens did not exhibit any audible or visible cracks before ultimate failure.

Specimens F and G, the strongest specimens, did not show any characteristic failure mode or deformation patterns. These were also the strongest types of the series and did not exhibit audible or visible cracks before ultimate failure.

Type H specimens exhibited matrix cracking in the outer 90° ply. Inspection of these cracks under a microscope showed them to extend through multiple plies. These cracks were also seen to propagate between plies for some distance before continuing through the matrix into other plies. No fiber failures (besides ultimate failure location) were seen. Figure 5 shows this matrix cracking.

Figure 5, Specimen H matrix cracking

Specimens I, J, K, and L

These laminates matched the laminate schedule of specimens A through D, but were made of IM7/5260 material. Results deviated from the tests of specimens A through D.

Specimen type I was compared to specimen A, which has the same laminate schedule. No evidence of the outer ply failure exhibited by type A specimens was seen. No indication of where failure was initiated could be found.

Type J specimens showed delamination of the 0° ply located under the 45° specimen, but did not show fiber failure similar to type B specimens. No visible cracks other than at the final failure location were found. Specimen J was slightly stronger than corresponding specimen B and was the second strongest of the series.

Specimens K and L showed no characteristic cracking or failure modes. Specimen K proved to be considerably weaker than corresponding specimen C. Specimen L continued to be the strongest of the series.

Summary

The tests show that failure is initiated by different modes in different series. Specimens A and B show fiber microbuckling failure in 0° plies in multiple locations, and at some time before ultimate failure. Perhaps specimens C and D fail by the same method with the fiber microbuckling failure immediately causing ultimate failure. The second set of tests show a different trend than the first, and failure seems to be initiated in a different manner. Specimen H showed matrix cracks that extend through multiple layers and delamination. Specimens E, F, and G could be failing by the same method, or perhaps by a combination of ply failures and matrix cracks. The third series of tests do not exhibit the fiber microbuckling failure shown by specimens A and B of the first series, demonstrating the effect of different material systems on the mode of failure.

ANALYSIS

Two analyses were performed. The first was calculations of stresses developed at the free edge of the specimen. The compressive fixture was designed to lessen the effects of free edges, but calculations were performed to confirm that the observed trend was not caused by free edge stresses.

The second and larger analysis consisted of a Finite Element Model of the loaded specimen. Non-linear analysis is used to calculate relative buckling strength of each laminate type. The model was also used to calculate interlaminar matrix stresses, as well as stresses in each ply.

Free Edge Analysis

The purpose of this analysis was to determine whether free edge stresses were the cause of the trends in the experimentally measured compressive strengths of the tested laminates described in the previous section.

Extensive research has been performed on the effects of a multiple ply laminate in tension [8], [9]. It has been shown that a loaded laminate will develop tensile forces between the plies at the free edge, and will tend to delaminate. To prove that free edge stresses were not initiating the failures in the study, two methods were used.

The first method was a mechanics of materials approach [10]. The method uses ply stresses solved by classical lamination theory (CLT) to calculate the moment developed at the free edge, see equation 1. The stresses between plies can then be estimated, see equation 2.

$$M_z = \sum_{n}^{plies} \sigma_y h_o \left(\frac{h_o}{2} + n h_o \right) \qquad \text{(eq. 1)}$$

$$\sigma_z = \frac{45 M_z}{14 h^2} \qquad \text{(eq. 2)}$$

The x-axis is the loading direction, the y-axis is across the plane of the laminate, and the z-axis is through the plies. The ply thickness is h_o, and the ply number is n.

The results showed no correlation between the experimentally observed results and those predicted if failure was caused by free edge stresses.

The second method used to predict free edge stresses was Finite Element Analysis. A 3-dimensional model with all 16 plies was created. A convergence study was performed using both element density and model aspect ratio to determine a reasonable model size. Before modeling the tested stacking sequences, laminates from other published studies (in tension) were modeled to verify this analysis. Results from the FEA model matched the published results.

The analysis of the experimentally tested 16 ply laminates predicted trends much different then those observed experimentally. However, the FEM analysis did show that large shear stresses were developed at the free edge between the ±45° plies of specimens E and H, which were the two weakest laminates. Later analysis also shows that large shear stresses develop in the clamp area of these same specimens.

In conclusion, it is assumed that free edge stresses are not the cause of the observed trend in compressive strength. Both the mechanics of materials approach and the Finite Element Model predicted that if free edge stresses initiated failure, a much different trend would be observed.

Laminate Analysis

A Finite Element Analysis (FEA) model was created in an attempt to analytically predict the experimentally observed trend in compressive strength. The observations with holographic interferometry gave indications that the failure could be either a local buckling phenomenon, interlaminar failure in the matrix between the plies, or in the plies themselves. The FEA model created allowed analysis of all three possible failure explanations. Using a combination of 2-dimensional shell elements and 3-dimensional brick elements, these models were used to solve for buckling load, interlaminar stresses, and for stresses within each ply. Non-linear analysis was used with ANSYS software.

The following discussion will show that calculated buckling load increases as the 0° plies are located toward the center of the laminate. This agrees with the experimental results of specimens A through D, but not E through H or I through L. The interlaminar stress analysis also shows that specimens E and H with adjacent ±45° plies develop large shear stresses, explaining their weak compressive strength compared to specimens F and G with separated 45° plies. Comparison of the maximum interlaminar and ply stresses between all laminate types also gives much insight to the compressive behavior of the specimens and will be discussed.

FEA Model Description

A combination of 2-dimensional shell elements and 3-dimensional brick elements were used to model the plies and interlaminar matrix regions in each specimen. Shell elements (2-dimensional) were used to model each unidirectional ply. Brick elements (3-dimensional) were then be used to model the resin-rich region between plies, creating a model of combined 2- and 3-dimensional elements.

Originally, a model entirely of brick elements was attempted, but the software limit in the number of elements prohibited using enough elements to accurately model the interaction between plies. Layered elements are also available in ANSYS, but would not calculate the interaction between plies. Combining the shell and brick elements allowed a reasonable size model that would allow calculation of both buckling shapes and interlaminar stresses.

All sixteen plies were modeled in a 4.90"L x 0.75"W x 0.0896"TH model. The ply thickness was 0.0056", but was modeled as 2-dimensional so plies had no thickness in the 3-dimensional model. The distance between plies was 0.0056". This created 16 plane areas (shell elements) separated by 15 volumes (brick elements). Clamped supports on each end and simple supports along each side were modeled. Actual constraints for the clamped end were in the x, y, and z direction, as well as y axis rotation. Simple supports were modeled with constraints in the z direction. The experimental test specimens were 4.90"L x 1.00"W x 0.090"TH, and the fixture clamps made the gage area 0.75"W.

Loads were applied in steps. Large steps were used to load the model to near the buckling load (for the buckling models). Small steps were then used to more accurately solve for the buckling load.

This method of modeling ply and matrix regions made it necessary to make some approximations. Each ply was represented by a 2-dimensional plane of elements located at the midplane of the ply. This caused each 3-dimensional matrix element to extend from midplane to midplane of each ply. Therefore, matrix layers are much thicker than in the actual laminate, causing a doubling of material in the areas where the ply thickness and the matrix elements overlap. In models used to calculate interlaminar stresses, the actual matrix properties were used. Relative stresses and out-of-plane displacements should show accurate trends. Matrix properties lower than the neat-resin properties were used for models calculating buckling shape to achieve more independent deformation between plies.

Models using full matrix properties had end loads slightly larger than measured. Models with lower matrix properties had end loads approximately equal to the measured loads.

Material Properties

Three different sets of material properties were used in the FEA models. Properties for IM7/K3B and IM7/5260 were provided by the Boeing Company and are shown in Tables 5 and 6. Values for tensile E_1 and compressive E_1 modulus of elasticity were averaged. Values for shear modulus G_{23} were calculated from the elasticity solution $G_{23} = E_2 / 2(1 + \nu_{matrix})$. These values were used for models calculating displacements and stresses. For models analyzing buckling strength, matrix properties were reduced until end loads approximately matched the experimental tests and buckling loads approximately matched ultimate failure levels. The resulting properties are shown in Table 7. Note specimens A, B, C, and D were made of IM7/K3B. Specimens E, F, G, and H were made of IM7/5260, and specimens I, J, K, and L were also made of IM7/5260.

Table 5: IM7/K3B Ply Properties

E_1	22.0 Msi
$E_2=E_3$	1.28 Msi
$\nu_{12}=\nu_{23}=\nu_{13}$	0.39
$G_{12}=G_{13}$	0.82 Msi
G_{23}	0.481 Msi
Matrix E	0.49 Msi
Matrix ν	0.33
Matrix G	0.184 Msi

Table 6: IM7/5260 Ply Properties

E_1	22.5 Msi
$E_2=E_3$	1.34 Msi
$\nu_{12}=\nu_{23}=\nu_{13}$	0.34
$G_{12}=G_{13}$	0.92 Msi
G_{23}	0.504 Msi
Matrix E	0.66 Msi
Matrix ν	0.33
Matrix G	0.248 Msi

Table 7: Buckling Model Properties

E_1	20.0 Msi
$E_2=E_3$	2.1 Msi
$\nu_{12}=\nu_{23}=\nu_{13}$	0.39
$G_{12}=G_{23}=G_{13}$	0.82 Msi
Matrix E	20 Ksi
Matrix ν	0.33

Results from the analysis show that changing the material properties affected the stress values and numerical results, but did not change any trends or relative strengths between specimens.

FEA Results

The remaining discussion describes the approaches used to predict the measured experimental trends. Applicable results from the FEA models (including buckling analysis, out-of plane displacements, interlaminar stresses and ply stresses) are summarized in the following sub-sections:

1. Buckling Strength
2. Exponential Growth of Matrix Stresses
3. Out-of-Plane Displacements
4. High Shear Stresses in Adjacent ±45° Plies
5. Matrix von Mises Stress
6. Ply Stresses

1. Buckling Strength

Non-linear analysis in ANSYS was used to solve for buckling loads. To solve for buckling load, an end displacement load was applied in even increments until a buckling mode shape was found. For post-buckling loads, the computer would not converge on a solution in a reasonable amount of iterations. The last pre-buckling load was considered to be the buckling load. A post-buckling mode shape could be observed, but stress and displacement results of the non-converged solution were not considered accurate. Figure 6 shows a typical post buckling mode shape.

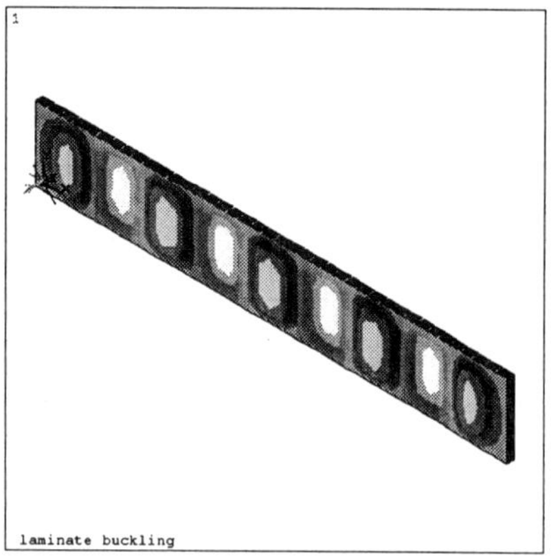

Figure 6: Specimen E Post Buckling z-displacement

Table 8 summarizes the buckling load for laminates A through D and I through L. It can be seen that the weakest laminate A has the lowest buckling load. Laminates B, C, and D buckle at the same end displacement. This analysis does not differentiate between laminates A through D and laminates I through L because the properties are modified to induce buckling (see Buckling Model Properties, Table 7).

Table 8: Buckling Load, Laminates A through D and I though L

Laminate Type	Stacking Sequence	Buckling End Displacement (in)
A or I	$[0/45/90/-45]_{2s}$	0.082
B or J	$[45/0/-45/90]_{2s}$	0.083
C or K	$[90/45/0/-45]_{2s}$	0.083
D or L	$[45/90/-45/0]_{2s}$	0.083

Table 9 lists the buckling load of laminates E through H. Laminate E (with the outer 0° plies) has the lowest buckling load and is the weakest experimentally. Laminate H has the highest buckling strength, but is the second weakest laminate of the series. (Later sections will note that large interlaminar shear stresses are developed in laminates E and H). Buckling analysis does not predict the observed compressive strength trend of laminates E, F, G, and H. (Recall observed trend is: E, H, G, and F in order of increasing strength.)

Table 9: Buckling Load, Laminates E through H

Laminate Type	Stacking Sequence	Buckling End Displacement (in)
E	$[0_2/45_2/-45_2/90_2]_s$	0.082
F	$[45_2/0_2/90_2/-45_2]_s$	0.0835
G	$[45_2/90_2/0_2/-45_2]_s$	0.0835
H	$[90_2/45_2/-45_2/0_2]_s$	0.1035

2. Exponential Growth of Matrix Stresses

Reviewing the matrix region stresses developed in the buckling models discussed above reveals that certain interlaminar stresses show exponential growth as the load is applied. Plotting all matrix stresses in a laminate show that tensile σ_z stresses grow exponentially as the load nears the buckling load. Figure 7 shows the maximum σ_z stress and location for specimens A through D. The stress values shown are for the matrix region between the plies referenced on the figure. Note that stresses in all matrix regions were recorded, but only the region with the maximum stress was plotted for each specimen type (A, B, C, and D).

The exponential behavior of the maximum σ_z stress can cause a different strength hierarchy to be predicted at different end displacement loads. Reviewing the stress levels in Figure 7 at a displacement of 0.08" gives a strength hierarchy D, A, C and B from weakest to strongest, respectively. Noting the slope of the σ_z stress in the B type laminate shows that it would surpass other laminates if the applied load was continued. The actual stress values are very low (approximately 350 PSI), but exponential stress growth near the buckling load could be causing stress levels large enough to initiate failure. Also, the Buckling Model Properties (Table 7) used have a low modulus that will cause lower stresses.

Most other matrix stresses exhibit linear behavior with increasing load, including all σ_z, τ_{yz}, and τ_{xz} stresses. The τ_{xy} stress shows a slight amount of exponential growth, but stress levels are very low. One or two ply interfaces show σ_y stresses that exhibit exponential growth under load, but the stress levels are also low.

Note that this exponential stress growth under load is observed only in the buckling models (with Buckling Model Properties, Table 7). Review of σ_z stresses in models with IM7/K3B and IM7/5260 properties (Table 5 and 6) show a very slight deviation from linear.

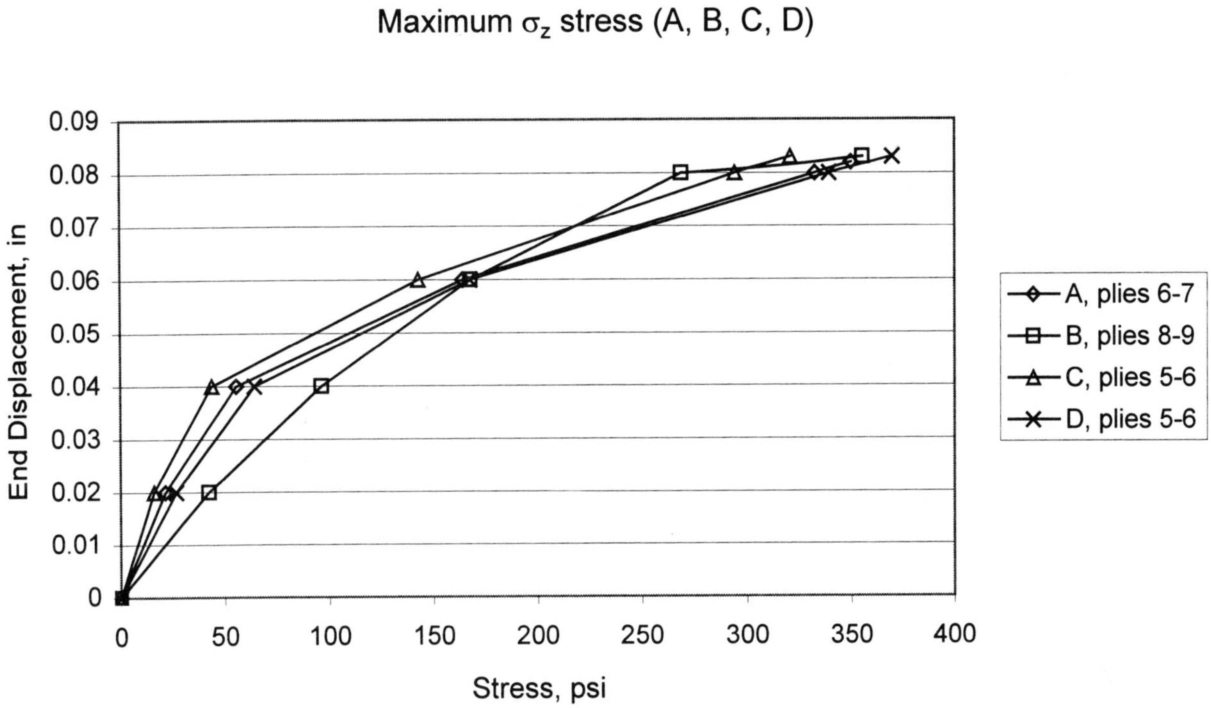

Figure 7: Maximum Tensile σ_z Stress (A, B, C, D)

3. Out of Plane Displacement

An attempt to correlate out-of-plane displacement with compressive strength was also made. As shown in Figure 8, the maximum displacement in the outer ply for specimens A through D does not correlate with the measured experimental results of increasing strength: A, B, C, to D. Specimens I through L show the same trend in maximum outer ply displacement. Similarly, specimens E through H have a different trend in max outer ply displacement, which does not correlate with the experimental order of increasing strength: E, H, G, to F.

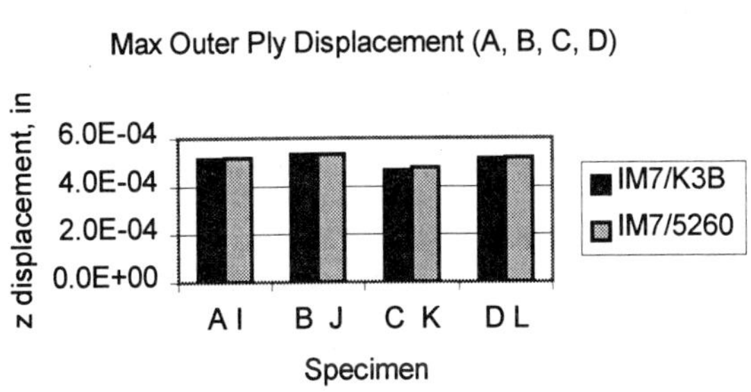

Figure 8: Max Outer Ply Displacement (A-D, and I-L)

The displacements between each ply were also recorded for each specimen type. It was interesting to note that the ply interface with the largest displacement did not always coincide with the highest σ_z stress. This is explained because the maximum σ_z stress can be located at stress concentrations caused by the constraints, whereas the maximum displacement will always be located away from the clamps. Figure 9 shows that maximum inter-ply displacement for specimens A - D roughly coincides with the trend in displacement for the outer ply of the specimens (Figure 8). The location of maximum displacement occurs at the same ply interface for corresponding specimens in series A-D and series I-L. End displacement is .009 inches for both graphs.

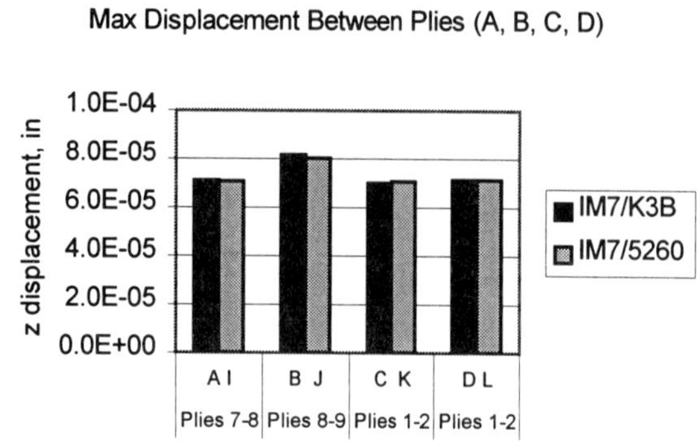

Figure 9: Max Displacement Between Plies (A-D, and I-L))

Plots of displacement from models with Buckling Model Properties show that out-of -plane displacement grows exponentially, similar to the σ_z stress in these models. (Displacement in models with IM7/K3B and IM7/5260 properties is nearly linear.) Also, as buckling loads were approached, the location of maximum displacement in each ply did not always coincide. Different plies in a laminate may take on different deformation mode shapes as they begin to buckle. This made it difficult to calculate accurate displacement between plies for near and post-buckling conditions using the Buckling Model Properties.

4. High Shear Stresses In Adjacent ±45° Plies

Both specimens E and H have ±45° plies located adjacent to each other. All other tested specimens have 45° angle plies separated by 0° and 90° plies. Results from the FEA model show that locating the ±45° plies together causes high shear stresses to develop. Figure 10 shows the maximum shear stress and the location between plies. It is interesting to note that this maximum τ_{yz} shear does not occur between the ±45° plies. The figure shows how specimens E and H have τ_{yz} shear stresses differing as much as 3,350 psi from specimens F and G. In comparison, shear values for Specimens A though D are much lower, with the largest increase at 1,650 psi, see Figure 11.

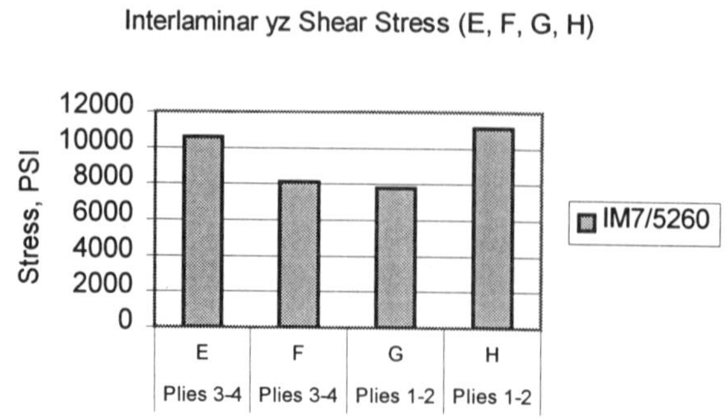

Figure 10: Maximum τ_{yz} Shear (E, F, G, H)

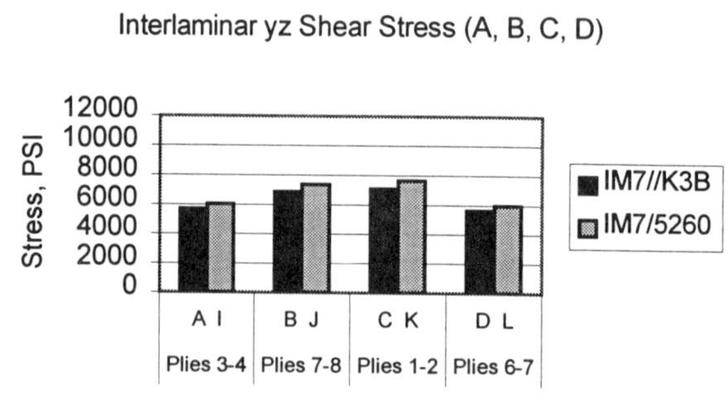

Figure 11: Maximum τ_{yz} Shear (A, B, C, D)

5. Von Mises Stresses

As a final effort to associate matrix stress to compressive strength, the maximum von Mises stress was plotted for each specimen. Figure 12 shows that there is little change in maximum von Mises stress between the A through D and I through L specimens. Figure 12 shows that specimens F and G actually have the highest von Mises stress, opposite to the τ_{yz} trend observed above.

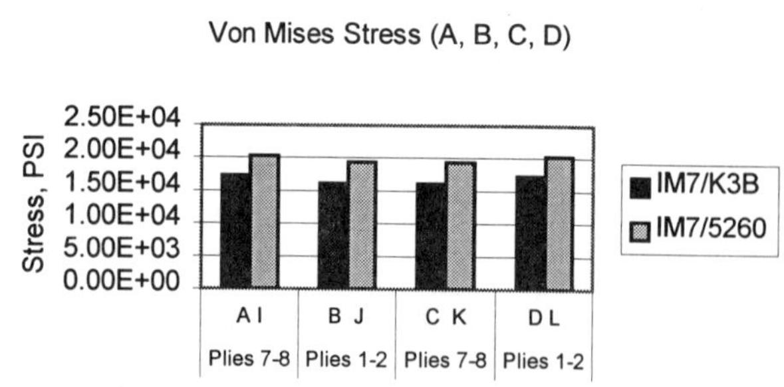

Figure 12: Von Mises Stress (A-D, and I-L)

6. Ply Stresses

Ply stresses were easily recorded from the 2-dimensional shell elements. Three stresses, σ_x, σ_y, and τ_{xy}, were tabulated for each ply. These were transformed into stresses aligned with the fibers of each ply (σ_1, σ_2, and τ_{12}) based on the angle of the ply. Transformation matrices were used for calculating σ_1 and σ_2 stresses in 45° and -45° plies. Aligned stress values for 90° plies were found by equating σ_y to σ_1 and σ_x to σ_2. Figure 13 shows that stress values are large, but maximum values are consistent in each laminate. All maximum ply stress are nearly identical for all specimens in each series, even though maximum stresses are located in different plies in each laminate.

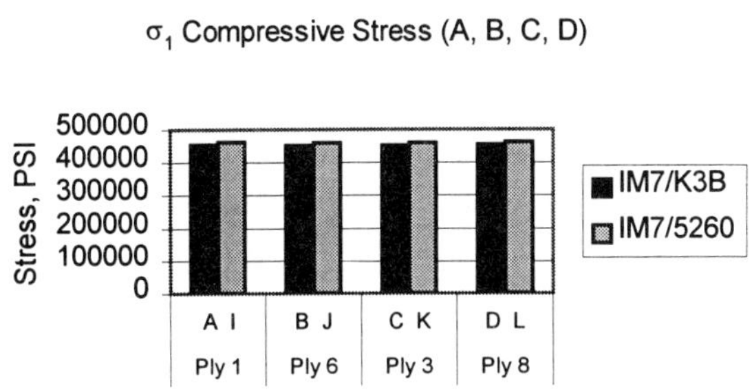

Figure 13: Ply σ_1 Stress (A, B, C, D)

Buckling and Stress Analysis Summary

In conclusion, neither analysis of buckling load, interlaminar stresses, or ply stresses gives a clear prediction of ultimate compressive strength of the tested laminates. The analysis of with FEA models gives much insight into the interactions between the plies in a multi-ply laminate. However, the results do not provide a clear explanation of the experimental trend in compressive strength.

CONCLUSIONS

The series of experimental compressive tests to ultimate failure demonstrated that there is not a universal trend in compressive strength based on the location of the 0° plies. Observation of out-of-plane deformation with holographic interferometry and inspection of loaded specimens showed that different modes of damage were occurring before ultimate failure in different specimens. Some specimens indicated that fiber failure initiated ultimate failure, while others indicated matrix failure was the cause.

Specimens A, B, C, and D

This test series showed a strong correlation between the location of the 0° plies and the ultimate compressive strength. Specimens with 0° located toward the midplane of the laminate were much stronger than those with 0° plies at the surface of the laminate. Specimens were made of IM7/K3B material.

Specimen A was the weakest specimen in the test series. All type A specimens exhibited audible and visible cracks on the surface of the laminate before ultimate failure. Inspection of the cracks found them to be fiber failure in the outer 0° ply of the specimen.

Specimen type B was the second weakest specimen type. Type B specimen also exhibited audible cracking, but cracks were not visible in all tests. Inspection of the specimens show that the cracks were fiber failures of the second ply from the surface (0°).

Specimens C and D showed no characteristic deformation shape between tests. These were the strongest specimen types, and most tests did not exhibit any audible or visible cracking before failure.

Specimens E, F, G, and H

This test series did not show any correlation between the location of the 0° plies and the ultimate compressive strength. Similar plies in the laminate were paired together in an effort to magnify the trends in the first test. The order of plies was slightly different from specimens A through D so there was no directly comparable specimens. These specimens were made of IM7/5260 material.

Type E specimens showed splintering of the outer 0° plies after failure, but did not exhibit the pre-failure cracking of the outer ply like specimen type A. Many specimens exhibit audible cracking before ultimate failure, but no visible cracks were seen.

Specimens F and G, the strongest specimens of the series, did not show any characteristic failure mode or visible damage before failure. Also, most did not exhibit audible cracks before ultimate failure.

Type H specimens exhibited matrix cracking in the outer 90° ply. Inspection of these cracks under a microscope showed them to extend through multiple plies. These cracks were also seen to propagate between plies for some distance before continuing through the matrix into other plies. Delamination cracks were also seen to propagate between various ply interfaces. This can be explained by the high shear stresses developed in this type of specimen. It is not clear whether the shear stresses are developed at the free edge, or near the clamped regions. No fiber failures (other than ultimate failure location) were seen.

Specimens I, J, K, and L

These specimens match the laminate schedule of specimens A through D, but were made of IM7/5260 material. Results did not match those of specimens A though D. These specimens were not observed with holographic interferometry.

Specimen I was again the weakest specimen of the series, but showed none of the outer ply fiber failure of specimen type A. All Type I specimens exhibited audible cracking before failure, but no damage was identified away from the final failure site.

Type J specimens exhibited delamination of the second ply (0°) at ultimate failure, but showed none of the fiber failure similar to specimen type B. Type J specimens were slightly stronger than type B. Note that other IM7/5260 specimens were weaker than their corresponding specimen types.

Type K specimens were the second weakest of the series, and much weaker than corresponding type C. No characteristic damage or failure mode was observed.

Type L was the strongest of the series, and no characteristic damage or failure mode was observed.

It is concluded that failure in specimens is initiated by different methods. Some specimens fail by fiber failure in the outer ply, and some fail by matrix cracking and delamination. The tests show that material system properties have a large effect on this mode of failure, based on the comparisons between specimens A through D and I through L. No one particular guideline or rule of thumb is applicable to the compressive strength of all quasi-isotropic laminates.

FEA Buckling and Stress Analysis Conclusions:

The Finite Element model was created in an attempt to analytically predict the experimentally observed trend in compressive strength. The observations with holographic interferometry gave indications that the failure could be either a local buckling phenomenon, interlaminar failure in the matrix between the plies, or in the plies themselves. The FEA model created allowed analysis of all three possible failure explanations. The following approaches were used to predict the experimentally observed trend in compressive strength:

- Buckling strength calculations showed that buckling load generally increased as the 0° plies moved toward the midplane of the laminate. This agreed with experimental tests of specimens A through D, but not E though H or I though L.
- It was observed that interlaminar σ_z stresses in the matrix region between plies grew at an exponential rate as the model approached buckling load. However, the strength trend calculated by this method still did not predict the observed strength trends.
- Out-of-plane displacement of the outer ply of specimens, as well as between plies, was calculated. Much insight was gained about the preferred deformation shape of the specimens, but could not be correlated with the holographic interferometry observations from the experiments.
- High τ_{xz} stresses were found to exist between the ±45° plies in the weak E and H specimens. This could explain why the trend in specimens E through H differed from the trend in specimens A through D.
- Von Mises stresses in the matrix regions were calculated. The predicted trend did not match with any observed trends.
- Stresses along the fiber direction were calculated for each ply in all specimens. It was found that the theoretical maximum σ_1, σ_2, and τ_{12} stresses were the same in each specimen, independent of stacking sequence or ply location. However, the tests demonstrate that ply location does affect ply strength.

In conclusion, results prove that predicting the compressive strength of composite laminate is a very complex problem. Failure occurs in different modes for different stacking sequences and material systems, requiring multiple criteria for failure.

The material systems used in this study differ from the woven resin-rich systems used in the marine industry. However, the theory and methods used to analyze these specimens is the same for any fiber-reinforced laminate composed of plies with differing 0° and 90° properties, including fiberglass with unbalanced bi-directional plies.

Examples of marine structures include single skin hull construction and deck equipment on racing vessels. Both high performance sailboats and powerboats are increasingly using unidirectional and bi-directional plies in hulls, superstructures, and equipment. Decks and bulkheads are also examples of compressively loaded structure (but are often sandwich type construction).

The results shown here give insight to how multiple-ply laminates behave in compression. Also, these types of laminates will be encountered more as marine structures incorporate aerospace type laminates using unidirectional fibers and low resin ratios.

REFERENCES

[1] SCOTT, R. *Fiberglass Boat Construction*, SNAME 1996.

[2] ELLIOT, H., “Switching to Pre-Pregs”, *Professional Boatbuilder*, No. 63, Feb/March 2000, pp.151-161.

[3] HALVERSON, L. “Effect of Stacking Sequence on the Compressive Strength of Composite Laminates”, MSME Thesis, University of Washington, 1999.

[4] GABOR, D., "A New Microscopic Principle", *Nature*, Vol. 161, pp. 777-778, 1948.

[5] LEITH, E.N., and UPATNIEKS, J., "Reconstructed Wavefronts and Communication Theory", *Journal of the Optical society of America*, Vol. 52, No. 10, 1962, pp. 1123-1130.

[6] HAINES, K.A., and Hildebrand, B.P., "Surface-Deformation Measurement Using the Wavefront Reconstruction Technique", *Applied Optics*, Vol. 5, No. 4, 1966, pp. 595-602.

[7] RANSON, W.F., SUTTON, M.A., and PETERS, W.H., "Holographic and Laser Speckle Interferometry", *Handbook on Experimental Mechanics*, Chpt 5, Prentice-Hall, 1987.

[8] PIPES. R.B. and PAGANO, N.J., "Interlaminar Stresses in Composite Laminates Under Uniform Axial Extension," *Journal of Composite Materials*, Vol. 4, Oct. 1970, pp. 538-548.

[9] HERAKOVICH, C.T., "On the Relationship Between Engineering Properties and Delamination of Composite Materials," *Journal of Composite Materials*, Vol. 15, July 1981, pp. 336-348.

[10] PAGANO, N.J., and PIPES, R.B., "Some Observations on the Interlaminar Strength of Composite Laminates," *International Journal of Mechanical Science*, Vol. 15, 1973, pp. 679-688.

Discussion

Jay N. Miner, Member

This paper demonstrates very well why it is notoriously difficult either to predict or to measure the in-plane compressive strength of a composite laminate; difficult to predict due to the multitude of possible failure modes, difficult to measure due to the sensitivity of results to test procedures and specimen geometry. These problems are difficult enough for a unidirectional laminate; adding angled lamina and raising questions about laminate stacking sequence adds layer upon layer of complexity.

The results of this paper suggest that some form of buckling is an important determinant of failure modes in these tests. Figure 3 of the paper is an interferometry image of the deflected shape, showing out of plane displacements in the form of bulges with length scales that appear to be set by the width of the specimen. The authors interpret these bulges as ply buckling. Figure 6 shows a buckling configuration from the FEA analysis that shows similar length scales. One would expect a different deflection pattern, and perhaps quite different results from samples with a different aspect ratio, or different boundary conditions. Thus one should be hesitant to make general conclusions about optimal laminate stacking sequence based on tests and analysis of one particular geometry. However, it would certainly seem sensible to avoid placing the compressive load bearing fibers on the surface, in order to inhibit fiber microbuckling and to improve damage tolerance of the critical fibers.

The geometry of the test specimens is very different from what is typically seen in the marine field. A typical marine application of a laminate of this thickness would be as one face of a sandwich laminate. The typical failure mode is much more likely to be global laminate buckling (or buckling of one skin of a sandwich laminate) rather than the ply or fiber microbuckling that seems to be seen in this study. For these more typical conditions, the standard ASTM tests, with their much smaller gauge lengths, would be more useful, being a clearer measure of laminate properties rather than the buckling properties of a particular sample geometry.

It is unfortunate that the authors were so limited in the number of elements they could use in their FEA model. In order to use shell elements for the plies, they have had to make the interlaminar regions much thicker than reality, which then required reducing the matrix E in order to approximate the real global stiffness of the laminate. As a result, the "spring constant" between one lamina and the next is too low by perhaps a couple of orders of magnitude, making interpretation of the results problematic at best.

It would be helpful if the standard deviations of the test results in Tables 2 through 4 were provided. Otherwise it is hard to know whether the differences between specimen types are statistically significant. It would also be useful to know the fiber volume fraction of the samples.

Authors' Closure

This discussion reiterates the difficulties in predicting the compressive strength of composite laminates.

As was pointed out, the study tested limited boundary conditions and aspect ratios. It would be difficult to apply the numerical results from the paper to actual applications. Instead, the results are presented to demonstrate the factors involved in compressive strength analysis, as well as to indicate the magnitude of possible strength trends.

The FEA model used in the study was chosen to simplify the analysis and to magnify the observed trends. Unfortunately, reducing the matrix modulus made the model inaccurate for predicting actual stresses in the analyzed specimen. However, the model did provide an indication of the interaction between plies.

Standard deviations of the test results are included in the following tables, corresponding to tables 3, 4, and 5.

Table 3 Compressive Test Results, Specimens A–D

Type	Ult Comp Strength	Std Dev	Sample Size
A	73,900 PSI	3,625	8
B	79,750 PSI	6,025	8
C	87,300 PSI	3,550	8
D	102,000 PSI	7,350	8

Table 4 Compressive Test Results, Specimens E–H

Type	Ult Comp Strength	Std Dev	Sample Size
E	66,800 PSI	3,350	4
F	95,400 PSI	5,850	4
G	87,600 PSI	7,700	4
H	81,500 PSI	7,600	4

Table 5 Compressive Test Results, Specimens I–L

Type	Ult Comp Strength	Std Dev	Sample Size
I	71,100 PSI	3,800	4
J	86,000 PSI	3,275	4
K	73,500 PSI	850	4
L	95,400 PSI	4,725	4

SNAME *Transactions*, Vol. 108, 2000, pp. 213–225

Technical and Metallurgial Developments in Bulb Flats and Other Hot Rolled Special Steel Profiles for Improved Ship Production

Iain Braidwood, Member, **Geoff Lofthouse,** Visitor, Corus Special Profiles, and **George Bruce,** Member, University of Newcastle upon Tyne

ABSTRACT

Special steel profiles are increasingly being used for commercial and warship construction to reduce production and life cycle costs. In particular, bulb flats (Holland profiles), manufactured by special profiles businesses, are increasingly being used for plate stiffening in warships and naval auxiliaries. Examples are the UK's Type 23 Frigates, the LPH HMS Ocean and the Dutch LPD HrMs Rotterdam, as well as, looking to the future, the UK Type 45 Destroyer.
Some comparison will be made of naval vessel construction on both sides of the Atlantic, particularly looking at the advantages that bulb flats can bring to United States Naval Vessel construction. This paper will discuss the challenge of introducing new shapes, examine the state of the art in the production of special steel shapes and describe metallurgical developments that enable the delivery of shapes with required structural performance and suitability for production.

INTRODUCTION

Hot rolled special steel profiles, available in standard and custom shapes, offer structural designers and fabricators opportunities to improve structural performance, minimise through life costs and reduce production expenditure. The industry has demonstrated its ability to produce a wide range of hot rolled steel special profiles that have material properties that support demanding structural performance requirements and it continues to improve on this capability.

The construction of Naval vessels in the USA is being driven towards a more competitive, commercial type of approach. This approach can include hot rolled steel profiles. Perhaps the first United States naval type ship built to commercial standards, in terms of material design at least was LPD17 [1] where a large proportion of the steel material is EH36, a readily available commercial grade. The steel grades HY80 and HY100, traditionally accepted in the USA as being used in potentially highly stressed parts of the hull, do not feature in the design. Another commercial practice, use of bulb flat plate panel stiffeners, could offer benefits in ship production and maintenance [2]. Future United States Naval Vessels that could be considered for bulb flats are DD21 as well as, for the United States Coast Guard, the Deepwater ships.

Bulb flats have particular advantages over competitor sections in terms of suitability for shipyard processes. These advantages show themselves particularly in terms of ease of application of coatings and also ease of fabrication. Bulb flats are in common use in commercial shipbuilding where their value is recognised. This paper will outline recent progress in the development of high yield strength bulb flats that are particularly suitable for naval vessel construction and may also see application in novel high speed ships. In addition, the industry is capable of producing shapes to suit unique design or production requirements. Research is ongoing into producing bulb flats and other special profiles with improved metallurgical properties, especially those that are suitable for laser welding. Laser trials are under way to evaluate the chemical and mechanical properties necessary to produce bulb flats for satisfactory laser welding. The trials involve both NdYAG and CO_2 lasers and are being partly carried out with the co-operation of shipbuilders in the United Kingdom. Special shapes may be produced to support adhesive bonding in shipbuilding. This paper will briefly discuss the research in those areas.

Availability

The manufacturing facilities of special profiles businesses are based primarily in Europe. Bulb flats (Holland profiles) are produced for ship and bridge building. Steel long-bar in complex cross section is produced for a variety of end users ranging from the construction industry and mining to the automotive and earth moving industry.

Standard sizes of bulb flats are readily available. Some manufacturers are capable of rolling every size at a minimum of six week intervals. Small quantities of bulb flats, for example, can be rolled on short notice, if there is available capacity. Custom special profiles can be produced from enquiry to first rolling in as short a period as six months.

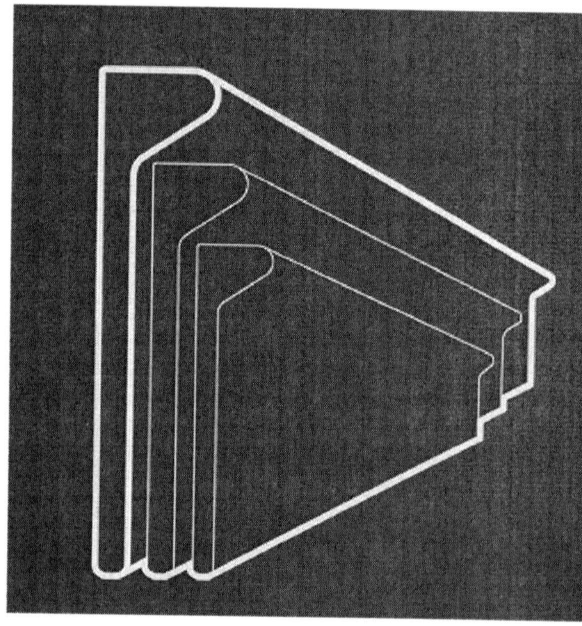

Figure 1. Typical cross sections of bulb flats

THE ADVANTAGES OF BULB FLATS

It is worth stating the advantages of bulb flats, as perceived by the commercial shipbuilding industry around the world and now increasingly by the Naval vessel industry.

Description

Bulb flats are an asymmetric, tailor made plate stiffening profile (Figure 1) manufactured in a range of sizes and grades. A bulb flat is a "ready to use" plate stiffener. Compared with a welded or fabricated stiffener, it eliminates the need for flange stripping, in the case of producing tees from I-Beams or for welding in the case of producing angles from flat bar. Bulb flats are available in a range of sizes dependent on the manufacturer, for example, from 120 by 6mm (4.7 X 0.24 inches) at the light end to 430 by 20mm (16.9 X 0.8 inches) at the heavy end. They are manufactured from steel using a hot rolling process and can be readily provided in steels with yield strength of up to 355 N/mm^2 (51ksi). Higher yields are possible, but are subject to approval by an appropriate classification society. For example, bulb flats of Grade D420 (61ksi) steel have recently received provisional approval from Germanischer Lloyd.

Structural properties

The unique shape of the bulb flat distributes the steel to increase its resistance to buckling. This results in a more efficient strength to weight ratio when compared with other forms of stiffener such as flat bar and structural angles. Bulb flats may allow a weight saving of up to 64% [3] when compared with flat bar stiffeners that offer structural equivalence in terms of buckling of the effective stiffener cross section, but are less structurally efficient than tees cut from I-Beams or welded from flat bars in resisting buckling.

Coatings and production

The compact shape of the bulb offers better access for welding and painting. The access benefits are felt on the panel line and also where join up butts need to be made at the berth. In addition the rounded profile of the bulb, with no sharp corners, assists effective and efficient coating, partly by reducing the shadow areas that are found under the face flats or flanges of tees and angles. Furthermore, bulb flats have a smaller surface area than other stiffeners of equivalent section modulus. The curved surface of a bulb is less likely to trap moisture or pollutants that can initiate corrosion. Through life cost benefits can thus be obtained. Leading shipbuilders recognise the advantages of bulb flats in their search for ways to minimise the surface area to be coated. Coatings are seen as an important and costly part of ship construction and minimisation of surface area is important. The rounded edges provided by bulb flats minimise sharp edge coverage problems. It is the sharp edges that can induce coating breakdown and corrosion leading to expensive repairs.

The rounded edge of the bulb flat eliminates the need for edge grinding before coating, a process that is labour intensive. It has been estimated [4] that elimination of edge grinding can save one to two man-years per vessel. In addition three pass grinding may not be particularly effective and may just result in more sharp edges if not properly carried out. Elimination of stripe coating would provide further savings.

The importance of coatings and corrosion in the US Navy can be seen from a recent report. The Commander in Chief Pacific and Atlantic Fleet Maintenance Officers have stated that about 25% of their annual maintenance budget is directed towards the prevention or correction of corrosion generated failures. Tank overhaul has been identified as the most common corrosion problem.

In the same report the annual expenditure of Fleet maintenance budget for seawater ballast tank preservation has been estimated by Kikuta [5] (1997 figures) at US$64million. Kikuta's report emphasises the importance of rounded edges in the establishment of a seawater ballast tank preservation protocol with an expected life of twenty years. Bulb flats are particularly mentioned as being used in European shipyards, (Kvaerner Masa and Danyard are quoted, although almost any European commercial shipyard would have given the same results).

Ballast tank corrosion is recognised by all navies as one of the biggest maintenance burdens for ships' Commanding Officers. The drive to reduce crew aboard navy ships to reduce operating costs will mean that there will be fewer crew available for tank inspection and repair. As a consequence the majority of such work may be deferred until scheduled maintenance periods, enhancing the importance of good long-term coatings performance. The problems in coatings performance are principally seen as being connected with poor coating coverage on tank angles, stiffeners and welds.

A new coating system has been developed by the US Navy for ballast tanks. A cost analysis was carried out in co-operation with Pacific Fleet Maintenance and it was determined that the Navy could save at least $590million in ballast tank applications [6] although no time scale is stated for the saving. The coating preservation technology developed by the US Navy is being developed to include all Navy tanks, where it will provide a cost saving of some $2billion.

It is generally accepted when making cost savings or efficiency improvements that where big savings are predicted, as in the case of the US Navy developed system, then it is the saving of the last few per cent of the cost that can be difficult. The provision of rounded edges in ballast tanks will increase the cost saving.

Looking at future United States nuclear powered aircraft carriers in particular it can be shown that maintenance represents a large proportion (31%) (http://www.nns.com/) of life cycle costs. It may be that around a quarter of that maintenance is due to corrosion related failures and, within that, tank coatings will form a significant part of the maintenance burden.

There are many features to be considered as ship designers move through the "design spiral" from concept to completion. Ship structures need to be designed to provide adequate local and global strength, to maintain structural integrity in shock and blast and to provide fire and watertight integrity. There is also an element of structural design for improved operability and ease of production. Coatings installation needs to be considered in the design and some simple statistics, relating to commercial shipbuilding, put coatings in perspective in the construction phase of a ship. It is known [7] that the cost of materials coatings for new building represents 1% of the new build price, however the cost of application represents 7% of the ship's price. For coating application the total man hours are 20% of the total ship hours, however rework is a minimum of 30% of the coatings' man hours. These levels of rework would not be tolerated in other areas of the build programme. A part of the shipyard coating strategy must include the development of the steel structure to aid the application of coatings and bulb flats are an accepted way of enhancing a coating strategy. It is this investment at new build that can pay dividends in reducing the through life cost of the ship.

Construction sequence and labour

The principal construction benefit of the bulb flat is in the avoidance of the need to fabricate tee sections. The implication of this is that bulb flats can be injected into the production process at the panel line stage. As the bulb flats can be provided in a blasted and primed condition several shipyard steps can be eliminated. A simplified example of NASSCO's steel fabrication and sub assembly areas illustrates the point. Bulb flats could be injected in a just in time basis at the panel line so eliminating much of the profile and T-beam fabrication.

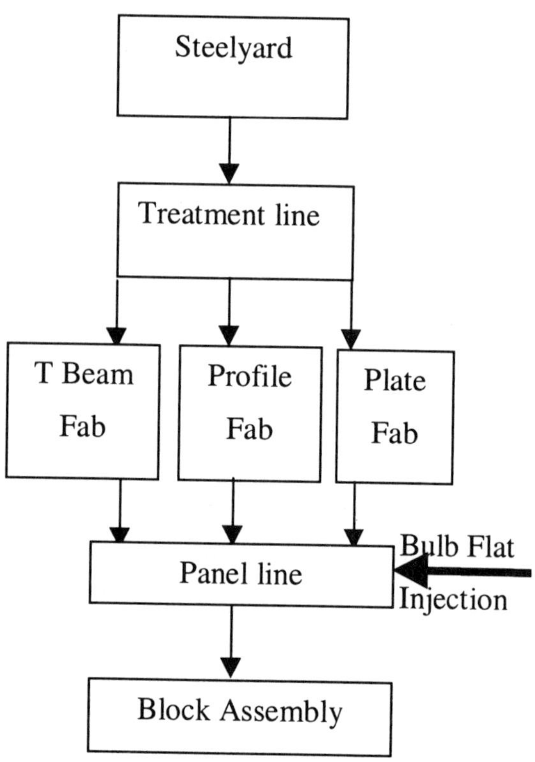

Figure 2. Simplified NASSCO Assembly Line [8]

The potential production benefits of using rolled profiles, particularly bulb flats, as an alternative to fabricated or stripped sections can be illustrated by a simple example. Taking two ship panel structures, identical except for the stiffening, one using bulbs and the other using fabricated sections.

Fabricated sections

For fabricated sections, the sequence of operations would be:

Receive material into storage.
Transport to cleaning.
Transport to cutting into flat bars.
Straighten as required
Transport to fabrication equipment.
Fabricate sections
Straighten as required.
Complete paint coating.
Transport to cutting equipment for mouseholes and cutting to length.
Deliver to panel line buffer storage.
Feed onto panel line in sequence.

Using conventional handling and manipulating equipment, the labour requirements would typically be:

Two workers to receive and store material.
One worker to manage cutting of flat bars
One worker to manage the transport through straightening
One worker to manage the cleaning and coating.
Automatic fabrication, with one worker to manage process.
Automatic (robotic) cutting.
One worker to manage the transport.
One worker to manage buffer and feed to panel line.

Bulb flats

Two modes are possible. If the shipyard carries out its own preparation, the sequence would be:

Receive material into storage.
Transport to section preparation.
Complete cleaning, straightening and paint coating.
Transport to cutting equipment for mouseholes and cutting to length.
Deliver to panel line buffer storage.
Feed onto panel line in sequence.

Using conventional handling and manipulating equipment, the labour requirements would typically be:

Two workers to receive and store material.

One worker to manage automatic straightening, cleaning and coating.
One worker to manage the transport.
Automatic (robotic) cutting.
One worker to manage buffer and feed to panel line.

If the bulbs are prepared by the supplier and then received at the shipyard ready for use, further operations can be removed from the shipyard. The sequence would now be:

Receive material into storage.
Deliver to panel line buffer storage.
Cut required mouseholes.
Feed onto panel line in sequence.

Using conventional handling and manipulating equipment, the labour requirements would typically be:

Two workers to receive and store material.
Automatic (robotic) cutting.
One worker to manage buffer and feed to panel line.

Savings from use of bulb flats

If the shipyard carries out its own preparation, the savings would be:

Three man-years (no flat bar cutting or fabrication)
No equipment for fabrication of stiffeners from flat bars OR no equipment to strip a flange from an H beam.

If the bulb flats are received by a shipyard ready prepared (blasted and coated), the savings become:

Five man-years
Equipment for fabrication and preparation
Additional straightening may be required for fabricated sections, both before and after the welding process.

If transport between the workstations is not by conveyor, then several additional crane movements would be required for shipyard fabrication of stiffeners:

in and out of flat cutting, via buffer storage,
in and out of fabrication via buffer storage.

The above gives some idea, in real terms, of the savings that can accrue on the production line, simply from the removal of the need for fabrication and further straightening and the handling that goes with those processes. Additional savings may result from the removal of maintenance intensive equipment from the shipyard.

BULB FLATS IN MILITARY VESSELS

Royal Navy warships have traditionally been stiffened using "long stalk tees" however this will change shortly. The Royal Navy's next twelve ship destroyer series, the Type 45 AAW Destroyer, will be stiffened with bulb flats. The Type 23 Duke Class frigates, where building started in the eighties are stiffened using bulb flats. Previous naval practice of using all teebars has been replaced by new practices that allow sections with moderate asymmetry, such as the bulb flat. The construction will feature a mix of commercial and naval styles. The Type 23s have seen front line service, most recently in the Kosovo crisis, and there is service experience stretching back to the late 80s. The Royal Navy is looking at a possible Service Life Extension Programme (SLEP) for the Type 23, at which time the benefits of the bulb flats will become particularly apparent

Other examples of military vessels where bulb flats are used are, in the Royal Navy, the LPH ***HMS Ocean*** and in the Dutch Navy the LPD ***HrMs Rotterdam***. Bulb flats are also used on the US Navy's Watson Class 24-knot strategic sealift vessels, NASSCO in San Diego has orders for seven of these ships.

A recent comparison between commercial and naval ship structure was made in the case of LSD41 (USA) and TCD ***Foudre*** (France) [9] The ships are broadly similar in their roles and operational requirements. The structure on TCD ***Foudre*** is lighter than on LSD41, 5158.3 metric tonnes as opposed to 6599.1 metric tonnes. There are several reasons for this according to the authors. The French ship is shorter and deeper and so has a more efficient hull girder. French vehicles (much of the "cargo" in this case) are lighter than US vehicles so secondary loads are reduced and, most importantly, Foudre is designed with a mix of commercial (bulb flat) and military standards (tees) and is not fully shock resistant. However it is principally the foundation weights for LSD 41 that add weight in the field of shock protection; there are many more foundations needed on LSD 41 than on ***Foudre***.

Afanasieff and Malory in their paper [10] debate the conceptual design of the FF21 Multi Mission Frigate, in particular its producibility and state that this is enhanced by the use of typical European stiffeners (bulb flats). However absolute producibility optimisation is tempered by the need for restraint of weight growth. Their hull structure estimate of weight is 1057 tonnes for a 100 metre frigate

The next generation of Sealift ships, the T-ADC(X) program, where design contracts for, reportedly, US$1.5million per yard were awarded in August 1999, to Halter Marine, NASSCO, Ingalls and

Avondale, represents the next big shipbuilding program in the USA where commercial practices are expected to continue.

DESIGN DRIVERS

There is a need to drive down costs in warship construction and a prime candidate for the focus of cost reduction is the platform. Allied with platform cost reduction is a need to minimise platform maintenance, reducing through life costs, and also to minimise steelweight, so increasing payload.

However all this needs to be achieved without increasing vulnerability or reducing structural integrity, particularly damage structural integrity.

It can be argued that platform cost is the one area where shipbuilders can compete effectively as much of the hardware and weapons fit is sold at comparable prices to competing builders. It is this competition aspect that encourages warship builders to look to the commercial field.

Bulb flats with their ready international availability and established commercial pedigree can be a straightforward cost reducing measure when set against fabricated sections or rolled tees. They are easily available both for building and repair purposes.

The whole area of acquisition reform was addressed by Mackey *et al* [11]. In summary they identified the need for cost avoidance, simplification, standardisation and commonality as well as designing for producibility. Moving towards a greater use of commercial shipbuilding techniques sits well with Mackey's paper.

BLAST AND SHOCK CONSIDERATIONS

Blast can be divided into two types, internal and external.

Internal blast

Internal blast is characterised by a quasi static over pressure phase. This would arise from a warhead explosion within the ship. The quasi static overpressure loads the decks and bulkheads laterally.

When it comes to the question of bulkhead collapse then, for a warship, collapse strength comes from membrane properties rather than stiffener properties, so in that case the stiffener shape is less important. Thus, bulb flats are an appropriate shape to specify for naval ship bulkhead stiffeners.

External blast

External blast effects can come from explosion of a weapon away from the ship. The first part of the above blast waves can be termed shock, since in a strictly engineering sense this is what it is, an abrupt, finite pressure change. This occurs over a distance of $0.3\mu m$ in air.

Shock

However the term shock is generally reserved to describe the transmission of energy through a ship's structure, from an underwater blast. Shock mountings can be used to isolate machinery and other equipment, but hull structure cannot be isolated in the same way.

There are no current shock (or blast) specifications for UK warship construction; instead ships are built in accordance with Naval Engineering Standards (NES). However there is a move away from NES with the introduction of Lloyd's Register Naval Ship Rules, which do allow for more flexibility and the incorporation of bulb flats. This is allied with "smart procurement", which enables UK Ministry of Defence to specify the performance and then it is up to the shipyards to provide an acceptable vessel that meets those performance requirements.

It has long been known that underwater explosions can cause buckling failure in ship's hull structure. [12] It was shown, using Liberty ships, that hull whipping could cause the hull girder strength to be exceeded. Failure usually occurred when the ship's bottom structure was compressed beyond its buckling strength.

Even though, as noted above, bulb flats have less lateral stability than symmetrical tee sections, equivalent resistance can probably be obtained by using slightly heavier sections. There would be some increase in weight and material cost, but it is expected that these would be more than offset by reductions in the construction costs associated with assembling stiffened plate panels and with painting. In addition, there will be substantial reductions in maintenance costs.

An alternative design approach to improve the lateral stability of bulb flats, particularly in long spans, is the incorporation of flat bars fitted between alternate pairs of stiffeners, perpendicular to the axis of the long span at half span position.

It is known that many NATO navies, including the UK are working on shock and blast resistance of ships' structures, with the eventual aim of calculating behaviour of structure from theory, however this is clearly not a straightforward problem. Indications from recent experiments are that bulb flats certainly perform comparably with alternative stiffener arrangements in shock and blast

It should be understood that the drive towards commercialisation of platform structure must not lead to a more vulnerable, less survivable warship. Vulnerability is a measure of damage control, recoverability and resistance to weapon effects. Improved survivability can come partly from passive vulnerability reduction methods like optimised system layout, blast resistant structure and adequately shock protected equipment. Cost savings arising from using commercial products and techniques can be ploughed into improving shock and blast performance in other parts of the ship or increasing its size and weapons capability.[13]

SPECIALISED SHAPES

In addition to bulb flats, the industry can produce shapes that the end user needs to satisfy engineering design requirements while minimising total manufacturing cost. This can take the form of replacing a complex fabrication with a single profile or producing profiles that enhance manufacturability by, for example, providing profiles that are easy to jig or that come with rolled weld edge preparations. Some examples are provided below, firstly from a shipyard and secondly from other industries.

Shipyards

Some work has been undertaken by Corus Special Profiles with shipyards examining ways that a special profile can help the yard overcome particular challenges. One example is given here, looking at the way that Meyer Werft [14] from Papenburg in Germany connect their laser welded deck panels to primary structure.

Meyer Werft – Laser welded sandwich panel to ship structure connection

Cruise liners have a requirement for flat decks in order that floor coverings can be laid directly without the need for a composite layer to be spread over the steel. Traditional light steel structure has almost invariably formed a type of "hungry horse" appearance on decks and one solution to the problem, developed by Meyer Werft, is laser welded deck sandwich panels. These sandwich panels are constructed from two sheets of 2.5 or 3mm steel (10 X 2/2.5/ 3 metres) separated by longitudinal flat bars (10 metre by 40mm deep by 4mm thick at 120mm centres. There is still a need to connect the laser-welded panels to the main structure of the ship and to each other. Meyer Werft's current solution is to use an extruded section (Figure 3). A cold or hot rolled solution has been proposed. The hot rolled solution could be composed from two hot rolled steel profiles, subsequently welded at the steel mill on an in line submerged arc facility (Figure 4) This is a well established process and is most familiar from bridge expansion or movement joints. Equally, a different single hot rolled solution could be used, avoiding the hollowness of both the existing extrusion and also of the proposed two part hot rolled profile

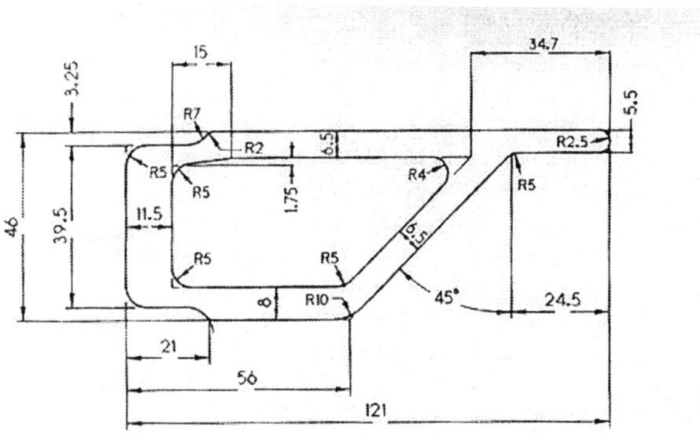

Figure 3. Current extruded section for connecting laser welded deck panels to ship structure

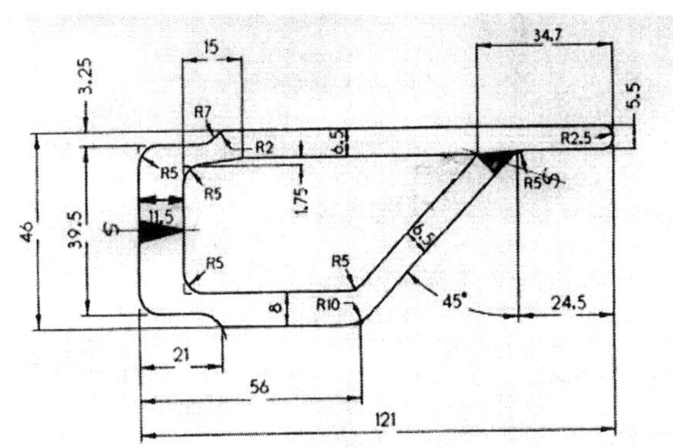

Figure 4. Proposed hot rolled and welded solution

Fabrication benefits of special profiles

The following two examples, although not from shipbuilding, give specific examples where improvements to manufacturability have been achieved by using special profiles.

Fabrication of Welded Box Section

The figure below shows a comparison between the conventional and special profiles approach to box sections.

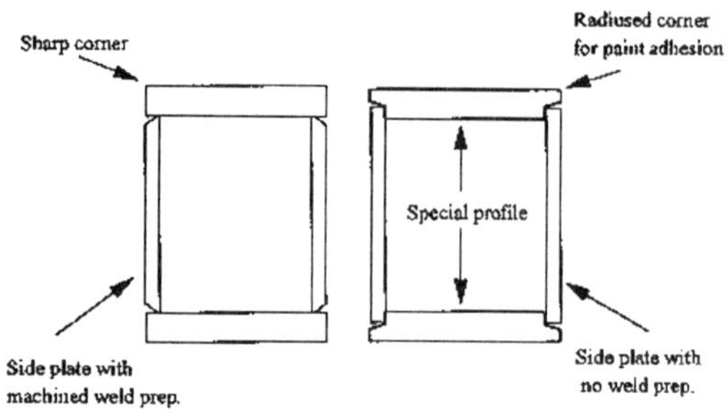

Figure 5. Box section fabrication

The special profile approach means that, for the fabricator, a weld preparation is already done and the box section is easily jigged for robotic welding. The final product has radiused corners so that the coating operation is facilitated.

Elimination of Machining

In the UK Ministry of Defence "DROPS" system for distributing palletised stores, the initial structural design was based on standard structural sections, including the use of a 254x89mm channel. The actual design required one flange to be shorter than the other by nine millimetres and initially this was achieved by machining the end of one flange. This was a fairly high cost procedure and also meant that throughput was slow. A special profile consisting of a 254x89x80 channel was developed to eliminate the machining operation. The cost saving over the full project more than paid for the tooling cost for the introduction of the special profile.

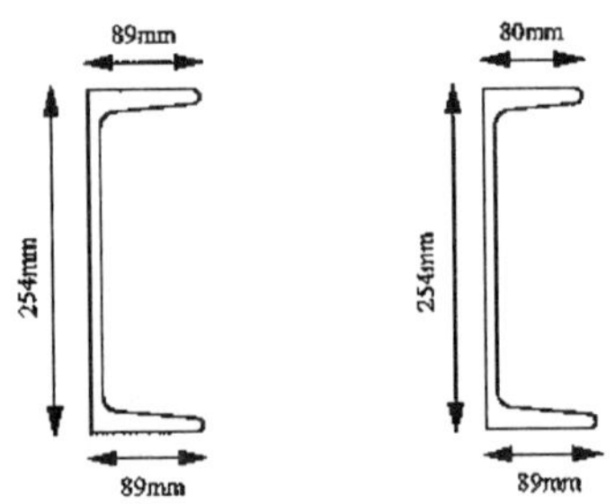

Figure 6. Standard and Special Profile Channel Sections

ADDITIONAL OPPORTUNITIES

There are other possible uses for hot rolled special profiles arising from the need to replace complex fabrications, notably on bulk carriers or multi hull steel vessels, such as trimaran warships, as well as joining sandwich structural panels. Flensburger shipyard has investigated use of a three armed hot rolled special shape to manufacture the complex tank top/longitudinal side girder/hopper plate joint on bulk carriers (Figure 7). [15]

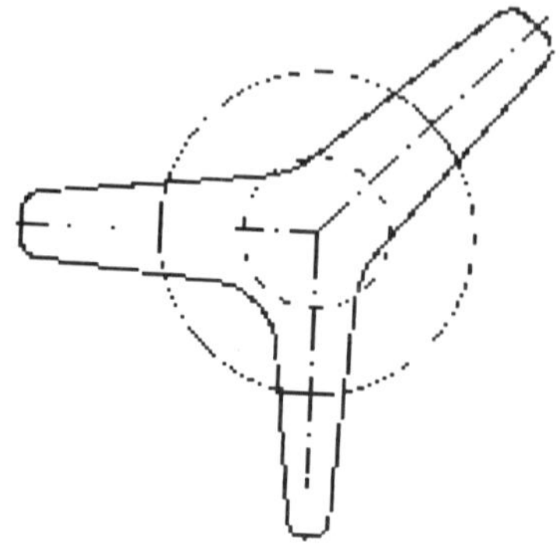

Figure 7. Proposed hot rolled solution for complex fabrication in a double hull bulk carrier

Similar solutions can be imagined for some of the complex structure in a steel multi-hull vessel, particularly at wet deck connections to hulls.

For laser welded aluminium sandwich deck panel, the challenge is similar to the Meyer Werft laser welded steel deck in joining main structure to the deck panel. Outlines of possible solutions are shown below.

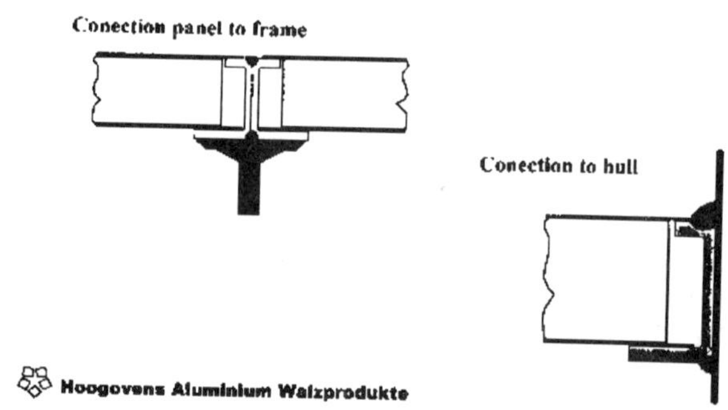

Figure 8. Laser welded aluminium sandwich deck panel connection to frame and hull

SPECIALISED GRADES

Bulb flats and other special shapes are currently available in a range of higher strength steels. Research continues to improve the strength and toughness of hot rolled special profiles

Higher Yield Notch Tough Bulb Flats

Commercial vessels are for the most part constructed from Grade A (235N/mm^2 (34 ksi) yield stress) and Grade AH36 (355N/mm^2 (51 ksi) yield stress). Notch tough plate is used in areas such as the deck, sheer strake and bilge radius where their special benefits are required. However more specialised grades have been developed for some applications, including bulb flats used for plate stiffeners in bridge building (such as the Oresund link between Denmark and Sweden). Given below are details of grades that are either currently manufactured or could be used for bulb flats, together with some information about process routes. The processes and grades described are specific to Corus Special Profiles, however it is believed that other special profile manufacturers will have similar, although not directly comparable, processes.

DH36/EH36

Lloyds and DNV has approved the manufacture of Grade DH36. Limited approval has been given for small orders of EH36 using the same steel composition and rolling procedure (up to 12mm thick with ABS). For these Grades a C-Mn-V-Al steel has been used. A very low S content (<0.003% aim) has been specified to enhance transverse toughness. The steel is treated with Calcium to provide inclusion shape control. All tensile tests satisfied the minimum yield stress required by DH36. The Charpy V impact tests were outstanding for a steel which has not been normalised - more than 100J longitudinal at -40°C (for full size test pieces), and generally more than 100J transverse at -20°C, easily meeting the toughness requirement of EH36. The composition of this Grade is designed so that the Carbon Equivalent Value (CEV) is not more than 0.41 ensuring that in normal commercial shipbuilding the steel can be welded without preheat, dependent on joint geometry.

EN 10113, Grade S420

This steel was especially developed for bulb flats used as plate stiffeners in the construction of the Oresund link bridge (between Denmark and Sweden). The steel has a minimum yield stress of 420N/mm^2 (61,000psi) and a minimum Charpy V notch impact requirement of 40J longitudinal at -20°C. Again this steel comfortably met the requirements, particularly the Charpy impact, which was generally above 100J even on a 10x7.5mm test piece. The CEV of this steel is slightly higher than the DH36 type described above (up to 0.43%). As transverse Charpy properties were not required, this steel had a S content of typically 0.020%. A low sulphur version of this steel has been granted provisional approval by Germanischer Lloyd.

A modification of the S420 steel has been developed for DH40 application. DNV Approval has been granted to this bulb flat type. DH40 requires a minimum yield stress of 390N/mm^2 (56,000psi). This allowed a reduced alloy content to bring the CEV down to 0.41% max. As transverse Charpy impact tests were specified, the S content was reduced to 0.010% max. which was sufficient to achieve the required values. The S content could be reduced further if required (0.003% max. as per DH36).

HC90

HC90 high strength steel special shapes have been produced for mast sections of fork lift trucks. Many of these require high strength steels to withstand the high contact roller loads. Mast sections are generally thicker than bulb flats so bulb flats made from this same composition could have marginally higher strength. The steel, a C-Mn-V-Al-N type, guarantees a minimum yield stress of 483N/mm2 (70,000psi), however this is achieved at the expense of CEV, which could be as high as 0.50%. Nevertheless, for forklift truck applications this steel is welded by MIG without preheat. Crack tip opening displacement fracture toughness properties show that toughness properties are not so good as the DH steels described above

Future steels

It may be that any future steel for bulb flats for United States' Naval Vessels would need to be of a higher yield than is currently available in the bulb flat range. Any new steel would require a development programme, probably including the production of laboratory melts, followed by laboratory rolling to simulate a bulb flat rolling schedule and finally the optimisation of any subsequent heat treatment that was needed. Production trials would then be needed.

The achievement of strength levels of over 480N/mm^2 (70 ksi) would mean moving away from conventional ferrite-pearlite steels supplied in the as rolled condition. The process could be based around the development of the ASTM A710 Cu precipitation hardened steels. There are two initial concerns to address.

Firstly, the manufacture of the steel blooms needed as the start point of the bulb flat rolling. Cu treated steels are known to be difficult to continuous cast, and on the bloom caster in use there is no experience of Cu contents around the typical 1% level required in A710. A recent evaluation exercise indicated that it may be possible to produce steels with around 0.5% Cu.

Secondly, heat treatment of long products, including bulb flats, is always potentially problematic due to the distortion that can take place, frequently meaning that re-straightening is needed. This can make the process expensive. However, for bulb flats we would envisage a process route being as-rolled plus aged, rather than the conventional normalised/quench and aged. This would minimise the risk of unacceptable distortion. However the exact strength level that could be achieved is uncertain, it may not be any higher than HC90, although the toughness and weldability would be better.

Successful development of new steel grades will certainly require cooperation between steelmakers and ship builders.

RESEARCH AREAS

Two principal research areas are laser welding and the use of adhesives. Additionally a computer program has been developed to assist bridge builders in their use of bulb flats [16].

Laser welding

Laser welding poses two challenges to the steel maker and profile producer, firstly chemistry and secondly tolerance. There is a need to work with those shipbuilders installing laser welding machines in order that we have the appropriate chemical composition and can produce profiles with the appropriate tolerances to enable efficient laser processing.

Laser welding provides benefits to the shipbuilder in terms of reduced distortion, reduced cost and possible higher welding speeds. Cost reduction comes from the need for tight dimensional control throughout the shipyard, leading to reduction in rework. It is known that rework is an expensive feature of shipbuilding and it has been estimated that rework on the berth can be ten times as expensive as the original job.

The principal laser welding shipyards in Europe are Fincantieri in Italy, Odense in Denmark and Meyer Werft in Germany. It is noted that successful application of laser welding depends on close co-ordination between the shipyard and the steel manufacturer.

The chemistry challenge in laser welding is to reduce solidification cracking. Control of phosphorous and sulphur is important and the Crack Index formula

$$P + \frac{S}{Mn} \leq 0.02$$

is commonly used. For Grade A steels with a manganese level of around 1% this equation implies that phosphorous should be at a maximum level of 0.015% and sulphur at a maximum of 0.005%. Control of the chemistry to produce the appropriate chemistry does give a cost penalty to the steelmaker. There is also a need to keep carbon to 0.12% maximum for hardness reasons, this has much less impact on the steelmaker.[17] February 2000 saw the first successful CO_2 laser welds of bulb flats to a plate panel. Further testing will compare the performance of a conventionally welded stiffened panel with a similar laser welded panel. The panel is intended to represent part of a double bottom grillage.

In terms of tolerances, laser welding works best and most economically when the bulb flat is absolutely flat to the plate. This means that filler wire is not required. A roller straightening machine is being developed so that the required tolerances can be provided. Laser welding prefers a 90° angle at the fillet connection, conventional bulbs have a slight radius. This challenge is now being addressed.

Laser welding also allows opportunities for expanding the component range available to shipbuilders, the LASCOR panel is an example. Similar thin sandwich structures, that could not be constructed by conventional welding, are in use in Europe. There are also opportunities to use laser welding within the manufacture of special profiles. For example some profiles cannot be rolled in one piece and so are rolled as a two piece and subsequently welded. This welding could be done by laser and with the use of stake welding the available profile range can be expanded. Some interesting applications for high speed ships might be developed in this way.

Adhesive bonding

Thin composite panels may also be stiffened, in hybrid construction, with steel stiffeners. Adhesive bonding can be used to join the stiffeners to the panel [18]. These stiffeners may also be suitable for manufacturing by hot rolling. The sketch shows indicative stiffener types.

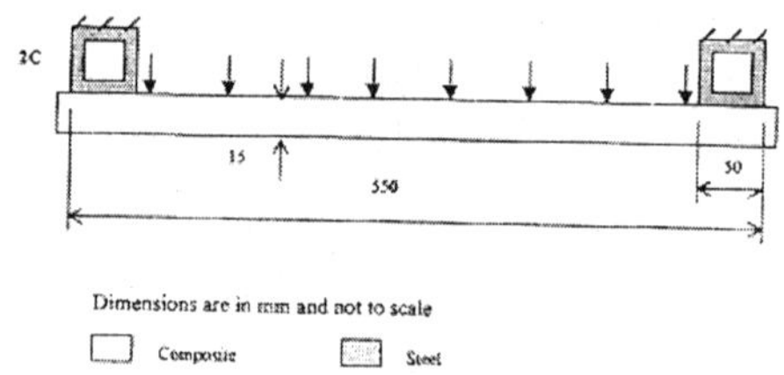

Figure 9. Schematic of composite bulkhead with steel stiffeners (stiffener type indicative only)

An additional research proposal [19] has been investigated. This work, should it come to fruition, will be a generic investigation of the design, joining processes and properties of joints that combine dissimilar materials such as steel, aluminium and composites. Generic property modelling of joints will allow research results to be applied in structures as diverse as vehicles and ships. The proposal developed from earlier work such as LIVEMAN (Advanced Joining Processes for Lightweight Vehicle Manufacture). Trials for validation of theoretical results would be carried out and guidelines could be developed for the increased use of adhesives in marine structures.

The preceding sketches give some indication of new developments in ship structures where hot rolled special profiles may have a role to play. Additional details can be found in Ref. [20].

QUALITY

Finally, running through the whole process is the vital quality issue. Quality is controlled throughout the steel making, hot rolling and delivery process by a defined and documented quality plan. This quality plan can be tailored to the needs of particular customers. Quality is particularly stressed at the enquiry stage, when a potential new steel section is being discussed with customers' engineers. This is the point at which quality can be designed into the end products. Implementation of the quality plan identifies process capability and compares it with product requirements. This ensures that quality is not simply a matter of inspection but that quality is assured before steel ordering commences.

CONCLUSIONS

This paper has demonstrated the suitability of bulb flat for plate stiffening in Naval Vessels. Some of the advantages have been outlined and in particular note has been made of the cost savings that can accrue by using a commercial style of build for naval vessels. The principle of increased commercial style in Naval vessels fits well with the need for acquisition reform within Naval circles in the USA. Additionally, it has been shown that steel hot rolled special profiles can perform structural optimisation roles and although these are not currently used in shipbuilding there is potential, especially in high speed vessels. Closer links between shipbuilders and steel profile manufacturers, particularly in the field of higher yield bulb flats, laser welding and adhesive bonding will enable these processes and products to develop successfully.

ACKNOWLEDGEMENTS

This paper has been a team effort and would not have been possible without contributions from a range of people on both sides of the Atlantic. Our colleagues from Special Profiles have also been helpful, in particular Phil Knowles and his team and also Barry Atkinson for his thoughts on quality.

REFERENCES

[1]SIEVE, Mike "LPD 17, Material Selection Process, Design Considerations" Second Annual NCEMT Optimised Commercial Steels Project. Users Group Meeting, September 21-22 1999, Arlington, VA

[2]BRAIDWOOD, I.T. "Bulb Flats-To Go In Harm's Way" Modern Shipbuilding Technologies Demonstration and Information Exchange, NCEMT, September 21-22 1999, Arlington, VA

[3]HAYWARD, A.G.C. "Application of Bulb Flats in Steel Bridges" Cass Hayward and Partners, 1994, Confidential Internal Document

[4]KATTAN, Raouf, Private Communication, 13th August 1998

[5]KIKUTA, Miles "A New Preservation Protocol for Seawater Ballast tanks in Naval Applications" in Marine Technology Volume 36, Number 1 pp 11-21, January 1999

[6]PARKS, Richard "Stopping Ballast Tank Corrosion" NAVSEA's Deckplate on the Web, Issue No 31, May-June-July 1999

[7]KATTAN, M. Raouf "The Integration of the Coating Process at New Building" PCE Europe, Genoa, March 2000

[8]NSRP 0553 N1-96-6 Automated Process Application in Steel Fabrication and Subassembly Facilities; Project Summary

[9]FERREIRO, Larrie and AUTRET, Gerard "A Comparison of French and US Amphibious Ships, the TCD Foudre and the LSD 41" Naval Engineers Journal, May 1995

[10]AFANASIEFF, Leonid and MALORY, John P. "The Design of the FF21 Multi Mission Frigate", Naval Engineers Journal Vol. 106 No 3 May 1994 pp 150-162

[11]MACKEY, Thomas P., BRESNAHAN, Aaron G., GORTON, G.John, KENDRICK, Andrew M. "Commercialisation, Standardisation and Acquisition Reform", Trans SNAME Vol 103 1995, pp 65-82

[12]KEIL, A.H. "The Response of Ships to Underwater Explosions" Trans SNAME 1961, Vol. 69 pp 366-410

[13]MARTIN, Andrew "The Place of Survivability in the Design of Future Surface Warships", in Proceedings of Warship 98 –RINA, International Symposium on Surface Warships: The Next Generation, London, 11-12 June 1998

[14]ROLAND, Dr. Frank "Laser Welded Sandwich panels for the Shipbuilding Industry" in Lightweight Construction - Latest Developments 24th and 25th February 1999

[15]BOHLMANN, Berend "Innovative Konstruktion von Massengutschiffen in Doppelhüllenbauweise", Stahl und Eisen 118 (1998) Nr. 8 pp 39-43

[16]BRAIDWOOD, I.T. and KELLY, R.B. "A Comparison of Theory and Practice in Plate Stiffener Behaviour Under Axial Loading" in Proceedings of the International Conference on Current and Future Trends in Bridge Design, Construction and Maintenance, Singapore, 4-5 October 1999

[17]PRICE, D. "Laser Processing at British Steel" in Lasers in Shipbuilding and Fabricated Structures, TWI and Make it with Lasers, Southampton 27th May 1999

[18]HASHIM, S.A. and KNOX, E.M. "Structural Performance of Thick Adherend Steel Composite Adhesive Joints", Proc Inst. Mech. Engrs 1999 Vol. 213 Part L

[19]Properties of Bonded Joints Between Dissimilar Materials" PROBOMAT Project Proposal, TWI, Abington, Cambridge 1998

[20]BRAIDWOOD, I.T. "Latest Developments in Special Steel Sections for Lightweight Ship Construction" in Proceedings of the International Conference on Lightweight Construction-Latest Developments, London, 24th and 25th February 2000.

Discussion

Authors' Closure

[No prepared discussions received. Author's responses are to oral comments from the floor.]

I thank Ed for his supportive comments and his highlighting of the range of benefits that bulb flats can bring to shipbuilders.

Robert vom Saal—Robert is right to draw attention to the difficult issue of weight reduction and bulb flats. Simply replacing a tee with a bulb is unlikely to bring weight savings. However, a total re-engineering approach, using a largely commercial ship structure in a naval vessel, can be expected to bring weight savings. Bulb flats are available in a wide range of sizes, sometimes with half millimeter increases in web thickness so allowing for a near optimal stiffener solution to be provided.

Weight saving and bulb flats is not a straightforward issue; however ,experience shows that it is possible to achieve a weight saving by going for a more commercial structure. It has also been suggested that even where there is a perceived weight penalty that the cost and manufacturing benefits of bulb flats will be such as to at least partially offset it.

There was also a comment from Electric Boat but I don't have the discusser's name. Basically his question was "Is it possible for naval shock data from the UK to be transferred to the USA?" The answer is yes on a government to government level, say DERA to NAVSEA, and in fact this is already being done.

I would also like to thank SNAME for giving me the opportunity to present our paper and also thank those who attended the presentation and contributed to the discussion.

SNAME *Transactions*, Vol. 108, 2000, pp. 227–248

Developing Design Criteria for Connections Around Cutout (Slot) Openings

Kai-tung Ma, Member, **Sanjay Srinivasan,** Visitor, **Hanqing Zhang,** Visitor, **Brian Healy,** Visitor, and **Hua Peng,** Visitor, American Bureau of Shipping, Houston

In ships such as tankers and bulk carriers, local connections near cutout openings in way of longitudinal stiffeners and transverse web frames or floors have been crack-prone. Periodic surveys of such ships reveal a large number of cracks to the extent that they pose a maintenance problem for ship owners. Until recently, not much guidance or design criteria have been given to assess these local connections.

In this paper, simple criteria are derived to evaluate typical connections. It is assumed that the pressure load on a hull panel is transferred through a flat bar and collar plates into the transverse web frames. These components are modeled as springs while the longitudinal stiffeners are assumed to be rigid. Formulas for estimating the mean stress in flat bars and collar plates are derived. Finite element analyses are performed to calibrate these formulas. Design criteria are proposed to check the local connections near cutout openings.

Using the stress equations to obtain nominal stresses, a set of stress concentration factors is created after hot spots in some fine-mesh finite element models have been determined. Fatigue life estimates are made for these hot spots and the fatigue performance of different design arrangements is discussed.

1. INTRODUCTION

Cutout openings serve the purpose of allowing longitudinal stiffeners to go through transverse web frames or floors. These cutouts, which are typically reinforced by welded stiffeners such as flat bars and brackets (structural details), have traditionally been crack-prone. Although past studies have provided useful information about the stress contours around these cutouts, little has been published in terms of developing criteria to evaluate them in the design stage. Questions often arise as to whether a collar plate and backing brackets are necessary to reinforce the connections during new construction. Unless a detailed finite element analysis (FEA) is performed, the question cannot be easily answered. Without an effective way to prevent local cracking, many ships experience hundreds of cracks after some years of service. In some cases, cracks are found after less than two years of service. Although most of these cracks are not considered a major threat to the overall safety of the hull, the ships' owners incur a large cost in repairing them.

In this paper, simple formulas are developed to check the design of such connections. The formulas are calibrated using FEA of eight sample details.

As illustrated in Figure 1, cracks around cutout openings (also known as slots) are often seen in many types of large ships. Bea, et al. (1995) concluded that single-hull tankers experienced most of these cracks in the side shell and bottom shell areas due to cyclic wave pressure. In double-hull tankers, however, the double bottom seems to be the main problem area due to a very high differential pressure between laden and ballast

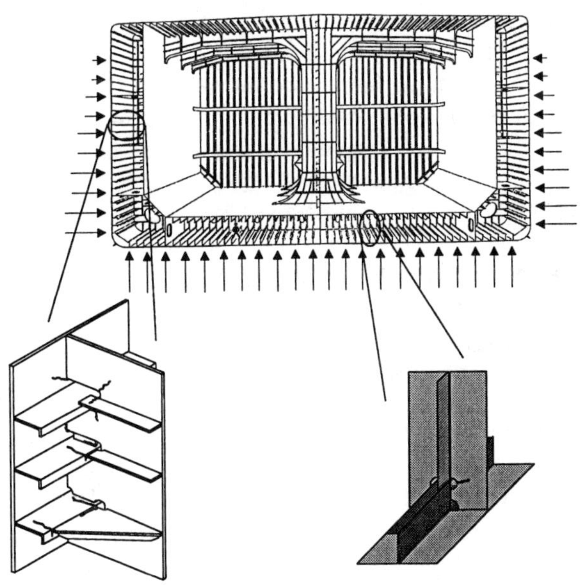

Figure 1: Typical cracking problem at end connections

NOMENCLATURE

A_1 flat bar footprint area
A_2 direct connection area
A_3 collar connection area
c_s coefficient related to longitudinal stiffeners
E Young's modulus
F_1, F_2, F_3 - forces transferred to flat bar, direct collar and optional collar plate, respectively
F_l total force transferred by longitudinal stiffener
F_T total force due to the pressure acting on a unit panel
F_w total force transferred by web frame
G shear modulus
h_L height of the bracket at the lower end of the side transverse
K_L empirical quantity
k_t total spring constant
L normalized fatigue life
l panel length
p static design pressure
s panel width
S_m section modulus of the longitudinal stiffener assuming full effectiveness of the side shell between longitudinals
α ratio of forces transferred to the web frame
β' coefficient calibrated for stress analysis
β_1, β_2, β_3 - factors that correct the bias of spring models
δ relative displacement between web frame and longitudinal stiffener
ν Poisons's ratio
σ_a allowable normal stress
σ_{fb} mean normal stress in flat bar
σ_{nom} nominal stresses
σ_{hss} hot spot stress from FEA
τ_a allowable shear stress
τ_{dc} mean shear stress at collar plate

conditions. Cheung and Slaughter (1998) have reported a large number of cracks in the inner bottom structure of several double-hull tankers. Many cracks occur along the flat-bar weld, between frame vertical stiffeners and inner bottom longitudinals.

Similarly, many survey reports show that cracks also occur in way of the connection of longitudinals to transverse floors inside the double bottom and hopper of bulk carriers (IACS, 1994). Again, the most common damage is cracking at the end of flat bar stiffeners. One of the main reasons for this type of damage is that high stresses develop in the connection area that is too small. Note that bulk carriers have more serious types of cracks in other locations, such as side frames and stools—these are not considered in this paper.

The flat bar appears to be the weakest link in the connections between longitudinals and transverse web frames and between longitudinals and floors. Ma (1998) and Bea, et al. (1995) report hundreds of flat bar failures in a single vessel. This subject was investigated by Glasfeld et al. (1977) who concluded that approximately 75 percent of the total number of cracks found around cutouts are at flat bars.

Cracking around end connections typically follows a sequence. The first crack normally appears along the footprint of the flat bar on the flange of the longitudinal (type B crack in Figure 2). The extensive corrosion, commonly observed at these cracks indicates that the crack growth rate is slow. As the flat bar cracks grow with time, the loads redistribute to the collar plates. Once the flat bar has cracked through, it loses its load-carrying capability completely and the entire load transfers to the remaining one or two collar connections. If this broken flat bar is not found and rectified, a second crack will start at the radius of the cutout (type D crack in Figure 2) and a third crack eventually occurs at the fillet weld on shell plate (type C or C1 failure in Figure 2). Many survey reports and field observations have confirmed this sequence of crack development. These surveys show that a cutout radius crack is only found when the flat bar has cracked completely through.

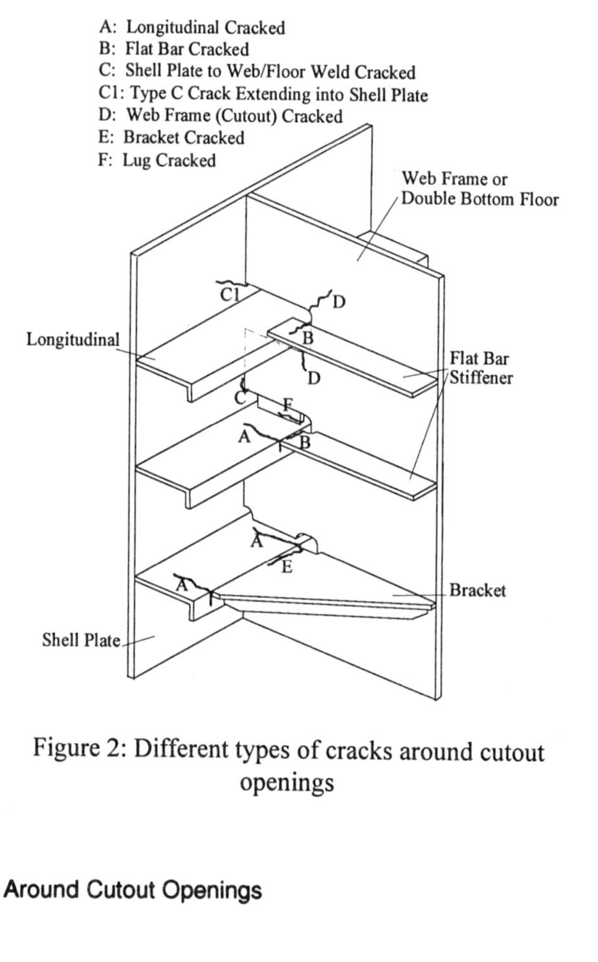

Figure 2: Different types of cracks around cutout openings

2. PROPOSED CRITERIA

In this paper, simple criteria are developed that will allow ship designers to perform a quick check of their design of end connections. The following criteria are preliminary in their form and are subject to extensive reviews and verification.

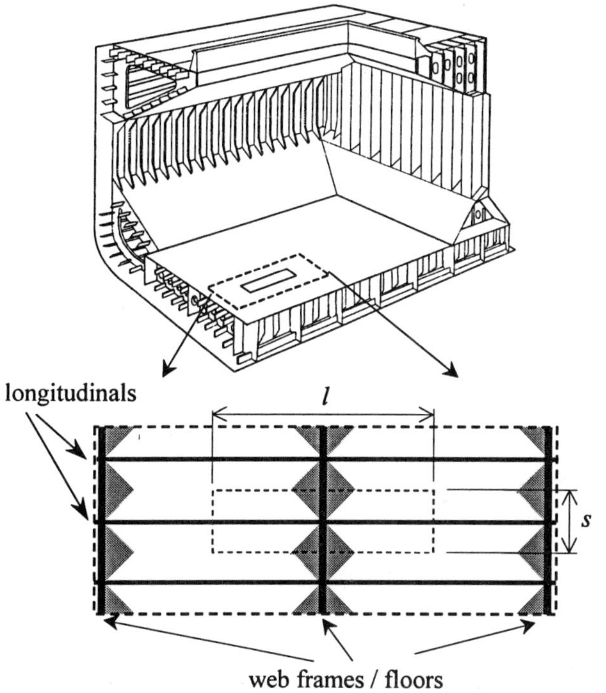

Figure 3: Unit panel within a sampled inner bottom plate (web frames take the pressure over shaded areas)

The criteria require that two checks be performed for each design of an end connection. First, the calculated mean normal stress in flat bar, σ_{fb}, should be less than an allowable value, σ_a:

$$\sigma_{fb} = \frac{ps(l - 0.7s)}{A_1 + 0.33c_s(A_2 + A_3)} < \sigma_a. \qquad (1)$$

Second, the calculated mean shear stress in the collar plate, τ_{dc}, should also be less than its allowable value, τ_a:

$$\tau_{dc} = \frac{ps(l - 0.7s)}{\dfrac{3.0}{c_s}A_1 + (A_2 + A_3)} < \tau_a. \qquad (2)$$

Here p, s and l represent static design pressure, panel width and panel length, respectively (see Figure 3). A_1, A_2 and A_3 are the flat bar footprint area, direct connection area, and collar connection area, respectively (see Figure 4). Units are in millimeters, newtons and their combinations. The coefficient, c_s, is determined as follows:

c_s = 1.0 for symmetrical longitudinal stiffeners,

c_s = 1.41 for unsymmetrical longitudinal stiffeners with one-sided support,

c_s = 1.12 for unsymmetrical longitudinal stiffeners with two-sided support.

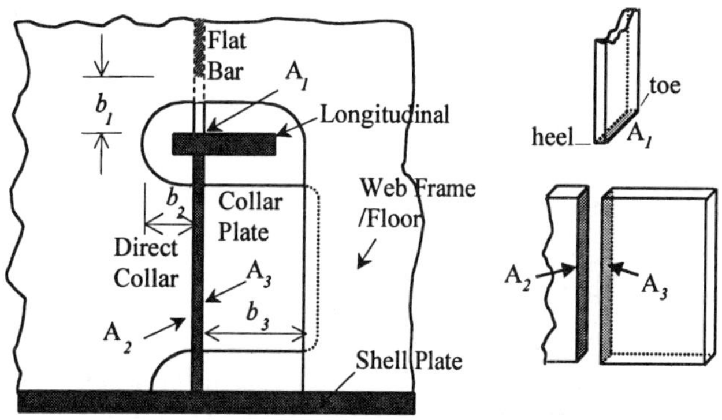

Figure 4: Definitions of geometry parameters

2.1 Derivation

Formulas for estimating stresses are presented in the preceding section. In this section, their derivations are explained in detail. The goal here is to formulate mean stresses in a flat bar and a collar plate as functions of scantling and material parameters. In a previous study, Chen and Wan (1999) performed a similar analysis. However, they focused more on defining stress concentration factors. The scope of the present paper is limited to conventional cutout designs. Specifically, the design is assumed to have a flat bar connected to the flange of a longitudinal by fillet weld. In addition, the web frame is welded to one side of the longitudinal stiffener by butt weld, while an optional collar plate may be fitted on the other side.

2.1.1 Assumptions

The ideal way to obtain the stresses in flat bar and collar plate is to perform a global and then a local FEA. The global FEA provides boundary conditions for the local one. This process is both resource demanding and time consuming. Clearly, it is not suitable for the purpose of a design check. A simple formula is needed to predict the stresses reasonably well. To achieve this, several assumptions have to be made to simplify the problem.

The first assumption is that pressure acting on a hull panel is the dominant load component causing the cracking problem. Shear forces in the web frame can be

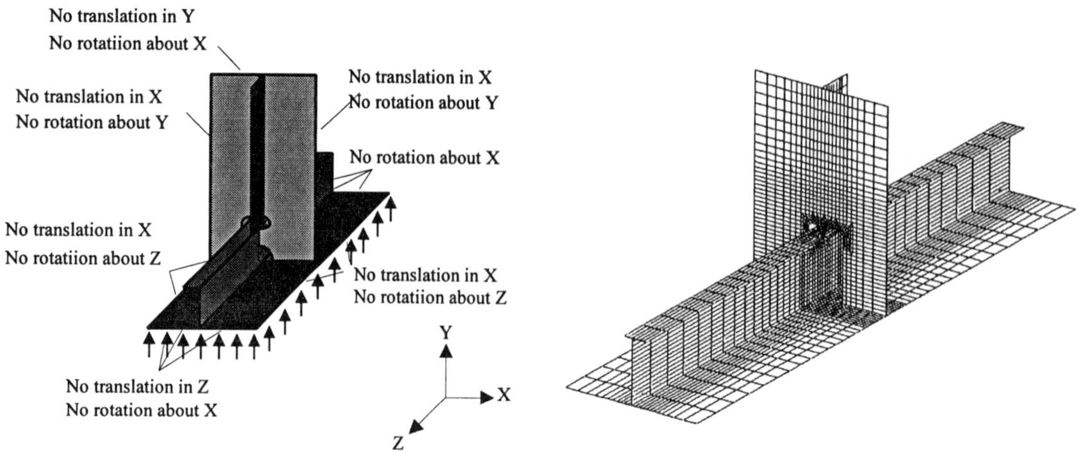

Figure 5: Finite element model and boundary conditions

neglected in this derivation, but will be accounted for by a larger factor of safety. Figure 3 shows an example of unit stiffened panels to be analyzed. The boundary conditions to be applied on it are defined in Figure 5. The longitudinal stiffener is assumed to be rigid. It is connected to the web frame by a flat bar, a direct collar and an optional collar plate as illustrated in Figure 4. These three components are modeled as springs holding the stiffener and web frame together.

The total force, F_T, due to the pressure acting on a unit panel is defined as:

$$F_T = psl, \tag{3}$$

where p is the pressure acting on unit panel, l is the web frame/floor spacing, and s is the longitudinal stiffener spacing.

F_T may be divided into two components, F_l and F_w, representing the portions of the total force transferred by the longitudinal stiffener and the web frame, respectively. Assume that the pressure on the shaded area of the panel in Figure 3 is transferred to the web frame. The longitudinal stiffeners take the rest of the load, i.e., the pressure on the unshaded area. The shaded area within a unit panel can be defined as αs^2, where α depends on the section modulus of the web frame and longitudinal stiffener. F_w and F_l can then be expressed by the following:

$$F_w = \alpha p s^2 \tag{4}$$

and

$$F_l = pls - \alpha p s^2. \tag{5}$$

If the section modulus of the transverse members is close to that of the longitudinals, α is taken as 0.5. For typical stiffened panels used in ships, this coefficient will be shown to be 0.7.

The force taken by the longitudinal stiffener, F_l, is transferred to the web frame through the flat bar and collar plates. That is,

$$F_l = F_1 + F_2 + F_3, \tag{6}$$

where F_1, F_2 and F_3 are the forces transferred by the flat bar, direct collar and optional collar plate, respectively.

2.1.2 Spring Model

Assume that the force transfer mechanism from the longitudinal stiffener to the web frame can be modeled by three springs, one each for the flat bar, direct collar and optional collar plate. The force-displacement relation for the three springs is expressed as:

$$F = k_i \delta \qquad i = 1, 2, 3, \tag{7}$$

where $k_1 = \beta_1 E A_1 / b_1$ is the spring constant of the flat bar, $k_2 = \beta_2 G A_2 / b_2$ is the spring constant of the direct collar, $k_3 = \beta_3 G A_3 / b_3$ is the spring constant of the optional collar plate, A_1, A_2, A_3 are the connection areas between web frame and longitudinal shown in Figure 4, b_1, b_2, b_3 are the working distances (spring lengths) shown in Figure 4, β_1, β_2, β_3 are factors that correct the bias of spring models, E is Young's modulus, G is the shear modulus, and δ is the relative displacement between the web frame and longitudinal. Note that A_1 is the total footprint area of the flat bar and backing bracket, where fitted. However, forward brackets on the toe side of the flat bar are not taken into account. The reason is explained later in Section 3.3.

The total spring constant, k_t, is the sum of the three component spring constants. The correction factors, β_3 and β_2, may be assumed equal, since they represent the same type of bias, i.e. the bias in modeling shear stiffness. Resulting in:

$$k_t = k_1 + k_2 + k_3 = \beta_1 \frac{EA_1}{b_1} + \beta_2 \left(\frac{GA_2}{b_2} + \frac{GA_3}{b_3} \right). \quad (8)$$

2.1.3 Solutions

The relative displacement between the web frame and the longitudinal stiffener, δ_0, can be calculated as:

$$\delta_0 = \frac{F_l}{k_t}. \quad (9)$$

Forces in the three components of interest can be expressed as:

$$F_i = k_i \delta_0 = \frac{k_i}{k_t} F_l, i = 1,2,3. \quad (10)$$

The above equation reveals the load distribution between the flat bar stiffener and the two collar plates. Once these forces are available, the mean stresses at the flat bar, direct collar and optional collar plate can be obtained by combining Equations 8 and 10, yielding:

$$\sigma_{fb} = \frac{F_1}{A_1} = \frac{F_l}{A_1 + \frac{0.5\beta}{1+\upsilon}\left(\frac{b_1}{b_2} A_2 + \frac{b_1}{b_3} A_3 \right)}, \quad (11)$$

$$\tau_{dc} = \frac{F_2}{A_2} = \frac{F_l}{2(1+\upsilon)\frac{b_2}{\beta b_1} A_1 + A_2 + \frac{b_2}{b_3} A_3}, \quad (12)$$

and

$$\tau_{cp} = \frac{F_3}{A_3} = \frac{F_l}{2(1+\upsilon)\frac{b_3}{\beta b_1} A_1 + \frac{b_3}{b_2} A_2 + A_3}, \quad (13)$$

where $\beta = \beta_2 / \beta_1$ and ν is Poison's ratio.

The above equations are simple mathematical expressions that can be used to calculate the three stresses. It is apparent that increasing the connection area reduces the mean stresses in the three structural members.

By studying the above equations, some observations can be made on the effectiveness of the second collar plate. Experience from field observation has shown that the likelihood of cracking and crack growth rate seems less for longitudinals fitted with two collars rather than with one. It is argued that the symmetry of load path and increased stiffness of the double collar plates reduce the stress in flat bar.

By examining Equation 11, the argument can easily be proven. In addition, the second collar plate is more effective when it is shorter in width. This is also obvious since its shear stiffness is inversely proportional to its width.

For practical use, Equations 11, 12 and 13, may be further simplified—they need to be as simple as possible. Since the ratios b_2/b_1, b_3/b_1 and b_3/b_2 are of the order of unity, they can be dropped from the formulas. Also, a new parameter, β', is introduced to merge β with its associated parameters. This results in the simplified formulas:

$$\sigma_{fb} = \frac{F_l}{A_1 + \beta'(A_2 + A_3)} \quad (14)$$

and

$$\tau_{dc} = \frac{F_l}{\frac{A_1}{\beta'} + (A_2 + A_3)}. \quad (15)$$

In conventional designs, the third equation (Equation 13) may be unnecessary because τ_{dc} is generally more critical than τ_{cp}. If the criterion for shear stress τ_{dc} is satisfied, this should normally guarantee that the shear stress criterion is met by τ_{cp}.

A calibration study will further defined β' in a later section. The final expressions for the stresses become:

$$\sigma_{fb} = \frac{ps(l - 0.7s)}{A_1 + 0.33c_s(A_2 + A_3)} \quad (16)$$

and

$$\tau_{dc} = \frac{ps(l - 0.7s)}{\frac{3.0}{c_s} A_1 + (A_2 + A_3)}, \quad (17)$$

Where, as before, c_s is 1.0 for symmetric longitudinal stiffeners, 1.41 for asymmetric longitudinal stiffeners with one - sided support, and 1.12 for asymmetrical

Table 1: Unity check of flat bar and collar plate

Detail ID		Bulker A	Bulker B	Bulker C	Bulker D(C2)	Tanker A	Tanker B	Tanker C	Tanker D
dead weight tonnage		48000	73500	170000	170000	na	105400	na	307000
longitudinal		250x90x 11/16 (IA)	310x10 + 125x30 (T)	310x150x 09/14 (T)	358x125x 09/12 (T)	400x10+ 150x16 (L)	450x150x 11.5/18 (IA)	500x150x 11/25 (T)	550x12 + 150x34 (T)
pressure head (meters)		14.8	17.8	21.7	20.0	22.5	18.7	21.1	28.8
pressure (N/mm2)	p	0.14	0.17	0.20	0.19	0.21	0.17	0.20	0.27
logitudinal spacing (mm)	s	850	835	840	850	880	830	850	860
frame/floor spacing (mm)	l	2700	3620	2580	2580	3700	3700	4500	4950
flatbar area (mm2)	A1	900	1500	2500	1500	2640	2275	2450	7312
direct collar area (mm2)	A2	2121	3264	2638	3338	4140	5395	6765	9486
flat bar unity check		94%	**117%**	72%	87%	89%	68%	100%	100%
collar unity check		88%	78%	48%	58%	84%	63%	66%	66%

longitudinal stiffeners with two-sided support.

These two formulas show that the stresses in the connection can be reduced by minimizing the spacing of the stiffeners (s and l) or maximizing the connection areas of the flat bar, direct collar and collar plate (A_1, A_2 and A_3).

2.2 Setting Allowable Stresses

Using Equations 16 and 17, design criteria for the connection details may be arranged as follows:

$$\sigma_{fb} < \sigma_a \tag{18}$$

and

$$\tau_{dc} < \tau_a, \tag{19}$$

where σ_a and τ_a are the allowable tensile and shear stresses, respectively. Based on service experience with four bulk carriers and four tankers, allowable stresses for flat bars and direct collars made of mild steel have been preliminarily determined to be:

$$\sigma_a = 140 \text{ N/mm}^2$$

and

$$\tau_a = 70 \text{ N/mm}^2.$$

The above criteria ensure that σ_{fb} and τ_{dc} are within acceptable levels. If these levels cannot be met, designers may be required to either increase the connection areas or reduce the load by using smaller stiffener spacing.

One of the four bulk carriers cited above has experienced extensive cracking in its flat bars. The allowable stress, σ_a, is so calibrated that this bulk carrier's design should fail the criterion of Equation 18. Table 1 lists the results of unity checks, σ_{fb}/σ_a and τ_{dc}/τ_a, for eight vessels. The vessel that would fail the criteria is listed in Table 1 as Bulker B. Bulker B is calculated to have a flat bar mean stress exceeding the allowable by 17 percent.

In addition to the unity check, Table 1 also lists the scantlings for four double hull tankers. Note that no single hull tankers were chosen since they will be gradually phased out. In their original design, these tankers have collar connections on both sides of their longitudinal stiffeners. However, the second collar plates are intentionally left out in the unity checks to see whether the proposed criteria can capture the under-design (A_3 is taken as zero for all eight structural details).

The unity checks give reasonable results. While they are able to pinpoint the design problem in Bulker B, the unity checks also show narrow safety margins for the four tankers with their second collar plates removed. In fact, two of them, namely Tanker C and Tanker D, just fail to meet the criteria.

Following the theory of shear failure, the allowable shear stress is set to be half of its normal counterpart. This allocation produces a result that all eight direct collar connections pass the shear unity checks. This agrees with the experience that flat bars in general are more susceptible to failure than collar plates.

2.3 Calibration of Coefficients

The coefficient α is the ratio of force transferred to the web frame to that transferred to the longitudinal stiffener. For a particular connection detail i, the parameter α can be defined as:

$$\alpha_i = \frac{F_{wi}}{p_i s_i^2}. \tag{20}$$

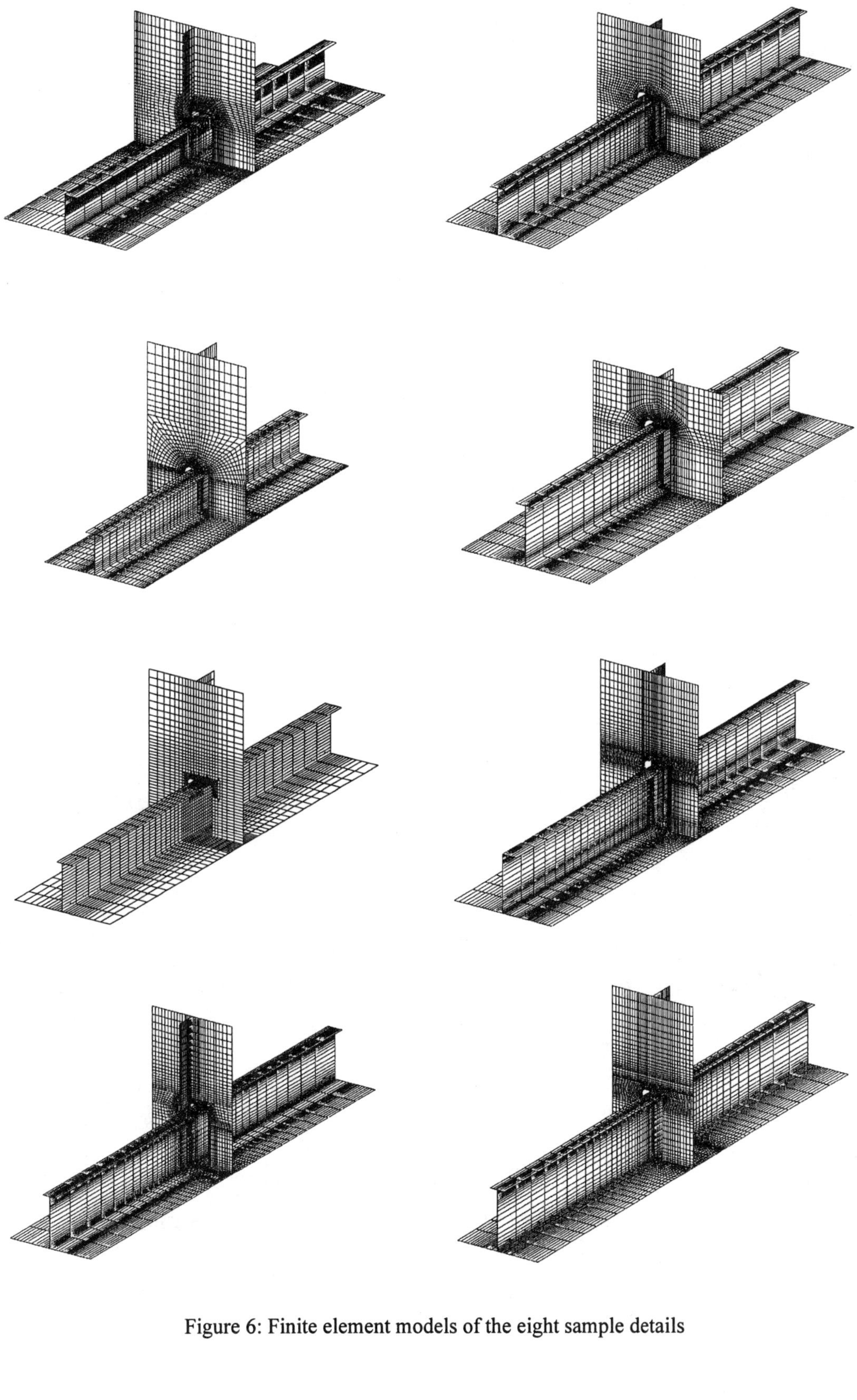

Figure 6: Finite element models of the eight sample details

Table 2: Scantlings of the eight sample details

Detail ID		Bulker A	Bulker B	Bulker C	Bulker D(C2)	Tanker A	Tanker B	Tanker C	Tanker D
dead weight tonnage		48000	73500	170000	170000	na	105400	na	307000
longitudinal		250x90x 11/16 (IA)	310x10+ 125x30 (T)	310x150x 09/14 (T)	358x125x 09/12 (T)	400x10 + 150x16 (L)	450x150x 11.5/18 (IA)	500x150x 11/25 (T)	550x12+ 150x34 (T)
pressure head (meters)		14.8	17.8	21.7	20.0	22.5	18.7	21.1	28.8
design pressure (N/mm2)	p	0.14	0.17	0.20	0.19	0.21	0.17	0.20	0.27
logitudinal spacing (mm)	s	850	835	840	850	880	830	850	860
floor spacing (mm)	l	2700	3620	2580	2580	3700	3700	4500	4950
flatbar connection area (mm2)	A1	900	1500	2500	1500	2640	2275	2450	7312
d.c. connection area (mm2)	A2	2121	3264	2638	3338	4140	5395	6765	9486
lug connection area (mm2)	A3	1575	2616	2638	3338	4140	3900	6765	9486
flatbar clearance (mm)	b1	68.0	49.5	50.0	30.0	30.0	84.0	102.0	90.0
d.c. width (mm)	b2	40.0	92.5	100.0	87.5	80.0	40.0	150.0	115.0
lug width (mm)	b3	125.0	135.0	100.0	87.5	150.0	155.0	169.0	115.0
lug height (mm)	bh	202.0	264.0	211.0	267.0	345.0	415.0	410.0	527.0
tran. web thickness (mm)	tw	10.5	12.0	12.5	12.5	12.0	13.0	16.5	18.0

Table 3: Sample values of α from sixteen models

Detail ID	Bulker A	Bulker B	Bulker C	Bulker D(C2)	Tanker A	Tanker B	Tanker C	Tanker D
p (N/mm2)	1.00	1.00	1.00	1.00	1.00	1.00	1.00	1.00
s (mm)	850	835	840	850	880	830	850	860
Details without Collar Plates								
Fw (MN)	0.58	0.65	0.55	0.58	0.59	0.50	0.67	0.69
F1 (MN)	0.76	1.39	1.10	1.00	1.59	1.27	1.71	2.01
F2 (MN)	0.96	0.98	0.52	0.61	1.08	1.30	1.48	1.56
F3 (MN)	na	na	na	na	na	na	na	na
α	**0.80**	**0.93**	**0.78**	**0.80**	**0.76**	**0.73**	**0.93**	**0.93**
Details with Collar Plates								
Fw (MN)	0.53	0.61	0.45	0.48	0.50	0.46	0.64	0.59
F1 (MN)	0.67	1.07	0.82	0.75	1.30	1.07	1.27	1.51
F2 (MN)	0.78	0.77	0.45	0.48	0.82	1.02	1.03	1.08
F3 (MN)	0.32	0.57	0.45	0.48	0.64	0.52	0.89	1.08
α	**0.73**	**0.87**	**0.64**	**0.66**	**0.65**	**0.67**	**0.89**	**0.80**

An average value of α for n connections can then be obtained as:

$$\bar{\alpha} = \frac{1}{n}\sum_{1}^{n}\alpha_i. \tag{21}$$

To calibrate α, structural detail samples are needed. Thus, sixteen finite element models (FEM) were established based on the eight details used in the previous section. Each detail has two variations: one with and the other without a collar plate. Scantlings and FEMs of the eight details are provided in Table 2 and Figure 6, respectively.

Linear analyses were performed using MSC/NASTRAN (MSC, 1998). Figure 5 shows the boundary conditions that are applicable to details in double bottoms or double sides. Also shown there is one example of the eight models. A unit pressure load is applied on the panel. Each model has about 2000 four-node shell (bending plate) elements. Some three-node shell elements also exist for smooth mesh transition. Though relatively fine meshes are used, the element size is not a concern since the objective of the analyses is merely to determine the load paths.

The values of F_w, F_1, F_2, and F_3 were manually computed from the stress results by summing up the element forces in the cross section of interest. Note that F_w is the share of load transferred by the web frame and F_1 is the portion of load that flows through the flat bar. Table 3 lists these forces and the calibrated coefficient. Sixteen sample values of α were obtained. The first eight values are from the models without collar plates; the

Table 4: Comparing predicted flat bar force percentages with actual data

	Detail ID	Bulker A	Bulker B	Bulker C	Bulker D(C2)	Tanker A	Tanker B	Tanker C	Tanker D
Details without Collar Plates									
Actual from FEA	$F_1/(F_1+F_2)$	0.442	0.586	0.679	0.621	0.596	0.494	0.536	0.563
Our prediction	$A_1/(A_1+0.33C_sA_2)$	0.477	0.582	0.742	0.577	0.578	0.475	0.523	0.566
Prediction error		8%	-1%	9%	-7%	-3%	-4%	-2%	1%
Details with Collar Plates									
Actual from FEA	$F_1/(F_1+F_2+F_3)$	0.379	0.444	0.477	0.439	0.471	0.410	0.398	0.411
Our prediction	$A_1/(A_1+0.33C_s(A_2+A_3))$	0.397	0.436	0.589	0.405	0.463	0.398	0.354	0.395
Prediction error		5%	-2%	24%	-8%	-2%	-3%	-11%	-4%

other eight values are from those models with collar plates. The values of α range from 0.64 to 0.93. Instead of using the mean value as defined in Equation 21, α is conservatively chosen to be 0.7.

As expected, α is larger than 0.5, since the transverse web frame/floor has larger section modulus than the longitudinal stiffener. In other words, the web frame is stiffer than the longitudinal, and thus absorbs more load than depicted in the shaded area of Figure 3. Similarly, the coefficient β' may be determined. β' is a tuning parameter used to accurately predict the load that is transferred into the flat bar. The purpose of the calibration is to derive an accurate, yet simple expression for β'.

The calibration process for β' is not as straight forward as that for α because several factors can have an effect. The degree of symmetry of a structural detail is probably the major one, the value of β' increases when the detail is less symmetrical. Physically, this means that an asymmetric longitudinal stiffener tends to twist (trip), and consequently collar connections take a larger share of F_l. It is easier to treat the symmetrical details first. It was determined that the base value for β' is approximately 0.33. An additional coefficient, c_s, is introduced here to adjust β' according to the degree of asymmetry, yielding:

$$\beta' = 0.33c_s, \tag{22}$$

c_s is 1.0 for symmetric longitudinal stiffeners, 1.41 for asymmetric longitudinal stiffeners with one-sided support, and 1.12 for asymmetrical longitudinal stiffeners with two-sided support. Chiou (2000) has also investigated the effect of asymmetry.

Table 4 shows that the given formula for β' provides reasonable predictions of the flat bar force, in cases both with and without collar plates. The worst prediction occurs in the case of Bulker C with a collar plate, which is fitted with an extra large flat bar. Bulker C's large A_1 causes the force F_1 to be over-predicted by 24 percent. Although this over prediction of the force is not desirable, the effect is alleviated later when F_1 is divided by A_1 to determine σ_{fb}. A new correction factor formulation could be introduced for oversized components; but this would make the criteria too complex.

3. SOME CONSIDERATIONS

This section discusses four different issues that are important to the derivation. First, the effect of web shear that has been ignored so far is studied. It is found that web shear has very little contribution to flat bar force. Second, nonlinear analyses are performed to verify whether the linear analyses have produced reasonable estimates of the force transfer percentages. Third, the effect of brackets is analyzed. Last, the limitations of the stress equations are reviewed.

3.1 Pressure Versus Shear

Stresses in a typical connection are induced by both secondary and tertiary deformation of the hull. As discussed in Section 2, the common practice for analysis of a local connection is a global-local FEA. However, this time-consuming FEA is unrealistic for the purpose of a design check. Consequently, in this paper only a local analysis of the connection is conducted. The loads on the connection may be idealized as a combination of fluid/cargo pressures on the shell/bulkhead and shear forces in the web plate. Both of these loads vary along the length, breadth, and height of the ship. Obviously, stresses in the connection are a function of the relative magnitudes of the two loads. Effects of each of these loads on the load transfer mechanisms in the connection must be investigated to develop proper design criteria.

In the previous section, analytical models have been developed for the force distribution in a typical connection with the shell/bulkhead subjected to uniform fluid pressures. In order to understand the response of a typical connection due to web shear, a linear FEA is conducted as follows. The connection is assumed to be at the side shell in the bilge region. Only connections without any collar plates were selected for this analysis. The ABS *Rules for Building and Classing Steel Vessel* (ABS 2000) were used to determine the magnitudes of shear stresses in the vicinity of the connection.

The shear stress in the web of the connection was estimated as follows. Referring to Section 5/1.4.11.7 of ABS (2000) for tankers, the shear force, F, in the side transverse in way of bilge is given as:

$$F = 1000\, ks\,[K_L\, l\,(P_U + P_L) - h_L P_L]$$

or

$$F = 350\; k\, s\, K_L\; l\,(P_U + P_L),$$

where $k = 1$, l is the span of the side transverse, s is the spacing of the side transverses, P_U and P_L are the nominal pressures at the upper and lower ends of the side transverse, h_L is height of the bracket at the lower end of the side transverse, and K_L is empirical quantity. For the selected connection in Tanker A, l = 20 m (assumed), s = 3.7 m, K_L = 0.3, h_L = P_U = 0, (assumed), and P_L = 98.1 KN/m^2. The web shear force is computed as F = 2175.6 KN and the web shear stress τ_{xy} is estimated to be 79.11 N/mm^2.

The displacement boundary conditions applied in the FEA are depicted in Figure 7. Stresses in the direct collar and the flat bar are presented in Figures 8a and 8b. As discussed by Liu and Bakker (1981), the web shear force induces significant stress concentrations at the web cut outs and locally in the flat bar. The stresses in the flat bar due to web shear are extremely localized. Consequently, the small force contributed by web shear in the flat bar may be neglected.

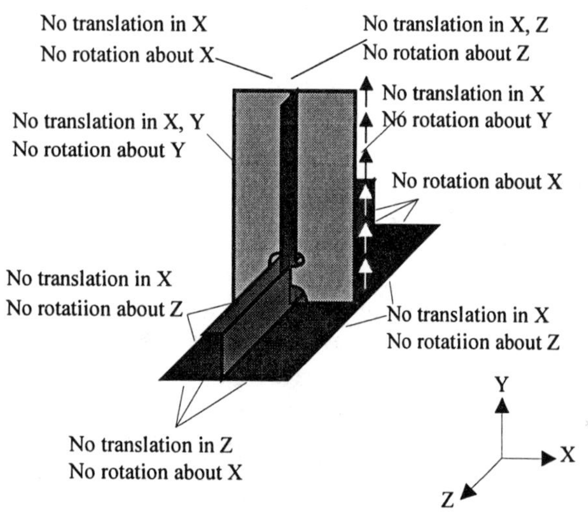

Figure 7: Boundary condition for analysis—web subjected to shear

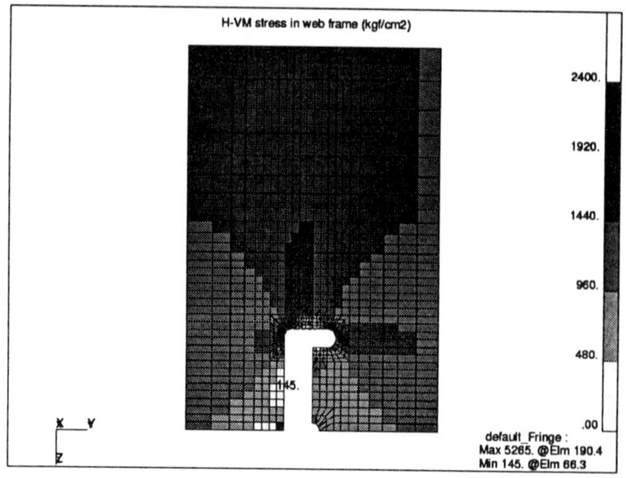

Figure 8a: Von Mises stress in the web plate in the connection in detail Tanker A

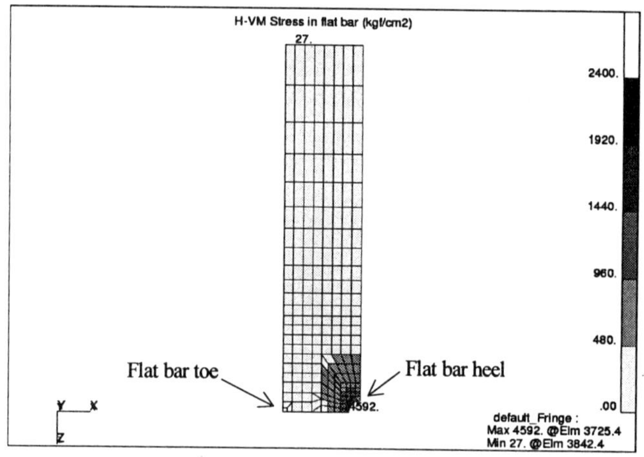

Figure 8b: Von Mises stress in way of the flat bar in the connection in detail Tanker A

3.2 Plastic Deformation

After a period of service, cracks are found in the local connections from web frames to longitudinal stiffeners. Cracks observed after short service duration may be due to yielding or low cycle fatigue. In either case, the local structure undergoes plastic deformation prior to failure. The elastic-plastic response of connections around cutout openings under fluid/cargo pressures has been investigated as part of this paper. The extent of the plasticity and the load redistribution in the adjoining structures during plastic deformation are simulated.

Linear elastic analysis of typical connections, subjected to nominal fluid pressure, described in the previous section, indicates that the linear stresses at the root (i.e., toe and heel) of the flat bar are above the yield stress. In order to quantify the effect of plastic deformation in the connection on the stress distribution a few elastic-plastic analyses were conducted, they are presented in Table 5.

Table 5: Load distribution considering effect of plastic deformation

Detail	Bulker B		Bulker C		Tanker B		Tanker C	
Applied load	Just yield	Beyond yield	Just yield	Beyond yield	Just yield	Beyond yield	Just yield	Beyond yield
	50%	100%	80%	100%	55%	100%	35%	100%
F_T (MN)	0.30	0.60	0.35	0.43	0.34	0.61	0.27	0.77
F_w (MN)	0.07	0.15	0.09	0.11	0.05	0.10	0.05	0.13
F_1 (MN)	0.14	0.27	0.17	0.22	0.14	0.25	0.12	0.33
F_2 (MN)	0.11	0.22	0.09	0.11	0.15	0.27	0.10	0.30
F_w/F_T	24%	24%	25%	25%	16%	16%	17%	17%
F_1/F_T	46%	44%	50%	50%	41%	41%	45%	44%
F_2/F_T	35%	36%	25%	25%	43%	44%	39%	40%
α	1.04	1.06	0.75	0.75	0.72	0.72	0.92	0.92

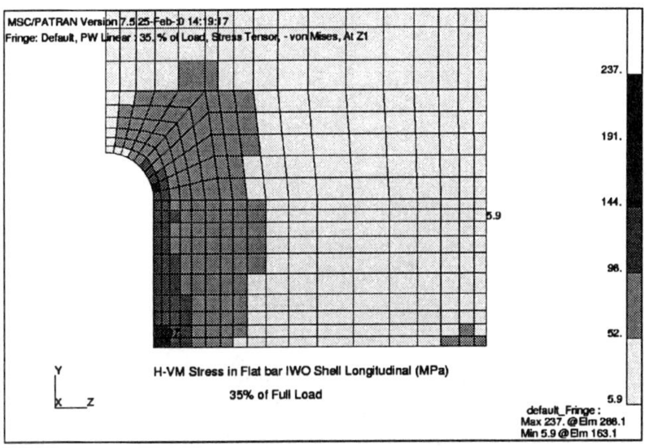

Figure 9a: Von Mises stress in way of the flat bar in the connection in detail Tanker C

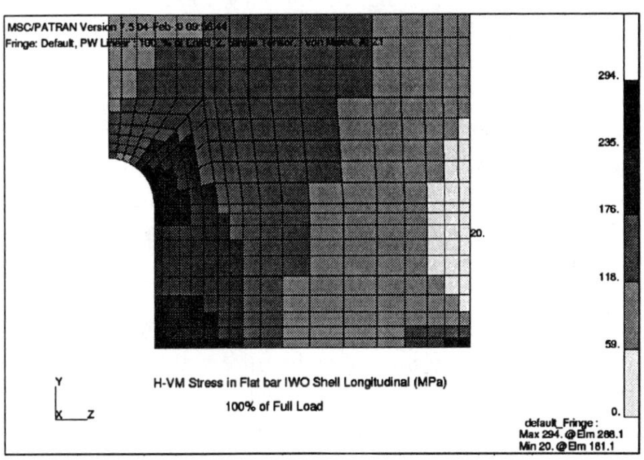

Figure 9b: Von Mises stress in way of the flat bar in the connection in detail Tanker C

The elastic-plastic analysis was conducted using MSC/NASTRAN. In the analysis the flat bar was assumed to be of mild steel with a yield stress 235 N/mm^2 and a hardening slope of E/30 where E is the elastic modulus of steel. A nominal uniform pressure of 0.2 N/mm^2 (full load) is applied on the shell. Yielding generally initiates at the root of the flat bar. The spread of plastic zone for two load magnitudes (a) at the initiation of yield in the flat bar and (b) at the final load, are illustrated in Figures 9a and 9b, respectively. For the magnitude of applied load, the plastic zone is limited to the root of the flat bar and the contour of the flat bar cutout. The force redistribution in the connection due to the plastic deformation in the flat bar is given in Table 5. As in Figures 9a and 9b, the forces are presented at two load stages, i.e., onset of yield (labeled 'Just yield') and applied full load (labeled 'Beyond yield'). For example, the column subheading "50%" for Bulker B in the table denotes the force distribution in the connection at the onset of yield, which is at fifty percent of the applied full load, i.e., at a pressure of 0.1 N/mm^2. Note that the maximum difference in the parameter α from that obtained through an elastic analysis is about 10 percent. This indicates that for nominal fluid pressures, local plastic deformation in the connection need not significantly modify the force distribution among the connection components.

3.3 Local Brackets

Heretofore, this paper has focused on connections with and without collar plates and with web stiffeners in the form of flat bars. Oftentimes, connection detail designs provide an additional bracket (Ma & Bea, 1995). Typically brackets are provided forward or aft of the flat bar for additional support to the shell longitudinal/web frame connection. The effect of an additional bracket on the force distribution in the connection is assessed in this section.

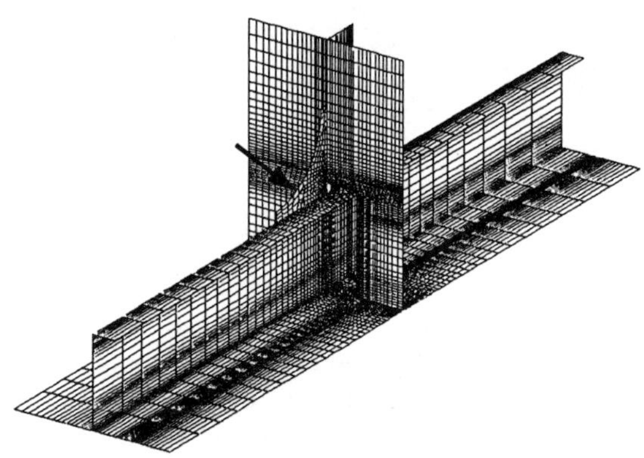

Figure 10: Connection detail with bracket aft of the flat bar on the web frame

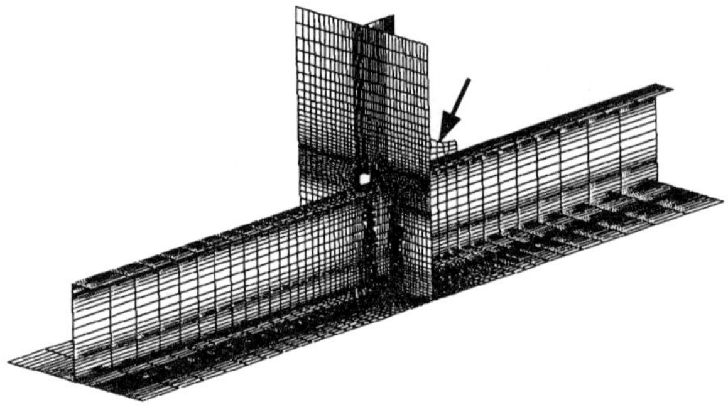

Figure 11: Connection detail with bracket forward of the flat bar on the web frame

Table 6: Load distribution in the case of bracket forward of the flat bar on the web frame

Detail	Tanker B		Tanker C	
	No Bracket	Bracket	No Bracket	Bracket
F_T (MN)	0.61	0.61	0.77	0.77
F_w (MN)	0.10	0.10	0.13	0.13
F_1 (MN)	0.25	0.25	0.34	0.34
F_2 (MN)	0.27	0.27	0.30	0.30
F_w/F_T	16%	16%	17%	17%
F_1/F_T	41%	41%	45%	44%
F_2/F_T	43%	44%	39%	38%
α	0.73	0.73	0.92	0.93

Table 7: Load distribution in the case of bracket aft of the flat bar on the web frame

Detail	Bulker B		Tanker B		Tanker C	
	No Bracket	Bracket	No Bracket	Bracket	No Bracket	Bracket
F_T (MN)	0.61	0.61	0.61	0.61	0.77	0.77
F_w (MN)	0.15	0.11	0.10	0.09	0.13	0.11
F_1 (MN)	0.28	0.41	0.25	0.41	0.34	0.52
F_2 (MN)	0.21	0.10	0.27	0.12	0.30	0.15
F_w/F_T	24%	18%	16%	14%	17%	14%
F_1/F_T	46%	67%	41%	66%	45%	67%
F_2/F_T	35%	17%	43%	20%	39%	19%
α	0.94	0.72	0.73	0.63	0.92	0.74

Figures 10 and 11 show two FEM that were constructed, one with a bracket forward of the flat bar and the other with a bracket aft of the flat bar. The shell was subjected to a constant normal pressure. The stress distribution in the connection details for Bulker A, Tanker B and Tanker C were computed as in the previous section and are presented in Tables 6 and 7. Here the force F_1 is the total force in the bracket and flat bar in way of the shell longitudinal. The additional bracket does not significantly modify the stress distribution in the connection when the bracket is forward of the flat bar (Table 6). However, when the bracket is placed aft (symmetrically about the web frame), the force resisted by the bracket and flat bar increases (Table 7). If the effective web stiffener foot print area, including the bracket, is denoted as A_1, then

$$A_1 = A_{fb} + \gamma A_b, \tag{23}$$

where γ is a parameter denoting the effectiveness of the bracket. Based on FEA, the parameter may be assumed as follows:

Bracket fwd of flat bar	$\gamma = 0$
Bracket aft of flat bar	$\gamma = 1.$

The above equation provides a method for estimating the force distribution in the connection.

3.4 Limitations

Although Equations 1 and 2 can be used for design check on flat bars and collar plates, they have some limitations.

For instance, Yoneya, et al. (1993) reported that one type of crack found in some second-generation VLCCs is different from the typical flat bar weld cracks. The cracks start in the flange of side longitudinals in way of the flat bar toe and propagate into their webs toward the side shell. While this type of cracks has a more serious consequence, the formulas unfortunately cannot predict it.

Another limitation of the stress equations relates to their application to non-conventional cutout designs. The formulas are only applicable to conventional cutout designs. Recently, ship designers and builders have investigated customized designs. One example is the apple-shaped slot designed by Kawasaki, another is a design that eliminates vertical flat bars but places a horizontal flat bar stiffener above the longitudinal stiffener. There are also designs that move flat bars to the side of a longitudinal stiffener. In these cases, the pressure load is transmitted solely by collar plates without the assistance of flat bars. While in general the formulas for shear stress may still work with these custom designs, the impact of special cutout shapes and flat bar arrangements needs to be investigated.

The present criteria does not distinguish the two types of flat bar joints, namely fillet weld and lap weld (see Figure 2). While the formulation is for a fillet weld on top of the longitudinal flange, the preferred way of attachment, a lap joint arrangement may use the same formulation with a lower allowable stress or other form of penalty. This is also needs further study.

4. CUTOUT FATIGUE PERFORMANCE

The emphasis of this paper is on the stress equations developed in Section 2.1. These equations were developed so that simple design level calculations could be performed that would mitigate potential fatigue problems. In this section, the fatigue performance of two representative side shell connections subjected to external pressure is examined for the purpose of correlation with the stress equations. To do so, the stress concentration factors (SCFs) characteristic of the connections is determined. The SCFs are defined as the ratios of the maximum local stress in the structural detail (the so called "hot spot stress") to the design-level average stress computed using the stress equations. The hot spot stresses are computed using FEA, in a way that is compatible with a hot spot stress S-N curve. The hot spot stresses computed for points along the flat bar load path are divided by the average direct stress, σ_{fb} to arrive at the SCFs. The hot spot stresses along the direct collar/collar plate load path are divided by the average shear stress, τ_{dc}.

The resultants SCFs are examined to gain insight into how the stress equations should correlate with fatigue performance. Toward this end, normalized fatigue life estimates are associated with the hot spot stresses. These normalized estimates result from the assumption that the nominal bending stress at the connection due to external pressure produces a life of 120 years when combined with a hot spot S-N curve that has a log-log slope of 3. The estimates serve to highlight the relative importance of the magnitude of the average stresses versus the SCFs when examined in the context of fatigue damage.

At each of the side shell connections examined herein, five alternative configurations are investigated. The first configuration investigated for each connection is the as-designed structural detail. Both as-designed structural details feature an overlapping collar plate and a direct connection of the flat bar to the longitudinal stiffener flange. In the second configuration the collar plate is removed, while in the third configuration the flat bar is offset from the longitudinal. To mitigate fatigue problems generated by the offset of the flat bar, the fourth configuration features symmetric offset brackets on each side of the web frame. In lieu of brackets, the fifth configuration features a collar plate in-line with the web frame and stress relieving profiles for the cutout, the collar plate, and the rat hole.

4.1 Description of Analyzed Details

Side shell connections from two vessels were selected for investigation. The vessels, designated Tanker D and Bulker C, were described in the earlier discussion. The connections themselves are located near the still-water-line of each vessel. This location is one that experience has shown to be fatigue-critical. As designed, each connection features a symmetric T-shaped longitudinal stiffener, an overlapping collar plate that does not extend to the side shell, and a direct connection of the flat bar to the longitudinal flange. The as-designed scantlings for Tanker D are provided in Figure 12, along with depictions of the cutout profiles and the flat bar connection. The same information for Bulker C is provided in Figure 13.

There are some noteworthy differences between the cutout profiles and flat bar connections designed for the respective vessels. For Tanker D the overlap of the collar plate has slanted corners, while those of Bulker C are circular. There is a rat hole in the web frame of Bulker C but not in that of Tanker D. Perhaps most significantly, there is a soft-toed backing bracket attached to the flat bar connection of Tanker D, while the connection of Bulker C is unsupported. Further, the flat bar rat hole in Tanker D has a stress-relieving soft-toed design while that of Bulker C is a simple circular arc with 90° attachments.

When the flat bar is offset from the longitudinal, it is sniped as illustrated in Figure 14. The symmetric backing brackets employed to mitigate potential fatigue problems caused by the offset of the flat bar are described in Figure 15. This bracket design has been employed successfully to mitigate problems in other fatigue studies (Healy and Thomsen, 2000); it is used here for the same purpose. The bracket design is not optimized for cost versus stress relief; it is included merely to demonstrate the effectiveness of brackets. The stress-relieving cutout and in-line collar plate profiles used to mitigate problems in lieu of brackets are presented in Figure 16.

4.2 Finite Element Modeling

In the FEA, a model of each side shell connection configuration was made that extends halfway to the adjacent transverse web frames, halfway to the adjacent longitudinals, and halfway to the inner hull. A unit pressure is applied to the side shell, and the boundaries are constrained in a way that mimics continuous beam behavior. These boundary conditions, illustrated in Figure 17, are akin but not identical to those used to derive the stress equations in Section 2.1. The two sets of boundary conditions differ in that the web frame boundary in Figure 17 is restrained against transverse motion (deflection in the Z direction). Restraint of the web frame in this way introduces more realistic web shear in the connection and leads to greater hot spot stresses.

The ABAQUS (1998) finite element program was employed. Each model consists of approximately 6000 8-noded bi-quadratic mid-surface thick shell elements with 100000 degrees of freedom. A typical mesh is

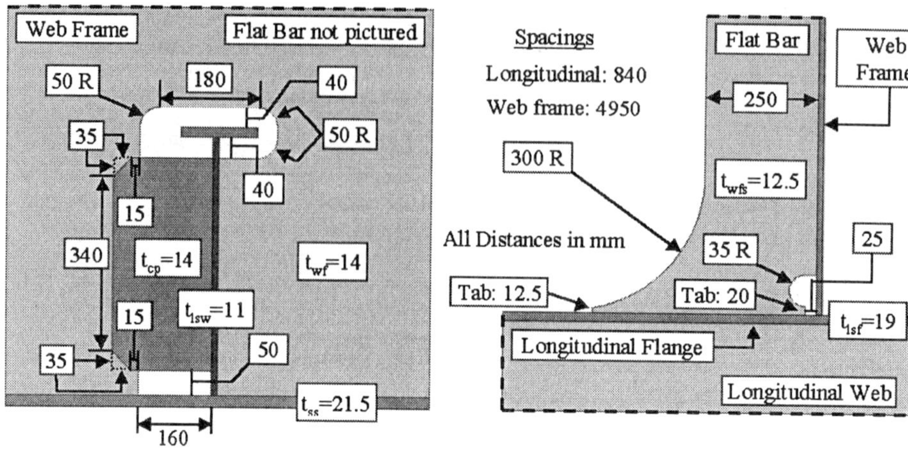

Figure 12: Scantlings for Tanker D

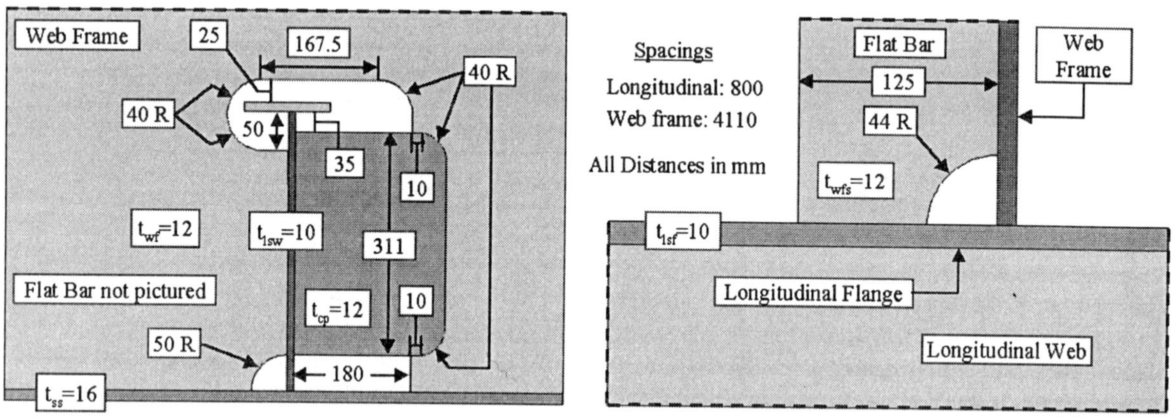

Figure 13: Scantlings for Bulker C

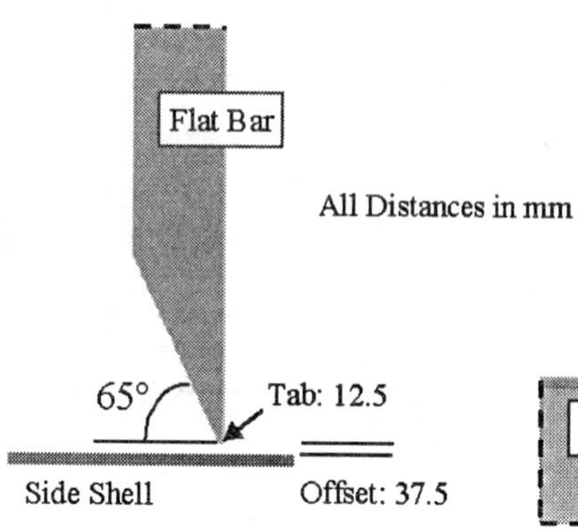

Figure 14: Snipe of the flat bar

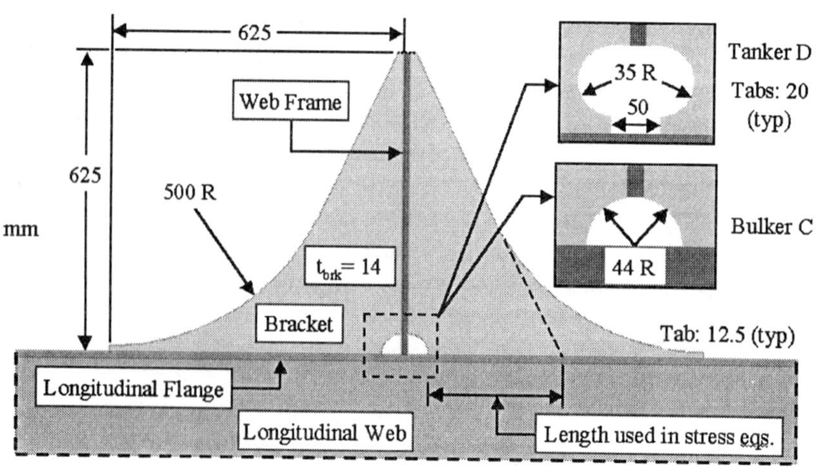

Figure 15: Particulars of symmetric brackets

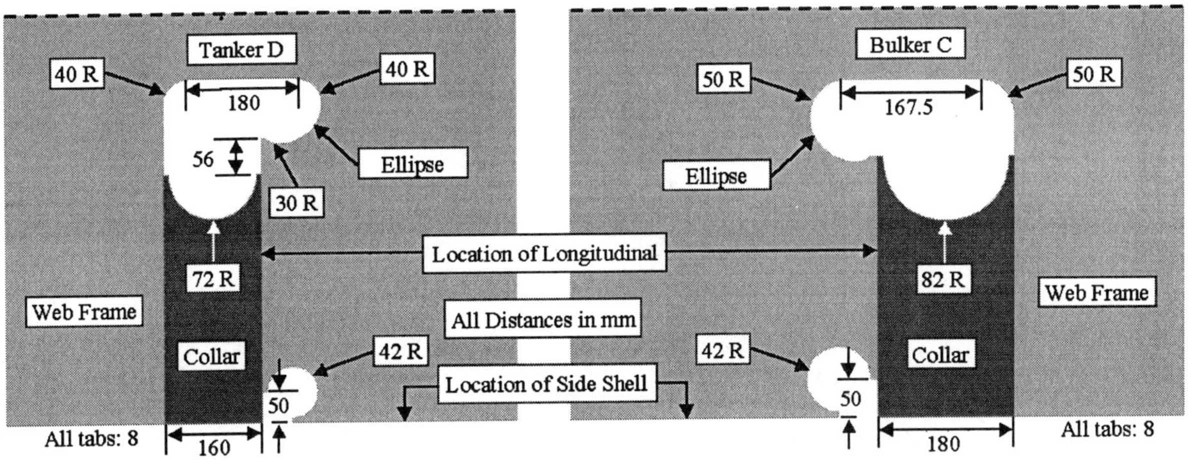

Figure 16: Stress relieving profiles

shown in Figure 18. In all expected hot spot areas, the mesh refinement is on the order of the plate thickness or smaller. The hot spot stress is defined as the principal stress—sampled at the material (integration) points in the top, middle, and bottom surfaces—with the greatest absolute value.

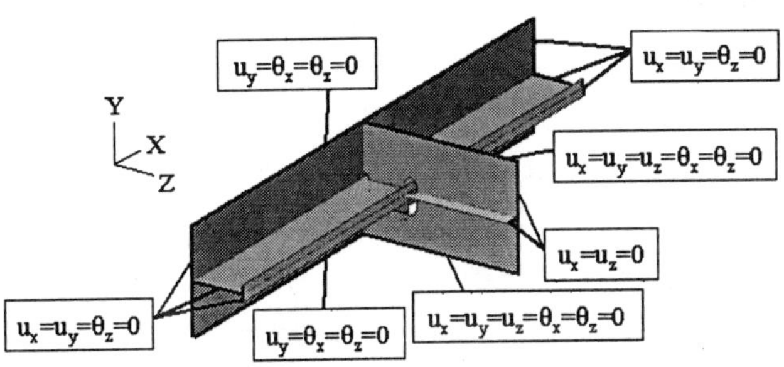

Figure 17: Boundary conditions on FE models

Ongoing research is dedicated to the derivation of S-N curves suitable for use with hot spot stresses (Lotsberg 2000). This research is also concerned with the procedure (element formulation, mesh density, extrapolation method, etc.) by which hot spot stresses are computed and the compatibility of the resulting hot spot stresses with the prospective S-N curves. While the findings from this research are not yet available, in the authors' experience the most accurate hot spot stresses result from a converged mesh of solid elements and linear extrapolation to the weld toe. The same experience indicates that a converged mesh of mid-surface shell elements and linear extrapolation to the mid-surface intersection produces very similar hot spot stresses in side shell connection details. If the elements and mesh densities employed in this paper are combined with linear extrapolation, the resulting hot spot stresses are accurate and conservative with respect to the converged solids/linear extrapolation technique. If linear extrapolation is forgone and the stresses are sampled as described above, the hot spot stresses remain reasonably accurate and useful in a comparative sense.

Where necessary, the overlap of the collar plate is modeled explicitly. To do so, shell elements with a thickness equal to the thicker of the connected plates are employed. It is important to model the overlap in order to capture fully the related hot spots. Without the overlap, localized bending effects associated with the eccentricity of the collar plate are lost. A close-up of the modeling used for the overlapped collars is also included with Figure 18.

4.3 Average Stresses from Stress Equations

The recommended stress equations (Equations 1 and 2) are applied to each of the five alternative configurations of each analyzed side shell connection. In general, the configurations featuring an offset of the flat bar lie outside the assumptions used to derive the stress equations. When symmetric brackets are added, however, the two brackets act as flat bar connections so that application of the stress equations is justified. Application of the stress equations is not justified for the other offset configurations, but for these the equations are intended to be instructive rather than practical.

The stress equations compute stresses that represent average values of the dominant stress acting along the longitudinal stiffener connections of the flat bar (σ_{fb}) and direct collar (τ_{dc}). In particular, the shear stress τ_{dc} may be thought of as the average stress considering both the direct collar and collar plate connections (when the latter is present). When σ_{fb} and τ_{dc} are compared with the hot spot stresses along their respective load paths, the resultant SCFs—in conjunction with the magnitude of the

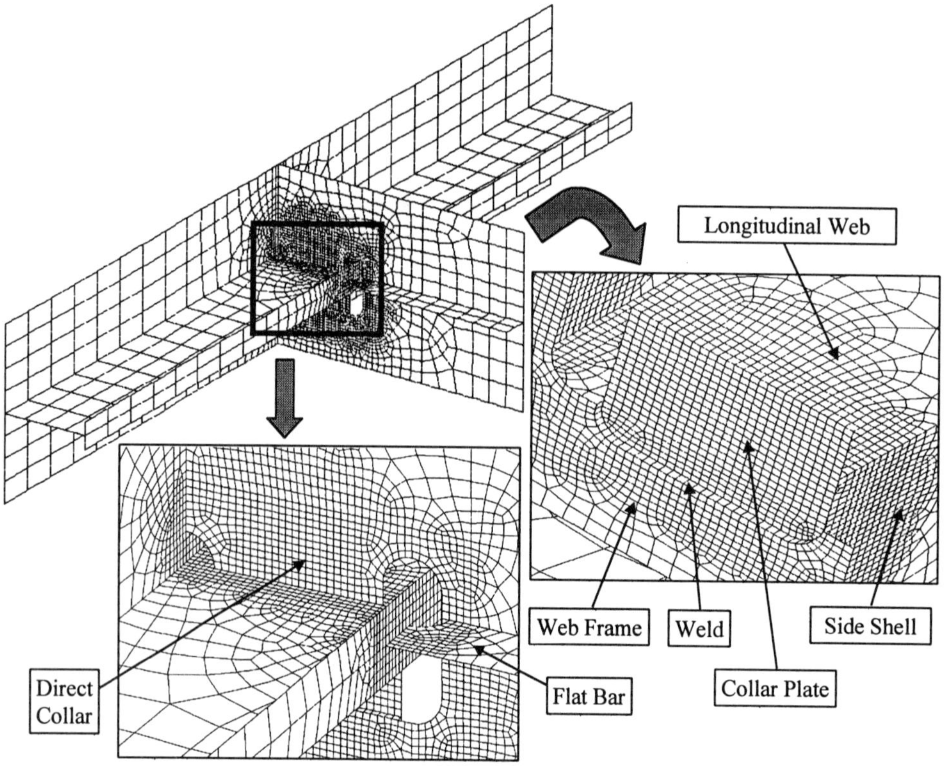

Figure 18: Typical FE mesh

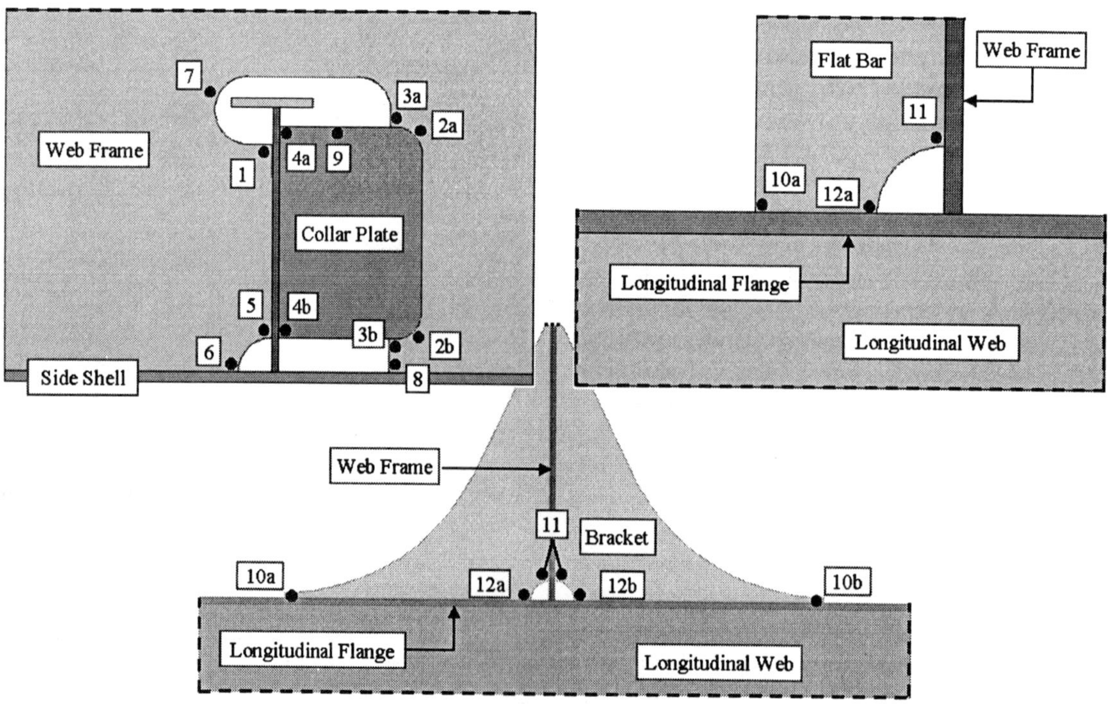

Figure 19: Hot spot locations

Table 9: Stress concentration factors from hot spot stress analysis

	As Designed		No Collar Plate		Offset Stiffener		Offset/Brackets		Offset/Profile	
HSL	Tnkr D	Blkr C	Tnkr D	Blkr C	Tnkr D	Blkr C	Tnkr D	Blkr C	Tnkr D	Blkr C
1	8.47	7.45	4.04	2.75	7.58	9.57	8.27	9.21	4.51	3.09
2a	7.31	4.74	n/a	n/a	5.94	4.44	6.60	5.61	n/a	n/a
2b	4.56	3.26	n/a	n/a	4.09	3.24	4.44	3.63	n/a	n/a
3a	8.86	6.24	n/a	n/a	6.98	5.59	7.56	6.43	2.86	2.52
3b	5.73	4.90	n/a	n/a	4.77	4.74	5.36	3.27	na	n/a
4a	8.55	5.41	n/a	n/a	7.39	3.87	8.31	6.69	2.50	3.09
4b	6.31	5.89	n/a	n/a	5.15	5.46	6.29	7.21	n/a	n/a
5	n/a	6.40	n/a	3.31	n/a	6.17	n/a	8.18	1.20	2.36
6	n/a	3.65	n/a	3.24	n/a	3.57	n/a	5.30	2.67	2.85
7	5.14	4.38	6.41	4.37	4.99	4.79	5.14	4.36	4.24	4.19
8	5.30	4.38	4.29	3.20	4.30	3.52	6.11	9.16	1.90	2.03
9	n/a	n/a	n/a	n/a	n/a	n/a	n/a	n/a	3.26	3.74
10a	1.32	2.76	1.03	1.87	n/a	n/a	1.27	1.65	n/a	n/a
10b	n/a	n/a	n/a	n/a	n/a	n/a	1.27	1.66	n/a	n/a
11	0.381	0.648	0.464	0.376	n/a	n/a	1.27	1.96	n/a	n/a
12a	3.85	4.10	3.16	3.32	n/a	n/a	1.89	1.42	n/a	n/a
12b	n/a	n/a	n/a	n/a	n/a	n/a	1.84	1.45	n/a	n/a

average stresses—should provide insight into how well the stress equations anticipate the fatigue performance of a connection.

The average stresses resulting from the application of the stress equations are given in Table 8. The tabulated stresses assume the units of the applied side shell pressure. The tabulated values of σ_{fb} for Tanker D take no account of the forward bracket attached to the flat bar when computing the footprint area. This is somewhat conservative, but the computations presented in Section 3 have shown and the hot spot stress analysis confirms that only a small part of the bracket footprint is effective in the flat bar load path. Likewise, the footprint area of the symmetric brackets does not include the full length of the brackets—the bracket is large enough and the toe is so soft that the full length does not participate in the load transfer. The length of the bracket footprint employed to derive the data in Table 8 is shown in Figure 15.

Table 8: Average stresses from stress equations, per unit side shell pressure

	Tanker D		Bulker C	
Configuration	σ_{fb}	τ_{dc}	σ_{fb}	τ_{dc}
As Designed	536	178	852	282
No Collar Plate	742	246	1350	448
Offset Stiffener	n/a	301	n/a	397
Offset/Brackets	322	107	292	97
Offset/Profile	n/a	353	n/a	440

4.4 Hot Spot Stresses

Given that there are numerous potential hot spots around the cutout area, some system is required to identify each for the purpose of reporting. Such a system is defined in Figure 19, where each hot spot location (HSL) is labeled. The SCFs at the hot spots are reported in Table 9. The reported SCFs refer to the greatest principal stress computed in any of the connecting parts.

4.5 Normalized Fatigue Life Estimates

The normalized fatigue life estimates are predicated on the assumption that the nominal bending stress due to external pressure results in a fatigue life of 120 years when used in conjunction with an appropriate hot spot stress S-N curve that has a log-log slope of 3. Doing so assumes that the design section modulus exceeds the section modulus required for a 20-year life by a factor of $(120/20)^{1/3}$, which is about 1.82. This factor is consistent with the authors' observations during previous design verification work, and the chosen reference life of 120 years is a convenient number. The nominal bending stress due to external pressure is the result of a continuous beam analysis, assuming a fixed support at the web frame:

$$\sigma_{\text{nom}} = psl^2/(12S_m), \tag{24}$$

where p is the external pressure, s is the longitudinal stiffener spacing, l is the web frame spacing, and S_m is the section modulus of the longitudinal stiffener assuming full effectiveness of the side shell between longitudinals.

Table 10: Normalized fatigue life estimates

	As Designed		No Collar Plate		Offset Stiffener		Offset/Brackets		Offset/Profile	
HSL	Tnkr D	Blkr C	Tnkr D	Blkr C	Tnkr D	Blkr C	Tnkr D	Blkr C	Tnkr D	Blkr C
1	32	12	110	59	5.5	2.0	160	150	120	43
2a	50	46	n/a	n/a	19	20	310	670	n/a	n/a
2b	120	140	n/a	n/a	59	52	1000	1900	n/a	n/a
3a	28	20	n/a	n/a	12	10	210	450	110	80
3b	60	41	n/a	n/a	37	16	570	3400	n/a	n/a
4a	19	31	n/a	n/a	9.9	30	160	400	160	43
4b	45	24	n/a	n/a	29	11	360	320	n/a	n/a
5	n/a	19	n/a	33	n/a	7.5	n/a	220	1400	98
6	n/a	100	n/a	36	n/a	39	n/a	820	130	55
7	(140)	(58)	(27)	(15)	(32)	(16)	(660)	(1500)	(33)	(18)
8	130	58	54	37	50	40	390	160	360	150
9	n/a	n/a	n/a	n/a	n/a	n/a	n/a	n/a	(72)	(25)
10a	310	8.5	280	6.8	n/a	n/a	1600	980	n/a	n/a
10b	n/a	n/a	n/a	n/a	n/a	n/a	1600	950	n/a	n/a
11	7500	650	2700	830	n/a	n/a	1600	580	n/a	n/a
12a	12	2.6	8.4	1.2	n/a	n/a	480	1500	n/a	n/a
12b	n/a	n/a	n/a	n/a	n/a	n/a	520	1400	n/a	n/a

The normalized fatigue life estimate, L, is then given in terms of σ_{nom} and σ_{hss} (the hot spot stress from FEA) as

$$L = 120\,(\sigma_{\text{nom}}/\sigma_{\text{hss}})^3. \qquad (25)$$

The exponent of 3 used in Equation 25 is taken from the log-log slope of hot spot S-N curves in the literature (IIW, 1996), as well as from a family of fatigue curves generally employed for welded connections (ABS, 2000). It should be emphasized here that the normalized fatigue life estimates are not intended to be indicative of the actual fatigue life experienced by the subject connections. As stated before, the estimates are merely intended to highlight the relative importance of the magnitude of the average stresses versus the SCFs when examined in the context of fatigue life.

The normalized fatigue life estimates resulting from the application of Equation 25 are given in Table 10. Normalized life estimates on the edge of a cut surface are reported in parenthesis; this is to reflect the well-known fact that an appropriate S-N curve for smooth cut surfaces is more favorable than an S-N curve applicable to welded connections. Consequently, the fatigue life estimate computed with a more appropriate curve would be greater. The C curve is often recommended for use with flame cut edges (ABS, 2000), but it has a slope of 3.5 and would be difficult to work into the current fatigue estimation method. Normalized life estimates in parenthesis may be loosely interpreted as pertaining to cut surfaces not free of defects (i.e. rough-cut surfaces that are not ground smooth).

4.6 Load Path Discussion

4.6.1 Flat Bar Load Path

For the as-designed configurations of both vessels, the lowest normalized fatigue life estimate occurs at the toe of the flat bar rat-hole connection to the longitudinal flange (location 12a). Since the as-designed SCF at location 12a (Table 9) is similar for both vessels (although it is slightly smaller for Tanker D due to the improved rat hole design), the lower fatigue life estimate for Bulker C (Table 10) is due to a higher average direct stress σ_{fb}. The lower σ_{fb} for Tanker D (Table 8) is due to the longer and thicker footprint of the flat bar.

At location 10a the situation is somewhat different. The as-designed SCF for Tanker D is half that for Bulker C, so there is a greater disparity between the normalized life estimates. The reduced SCF for Tanker D at location 10a is the result of the attached bracket with a soft toe. On the other hand, for Tanker D the attached bracket does not provide relief to the hot spot at location 12a.

When the collar plate is removed, more load passes through the flat bar and the average direct stress should increase. This is indeed the situation found in Table 8, where σ_{fb} increases relative to the as-designed configuration by a factor of about 1.4 for Tanker D and 1.6 for Bulker C. The SCFs actually decrease by a factor of about 1.2, but the normalized life estimates still drop.

Thus if a fatigue problem existed in the flat bar load path for the as-designed configuration, then the problem would be amplified by the removal of the collar plate. The improvement in the SCFs is probably due to the absence of the overlapping collar, which slightly unbalances the connection and aggravates the flat bar hot spots. Again, the attached bracket protects location 10a while doing little to protect location 12a.

When the flat bar is offset and symmetric brackets are added, the increased footprint area drives σ_{fb} down by factors of about 1.4 and 1.65 when compared to the as-designed configurations of Tanker D and Bulker C, respectively. The SCFs at location 12a are also reduced by factors of 2 or more, because the brackets balance the connection and relieve the inner hot spots. One would thus expect that if a fatigue problem existed in the flat bar load path for the as-designed configuration, then the problem would be all but eliminated by the addition of brackets. This is verified by the normalized life estimates in Table 10, where the improvement is particularly impressive for Bulker C since its σ_{fb} and SCF at location 12a decrease the most. Based on Healy and Thomsen (2000), the addition of brackets should lead to similar improvements in the fatigue lives along the flat bar load path when either the flat bar is offset and symmetric brackets are added, or the flat bar is connected to the longitudinal and double-sided brackets are added.

A potential drawback associated with double-sided brackets is highlighted by the increase in the SCFs at location 11 when such brackets are added. When compared with the as-designed configuration, the subject SCFs increase by a factor of 3 for each vessel. It is conceivable that a connection could exist where the original fatigue life at location 11 computed during a design assessment did not control, but was relatively low. Under this circumstance, the increased SCF resulting from the addition of double-sided brackets could offset the reduced σ_{fb} and cause the computed fatigue life at location 11 to become critical.

4.6.2 Direct Collar/Collar Plate Load Path

In the as-designed configuration for each vessel, the average shear stress τ_{dc} is relatively benign when compared to the direct stress σ_{fb}. Yet, a number of the hot spots along the direct collar/collar plate load path produce relatively low normalized fatigue life estimates. Among them are locations 1 and 5 (direct collar) and locations 2a, 3a, and 4b (collar plate). The reason for these relatively low fatigue life estimates is that the SCFs associated with the hot spots are quite severe. Table 9 reveals that these SCFs can reach as high as 8.9. These SCFs are high because the stress fields at the subject locations include direct stresses that are not accounted for in the stress equations. The direct stresses result from both membrane effects (the direct load into the web frame causing it to deform and pull away from the longitudinal) and localized bending (due to the eccentricity of the collar plate overlap). If the longitudinal were asymmetric, twisting of the stiffener would also introduce direct stresses at the subject locations.

For Tanker D, the average shear stress τ_{dc} increases as the collar plate is removed (by a factor of about 1.4 relative to the as-designed configuration), the flat bar is offset (factor of 1.7), and the offset of the flat bar is combined with a collar plate that is moved in-line and profiled (factor of 2). For Bulker C, the worst case occurs when the collar plate is removed (factor of 1.6). This is because the footprint area of the collar plate exceeds three times the footprint area of the flat bar, making its removal more significant than offsetting the flat bar. Regardless, removal of the collar plate or offsetting the flat bar should both have a deleterious effect on the fatigue performance of the direct collar/collar plate load path.

On the other hand, removal of the collar plate also removes all associated hot spots, and without the eccentricity of the overlap to drive localized bending the SCFs at locations 1 and 5 decrease relative to the as-designed configuration. The normalized life estimates at these locations thus improve despite the increased τ_{dc}. The down side to removal of the collar plate for the direct collar load path is lower fatigue life estimates along the web frame cutout at location 7 and at location 8, where the web frame meets the side shell. The latter is the more serious, as it could lead to cracks directly in the outer hull.

When the flat bar is offset, the increased τ_{dc} does lead to degraded fatigue performance. This is true despite the observation that along with the increased τ_{dc} come SCFs that are generally reduced relative to the as-designed configuration. The SCFs may decrease somewhat because the restraint of the flat bar no longer prevents the connection from adjusting to compensate for the eccentricity of the collar plate. Still, the increased τ_{dc} dominates and the normalized fatigue life estimates drop significantly.

When the flat bar is offset, the addition of symmetric brackets reduces τ_{dc} by factors of 1.7 for Tanker D and 2.9 for Bulker C when compared to the as-designed configuration. The latter factor is much larger than the former due to the much smaller as-designed flat bar area of Bulker C. In the presence of brackets, the SCFs along the direct collar/collar plate load path generally increase relative to the offset configuration. Again, this may occur because the restraint of the brackets prevents adjustment to the eccentricity of the collar plate, drawing extra load to the hot spots. And again, the change in τ_{dc} dominates, as the normalized fatigue life estimates increase significantly. Healy and Thomsen (2000) have shown that similar improvements in fatigue life along the direct collar/collar plate load path should result from either the addition of symmetric brackets when the flat bar is offset or the addition of double-sided brackets when the flat bar is connected to the longitudinal.

The normalized fatigue life estimates in Table 9 indicate that the use of an in-line, profiled lug plate and a profiled web frame cutout and rat hole are also effective in minimizing potential fatigue problems caused by offsetting the flat bar. This is true despite the fact that τ_{dc} is greater for the profiled configuration than it is for the offset configuration. The normalized fatigue life estimates improve relative to the offset configuration because the SCFs drop significantly. This drop occurs because of the stress-relieving profiles and, most importantly, because the collar plate overlap is eliminated. It also helps that the collar plate is extended to the side shell, eliminating the hot spots associated with the gap. The collar plate overlap leads to many problems unless much of the load is diverted elsewhere.

4.7 Summary of Fatigue Performance

In this section, the ability of the stress equations to predict the fatigue performance of a connection has been examined. When the flat bar is attached to the longitudinal, the analysis presented herein indicates that an SCF of 4 relative to σ_{fb} is sufficient to derive a quick and meaningful fatigue life estimate at the hot spots along the flat bar load path. When the flat bar is offset and symmetric brackets are employed, an SCF of 2 appears to be sufficient. The latter SCF should also pertain when the flat bar is attached to the longitudinal and double-sided brackets are added. In the absence of double-sided brackets, the fatigue life at location 11 need not be checked. When double-sided brackets are present, the fatigue life location 11 should be checked with an SCF of 2.

Along the direct collar/collar plate load path, the findings are not as optimistic. The average shear stress τ_{dc} does not account for the significant direct stresses present at the hot spots associated with the direct collar/collar plate load path. These direct stresses may result from in-plane membrane behavior, or they may result from localized bending. The latter is particularly true at the hot spots associated with an overlapped collar plate. Failure to account for the direct stresses leads to potentially high and widely scattered SCFs, although the situation is better when an in-line collar plate and stress-relieving profiles are employed. (The latter should be true even if the flat bar is not offset; then the primary causes of the direct stresses are eliminated or mitigated). Fortunately, when the flat bar is attached to the longitudinal, or when the flat bar is offset and symmetric brackets are present, much of the load is diverted to the flat bar load path so that the hot spots associated with σ_{fb} will control the fatigue life of the connection. However, if the flat bar is offset without the addition of brackets, hot spots along the direct collar/collar plate load path will control. For these cases, further development of the stress equations is important. In particular, the effects of an offset flat bar and an overlapped collar plate require investigation.

During the course of the study related to fatigue performance, a number of stress-relieving configurations other than those reported were investigated. The results of this additional work are not presented due to space restrictions. Nonetheless, one interesting trend that came out of that work is worth mentioning, even without documentation. When the ends of the direct collar connection to the longitudinal web line up with those of the collar plate (in-line or offset), a hard spot is created that drives up the hot spot stresses associated with both. When the ends of the direct collar connection lie inside those of the collar plate, the hard spots associated with the collar plate are relieved while those associated with the direct collar are heightened. When the ends of the direct collar connection lie outside those of the collar plate, the direct collar hot spots are relieved while those of the collar plate are heightened. The best solution seems to be to extend the collar plate to the side shell (adding a web frame rat hole if necessary), placing the other end of the collar plate beneath the upper end of the direct collar, and using stress-relieving profiles. This is the strategy adopted for Tanker D, where the collar plate is also moved in-line. For Bulker C, the resulting collar plate size was small enough to raise a concern about the hot spots associated with the web frame rat hole, especially at location 6 where a crack could form and propagate into the side shell. In this latter case, the collar plate was extended to line up with the direct collar end, but the toe of the collar plate is so soft that the hot spot stress at location 1 remains relatively low.

The study of fatigue performance was conducted considering only external pressure as a driver for fatigue. As such, symmetric (or double-sided if the flat bar is attached to the longitudinal) brackets have been seen to effectively combat fatigue problems that may occur in side shell connections at the still-water-line. However, the performance of brackets is less robust if the connection occurs further away from the vessel neutral axis. At the turn-of-the-bilge and along the bottom shell, for instance, Healy and Thomsen (2000) indicate that longitudinal stress due to vertical hull girder bending can lead to low fatigue lives at the bracket toes. Under these circumstances, the problem area is simply moved, not eliminated.

Another factor that has not been considered in either the study of fatigue performance or in the development of the stress equations is the effect of in situ boundary conditions. That is, the conditions resulting from deformation of the vessel that occur along the defined boundary of the connection. The relative deformations imposed by these conditions may lead to high web shear or twist of the longitudinal. Healy and Thomsen (2000) found that such deformations, particularly those leading to web shear, could seriously degrade fatigue performance. The hot spots associated with the direct collar and the collar plate are the most affected, especially when the collar plate is overlapped and the longitudinal is asymmetric. The side shell connections near the still-water-line are generally not as affected by relative deformations as the connections nearer the turn-of-the-bilge. The use of brackets at the latter locations can lead to more problems than they solve, both due to the longitudinal stresses previously described and to the inability of the brackets to adequately mitigate the original problems. Unfortunately, the stress equations cannot be used to estimate the negative impact that in situ boundary conditions can have on the fatigue performance. At some point in the design process the possibility of large local deformations needs to be assessed and, if deemed significant, steps taken to quantify their impact. In the SafeHull system (Horn, et al., 1999), one can use the zooming capability (elsewhere-called submodeling or telescoping) to isolate connections of interest and perform fatigue analysis including the effects of local deformations.

5. CONCLUSIONS

Simple formulas have been derived for estimating mean stresses in flat bars and direct collars around typical end connections. Under several assumptions, these formulas imply that the stresses are functions of geometric parameters and pressure load. The percentage share of the pressure load taken by the longitudinal stiffener can be estimated easily. The share is further transferred into the transverse web frame through the small connection areas of the flat bar, direct collar and optional collar plate.

The coefficients in the formula have been calibrated based on FEA results from eight sample details borrowed from existing tankers and bulk carriers. The stress equations may be used to estimate the stresses in the structural details of cutouts, and in conjunction with design criteria to develop designs that will not have local cracking around end connections. To reduce the stresses, the total connection area should be as large as possible.

Using the stress equation to obtain nominal stresses, a set of stress concentration factors is obtained, and the fatigue performance of different design arrangements is discussed. The present work can be integrated with most existing fatigue assessment procedures. The formulas can be used as a quick way to determine the nominal stress range in a fatigue analysis. Because of their simplicity, the formulas may also be used as a tool for fatigue screening.

ACKNOWLEDGEMENTS

The authors wish to thank those colleagues who provided guidance or comments. These include Dr. Donald Liu, Dr. Jack Spencer, Dr. Yung Shin and many others. Also, the technical review and guidance on writing from the review committee have greatly improved the quality and readability of this paper. In particular, the authors wish to thank Dr. Arthur Reed for his time and effort. The views expressed in this paper are those of the authors and are not necessarily those of the American Bureau of Shipping.

REFERENCES

ABAQUS (1998) ABAQUS/Standard User's Manual, Version 5.8. Hibbit, Karlsson, and Sorenson Inc.

ABS (2000) *Rules for Building and Classing Steel Vessels*, American Bureau of Shipping.

Bea, R. G., Cramer, E., Schulte-Strauthaus, R., Mayoss, R., Gallion, K., Ma, K. T., Holzman, R. and Demsetz, L. (1995) "*Ship's Maintenance Project*," Conducted at University of California, Berkeley for U.S. Coast Guard/Ship Structures Committee (SSC), Ship Structures Committee Report SSC-386.

Chen, Y. K. and Wan, Z. Q. (1999) "Re-evaluation of SVR 5/3 AA.4.2.2, Fatigue Assessment of Cutouts for Longitudinals," ABS Report.

Cheung, M. C. and Slaughter, S. B. (1998) "Inner Bottom Design Problems in Double-Hull Tankers," *Marine Technology,* **35**(2):65–73.

Chiou, J. W. (2000) "Stress Analysis of Transverse Cutouts for Longitudinal Stiffeners of Tankers and Verifications of ABS/Rules Section 5/2 AA.4.2.2 and Newly Modified Formulas for Fatigue Assessment," ABS Report No. 6.

Glasfeld, R., Jordan, D., Kerr, M., and Zoller, D. (1977) "*Review of Ship Structural Details,*" Ship Structure Committee Report SSC-226.

Healy, B. and Thomsen, T. (2000) "Comparative Fatigue Performance of Alternative Outer Hull Connection Details," Parts 1 and 2, Draft accepted by ISOPE 2000, Seattle, Washington.

Horn, G. E., Chen, Y. K., and Chen, J. M. (1999) "SafeHull FEM Fatigue Assessment of Ship Structural Details," ICCAS 99, Cambridge, Massachusetts. **2**: 295–309

IACS (1994) "*Bulk Carriers—Guidelines for Surveys Assessment and Repair of Hull Structure*," International Association of Classification Societies.

IIW (1996) *Fatigue Design of Welded Joints and Components*, International Institute of Welding, Abingdon Publishing, Cambridge.

Liu, D. and Bakker, A. (1981) "Practical Procedures for Technical and Economic Investigation of Ship Structural Details," *Marine Technology,* **18**(1)**:**51–68.

Lotsberg, I. (2000) "Background and Status of the FPSO: Fatigue Capacity JIP," OTC 12144.

Ma, K. T. and Bea, R. G. (1995) "A Repair Management System for Fatigue Cracks in Ships," *Transactions of SNAME*, **130**:343–69.

Ma, K. T. (1998) "Tanker Inspection and a Risk-Based Approach," Proceedings of ISOPE98, International Offshore and Polar Engineering Conference, Montreal. **IV**: 504–512.

MSC (1998) MSC/NASTRAN Quick Reference Guide, Version 70.5. The MacNeal-Schwendler Corporation.

Yoneya, T., Kumano, A., Yamamoto, N. and Shigemi, T. (1993) "Hull Cracking of Very Large Ship Structures," Integrity of Offshore Structures–5.

SNAME *Transactions*, Vol. 108, 2000, pp. 249–290

On Advanced Buckling and Ultimate Strength Design of Ship Plating

Jeom Kee Paik, Member, Pusan National University, Pusan, Korea, **Anil Kumar Thayamballi,** Life Member, Chevron Shipping, San Ramon, California, **Ge Wang,** Member, American Bureau of Shipping, Houston, and **Bong Ju Kim,** Student Member, Pusan National University

ABSTRACT

This paper is a summary of recent research and development in areas related to advanced buckling and ultimate strength design of ship plating, jointly undertaken by the American Bureau of Shipping and the Pusan National University. The behavior of ship plating normally depends on a variety of influential factors, namely geometric / material properties, loading characteristics, initial imperfections, boundary conditions and deterioration arising from corrosion, fatigue cracking and accidental dents. In achieving a more advanced buckling and ultimate strength design of ship plating, we are still confronted with a number of problem areas to be more completely solved and these would need methods that are more sophisticated than most existing simplified approaches. In this regard, this paper focuses on the following five subjects which have been studied by the authors theoretically, numerically and experimentally: mathematical modeling of fabrication related imperfections (i.e., initial deflections and residual stresses), characteristics of the plate buckling with elastically restrained edge conditions, capacity equations based on buckling and ultimate strength under combined loads including biaxial loads, edge shear and lateral pressure, collapse strength characteristics under axial compressive dynamic loads, and design equation of the plate capacity under impact lateral pressure loading. Useful results, important insights and conclusions developed from the studies are summarized and recommendations are made with respect to both technologically improved design procedures, and also needed future research.

NOMENCLATURE

a = plate length

b = plate breadth

b_{fx}, b_{fy} = flange breadth of longitudinals or transverses

D = $\dfrac{Et^3}{12(1-\nu^2)}$

E = Young's modulus

h_{wx}, h_{wy} = web height of longitudinals or transverses

p = average net lateral pressure

t = plate thickness

t_{fx}, t_{fy} = flange thickness of longitudinals or transverses

t_{wx}, t_{wy} = web thickness of longitudinals or transverses

β = $\dfrac{b}{t}\sqrt{\dfrac{\sigma_o}{E}}$

ν = Poisson's ratio

σ_o = yield stress

σ_{rcx}, σ_{rcy} = compressive (negative) residual stress in the x or y direction

σ_{xav}, σ_{yav} = average longitudinal or transverse axial stress (compression: negative, tension: positive)

σ_{xE}, σ_{yE} = elastic longitudinal or transverse compressive buckling stress

σ_{xB}, σ_{yB} = buckling based capacity for σ_{xav} or σ_{yav}

σ_{xu}, σ_{yu} = ultimate strength based capacity for

	σ_{xav} or σ_{yav}
τ_{av}	= average edge shear
τ_o	$=\dfrac{\sigma_o}{\sqrt{3}}$
τ_B	= buckling based capacity for τ_{av}
τ_u	= ultimate strength based capacity for τ_{av}

INTRODUCTION

The overall failure of a ship hull girder is normally governed by buckling and plastic collapse of the deck, bottom or sometimes the side shell stiffened panels. Therefore, the relatively accurate calculation of buckling and plastic collapse strength of stiffened plating of the deck, bottom and side shells is a basic requirement for the safety assessment of ship structures. In stiffened panels, local buckling and collapse of plating between stiffeners is a primary failure mode, and thus it would also be important to evaluate the buckling and collapse strength interactions of plating between stiffeners under combined loading.

The behavior of ship plating normally depends on a variety of influential factors, namely geometric / material properties, loading characteristics, initial imperfections (i.e., initial deflections and residual stresses), boundary conditions and existing local damage related to corrosion, fatigue crack and denting.

The geometry of plating found in ship and offshore structures is normally rectangular and the material used is usually mild or high tensile steel (Note that the use of aluminum alloys is increasing in the design and fabrication of high speed vessel structures). The boundary condition for the rectangular plate elements making up steel plated structures is normally assumed to be simply supported or sometimes clamped for practical purposes of analysis. In real ship plating, however, such ideal edge conditions may never occur due to rotational restraint by support members along the plate edges.

The ship plating is generally subjected to combined in-plane and lateral pressure loads. In-plane loads include biaxial compression / tension and edge shear, which are mainly induced by overall hull girder bending and / or torsion of the vessel. Lateral pressure loads are due to water pressure and / or cargo. The extrema of such load components may not occur simultaneously, and more than one load component may normally exist and interact. Hence, for more advanced design of ship structures, it is of crucial importance to better understand the characteristics of the buckling and ultimate strength for ship plating under combined loads.

Since the post-weld initial imperfections in the form of initial deflections and residual stresses exist in ship steel plating and can affect significantly the strength, such welding induced initial imperfections should be included in the strength calculations as parameters of influence.

When a perfectly flat plate (i.e., without initial imperfections) is subjected to predominantly compressive loads, buckling (bifurcation) can occur if the applied compressive stress reaches a critical bifurcation stress, see Fig.1. However, the in-plane stiffness of plating with initial imperfections decreases from the very beginning as the compressive loads increase. In this more general case, it is not possible to define a bifurcation point for buckling.

The phenomenon of buckling may be categorized by plasticity considerations into three classes, namely elastic buckling, elastic-plastic buckling and plastic buckling, the last two being called inelastic buckling. The first class (i.e., elastic buckling) typically indicates that buckling occurs solely in the elastic regime. This class of buckling is often seen in very thin steel plates. The second (i.e., elastic-plastic buckling) normally represents the case wherein buckling occurs after plastification has occurred in a local region in the plate. The third (i.e., plastic buckling) indicates that buckling occurs in the regime of gross yielding, i.e., after the plate has yielded over large areas. Relatively thick plating may exhibit either elastic-plastic or plastic buckling.

Unlike columns, plating can normally sustain additional applied loads even after elastic buckling occurs since membrane tension develops along the plate edges resists any abrupt increase in lateral deflection. A plate buckled in the elastic regime will eventually collapse by a rapid decrease of in-plane stiffness (or an abrupt increase of lateral deflection) as the yield zone inside the plate is expanded. On the other hand, if buckling occurs in the elastic-plastic or plastic regime the plating normally immediately reaches the ultimate limit state.

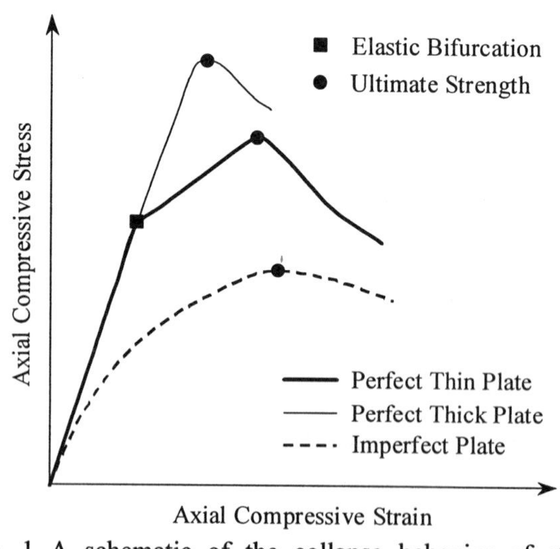

Fig. 1 A schematic of the collapse behavior of steel plating under predominantly compressive loads

From the view point of a structural designer, it can be said with reasonable certainty that the buckling and ultimate strength problem for ship plating under a single load application and common idealized edge conditions (e.g., simply supported along four edges) has been almost completely solved. In the more general case, however, we are still confronted with a number of problem areas that remain unsolved due to the various influential factors previously mentioned. In the following, a literature review of selected studies related to the buckling and ultimate strength of plating is now made.

Depending on the rotational restraints and torsional rigidity of support members along the plate edges, the common ideal edge conditions (i.e., the assumption that the plate edges are simply supported or clamped) may or may not be appropriate to apply (Bleich 1952, Timoshenko & Gere 1963). The plate element is normally subjected to combined loads and the buckling mode depends on the interaction of these load components. Therefore, the plate buckling strength should in principle be evaluated by taking into account the effects of boundary condition and load component interactions among other factors.

Williams (1976) investigated the buckling strength characteristics of plate elements varying torsional rigidity of support members along their edges. Paik et al. (1993) surveyed the bending and torsional rigidities of support members for plate elements in merchant vessel structures. Based on the survey results, they concluded that due to the rotational restraint by support members at plate edges the plate edge condition would be in an intermediate situation, i.e., between a simply supported and a clamped condition. Most recently, Paik & Thayamballi (2000) investigated the buckling strength characteristics of steel plating elastically restrained at their edges and developed simple design formulations for buckling strength as function of the torsional rigidity of support members that provide the rotational restraints along either one set of edges or all (four) edges.

Mansour (1976) developed charts for predicting the buckling and post-buckling behavior of simply supported plates under combined in-plane and lateral pressure loads. Steen & Valsgard (1984) developed a simplified buckling and ultimate strength equation for plates under biaxial compression and lateral pressure loads. They define a pseudo-buckling (non-bifurcation) strength for initially deflected plating. Ueda et al. (1985) developed elastic buckling interaction equations for simply supported plates subject to five load components, namely biaxial compression, biaxial in-plane bending and edge shear. Paik et al. (1992a) developed the elastic buckling interaction equation for simply supported plates under biaxial compression, edge shear and lateral pressure loads. The post-weld residual stresses were also later incorporated in the plate buckling design formula (Paik et al. 1992b). To appropriately include the effects of post-weld initial imperfections in the strength calculations, an idealized model representing the distribution of the post-weld initial imperfections is used. Mazzolani et al. (1998) studied the effect of welding on the local buckling of aluminum thin plates. The influence of welding induced initial deflection and residual stresses on the buckling and ultimate strength of plating under uniaxial compression and lateral pressure was studied by Yao et al. (1998).

Most design rules of classification societies approximately calculate the inelastic buckling strength of plate elements by a correction for plasticity applied to the elastic buckling strength, using the so-called Johnson-Ostenfeld formula. This approach normally tends to underestimate the buckling strength for one single stress component loading, but in some cases for combined loading it can overestimate the buckling strength. Paik et al. (1992b) and Fujikubo et al. (1997) have derived newer empirical formulations of the plasticity correction by curve fitting based on nonlinear finite element solutions.

Following von Karman et al. (1932), the concept of effective width has been recognized as an efficient device for characterizing the post-buckling strength behavior of a plate in compression. For collapse strength prediction of steel plates, the effective width concept has also been widely used (Faulkner 1975). For such use, the reduction of in-plane stiffness of a buckled plate is evaluated by using the effective width concept, and it is assumed that the plate reaches the ultimate limit state if the normal stress components within the plate field satisfy certain predefined ultimate strength criteria.

An extensive review of a number of studies for the derivation of the effective width formulae for plates, undertaken until the early 80s, has been made by Rhodes (1984). Since then, Ueda et al. (1986a) derived the effective width formula for a plate under combined biaxial compression and edge shear taking into account the effects of initial deflections and welding induced residual stresses. Usami (1993) studied the effective width of plates buckled in compression and in-plane bending.

While the concept of effective width is aimed at the evaluation of in-plane stiffness of plate elements buckled in compression, Paik (1995) suggested a new concept of the effective shear modulus to evaluate the effectiveness of plate elements buckled in edge shear. The effective shear modulus concept is useful for computation of the post-buckling behavior of plate girders under predominant shear forces.

Regarding the ultimate strength interaction equations for plate elements under combined loads, a number of studies have also been undertaken in the past, e.g., for uniaxial compression and shear (Fujita et al. 1979), for in-plane compression and tension (Smith et al. 1987), for uniaxial compression and lateral pressure (Aalami & Chapman 1972, Aalami et al. 1972, Okada et al. 1979, Paik & Kim 1988), for biaxial compression (Dier & Dowling 1983, Ohtsubo & Yoshida 1985), for biaxial compression and lateral pressure (Dowling & Dier 1978, Soreide & Czujko 1983, Steen & Valsgard 1984, Davidson et al. 1991, Soares & Gordo 1996, Wang & Moan 1997), for biaxial compression and shear (Ueda et al. 1984, 1995, Davidson et al. 1989), for biaxial compression, shear and lateral pressure (Ueda et al. 1986b), among others. Some of the methods mentioned above approximately accommodate post-weld initial imperfections, but others neglect them.

For safety assessment of aging ship structures, it is necessary to better understand the influence of local damage related to corrosion, fatigue cracking and dents on the strength. Smith & Dow (1981) review structural damage in a ship's bottom or side shell as may be caused by collisions, grounding, hydrodynamic impact or explosions, with particular reference to the influence of such damage on hull girder bending strength. Paik et al. (1998a) proposed a probabilistic corrosion rate estimation model of ship plating. They also studied the ultimate strength reliability of ship structures related to corrosion damage (Paik et al. 1998b, 1998c). Mateus & Witz (1997, 1998) studied the buckling and post-buckling behavior of corroded steel plates using the nonlinear finite element method.

The response of ship plating is dynamic in principle since ships are subjected to both low and high frequency dynamic loads induced by waves. It is recognized that the strength characteristics of structural elements under dynamic loading can be quite different from those under a static or quasi-static loading situation (Jones 1989). An extensive survey of the studies related to dynamic plastic behavior of marine structures that have been published over the last two decades was made by Jones (1997). According to his review, most previous studies related to the dynamic plastic behavior of structural members are limited to either beam members or plates under lateral pressure loading, and the studies for plate elements subject to dynamic in-plane loads are limited to crushing and are not seen for dynamic ultimate strength. Also, the ship bottom plating can be subjected to impact pressure loads due to slamming and can collapse (Jones 1973). For advanced strength design of ship structures, it would therefore be important to better understand the collapse strength characteristics of ship plating under dynamic / impact in-plane or lateral pressure loads.

Based on the literature surveys mentioned above, it is evident that some of the issues that need to be studied further for facilitating more refined buckling and ultimate strength of ship plating are as follows

- Modeling of the fabrication related initial imperfections (i.e., initial deflections and residual stresses) and their effect,
- Effects of rotational restraints and torsional rigidity of support members,
- Ultimate strength interaction characteristics under any combination of potential load components, including biaxial compression / tension, edge shear and lateral pressure loads,
- Dynamic collapse strength characteristics under dynamically loaded in-plane forces,
- Dynamic collapse strength characteristics under lateral impact pressure loading, and
- Effects of structural deterioration such as due to corrosion, fatigue cracking and local dents.

This paper is a summary of recent research and development in the above areas, jointly undertaken by the Pusan National University and the American Bureau of Shipping. The paper focuses on the following five subjects which are studied theoretically, numerically and experimentally: modeling of post-weld initial imperfections (i.e., initial deflections and residual stresses) and their effects, influence of rotational restraints and torsional rigidity of support members on the plate buckling strength, ultimate strength design equations under combined loads including biaxial compression / tension, edge shear and lateral pressure loads, and dynamic collapse strength characteristics under dynamic axial compressive loads or slamming induced impact lateral pressure loading. Selected useful results and insights developed are summarized, and recommendations are made with respect to related enhancements in the advanced ship structural design and also needed future research.

BUCKLING / ULTIMATE STRENGTH DESIGN PROCEDURE

Figure 2 shows a schematic of the typical steel plated structure. The response of such a structure can be classified into three levels, namely the bare plate element level, the stiffened panel level and the entire plated structure level. This paper is concerned with the design for the first level (i.e., the plating between longitudinals and transverses). In such a case, the structure is to be

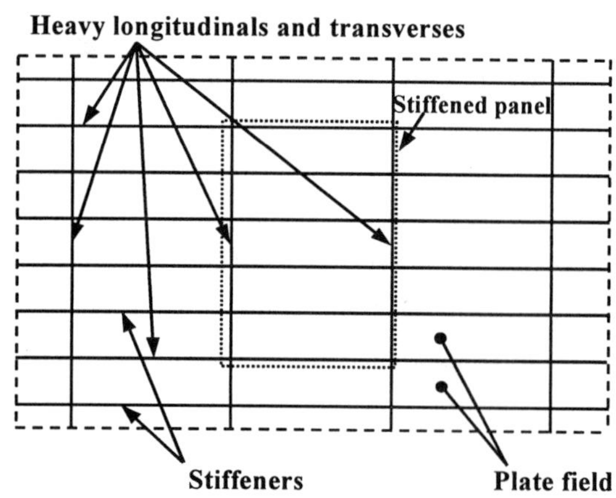

Fig.2 A typical stiffened plate structure in a ship

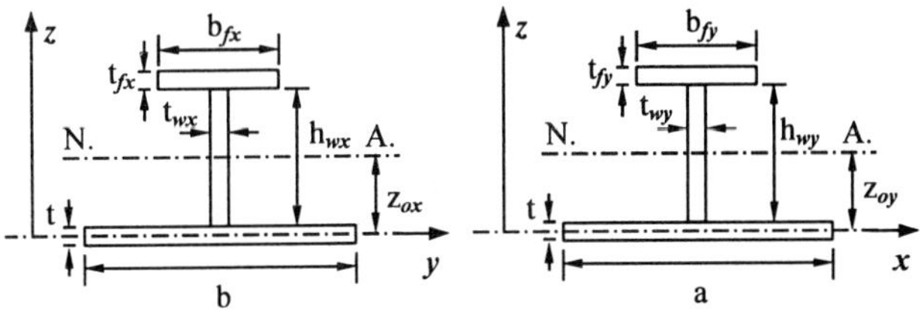

(a) x-stiffener (b) y-stiffener

Fig. 3 Typical geometry for the longitudinals and transverses

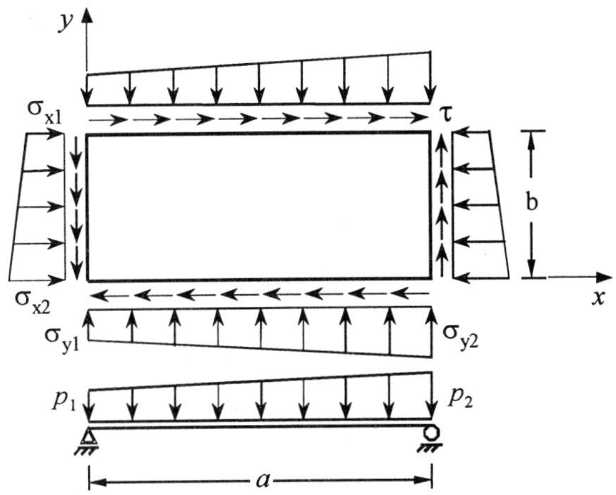

Fig. 4 The plating under a general pattern of combined external loads

designed so that the capacity (resistance) with allowable usage factor should not be less than the corresponding applied loads. To prevent the structure from failure (instability) under applied loading, therefore, the following criterion is to be satisfied:

$$\frac{\sigma}{\sigma_c} \le \eta \tag{1}$$

where σ = applied load (stress), σ_c = structural capacity (stress), and η = allowable usage factor which is the inverse of the conventional factor of safety.

The applied load (stress) components are to be determined using any acceptable method such as the finite element approach. The structural capacity is normally determined based on either buckling or ultimate strength. This paper focuses on the advanced design equations for the capacity based on both buckling and ultimate strength.

GEOMETRIC AND MATERIAL PROPERTIES

The length and breadth of plating are a and b, respectively. The long direction is taken as the x axis and the short direction is taken as the y direction, that is, $a/b \ge 1$. The thickness of plating is t. The Young modulus and Poisson ratio are E and ν, respectively. The yield stress of material is σ_o. The plating is supported by longitudinals and transverses. Figure 3 shows a typical geometry of the supporting members in the x and y directions. The rotational restraint parameters for the boundary longitudinals and transverses are defined as follows

$$\zeta_L = \frac{GJ_L}{bD}, \quad \zeta_S = \frac{GJ_S}{aD} \tag{2}$$

where ζ_L, ζ_S = rotational restraint parameters for the longitudinals and transverses, respectively, with $G = \dfrac{E}{2(1+\nu)}$, $J_L = \dfrac{h_{wx}t_{wx}^3 + b_{fx}t_{fx}^3}{6}$, $D = \dfrac{Et^3}{12(1-\nu^2)}$, $J_S = \dfrac{h_{wy}t_{wy}^3 + b_{fy}t_{fy}^3}{6}$.

For a simply supported condition, ζ_L and ζ_S are set to be zero, while their values will become infinity for a clamped edge condition. For practical purposes, the value of the rotational restraint parameter for clamped edges may be considered to be 20.

LOAD (STRESS) APPLICATION

Figure 4 shows a general loading condition on the plating between longitudinals and transverses. For the plate capacity calculations, the distribution of applied loads is often idealized by their average values, similar to that shown in Fig.5. The compressive stress is taken as negative and the tensile stress is taken as positive. The average values of the applied stresses (loads) are defined as follows

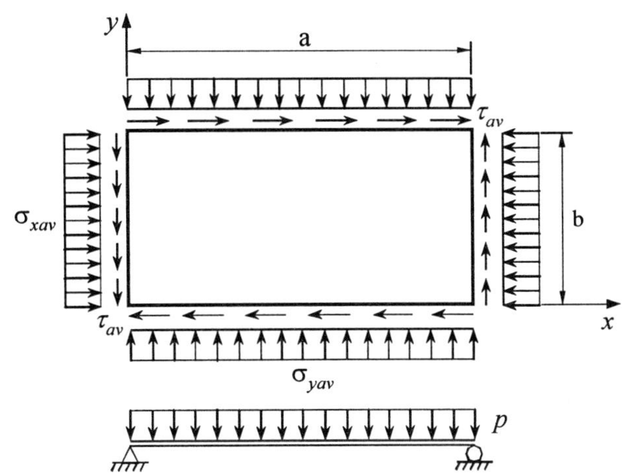

Fig.5 Idealized load application for the plating under uniform biaxial, edge shear and lateral pressure loads

$$\sigma_{xav} = \frac{\sigma_{x1} + \sigma_{x2}}{2}, \qquad \sigma_{yav} = \frac{\sigma_{y1} + \sigma_{y2}}{2}, \qquad \tau_{av} = |\tau|,$$

$$p = \left| \frac{p_1 + p_2}{2} \right| \tag{3}$$

where σ_{xav} = average axial stress in the x direction, σ_{yav} = average axial stress in the y direction, τ_{av} = average edge shear stress, and p = average net lateral pressure.

The effect of in-plane bending stress in the x or y direction is included in the buckling based capacity analysis. The in-plane bending stresses are defined as follows (For the symbols, see Fig.4)

$$\sigma_{xb} = |\sigma_{x1} - \sigma_{xav}| = |\sigma_{x2} - \sigma_{xav}|$$
$$= \begin{cases} \left| \dfrac{1-\phi_x}{1+\phi_x} \sigma_{xav} \right|, \ \phi_x = \dfrac{\sigma_{x2}}{\sigma_{x1}} & if\ \sigma_{x1} \neq 0, \phi_x \neq -1 \\ |\sigma_{x1}| = |\sigma_{x2}| & if\ \sigma_{x2} = -\sigma_{x1} \\ |\sigma_{xav}| = \left| \dfrac{\sigma_{x2}}{2} \right| & if\ \sigma_{x1} = 0 \end{cases} \tag{4.a}$$

$$\sigma_{yb} = |\sigma_{y1} - \sigma_{yav}| = |\sigma_{y2} - \sigma_{yav}|$$
$$= \begin{cases} \left| \dfrac{1-\phi_y}{1+\phi_y} \sigma_{yav} \right|, \ \phi_y = \dfrac{\sigma_{y2}}{\sigma_{y1}} & if\ \sigma_{y1} \neq 0, \phi_y \neq -1 \\ |\sigma_{y1}| = |\sigma_{y2}| & if\ \sigma_{y2} = -\sigma_{y1} \\ |\sigma_{yav}| = \left| \dfrac{\sigma_{y2}}{2} \right| & if\ \sigma_{y1} = 0 \end{cases} \tag{4.b}$$

where σ_{xb}, σ_{yb} = in-plane bending stress in the x or y direction, respectively.

For safety evaluation using equation (1), the measure of the applied stresses can be defined for combined loading, as follows

$$\sigma = \sqrt{\sigma_{xav}^2 + \sigma_{yav}^2 + \tau_{av}^2} \tag{5}$$

MODELING OF FABRICATION RELATED IMPERFECTIONS

To fabricate the ship stiffened plate structure, welding is normally used and thus the post-weld initial imperfections (initial deflections and residual stresses) develop in the structure. In advanced ship structural design, capacity calculations of ship plating should accommodate post-weld initial imperfections as parameters of influence. The characteristics of the post-weld initial imperfections are uncertain, and an idealized model is used.

Figure 6 shows a schematic of the post-weld initial deflections in ship stiffened plate structure. The measurements of welding induced initial deflection for plating in merchant ship structures reveal a complex multi-wave shape in the long direction and one half wave is found in the short direction (Carlsen & Czujko 1978, Antoniou 1980, Kmiecik et al. 1995). In this case, the plate initial deflection can approximately be expressed by

$$\frac{w_o}{w_{opl}} = \sum_{i=1}^{M} B_{oi} \sin\frac{i\pi x}{a} \sin\frac{\pi y}{b} \tag{6}$$

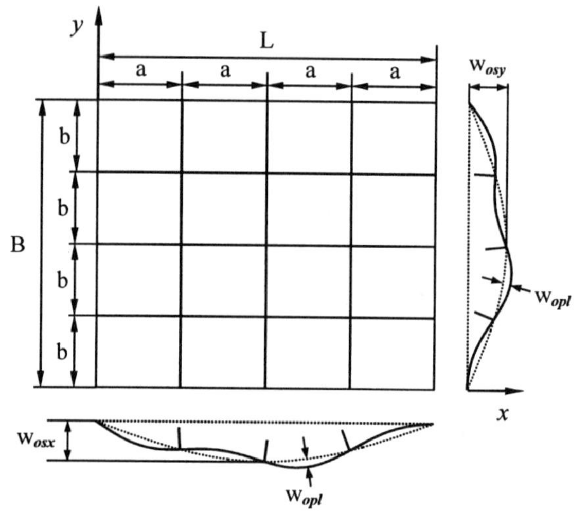

Fig.6 Fabrication related initial deflections in steel stiffened panels

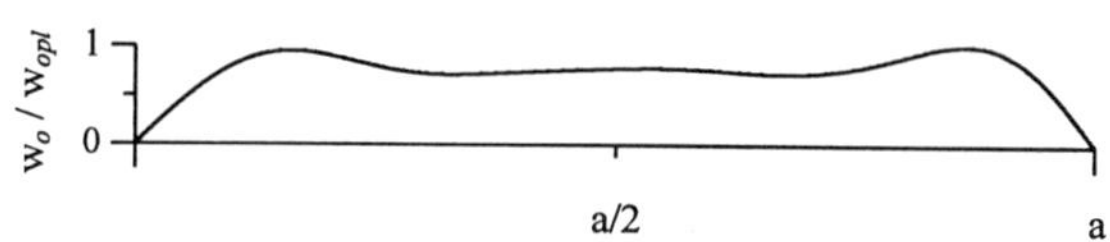

(a) Initial deflection shape #1

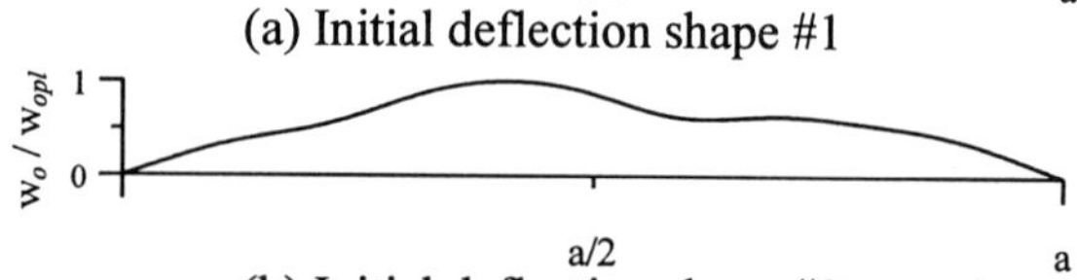

(b) Initial deflection shape #2

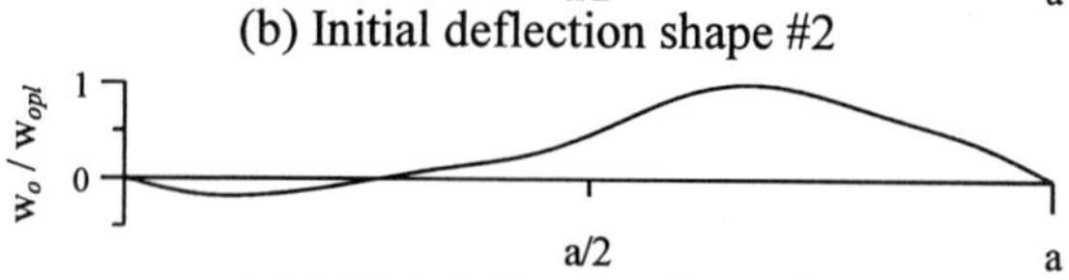

(c) Initial deflection shape #3

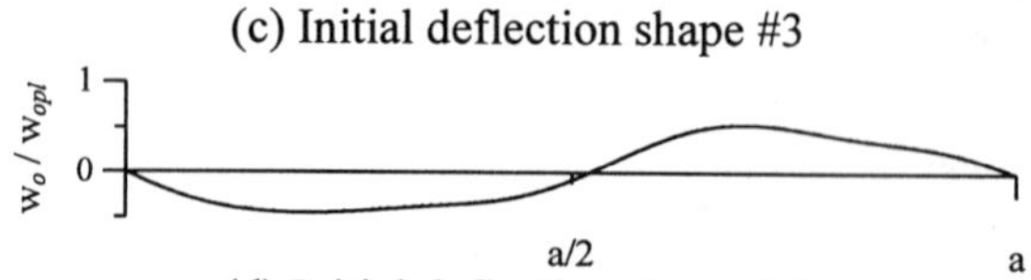

(d) Initial deflection shape # 4

Fig.7 Some typical patterns of welding induced initial deflection in ship plating

Table 1 Initial deflection amplitudes for various initial deflection shapes indicated in Fig.7

Initial Deflection Shape No.	B_{o1}	B_{o2}	B_{o3}	B_{o4}	B_{o5}	B_{o6}	B_{o7}	B_{o8}	B_{o9}	B_{o10}	B_{o11}
#1	1.0	-0.0235	0.3837	-0.0259	0.2127	-0.0371	0.0478	-0.0201	0.0010	-0.0090	0.0005
#2	0.8807	0.0643	0.0344	-0.1056	0.0183	0.0480	0.0150	-0.0101	0.0082	0.0001	-0.0103
#3	0.5500	-0.4966	0.0021	0.0213	-0.0600	-0.0403	0.0228	-0.0089	-0.0010	-0.0057	-0.0007
#4	0.0	-0.4966	0.0021	0.0213	-0.0600	-0.0403	0.0228	-0.0089	-0.0010	-0.0057	-0.0007

where w_{opl} = relative maximum initial deflection of the plating between stiffeners, and B_{oi} = initial deflection amplitudes normalized by w_{opl}.

Paik & Pedersen (1996) examined 33 sets of measurements and showed that equation (6) with $M = 11$ could reasonably model the measured initial deflections. For the shapes of initial deflection in ship plating shown in Fig.7, for instance, the coefficients B_{oi} are given as those indicated in Table 1. Smith et al. (1987) suggest the following maximum values of representative initial deflections for plating in merchant vessel structures which may be used to approximate w_{opl} in equation (6):

$$\frac{w_{opl}}{t} = \begin{cases} 0.025\beta^2 & for\ slight\ level \\ 0.1\beta^2 & for\ average\ level \\ 0.3\beta^2 & for\ serious\ level \end{cases} \tag{7}$$

The welding induced residual stress distributions can be idealized to be composed of tensile and compressive stress blocks, as shown in Fig.8. Along the welding line, tensile (positive) residual stresses are usually developed with magnitude σ_{rtx} in the x direction and σ_{rty} in the y direction since welding is normally performed in both

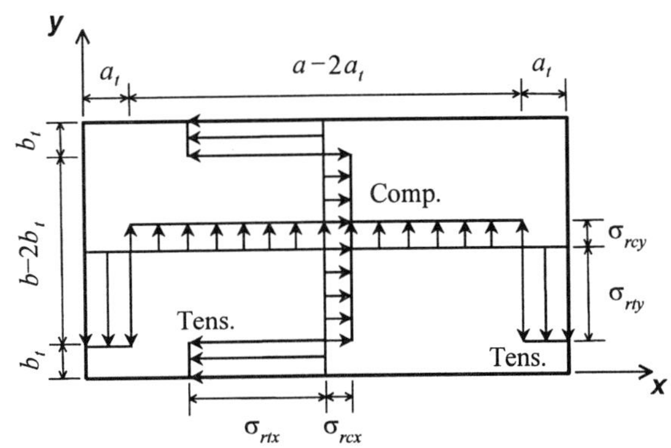

Fig.8 Idealization of welding induced residual stress distribution inside plating in the x and y directions

x and y directions. In order to obtain equilibrium, corresponding compressive (negative) residual stresses with magnitude σ_{rcx} in the x direction and σ_{rcy} in the y direction are developed in the middle part of plating. The breadths of the related tensile residual stress blocks in the x and y directions can be shown to be as follows:

$$\frac{2b_t}{b} = \frac{\sigma_{rcx}}{\sigma_{rcx} - \sigma_{rtx}}, \quad \frac{2a_t}{a} = \frac{\sigma_{rcy}}{\sigma_{rcy} - \sigma_{rty}} \tag{8}$$

where the tensile residual stress normally reaches the yield stress of material for mild steel plating (e.g., $\sigma_{rtx} = \sigma_{rty} \approx \sigma_o$), while it is usually somewhat less (approximately 80% of the material yield stress) for high tensile steel plating (e.g., $\sigma_{rtx} = \sigma_{rty} \approx 0.8\sigma_o$).

Once the magnitudes of the compressive and tensile residual stresses are known, breadths of the tensile residual stress blocks can be determined from equation (8). The residual stress distributions in the x and y directions may be approximated by

$$\sigma_{rx} = \begin{cases} \sigma_{rtx} & for\ 0 \le y < b_t \\ \sigma_{rcx} & for\ b_t \le y < b - b_t \\ \sigma_{rtx} & for\ b - b_t \le y \le b \end{cases} \tag{9.a}$$

$$\sigma_{ry} = \begin{cases} \sigma_{rty} & for\ 0 \le x < a_t \\ \sigma_{rcy} & for\ a_t \le x < a - a_t \\ \sigma_{rty} & for\ a - a_t \le x \le a \end{cases} \tag{9.b}$$

Smith et al. (1987) also suggest the following representative values of welding induced compressive residual stress in the longitudinal (x) direction:

$$\frac{\sigma_{rcx}}{\sigma_o} = \begin{cases} -0.05 & for\ slight\ level \\ -0.15 & for\ average\ level \\ -0.3 & for\ serious\ level \end{cases} \tag{10}$$

The magnitude of welding induced residual stresses

in the longer direction will normally be larger because the weld length is longer. Therefore, the transverse (plate breadth direction) residual stresses may be approximated as follows:

$$\sigma_{rcy} = \frac{b}{a}\sigma_{rcx} \tag{11}$$

By substituting equation (7) with Table 1 into equation (6) or equations (8), (10) and (11) into equation (9), the post-weld initial deflection and residual stress distribution can reasonably be defined for practical design purposes.

BUCKLING BASED CAPACITY

Design Equations

The basis of the plate capacity nominally adopted by most classification societies is buckling. For one single stress component loading, the buckling based capacity σ_B (i.e., σ_{xB} for σ_{xav}, σ_{yB} for σ_{yav} and τ_B for τ_{av}) is defined using the so-called Johnson-Ostenfeld equation (or sometimes called Bleich-Ostenfeld equation) to account for the effect of plasticity, as follows

$$\sigma_B = \begin{cases} \sigma_E & if \ \dfrac{|\sigma_E|}{\sigma_k} \le 0.5 \\ (-1)\sigma_k\left(1 - \dfrac{\sigma_k}{4|\sigma_E|}\right) & if \ \dfrac{|\sigma_E|}{\sigma_k} > 0.5 \end{cases} \tag{12}$$

where σ_E = elastic buckling stress for one single stress component, (i.e., σ_{xE} = as defined in eq. (26) for compressive σ_{xav}, σ_{yE} = as defined in eq. (27) for compressive σ_{yav} and τ_E = as defined in eq. (28) for τ_{av}), $\sigma_k = \sigma_o$ for either σ_{xav} or σ_{yav}, and $\sigma_k = \tau_o = \sigma_o / \sqrt{3}$ for τ_{av}. It is taken as $\sigma_{xB} = \sigma_o$ for tensile σ_{xav} and $\sigma_{yB} = \sigma_o$ for tensile σ_{yav}.

The elastic buckling stress equations suggested in our study accommodate the in-plane bending, lateral pressure, residual stress, and rotational restraints as necessary, but the effect of initial deflection is not included since clear bifurcation buckling is not defined for the initially deflected plating.

For combined stress component loading, the buckling based capacity component σ_B^* is obtained as a solution of the following equations (comp.:-, tens.:+)

(a) When both σ_{xav} and σ_{yav} are compressive:

$$\left(\frac{\sigma_{xav}}{\sigma_{xB}}\right)^2 + \left(\frac{\sigma_{yav}}{\sigma_{yB}}\right)^2 + \left(\frac{\tau_{av}}{\tau_B}\right)^2 = 1 \tag{13.a}$$

(b) When either σ_{xav}, σ_{yav} or both are tensile:

$$\left(\frac{\sigma_{xav}}{\sigma_{xB}}\right)^2 - \left(\frac{\sigma_{xav}}{\sigma_{xB}}\right)\left(\frac{\sigma_{yav}}{\sigma_{yB}}\right) + \left(\frac{\sigma_{yav}}{\sigma_{yB}}\right)^2 + \left(\frac{\tau_{av}}{\tau_B}\right)^2 = 1 \tag{13.b}$$

By taking σ_{xav} as the reference (non-zero) stress component, for instance, the solution of equation (13) with regard to σ_{xav} is given by

(a) When both σ_{xav} and σ_{yav} are compressive:

$$\sigma_{xB}^* = (-1)\frac{\sigma_{xB}\sigma_{yB}\tau_B}{\sqrt{\sigma_{yB}^2\tau_B^2 + C_1^2\sigma_{xB}^2\tau_B^2 + C_2^2\sigma_{xB}^2\sigma_{yB}^2}} \tag{14.a}$$

(b) When either σ_{xav}, σ_{yav} or both are tensile:

$$\sigma_{xB}^* = (s)\left|\frac{\sigma_{xB}\sigma_{yB}\tau_B}{\sqrt{\sigma_{yB}^2\tau_B^2 - C_1\sigma_{xB}\sigma_{yB}\tau_B^2 + C_1^2\sigma_{xB}^2\tau_B^2 + C_2^2\sigma_{xB}^2\sigma_{yB}^2}}\right| \tag{14.b}$$

where $s = -1$ if σ_{xav} is compressive and $s = 1$ if σ_{xav} is tensile. $C_1 = \dfrac{\sigma_{yav}}{\sigma_{xav}}$, $C_2 = \dfrac{\tau_{av}}{\sigma_{xav}}$.

For safety evaluation using equation (1), the buckling based capacity measure σ_{cB} of the plating under combined loading are therefore given by holding the loading ratio constant, as follows

$$\sigma_{cB} = \sigma_{xB}^*\sqrt{1 + C_1^2 + C_2^2} \tag{15}$$

A similar method can be applied to calculate the buckling based capacity measures for the cases in which either σ_{yav} or τ_{av} is taken as the reference stress.

Validity of the Johnson-Ostenfeld Equation

To account for the influence of plasticity on the buckling based capacity equation, the Johnson-Ostenfeld formula is used, as defined in equation (12). Figures 9 to 11 show the validity of the Johnson-Ostenfeld equation by comparing with the nonlinear finite element inelastic buckling (ultimate strength) solutions for the plating under one single stress component, varying the edge condition and the aspect ratio. It is seen that the Johnson-Ostenfeld equation generally has the tendency to underestimate the inelastic buckling strength.

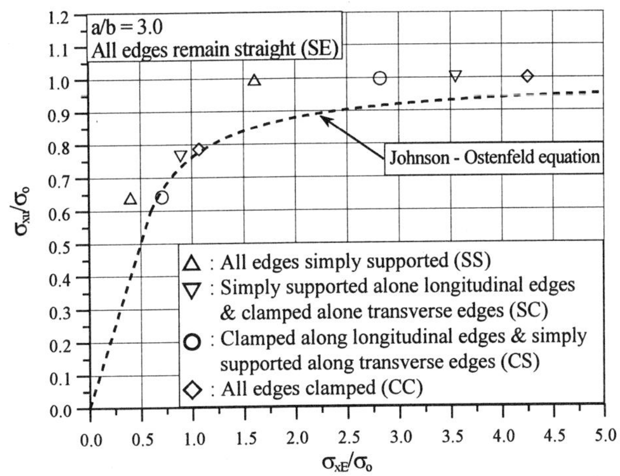

Fig.9 The ultimate capacity versus the elastic bifurcation buckling stress of plating under longitudinal compression alone, $a/b=3.0$ (symbol: FEA)

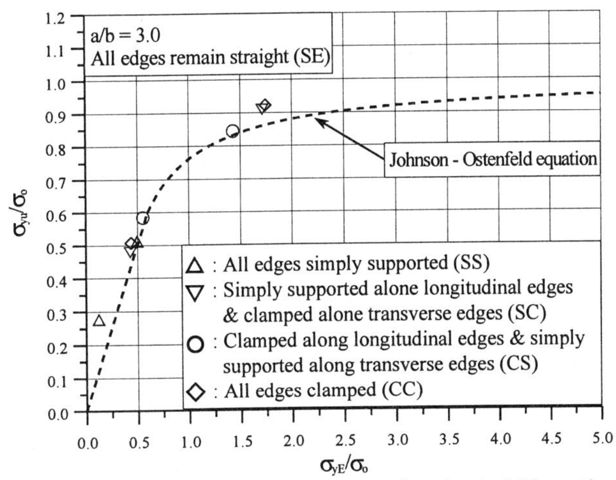

Fig.10 The ultimate capacity versus the elastic bifurcation buckling stress of plating under transverse compression along, $a/b=3$ (symbol: FEA)

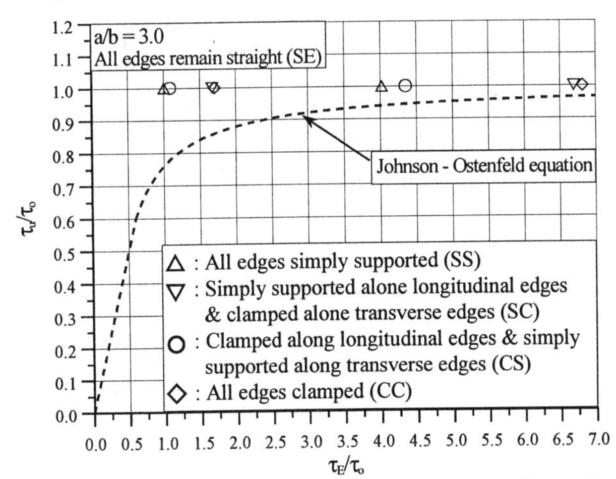

Fig.11 The ultimate capacity versus the elastic bifurcation buckling stress of plating under edge shear alone, $a/b=3$ (symbol: FEA)

Effect of Rotational Restraints

The rotational restraints of the support members are included in the elastic buckling equations for both σ_{xav} and σ_{yav} as parameters of influence. Ship plating is supported by various types of members along the edges, which have a finite value of the torsional rigidity. This is in contrast to the idealized simply supported boundary conditions often assumed for design purposes. Depending on the torsional rigidity of support members, the rotation along the plate edges will to some extent be restrained. When the rotational restraints are zero, the edge condition corresponds to a simply supported case, while the edge condition becomes clamped when the rotational restraints are infinite.

Most current practical design guidelines from classification societies for the buckling and ultimate strength of ship plating are based on boundary conditions in which all (four) edges are simply supported. In real ship plating, idealized edge conditions such as simply supported or clamped however may never occur because of finite rotational restraints.

According to the study of Paik et al. (1993) who investigated the bending and torsional rigidities of support members for deck, side and bottom plating in merchant ships, the magnitude of the rotational restraint parameter ζ_L at long edges (ship's longitudinal direction) is normally in the range of 0.05 to 3.0 (and usually not exceeding 5.0) while the amount ζ_S at the short edges (normal to the ship longitudinal direction) is normally in the range of 0.1 to 8.0 (and usually not exceeding 13.0). Thus, there is of course no case with zero or infinite rotational restraints in practice as long as support members exist at their edges, and the amount of the rotational restraints at one set of long or short edges is normally different from each other as well. It was also found from the same investigation that the bending rigidities of support members are usually sufficient enough so that the relative lateral deflection of typical members providing the support to plating at edges can be taken to be small.

For advanced design of ship plating against buckling, it is hence important to better understand the buckling strength characteristics of plating as a function of the rotational restraints of support members along the edges. This problem has of course been studied before, by a number of investigators. Lundquist & Stowell (1942) studied the effect of the edge condition on the buckling strength of rectangular plates subject to uniaxial compressive loads where the support along the unloaded edges was intermediate between simply supported and clamped. Bleich (1952) and Timoshenko & Gere (1963) discussed the buckling strength of plates with various boundary conditions that one set of edges is elastically restrained while the other set of edges is either simply supported or clamped. Gerard & Becker (1954) surveyed literature for the buckling of rectangular plates under various combinations of two or three types of loading under a number of edge conditions. Evans (1960) carried out an extensive experimental study on the buckling strength of wide plates with the loaded (long) edges elastically restrained while the unloaded (short) edges are simply supported. Based on the experimental results, he derived a closed-form expression of the compressive strength of wide plates taking into account the effect of rotational restraints along the loaded edges. McKenzie (1963) studied the buckling strength of plating under biaxial compression, bending and edge shear that is simply supported along short edges (at which bending is applied) and elastically restrained along long edges.

These various previous studies are quite useful for the buckling strength design of plating considering the rotational restraint effect along the edges. To the authors' knowledge, however, systematic investigations on the buckling strength of plating which is elastically restrained along both long and short edges appear to be difficult to come by and were thus needed. The aims of our study related to this issue (Paik & Thayamballi 2000) were to

- investigate the buckling strength characteristics of plating with the boundary conditions which are elastically restrained along the edges, and to
- develop simple buckling design formulations of plating taking into account the rotational restraints of support members along either one set of edges or all (four) edges.

The simplified formulations referred to are based on more exact solutions as obtained by directly solving the buckling characteristic equations for a variety of the torsional rigidities of support members and the plate aspect ratio. The characteristic equation for the buckling of plating with elastic restraints along either long or short edges while the other edges are simply supported is derived analytically. By solving the characteristic equation, the buckling strength characteristics of plating are investigated varying the plate aspect ratio and the torsional rigidity of support members. Based on the computed results, closed-form expressions of the plate buckling strength are obtained empirically by curve fitting. Simplified buckling design formulations for plating with all edges elastically restrained are also derived.

Figures 12 to 15 show some selected sets of the buckling coefficients as obtained by directly solving the theoretical characteristic buckling equation plotted against the plate aspect ratio and the torsional rigidity of support members along the plate edges. The accuracy of the proposed simplified equations obtained by curve fitting the more exact results may be verified by comparison with the exact theoretical solutions, see Figs. 12 and 16 to 18. The curve-fit design equations k_{x1}, k_{x2}, k_{y1} and k_{y2} are given in Appendix 1.

One of the useful insights developed herein is that the buckling coefficient for the plating elastically restrained at both long and short edges can be expressed by a relevant combination of the following three edge conditions, namely (a) elastically restrained at long edges and simply supported at short edges, (b) simply supported at long edges and elastically restrained at short edges, and (c) simply supported at all edges. Specifically it was noted that the following held approximately:

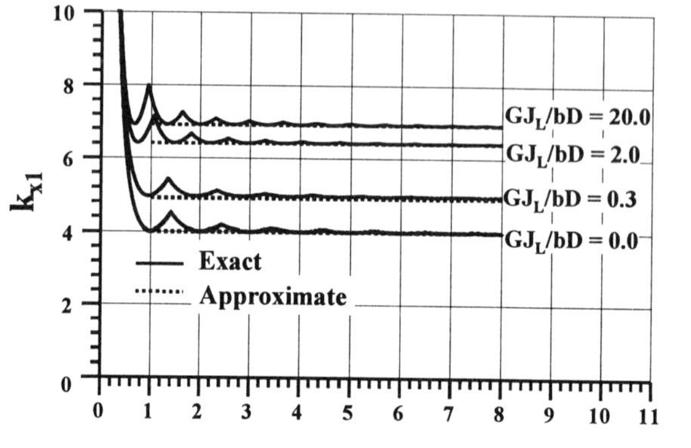

Fig. 12 Buckling coefficient k_{x1} for a plate under longitudinal compression, elastically restrained at the long edges and simply supported at the short edges as obtained by directly solving the buckling characteristic equation and by the proposed approximate equation

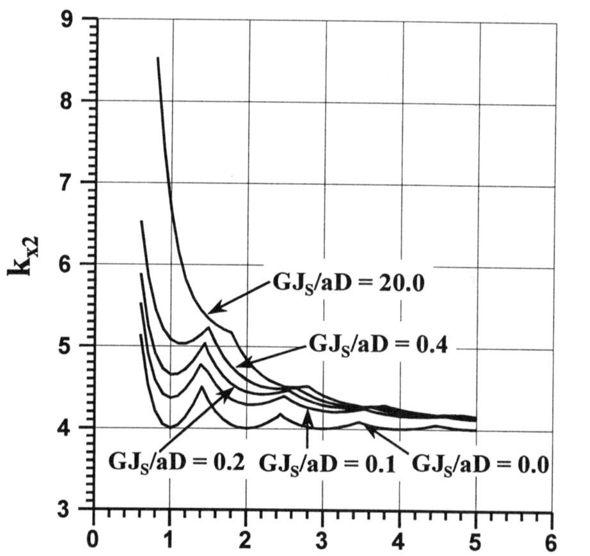

Fig. 13 Buckling coefficient k_{x2} for a plate under longitudinal compression, elastically restrained at the short edges and simply supported at the long edges as obtained by directly solving the buckling characteristic equation

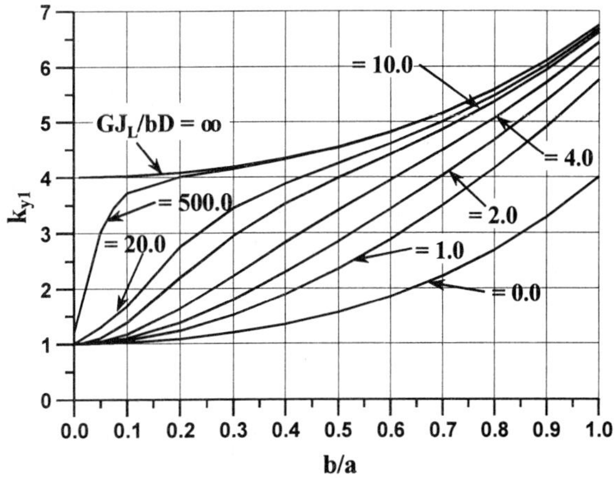

Fig. 14 Buckling coefficient k_{y1} for a plate under transverse compression, elastically restrained at the long edges and simply supported at the short edges as obtained by directly solving the buckling characteristic equation

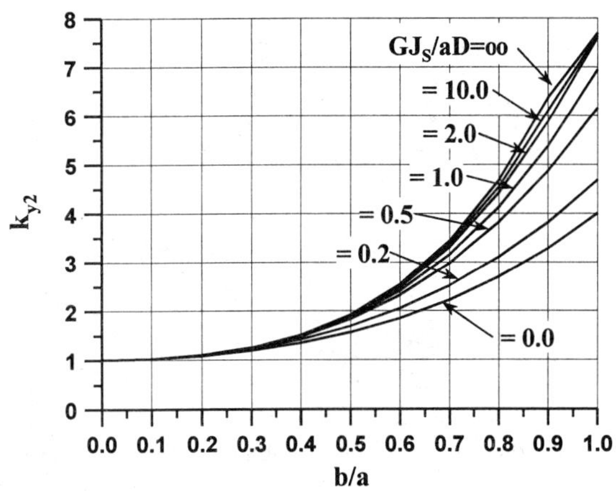

Fig. 15 Buckling coefficient k_{y2} for a plate under transverse compression, elastically restrained at the short edges and simply supported at the long edges as obtained by directly solving the buckling characteristic equation

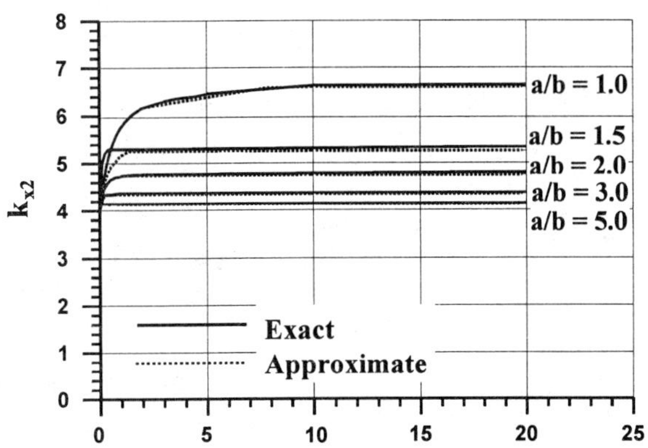

Fig. 16 Accuracy of the design equation for the buckling coefficient k_{x2}

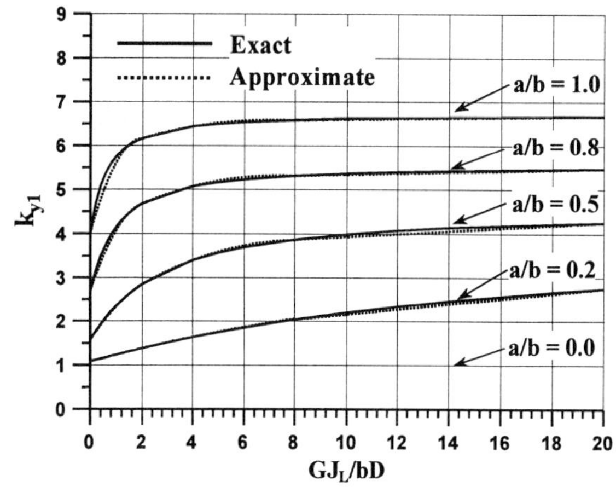

Fig. 17 Accuracy of the design equation for the buckling coefficient k_{y1}

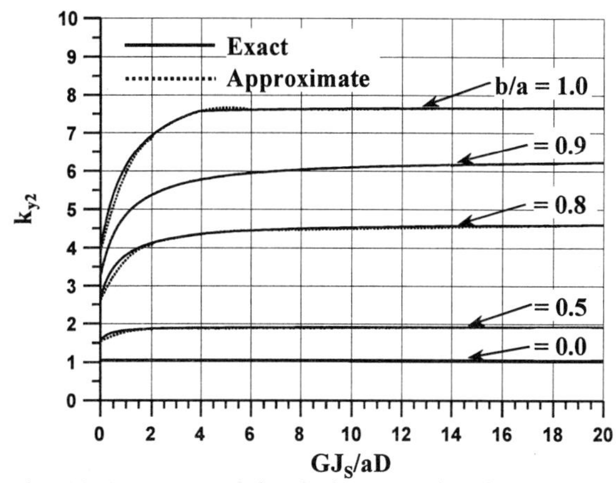

Fig. 18 Accuracy of the design equation for the buckling coefficient k_{y2}

$$k_x = k_{x1} + k_{x2} - k_{xo}\ ,\quad k_y = k_{y1} + k_{y2} - k_{yo} \tag{16}$$

where k_x = buckling coefficient of plating elastically restrained at both long and short edges for longitudinal compression, k_y = buckling coefficient of plating elastically restrained at both long and short edges for transverse compression, k_{xo} = buckling coefficient of plating simply supported at all edges for longitudinal compression which may be taken as $k_{xo} \cong 4.0$, and k_{yo} = buckling coefficient of plating simply supported at all edges which may be taken as $k_{yo} = \left\{1 + (b/a)^2\right\}^2$. k_{xo}, k_{x1}, k_{x2}, k_{yo}, k_{y1}, k_{y2} = as defined in Appendix 1.

It was also found that the buckling interaction equation of the plating elastically restrained along all edges and under combined loading can approximately take the same relationship as that with simply supported conditions at all edges, but by replacing the buckling stress components of the plating simply supported at all edges with the corresponding ones for the elastically restrained plating. As two specific cases of plating under combined biaxial compression or combined axial compression and edge shear, where the plate edges are all clamped, i.e., with infinite rotational restraints, Figures 19 and 20 show the elastic buckling interaction relations varying the plate aspect ratio where the theoretical predictions were obtained by the formulae for simply supported plates as given in Appendices 2 and 3 while FE solutions were calculated for clamped plates.

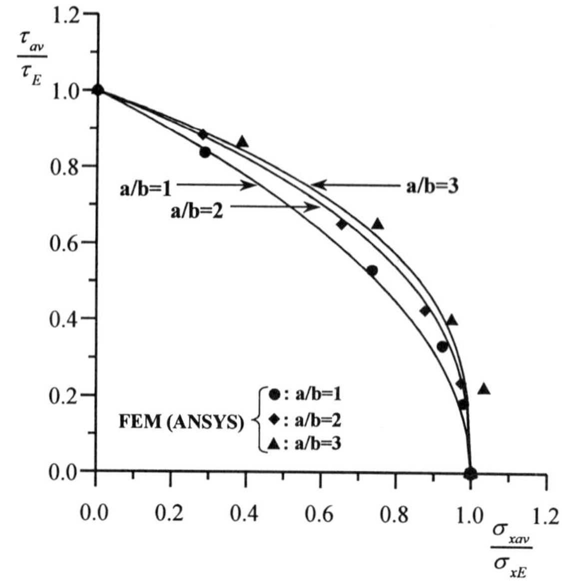

Fig. 20 Elastic buckling interaction relationships of plating under combined axial compression and edge shear (symbol: eigen value finite element solutions for plating clamped at all edges, line: design equation for plating simply supported at all edges)

Effect of Residual Stresses

The welding induced residual stress (compression: -, tension: +) is included in the elastic buckling equations for both σ_{xav} and σ_{yav} as a parameter of influence. The elastic buckling stress of simply supported plating under uniform axial compression (i.e., without in-plane bending) is given by (Paik et al. 2000) (For the symbols unless specified below, refer to the section on "modeling of fabrication related imperfections")

$$\sigma_{xE} = -\frac{\pi^2 D}{b^2 t}\left(\frac{mb}{a}+\frac{a}{mb}\right)^2 - \sigma_{rex} - \frac{a^2}{m^2 b^2}\sigma_{rey} \qquad (17)$$

where $\sigma_{rex} = \sigma_{rcx} + \frac{2}{b}(\sigma_{rtx}-\sigma_{rcx})\left(b_t - \frac{b}{2\pi}\sin\frac{2\pi b_t}{b}\right)$,

$$\sigma_{rey} = \sigma_{rcy} + \frac{2}{a}(\sigma_{rty}-\sigma_{rcy})\left(a_t - \frac{a}{2m\pi}\sin\frac{2m\pi a_t}{a}\right)$$

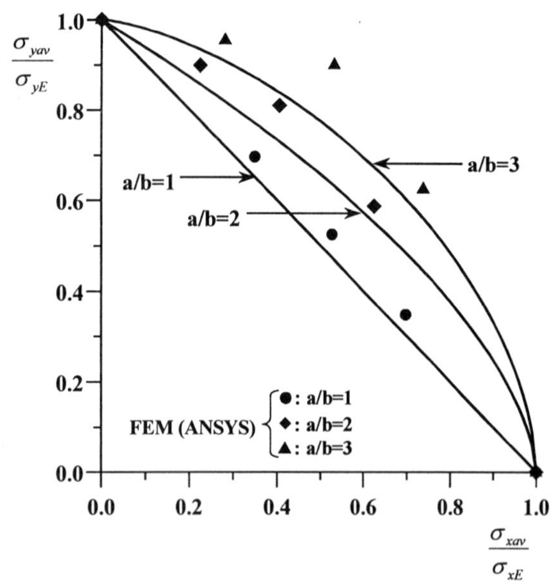

Fig.19 Elastic buckling interaction relationships for plating under combined biaxial compression (symbol: eigen value finite element solutions for plating clamped at all edges, line: design equation for plating simply supported at all edges)

The second and third terms of the right hand side of equation (17) reflect the effect of welding induced residual stresses on the plate compressive buckling stress. m is the buckling half wave number which is determined as a minimum integer satisfying the following equation

$$\frac{\pi^2 D}{b^2 t}\left(\frac{mb}{a}+\frac{a}{mb}\right)^2 + \frac{a^2}{m^2 b^2}\sigma_{rey} \le \frac{\pi^2 D}{b^2 t}\left[\frac{(m+1)b}{a}+\frac{a}{(m+1)b}\right]^2 + \frac{a^2}{(m+1)^2 b^2}\sigma_{rey} \qquad (18)$$

Without the post-weld residual stresses, i.e., $\sigma_{rex} = \sigma_{rey} = 0$, equation (18) simplifies to the well-known condition

$$\frac{a}{b} \le \sqrt{m^*(m^*+1)} \qquad (19)$$

where m^* is the buckling half wave number when the residual stresses do not exist.

In the similar way, the elastic buckling stress σ_{yE} of the simply supported plating subject to axial compression in the y direction can be given by

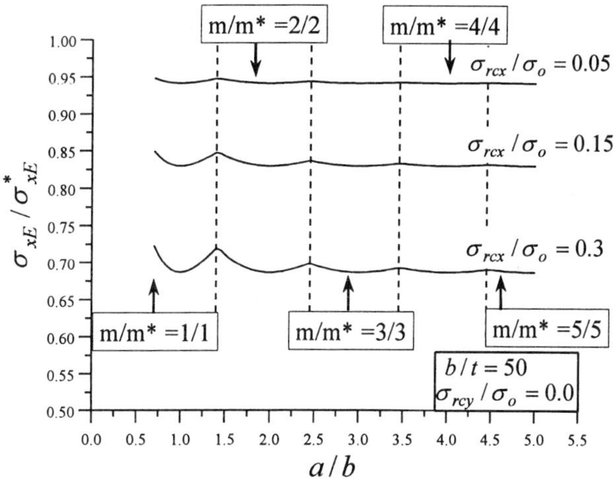

Fig.21.a Variation of the elastic compressive buckling stress (normalized by the elastic buckling compressive stress without residual stresses) as a function of the welding induced residual stress and the plate aspect ratio, $\sigma_{rcy} = 0$, $b/t = 50$, $\beta = 2.07$, $\sigma_o = 352MPa$, $\sigma_{rtx} = 0.8\sigma_o$ (σ^*_{xE} = buckling stress without residual stress)

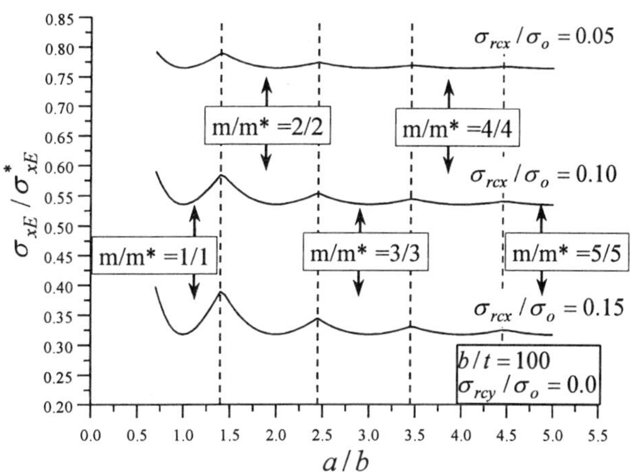

Fig.21.b Variation of the elastic compressive buckling stress (normalized by the elastic buckling compressive stress without residual stresses) as a function of the welding induced residual stress and the plate aspect ratio, $\sigma_{rcy} = 0$, $b/t = 100$, $\beta = 4.14$, $\sigma_o = 352MPa$, $\sigma_{rtx} = 0.8\sigma_o$ (σ^*_{xE} = buckling stress without residual stress)

$$\sigma_{yE} = -\frac{\pi^2 D}{b^2 t}\left(1 + \frac{b^2}{a^2}\right)^2 - \frac{b^2}{a^2}\sigma_{rex} - \sigma_{rey} \tag{20}$$

where σ_{rex} and σ_{rey} are defined as those in equation (17) but replacing by $m = 1$. The second and third terms of the right hand side of equation (20) reflect the effect of welding induced residual stresses.

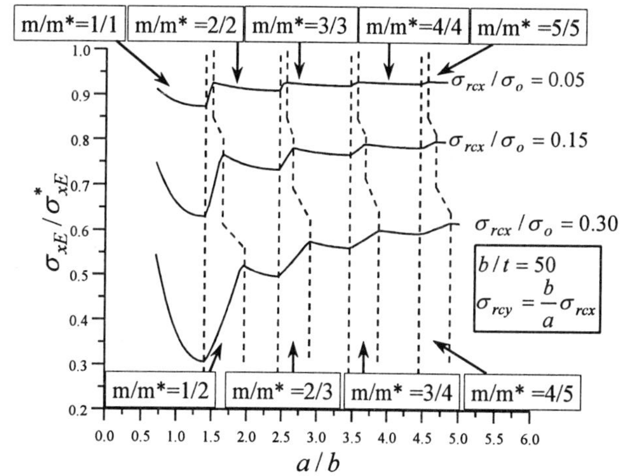

Fig.21.c Variation of the elastic compressive buckling stress (normalized by the elastic compressive buckling stress without residual stresses) as a function of the welding induced residual stress and the plate aspect ratio, $\sigma_{rcy} = \sigma_{rcx} b/a$, $b/t = 50$, $\beta = 2.07$, $\sigma_o = 352MPa$, $\sigma_{rtx} = 0.8\sigma_o$ (σ^*_{xE} = buckling stress without residual stress)

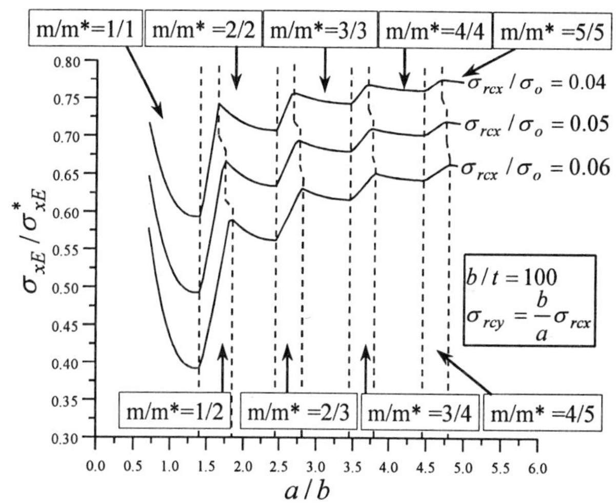

Fig.21.d Variation of the elastic compressive buckling stress (normalized by the elastic compressive buckling stress without residual stresses) as a function of the welding induced residual stress and the plate aspect ratio, $\sigma_{rcy} = \sigma_{rcx} b/a$, $b/t = 100$, $\beta = 4.14$, $\sigma_o = 352MPa$ $\sigma_{rtx} = 0.8\sigma_o$ (σ^*_{xE} = buckling stress without residual stress)

Figure 21 shows the influence of welding induced residual stress on the compressive buckling stress for the high tensile steel plating with the yield stress of $\sigma_o = 352MPa$. In the calculations indicated in Fig.21, the level of residual stresses and the plate slenderness ratio (i.e., b/t ratio) are varied. Two types of welding induced residual stresses in the y direction are presumed, namely one with zero residual stresses and the other with $\sigma_{rcy} = \frac{b}{a}\sigma_{rcx}$. It is in the analysis assumed that the magnitude of the tensile residual stresses is 80% of the yield stress, that is, $\sigma_{rtx} = \sigma_{rty} = 0.8\sigma_o$. It is evident from Fig.21 that the welding residual stresses can significantly reduce the compressive buckling stress of the plating. The reduction tendency of the buckling stress for thin plating is faster than that for thick plating, as expected. It is also noted from Figs.21.c and 21.d that the residual stresses in the y direction may change the longitudinal buckling half wave number of the plating.

Effect of Cut-outs

In ship plating, cut-outs are often located to make a way of access or to lighten the structure. These will reduce the capacity of the plating. The opening is included in the elastic buckling equations for σ_{xav} (compression), σ_{yav} (compression) and τ_{av} as a parameter of influence. For a circular type of opening, the buckling reduction factors are in our study suggested by curve fitting based on the eigen value finite element solutions, as follows

$$R_{cx} = R_{cy} = 1.0 - 0.57\left(\frac{d_c}{b}\right), \quad R_{cs} = 1.0 - 0.68\left(\frac{d_c}{b}\right) \qquad (21)$$

where R_{cx}, R_{cy}, R_{cs} = buckling reduction factor accounting for the effect of cut-outs under σ_{xav}, σ_{yav} and τ_{av}, respectively, and d_c = diameter of circular opening .

Figures 22 and 23 show the validity of equation (21) by comparing with the eigen value finite element solutions, varying the diameter of opening. It is seen that with an opening of which diameter is 50% of the plate breadth, the plate compressive buckling becomes 70% of the original strength. As the diameter of opening increases, the buckling reduction tendency for edge shear loading is slightly faster than that for axial compressive loading.

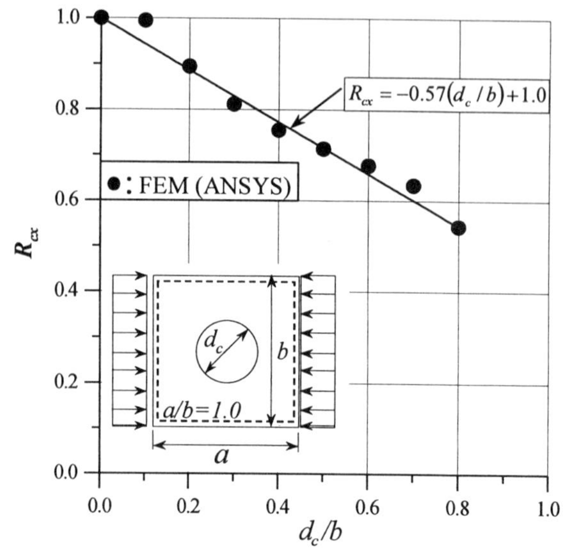

Figure 22 Buckling reduction factor accounting for the effect of cut-outs under axial compression

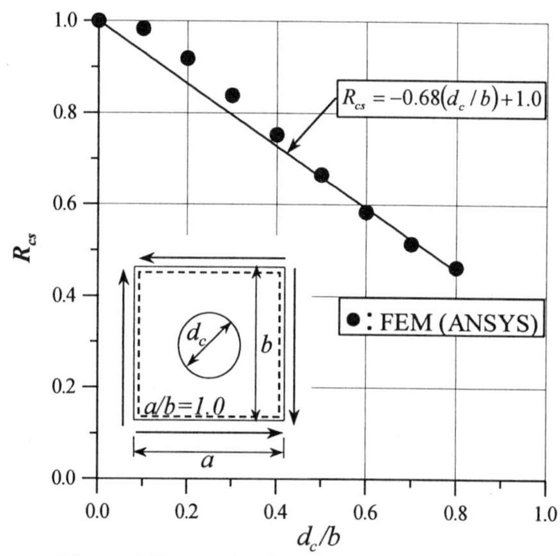

Figure 23 Buckling reduction factor accounting for the effect of cut-outs under edge shear

Effect of Lateral Pressure

The lateral pressure is included in the elastic buckling equations for both σ_{xav} and σ_{yav} as a parameter of influence. When the buckling half wave corresponds to the plate deflection pattern caused by lateral pressure alone, e.g., in a nearly square plate with $a/b \approx 1.0$ or a long plate under predominantly transverse compression (σ_{yav}), lateral pressure reduces the compressive buckling stress. However, lateral pressure loads increase the compressive buckling stress for a long plate under predominantly longitudinal compressive loads (σ_{xav}) since lateral pressure disturbs the occurrence of buckling in this case. The buckling correction factors accounting for the effect of lateral pressure are given by (Paik & Kim 1988)

$$R_{px} = \begin{cases} 1.0 - 0.056\left(\dfrac{pb^4}{Et^4}\right) & for\ 1 \le \dfrac{a}{b} \le \sqrt{2} \\ \left(1.0 + C_p\dfrac{pb^4}{Et^4}\right)^{1/2} & for\ \dfrac{a}{b} > \sqrt{2} \end{cases}$$

$$R_{py} = 1.0 - 0.056\left(\frac{pb^4}{Et^4}\right) \qquad (22)$$

where R_{px}, R_{py} = buckling correction factor accounting for the effect of lateral pressure under σ_{xav} and σ_{yav}, respectively, with

$$C_p = \begin{cases} 0.025\left(\frac{a}{b}\right)^2 - 0.155\left(\frac{a}{b}\right) + 0.360 \ for \ \sqrt{2} \le \frac{a}{b} \le 4 \\ 0.14 \ for \frac{a}{b} > 4 \end{cases}$$

Figure 24 shows the variation of the elastic longitudinal compressive buckling plotted against the magnitude of lateral pressure loads. It is seen from Fig.24 that for a square plate lateral pressure reduces the buckling, while buckling strength of a long plate under axial compressive loads in the longitudinal direction is increased by lateral pressure.

Effect of In-plane Bending

As shown in Fig.4, ship plating is often subjected to in-plane bending which affects the buckling capacity. The in-plane bending stresses are included in the elastic compressive buckling equations as parameters of influence. In general, the in-plane bending stresses can be given in terms of average axial stresses, as defined in equation (4). The elastic buckling interaction equations between σ_{xav} and σ_{xb} or between σ_{yav} and σ_{yb} may be approximated by

$$\frac{\sigma_{xav}}{\sigma^*_{xE}} + \left(\frac{\sigma_{xb}}{\sigma_{xbE}}\right)^2 = 1, \quad \frac{\sigma_{yav}}{\sigma^*_{yE}} + \left(\frac{\sigma_{yb}}{\sigma_{ybE}}\right)^2 = 1 \qquad (23)$$

where σ_{xbE}, σ_{ybE} = as defined in Appendix 4

In the above equation, σ^*_{xE} is the elastic buckling stress under uniform longitudinal axial compression (i.e., without in-plane bending) which is given considering the effects of rotational restraints, residual stress, lateral pressure and cut-outs and is defined, as follows

$$\sigma^*_{xE} = \left[-k_x \frac{\pi^2 E}{12(1-\nu^2)}\left(\frac{t}{b}\right)^2 - \sigma_{rex} - \left(\frac{a}{mb}\right)^2 \sigma_{rey} \right] R_{px} R_{cx} \qquad (24)$$

where k_x = as defined in Appendix 1, σ_{rex}, σ_{rey}, m = as defined in equation (17), R_{px} = as defined in equation (22), R_{cx} = as defined in equation (21)

σ^*_{yE} in equation (23) is the elastic buckling stress under uniform transverse axial compression (i.e., without in-plane bending) which is given considering the effects of rotational restraints, residual stress, lateral pressure and cut-outs and is defined, as follows

$$\sigma^*_{yE} = \left[-k_y \frac{\pi^2 E}{12(1-\nu^2)}\left(\frac{t}{b}\right)^2 - \left(\frac{b}{a}\right)^2 \sigma_{rex} - \sigma_{rey} \right] R_{py} R_{cy} \qquad (25)$$

where k_y = as defined in Appendix 1, σ_{rex}, σ_{rey} = as defined in equation (20), R_{py} = as defined in equation (22), R_{cy} = as defined in equation (21)

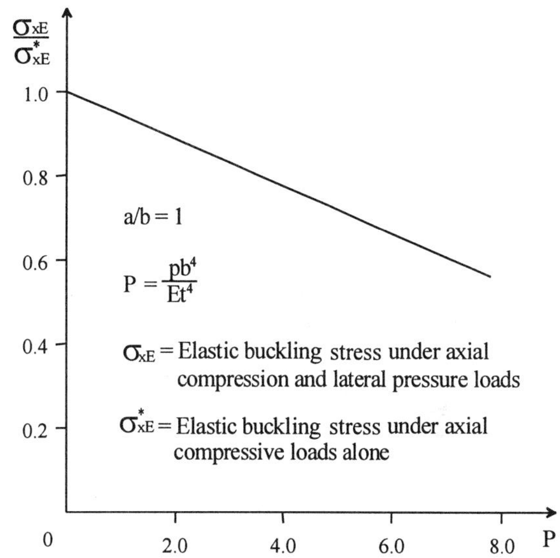

Fig.24.a Variation of the elastic compressive buckling for a square plate against the magnitude of lateral pressure loads

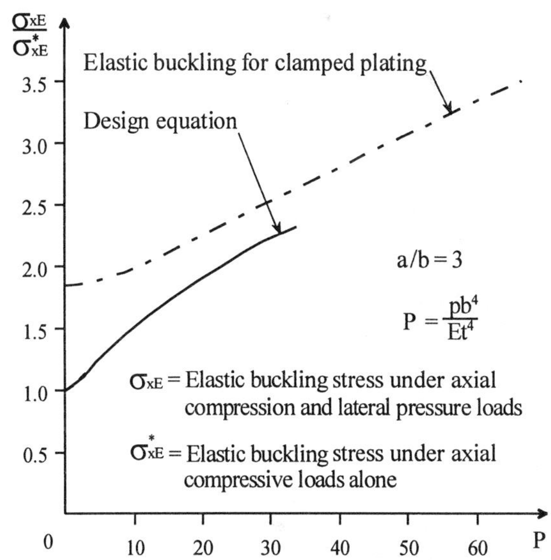

Fig.24.b Variation of the compressive buckling for a rectangular plate of a/b=3 against the magnitude of lateral pressure loads

Considering the relationship between σ_{xav} and σ_{xb} as defined in equation (23), the axial compressive buckling stress σ_{xE} taking into account the effect of in-plane bending stresses are given, as follows

(a) When $\sigma_{x1} = \sigma_{x2}$:

$$\sigma_{xE} = \sigma_{xE}^{*} \tag{26.a}$$

(b) When $|\sigma_{x1}| \neq |\sigma_{x2}|$:

$$\sigma_{xE} = \frac{-F_2 - \sqrt{F_2^2 + 4F_1}}{2F_1} \tag{26.b}$$

(c) When $\sigma_{x1} = -\sigma_{x2}$:

$$\sigma_{xE} = (-1)\sigma_{xbE} \tag{26.c}$$

where $F_1 = \dfrac{C^2}{\sigma_{xbE}^2}$, $F_2 = \dfrac{1}{\sigma_{xE}^{*}}$,

$$C = \begin{cases} \dfrac{1-\phi_x}{1+\phi_x}, \phi_x = \dfrac{\sigma_{x2}}{\sigma_{x1}} \ if\ \sigma_{x1} \neq 0, \sigma_{x1} \neq -\sigma_{x2} \\ 1.0\ if\ \sigma_{x1} = 0, \sigma_{x1} \neq -\sigma_{x2} \end{cases}$$

In the similar way, the axial compressive buckling stress σ_{yE} taking into account the influence of in-plane bending is calculated from considering the relation between σ_{yav} and σ_{yb} defined in equation (23), as follows

(a) When $\sigma_{y1} = \sigma_{y2}$:

$$\sigma_{yE} = \sigma_{yE}^{*} \tag{27.a}$$

(b) When $|\sigma_{y1}| \neq |\sigma_{y2}|$:

$$\sigma_{yE} = \frac{-G_2 - \sqrt{G_2^2 + 4G_1}}{2G_1} \tag{27.b}$$

(c) When $\sigma_{y1} = -\sigma_{y2}$:

$$\sigma_{yE} = (-1)\sigma_{ybE} \tag{27.c}$$

where $G_1 = \dfrac{C^2}{\sigma_{ybE}^2}$, $G_2 = \dfrac{1}{\sigma_{yE}^{*}}$,

$$C = \begin{cases} \dfrac{1-\phi_y}{1+\phi_y}, \phi_y = \dfrac{\sigma_{y2}}{\sigma_{y1}} \ if\ \sigma_{y1} \neq 0, \sigma_{y1} \neq -\sigma_{y2} \\ 1.0\ if\ \sigma_{y1} = 0, \sigma_{y1} \neq -\sigma_{y2} \end{cases}$$

Elastic Edge Shear Buckling

The elastic shear buckling stress taking into account the influence of cut-outs is defined as follows

$$\tau_E = k_s \frac{\pi^2 E}{12(1-\nu^2)}\left(\frac{t}{b}\right)^2 R_{cs} \tag{28}$$

where k_s = as defined in Appendix 5, R_{cs} = as defined in equation (21)

ULTIMATE STRENGTH BASED CAPACITY

Most classification society criteria and procedures for ship structural design are based on the first yield of hull structures together with buckling checks for structural components. While service proven, the traditional design criteria and associated linear elastic stress calculations do not necessarily define the true ultimate limit state which is the limiting condition beyond which a ship hull will fail to perform its function. Neither do such procedures help understand the likely sequence of local failure prior to reaching the ultimate limit state. It is of course important to determine the true ultimate strength if one is to obtain consistent measures of safety which can form a fairer basis for comparisons of vessels of different sizes and types. An ability to better assess the true margin of safety should also inevitably lead to improvements in related regulations and design requirements.

In the case of plate elements which constitute a significant portion of the hull and thus affect its weight and other design characteristics, it is now known that a single set of ultimate strength interaction equations will not successfully represent the ultimate limit state of ship plating under combined loads since collapse patterns significantly depend on the types and relative magnitudes of primary load component involved. The strength interaction relationship would thus be different depending on which load component is predominant. In this regard, the present authors have developed three sets of such equations considering each primary load component, namely longitudinal axial load, transverse axial load and edge shear, while lateral pressure is regarded as secondary. The ultimate strength interaction equation under all of the load components is derived by a relevant combination of the individual strength formulae (Paik et al. 1999a). In the following, the plate ultimate strength equations for plating under combined in-plane and lateral pressure loads are presented.

Ultimate Strength Equation for Combined Longitudinal Axial Load and Lateral Pressure

Figure 25 shows a typical example of the axial membrane stress distribution inside a plate element under predominantly longitudinal compressive loading, before and after buckling occurs. It is noted that the membrane

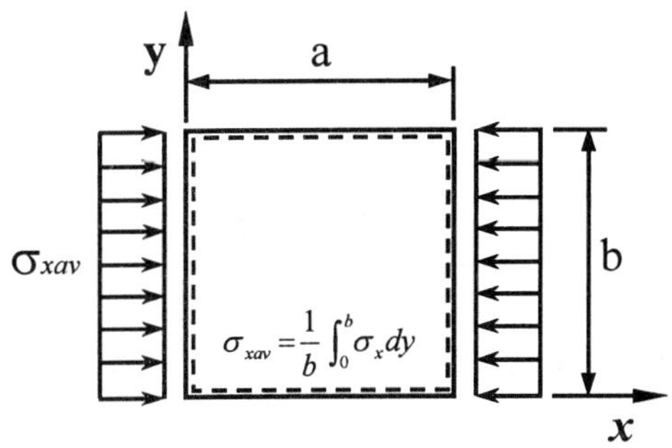

(a) Before buckling

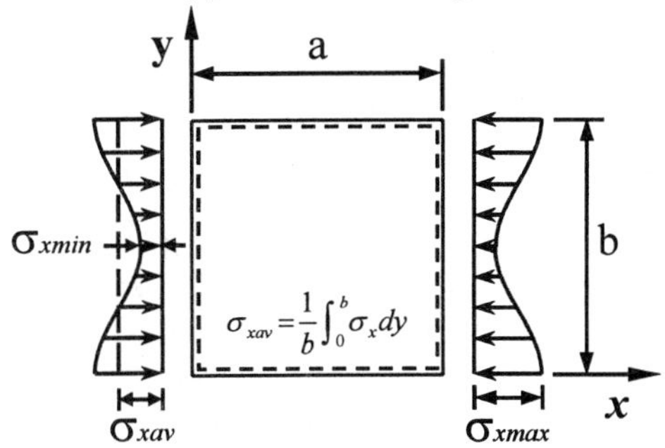

(b) After buckling, unloaded edges move freely in plane

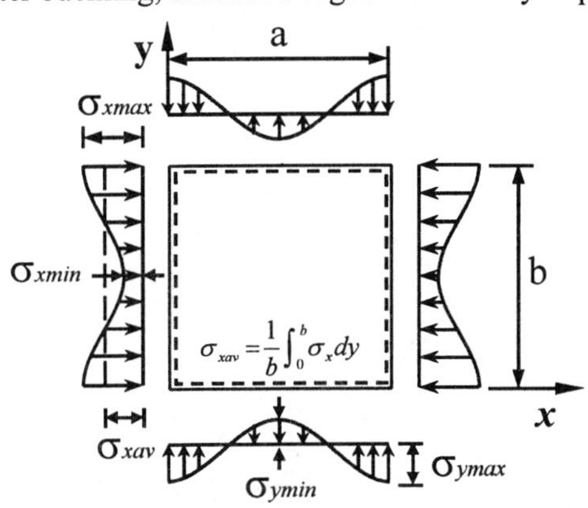

(c) After buckling, unloaded edges keep straight

Fig.25 Membrane stress distribution inside the plate element under longitudinal compressive loads

stress distribution in the loading (x) direction can become non-uniform as the plate element deflects (or buckles). The membrane stress distribution in the y direction also becomes non-uniform if the unloaded plate edges remain straight, while no membrane stresses will develop in the y direction if the unloaded plate edges move freely in plane. The maximum compressive membrane stresses are developed around the plate edges that remain straight, while the minimum (tensile) membrane stresses occur in the middle of the plate element where a membrane tension field is formed by the plate deflection since the plate edges remain straight.

With increase in the plate deflection, the upper and / or lower fibers in the mid-region of the plate element will initially yield by the action of bending. However, as long as it is possible to redistribute the applied loads to the straight plate boundaries by the membrane action, the plate element will not collapse. Collapse will then occur when the most stressed boundary locations yield, since the plating can not keep the boundaries straight any further, resulting in a rapid increase of lateral plate deflection (Paik & Pedersen 1995).

Hence the ultimate strength formulation for ship plating subject to uniaxial compression / tension and lateral pressure loads is in the present study derived under the somewhat pessimistic assumption that the plating collapses when initial plastic yield at the plate edges occurs.

The occurrence of yielding can be assessed by using the von Mises yield criterion. For longitudinal axial load and lateral pressure, the most probable yield locations will be found at longitudinal mid-edges where the maximum compressive stress in the x direction and the minimum tensile stress in the y direction develop, as shown in Fig.26.a. The resulting yield criterion is in this case expressed by (For the stress related symbols, see Fig.25.c)

$$\Gamma_u = \left(\frac{\sigma_{x\max}}{\sigma_o}\right)^2 - \left(\frac{\sigma_{x\max}}{\sigma_o}\right)\left(\frac{\sigma_{y\min}}{\sigma_o}\right) + \left(\frac{\sigma_{y\min}}{\sigma_o}\right)^2 - 1 = 0 \tag{29}$$

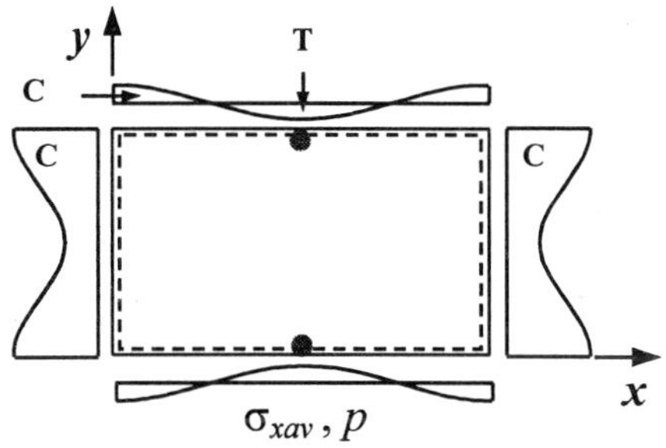

(a) Plasticity at plate longitudinal mid-edges

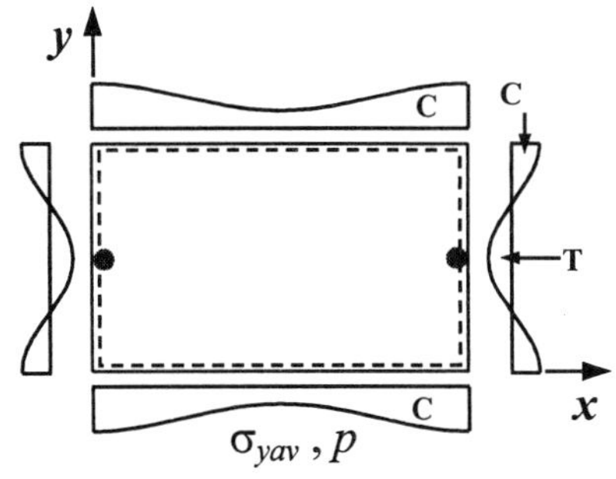

(b) Plasticity at plate transverse mid-edges

•: Expected Yielding Locations, T: Tension, C: Compression

Fig.26 Possible locations for the initial plastic yield at the plate edges under combined uniaxial load and pressure

where $\sigma_{x\max}$ and $\sigma_{y\min}$ are given in terms of σ_{xav}, p and post-weld initial imperfections (initial deflection and residual stresses), as defined in Appendix 6.

The ultimate strength based capacity σ_{xu} for longitudinal axial load is obtained as the solution of equation (29) with regard to σ_{xav}. As an approximation, σ_{xu} is taken as the initial (minimum) value at $\Gamma_u \geq 0$ by increasing σ_{xav} with the increment of 1% yield stress, i.e., $\Delta\sigma_{xav} = -0.01\sigma_o$ for compressive σ_{xav} and $\Delta\sigma_{xav} = 0.01\sigma_o$ for tensile σ_{xav}. Figure 27 shows the variation of the ultimate longitudinal compressive stresses plotted against the plate reduced slenderness ratio, as obtained from equation (29) and from nonlinear FEA. The compressive stresses at the initial yielding are also plotted. It is evident that equation (29) agrees very well with the more refined nonlinear FEA. Equation (29) shows mean bias = 0.978 and COV = 0.065 against FEA. Figure 28 shows the variation of the ultimate axial compressive stress plotted against the initial deflection with the shape #1 in Fig.7. Figure 29 shows the ultimate strength interaction relationship between longitudinal axial compression and lateral pressure. In these calculations, the plate edges are simply supported and kept straight.

Ultimate Strength Equation for Combined Transverse Axial Load and Lateral Pressure

In this case, the most probable yield location is found at the transverse mid-edges where the maximum compressive stress in the y direction and the minimum tensile stress in the x direction develop, as shown in Fig.26.b. The resulting yield criterion is then given by (For the stress related symbols, see Fig.25.c)

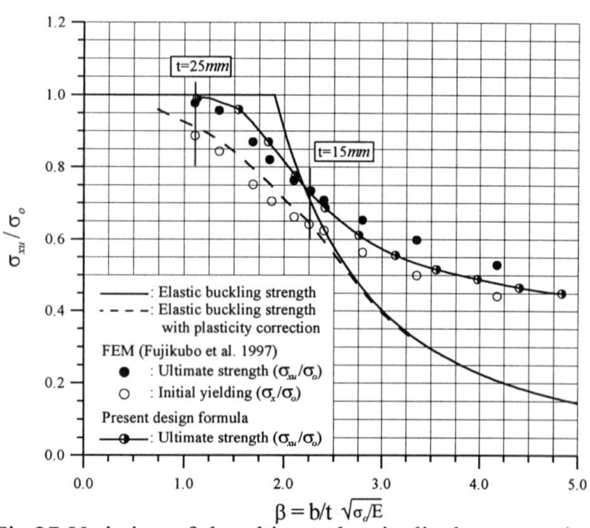

Fig.27 Variation of the ultimate longitudinal compressive strength of a long plating shown as a function of the reduced slenderness ratio, $a/b = 3$

$$\Gamma_u = \left(\frac{\sigma_{x\min}}{\sigma_o}\right)^2 - \left(\frac{\sigma_{x\min}}{\sigma_o}\right)\left(\frac{\sigma_{y\max}}{\sigma_o}\right) + \left(\frac{\sigma_{y\max}}{\sigma_o}\right)^2 - 1 = 0 \tag{30}$$

where $\sigma_{x\min}$ and $\sigma_{y\max}$ are obtained in terms of σ_{yav}, p and post-weld initial imperfections (initial deflection and residual stresses), as defined in Appendix 7.

The ultimate strength based capacity σ_{yu} for transverse axial load is obtained as the solution of equation (30) with regard to σ_{yav}. As an approximation, σ_{yu} is taken as the initial (minimum) value at $\Gamma_u \geq 0$ by increasing σ_{yav} with the increment of $\Delta\sigma_{yav} = -0.01\sigma_o$

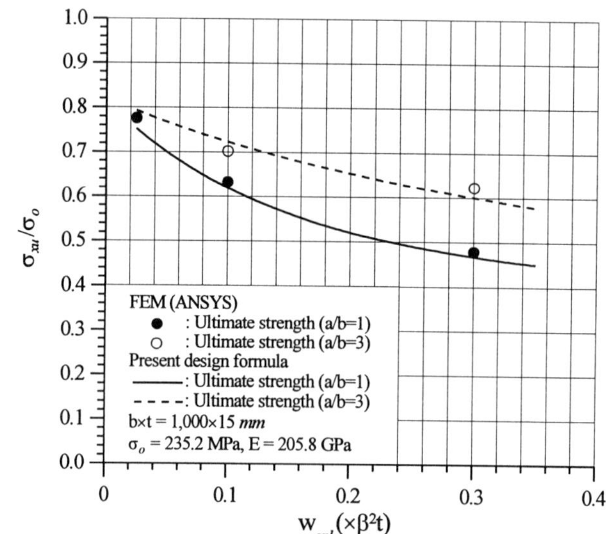

Fig.28 Effect of initial deflection on the plate ultimate compressive strength

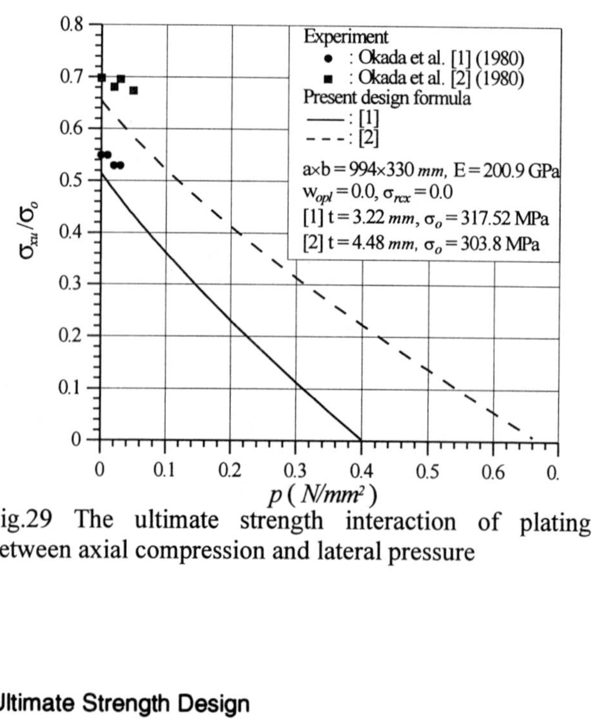

Fig.29 The ultimate strength interaction of plating between axial compression and lateral pressure

for compressive σ_{yav} and $\Delta\sigma_{yav} = 0.01\sigma_o$ for tensile σ_{yav}. Figure 30 shows the variation of the ultimate transverse compressive stress plotted against the plate slenderness ratio, as obtained from equation (30) and from nonlinear FEA. Equation (30) shows mean bias = 0.997 and COV = 0.078 against FEA.

Ultimate Strength Equation for Edge Shear

Based on a series of the nonlinear finite element calculations for simply supported plating varying the plate geometry and the aspect ratio, Paik (1999) derived the following empirical formula for the ultimate edge shear strength of plating, namely

$$\frac{\tau_u}{\tau_o} = \begin{cases} 1.324\left(\frac{\tau_E}{\tau_o}\right) & for\ \ 0 < \frac{\tau_E}{\tau_o} \le 0.5 \\ 0.039\left(\frac{\tau_E}{\tau_o}\right)^3 - 0.274\left(\frac{\tau_E}{\tau_o}\right)^2 + 0.676\left(\frac{\tau_E}{\tau_o}\right) & \\ \quad + 0.388 \ \ for\ \ 0.5 < \frac{\tau_E}{\tau_o} \le 2.0 & \\ 0.956 & for\ \ \frac{\tau_E}{\tau_o} > 2.0 \end{cases} \qquad (31)$$

where τ_E is the elastic shear buckling stress of the plating with $a/b = 1$, which is taken as

$$\tau_E = 9.34\frac{\pi^2 E}{12\left(1-\nu^2\right)}\left(\frac{t}{b}\right)^2.$$

Figure 31 shows the variation of the ultimate shear strength of ship plating against the elastic shear buckling stress. The nonlinear finite element solutions varying the magnitude of post-weld initial deflections are compared. The dotted line represents the elastic shear buckling strengths with plasticity correction made by the Johnson-Ostenfeld formula. It is noted that the influence of lateral pressure on the ultimate shear strength is normally small, and equation (31) can approximately be applied for the plating under combined τ_{av} and p as well. Equation (31) shows mean bias = 0.931 and COV = 0.075 against FEA.

Ultimate Strength Equation for Combined Biaxial Load, Edge Shear and Lateral Pressure

Based on the insights developed (Paik 1999), our studies suggest the following ultimate strength equations of plating under combined biaxial load, edge shear and lateral pressure (comp.:-, tens.:+)

(a) When both σ_{xav} and σ_{yav} are compressive:

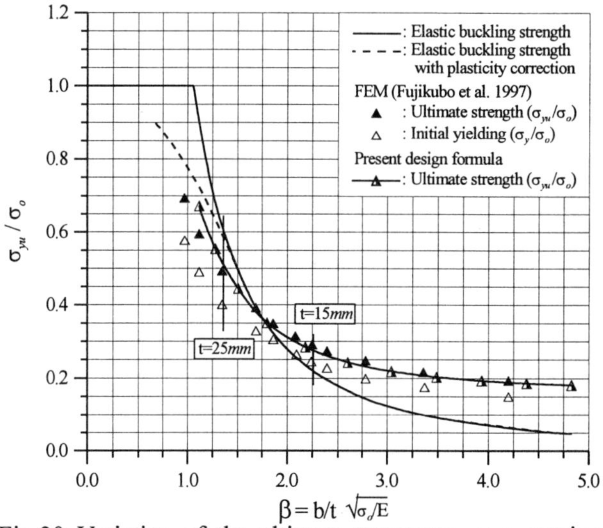

Fig.30 Variation of the ultimate transverse compressive strength of a long plating shown as a function of the reduced slenderness ratio, $a/b = 3$

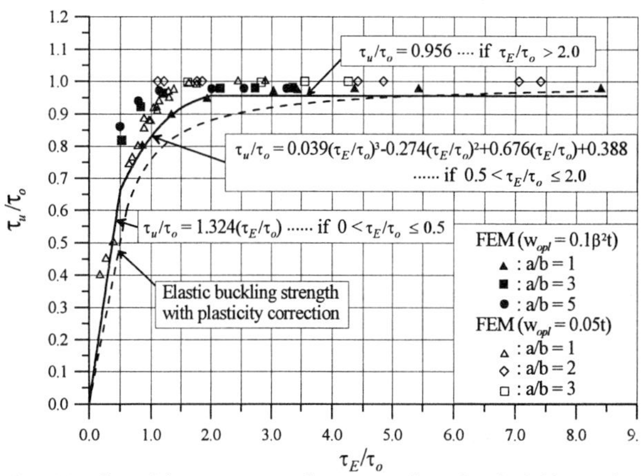

Fig. 31 The ultimate strength versus the elastic bifurcation buckling stress of plating under edge shear

$$\left(\frac{\sigma_{xav}}{\sigma_{xu}}\right)^2 + \left(\frac{\sigma_{yav}}{\sigma_{yu}}\right)^2 + \left(\frac{\tau_{av}}{\tau_u}\right)^2 = 1 \qquad (32.a)$$

(b) When either σ_{xav}, σ_{yav} or both are tensile:

$$\left(\frac{\sigma_{xav}}{\sigma_{xu}}\right)^2 - \left(\frac{\sigma_{xav}}{\sigma_{xu}}\right)\left(\frac{\sigma_{yav}}{\sigma_{yu}}\right) + \left(\frac{\sigma_{yav}}{\sigma_{yu}}\right)^2 + \left(\frac{\tau_{av}}{\tau_u}\right)^2 = 1 \quad (32.b)$$

where σ_{xu}, σ_{yu} = solution of equations (29) and (30), respectively, and τ_u = as defined in equation (31).

By taking σ_{xav} as the reference (non-zero) stress component, for instance, the solution of equation (32) with regard to σ_{xav} is given by

(a) When both σ_{xav} and σ_{yav} are compressive:

$$\sigma_{xu}^{*} = (-1)\frac{\sigma_{xu}\sigma_{yu}\tau_u}{\sqrt{\sigma_{yu}^2\tau_u^2 + C_1^2\sigma_{xu}^2\tau_u^2 + C_2^2\sigma_{xu}^2\sigma_{yu}^2}} \tag{33.a}$$

(b) When either σ_{xav}, σ_{yav} or both are tensile:

$$\sigma_{xu}^{*} = (s)\left|\frac{\sigma_{xu}\sigma_{yu}\tau_u}{\sqrt{\sigma_{yu}^2\tau_u^2 - C_1\sigma_{xu}\sigma_{yu}\tau_u^2 + C_1^2\sigma_{xu}^2\tau_u^2 + C_2^2\sigma_{xu}^2\sigma_{yu}^2}}\right| \tag{33.b}$$

where $C_1 = \dfrac{\sigma_{yav}}{\sigma_{xav}}$, $C_2 = \dfrac{\tau_{av}}{\sigma_{xav}}$, s = as defined in equation (14.b)

For safety evaluation based on the ultimate strength using equation (1), the plate capacity measure σ_{cu} is then given by holding the loading ratio constant, as follows

$$\sigma_{cu} = \sigma_{xu}^{*}\sqrt{1 + C_1^2 + C_2^2} \tag{34}$$

Figures 32 and 33 show the validity of equation (32) for the plating under combined biaxial compression by a comparison with the conventional nonlinear finite element solutions, varying the aspect ratio, the plate thickness and the level of the post-weld initial imperfections. For both FEA and design formula

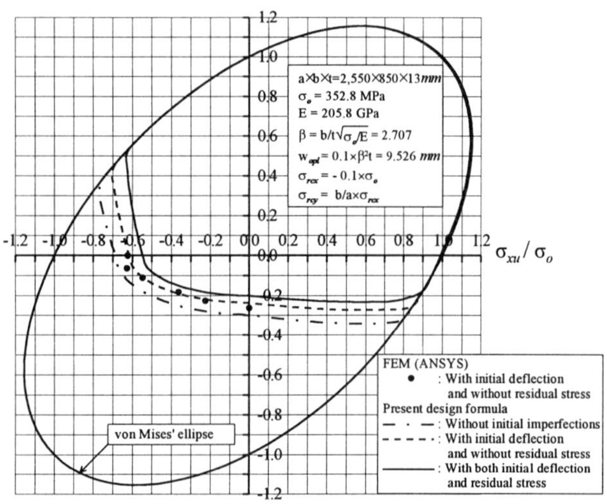

Figure 32.a Plate ultimate strength interaction between biaxial compression, $a/b = 3$, $t = 13mm$, initial deflection shape #1

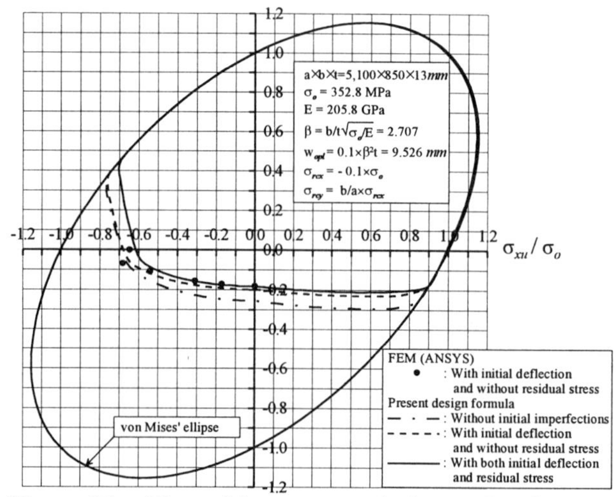

Figure 33.a Plate ultimate strength interaction between biaxial compression, $a/b = 6$, $t = 13mm$, initial deflection shape #1

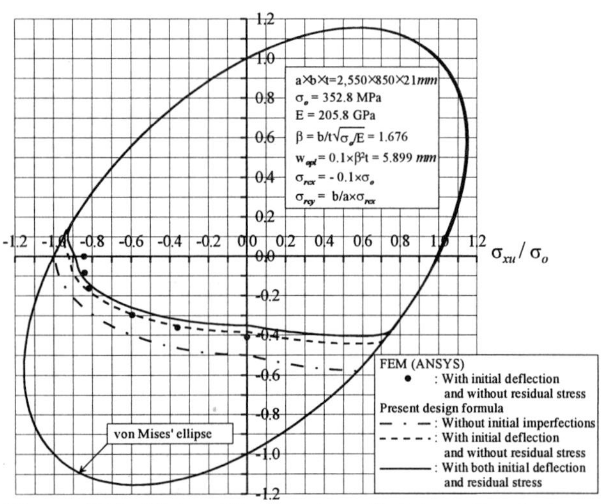

Figure 32.b Plate ultimate strength interaction between biaxial compression, $a/b = 3$, $t = 21mm$, initial deflection shape #1

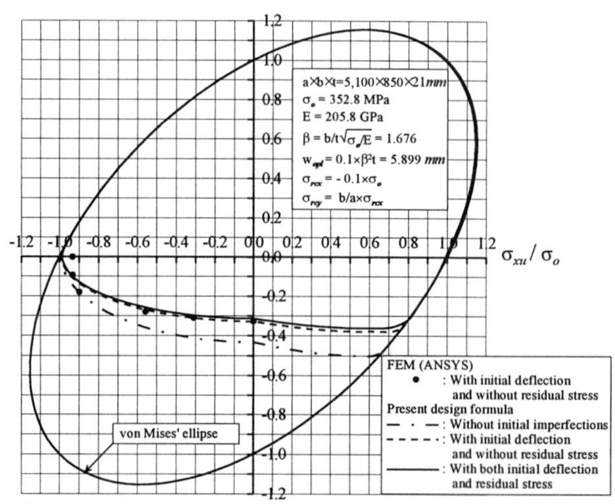

Figure 33.b Plate ultimate strength interaction between biaxial compression, $a/b = 6$, $t = 21mm$, initial deflection shape #1

predictions, the shape #1 of initial deflection as indicated in Fig.7 is presumed. All edges are simply supported and kept straight. Figure 34 shows the plate ultimate strength interaction between axial compression and edge shear, as those obtained by the present design formula and the FEA.

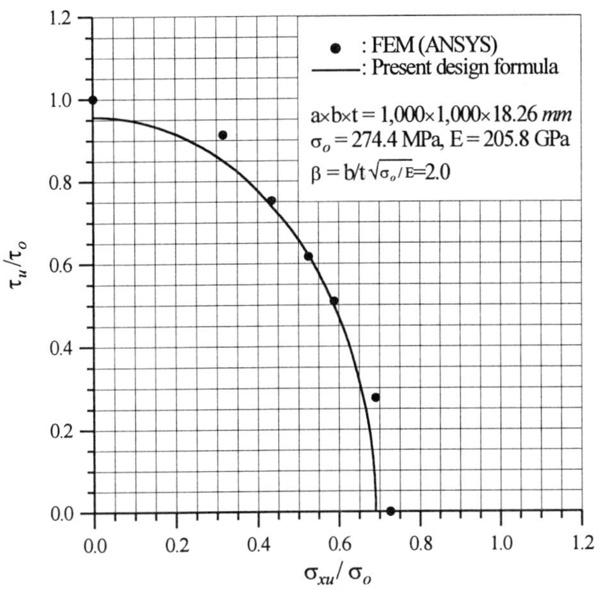

Fig.34.a Plate ultimate strength interaction between axial compression and edge shear, $a/b=1$ and $\beta=2$, $w_{opl}/t=0.1\beta^2$

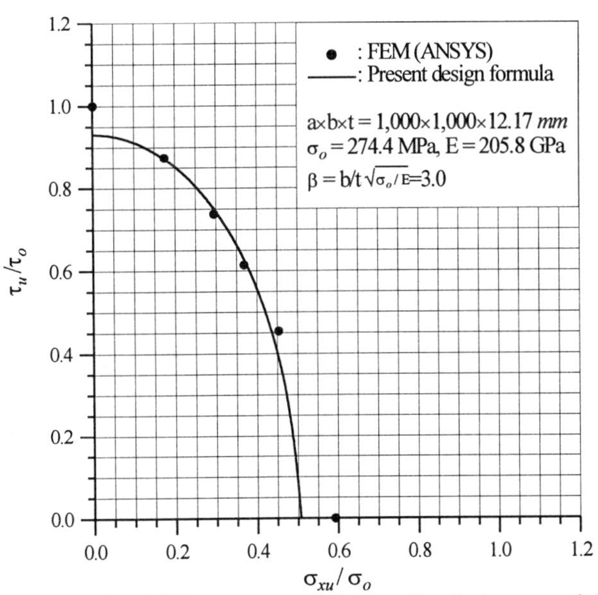

Fig.34.b Plate ultimate strength interaction between axial compression and edge shear, $a/b=1$ and $\beta=3$, $w_{opl}/t=0.1\beta^2$

COMPARISON BETWEEN BUCKLING AND ULTIMATE STRENGTH BASED CAPACITIES

Figures 35 and 36 compare the plate ultimate strength interactions between biaxial compression or tension, as those obtained by the FEA, the buckling or ultimate strength based capacity equations, varying the aspect ratio and the plate thickness. For FEA, average level of initial deflection is considered with the shape #1 as indicated in Fig.7, but no residual stresses are presumed. For the buckling formula predictions, no initial deflection is considered while a slight level of residual stresses is assumed. For the ultimate strength formula predictions, both initial deflection and residual stresses are presumed.

It is seen that for thin plating which buckles in the elastic regime the formula prediction based on the buckling is too pessimistic when compared against the ultimate limit state, while for relatively thick plating that buckles in the inelastic regime it provides good measures for the structural capacity. The capacity formula based on the ultimate strength gives excellent indications for both thin and thick plating.

ADVANCED ULTIMATE STRENGTH DESIGN OF SHIP PLATING UNDER IN-PLANE DYNAMIC LOADS

As previously noted, since ships are subjected to both low and high frequency dynamic loads induced by waves, the response of ship plating is in principle dynamic. With increase of the design speed of vessels, particularly for high speed craft, the speed of loading on ship plating is now tending to get faster than before. It is known that the strength characteristics of structural elements under dynamic loading can be quite different from those under a static or quasi-static loading situation.

According to the review of Jones (1997) who made an extensive survey of developments in the static and dynamic plastic behavior of structures as published over the last 20 years, most previous studies related to the dynamic plastic behavior of structural members are limited to either beam members or plates under lateral pressure loading, and studies relating to the buckling collapse of plate elements under dynamic in-plane loads are few in number. For more advanced design of ship structures, it is thus important to better understand the buckling and collapse strength characteristics of ship plating under dynamic in-plane loads.

In the present study, an experimental investigation of the buckling and collapse strength characteristics for ship plating under dynamic compressive loads is undertaken. For this purpose, a series of the dynamic buckling collapse tests were carried out on mild steel plates subject to axial compressive loads over a range of loading speeds.

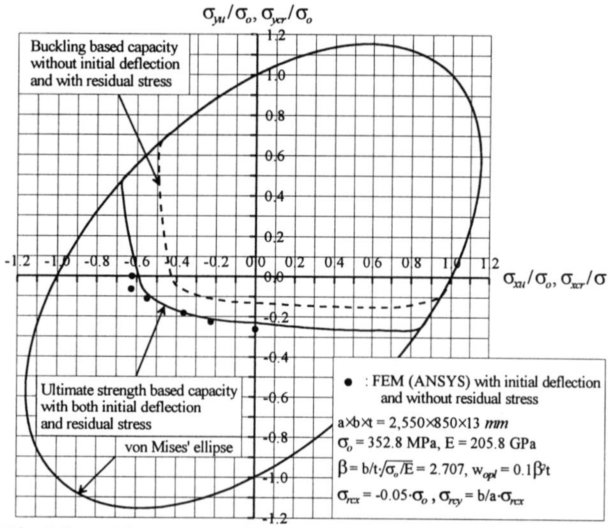

Fig.35.a Plate capacity interactions between biaxial compression as those obtained by FEA, buckling and ultimate strength based capacity formulae, $a/b = 3$, $t = 13mm$, initial deflection shape #1

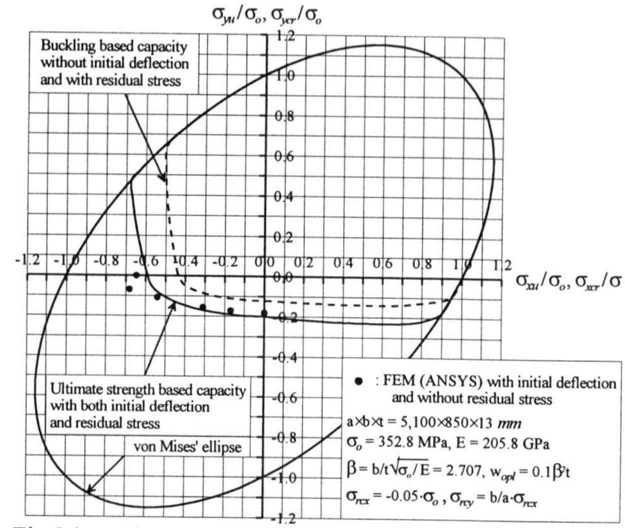

Fig.36.a Plate capacity interactions between biaxial compression as those obtained by FEA, buckling and ultimate strength based capacity formulae, $a/b = 6$, $t = 13mm$, initial deflection shape #1

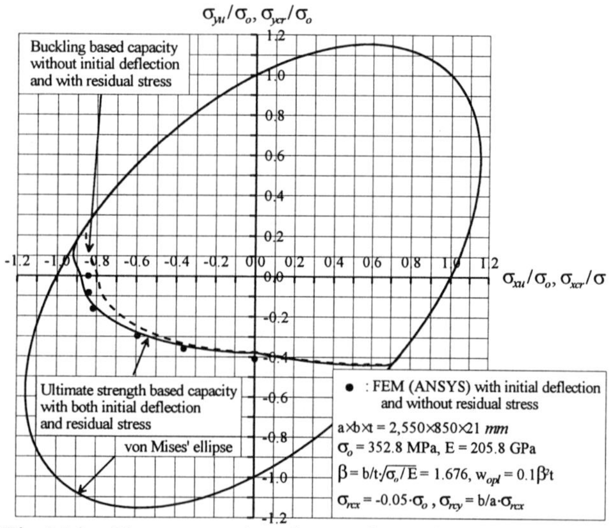

Fig.35.b Plate capacity interactions between biaxial compression as those obtained by FEA, buckling and ultimate strength based capacity formulae, $a/b = 3$, $t = 21mm$, initial deflection shape #1

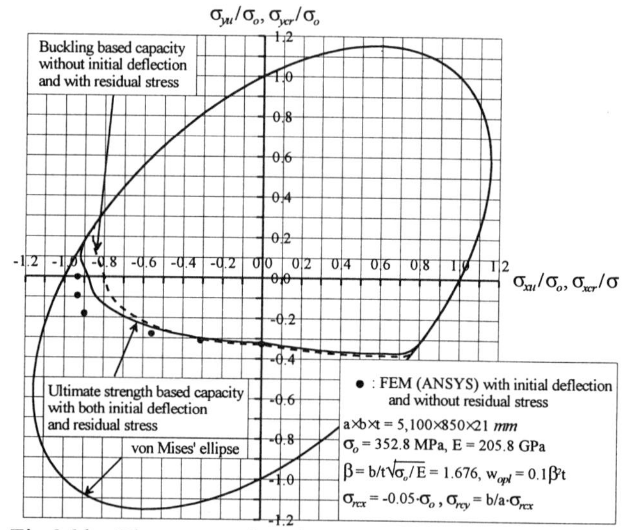

Fig.36.b Plate capacity interactions between biaxial compression as those obtained by FEA, buckling and ultimate strength based capacity formulae, $a/b = 6$, $t = 21mm$, initial deflection shape #1

The loading speed was varied from 0.05 to 400 mm/sec. Based on the test results, the effect of the loading speed on the ultimate compressive strength of mild steel plates is investigated. Also, a simple formula for predicting the buckling collapse strength of steel plates taking into account the strain rate effect is empirically derived by curve fitting the experimental results. A more detailed description of the study may be found in Paik et al. (1999b).

Figure 37 shows a selected set of the test results indicating the applied axial compressive loads versus lateral deflection relations for the plate test specimen. It is seen from Fig.37 that with increase in the speed of axial compressive loading the stiffness and ultimate compressive strength of steel plates both increase. This is because the strain rate for the material involved increases as the speed of loading increases. However, it should be noted that the value of axial displacements at the ultimate limit state is also increased as the speed of loading increases. Further, it appears that the lateral deflections at the ultimate limit state also increase with increase in the loading speed. The unloading pattern in the post-ultimate

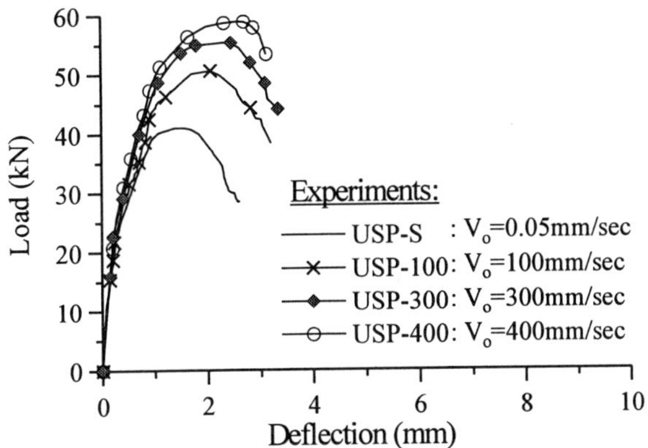

Fig. 37 Dynamic compressive load versus deflection behavior at center of the specimen varying the loading speed (USP stands for un-stiffened plate specimen)

strength regime tends to be more rapid as the speed of loading increases, meaning that the tendency for unstable plate behavior is greater as the speed of axial compressive loading becomes faster.

Based on such test results, relevant useful formulations for assessing the dynamic collapse strength characteristics of plates are empirically derived as a function the strain rate by curve fitting, resulting in

$$\frac{\sigma_{ud}}{\sigma_u} = 1.0 + \left(\frac{\bar{\varepsilon}}{5.41}\right)^{\frac{1}{2.21}} \tag{35}$$

where σ_u, σ_{ud} = ultimate compressive stress of plate elements under a quasi-static or dynamic condition, respectively. $\bar{\varepsilon}$ = strain rate which may be taken as $\bar{\varepsilon} = V_o / a$, where V_o = loading speed in m/sec. and a = plate length.

In a manner similar to the derivation of the buckling collapse strength formula, our studies indicated that the end-shortening and maximum deflection at the ultimate limit state of plating under dynamic axial compressive loads may be approximately predicted by

$$\frac{U_{ud}}{U_u} = 1.0 + \left(\frac{\bar{\varepsilon}}{0.74}\right)^{\frac{1}{3.16}} \tag{36.a}$$

$$\frac{W_{ud}}{W_u} = 1.0 + \left(\frac{\bar{\varepsilon}}{1.21}\right)^{\frac{1}{2.43}} \tag{36.b}$$

where U_u, U_{ud} = end-shortening at the ultimate limit state of plate elements under a quasi-static or dynamic loading condition, respectively, and W_u, W_{ud} = maximum deflection at the ultimate limit state of plate elements under a quasi-static or dynamic loading condition, respectively. The accuracy of equations (35) and (36.a) is seen in Fig. 38 by comparison with the plate test data.

ADVANCED ULTIMATE STRENGTH DESIGN OF SHIP PLATING UNDER IMPACT LATERAL PRESSURE LOADS

Ship structures can be subjected to impact loads induced by waves. The magnitude of such loads is likely to become larger as vessel speeds increase. In vessels carrying liquid cargo, structural members are internally subjected to sloshing induced impact loads resulting from roll or pitch motion of the vessel. The plating of ship's bow and flare parts may be externally subjected to slamming induced impact loads in rough sea state. Such hydrodynamic impact loading can in fact occasionally result in serious structural damage.

According to the ISSC (1991), more than 10% of structural damages for conventional vessel structures are caused by hydrodynamic impact loading, e.g., by slamming. High speed vessel structures fabricated of aluminum alloys, composites or high tensile steels may in theory experience even larger impact loads compared to

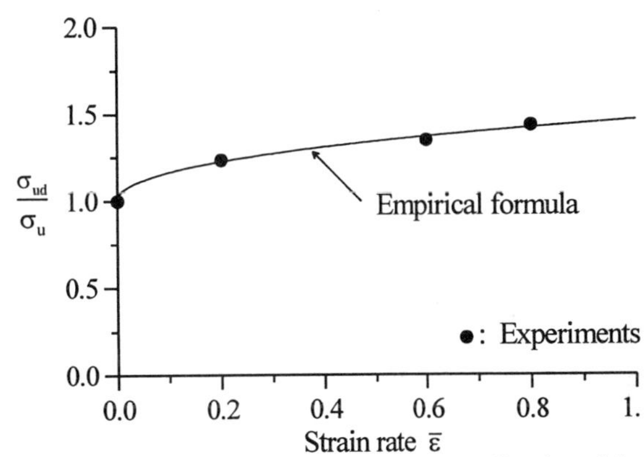

Fig. 38.a Variation of the normalized ultimate compressive stress against the strain rate

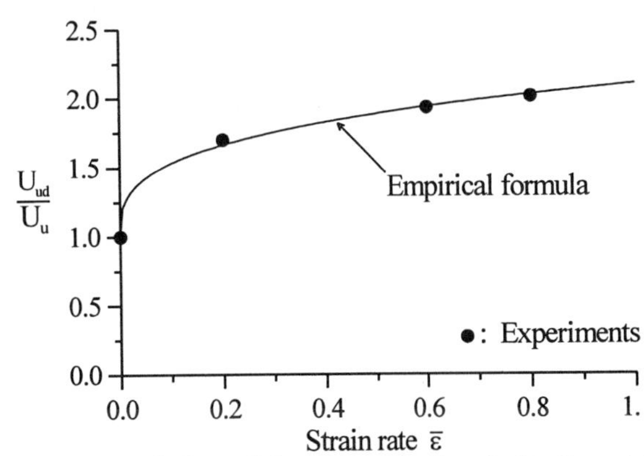

Fig. 38.b Variation of the normalized end-shorting at the ultimate limit state against the strain rate

the conventional vessels, and thus the amount and extent of structural damage they may be subjected to can potentially be more serious.

As previously noted, structural response in an impact loading condition is quite different from that in a static or quasi-static condition. For practical purposes, ship structural design for an impact load is usually carried out based on equivalently defined values of effective quasi-static loads instead of direct application of the impact loads themselves, the equivalence being based on certain characteristics of the likely actual response. Since it lacks a strict theoretical basis, it is possible that such an "equivalent quasi-static load approach" may in some cases lead to either unsafe or too heavy structures. For more rational design of modern ship structures, therefore, it is of importance to better understand the strength and response characteristics of ship plating under dynamic / impact loading.

The literature review made by Jones (1997) includes a number of useful studies related to the strength of structural members under impact loading. Selected literature related to the strength of plating under impact pressure loading is now reviewed. Chuang (1966) experimentally investigated the characteristics of the pressure distribution on ship plating subject to slamming induced impact loads. Jones (1973) derived the maximum permanent deflection equations for ship bottom plating under slamming. Mori (1976) investigated the collapse strength characteristics of aluminum alloy ship bottom plating under slamming induced impact pressures, both theoretically and experimentally. Recently, Caridis & Stefanou (1997) performed a nonlinear elastic / visco-plastic numerical simulation to investigate the strength characteristics of plating subject to wave impact loads.

The previously mentioned studies provide useful information. However, there are still a number of problem areas where the state of the art can be potentially improved. In particular, to more reasonably and precisely accommodate the impact load effects in the preliminary design stage of ship structures, we need to have better understanding of (a) hydrodynamic wave impact loading characteristics and (b) collapse strength characteristics of ship plating under impact pressure, among others.

The present authors (Paik et al. 1999c) aim to provide some new contributions to the design technology for ship structures considering the dynamic / impact load effects. For this purpose, the idealization of impact pressure distribution at the ship bottom plating subject to slamming induced impact loading is studied. Using a nonlinear finite element program that applies the dynamic approach, a series of collapse strength analyses for ship plating under lateral pressure loading are carried out varying the plate dimensions, aspect ratio and loading speed. Based on the computed results, the collapse strength characteristics of ship bottom plating subject to impact pressure loading are investigated. A simplified theoretical

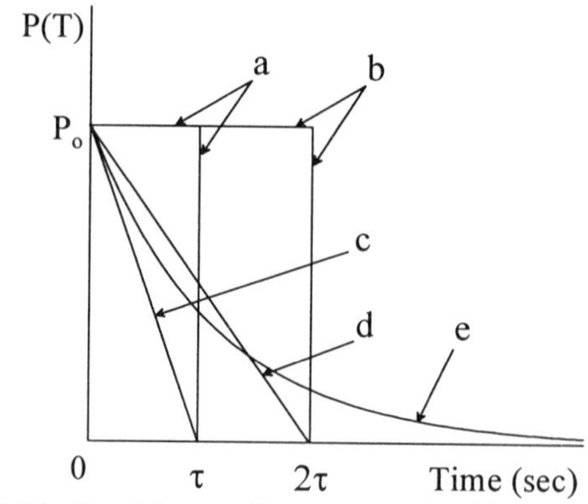

Fig. 39 Idealized types of pressure pulse history

formula for predicting the collapse strength of ship plating under impact pressure loads is then developed by including the strain rate sensitivity within a strength formula that is derived based on the rigid-plastic theory for plating under a quasi-static pressure loading.

Idealization of Pressure Distribution on Ship Plating under Hydrodynamic Wave Impact Loads

To analyze the impact load related response characteristics (i.e., magnitude and pattern of the response as a function of time) for structural members, it is important to first define the characteristics of the impact load versus time history. The characteristics of dynamic / impact loading for ship structural members can be quite complex. The developments presented in this study are, for reasons of convenience and efficiency, based on certain simplifications of typical load time histories. The nature of these simplifications and their relationship to the response under the actual (non-idealized) load versus time histories are studied. According to the experiments undertaken by Chan & Melville (1987), the pressure rise time is one order of magnitude smaller than the decay period and it is more relevant to idealize that the pressure increases instantaneously to the peak value (i.e., with zero rise time) and then decays following the exponential law.

Figure 39 indicates possible idealizations of the pressure impulse history on ship plating, where the curve "e" is the most realistic. As shown in the figure, in the limit, two types of idealizations, one being of a rectangular pattern and the other being of a triangular pattern, can be possible. Holding the amount of impulse the same, the former type (i.e., rectangular pattern) is depicted by curve "a" and the latter (i.e., triangular pattern) is given by curve "d". For comparison purposes, curves denoted "b" and "c", the former with two times impulse of curve "a" and the latter with a half impulse of curve "c", are also shown in the figure.

Figure 40 shows the variations of lateral deflection with time for an example rectangular steel plate subject to impact lateral pressure loads varying the pressure impulse histories as indicated above. The results of Fig.40 in the

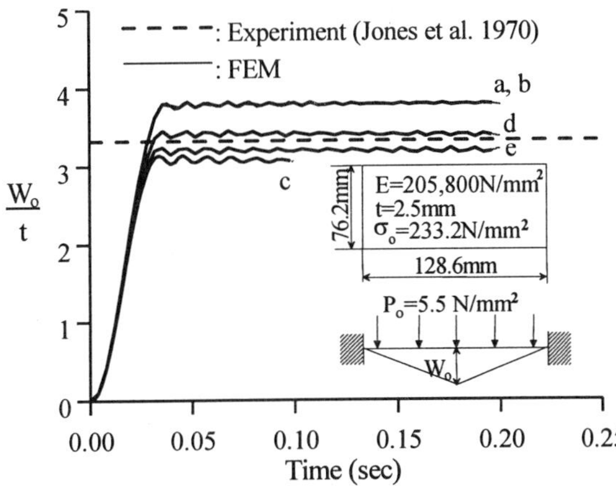

Fig. 40 Deflection-time history at center of the all edged clamped plate subject to various types of pressure pulse

present study are obtained by using a nonlinear finiteelement (FE) program based on a dynamic formulation, STARDYNE (1996). For these analyses, the boundary conditions of the plating are modeled as clamped at all (four) edges. Figure 40 also shows relevant experimental data for the maximum permanent deflection for the same case, as obtained by Jones et al. (1970) using a pressure versus time history of the more realistic "e" type.

It is of interest to note that when using the rectangular type of pressure impulse, as in either curve "a" or "b" of Fig.39, the responses obtained are almost same in either case even if the duration of impact impulse is different (although the maximum permanent deflection is overestimated compared to the experimental results). As evident from Fig.40, curves "a" and "b" have the same initial maximum impact pressure (and the same impulse history at the early impact stage). That is, the rectangular type of impulse may in such cases give almost same maximum structural response regardless of duration of the pressure impulse (or the impulse amount) as long as the initial maximum impact pressure is the same.

Unlike the rectangular type of pressure impulse, it is seen from Fig.40 that the triangular type may produce different results if the duration of impact pulse is different (although the initial maximum impact pressure is the same). In some cases (e.g., if the duration of impact pulse is assumed to be small), the calculation using the triangular type of impulse history will underestimate the permanent deflection of plating as well. The differences in the permanent plate deflection between curves "c" and "d" using the triangular type of impact pulse history is due to the different duration of impulse even though the initial maximum impact pressure is the same in either case.

An advantage of using the rectangular type of impulse history for simplified analysis or design is that the related calculations result in similar impact responses with only the initial maximum pressure known, i.e., regardless of the duration of pulse (or impulse amount). For simplicity, therefore, the present theoretical calculation idealizes the impulse history for ship plating under impact pressure loads as a rectangular type with the same initial maximum impulse as the actual and more complicated impact load history. Of course, if one knows the values of both duration of pulse (or impulse amount) and initial maximum pressure, an idealization by way of the triangular type of impulse history would be desirable in the sense of accuracy. For illustrative examples (where both duration of pulse and initial maximum pressure are known), therefore, the various finite element numerical analyses of this study are carried out using a triangular type of impact pressure pulse history which is assumed to represent the same amount of impulse as in the actual more complicated impact load history which in reality may take an exponential form in its latter phase. The numerical finite element results are then compared with the simplified theoretical solutions, i.e., using the rectangular type of pressure pulse with only the maximum impact pressure known.

It is known that duration of impact pulse for ship bottom plating subject to slamming induced impact loads is in the range of 0.025~0.25 seconds (Ochi 1967, Wheaton et al. 1970). Hence the duration of an impact pressure pulse may be taken as 0.1 second for the interests of illustrative examples. The study methodology can of course be applied to any other duration value equally well.

Prediction of the Initial Maximum Impact Pressure

It is necessary to predict the initial maximum impact pressure to analyze the behavior of ship plating subject to pressure impact. When ship bottom plating strikes water surface, e.g., in slamming event of vessels, air cushioning between bottom plating and water surface makes the time history of impact pressure pulse more complex. Since it is not an easy task to theoretically study the influence of air cushioning on the behavior of plating under hydrodynamic wave impact loads, most previous studies related to this problem were undertaken experimentally. According to such previous studies (for instance, Chuang 1966, Lewison & Maclean 1968, Verhagen 1967), it is known that the magnitude of the initial maximum impact pressure felt by the structure becomes smaller if there exists more of an air cushioning effect, because air particles between plating and water surface which are more compressible than water absorb the impact energy to a greater extent.

Based on his own experimental results, Chuang (1966) suggested an empirical formula for predicting the initial maximum impact pressure of plating which strikes the water surface at a right angle with the initial impact speed of V_0, taking into account the influence of air cushioning, as follows:

$$P_o = \frac{1}{32}\left[\frac{1.4^2 + \pi^2}{e^{-1.4} + 1.0}\right]\rho_f C_a V_o \quad (37)$$

where P_o = peak impact pressure on the plating, ρ_f = density of sea water (= 1,025 kg/m^3), C_a = velocity of sound in the air (=342.9 $m/\mathbf{sec}$), and V_o = initial impact velocity between plate and water surface.

For simplicity, the present study uses equation (37) to calculate the magnitude of the initial maximum impact pressure for the various theoretical and numerical analysis of ship bottom plating under slamming induced impact loads. Actual values from realistic cases may also be readily used for the same purposes where available.

Ultimate Strength Design Formulation of Ship Plating under Impact Pressure Loads

Based on rigid-plastic theory, a simplified formula for predicting the collapse strength of ship plating under impact pressure loads is now theoretically derived. For convenience, the process of the formula derivation is split into two steps, namely the quasi-static loading step and the dynamic / impact loading step. The accuracy of the formula so derived is then verified by comparing with the corresponding nonlinear FE solutions. The boundary conditions for the plate are assumed to be clamped at all (four) edges. The plastic collapse mode of the plate under equilibrium condition can be assumed as shown in Fig.41. In the figure, the dotted lines indicate the plastic hinge lines which are in a gross yielding condition. Except for the plastic hinge lines, each plate region is assumed to behave as a rigid body. The collapse load p_c of the plating under static pressure is given as follows

$$p_c = \frac{48M_p}{b^2\left(\sqrt{(b/a)^2 + 3} - b/a\right)^2} \quad (38)$$

where $M_p = \dfrac{\sigma_o t^2}{4}$.

The relationship between the applied static pressure p and the maximum (permanent) deflection W_o for the plating can be shown as follows

$$\frac{p}{p_c} = 1 + \frac{1}{3}\left(\frac{W_o}{t}\right)^2 \frac{\xi_o + (3 - 2\xi_o)^2}{3 - \xi_o} \quad \text{if} \quad \frac{W_o}{t} \le 1 \quad (39.a)$$

$$\frac{p}{p_c} = \frac{2W_o}{t}\left\{1 + \frac{\xi_o(\xi_o - 2)\left(1 - \dfrac{t^2}{3W_o^2}\right)}{3 - \xi_o}\right\} \quad \text{if} \quad \frac{W_o}{t} > 1 \quad (39.b)$$

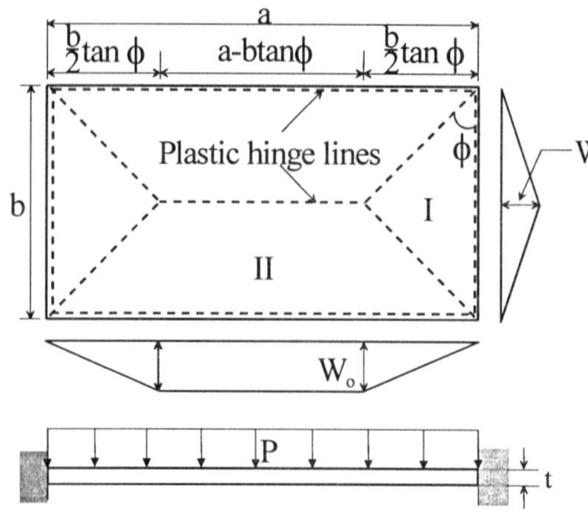

Fig.41 Plastic hinge mechanism for the all edges clamped plate subject to uniformly distributed static pressure loads

where p_c = as defined in equation (38), $\xi_o = \dfrac{b}{a}\left(\sqrt{\left(\dfrac{b}{a}\right)^2 + 3} - \dfrac{b}{a}\right)^2$.

The structural response in the dynamic / impact condition is in principle different from that in the quasi-static condition. Three related aspects of a dynamic / impact loading situation are possibly relevant, namely material strain rate sensitivity, inertia effects and dynamic frictional effects. For steels, with increase in the strain rate, the yield strength of the material can increase and rupture strain can decrease. Due to inertia effects, deformation patterns may be varied as well. It is also known that as the speed of dynamic loading increases, the coefficient of friction becomes lower.

However, in most practical marine cases of interest where dynamic / impact loads are applied, the structural response is mainly affected by the material strain rate sensitivity (Jones 1989, Paik & Wierzbicki 1997). Therefore, the present study approximately develops the dynamic collapse strength formula by including the strain rate effect alone into the static collapse strength formula. The validity of this approximation is to some extent verified by comparing with appropriate numerical and experimental results.

As previously noted, the pressure versus time history during hydrodynamic wave impact loading can be simplified using the rectangular type of pressure pulse. If the duration of impact pulse is relatively short, the relationship between initial velocity V_o and maximum pressure pulse P_o can then be given by

$$\mu V_o = P_o T \quad (40)$$

where T = duration of pressure pulse, μ = mass per unit area of the impacted plate, P_o = peak impact pressure on the plate, and V_o = initial impact velocity between plate and water surface.

Perrone & Bhadra (1984) formulated the strain rate $\bar{\varepsilon}$ of plate elements under impact pressure loading in terms of initial impact velocity V_o and maximum permanent deflection W_o as follows

$$\bar{\varepsilon} \approx \left(\frac{2W_o}{3}\right)\left(\frac{V_o}{\sqrt{2}}\right)\left(\frac{2}{b^2}\right) \tag{41}$$

Using the initial impact velocity V_o calculated from equation (40), the strain rate $\bar{\varepsilon}$ can then be expressed by equation (41) in terms of the plate dimensions and the maximum deflection. To estimate the dynamic yield strength of the material, σ_{od} from the static yield stress σ_o, with $\bar{\varepsilon}$ known, the Cowper & Symonds equation has been widely used, namely

$$\frac{\sigma_{od}}{\sigma_o} = 1.0 + \left(\frac{\bar{\varepsilon}}{H}\right)^{1/q} \tag{42}$$

where H and q are coefficients to be determined based on test data. For example, the parameters for the initial yield of mild steel under dynamic loading are $H = 40.4\,\text{sec}^{-1}$ and $q = 5$ (Cowper & Symonds 1957). Recently, Paik et al. (1999d) found that equation (42) is also applicable to high tensile steel material, but with coefficients that are different from those of mild steel. Based on existing test data for high tensile steel materials, they determined a sample set of the Cowper-Symonds coefficients as $H = 3200\,\text{sec}^{-1}$ and $q = 5$. For aluminum alloys, Jones (1974) suggests that the relevant coefficients can be $H = 1288000\,sec^{-1}$ and $q = 4$.

Substituting equation (41) into equation (42), the dynamic yield stress can be given by

$$\frac{\sigma_{od}}{\sigma_o} = 1.0 + \left(\frac{V_o W_o}{12\sqrt{2}\cdot Hb^2}\right)^{1/q} \tag{43}$$

Therefore, it is proposed that the collapse behavior of plating under impact pressure loading can be assessed by the formula developed for a quasi-static condition, i.e., equation (39), but with the use of the dynamic yield stress σ_{od} defined in equation (43) in place of its static counterpart σ_o.

Figure 42 shows a selected set of the illustrative examples representing the variation of the maximum permanent deflection for the all edges clamped steel plate under impact pressure as a function of the loading speed, as obtained by the above developed theoretical formula, and also FE calculations and experiments. The theoretical solutions are compared with the nonlinear FE results (with 5% of the strain hardening) and corresponding test data as obtained by Jones (1970). It is seen from Fig.42 that both theoretical and numerical solutions agree well with the experimental data.

Figure 43 shows the influence of the plate aspect ratio and slenderness ratio of the maximum lateral deflection. With a constant speed of loading, the influence of the plate aspect ratio can be ignored for large aspect ratios, e.g., here when $a/b > 2$. With increase in the plate slenderness (or with decrease in the plate thickness)

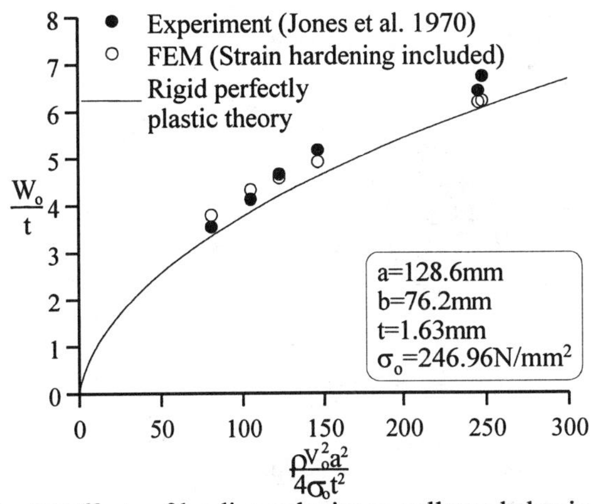

Fig. 42 Effects of loading velocity on collapse behavior of the all edges clamped rectangular plate subject to impact uniform pressure load

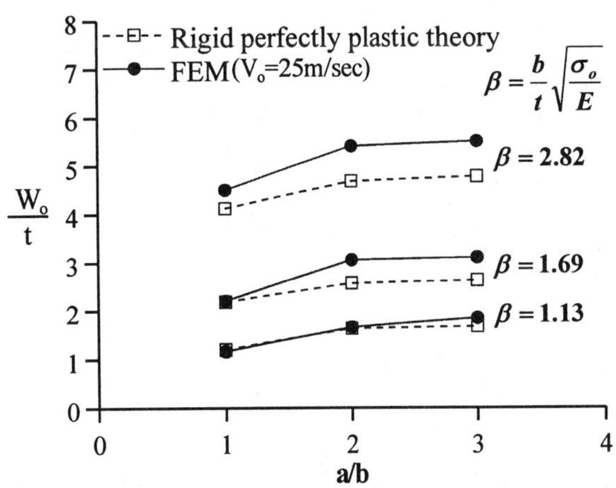

Fig.43 Effects of aspect and slenderness ratio on the maximum deflection of clamped plating under impact lateral pressure loads, as those obtained by FEA and the present design formula

the maximum deflection is however seen to increase significantly.

Based on the insights developed above, it may be concluded that the collapse behavior of steel plating under impact pressure loading is mostly affected by the loading speed and the plate slenderness ratio, and to much less extent by the plate aspect ratio. In addition, some important insights developed in the present study are to be noted, as follows: The strain hardening effect of material may vary with the magnitude of plate deflection and it can not be neglected in the regime of large deflection that produces membrane effects. Further investigations are needed in this regard. Also, it is evident that as the loading speed increases the plate maximum deflection increases remarkably. It is also evident that the theoretical formula developed in the present study is useful for assessment of collapse strength characteristics for ship plating under impact pressure loading.

CONCLUDING REMARKS

The behavior of ship plating normally depends on a variety of influential factors, namely geometric / material properties, loading characteristics, initial imperfections, boundary conditions and also local deterioration related to corrosion, fatigue cracking and dents. To achieve the advanced buckling and ultimate strength design of ship plating, we would need more sophisticated methods than existing simplified approaches. The aim of the present study has been to develop more advanced buckling and ultimate strength design technology for ship plating. The present paper focuses on the following five areas which were studied theoretically, numerically and experimentally as appropriate:

- Mathematical modeling for fabrication related imperfections (i.e., initial deflections and residual stresses),
- Characteristics of the plate buckling with elastically restrained edge conditions,
- Strength equations for ship plating under combined static loads including biaxial compression / tension, edge shear and lateral pressure,
- Characteristics of the plate capacity under in-plane dynamic loads, and
- Characteristics of the plate capacity under slamming induced lateral impact pressure loads.

Some collected results and conclusions developed in the present study are as follows:

(1) During fabrication of ship structures, the initial imperfections (initial deflection and residual stresses) inevitably develop and can significantly affect the structural capacity. The characteristics of the fabrication related imperfections are uncertain, and an idealized model was proposed in the present study.

(2) The buckling strength of plating can be affected significantly by the torsional rigidity of support members as well as the plate dimensions. The proposed buckling based capacity formula accommodates the torsional rigidity of support members as a parameter of influence.

(3) Ship plating is generally subjected to combined in-plane and lateral pressure loads and the plate capacity should be evaluated taking into account the effect of combined loads. Two types of the capacity formulations, namely one based on the buckling and the other based on the ultimate strength, were proposed by accommodating the combined loads (i.e., biaxial compression / tension, edge shear and lateral pressure) and fabrication related imperfections. For thin plating which buckles in the elastic regime, the capacity formula based on the buckling is too pessimistic against the ultimate limit state, while it provides a good measure of the structural capacity for relatively thick plating which buckles in the inelastic regime. The capacity formula based on the ultimate strength provides a good indication for both thin and thick plating against collapse.

(4) The in-plane stiffness and ultimate compressive strength of steel plating both tend to increase with increase in the speed of loading. This is due to the fact that the strain rate of the material involved increases as the loading speed increases. However, it is also noted from the tests that the values of axial displacements and lateral deflections until the plate reaches the ultimate limit state increases rapidly as the loading speed increases. Also, the unloading pattern in the post-ultimate strength regime is more rapid with increase in the loading speed. This would mean that as the speed of loading increases the response of steel plates will become more unstable after the ultimate strength is reached. For structural design of ship plating, while it is recommended that the increase of the ultimate strength due to the dynamic loading effect should not be overestimated, it is also recommended that the potential for dynamic instability of steel plates in the post-ultimate strength regime should not be underestimated. The related phenomena depend of course on the rates of loading involved, as the data presented herein clearly indicate.

(5) The collapse behavior of steel plating under impact pressure loading is in principle affected significantly by the impact pressure versus time history. Starting some time after the structure is initially impacted, the impact pulse usually decreases in a way of the exponential function with time. When only the initial maximum impact pressure is known, the rectangular type of idealization of an impact pulse history would be useful for practical purposes. The maximum permanent deflection for ship plating under impact pressure loading increases with increase in the loading speed. For a constant loading speed, the maximum deflection for a rectangular plate is larger than that of a square plate, but the effect of the

plate aspect ratio appears to become smaller as aspect ratios get to be greater than 2. Also, the maximum deflection of ship plating increases significantly as the plate slenderness ratio increases, as expected. To prevent ship plating under impact pressure loading from dynamic buckling collapse, one consideration of importance is of course to limit the slenderness ratio under a critical value.

ACKNOWLEDGEMENTS

The present research study was undertaken with support from the American Bureau of Shipping, the Korea Research Foundation (for Faculty Research abroad), the Brain Korea 21 project and the Pusan National University who are thanked for this support.

REFERENCES

Aalami, B. and Chapman, J.C. (1972). Large deflection behavior of ship plate panels under normal pressure and in-plane loading, Trans. RINA, Vol.114, March.

Aalami, B., Moukhtarade, A. and Mahmudi-Saati, P. (1972). On the strength design of ship plates subjected to in-plane and transverse loads, Trans. RINA, Vol.114, November.

Antoniou, A.C. (1980). On the maximum deflection of plating in newly built ships, J. of Ship Research, Vol.24, No.1, pp.31-39.

Bleich, F. (1952). Buckling strength of metal structures, McGraw-Hill Book Co., New York.

Caridis, P.A. and Stefanou, M. (1997). Dynamic elastic / visco-plastic response of hull plating subjected to hydrodynamic wave impact, J. of Ship Research, Vol.41, No.2, pp.130~146.

Carlsen, C.A. and Czujko, J. (1978). The specification of post-welding distortion tolerance for stiffened plates in compression, The Structural Engineer, Vol.56A, No.5, pp.133~141.

Chuang, S.L. (1966). Experiments on flat-bottom slamming, J. of Ship Research, Vol.10, No.1, pp.10~17.

Cowper, G.R. and Symonds, P.S. (1957). Strain hardening and strain rate effects in the impact loading of cantilever beams, Brown University Division of Applied Mathematics, Report No.28, September.

Davidson, P.C., Chapman, J.C., Smith, C.S. and Dowling, P.J. (1989). The design of plate panels subjected to in-plane shear and biaxial compression, Trans. RINA, Vol.132, pp.267-286.

Davidson, P.C., Chapman, J.C., Smith, C.S. and Dowling, P.J. (1991). The design of plate panels subjected to biaxial compression and lateral pressure, Trans. RINA, Vol.134, pp.149-160.

Dier, A.F. and Dowling, P.J. (1983). The strength of plates subjected to biaxial forces, Proc. of Conference on Behavior of Thin-Walled Structures, University of Strathclyde, Glasgow, pp.329-353.

Dowling, P.J. and Dier, A.F. (1978). Strength of ship's plating under combined lateral loading and biaxial pressure, CESLIC Report SP6, Imperial College, London, September.

Evans, J.H. (1960). Strength of wide plates under uniform edge compression, Trans. SNAME, Vol.68, pp.585-621.

Faulkner, D. (1975). A review of effective plating for use in the analysis of stiffened plating in bending and compression, J. of Ship Research, Vol.19, No.1, March, pp.1-17.

Fujikubo, M., Yao, T. and Varghese, B. (1997). Buckling and ultimate strength of plates subjected to combined loads, Proc. of the 7th International Offshore and Polar Engineering Conference, Vol. IV, Honolulu, pp.380-387.

Fujita, Y., Nomoto, T. and Niho, O. (1979). Ultimate strength of rectangular plates subjected to combined loading (1st report) –Square plates under compression and shear, J. of the Society of Naval Architects of Japan, Vol.145, pp.194-202 (in Japanese).

Gerard, G. and Becker, H. (1954). Handbook of structural stability, Part I. Buckling of flat plates, NACA Technical Note, No.3781.

ISSC (1991). Dynamic load effects, Report of Committee II.2, The 11th International Ship & Offshore Structures Congress, Wuxi, China, pp.239~318.

Jones, N. (1973). Slamming damage, J. of Ship Research, Vol.17, No.2, pp.80-86.

Jones, N. (1974). Some remarks on the strain rate sensitive behavior of shells, Problems of Plasticity, Vol.2, pp.403~407.

Jones, N. (1989). Structural Impact, Cambridge University Press.

Jones, N. (1997). Dynamic plastic behavior of ship and ocean structures, Trans. RINA, Vol.139, Part A, pp.65~97.

Jones, N., Uran, T.O. and Tekin, S.A. (1970). The dynamic plastic behavior of fully clamped rectangular plates, Int. J. of Solids and Structures, Vol.6, pp.1499~1512.

Kmiecik, M., Jastrzebski, T. and Kuzniar, J. (1995). Statistics of ship plating distortions, Marine Structures, Vol.8, pp.119-132.

Lewison, G. and Maclean, W.M. (1968). On the cushioning of water impact by entrapped air, J. of Ship Research, Vol.12, No.2, pp.116~130.

Lundquist, E. and Stowell, E.Z. (1942). Critical compressive stress for flat rectangular plates – Elastically restrained, NACA Technical Note, No.733.

Mansour, A.E. (1976). Charts for the buckling and post-buckling analyses of stiffened plates under combined loading, Technical and Research Bulletin, No.2-22, SNAME, July.

Mazzolani, F.M., Landolfo, R. and De Matteis, G. (1998). Influence of welding on the stability of aluminum thin plates, Stability and Ductility of Steel Structures,

Elsevier Science, pp.225-232.

Mateus, A.F. and Witz, J.A. (1997). Post-buckling of corroded steel plates: an assessment of the design codes, Proc. of the 8th International Conference on the Behavior of Offshore Structures (BOSS'97), Elsevier Science, Delft, July.

Mateus, A.F. and Witz, J.A. (1998). On the post-buckling of corroded steel plates used in marine structures, Trans. RINA, Vol.140, Part C, pp.165-183.

McKenzie, K.I. (1963). The buckling of a rectangular plate under combined biaxial compression, bending and shear, The Aeronautical Quarterly, August.

Mori, K. (1976). Response of the bottom plate of high speed crafts under implusive water pressure, J. of the Society of Naval Architects of Japan, Vol.142, pp.297~305.

Ochi, M.K. (1967). Ship slamming hydrodynamic impact between waves and ship bottom forward, Fluid-Solid Interaction, ASME, Edited by J.E. Greenspon, pp.223~240.

Ohtsubo, H. and Yoshida, J. (1985). Ultimate strength of rectangular plates under combination of loads (part 2) – Interaction of compressive and shear stresses, J. of the Society of Naval Architects of Japan, Vol.158, pp.368-375 (in Japanese).

Okada, H., Oshima, K. and Fukumoto, Y. (1979). Compressive strength of long rectangular plates under hydrostatic pressure, J. of the Society of Naval Architects of Japan, Vol.146, pp.270-280 (in Japanese).

Paik, J.K. and Kim, C.Y. (1988). Simple formulae for buckling and ultimate strength estimation of plates subjected to water pressure and uniaxial compression, J. of the Society of Naval Architects of Korea, Vol.25, No.4, pp.69-80 (in Korean).

Paik, J.K., Ham, J.H. and Kim, U.N. (1992a). A new plate buckling design formula, J. of the Society of Naval Architects of Japan, Vol.171, pp.559-566.

Paik, J.K., Ham, J.H. and Ko, J.H. (1992b). A new plate buckling design formula (2nd report) –On the plasticity correction, J. of the Society of Naval Architects of Japan, Vol.172, pp.417-425.

Paik, J.K., Kim, J.Y., Kim, W.S. and Lee, B.H. (1993). Development of an efficient and accurate buckling design system for ship structures (II), Final Report to the Hyundai Heavy Industries, Ulsan, by the Department of Naval Architecture and Ocean Engineering, Pusan National University, Pusan, August.

Paik, J.K. (1995). A new concept of the effective shear modulus for a plate buckled in shear, J. of Ship Research, Vol.39, No.1, pp.70-75.

Paik, J.K. and Pedersen, P.T. (1995). Ultimate and crushing strength of plated structures, J. of Ship Research, Vol.39, No.3, Sept., pp.250-261.

Paik, J.K. and Pedersen, P.T. (1996). A simplified method for predicting the ultimate compressive strength of ship panels, International Shipbuilding Progress, Vol.43, No.434, pp.139-157.

Paik, J.K. and Wierzbicki, T. (1997). A benchmark study on crushing and cutting of plated structures, J. of Ship Research, Vol.41, No.2, pp.147-160.

Paik, J.K., Kim, S.K. and Lee, S.K. (1998a). A probabilistic corrosion rate estimation model for longitudinal strength members of bulk carriers, Ocean Engineering, Vol.25, No.10, pp.837-860.

Paik, J.K., Thayamballi, A.K., Kim, S.K. and Yang, S.H. (1998b). Ultimate strength reliability of corroded ship hulls, Trans. RINA, Vol.140, pp.1-18.

Paik, J.K., Thayamballi, A.K., Kim, S.K. and Yang, S.H. (1998c). Ship hull ultimate strength reliability considering corrosion, J. of Ship Research, Vol.42, No.2, June, pp.154-165.

Paik, J.K. (1999). Buckling and ultimate strength design equations of plates and stiffened panels under combined loads, Final Report to ABS New York.

Paik, J.K., Thayamballi, A.K. and Kim, B.J. (1999a). Advanced ultimate strength design equations for ship plating subject to combined biaxial compression / tension, edge shear and lateral pressure loads, to be published.

Paik, J.K., Thayamballi, A.K. and Chung, J.Y. (1999b). Experimental investigation of the dynamic collapse strength characteristics for ship plating under axial compressive loads, to be published.

Paik, J.K., Chung, J.Y. and Paik, Y.M. (1999c). On collapse strength characteristics of ship bottom plating subject to slamming induced impact pressure loads, J. of the Society of Naval Architects of Korea, Vol.36, No.2, pp.77~93 (in Korean).

Paik, J.K., Chung, J.Y. and Paik, Y.M. (1999d). On dynamic / impact tensile strength characteristics of thin high tensile steel materials for automobiles, Trans. the Korea Society of Automotive Engineers, Vol.7, No.2, SAE No. 99370060, pp.268~278 (in Korean).

Paik, J.K. and Thayamballi, A.K. (2000). Buckling strength of steel plating with elastically restrained edges, Thin-Walled Structures, Vol.37, Issue 1, pp.27~55.

Paik, J.K., Thayamballi, A.K. and Kim, B.J. (2000). Ultimate strength and effective width equations for ship plating under combined axial load, edge shear and lateral pressure, to be published.

Perrone, N. and Bhadra, P. (1984). Simplified large deflection mode solution for impulsively loaded, viscoplastic, circular membranes, J. of Applied Mechanics, Vol.51, pp.505~509.

Rhodes, J. (1984). Effective widths in plate buckling, Chapter 4, Developments in Thin-Walled Structures, Edited by J. Rhodes and A.C. Walker, Applied Science Publishers, London, pp.119-158.

Smith, C.S., Davidson, P.C., Chapman, J.C. and Dowling, P.J. (1987). Strength and stiffness of ships' plating under in-plane compression and tension, Trans. RINA, Vol.129, pp.277-296.

Smith, C.S. and Dow, R.S. (1981). Residual strength

of damaged steel ships and offshore structures, J. of Constructional Steel Research, Vol.1, No.4, pp.2-15.

Soares, C.G. and Gordo, J.M. (1996). Compressive strength of rectangular plates under biaxial load and lateral pressure, Thin-Walled Structures, Vol.24, pp.231-259.

Soreide, T.H. and Czujko, J. (1983). Load carrying capacities of plates under combined lateral load and axial/biaxial compression, Proc. of the Second International Symposium on Practical Design in Shipbuilding (PRADS'83), Tokyo and Seoul, October.

STARDYNE (1996). User's manual (version 4.4), Research Engineers, Inc.

Steen, E. and Valsgard, S. (1984). Simplified buckling strength criteria for plates subjected to biaxial compression and lateral pressure, Proc. of the Ship Structure Symposium, SNAME, Arlington, October, pp.257-272.

Timoshenko, S.P. and Gere, J.M. (1963). Theory of elastic stability, McGraw-Hill Book Co., London.

Ueda, Y., Rashed, S.M.H. and Paik, J.K. (1984). Buckling and ultimate strength interactions of plates and stiffened plates under combined loads –In-plane biaxial and shearing forces, J. of the Society of Naval Architects of Japan, Vol.156, pp.377-387 (in Japanese).

Ueda, Y., Rashed, S.M.H. and Paik, J.K. (1985). New interaction equation for plate buckling, Trans. JWRI, Vol.14, No.2, Welding Research Institute of Osaka University, pp.159-173.

Ueda, Y., Rashed, S.M.H. and Paik, J.K. (1986a). Effective width of rectangular plates subjected to combined loads, J. of the Society of Naval Architects of Japan, Vol.159, pp.269-281 (in Japanese).

Ueda, Y., Rashed, S.M.H. and Paik, J.K. (1986b). Plate and stiffened plate units of the idealized structural unit method (2nd report) –Under in-plane and lateral loading considering initial deflection and residual stress, J. of the Society of Naval Architects of Japan, Vol.160, pp.321-339 (in Japanese).

Ueda, Y., Rashed, S.M.H. and Paik, J.K. (1995). Buckling and ultimate strength interaction in plates and stiffened panels under combined inplane biaxial and shearing forces, Marine Structures, Vol.8, pp.1-36.

Usami, T. (1993). Effective width of locally buckled plates in compression and bending, J. of Structural Engineering, ASCE, Vol.119, No.5.

Verhagen, J.H.G. (1967). The impact of a flat plate on a water surface, J. of Ship Research, Vol.11, No.4, pp.211~223.

von Karman, T., Sechler, E.E. and Donnell, L.H. (1932). Strength of thin plates in compression, Trans. ASME, Vol.54, No.5.

Wang, X. and Moan, T. (1997). Ultimate strength analysis of stiffened panels in ships subjected to biaxial and lateral loading, Int. J. of Offshore and Polar Engineering, Vol.7, No.1, pp.22-29.

Wheaton, J.W., Kano, C.H., Diamant, P.T. and Bailey, F.C. (1970). Analysis of slamming data from the S.S. Wolverine State, Ship Structure Committee, Report No. SSC-210.

Williams, D.G. (1976). The influence of the torsional rigidity of plate stiffeners on plate effectiveness, International Shipbuilding Progress, Vol.23, No.268, pp.355-360.

Yao, T., Fujikubo, M., Yanagihara, D., Varghese, B. and Niho, O. (1998). Influences of welding imperfections on buckling / ultimate strength of ship bottom plating subjected to combined biaxial thrust and lateral pressure, Thin-Walled Structures, Research and Development, 2nd International Conference on Thin-Walled Structures, pp.425-432.

APPENDIX 1: BUCKLING COEFFICIENTS FOR SHIP PLATING WITH ELASTICALLY RESTRAINED EDGE CONDITIONS

$$k_{x0} = \left(\frac{mb}{a} + \frac{a}{mb}\right)^2, \quad k_{x1} = \begin{cases} 0.396\zeta_L^3 - 1.974\zeta_L^2 + 3.565\zeta_L + k_{x0} & for\ 0 \le \zeta_L < 2 \\ 6.951 - \dfrac{0.881}{\zeta_L - 0.4} & for\ 2 \le \zeta_L < 20 \\ 7.025 \quad for\ 20 \le \zeta_L \end{cases}$$

$$k_{x2} = d_1\zeta_S^4 + d_2\zeta_S^3 + d_3\zeta_S^2 + d_4\zeta_S + d_5$$

$$d_1 = \begin{cases} -1.010(a/b)^4 + 12.827(a/b)^3 - 52.553(a/b)^2 + 67.072(a/b) - 27.585 \ for\ 0 \le \zeta_S < 0.4 \\ 0.047(a/b)^4 - 0.586(a/b)^3 + 2.576(a/b)^2 - 4.410(a/b) + 1.748 \ for\ 0.4 \le \zeta_S < 0.8 \\ -0.017(a/b)^2 + 0.099(a/b) - 0.150 \ for\ 0.8 \le \zeta_S < 2 \\ 0.0 \ for\ 2 \le \zeta_S \end{cases}$$

$$d_2 = \begin{cases} 0.881(a/b)^4 - 10.851(a/b)^3 + 41.688(a/b)^2 - 43.150(a/b) + 14.615 & for\ 0 \le \zeta_S < 0.4 \\ -0.123(a/b)^4 + 1.549(a/b)^3 - 6.788(a/b)^2 + 11.299(a/b) - 3.662 & for\ 0.4 \le \zeta_S < 0.8 \\ 0.138(a/b)^2 - 0.793(a/b) + 1.171 & for\ 0.8 \le \zeta_S < 2 \\ 0.0 & for\ 2 \le \zeta_S \end{cases}$$

$$d_3 = \begin{cases} -0.190(a/b)^4 + 2.093(a/b)^3 - 5.891(a/b)^2 - 2.096(a/b) + 1.792 & for\ 0 \le \zeta_S < 0.4 \\ 0.114(a/b)^4 - 1.412(a/b)^3 + 5.933(a/b)^2 - 8.638(a/b) + 0.224 & for\ 0.4 \le \zeta_S < 0.8 \\ -0.457(a/b)^2 + 2.571(a/b) - 3.712 & for\ 0.8 \le \zeta_S < 2 \\ 0.0 & for\ 2 \le \zeta_S \end{cases}$$

$$d_4 = \begin{cases} 0.004(a/b)^4 - 0.007(a/b)^3 - 0.243(a/b)^2 + 0.630(a/b) + 3.617 & for\ 0 \le \zeta_S < 0.4 \\ -0.021(a/b)^4 + 0.184(a/b)^3 - 0.126(a/b)^2 - 2.625(a/b) + 6.457 & for\ 0.4 \le \zeta_S < 0.8 \\ 0.822(a/b)^2 - 4.516(a/b) + 6.304 & for\ 0.8 \le \zeta_S < 2 \\ -0.106(a/b) + 0.176 & for\ 2 \le \zeta_S < 20 \\ 0.0 & for\ 20 \le \zeta_S \end{cases}$$

$$d_5 = \begin{cases} k_{x0} & for\ 0 \le \zeta_S < 0.4 \\ -0.001(a/b)^4 + 0.033(a/b)^3 - 0.241(a/b)^2 + 0.684(a/b) + 3.539 & for\ 0.4 \le \zeta_S < 0.8 \\ -0.148(a/b)^2 + 0.596(a/b) + 3.847 & for\ 0.8 \ge \zeta_S < 2 \\ -1.822(a/b) + 7.850 & for\ 2 \le \zeta_S < 20 \\ 0.041(a/b)^4 - 0.602(a/b)^3 + 3.303(a/b)^2 - 8.176(a/b) + 12.144 & for\ 20 \le \zeta_S \end{cases}$$

In calculating k_{x2}, the following conditions must be satisfied in order for the approximations to hold: If $4.0 < a/b \le 4.5$ and $\zeta_S \ge 0.2$ then $\zeta_S = 0.2$, (2) If $a/b > 4.5$ and $\zeta_S \ge 0.1$ then $\zeta_S = 0.1$, (3) If $a/b \ge 2.2$ and $\zeta_S \ge 0.4$ then $\zeta_S = 0.4$, (4) If $a/b \ge 1.5$ and $\zeta_S \ge 1.4$ then $\zeta_S = 1.4$, (5) If $8 \le a/b \le 20$ then $\zeta_S = 8$, and (6) If $a/b \ge 5$ then $a/b = 50$.

$$k_{y0} = \left(1.0 + \frac{b^2}{a^2}\right)^2, \quad k_{y1} = e_1\zeta_L^2 + e_2\zeta_L + e_3, \quad k_{y2} = f_1\zeta_S^2 + f_2\zeta_S + f_3$$

$$e_1 = \begin{cases} 1.322(b/a)^4 - 1.919(b/a)^3 + 0.021(b/a)^2 + 0.032(b/a) & for\ 0 \le \zeta_L < 2 \\ -0.463(b/a)^4 + 1.023(b/a)^3 - 0.649(b/a)^2 - 0.073(b/a) & for\ 2 \le \zeta_L < 8 \\ 0.0 & for\ 8 \le \zeta_L \end{cases}$$

$$e_2 = \begin{cases} -0.179(b/a)^4 - 3.098(b/a)^3 + 5.648(b/a)^2 - 0.199(b/a) & for\ 0 \le \zeta_L < 2 \\ 5.432(b/a)^4 - 11.324(b/a)^3 + 6.189(b/a)^2 - 0.068(b/a) & for\ 2 \le \zeta_L < 8 \\ -1.047(b/a)^4 + 2.624(b/a)^3 - 2.215(b/a)^2 + 0.646(b/a) & for\ 8 \le \zeta_L < 20 \\ 0.0 & for\ 20 \le \zeta_L \end{cases}$$

$$e_3 = \begin{cases} 0.994(b/a)^4 + 0.011(b/a)^3 + 1.991(b/a)^2 + 0.003(b/a) + 1.0 & for\ 0 \le \zeta_L < 2 \\ -3.131(b/a)^4 + 4.753(b/a)^3 + 3.587(b/a)^2 - 0.433(b/a) + 1.0 & for\ 2 \le \zeta_L < 8 \\ 20.111(b/a)^4 - 43.697(b/a)^3 + 30.941(b/a)^2 - 1.836(b/a) + 1.0 & for\ 8 \le \zeta_L < 20 \\ 0.751(b/a)^4 - 0.047(b/a)^3 + 2.053(b/a)^2 - 0.015(b/a) + 4.0 & for\ 20 \le \zeta_L \end{cases}$$

$$f_1 = \begin{cases} 0.543(b/a)^4 - 1.297(b/a)^3 + 0.192(b/a)^2 - 0.016(b/a) \;\; for\; 0 \le \zeta_S < 2 \\ -0.347(b/a)^4 + 0.403(b/a)^3 - 0.147(b/a)^2 + 0.016(b/a) \;\; for\; 2 \le \zeta_S < 6 \\ 0.0 \;\; for\; 6 \le \zeta_S \end{cases}$$

$$f_2 = \begin{cases} -1.094(b/a)^4 + 4.401(b/a)^3 - 0.751(b/a)^2 + 0.068(b/a) \;\; for\; 0 \le \zeta_S < 2 \\ 2.139(b/a)^4 - 1.761(b/a)^3 + 0.419(b/a)^2 - 0.030(b/a) \;\; for\; 2 \le \zeta_S < 6 \\ -0.199(b/a)^4 + 0.308(b/a)^3 - 0.118(b/a)^2 + 0.013(b/a) \;\; for\; 6 \le \zeta_S < 20 \\ 0.0 \;\; for\; 20 \le \zeta_S \end{cases}$$

$$f_3 = \begin{cases} 0.994(b/a)^4 + 0.011(b/a)^3 + 1.991(b/a)^2 + 0.003(b/a) + 1.0 \;\; for\; 0 \le \zeta_S < 2 \\ -2.031(b/a)^4 + 5.765(b/a)^3 + 0.870(b/a)^2 + 0.102(b/a) + 1.0 \;\; for\; 2 \le \zeta_S < 6 \\ -0.289(b/a)^4 + 7.507(b/a)^3 - 1.029(b/a)^2 + 0.398(b/a) + 1.0 \;\; for\; 6 \le \zeta_S < 20 \\ -6.278(b/a)^4 + 17.135(b/a)^3 - 5.026(b/a)^2 + 0.860(b/a) + 1.0 \;\; for\; 20 \le \zeta_S \end{cases}$$

ζ_L, ζ_S = rotational restraint parameters as defined in equation (2), m = as defined in equation (18)

APPENDIX 2: ELASTIC BUCKLING INTERACTION EQUATION FOR SIMPLY SUPPORTED PLATING BETWEEN BIAXIAL COMPRESSION

$$\left(\frac{\sigma_{xav}}{\sigma_{xE}}\right)^{\alpha 1} + \left(\frac{\sigma_{yav}}{\sigma_{yE}}\right)^{\alpha 2} = 1 \text{ where } \alpha_1 = \alpha_2 = 1 \text{ for } \frac{1}{\sqrt{2}} \le \frac{a}{b} \le \sqrt{2},$$

$$\left.\begin{aligned} \alpha_1 &= 0.0293\left(\frac{a}{b}\right)^3 - 0.3364\left(\frac{a}{b}\right)^2 + 1.5854\left(\frac{a}{b}\right) - 1.0596 \\ \alpha_2 &= 0.0049\left(\frac{a}{b}\right)^3 - 0.1183\left(\frac{a}{b}\right)^2 + 0.6153\left(\frac{a}{b}\right) + 0.8522 \end{aligned}\right\} \text{ for } \frac{a}{b} > \sqrt{2}$$

APPENDIX 3: ELASTIC BUCKLING INTERACTION EQUATION FOR SIMPLY SUPPORTED PLATING BETWEEN AXIAL COMPRESSION AND EDGE SHEAR

$$\frac{\sigma_{xav}}{\sigma_{xE}} + \left(\frac{\tau_{av}}{\tau_E}\right)^{\alpha_{11}} = 1 \text{ where } \alpha_{11} = \begin{cases} -0.160\left(\frac{a}{b}\right)^2 + 1.080\left(\frac{a}{b}\right) + 1.082 & for\; 1 \le \frac{a}{b} \le 3.2 \\ 2.90 & for\; \frac{a}{b} > 3.2 \end{cases}$$

APPENDIX 4: ELASTIC BUCKLING STRESS FOR SIMPLY SUPPORTED PLATING UNDER IN-PLANE BENDING

$$\sigma_{xbE} = 23.9\frac{\pi^2 E}{12(1-\nu^2)}\left(\frac{t}{b}\right)^2, \quad \sigma_{ybE} = k_{yb}\frac{\pi^2 E}{12(1-\nu^2)}\left(\frac{t}{b}\right)^2, \text{ where } k_{yb} = \begin{cases} 23.9 & for\; 1 \le \frac{a}{b} \le 1.5 \\ 15.87 + 1.87\left(\frac{a}{b}\right)^2 + 8.60\left(\frac{b}{a}\right)^2 & for\; \frac{a}{b} > 1.5 \end{cases}$$

APPENDIX 5: ELASTIC PLATE SHEAR BUCKLING COEFFICIENTS

(a) For all edges simply supported: $k_s = 4.0\left(\frac{b}{a}\right)^2 + 5.34$

(b) For short edges simply supported and long edges clamped: $k_s = 2.4\left(\frac{b}{a}\right)^2 + 1.08\left(\frac{b}{a}\right) + 9.0$

(c) For short edges clamped and long edges simply supported:

$$k_s = \begin{cases} 2.25\left(\dfrac{b}{a}\right)^2 + 1.95\left(\dfrac{b}{a}\right) + 5.35 & if\ 0 < \dfrac{b}{a} \le 0.4 \\ 22.92\left(\dfrac{b}{a}\right)^3 - 33.0\left(\dfrac{b}{a}\right)^2 + 20.43\left(\dfrac{b}{a}\right) + 2.13 & if\ 0.4 < \dfrac{b}{a} \le 1.0 \end{cases}$$

(d) For all edges clamped: $k_s = 5.4\left(\dfrac{b}{a}\right)^2 + 0.6\left(\dfrac{b}{a}\right) + 9.0$

APPENDIX 6: MAXIMUM AND MINIMUM MEMBRANE STRESSES IN SHIP PLATING UNDER COMBINED LONGITUDINAL AXIAL LOAD AND LATERAL PRESSURE

$$\sigma_{x\max} = \sigma_{xav} + \sigma_{rcx} - \frac{m^2\pi^2 E A_m (A_m + 2A_{om})}{8a^2}\cos\frac{2\pi b_t}{b}, \quad \sigma_{y\min} = \sigma_{rcy} + \frac{\pi^2 E A_m (A_m + 2A_{om})}{8b^2}$$

where $A_m = \left|\frac{1}{2}\left(-Y_1 + \sqrt{Z_1}\right)^{1/3} - \frac{1}{2}\left(Y_1 + \sqrt{Z_1}\right)^{1/3} - \frac{S_1}{3}\right|$, $X_1 = \frac{S_2}{3} - \frac{S_1^2}{9}$, $Y_1 = S_3 - \frac{S_1 S_2}{3} + \frac{2S_1^3}{27}$, $Z_1 = 4X_1^3 + Y_1^2$,

$S_1 = \frac{C_2}{C_1}$, $S_2 = \frac{C_3}{C_1}$, $S_3 = \frac{C_4}{C_1}$, $C_1 = \frac{\pi^2 E}{16}\left(\frac{m^4 b}{a^3} + \frac{a}{b^3}\right)$, $C_2 = \frac{3\pi^2 E A_{om}}{16}\left(\frac{m^4 b}{a^3} + \frac{a}{b^3}\right)$,

$$C_3 = \frac{\pi^2 E A_{om}^2}{8}\left(\frac{m^4 b}{a^3} + \frac{a}{b^3}\right) + \frac{m^2 b}{a}\left(\sigma_{xav} + \sigma_{rex}\right) + \frac{a}{b}\sigma_{rey} + \frac{\pi^2 D}{t}\frac{m^2}{ab}\left(\frac{mb}{a} + \frac{a}{mb}\right)^2,$$

$C_4 = A_{om}\left\{\frac{m^2 b}{a}\left(\sigma_{xav} + \sigma_{rex}\right) + \frac{a}{b}\sigma_{rey}\right\} - \frac{16ab}{\pi^4 t}p$, b_t, m, p = as defined in equations (8), (18) or (3), respectively, σ_{rex}, σ_{rey} = as defined in equation (17), σ_{rcx}, σ_{rcy} = as defined in Fig.8, $A_{om} = \left|B_{om}\right| w_{opl}$, w_{opl} = as defined in Fig.6, B_{om} = as defined in equation (6) when $i = m$

APPENDIX 7: MAXIMUM AND MINIMUM MEMBRANE STRESSES IN SHIP PLATING UNDER COMBINED TRANSVERSE AXIAL LOAD AND LATERAL PRESSURE

$$\sigma_{y\max} = \sigma_{yav} + \sigma_{rcy} - C_f\frac{\pi^2 E A_1 (A_1 + 2A_{o1})}{8b^2}\cos\frac{2\pi a_t}{a}, \quad \sigma_{x\min} = \sigma_{rcx} + C_f\frac{\pi^2 E A_1 (A_1 + 2A_{o1})}{8a^2}$$

where $A_1 = \left|\frac{1}{2}\left(-Y_1 + \sqrt{Z_1}\right)^{1/3} - \frac{1}{2}\left(Y_1 + \sqrt{Z_1}\right)^{1/3} - \frac{S_1}{3}\right|$, $X_1 = \frac{S_2}{3} - \frac{S_1^2}{9}$, $Y_1 = S_3 - \frac{S_1 S_2}{3} + \frac{2S_1^3}{27}$, $Z_1 = 4X_1^3 + Y_1^2$,

$S_1 = \frac{C_2}{C_1}$, $S_2 = \frac{C_3}{C_1}$, $S_3 = \frac{C_4}{C_1}$, $C_1 = \frac{\pi^2 E}{16}\left(\frac{b}{a^3} + \frac{a}{b^3}\right)$, $C_2 = \frac{3\pi^2 E A_{o1}}{16}\left(\frac{b}{a^3} + \frac{a}{b^3}\right)$

$C_3 = \frac{\pi^2 E A_{o1}^2}{8}\left(\frac{b}{a^3} + \frac{a}{b^3}\right) + \frac{b}{a}\sigma_{rex} + \frac{a}{b}\left(\sigma_{yav} + \sigma_{rey}\right) + \frac{\pi^2 D}{t}\frac{1}{ab}\left(\frac{b}{a} + \frac{a}{b}\right)^2$, $C_4 = A_{o1}\left\{\frac{b}{a}\sigma_{rex} + \frac{a}{b}\left(\sigma_{yav} + \sigma_{rey}\right)\right\} - \frac{16ab}{\pi^4 t}p$,

$C_f = \begin{cases} 1.0\ if\ 1 \le a/b \le \sqrt{2} \\ 0.2\left(a/b - \sqrt{2}\right) + 2.0\ if\ a/b > \sqrt{2} \end{cases}$, a_t, p = as defined in equations (8) or (3), respectively, σ_{rex}, σ_{rey} = effective residual stress as defined in equation (20), σ_{rcx}, σ_{rcy} = as defined in Fig.8, $A_{o1} = \left|B_{o1}\right| w_{opl}$, w_{opl} = as defined in Fig.6, B_{o1} = as defined in equation (6) when $i = 1$

Discussion

E. Steen, T.K. Østvold, and S. Valsgård, Members

The authors present a comprehensive paper on the buckling strength of plates and a lot of information is gathered from relevant research taken place over the last decades. The paper is exclusively devoted to the strength assessment of unstiffened plating between longitudinal stiffeners and transverse frames in ship structures. Load interaction from bi-axial compression/tension loads combined with shear and lateral pressure, and effects from geometrical imperfections and residual stresses are of main interest. The less explored field of impact loading is also dealt with.

Five different topics are presented:

1. Modeling of geometrical imperfections and residual stresses.
2. Rotational restraints and torsional rigidity of stiffeners.
3. Ultimate strength design equation under combined bi-axial compression/tension, edge shear and lateral pressure loads.
4. Dynamic collapse strength characteristics, dynamic axial loads.
5. Dynamic collapse strength characteristics, dynamic slamming induced lateral pressure loads.

We have read the paper with great interest and we have some comments, as follows.

1. The authors show a significant knockdown effect from the residual stresses on the ideal elastic buckling stress. The knockdown effect is considered as a type of pre-stress effect for geometrically perfect flat plates, and it would have been interesting to see verifications of these values, using FE analysis. Another matter is whether it is of any interest to study the effect of residual stresses in this way, isolated from the simultaneous effect of the real geometrical imperfections and nonlinear post-buckling behavior. These elastic buckling stresses do not enter the ultimate strength assessment presented later in the paper and it would be interesting to hear the authors' meaning on the field of application for these parameters.
2. The authors document that the boundary conditions for plate element between stiffeners is found to vary between simply supported and clamped conditions, and a comprehensive study is presented on how the torsional restraints influences the ideal elastic buckling stress of the plating. We agree that this range of boundary conditions normally applies, but we have also observed that slender stiffeners, e.g., flat bar profiles prone to torsional buckling, may even destabilize the plating. Thus the plate and stiffener seen together as a unit may then even buckle at a nominal stress level below that for a simply supported plate. The fact that the stiffeners and girder members also have a buckling problem (in interaction with the plate) makes it difficult to assess the level of rotational restraint and it will be interesting to hear the authors' comments on this. It is not clear how the rotational restraint effects influence the ultimate strength assessment presented later in the paper.
3. The authors' ultimate strength procedure is separated from the buckling strength calculation procedure, in particular we cannot see that the rotational restraint effects influence the ultimate strength values. Is the buckling strength procedure only meant for serviceability type of criteria and thus not to be applied for ultimate design loading at all? The ultimate strength solution is formulated as an implicit solution of a limit state based on von Mises yield criterion in certain points along the supported edges. This is as sound approach as we see it. The authors document a significant knockdown effect from the presence of lateral pressure (Fig. 29) which we cannot see has been verified by nonlinear FE analysis. The opposite effect has been documented for the elastic buckling stress of rectangular plates (a/b=3) as shown in Fig. 24b. Have the authors considered the effect of boundary conditions in the assessment of the ultimate strength assessment, i.e., the fact that the lateral pressure will act over several bays in all directions? Presently classification societies consider the lateral pressure effect not to be critical for the in-plane capacity of plate-elements between stiffeners and girders and it is neglected in this respect. However, pressure effects are more important for the strength assessment of the stiffeners. We will be interested to hear if the authors have any comments on such a basic philosophy.

4, 5. The authors present the cases of impact loading and the results clearly illustrate the effect of dynamic effects on the capacity. This field is not as extensively explored as the case of static loads, but the results presented are interesting in a general context. For practical applications to ship structures, say to vessels with a large bow flare, studies like the present may prove to be useful for designing the plating such that not too large permanent deflections or rupture will occur. This is a very complex area and more studies are needed before dynamic effects can be utilized in the strength assessment of both ship skin and in the dimensioning of the internal supporting stiffener and frame structures.

We conclude that the paper contains a lot of interesting results and descriptions of new design methods. The philosophy is clearly a step in the direction of applying more advanced methods in the design of ship structures, a philosophy the present discussers fully support.

Philippe Rigo, Member

First of all, I would like to congratulate Professor J.K. Paik and the co-authors for the very thorough and extensive study carried out. In my discussion, I would like to make some comments and ask for additional information. These comments and questions are not based on criticisms but should be taken positively as being a guide for the authors for further improvement of this already valuable world-class contribution.

1. With regards to yielding limit state, use of equations (1) and (5) does not allow for consideration of the effect of the in-plane bending moment. Have the authors assessed the effect on the safety of neglecting these in-plane bending stresses?
2. The authors say that "*the Johnson-Ostenfeld equation has a tendency to underestimate the inelastic buckling strength.*" This statement is obvious, as this equation is only valid to account for the effect of plasticity of stocky plate and not to evaluate ultimate strength. Nevertheless, their analysis is interesting as:

- They point out the fact that the underestimation is particularly large with plating under shear load; and
- They provide a quantitative evaluation of this underestimation (about 5% for in plane compressive stress).

3. The authors mention that the rotational restraint parameter at long edges is normally in the range of 0.05 to 3.0. This is true for tankers with plating over 10 mm. For lightship, having thinner plates, lower values are usually obtained.

Based on this remark, it should be valuable to know if the proposed design equations based on fitting curves have some limits (plate thickness, plate slenderness, etc.)? What is the validity of the proposed buckling and ultimate strength equations for aluminum plates?

4. From Figs. 19 and 20 the authors conclude that the buckling interaction equation of the plating with simply supported edges is also valid for elastically restrained edges but by replacing the right buckling stress components. Have the authors some data to assess this statement? It should be interesting if they could superpose on these figures the two design relationships (simply supported and clamped).
5. The effect of cut-outs, Equation (21) should have a validity limit. Based on Figs. 22 and 23, it should probably be $0 < d_c/b < 0.8$.
6. Equation (23) allows to consider in-plane bending effect. For this purpose σ_{xbE} and σ_{ybE} are defined. We feel uncomfortable to see that these two stresses do not consider effects of rotational restraints, residual stress, lateral pressure and cut-outs. Why not define reduced stresses called σ^*_{xbE} and σ^*_{ybE} as it is done with σ^*_{xE} and σ^*_{yE} (Equations 23 and 24) which are reduced σ_{xE} and σ_{yE} elastic buckling stress (equations 17 and 20)?

   Note that the k_x and k_y coefficients used in Equations (24) and (25) are defined in equation (16) and not in Appendix 1 (as mentioned in the text). Only k_{xo} and k_{yo} are defined in Appendix 1.
7. Has the lateral pressure an effect on the elastic shear buckling stress defined by Equation (28)? Have the authors any quantitative data about that?
8. The proposed design equations are mainly parametric equations based on fitting adjustment (Appendix 1 to 7). For that reason, a potential user does not have the possibility to check the proposed equation. He has to thrust the provided equations and does not have the opportunity to find any eventual typing errors (in the paper itself or in the code). A series of simple benchmark tests (with data and results) would be helpful to solve this uncertainty.
9. It is not clear if the rotational restraint effect and the in-plane bending effect can be considered by the proposed ultimate strength.
10. Ship structures are composed mainly of stiffened panels and ship plates, studied here, are a component of these stiffened panels. Have the authors some plan to extend their study to the ultimate strength of stiffened panels?
11. Ultimate strength of stiffened panels and plating are the keys of a reliable tool to assess the ultimate bending moment of hull girders. Nowadays, many methods dedicated to perform ultimate bending moment analysis are based on a progressive collapse model using an incremented curvature. For that purpose, "average stress–average strain curves" of elements (panels, plating, beam-column, etc.) are required. The design equations presented today give the ultimate strength but do not provide the stress-strain curves. Is it possible to extend this study in order to provide the ultimate strain associated with the ultimate stress?

Paul A. Frieze, Member

The authors are to be congratulated on the production of a paper that comprehensively deals with ship plate panels. The breadth of the material included should solve nearly all a designer's problems although they will have to spend some time coding the equations in the Appendices if they wish to exploit the material effectively. Or perhaps, many of the results are already incorporated in SafeHull and that may be the most straightforward and perhaps reliable way to use the results.

However, as with most summary papers, some of the details have necessarily been omitted which raises questions regarding some of the findings and conclusions. I would appreciate the authors answering the following so that my mind, at least, can be put to rest.

The buckling phenomenon illustrated in Figure 1 is correct in principle. However, everything else being equal, elastic buckling of a perfect thin plate will always occur at a lower stress than that of a thick plate in contrast to that shown in Figure 1.

In discussing welding residual stress magnitudes, the authors note immediately after equation (8) that the tensile residual stress may only be some 80% of yield in high strength steel whereas it normally reaches yield in mild steel plating. Many of us will be familiar with the latter but not the former. Can the authors please cite the background material from which this former finding was derived?

In dealing with both Buckling Based Capacity (page 8 and onwards) and Ultimate Strength Based Capacity (page 16 and onwards), the authors widely assume the plate in-plane condition of straight edges. I believe this condition was possibly first introduced by Frieze et al. in 1977. This sought to take more realistic account of the effect of in-plane continuity of plating in place of the free in-plane conditions that had normally been adopted. However, the experimental results under combined in-plane loading and lateral pressure of Dean & Dowling (1977) coupled with the findings of Jackson & Frieze (1981) suggest that this straight edge condition may not always be appropriate. Shear continuity seems important in this context and, as well, the edges clearly do not remain straight. One simple way around this is to extend the plate model to include surrounding plate elements as adopted by Jackson and Frieze. This has the major advantage that all continuity effects are correctly represented, i.e., transverse, longitudinal and flexural, so that no assumptions have to be made regarding these. Can the authors please comment on what impact these more realistic boundary representations might have on their results, recognizing of course that this approach does not seek in any way to reflect interaction with stiffeners.

The investigation into the effects of rotational restraints (page 9) leads to some interesting findings. However, for practical applications such as plating supported on stiffeners and subject to uniaxial or biaxial compression, can the stiffeners torsional stiffness simply be divided in two and one half assumed to be allocated to each plate panel?

When discussing the accuracy of ultimate strength capacity strength solutions, the authors draw comfort from comparisons with FEA solutions. For the purposes of their study, this is reasonable since the input data to FEA solutions can be defined precisely in contrast with experimental solutions where nearly all aspects of modeling are uncertain to a lesser or greater degree. It would be more comforting to the designer to see the strength equation comparisons extended to include experimental results. Perhaps the authors have already done this. I would welcome seeing such comparisons.

The results for dynamic axial loading are extremely useful. However, in summarizing the results thereof, the authors have omitted just one or two too many details so it is not possible to put the results in perspective. For example, V_o is defined in equation (35) to be in m/sec. Presumably 'a' must also be in meters. Also, how do the strain rates examined, up to some 0.8, relate to strain rates likely to be experienced in typical ship structures including high speed craft? Does not even the slowest dynamic result reported (= 0.2) correspond to a ship in waves achieving full bending moment (from zero) in 5 seconds? Further, does the equation take account of strain-rate effects in steel? Introducing typical steel tensile coupon test rates into equation (35) leads to strength increases very similar to those found from equations by Cowper and Symonds (cited later by the authors), Reckling (1976) and Kaminski (1992) that quantify such strain-rate effects. Thus the recorded increase in strength may be entirely due to the strain-rate effect increase in yield strength rather than anything else.

Have the authors satisfied themselves that these dynamic test results can be nondimensionalized in the same way as similar quasi-static tests which is what the presented results seem to suggest. Frieze (1975) confirmed that under quasi-static conditions, identical nondimensional results are obtained for uniaxial compressed plates of different sizes and material properties (but with the same shape of material stress-strain curve) if they are normalized using (adopting the same symbols as the authors) β for slenderness, σ_{xav}/σ_o for stresses, σ_{rx}/σ_o for welding residual stresses and w/t for out-of-plane deflections where w is the magnitude at the center of the plate. However, with increasing axial loading rate, changes in buckling mode are expected, culminating in local buckling immediately under the point of load application. Perhaps the reported tests were not subject to loading rates that changed the buckling mode. Nevertheless, the question of nondimensionalization still arises. This is because the axial strain rate might be the same for two plates of similar proportions but different thickness, but the strain-rate in the extreme fibers of the thicker plate will be greater than those in the thinner plate. Consequently the effective yield stress in the extreme fiber of the thicker plate will be higher so it might be expected to demonstrate greater strength.

The authors examine dynamic lateral loading of plates from page 23 onwards. The reference by Biggs (1964) is well worth consulting to gain insight to this problem although Biggs concentrates on beams rather than plates. Nevertheless, a study of Biggs would give considerable appreciation into the effect of different forms of pulse impact loading. It might also discourage the authors from generalizing as they do at the end of the first complete paragraph on page 25 in relation to consequences of rectangular pulses of different durations which they then go on to exploit. When discussing such effects, it is extremely important to be aware of the natural frequency of the component in the mode into which it will deform when loaded.

Equation (38) appears to be a function of flexural hinge capacity only. Under the large deflections experienced by plates under lateral loading, membrane stresses will develop so that, along hinge lines, in-plane axial and shear stresses will also be present. Does the hinge line theory used in the authors' work include allowances for such axial and shear stresses? A proper analysis including such effects would also have to account for the normality rule of plastic flow in order to ensure the correct kinematics in the response.

In closing, and looking forward to the replies by the authors, again I commend the authors on their comprehensive and useful publication.

J. Roberts, Member

As requested at the AGM, this is a followup to the verbal discussion that I presented. I have read your paper in detail and consider it well presented with useful results. I was particularly encouraged that your references included one of the definitive aerospace works on the subject (Gerard and Becker, NACA Technical Note 3781 Handbook of Structural Stability Part I—Buckling of Flat Plates). I mention this because in many respects ship and aircraft structure are similar. Both make extensive use of stiffened panels where buckling is a concern.

One of the 'bibles' for aerospace structural engineering is: Bruhn, E.F., Analysis and Design of Flight Vehicles Structures Published and distributed by: Richard Jacobs Company, Indianapolis, Indiana 317-815-9400, Copyright 1973. Although 'Bruhn' is now somewhat dated it is still in use. There are several chapters on buckling. What makes 'Bruhn' of particular interest is that it references much of the earlier, definitive work such as that of Gerard and Becker. Aerospace addresses the issue of local buckling failures (e.g., crippling) that so far have not been applicable to ship structure. However, I suspect that the introduction of very thin high strength steel could result in local failures.

I have included a description of our experience. You will see that a good portion of our work is aerospace. We have witnessed opportunities for technology transfer between both industries—and in both directions. Also included is a paper that I gave at the FAST' 99 conference in Seattle, Washington. The paper gives an example of how aerospace technology might be adapted to high speed vessels. Clearly, there are differences between aerospace and ship structure including the important effects of initial deformation and residual stresses.

The issue that I raised was how can we manage nonlinear finite element analysis (FEA) in an appropriate engineering manner. Nonlinear FEA can be categorized into the following three general areas:

1. Geometric nonlinearity, for example: "tension only" cables or "surface-to-surface" contact.
2. Large strain, for example: secondary effects due to displacement such as the development of "membrane stresses" in a flat panel subject to transverse pressure load.
3. Material nonlinearity, or plasticity.

It is my opinion that 1 and 2 above have already been well addressed by FEA software. An exception might be excessive distortion of a finite element mesh that produces 'ill mannered' elements but at least the problem is clearly understood. It is the issue of material nonlinearity that needs careful thought as to how we can proceed with *engineering solutions*. I make the distinction between *engineering* and *science* because engineers need a more general (hence, less rigorous) approach to problem solving. One of the reasons this is a problem is the rapid advancement of engineering tools such as FEA. Nonlinear capability is now common within FEA software; consequently, engineers are making more use of it. I believe that we do so with only a limited understanding of what the analysis means. Engineering traditionally deals with simplistic models and this is fine as long as the model works for the intended application. For example, in a uniaxial stress condition, assuming a piece-wise linear *engineering* (as opposed to *true*) stress-strain curve is probably quite acceptable. I think you can easily see, however, that a three-dimensional state of stress is not so easily described. I don't believe that engineers in general receive adequate training in materials science to clearly understand multi-dimensional yielding within a volume. Obviously I cannot go into great detail on this complex subject but let's just consider a few simple points. First, there are actually three types of nonlinear material behavior:

1. Nonlinear elastic
2. Inelastic behavior (hysteresis)
3. Linear elastic—plastic with nonlinear strain hardening

Material nonlinearity studies (yielding) require the following to be addressed:

1. Strain decomposition to elastic and plastic parts
2. Yield criterion to define initiation of yielding (such as Tresca or von Mises)
3. Flow rule (to define growth of plastic strain)
4. Hardening rule

Strain decomposition is not as simple as expected. Material data are usually characterized in terms of engineering strain. However, plasticity usually deals with *Green's strain* or *Almansi strain* (as two examples). Two apparently common hardening rules are *isotropic* and *kinematic*. In isotropic hardening the yield surfaces are a uniform expansion of the original yield surface without translation. In kinematic hardening the yield surface translates without change in shape or size. We need to understand how the physical material behaves relative to various models and then understand what 'magic' the finite element software is performing. It would be natural to expect that we can test how realistic the FEA is to the physical material but this is apparently not quite that easy. Consider the points that you made regarding the sensitivity of some materials to strain rates. Let's ignore the more esoteric effects of varying strain rates, visco-elastic and visco-plastic behavior and consider only quasi-static analysis. The FEA software we use (Algor) offers two formulations for nonlinear analysis:

1. Total Lagrangian, formulated in terms of *2nd Piola-Kirchoff stresses* and *Green-Lagrangian strains*
2. Updated Lagrangian, formulated in terms of *Cauchy stresses* and *Almansi strains*

Don't be concerned if you are not familiar with the definition of the above stresses and strains as this only reinforces my point. Where does the practicing engineer get material data characterized in terms of such stresses and strains? I suspect that material data are one of the more important issues for material non-linear analysis. I hope that I am correctly interpreting Fig. 11 of your paper. If so, then I find it very interesting how the FEA results greatly differ with the Johnson-Ostenfeld equation. We have tried duplicating empirical results for panels loaded in shear using FEA but with poor results—even for elastic buckling. I suspect that a significant problem with shear loading is the correct balance of loads and boundary conditions. I certainly agree with your statement (just before the Concluding Remarks) "The strain hardening effect of material may vary with the magnitude of plate deflection. . . ." This statement is the essence of my discussion. Our main application for nonlinear FEA is service failure analysis. As opposed to design work, service failure analysis cannot rely on "conservative" assumptions. Instead, we are trying to duplicate—as exactly as possible—a material behavior. If you have any suggestions on how engineers can improve their use of nonlinear FEA then I would be happy to hear them.

Authors' Closure

First of all, the authors would like to thank all discussers for their valuable discussions and comments. The authors will try to reply for the discussers one by one.

Dr. Steen and his colleagues from the DnV raise a question about the influence of welding residual stress on the elastic buckling and ultimate strength stress. While some researchers (e.g., Ueda & Tall 1967) previously investigated the reduction characteristics of elastic and elastic-plastic compressive buckling strengths of plates due to welding residual stress in the loading direction, the added contribution of the present paper is to study the expected knockdown effect when residual stresses exist in the direction normal to the loading direction as well as in the loading direction. While the buckling strength formulations presented in our paper are based on analytical solutions to take into account the influence of residual stresses, comparisons with FE analysis are being planned. The buckling formulations including residual stress as a parameter of influence can be used as the basis of the design for serviceability limit state. Even though the elastic buckling stresses are not directly involved in the proposed design procedure for the ultimate limit state, the post-weld initial imperfections (i.e., both initial deformations and residual stresses) are included in the ultimate strength assessment as parameters of influence.

Regarding the rotational restraints along the plate-stiffener intersection, we in a general sense agree that for slender stiffeners, including flat bar profiles prone to torsional buckling, the torsional restraints may be small or even the stiffeners may destabilize the plating. However, our treatment is based on the normal presumption that stiffeners and supporting members have been properly designed so that their local instability will not occur prior to the failure of plating. When the stiffeners are very weak, they can buckle together with plating as part of what may be called overall buckling. The design procedure for the overall buckling of stiffened panels is outside the scope of the present paper. As Paik & Thayamballi (2000) previously suggested, however, the rotational restraint coefficients (i.e., ζ_L or ζ_s) may be approximately reduced to take into account the effect of local buckling of stiffeners. In spite of the rotational restraint effects on the elastic buckling, we do approximate that the ultimate strength may not be affected significantly. This is because if plasticity occurs earlier along the edges where the larger bending moments are developed, the rotational restraints at the yielded edges will then be lessened as the applied loads increase.

We agree that in our treatment, the ultimate strength procedure as presented is separated from the buckling strength calculation procedure, with the latter being considered as a basis of the serviceability limit state and the former being employed for the assessment of the ultimate limit state. Perhaps in the future, a more unified treatment might be possible.

Figure 24(b) shows the influence of lateral pressure on the elastic buckling strength of a rectangular plate with aspect ratio $a/b = 3$, indicating that lateral pressure can disturb occurrence of buckling in the elastic regime. On the other hand, Figure 29 represents the effect of lateral pressure on the plate ultimate strength when simply supported plate edges are presumed. In our study, the behavior of bare plate elements between stiffeners is being considered. For the design of stiffened panels under lateral pressure, the authors agree with the procedure of classification societies that stiffeners with associated plating will resist the lateral pressure loading. However, the authors point out that the effective breadth of associated plating can be reduced by a shear lag effect due to lateral pressure, a factor that is sometimes neglected.

Regarding Dr. Steen's comments about dynamic effects on the collapse behavior of ship plating, the authors fully agree that this subject is quite complicated to deal with, and that further studies are required in order to establish a more refined design procedure.

Regarding *Dr. Rigo*'s discussion to the influence of in-plane bending, the buckling based capacity formulations include in-plane bending as a parameter of influence, while the same is not considered for the ultimate strength based capacity formulations. The difficulty in considering in plane bending effects on ultimate strength arises because some part of in-plane bending results in axial tension while axial compression is applied by the other part. The tensile part works toward stabilizing the plate, and it is difficult to properly account for its effects until the ultimate strength is reached. It of course helps that in most real cases, in-plane bending effects tend to be small, but this is not always so, which is a fact that one needs to bear in mind.

The authors agree with Dr. Rigo's comment that the Johnson-Ostenfeld equation is intended to account for the effect of plasticity which is of course important for stocky plates.

Dr. Rigo had a question about applicable range of the rotational restraint parameters included in the elastic buckling formulations. We note in this regard that there are basically no range limitations assumed for the parameters; that is, they can have values that may range from zero to very large values. From a theory point of view, it would appear that the buckling and ultimate strength formulations presented in the paper can be directly applied to aluminum plates as well; verification studies in this regard are being planned.

Paik & Thayamballi (2000) suggested and provided calculations that appear to support that the buckling interaction relationship for simply supported plates can also be valid for plates with elastically restrained edges by using the corresponding buckling stress components. Figs. 19 and 20 show results for a limiting extreme case of plates with infinite rotational restrained edges, which might be a worse situation. The validity of the above suggestion appears to be confirmed by Figs. 19 and 20.

The authors agree with Dr. Rigo's characterization of the upper limit for the formulations of the buckling reduction factors due to the cut-outs presented in equation (21), which should cover the majority of practical cases. Related to this, equation (21) might be valid only for a square plate since the plate aspect ratio is of course an influential parameter on the buckling strength. Figures A.1–A.3 show the effects of the plate aspect ratio on the plate buckling strength varying the size of cut-outs and loading condition, as those obtained by ANSYS eigenvalue analysis, where all edges are simply supported and kept straight.

Dr. Rigo raises a question why the buckling strength in in-plane bending does not take into account the influence of residual stresses. This is because the strength formulations presented are concerned with the failure of plating primarily due to in-plane axial compressive loading, and thus equation (23) attempts to include the in-plane bending stress in the axial compressive buckling formulations as a parameter of influence. As described in the text of the paper, Appendix 1 defines the plate buckling coefficients for a variety of elastically restrained edge conditions.

Since the influence of lateral pressure on the buckling and ultimate strength of plating under shear is normally small in most practical cases, the shear strength formulations presented do not at this stage include lateral pressure as a parameter of influence.

While the elastic buckling coefficients for plates with elastically restrained edge conditions and the ultimate shear strength formulations are derived by curve fitting based on a series of analytical or numerical computations, the primary strength formulations for axial loads are based on analytical solutions as those obtained by solving the nonlinear governing differential equations of large deflection plate theory. The validity of the developed strength formulations are to some extent undertaken by a comparison with either the nonlinear finite element solutions or the mechanical model tests in the paper.

As described in our reply to the discussion of Dr. Steen and his colleagues as well as in the text of the paper, the effects of either rotational restraints along the plate edges or in-plane bending are not considered in the plate ultimate strength formulations.

As indicated in the section "Buckling/Ultimate Strength Design Procedure," the present paper is concerned with the strength behavior of the bare plate element level alone where local buckling and collapse of plating between stiffeners is a primary failure mode in stiffened panels. The authors have separately developed the advanced ultimate strength formulations for ship stiffened panels and grillages, and plan to present them in near future.

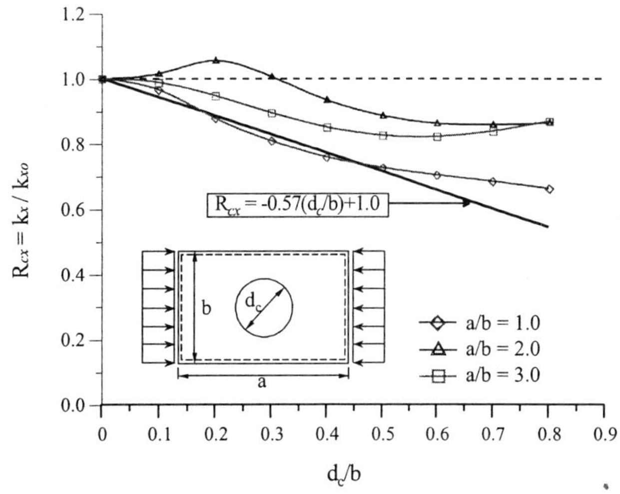

Figure A.1 Buckling reduction factor accounting for the effect of cut-outs under longitudinal compression

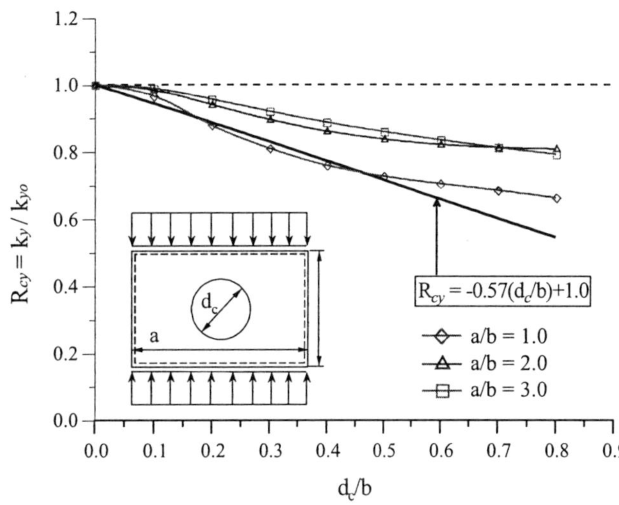

Figure A.2 Buckling reduction factor accounting for the effect of cut-outs under transverse compression

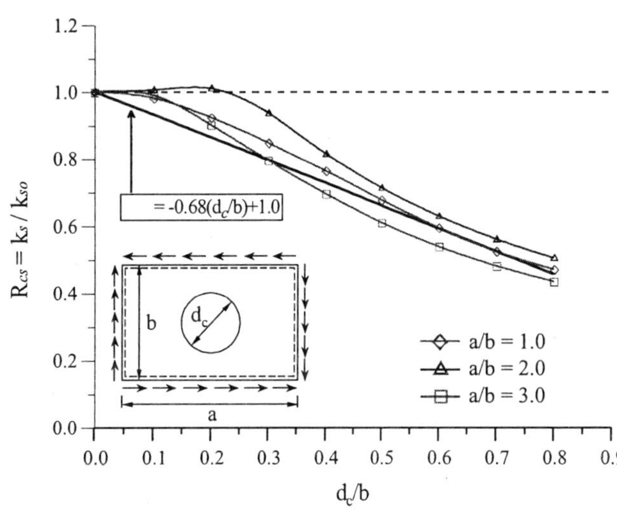

Figure A.3 Buckling reduction factor accounting for the effect of cut-outs under edge shear

To analyze the progressive collapse behavior of ship hulls under vertical hull girder bending, the stress-strain relationship of plating in the pre-buckling, buckling, post-buckling, ultimate strength and post-ultimate regime are needed. While Ueda et al. (1984) and Paik (1995) present the stress-strain formulations of simply supported plates, the present paper develops the strength formulations for buckling and plastic collapse. In the following, the relationships between average stress and average strain of imperfect plates under predominantly longitudinal axial compressive loading are suggested, consistent with the developments presented in this paper. Note that the imperfect plates do not show a bifurcation buckling, and that the in-plane stiffness decreases from the beginning as the axial compressive stresses σ_{xav} increase. In this case, therefore, the stress-strain formulations can be divided for convenience into three regimes, namely pre-ultimate strength, ultimate strength and post-ultimate strength.

(1) Pre-ultimate strength regime

As long as the plate edges keep straight, the average strains can be obtained in terms of maximum membrane stresses which are a function of average axial stress σ_{xav} as well as post-weld initial imperfections, as follows

$$\varepsilon_{xav} = \frac{1}{E}\left(\sigma_{x\max} - \nu\sigma_{y\max}\right) \tag{A.1}$$

where

ε_{xav} = average strains in the direction

$\sigma_{x\max}$ = as defined in Appendix 6 of the present paper

$$\sigma_{y\max} = \sigma_{rcy} - \frac{\pi^2 E A_m (A_m + 2A_{om})}{8b^2}\cos\frac{2m\pi a_t}{a}$$

The above relationship can be rewritten in the incremental form, as follows

$$\Delta\varepsilon_{xav} = \frac{1}{E}\left(\frac{d\sigma_{x\max}}{d\sigma_{xav}} - \nu\frac{d\sigma_{y\max}}{d\sigma_{xav}}\right)\Delta\sigma_{xav} \tag{A.2}$$

(2) Ultimate strength regime

The plate reaches the ultimate strength when $\sigma_{xav} = \sigma_{xu}$, where σ_{xu} is obtained as the solution of equation (29) with regard to σ_{xav}.

(3) Post-ultimate strength regime

Paik (1995) derived an analytical expression of the average axial stress-strain relationship of the plate in the post-ultimate strength regime, as follows

$$\sigma_{xav} = \frac{1}{2}\left(1 + \frac{\sigma_{xE}}{E\varepsilon_{xav}}\right)\sigma_{xu} \tag{A.3}$$

where σ_{xE} = as defined in equation (17).

The incremental form of equation (A.3) can be given by

$$\Delta\sigma_{xav} = -\frac{\sigma_{xu}}{2}\frac{\sigma_{xE}}{E\varepsilon_{xav}^2}\Delta\varepsilon_{xav} \tag{A.4}$$

Figure A.4 demonstrates the validity of equations (A.1) to (A.4) by a comparison with the nonlinear finite element analysis in the case of square plating with initial deflection but without residual stresses. It is evident that the proposed stress-strain formulations fairly well correspond to the non-linear FEA. For combined loading, the stress-strain formulations can be derived by a method similar to that of the uniaxial load case.

Dr. Frieze has a comment about Figure 1. In this figure, the authors attempt to represent that thick plates will not buckle in the elastic regime while thin plates do.

Dr. Frieze raises a question about idealization of welding induced residual stress distribution for high tensile steel plates. According to the measurements of welding induced residual stresses in plating between stiffeners (Kmiecik 1970, Cheng et al. 1996), the residual stress distribution for mild steel plates is somewhat different from that for high tensile steel plates, as those indicated in Figure A5, while the maximum tensile residual stress well reaches the yield stress for both types of material. In evaluating the effect of residual stresses on the compressive buckling strength, a somewhat reduced tensile residual stress (e.g., 80% of the yield stress) has been used for idealization of the welding residual stress distribution for high tensile steel plates as long as the actual breadth of the tensile residual stress block is applied, as that indicated by the dotted line in Figure A.5b. Whether this type of approximation is appropriate in a general sense is worth further investigation, as there is no reason to think that residual stresses are a function of base material yield strength alone. We thank Dr. Frieze for pointing this out.

Regarding the comments on the straight edge condition of plating between stiffeners, the authors agree that the plate edges may not remain straight in some cases, specifically under large lateral pressure loading. In these cases, however, assuming that the stiffeners are slender, the plate will buckle together with stiffeners showing the overall buckling of the entire stiffened panel, and the behavior of the panel may be analyzed by a different approach, e.g., large deflection orthotropic plate theory. As long as the stiffeners are strong enough so that they do not fail prior to buckling of the plate, which is the case that the present study is concerned with, the plate will fail locally. In a continuous plated structure where such a hypothesis can be accepted, the edges of individual plate elements will remain almost straight due to the relative structural response to the adjacent plate elements even after the plate deflects.

Dr. Frieze raises a question as to whether the torsional stiffness of a stiffener can be simply divided into two, to be allocated to each adjacent plating. As long as the plate deflection pattern is unsymmetrical with regard to a stiffener, the authors approximate that the contribution of the stiffener to the rotational restraints along the plate-stiffener intersection may approximately be equal for two adjacent plates. The validity of this approximation will be looked into in the future.

Dr. Frieze would like to see some comparisons of the proposed ultimate strength formulations with the mechanical model test results. While a limited amount of such effort has been made for plates under combined axial compression and lateral pressure as indicated in Fig. 29, we certainly feel that it is more worthwhile to perform the comparisons with nonlinear FEA, the FEA procedure being validated by the admittedly limited number of actual mechanical model tests.

In equation (35), the dimension of plate length is considered to be mm (or m) when the loading speed is in mm/sec (or m/sec) since the strain rate is in sec^{-1}. Also, this equation is valid for typical mild steel. It is realized that the strain rate effects for high tensile steel might be different from that for mild steel (Paik et al. 1999d, and others). There is usually a difficulty to theoretically estimate the strain rates of the structural components making up ship hulls, even though a monitoring system may be used to measure the strain rates at individual locations. While the Cowper-Symonds type equations represent

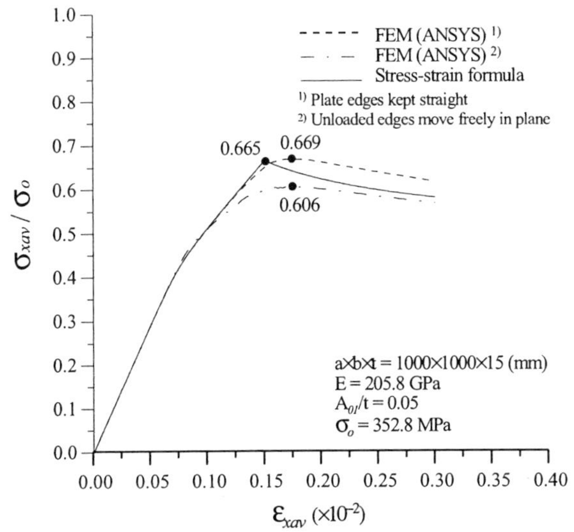

Figure A.4 The average stress-strain curves for a simply supported square plating under axial compressive loads

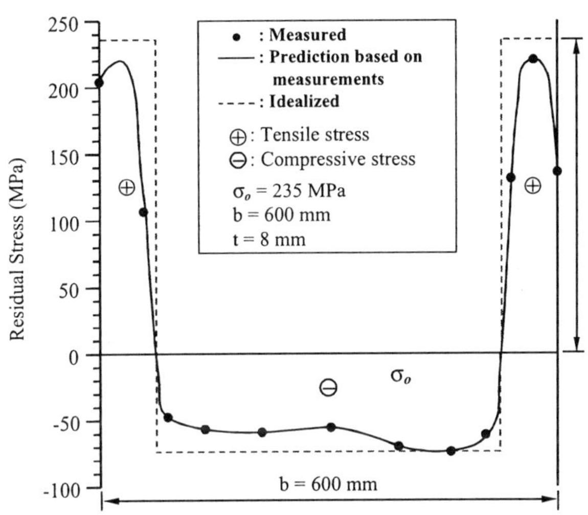

Figure A.5(a) Idealization of residual stress distribution in the mild steel plate between stiffeners based on the measurements of Kmiecik (1970)

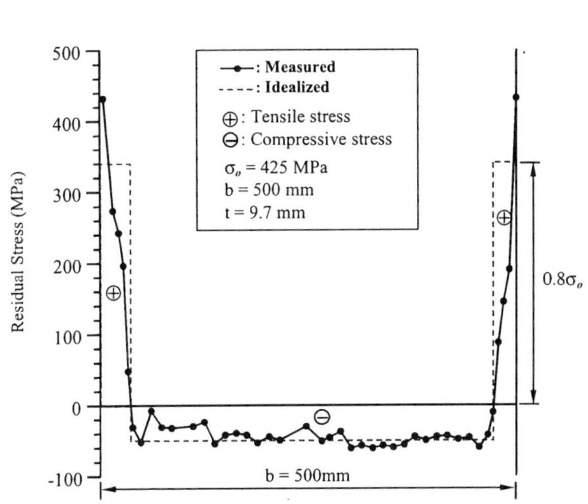

Figure A.5(b) Idealization of residual stress distribution in the high tensile steel plate between stiffeners based on the measurements of Cheng et al. (1996)

the effect of strain rates on the dynamic yield stress, equations (35) or (36) attempt to investigate the strain rate effects on the plate ultimate compressive stress.

The authors appreciate the comments and insights of Dr. Frieze on the nondimensionalization for the dynamic collapse test results. As previously noted, equations (35) or (36) represent the dynamic (strain rate) effects on the ultimate compressive strength and the authors approximate that the dynamic ultimate strength normalized by the quasi-static ultimate strength can be approximately predicted as a function of only strain rates, i.e., without the geometric and material properties which will be included in the calculations of the latter (i.e., the quasi-static ultimate strength). However, we agree that more data are needed to study the degree of approximation involved in the particular case of ultimate strength design of ship plating under dynamic in-plane loading.

Regarding collapse of the plating under dynamic lateral loading, the authors appreciate Dr. Frieze for the information on the effect of different forms of pulse impact loading. As that shown in Fig.39, various types of pressure pulse history can usually be considered. However, our conclusion based on some numerical computations is that idealization of impact pulse with the rectangular type may be appropriate for practical purposes.

Equation (38) based on the hinge line theory may not properly consider the effects of membrane stresses which are neglected, specifically when the plate lateral deflection is large. Hence we agree that the treatment only approximately takes into account the plate large deflection effects.

We thank *Mr. Roberts* for discussing the difficulties on the finite element analysis of material nonlinearity. He raises a question as to why there is a large difference between FEA and the Johnson-Ostenfeld formula in Fig. 11. This is because the FEA has in Fig. 11 calculated the plastic collapse strength of plating while the Johnson-Ostenfeld formula is based on initial yielding. Regarding the difficulty for plate shear strength analysis, we got the FE solutions applying the displacement control method when all edges are simply supported and kept straight. Mr. Roberts questions how engineers can improve their use of nonlinear FEA. This is a tough question while the FE technology has rapidly advanced in the last decade. Certainly the FE solutions significantly depend on how to model the object structures. A basic response to this question is that considering both accuracy and efficiency of FEA the FE modeling should be made so that all primary parameters affecting both geometric and material nonlinear behavior of the structures are more precisely and effectively handled.

Finally, the authors would again like to thank the various discussers for their time and effort, which have helped clarify the design oriented technology for ship plated structures against buckling and plastic collapse developed in this paper. In closing, the authors would like to thank the SNAME Papers Committee members for their guidance related to this paper, and the Society itself for having given the authors the opportunity to present this work.

Additional references

Biggs, J. M. (1964). Introduction to Structural dynamics, McGraw-Hill.

Cheng, J. J. R., Elwi, A. E., Grodin, G. Y. and Kulak, G. L. (1996). Material testing and residual stress measurements in a stiffened steel plate, Appendix of the Ship Structure Committee Report "Strength and stability of stiffened plate components", SSC-399. To be published.

Dean, J. A. and Dowling, P. J. (1977). Ultimate load tests on three stiffened plates under combined in-plane and lateral loading, Steel Plated Structures, Eds. Dowling, P. J., Harding, J. E. and Frieze, P.A., Crosby Lockwood Staples, London.

Frieze, P. A. (1975). Ultimate load behavior of steel box girders and their components, Ph.D. Thesis, University College, of London.

Frieze, P. A., Dowling, P. J. and Hobbs, R. E. (1977). Ultimate load behavior of plates in compression, Steel Plated Structures, Eds. P. J. Dowling, J. E. Harding and P. A. Frieze, Crosby Lockwood Staples, London, pp. 24–50.

Jackson, R. I. and Frieze, P. A. (1981). Design of deck structures under wheel loads, *Trans. RINA,* Vol. 123, pp. 119–144.

Kaminski, M. L. (1992). Cyclic compression of imperfect steel plates, Ph.D. Thesis, Delft University of Technology, Delft.

Kmiecik, M. (1970). The load carrying capacity of axially loaded longitudinally stiffened plates having initial deformation, Ship Research Institute, Report No. R80, Trondheim.

Paik, J. K. (1995). Advanced idealized structural units considering excessive tension-deformation effects, *J. of Hydrospace Technology,* Vol.1, No.1, pp. 125–145.

Reckling, K. A., (1976). Beitrag der Elasto- und Plastomechanik zur Untersuchung von Schiffskollisionen', *Jahrbuch der Schiffbautechnischen Gesellschaft,* 70. Band, pp. 443–464.

Ueda, Y., Rashed, S. M. H. and Paik, J. K. (1984). Plate and stiffened plate units of the idealized structural unit method (1st report), *J. of the Society of Naval Architects of Japan,* Vol. 156, pp. 389–400.

Ueda, Y. and Tall, L. (1967). Inelastic buckling of plates with residual stresses, *Proc. of International Association for Bridge and Structural Engineering,* pp. 212–254.

SNAME *Transactions*, Vol. 108, 2000, pp. 291–298

Corrosion Protection Strategies for High Speed Craft

Daniel Engel, Visitor, and **John Holland,** Associate Member, Germanischer Lloyd AG, Hamburg

ABSTRACT

The introduction of car carrying high speed craft in the early nineties lead to a worldwide revolution in short sea passenger transport. Although steel is employed in some types and sizes of HSC, aluminum has become the material of choice for most designs. The corrosion protection strategy applied to conventional ships does not apply to HSC for numerous reasons. Therefore special attention for both lightweight steel and aluminum structures are required to insure a long service life. The paper describes the often conflicting requirements involved in corrosion protection of HSC craft and offers solutions which satisfy the designer's, owner's and society's concerns. Strategies for protection of lightweight steel and aluminum structures against corrosion are explored.

INTRODUCTION

One of the factors influencing the service life and reliability of marine structures exposed to seawater and sea atmosphere is corrosion. High corrosion rates of uncoated or improperly coated steel surfaces may significantly reduce the scantlings of structural elements, so that within a relatively short period the required minimum wall thickness is no longer present. Other materials such as stainless steels, aluminum alloys and copper or nickel alloys may also be attacked by marine corrosion under particular conditions. Extensive knowledge about the corrosion characteristics of the materials and service conditions of the structures is required to avoid deterioration.

Fig.1 Corrosion in a coated ballast water tank

Today much greater attention is paid to the corrosion protection [1] especially for oil tankers and bulk carriers. Reliable corrosion protection, particularly in the critical areas such as ballast tanks and holds, plays a key role in ensuring economical and safe operation. Figure 1 shows the condition of a ballast tank after coating failure.

Lightweight construction has always been of great importance in shipbuilding. For high speed craft weight saving is absolutely essential. Reducing the weight of ship structures requires increased effort in the selection and application of sophisticated materials and manufacturing processes. For ship operation and survey activities it must be remembered that reducing the weight increases the risk of under dimensioning with respect to possible failure modes such as fatigue, buckling and corrosion. For high tensile steels or aluminum alloys applied for lightweight construction, it is important to realize that traditional strategies for design and survey should be adapted to the characteristics of the material in question.

Conventional ships like container ships and oil tankers are normally designed for a lifetime of about 30 years. Safe ship operation with regard to classification societies' requirements can be achieved through out the life of the ship by applying a complex set of measures. These measures are based on the material properties, careful engineering and fabrication technology. Additionally, such measures must be combined with periodic surveys which take into consideration ship type, age, structural condition etc. In this context the corrosion allowance presents a significant gain in the safety margin of the structural design.

High speed craft typically operate in considerably different conditions compared to conventional vessels. Hull, machinery and equipment are highly loaded and tight operating schedules give limited opportunity for maintenance. The planned operating life of high speed craft (HSC) is usually much shorter than for conventional ships. In addition, the operating profile is more often changed to meet changing market conditions. This may result in more intense operation and a shorter operational lifetime. Additionally HSC are build to special classification rules [2] which take into account the needs and demands for reliable construction and safe ship operation.

Fig. 2 Light weight structure

LIGHTWEIGHT STRUCTURES

Lightweight construction has always been an important issue in shipbuilding, but is increasingly significant. The reasons for this are obvious:

1. a reduced lightship weight permits increased deadweight for a given displacement, and/or
2. operating costs are reduced due to lower fuel consumption at a given speed.

As defined in the SOLAS regulations a High Speed Craft (HSC) is a craft capable of a speed equal to or exceeding 3.7 V $^{0.1667}$ (m/s). As an example Fig. 2 shows the typical construction of an HSC with a cruising speeds in excess of 40 knots.

Reduction in light-ship weight can be achieved by optimization of the scantlings, by innovative design and by the use of higher strength to weight ratio structural steels, aluminum alloys or composite materials (fiber-reinforced plastics, FRP). In the endeavor to obtain structures of increasingly lighter weights adequate strength must, of course, be ensured with respect to possible failure modes. Optimized scantlings are obtained using highly sophisticated design and analysis tools. The corrosion allowances normally applied to conventional ships are not used for HSC due to weight constraints. Therefore the margin of structural safety at delivery is considerably less than that for conventional vessels.

MATERIAL APPLICATIONS IN HSC's

Ordinary and higher-tensile steels (HTS), materials traditionally employed in shipbuilding, are also frequently used in present-day lightweight structures. For seagoing ships of approx. 150 m in length and over, higher-tensile hull structural steels with minimum yield strength of up to 390 N/mm² are employed. This permits the amount of required material to be considerably reduced and also avoids forming and welding problems associated with excessive plate thickness. HTS is generally applied only in larger high speed craft where the minimum thickness criteria for steel is at least 4mm.

Aluminum alloys are currently the most commonly employed materials for high speed craft. All aluminum vessels of over 100 meters in length with cruising speeds of over 40 knots are becoming common. In shipbuilding, aluminum alloys of the 5000 series (AlMg alloys) and the 6000 series (AlMgSi alloys) are employed both for ships' hulls and superstructures. These aluminum alloys are characterized by good weldability and corrosion resistance. For superstructures such as the wet decks of catamarans coatings and/or cathodic protection are not necessarily required. However if aluminum alloys are in direct contact or immersed in seawater for a prolonged period, corrosion protection measures are absolutely necessary to avoid critical corrosion attack. The so-called fully seawater resistant aluminum alloys which are often mentioned have yet to be developed

When employing higher-tensile steel and aluminum alloys in lightweight ship construction, special attention must be paid to the particular characteristics of these materials regarding fatigue strength, buckling and corrosion behavior.

STRENGTH BEHAVIOR OF HSC's

Fatigue strength in welded structures has a major influence on structural design. Fatigue problems for high tensile steel are well known, however, fatigue is also especially important for aluminum alloys. In view of the heavy dynamic loads acting on fast vessels and the magnitude of load cycles it becomes particularly important to ensure adequate fatigue strength in the design phase.

Fig. 3 Construction detail

The design of intersecting structural members employing fillet welds and the welded connections of highly stressed transverse and longitudinal members must be considered with respect to fatigue. Figure 3 shows the fatigue optimized connection of a stanchion supporting an extruded aluminum deck. Smooth transitions of transverse joints, soft brackets, single-bevel and/or double-bevel welds with full root penetration and well rounded free edges all improve the fatigue strength properties of highly stressed structures. For special structures, such as the waterjet inlet ducts, analysis of structural details on the basis of finite element models is required to ensure adequate fatigue strength.

CORROSION BEHAVIOR AND PROTECTION MEASURES

High Tensile Steel

Unprotected carbon steels such as high tensile steels in direct contact with seawater are normally attacked by uniform corrosion. No significant difference in corrosion rates can be observed between mild and high tensile steels. Under the worst conditions corrosion rates of 1,5 to 2 mm/year may be assumed. Since such corrosion can ultimately result in a reduction of the structural strength or induce leakage in watertight boundaries, organizations like IMO and classification societies have introduced measures to deal with this problem. Regulations and rules have been adopted providing suitable corrosion control systems such as protective coatings and/or cathodic protection. In addition surveys have been adjusted to account for structural deterioration due to corrosion. Figure 4 shows the relatively uniform corrosion of an uncoated steel ballast tank.

Fig. 4 Uniform corrosion in a ballast tank

In the design phase structural scantlings are calculated under consideration of classification requirements. A corrosion allowance is required when following Germanischer Lloyd's Rules for Classification and Construction of Seagoing Ships,. Plate thicknesses up to 10 mm require a corrosion allowance of t_k=1,5 mm. For plate thickness above 10 mm, the corrosion allowance is determined by $t_k = 0,1 \times t'/ k^{-2} + 0,5$ or a maximum of 3 mm. The t′ is the thickness required to provide adequate strength. Introducing a corrosion allowance is equivalent to increasing the time frame available for safe ship operation.

Coated carbon steels behave much like passive materials such as stainless steels and aluminum alloys. The coating system or rather the passive layer is normally able to protect the material from uniform corrosion. When the passive layer is damaged and not recoated, localized corrosion will attack the base metal. Such attacks are characterized by being extremely localized and the penetration is deep in relation to the area attacked. Corrosion rates in such situations will depend on factors such as the size of the breakdown and the availability of oxygen. It can generally be assumed that significantly increased corrosion rates compared to uniform corrosion will result.

In such cases a corrosion allowance will not enhance longevity. Because it is impractical to apply a perfect coating without some defects and given the harsh environmental conditions and service which ships operate in, it is unlikely that local corrosion attack can be totally prevented. Therefore coated carbon steels submerged in seawater need additional measures to protect the material from localized corrosion. A cathodic protection system is able to provide suitable protection against pitting corrosion in cases where the coating has broken down.

A detailed discussion of the principles of cathodic protection are beyond the scope of this paper. Briefly one method of cathodic protection is the use of sacrificial anodes made of Zinc or aluminum alloys. Another method or protection is through an active impressed current system. Details for design and layout of cathodic protection systems can be found in [2] and in [3].

Aluminum Alloys

Aluminum is a reactive metal with a high affinity for oxygen. The corrosion of aluminum is dependent on the pH value of the immersing electrolyte. Since the pH value of seawater is close to neutral, the natural oxide film which forms on unprotected aluminum is normally stable. The exact chemical composition of the aluminum alloy and the related presence and absence of other elements which are used to produce the alloy also determines the way in which aluminum will suffer corrosion. The very limited copper content combined with magnesium of the 5000 and 6000 series alloys provides the good corrosion resistance in the marine environment. However, where no corrosion measures are taken aluminum alloys may be attacked by pitting corrosion when permanently immersed in seawater. The susceptibility to corrosion is mainly related to the time of exposure, the alloy's chemical composition and its grain structure.

According to the galvanic series in seawater zinc and aluminum are the least noble materials. This means aluminum alloys combined and in contact with other more noble materials such as carbon steel or stainless steels will suffer bimetallic corrosion. The corrosion which can be observed can be described usually as enhanced pitting or swallow pitting corrosion. This phenomenon is most visible in areas where stainless steels or copper alloys are present. Figure 5 shows the bimetallic corrosion of a waterjet flange welded to the transom of a HSC. The waterjet has a high percentage of stainless steel components.

If aluminum alloys are utilized in a marine environment, special corrosion protection measures must be applied. Coatings applied to aluminum ships

Fig. 5 Bimetallic corrosion of aluminum

are frequently used for corrosion protection as well as for decorative purposes. Products such as epoxy polymer and epoxy glass flake coatings can be used to reduce the risk of pitting corrosion. These coatings can also be used to provide an electrically non-conductive surface. The application of coatings can also be used to achieve an electrical isolation between aluminum and, for example, stainless steels in order to avoid bimetallic corrosion. Of course the specified coating procedure must be closely observed. Usually surfaces on which epoxy coatings are applied must be blasted and the coating should be applied in two or possibly three coats to achieve a long-standing protection against bimetallic and pitting corrosion. If coatings are applied on aluminum it is important to realize that any coating failure will concentrate the corrosion on the exposed corrosion cell, leading to deeper and faster penetration of the pit. Therefore during construction and maintenance considerable attention must be paid to the coating application. It is also essential that any mechanical damage to the coating during construction be repaired before launching.

Aluminum protected by coatings and immersed in seawater should always be combined with cathodic protection to avoid pitting corrosion where the coating has broken down. Cathodic protection by means of sacrificial anodes made from aluminum or zinc can be used. Additionally, bimetallic corrosion can be avoided with anodes. The protective potential applicable for AlMg and AlMgSi alloys has to be limited to U_H= -0,55 V versus the standard hydrogen electrode. Any shift in the protective potential may result in an altered pH- value leading to accelerated corrosion rates. If anodes are used to protect aluminum against bimetallic corrosion, for example in the presence of large stainless steel components, the design of cathodic protection must be thoroughly investigated.

Impressed current systems (ICS) have so far not been used for the corrosion protection of aluminum ship's hulls. As mentioned above the electrochemical potential necessary for the corrosion protection of aluminum has to be precisely adjusted. Traditionally designed impressed current systems with one or maximum two impressed anodes will not be able to achieve or adjust to an acceptable current distribution. Thus, unacceptable potentials may result in corrosion being accelerated by the ICS.

CORROSION PROTECTION STRATEGIES

Unlike conventional sea going vessels, corrosion protection for lightweight structures in high speed craft needs to be adapted according to the material employed, its intended service and its environment.

Protection of Steel in HSC's

All surfaces of high speed craft built of carbon steel should be protected against uniform corrosion.

This includes high tensile steels and is applicable for all surfaces in direct contact to seawater and to sea atmosphere. The protective coatings employed are traditionally epoxy and polyurethane based. Surfaces immersed in seawater such as ballast tanks and the underwater hull are to be protected by coatings with a dry film thickness of at least 300 μm before antifouling. For these areas cathodic protection is additionally required to avoid corrosion in case of a coating break down. If coatings are combined with cathodic protection their compatibility must be checked. For ballast tanks sacrificial anodes (zinc or aluminum) can be used. Impressed current systems are not permitted in confined spaces such as ballast tanks due to the possible development of hydrogen gas. As with ballast tanks, the underwater hull should be protected by cathodic protection. Sacrificial anodes (zinc or aluminum) or an impressed current system may be utilized.

Insulated interior areas exposed to dry heat such as passenger spaces normally receive only shop primer. Areas susceptible to sweating should receive a protective coating in addition to the shop primer.

For newbuildings the corrosion protection specification has to be submitted to class for approval. The specification must contain information about the intended surface preparation before coating application, coating types (name and manufacturer), number of layers, film thickness, etc. Coatings used for immersed areas are to be type approved. The design and layout of cathodic protection measures must be specified to obtain approval. The specification of the cathodic protection system should follow the requirements given in [2]. For surfaces immersed in seawater the application of the corrosion protection system has to be documented. Surface preparation as well as activities connected to coating application must be fixed in a protocol. If applicable, a functional test of the impressed current system has to be undertaken during sea trials.

Protection of Aluminum in HSC's

For HSC built of aluminum the corrosion protection measures are divided into two broad categories. The first category is for areas exposed to sea and interior atmospheres; the second category is for surfaces which are immersed in seawater.

Surfaces Exposed to the Atmosphere

Surfaces which are exposed to sea and interior atmospheres do not necessarily need corrosion protection by means of coating systems. The corrosion behavior of aluminum alloy employed on HSC's has demonstrated no uniform corrosion. In some cases, local corrosion attack may be observed as an uneven appearance of the surface. However, the depth of the corrosion penetration is rather limited and the corrosion rates are slowed through the formation of an oxide layer. Current practice is to coat all visible surfaces above the waterline on HSC to improve the visual appearance for passengers. The coatings are usually well maintained thus avoiding local corrosion. Suitable coatings are generally epoxy or polyurethane based. A phosphate wash primer with a polyurethane top coat or similar makes a durable and very lightweight coating system for such areas.

If aluminum is used in combination with other materials in a sea atmosphere it is recommended that stainless steels such as 316 grades be used. Carbon steel should not be combined with aluminum.

Surfaces Immersed in Seawater

For the corrosion protection of aluminum surfaces immersed in seawater suitable coating systems combined with cathodic protection is required. The corrosion protection of ballast tanks on HSC's is not as critical as for conventional ships as these craft rarely use water ballast. Therefore the corrosion protection strategies should be focused on the protection of the underwater hull to avoid local and bimetallic corrosion.

The requirements applicable for HSC's using aluminum are basically similar to those for using steel. For coating systems on underwater surfaces of aluminum the dry film thickness required is 300 μm without antifouling. The coatings are to be type approved. As the coatings are required to be combined with cathodic protection, the coating must be compatible for the chosen cathodic protection system. Vinyl epoxies or similar are generally suitable for this type of service.

The corrosion protection specification must be submitted for approval to class for each newbuilding project. Information about the intended surface preparation, coating types (name and manufacturer), number of layers, film thickness, etc. must be provided. The application of the coating system has to be documented. The protocol must include the surface preparation, environmental conditions as well as complete activities connected with the coating application.

The cathodic protection of aluminum can be assured by using zinc or aluminum sacrificial anodes. Due to the risk of overprotection magnesium anodes should not be used. The anodes are generally flush mounted to the surface to avoid increased resistance. In the presence of stainless steel components such as waterjets a calculation of anode mass is required to avoid bimetallic corrosion. The calculation has to consider the exposed area of the stainless steel as well as water velocities, alloy characteristics, coating system, etc. The distribution and size of the anodes should be laid out for a protective potential of U_H= - 0,55 V. For the approval of cathodic protection measures the design and layout of the system has to be specified and approved.

The use of sacrificial anodes for waterjet propulsion systems is rather limited due to the very high water velocities and limited space for attaching anodes. Recently, impressed current protection systems have been used to protect aluminum against bimetallic corrosion for such systems. These areas are highly susceptible to bimetallic corrosion as the propulsion systems are normally manufactured from stainless steel. Such impressed current systems use a stable anode manufactured from an inert material coated with an active layer of platinum or metal oxide. The surfaces which are to be protected are connected to the negative supply while the positive supply is connected to the anode, thus forming an electrolytic cell. Potentially harmful corrosion is prevented by introducing a negative charge onto the aluminum surface thus returning the potential to nearly the passive potential.

The potential and current flow has to be adjusted by a control system using reference electrodes cells. This control system is similar to those for impressed current systems used to protect steel structures. It is absolutely essential that the correct potential be maintained to obtain sufficient protection on the one hand and avoid overprotection on the other hand. Any shift in the potential generated on the aluminum surface may introduce areas of over- or under protection resulting in rapid corrosion attack. The protective potentials which have to be achieved to avoid bimetallic as well as pitting corrosion in the presence of highly divergent velocities were recently investigated. The final conclusions of this study are still outstanding. However, recent experience has improved our understanding of the mechanics involved and the future application of impressed current systems for propulsion units looks promising.

Inspection and Maintenance

In addition to corrosion protection systems, the inspection procedures by the owner and classification society have to be adapted for the service and operating conditions of high speed craft. The lack of a corrosion allowance and the highly stressed structure of HSC's impose the need for owners and classification societies to pay increased attention to localized corrosion attack.

Since high speed craft are usually operated as passenger vessels, the annual inspections are undertaken in dry dock. It is imperative that a detailed, thorough inspection of the vessel be carried out together with the classification society. Increased attention has to be paid to highly loaded structural components to detect possible failure modes such as cracks, buckling and corrosion. The time required for thorough inspections has to be adapted to the type of details and scope of surface area to be inspected.

Due to the localized nature of HSC corrosion, more time is needed than for conventional vessels where uniform corrosion takes place. Special consideration must be given to the characteristics of the affected material, its composition and its suitability for welding. Pitting corrosion can also take place below the surface of the coating in the absence of oxygen. In such cases there may be no visible evidence or perhaps only a slight swelling of the coating surface. Where corrosion is evident repair strategies have to be discussed with class.

Areas of increased probability of fatigue failure should also be regularly inspected during periodic surveys and special inspection plans are to be scheduled. Relevant detection technology and documentation systems must be provided.

Due to the rapid development of pitting corrosion it is recommended that operators implement regular inspection between class surveys for the underwater and susceptible areas of the hull and equipment. The findings should be documented with reports, photos and/or videos. Such an inspection plan need not be approved by class. It will however prove helpful in reducing maintenance costs and unscheduled time out of service. In addition, the longevity and resale value of the vessel can be enhanced when such measures are well documented.

CONCLUSION

High speed craft are built under consideration of rules specially developed for their design. In order to reduce weight and improve performance their structures are dimensioned solely on the basis of strength regardless of the corrosion resistance of the material employed. A corrosion allowance is not usually considered. This imposes the demanding requirement for a highly reliable corrosion protection system as well as thorough inspection and supervision of the vessel and its protection system. Additionally, it is important for the owner and operator to realize that the increased safety margins due to a corrosion allowance are not present. Where local corrosion is discovered, immediate action to repair the damage and mediate the cause of the corrosion is imperative.

REFERENCES

[1] Germanischer Lloyd "Germanischer Lloyd Rules for Classification and Construction, I – Ship Technology, Part 1 – Seagoing Ships, Chapter 1 – Hull Structures, Section 35 – Corrosion Protection", 1998

[2] Germanischer Lloyd "Germanischer Lloyd Rules for Classification and Construction, I – Ship

Technology, Part 1 – Seagoing Ships, Chapter 5 – High Speed Craft", 1998

[3] Weber, S. and B. Richter "Aluminium Anwendungen im Schiffbau", Gesellschaft für Korrosionschutz, Frankfurt - Jahrestagung 1998 – 21-22 October 1998

SNAME *Transactions*, Vol. 108, 2000, pp. 299–325

Development of a Large, High-Speed Ferry

Seiko Ogiwara, Member, **Isao Kawabe,** Visitor, **Hiroshi Nogami,** Visitor, **Tetsuya Omatsu,** Visitor, **Satoshi Toyama,** Visitor, Ishikawajima-Harima Heavy Industries Co., Ltd., **Akitoshi Ando,** Visitor, I.E.M. Co., Ltd., and **Kiyoshi Takaoka,** Visitor, Shin-Nihonkai Ferry Co., Ltd.

Large, high-speed ferries "Suzuran" and "Suisen"

ABSTRACT

Ishikawajima-Harima Heavy Industries Co., Ltd. (IHI) developed and built two large, high-speed ferries, "Suzuran" and "Suisen," for delivery to Sinnihonkai Ferry Co., Ltd. in response to the demand for high-speed distribution by ferries along the Japanese coast. The ferries have the highest speed for large, long-distance ferries (29.4 knots), and make one round trip in two days over a distance of 552 sea miles between Otaru and Tsuruga, and daily service is made possible with two vessels. To satisfy the requirements for the vessels, versatile studies including principal dimension optimization, the improvement of the main engine, hull form and propeller optimization, the reduction of hull weight, and a review of arrangement plans have been conducted. To maintain the on-schedule service required for the ferries, speed performance in actual service were predicted and the performance was monitored through sea trials and actual service to confirm the vessels' satisfactory high-speed performance. The development of these vessels has attracted attention as a technical development that not only changes the conventional form of ferry transportation, but also promotes a modal shift from land transportation to marine transportation.

INTRODUCTION AND PROJECT BACKGROUND

Recently, to deal with environmental issues such as air pollution, a smaller workforce for the land transportation industry in the future, and traffic congestion concurrent with land transportation in Japan, businesses, government bodies, and universities together began studies. Movements that promote a so-called "Modal Shift" to change the principal domestic form of transportation from land transportation by truck to marine transportation and railway transportation become active. As part of this effort, the development of the Techno-Super Liner (TSL) has been conducted under the Ministry of Transportation's guidance. To accomplish the modal shift, naval architects have sought innovations in shipbuilding technology through consideration of economics, convenience, and environmental effects as they relate to the overall distribution system.

Meanwhile, the present long-distance ferries for cars travel more than 400 sea miles in Japan and are incorporated into an overall transportation system that ranges from cargo collection to delivery. Many of these ferries are used for marine transport services to make one round trip every three days. In the marine distribution industry, the demand for improvement of the speed of marine transportation is gradually increasing. An incremental increase in transportation speeds has also been sought, going from three days to two days or from two days to one day; thus, gradual increases in speed have no real meaning.

Such incremental increases in speed, that is, an increase in speed by a large margin at one stroke, involve many economical and technical issues, such as a rapid increase in fuel costs. Thus, most long-distance ferries operate at ship speeds of less than 23 knots, and the service frequency or the number of vessel changes according to the service distance. This is illustrated by the following figures:

Service distance	Vessels necessary for one departure per day
About 400 sea miles	2 vessels (one round trip in 2 days)
About 600 sea miles	3 vessels (one round trip in 3 days)

Under these circumstances, the demand for faster ferries in the transportation industry has become concrete, and there is a progressive tendency toward daily service by shortening one voyage that takes three days for a round trip to two days and another voyage that takes one day and one night for a round trip to a half-day trip. The market for faster ferries and RO/RO vessels is expanding.

The route of Sinnihonkai Ferry Co., Ltd. between Tsuruga and Otaru (552 sea miles) has been developed mainly to transport fresh vegetables, fresh fishes and livestock products from Hokkaido to the Kansai area and also to transport miscellaneous goods for daily use from the Kansai area to Hokkaido. As illustrated in Fig.1, in the case of transportation of cargoes between these two locations in the islands of Japan, marine transportation has an advantage of a shorter distance compared with land transportation.

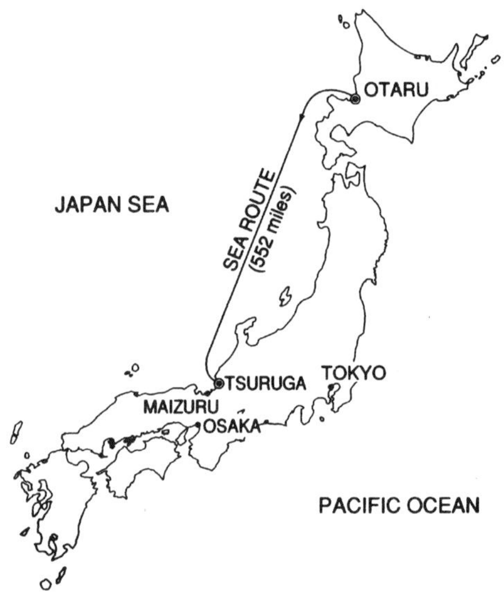

Figure 1 The sea route between Tsuruga and Otaru

For conventional services on this route, it took 30 hours one way and three days for a round trip, however, the high-speed sister ferries "Suzuran" and "Suisen" that were developed and built for Sinnihonkai Ferry Co., Ltd. are 5,440 tons in deadweight and travel the route in 21 hours, shortening a round trip to two days. Having a two vessel-operation system tends to improve distribution services greatly with ferries sailing from both ports at the same time every day. As illustrated in Fig.2, conventional services took four calendar days from cargo collection to market delivery; however, high-speed services could reduce that time to three days.

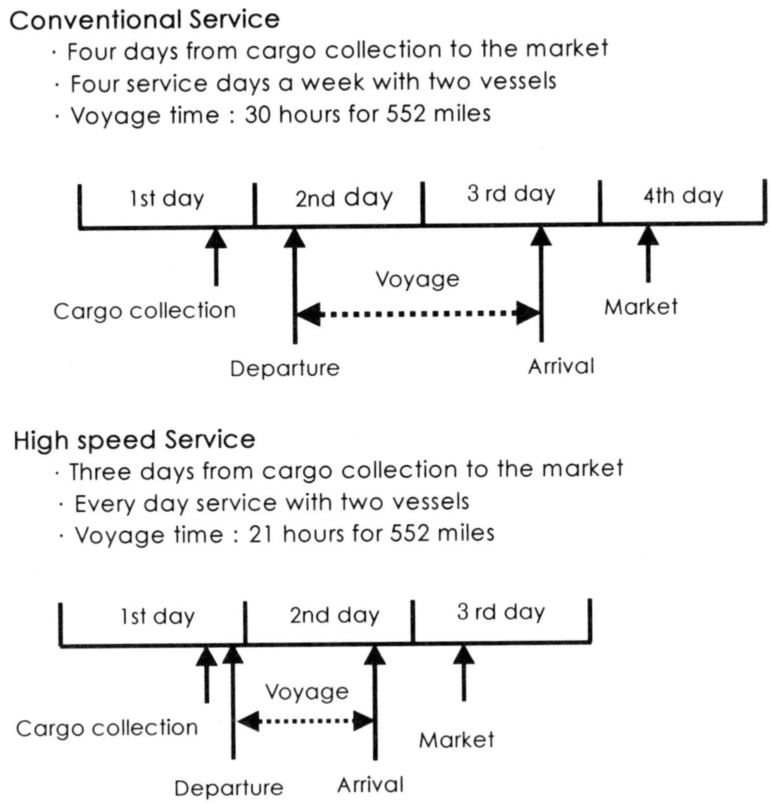

Figure 2 Improvement of long distance ferry service

After receiving an inquiry about large, high-speed ferries, Ishikawajima-Harima Heavy Industries Co., Ltd. (IHI) needed two years to develop and design the vessels. Building started in 1995 and the vessels were put into service in June of 1996. This report describes the methods employed for the optimization of the hull dimensions and hull form, the improvement of the main engine and propulsion system, and the design of a hull structure to satisfy the basic specification requirements for development of a large, high-speed ferry. Also, the propulsive performance of the vessels in actual sea areas after commissioning has been discussed.

DESIGN REQUIREMENTS AND OPTIMIZATION OF SHIP DIMENSIONS

Design requirements

Requirements were for a service speed to travel a distance of 552 sea miles between Otaru and Tsuruga in 21 hours or less with a full cargo load of 4,900 tons or more. The main engines had to be mounted to have a margin of output power to maintain on-schedule service for the vessels even in rough conditions in the Japan Sea during the winter and early spring seasons. A medium-speed diesel engine that maintains a good level fuel efficiency even with an increased output of the main engine and that is lightweight was employed.

The ship's length overall and draft were restricted to 200 m max. and 7.5 m max., respectively, due to harbor conditions. To improve the maneuverability of the vessels in harbors, controllable pitch propellers (CPP) were employed. The basic design requirements for the vessels are:

Service speed:	29 knots or over
Loading capacity:	4,900 tons or over
	(130 12-m trailers and about 500 passengers)
Length overall:	200 m max.
Draft:	7.5 m max.
Main engine:	Medium speed diesel 2 sets
Propeller:	CPP 2 sets

The reason for restriction of the vessel's length overall to 200 m was due to the fact that a ferry or passenger vessel operating in regular service along the Japanese coast need not significantly exceed 200 m for overall length because of transportation lots. Furthermore, the vessel had to meet a maximum of 200 m in overall length due to the fact that a vessel over 200 m that navigates in certain traffic-congested sea routes is imposed severe restrictions under Maritime Traffic Safety Laws in Japan, and piers of ferry terminals in

Japan are mostly a maximum of 200 m.

A loading capacity requirement of 4,900 tons was determined by the total weight of 130 trailers of 30 tons in weight; that is, a payload of 3,900 tons, fresh water of 150 tons, a constant of 300 tons and the weight of fuel. Fuel consumption for one round trip was about 360 tons according to results of performance examinations in the initial design. By adding the fuel consumption of generators to these figures, a total fuel weight of 550 tons was determined. By this analysis, the loading capacity minimum requirement turned out to be 4,900 tons.

Optimization of ship dimensions

After ensuring the satisfaction of requirements for a passenger ferry, that is, sufficient stability, safety, loading capacity, and comfortable accommodations, basic planning to achieve the required speed proceeded.

A design speed of 29.4 knots (90% MCR, 15% sea margin) was determined by taking the sea margin into consideration. Principal dimensions of the vessel were determined by conducting an iteration study for optimization using the design speed, stability, and load capacity within the limitations of ship length and draft as parameters.

(1) Ship length

The Froude number of the vessel is over 0.35 and it is the speed range that resistance abruptly increases due to an increase of wave resistance. Optimization of ship length was studied from wave resistance point of view, and the vessel's overall length was designed to be as long as possible by compromising a decrease in wave resistance and an increase in friction resistance caused by extending the vessel length.

(2) Ship breadth

To avoid an increase in wave resistance, the ratio of the length and breadth should be as large as possible. On the other hand, sufficient stability must be maintained by increasing the breadth. Iteration studies were carried out to obtain optimum breadth through calculations of ship resistance and stability for a series models of systematically changing ship dimensions. Furthermore, the breadth of a ferry should be selected from multiples of the trailer width, that is, a trailer width of 2.5 m and clearance of 0.7 m between trailers. The breadth of 25.0 m was employed for this project in comparison to a former ferry width of 29.4 m.

(3) Draft

The required draft was obtained by calculating the optimal block coefficient and prismatic coefficient based on the vessel length, breadth, predicted displacement, and designed speed of 29.4 knots. A maximum draft of 7.23 m and a designed draft of 7.08 m were determined given the maximum depth at berth for the vessels in the ports at Otaru and Tsuruga.

(4) Longitudinal center of buoyancy

A chart in Fig.3 was prepared by studying the relation between longitudinal center of buoyancy (l_{cb}), prismatic coefficient (Cp) and the wave resistance [1]. In order to reduce the wave resistance as much as possible, the longitudinal center of buoyancy was positioned at approximately the middle of the hull.

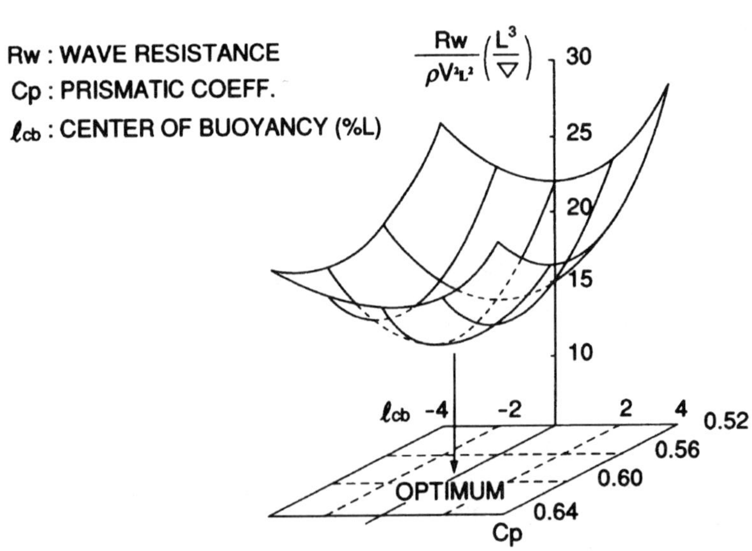

Figure 3 Optimization of lcb and Cp

(5) Depth at car-deck

To increase stability, the center of gravity of the vessel must be positioned as low as possible. To that end, an effective solution is for the car-deck that stores heavy cargoes to be as low as possible. The midship coefficient is rather smaller than conventional ferries by taking high-speed performance seriously, the bottom amidships is almost a round shape. Therefore, the mounting level of the main engine is a distance 1 m higher than the level in the conventional ship. This will raise the car-deck by 1 m; however, the car-deck height could be restricted to be only 0.5 m higher than the car-deck height of a conventional vessel by improving the shape of the main engine foundation, exhaust pipe size, and its layout, with the car-deck height of the vessel becoming 9.5 m.

(6) Payload

The minimum payload requirement is 4,900 tons; however, operational profits could be increased by loading more trailers. Therefore, a maximum payload that matches the service speed of 29.4 knots was examined. The loading capacity upon completion of the vessel was 5,440 tons, with a loading space that can store 159 trailers 12 m in length.

The main characteristics of the hull as determined by optimization of the ship dimensions are listed in Table 1. Also, the characteristics of a conventional vessel operating in the same route are shown in Table 2 for comparison.

Table 1
Principal dimensions of the "Suzuran" and "Suisen"

Certification		JG, Class 2
Overall length		199.45 m
Length between perpendiculars		187.00 m
Breadth (molded)		25.00 m
Depth (molded)	Up to C deck	14.80 m
	Up to D deck	9.50 m
Designed draft (molded)		7.08 m
Full load draft (molded)		7.23 m
Dead weight (designed draft)		5,440 tons
Gross tonnage		17,300 tons
Displacement tonnage		16,420 tons

Table 2
Compared performance of the "Suzuran", "Suisen," and a conventional ferry "Lavender"

Items / Ship Name	"Suzuran" and "Suisen"	"Lavender"
Overall length (m)	199.5	192.9
Breadth (m)	25.0	29.4
Depth (m)	9.5	9.0
Designed draft (m)	7.08	6.50
Dead weight (ton)	5,440	7,689
Main engine	18PC4-2B	8PC40L
Main engine output (KW)	23,850 x 2	9,710 x 2
Service speed (knots)	29.4	21.8
Route	Otaru - Tsuruga	Otaru - Maizuru
Distance of the route (sea miles)	552	580
Delivery	June 1996	Sept. 1991

IMPROVEMENT OF THE MAIN ENGINE

Development of a medium speed diesel engine

Of the basic requirements for the vessel, a high speed is extremely important, and a gas turbine engine could have been considered for the main engine prior to this project. However, a medium speed diesel engine was employed in light of fuel economy as the first priority.

S.E.M.T Pielstick and Diesel United, Ltd. (DU), a related company of IHI, developed the DU-S.E.M.T Pielstick 18PC4-2B model diesel engine that had the largest output in the world as a single medium speed diesel engine in response to the trend for high speeds for large car-ferries and RO/RO ships. IHI employed this engine model, mounting two sets of the engine on the vessel.

This engine was developed following the reliable structure of the PC4-2 model and the PC40 model, which hold a number of operation records as a marine main engine. The engine achieved the largest outputs among medium speed diesel engines, which is a rated output of 23,850 KW with 18 V-type cylinders with an output of 1,325 KW per cylinder as same as PC40 model

Principal characteristics of the 18PC4-2B model are shown in Table 3. Also, a section plan and photograph at the shop test are shown in Figs. 4 and 5, respectively.

Table 3
Principal characteristics of the
DU-S.E.M.T Pielstick 18PC4-2B Engine

Maximum Continuous Rating	23,850 kw x 410 rpm
Normal output	21,465 kw x 410 rpm
Rated revolution	410 rpm
Cylinder bore	570 mm
Piston stroke	660 mm
Ratio of stroke and bore	1.15
Mean piston speed	9.0 m/s
Mean effective pressure	2.3 Mpa
Maximum combustion pressure	15.5 Mpa
Specific fuel consumption at service output on shop test	178 gr/kwh

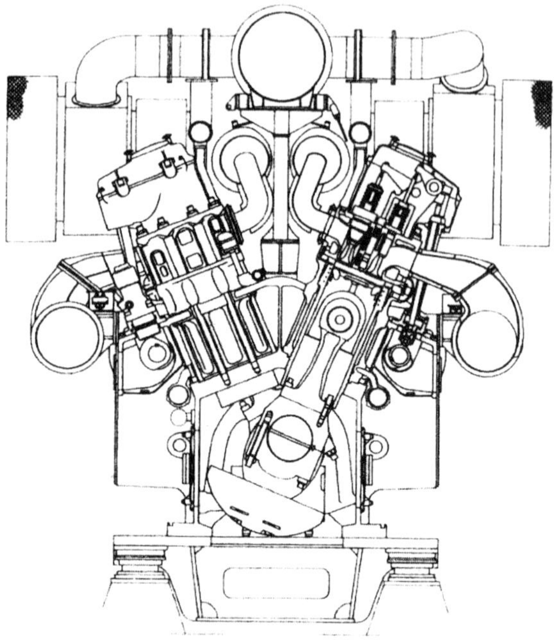

Figure 4 Section plan of the main engine

Figure 5 The main engine:
DU-S.E.M.T Pielstick 18PC4-2B

Taking space saving in the engine room into account, a cylinder bore of 570 mm as same as that of the PC4-2 and the PC40 models and 660 mm of the piston stroke, which is between 620 mm of the PC4-2 model and 750 mm of the PC40 model ,was employed. For the dimension between cylinders, 980 mm was employed as with the dimension of the PC4-2 model, and the almost same engine mounting space as that of the PC4-2 model with an output of 1,215 KW per cylinder was maintained. Meanwhile, for reliability, the mean effective pressure and the rated revolution were maintained as the same levels as those of experienced engines with confirmed reliability; however, the new engine could produce an increase in output per cylinder of 9% in comparison with the PC4-2 model. Furthermore, structural reliability has been increased by improvement of the bearing performance with increments of the crank shaft journal diameter and also improvement of the piston and piston rod.

The propulsion plant consists of two engines and two shafts, which drive controllable pitch propellers (CPP) via Geislinger couplings and planetary type reduction gears. A combination control combining the revolution and propeller pitch for the main engine and CPPs was employed and Joystick control is also possible for better maneuverability in harbors.

Measures for reducing vibrations and noise

To ensure comfortable accommodations as a passenger vessel, an extremely important aspect is to isolate vibrations and noise caused by the engines to whatever extent possible. Elastic supports with rubber units are employed on the vessel for the main engines and generator engines to interrupt vibration. In particular, the total weight of the main engines, 385 tons per engine, is supported with 24 rubber units. The arrangement of the rubber units and spring constants was determined after accounting for the exciting force of the engines and vessel motion so that harmful resonance will not occur in the range of the engine revolution and also, in light of vibration-isolating properties and deformation due to deterioration with age. Furthermore, to reduce engine vibrations, the improvement of the rigidity of the main engine columns and oil sump and the arrangement of the turbocharger on the engine columns proceeded as the results of natural frequency analysis by a finite element method. To reduce vibrations and noise for accommodations, elastic supports for the engine exhaust gas pipe, a high performance silencer for exhaust gas, vibration suppression steel plates to engine casing walls, and a low-noise tooth profile for reduction gears have been employed.

HULL FORM OPTIMIZATION

Sectional area curve and bulbous bow

It is impossible to satisfy the speed requirement just by decreasing the displacement of a conventional ferry that travels the same route to produce an equivalent level for this vessel. The main reason for this is due to the fact that the Froude number of this vessel at the designed speed is as high as exceeding 0.35, positioning the vessel in a speed range where the ratio of the wave resistance component in the total resistance increases. Thus, it is essential to keep the wave resistance to a minimum level by improving the sectional area curve and bulbous bow form.

First, as a measure for improving the sectional area curve, a rough hull form was prepared by optimizing calculations for the sectional area curve and the size of bulbous bow with first order analysis like a linearized wave resistance theory [1].

Using this as a prototype, four kinds of hull forms with changes in the sectional area curve and the bulbous bow form were prepared. The performance of each of these hull forms was checked by proceeding with more precise numerical calculations and towing tests, and optimization of the sectional area curve and bulbous bow form was conducted so that the minimum wave resistance would be obtained. The towing tests were conducted in the IHI Ship Model Basin. On the basis of this analysis, a considerably large bulbous bow as compared to a conventional ferry was employed (see Fig.6). For numerical analysis to predict the wave resistance, a nonlinear Rankine source method [2] was used in potential flow analysis and this was utilized for local improvement of the hull form by calculating the waves generated by the hull and the pressure distribution on the hull. The panel arrangement on the hull surface used for the Rankine source method is illustrated in Fig.7. The computed wave pattern and the pressure distribution on the hull are illustrated in Figs. 8 and 9 respectively.

By this method, as illustrated in Fig.10, the ratio of the wave resistance component in the total resistance decreased to about 30%.

Figure 6 The bulbous bow of the "Suzuran"

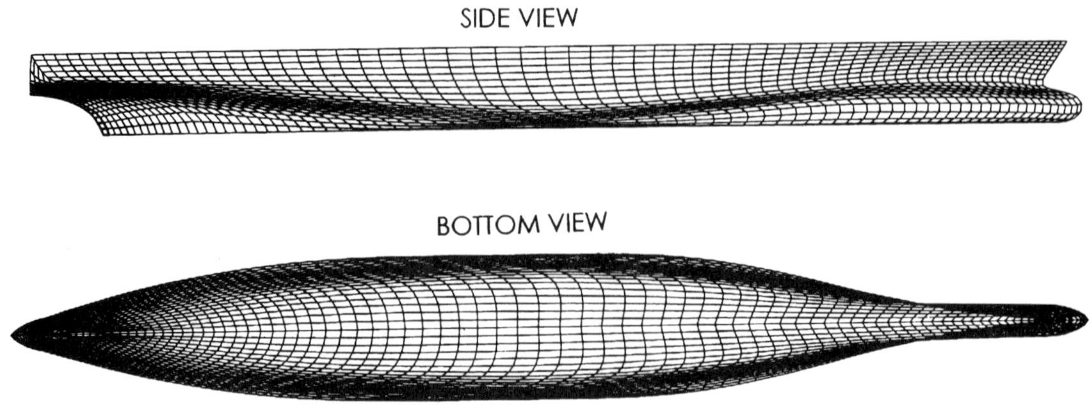

Figure 7 Hull surface panel arrangement for wave resistance calculation

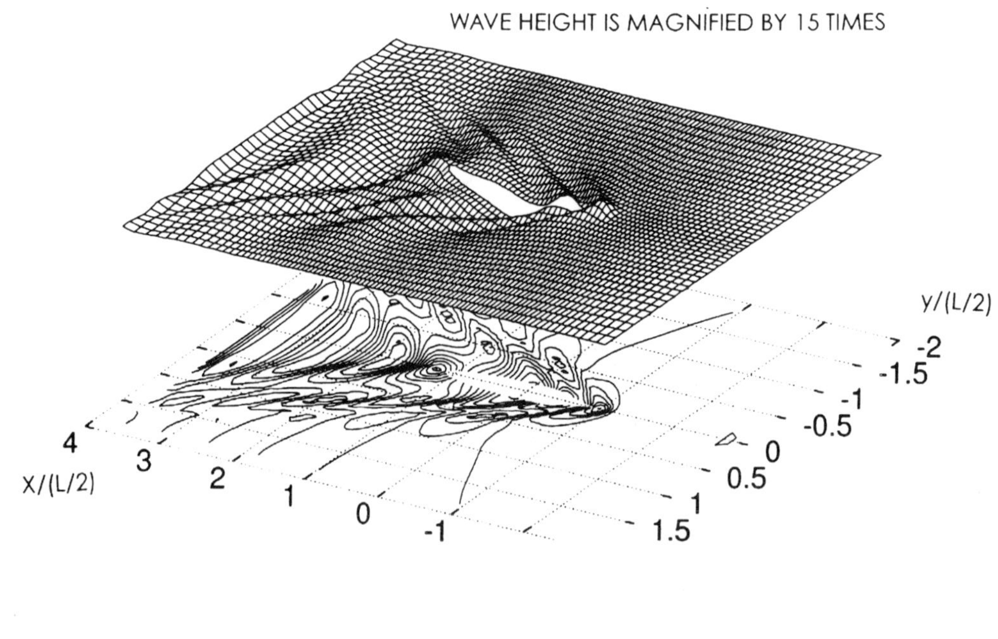

Figure 8 Computed wave pattern at 29.4 knots

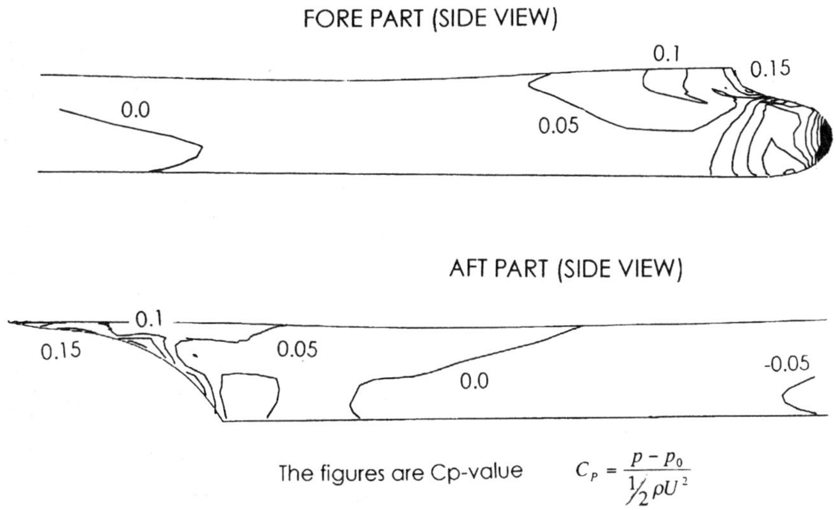

$$C_P = \frac{p - p_0}{\frac{1}{2}\rho U^2}$$

Figure 9 Computed pressure distribution on the hull surface at 29.4 knots

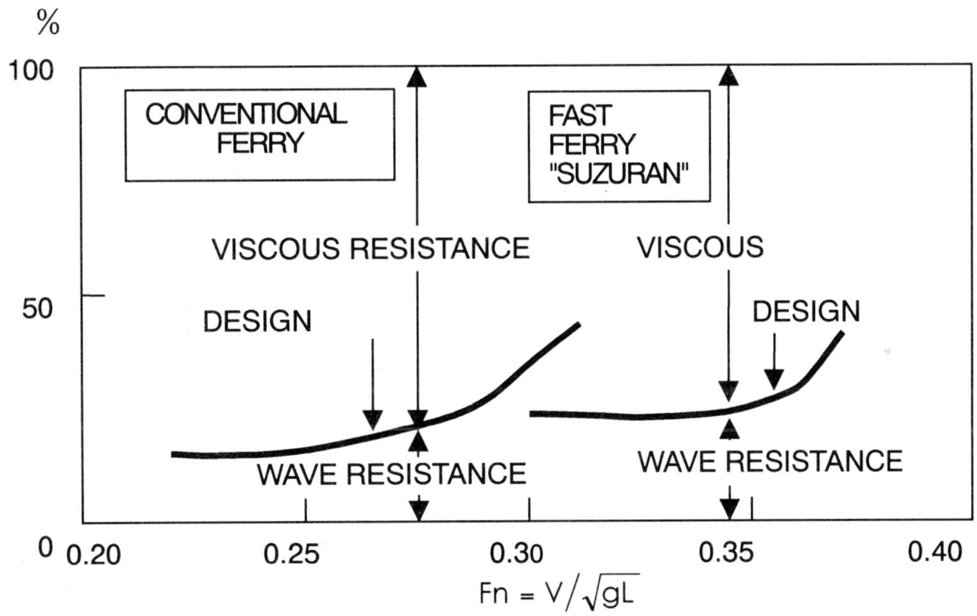

Figure 10 Ratio of resistance components

Appendages

The vessel is equipped with supporting members for two propeller shafts and other appendages such as a bilge keel and rudder. Due to the large output of the main engine, the resistance caused by the supporting members for the propeller shafts is significant, and the resistance from these appendages reaches 30% of the residual resistance during operation at the designed speed. Therefore, a design must be made so that the resistance caused by the appendages drops as much as possible. Results of towing tests indicated that if a shaft bracket that is commonly used is employed, a bracket with a large section will be required because of the large shaft horse power, which causes an increase in resistance and also harms the propeller performance. Thus, towing tests were conducted with both a shaft bracket and bossing as supporting members, and test results confirmed that the bossing showed less viscous resistance and improved the propulsive performance. Bossing was employed according to these test results. The mounting angle of the bossing against the hull surface was examined using computational fluid dynamics so that the resistance would be slight [3]. The stern form is illustrated in Fig.11. The towing tests confirmed that by employing bossing the appendage drag was reduced by 15% in comparison to a conventional bracket.

The form and mounting location of bilge keels were designed using numerical analysis of the flow around the hull. The number of rudder and size were designed using a simulation of maneuverability performance, and the determination was made to install one rudder on the centerline at the stern, which gives satisfactory maneuverability even under the harbor conditions of Tsuruga and Otaru.

The vessel is equipped with two thrusters at the stern and one thruster at the bow to improve maneuverability in the harbors. The increment of resistance caused by thruster holes was reduced as much as possible by tapering the hole edges of down stream side.

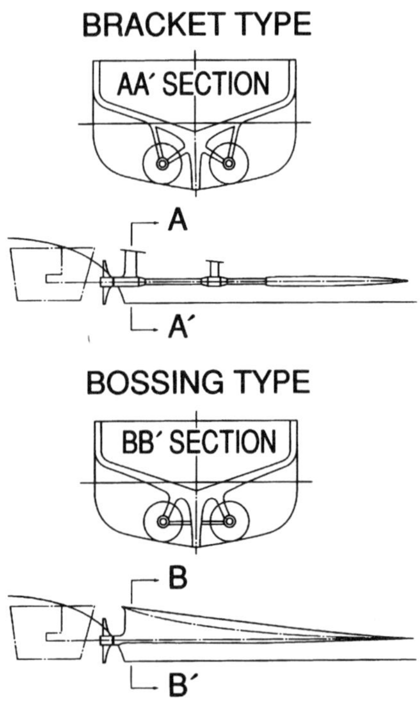

Figure 11 Stern form for the high-speed ferry(below) and a conventional ferry(above)

Evaluation of the sea margin

The vessel is scheduled to travel one voyage in 21 hours or less. To maintain on-schedule service for one round trip in two days, the vessel must have performance that enables travel even in rough sea conditions. Many days of rough sea conditions are expected in the Japan Sea from winter to early spring, which may result in a reduction of ship speed. Thus, the propulsive performance in calm seas as well as in heavy seas, the "sea margin," was evaluated.

As factors involved in the sea margin, added resistance in waves, wind drag, and resistance due to fin stabilizer operation were examined. For the analysis of the sea margin, the logbooks of a conventional ferry operating the same route were analyzed and weather and sea data in a year was utilized.

Added resistance in waves was predicted by model tests with regular and irregular waves in the IHI Ship Model Basin. The results of the model test for the added resistance in regular head waves are illustrated in Fig.12. The largest added resistance in regular head waves occurred at a ratio of wave length and ship length (λ / L) =1.1, which was 29% of the resistance in calm seas. The waves most frequently encountered are Beaufort scale=6~7 and λ / L=0.6 or less, and added resistance in this case was 8% of the resistance in calm seas.

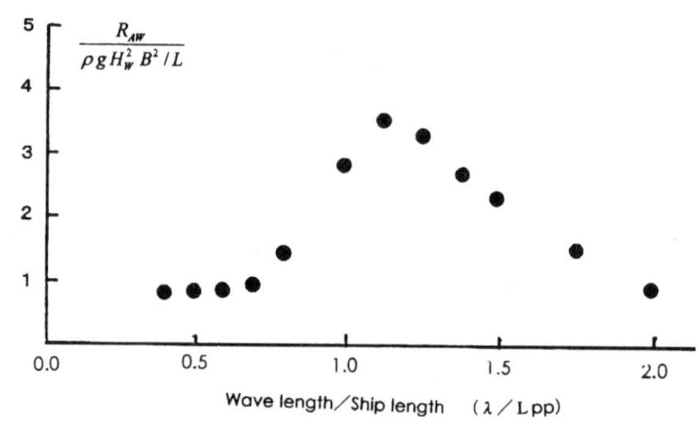

Figure 12 Response function of added resistance in regular head waves at Fn=0.353

Wind drag was predicted using the empirical chart of the wind drag for ships. A wind pressure resistance coefficient of 0.7 was employed by the actual records of the ships that were built up so far. Accounting for the probability of wind occurrence in the Japan Sea, the wind drag of the vessel was obtained using statistical data for wind velocity and wind direction collected from conventional ships' logbooks.

To estimate the resistance due to fin stabilizer operation, the resistance coefficient of the fin was calculated in reverse by measuring a speed reduction in cases where a similar fin stabilizer installed on a conventional ship is and is not in operation. On the basis of these results, the resistance when the fin stabilizer of the vessel is in operation was estimated. The resistance due to fin stabilizer changes depend on the attack angle, which is 4~8% of the hull resistance in calm seas. When all of the forces of resistance from these factors are added, the sea margin of the vessel in the Japan Sea can be estimated at 8%~35%, varying greatly with the season.

PROPELLER DESIGN AND POWERING

Propeller Design

Design of a high-power, high-speed ship propeller requires suppression of propeller cavitation that causes vibrations, noise, and erosion; the efficiency must be improved as well. The design point of the CPPs mounted on this twin-screw vessel were greatly deviated from design points of other merchant ships as illustrated in Fig.13. Therefore, it seems necessary to increase the expanded area of the blade to suppress cavitation, and also decrease of the propeller efficiency can't be avoided. To resolve these issues, the propeller design proceeded combining with improvement of flow into the propeller.

(1) Improvement of the flow at the stern

To avoid flow separation and cavitation originated from the shaft brackets and improve the flow field around the propeller shafts during high-speed operation, a method that covered the propeller shaft with bossing was employed instead of a conventional bracket-supporting method as illustrated in Fig.11, and

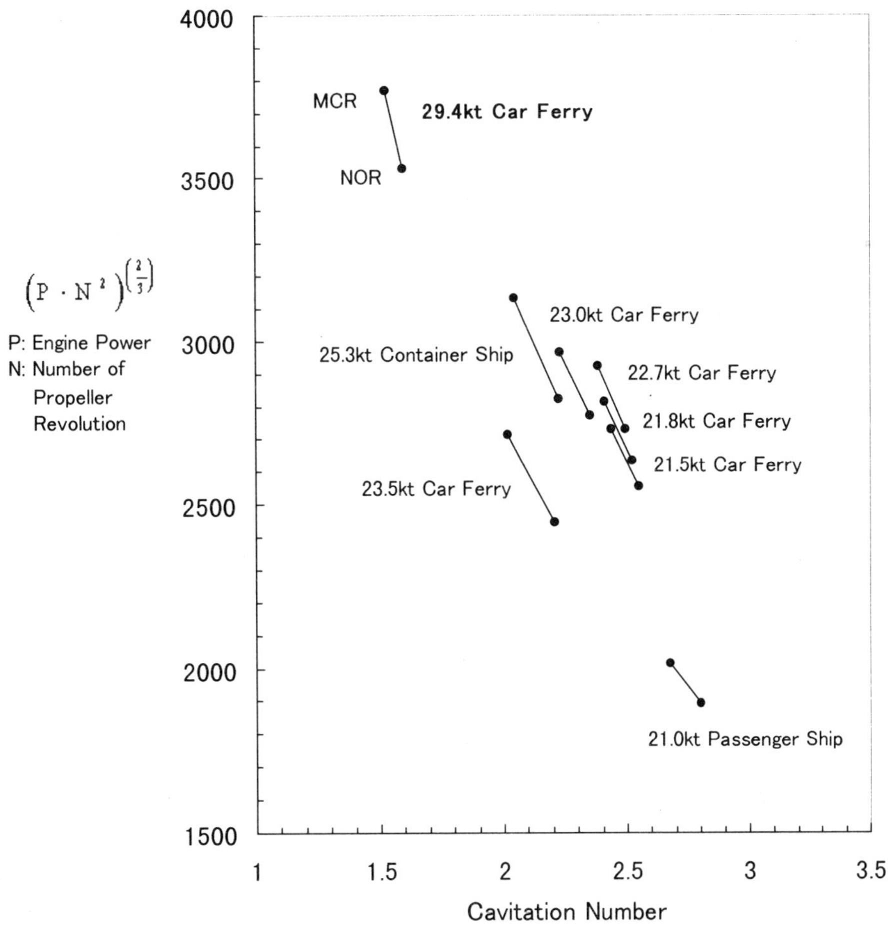

Figure 13 Design point of CPP of twin-screw merchant ships (built by IHI)

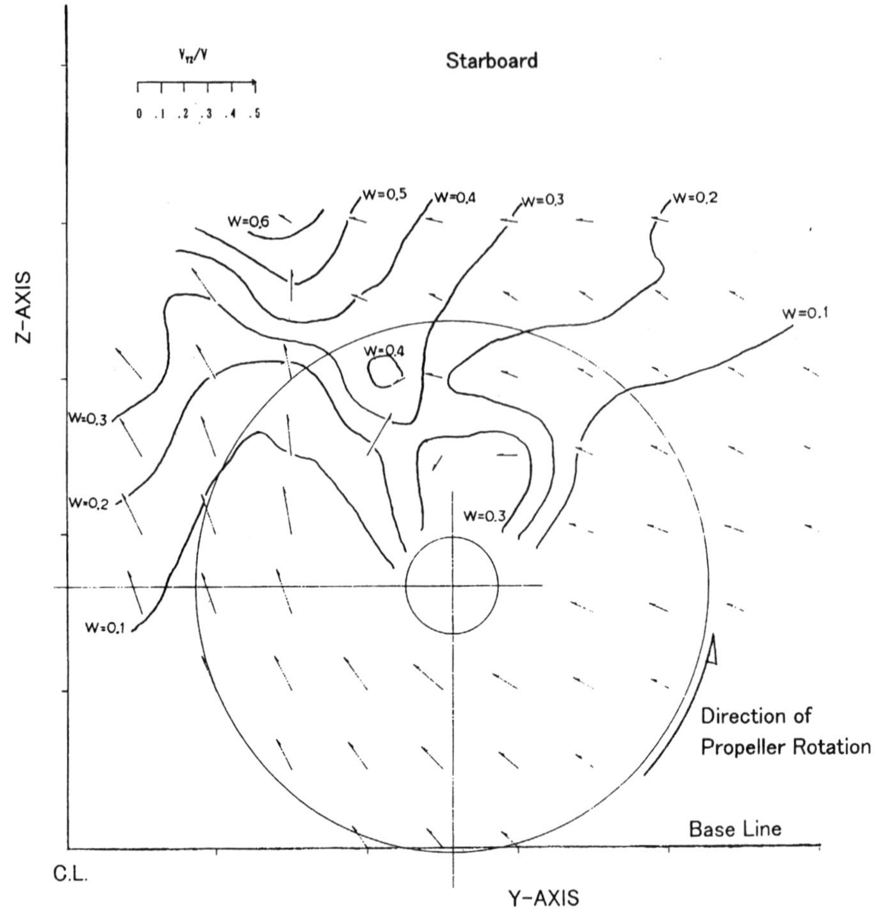

Figure 14 Wake distribution on the propeller plane (Looking forward from the stern)

improvement of the wake distribution at the plane of the propeller and a decrease of the form drag of shaft supporting system were achieved. With improvement of the wake distribution, an occurrence of severe propeller cavitation was alleviated and an increase in the expanded area of the propeller blade was suppressed; thus, resulting propeller efficiency improved. Measurement of the wake distributions in the plane of the propeller of the ship model were conducted in the towing tank, and the result is illustrated in Fig.14.

By employment of this bossing, self-propulsion factors, that is an interaction of the hull and propeller, also improved. The vessel achieved good propulsive performance, which assisted in the work to meet requirements for high speeds and comfort at the same time.

(2) Selection of propeller characteristics

In preparation of designing the propeller, research work of existing propellers mounted on high speed vessels were conducted. Propeller model tests including open water tests, cavitation tests and fluctuating pressure measurements were conducted. Computation using the lifting surface theory[4] were made to predict performance and study the modification of propeller characteristics. Also, accounting for propeller efficiency, blade strength, and cavitation erosion as well as hull vibration / strength and vibration of the shaft system along with performance and vibration of the main engine, propeller characteristics were selected so that an optimized propeller was obtained.

The propeller diameter was examined with normal output of main engine as a design point. Employing four blades for the CPP, an optimum propeller diameter at the design point was selected using a Troost Bp-δ chart and Burrill's cavitation chart. However, the Troost chart is made based on the data of propeller open water test data and when the propeller is mounted at the stern, the optimal diameter is said to be a value slightly smaller than the optimal value obtained from the chart. Therefore, by determining a reduction rate based on past experience, a propeller diameter of 5.2 m was employed.

An expanded area ratio, a pitch distribution, and camber distribution were examined with maximum output of main engine as a design point. Since the design condition of the propeller was much more severe than usual, as the first step of design study, two kinds of prototype propellers were designed and model propellers of 250mm in diameter were manufactured as shown on Table 4 . One is a propeller (Type-A) with priority on cavitation performance as the design point and the other is propeller (Type-B) with priority on propeller efficiency. Various characteristics obtained from the model tests in the cavitation tunnel were compared and examined to determine essential characteristics.

Table 4
Characteristics of prototype and the final propeller

	Prototype		Final
	Type-A	Type-B	Propeller
Pitch ratio (0.7R)	1.451	1.376	1.383
Expanded area ratio	0.710	0.680	0.680
Efficiency	0.667	0.678	0.682

When focusing on characteristics, special consideration should be given to:

- Restraining bubble cavitation that occurs around the leading edge on the face side of the propeller blade at blade root which is typical to CPP.
- Restraining tip vortex cavitation that occurs on the blade tip.
- Stabilizing fluctuation of sheet cavitation on the back side of the propeller blade.
- Improving propeller efficiency.

Meanwhile, the blade was optimized by improving pitch, camber and the blade width distribution in the direction of the radius using theoretical calculations. For the pitch distribution, having considered the bearing force in the wake, the reference pitch was determined by reducing the circulation intensity at blade roots and blade tips. For camber distribution, when an extremely large camber is employed and if it is operated with a reduced pitch angle while maintaining high revolutions with a low load, there is a risk of face cavitation resulting in erosion. Therefore, a camber distribution with a slightly reduced camber at the blade tips was employed. For blade width distribution, as with the CPP, the blade width at the root is restricted and the boss ratio is large; thus, it is impossible to use the same blade width distribution as that of a fixed pitch propeller obtained from the design chart. From empirical knowledge of various CPP designs, equivalent cavitation performance to the chart-based design can be obtained by increasing the blade width of the CPP from around 0.7R (R: propeller radius) to the blade tip, and thus, the blade width distribution was determined

(3) Confirming performance by model tests

Results of cavitation tests for the final CPP are shown in Figs.15 (photograph) and 16 (sketch). These are cavitation patterns on the backside of the propeller blade. Tests were conducted at maximum output condition (Thrust Coefficient Kt=0.195, Cavitation Number σn = 1.493, Propeller Revolution n=21.5rps). Fig.15 shows the results of the final propeller, Type-A and Type-B for comparison. From these results, harmful cavitation like the face cavitation and bubble cavitation that may cause severe erosion did not occur. The results of measurements for pressure fluctuations are also illustrated in Fig.17. When the peak pressure is converted for an actual ship, it is 4.7 Kpa, which is the value in a range that does not produce a hull vibration problem when refered to other vessels built by IHI. 4 sets of the full scaled CPP were manufactured by Kawasaki Heavy Industries Co. Ltd., and mounted to "Suzuran" and "Suisen".

Powering

Ensuring the service speed of the vessel was an extremely important requirement. The speed of the actual vessel was predicted using hull resistance and self-propulsion factors obtained by tank tests with a 6-m model ship along with the propeller open characteristics. Results confirm that the vessel can achieve the designed speed of 29.4 kt with sea margin of 15%. Measures to improve speed as applied to the conventional ferry "Lavender" with a speed of 21.8 knots and results are summarized as follows:

- Increase of the main engine output (19,420 kw→47,700 kw) 21.8 knots →25.9 knots
- Achieving hull slenderness while ensuring stability →28.5 knots
- Decrease in the wave resistance by improving the bow form →29.0 knots
- Improvement of propulsive performance by employing the bossing →29.2 knots
- Improvement of the blade shape of CPP →29.4 knots

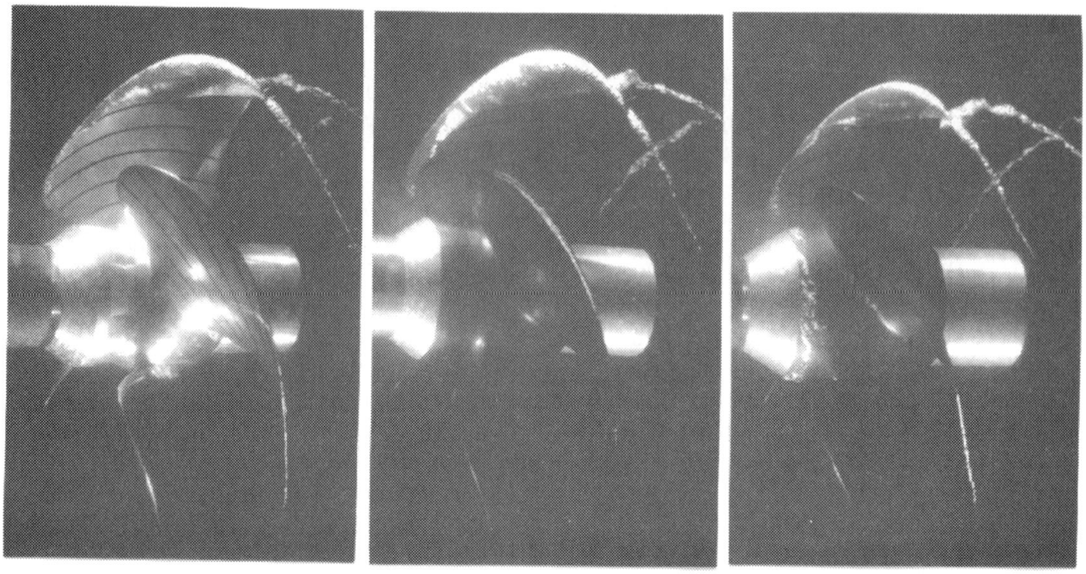

The final propeller | Prototype propeller Type-A | Prototype propeller Type-B

Figure 15 Comparison of cavitation pattern at θ =10deg.

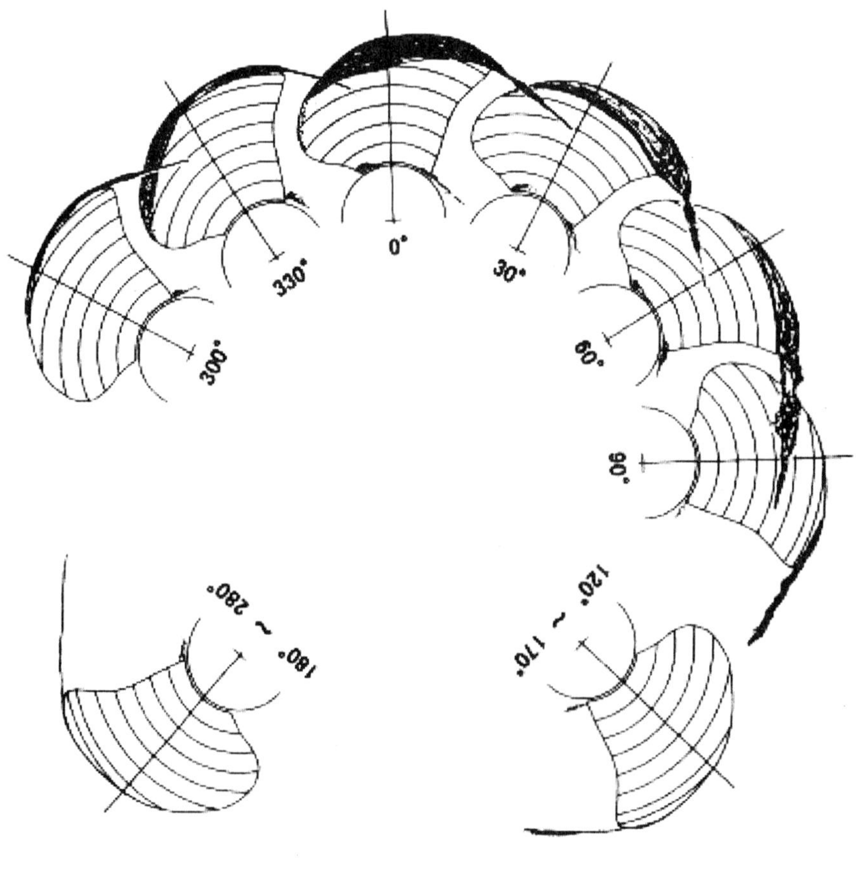

Figure 16 Cavitation pattern of the final propeller

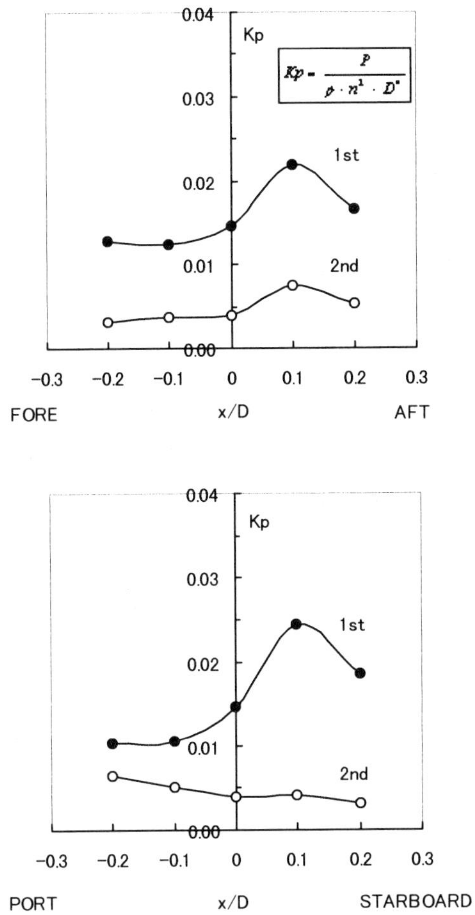

Figure 17 Fluctuating pressure of the final propeller

HULL STRUCTURE DESIGN

Design policy for the hull structure

Since the vessel has speed and power that are greater than conventional ferries built by IHI, the hull structure design proceeded by placing great importance on several points.

- Weight reduction for the hull structure
- Measures to withstand the wave impact load
- Measures to minimize vibration

Focusing on these points, an outline of the vessel's hull structure design is provided in detail.

Weight reduction for the hull structure

To reduce the weight of the hull structure, more high-tensile steel was used than in conventional ferries. High-tensile steel with a yield stress of 315 N/mm^2 has been used for strength decks in the main hull, car decks, the shell, and the longitudinal frame. For the bottom plating and double bottom, mild steel has been used so as to lower the center of gravity. The ratio of high-tensile steel components to the hull steel weight is about 30%.

It is possible to use high-tensile steel with yield stresses that are higher than 355 N/mm^2; however, there is not a significant advantage in using such steel because the plate thickness may not be reduced due to the criteria for buckling and other factors with this size of ship, and subsequently high-tensile steel with a yield stress of 315 N/mm^2 was used. For instance, the strength deck for the longitudinal strength of the vessel is B-deck, and its plate thickness is 9.5 mm, which is determined by buckling due to a wave longitudinal bending moment with a sagging condition. The same plate thickness cannot be reduced even through the use of higher strength steels unless the longitudinal fame space narrows.

The frame and longitudinal frame spaces were determined so as to reduce the car-deck weight to a minimum. Scantlings of car-decks are determined by the class rule based on the design wheel load of loaded vehicles; however, since the load is a concentrated load, narrowing the frame space does not always reduce the weight. With this vessel, the frame and longitudinal spaces of 750 mm and transverse deck girder space of 3000 mm present a minimum weight and thus these dimensions were employed.

Vessel car-decks have a line of pillars nearly on the hull center line. Although a structure without pillars is preferable for car loading operations, but a structure without pillar requires an increase in the strength of the transverse deck girders and the weight increases due to enlarged transverse deck girder dimensions. The depth of the transverse deck girders deepens, increasing the depth of the vessel and subsequently raises the center of the gravity. The vessel was designed so that the depth of the girders could be as shallow as possible by providing pillars on the car-decks and support for the girders. By designing the longitudinal deck girders supporting the transverse deck girders to be the same depth as the transverse deck girders and arranging the pillars at interval of nearly 3 spaces of the transverse deck girder, the weight was reduced and the center of gravity was lowered.

Since the vessel's hull form is fine, the ship's transverse girders under the waterline have a curved form. Furthermore, the double bottom also has a shallow form in the ship's side, and the vessel's structure is considerably different from that of a conventional ferry. By taking out the portion between the transverse bulkheads amidships and conducting FEM analysis on the portion, the strength to withstand wave loads and car loads was confirmed. Since the stern section consists of a thin skeg structure and a shallow stern overhang structure, the strength to undergo hydrodynamic force and rudder reaction force was confirmed by FEM analysis of the respective structures. Amidships are shown in Fig.18.

Measures to withstand wave impact loads

Due to the vessel's high speeds, its relative speed between hull and wave is high and the vessel is liable to suffer wave impact loads. In particular, the bow and stern sections have a high relative speed and are likely to move above the water surface. Therefore, these sections must be designed to withstand wave impact with special precautions.

First, the hull motion calculations were conducted by using the strip method, and the relative speed between the subjected hull section and water surface was obtained, and then, the impact loads were calculated by a theoretical formula for water surface impact using the relative speed obtained. That is, by predicting the hull motion of the vessel in the long term by a strip method using wave data for the Japan Sea, the maximum value in 20 years of the relative speed between water surface and respective hull sections was calculated.

For example, the impact pressure on the bottom section was calculated as about 1200 kN/m^2 around the bow and about 850 kN/m^2 around ordinate 8. The structural dimensions of each hull section, such as the bow section structure that is liable to suffer impact loads, were determined using the impact loads obtained by the process mentioned. For the structure of the bow section, since the vessel is a slender form when compared to conventional ferries, the strength to withstand wave impact loads was confirmed by conducting FEM analysis not only on local members such as plates and frames, but also on the whole bow structure.

Measures to minimize vibration

Generally, high-power ships like this vessel have a large exciting force and will likely encounter vibration problems. The vessel is a ferry to carry passengers and thus requires lower vibrations. Therefore, vibrations were examined during design as an extremely important issue. The propellers and the main engine are the major exciting sources that cause hull structure vibration. During design, measures to minimize vibration were examined so that the natural frequencies of the hull structure would avoid resonance with the excitation frequency of the exciting source.

The vessel's characteristics concerning the exciting force are:

- The exciting force of the vessel's 18-cylinder main engine is less than an 8-cylinder to 9-cylinder main engine on a conventional ferry.
- By employing elastic supports for the main engine, a greater vibration interrupting effect should be achieved.
- The exciting force of the propeller will be increased due to high-speed and high-power operation when compared to conventional ferries.

Therefore, as a structural measure to withstand the exciting force of the main engine, resonance with

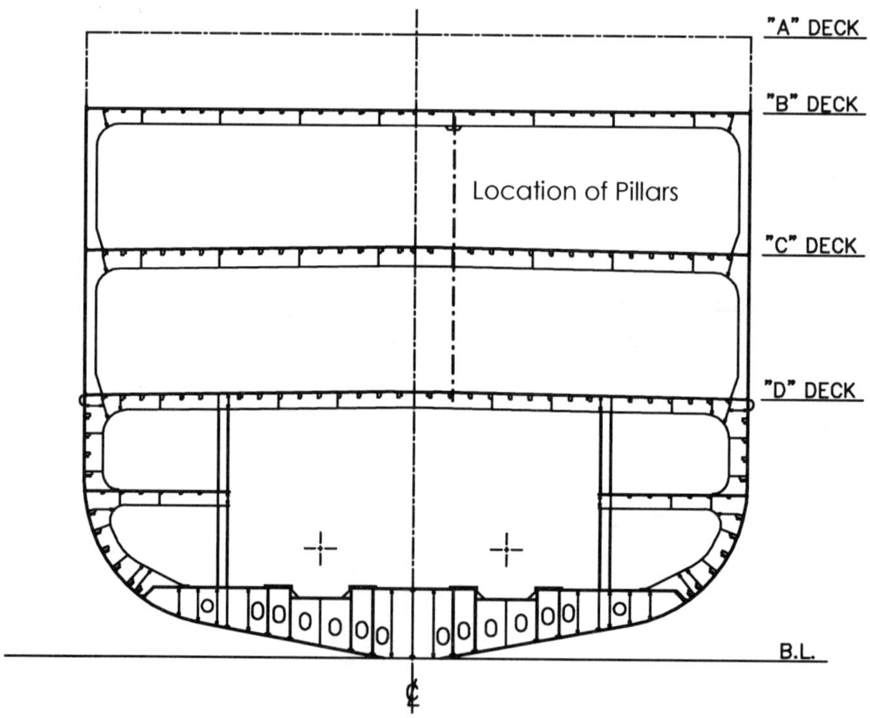

Figure 18 The midship section of "Suzuran" and "Suisen"

structural members was avoided only in the sections around the engine room by focusing only on the 1.5th X type and 4.5th H type caused by the guide force that is significant among exciting forces. As for exciting force of the propeller, resonance with the 1st exciting force of the propeller was avoided in whole sections of the vessel. Normal revolution numbers of the main engine and propeller are 410 rpm and 160.8 rpm respectively, and considering the operating conditions, the following revolution ranges that have a high frequency of use were selected as the ranges where resonance would be avoided.

Main engine: 348-410 rpm, Propeller: 136-160.8 rpm

As for hull girder vibration, on the basis of the vibration measurements for conventional ferries, the natural frequency of the vessel was predicted using the bending rigidity as well as weight and length of the hull as parameters. Predicted results indicated that the natural frequency range of hull girder vibration was about 80~350 cpm from 2nd to 8th-node vibration. Meanwhile, both the excitation frequencies of the 2nd unbalance moment of the main engine and the exciting force of propeller that causes hull girder vibration are higher than 500 cpm. The frequency of hull girder vibrations that produce resonance with these excitation frequencies will be of high order, and therefore this excitation frequency range will be dominated by local vibration, not by hull girder vibration. The analysis of the natural frequency was conducted for every local structure in each hull section, and the hull structure was designed so that resonance with the excitation frequency has been avoided.

The analysis of the natural frequency of local structures in each hull section was conducted with FEM. The local structures subjected to the analysis were the accommodation deck, car-deck, double bottom of engine room, stern skeg section, and stern overhang section. As for the stern overhang section, the results of analysis indicated the possibility of resonance with the exciting force of the propeller in the initial structural design, and resonance was avoided by increasing the natural frequency with an additional longitudinal bulkhead.

Furthermore, there was a possibility of resonance with the exciting force of the propeller in relatively wide deck structures for several accommodations, and reinforcement was prepared to increase the rigidity of the girders. As for stern skeg section and stern overhang section, a large forced vibration caused by the exciting force of the propeller was expected even if the resonance was avoided. Results confirmed that the response value was within the allowable amount in the revolution range of use after FEM vibration response analysis using the surface force and the bearing force predicted by the lifting surface method of propeller performance.

GENERAL ARRANGEMENT

The first examination of the arrangement of the hull under the car-decks was to determine the longitudinal location of the main engine in the hull. By optimization of the center of buoyancy, longitudinal location of the center of buoyancy came closer to the midship compared to conventional ferries. Therefore, the position of a heavyweight main engine was determined to be amidship.

Meanwhile, since the center of the car-deck area is in aft of the middle of the vessel due to the bow mooring deck, the vessel tends to be trim by the stern with a full load of cars. To compensate the stern trim moment due to cars, fuel tanks and fresh water tanks were arranged forward. It is better to locate the fuel tanks as close to the main engine as possible; however, for the vessel, the trim was adjusted by locating the fuel tanks forward by a considerable distance.

To compensate the heel produced by an unexpected weight unbalance of cars in the direction of the ship's width or a side wind, a heel adjusting tank and pump were located amidships.

When the initial design of this arrangement was prepared, the length of each compartment was determined by examining damage stability requirements after two compartments flooded. With the results of this examination, a cross flooding system was employed for the heel tank.

The engine casing is located in the car-deck area as well, however, in order to increase number of loading trailers and to improve the cargo handling efficiency, the total length of car loading could reach 1,960 m by improving the arrangement of the exhaust pipes and other parts inside the casing to make the casing width equivalent to the trailer width, that is, 3.2 m or less. The loading capacity and main equipment are described below.

Loading capacity for cars		
C deck	12 m trailer	82
D deck	12 m trailer	77
Passenger capacity		507 persons
Crew capacity		50 persons
Cargo handling equipment		
Stern lamp		1 set
Stern side lamp		1 set
Turn table		2 sets
Car-loading door		1 set
Fin stabilizer		1 set
Bow thruster		1 set
Stern thruster		2 sets
Elevator		2 sets
Liftable car deck		1 set

The vessel has two tiers of car decks and three tiers of passenger decks. Up to 159 12-meter trailers can be loaded on the car decks, with liftable car decks for a capacity of 40 passenger cars. The breakdown of the passenger capacity of 507 persons is as follows:

Suite (capacity of 2): 4 rooms
Special-class Japanese-style room (capacity of 2): 6 rooms
Special-class Wester- style room (capacity of 2): 14 rooms
First-class Western-style room (capacity of 2): 26 rooms
First-class Japanese-style room (capacity of 2 and 3): 5 rooms for each capacity
First-class Western-style room (capacity of 4): 34 rooms
Second-class cabin with berths (total capacity of 180): 10 rooms
Second-class Japanese-style room (total capacity of 34): 2 rooms
Driver's room (capacity of 8): 4 rooms

The general arrangement of the completed vessel is shown in Fig.19.

For cabin interiors, "Suzuran" uses a color theme, with color contrasts for the walls and furniture. "Suisen" uses a light theme with elaborate lighting fixtures. These themes are used throughout, not only in the three-tier entrance and restaurant, but also in the cabins. The interiors are luxurious like a passenger ship as shown by the entrance space of the "Suisen" in Fig.20.

ELECTRICAL SYSTEMS

Generators

The vessel has three main diesel generators with a capacity of 1,200kW each, two shaft-driven generators with respective capacities of 1,600kW and 1,300kW, and 1 emergency diesel generator with a capacity of 190kW. The shaft-driven generators are utilized for each thruster as a dedicated power supply, and are combined with a constant revolution of main engine and CPP system.

During thruster operation at leaving and arriving port, the 1,500kW bow thruster is power supplied from the 1,600kW shaft-driven generator, and to be started by a constant-current starter. The two sets of stern thrusters are power supplied from the 1,300kW shaft-driven generator as a same method. In the normal operating conditions, each thruster is power supplied from the two shaft-driven generators, but in the emergency condition each thruster is power supplied from main diesel generators alternatively.

Receptacles for refrigerated containers

On car decks, there are 110 receptacles of AC 220 V three-phase for refrigerated containers.

Safety device

To ensure safe navigation of the vessel and to respond to emergencies, there is a fire detector and alarm system, a remote-control system for fire-proof doors, a closing indication system for entrances to car areas, a closing indication system for cargo loading doors, a water-tight sliding door control system, a car-deck water leakage detection system and a car-deck monitoring television system. Centralized monitoring and control of these systems is possible the wheelhouse.

Furthermore, a data logger and engine room monitoring television system with an oil-leakage detection function are located in the engine room, allowing quick detection of any abnormalities in the engine control room.

Radio equipment

Other than radio equipment satisfying the rule and regulation for JG passenger ship, the vessel has INMARSAT-M to serve as a ship earth station for maritime satellite communications. The INMARSAT-M is small and lightweight compared to INMARSAT-A and INMARSAT-B, which were mainly used for satellite communications telephone equipment on ships.

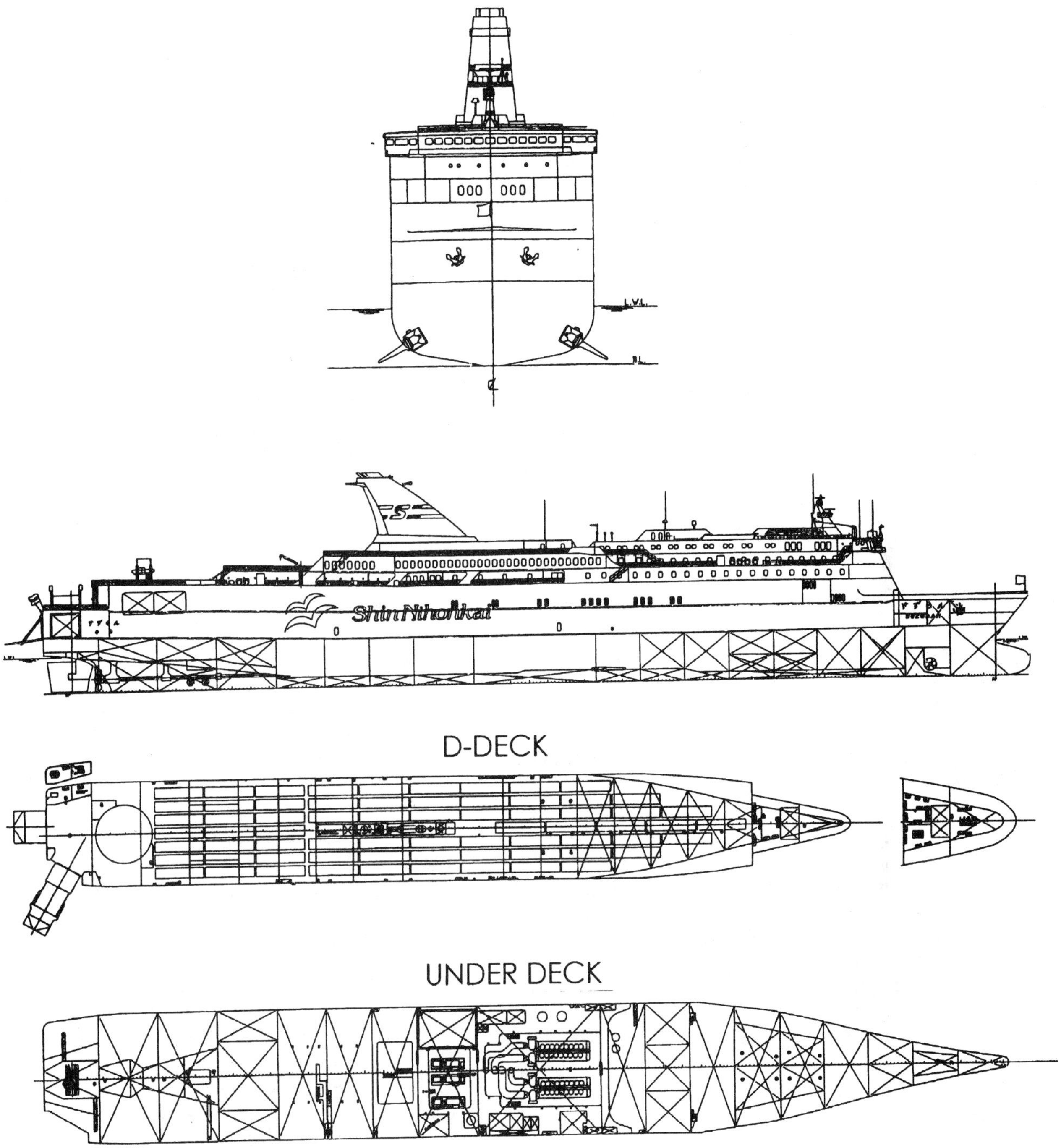

Figure 19 The general arrangement of the “Suisen” and “Suzuran”

Figure 20 The entrance space of the "Suisen"

TRIAL AND MONITORING OF PERFORMANCE

Sea trials

Sea trials for the "Suzuran" and "Suisen" were conducted in February and May 1996 respectively off of Hinomisaki. The water depth in the sea trial area was about 330 m. The wind velocity and tidal current during speed trials for the "Suzuran" were 5~12 m/s and 0.5 knot respectively, and 2~7m/s and 0.1 knot for "Suisen." Wave conditions were relatively calm. The speed trial was conducted by navigating back and forth with a sufficient approach run in a specified course, and the main engine output (BHP), the propeller revolution, wind velocity and ship speed were measured. The ship speed was measured by a radio wave log during a run of a distance of one sea mile. The propeller revolution and BHP were measured by keeping the CPP blade angle constant. After making adjustments for the wind and tidal current, the BHP, propeller revolution, and ship speed in windless conditions were analyzed. The ship speed at MCR power of the main engine is shown in Table 5. While guarantee speed at MCR is 30.8 knots, a ship speed of 31.4 knots was achieved, which was a satisfactory result.

For maneuvering tests at sea, a turning test, stopping test, zig-zag maneuvering test, and spiral test were conducted.

On the turning test, the advance was 790 m (4.2 Lpp) and the tactical diameter was 1138 m (6.1 Lpp) with an initial speed of 29.7 knots. On the stopping test, a satisfactory stopping ability was confirmed by a track reach of 1493 m (8.0 Lpp) with an initial speed of 30.6 knots. On the zig-zag maneuvering test, the first overshoot angle was 1.6° and the second overshoot angle was 4.3° at a helm angle 20° to -20° with an initial speed of 29.5 knots. Spiral test results were satisfactory, then the yaw check and course-keeping ability were extremely good. Similar maneuvering test results were obtained for both vessels

Table 5
Results of speed trials for the "Suzuran" and "Suisen"

	"Suzuran"	"Suisen"
Lpp (m)	187.0	
B (m)	25.0	
Aft draft (m)	6.74	6.83
Fore draft (m)	5.22	5.14
Displacement (ton)	13,042	13,101
BHP (MCR : KW)	23,850	23,850
Propeller rev. (rmp)	163	162
Speed (knot)	31.5	31.4

Monitoring propulsive performance

The "Suzuran" and "Suisen" were commissioned in June 1996 for the Tsuruga-Otaru route with a service frequency of one departure per day. Wind and wave conditions in the Japan Sea are very severe from winter to early spring, and on-schedule service of a ferry must be ensured during this period.

Data of service speed and the main engine horsepower of the "Suzuran" and "Suisen" were collected and analyzed from the logbooks for the period from February to May in 1997 after commissioning. The speed is that over the ground and the horsepower

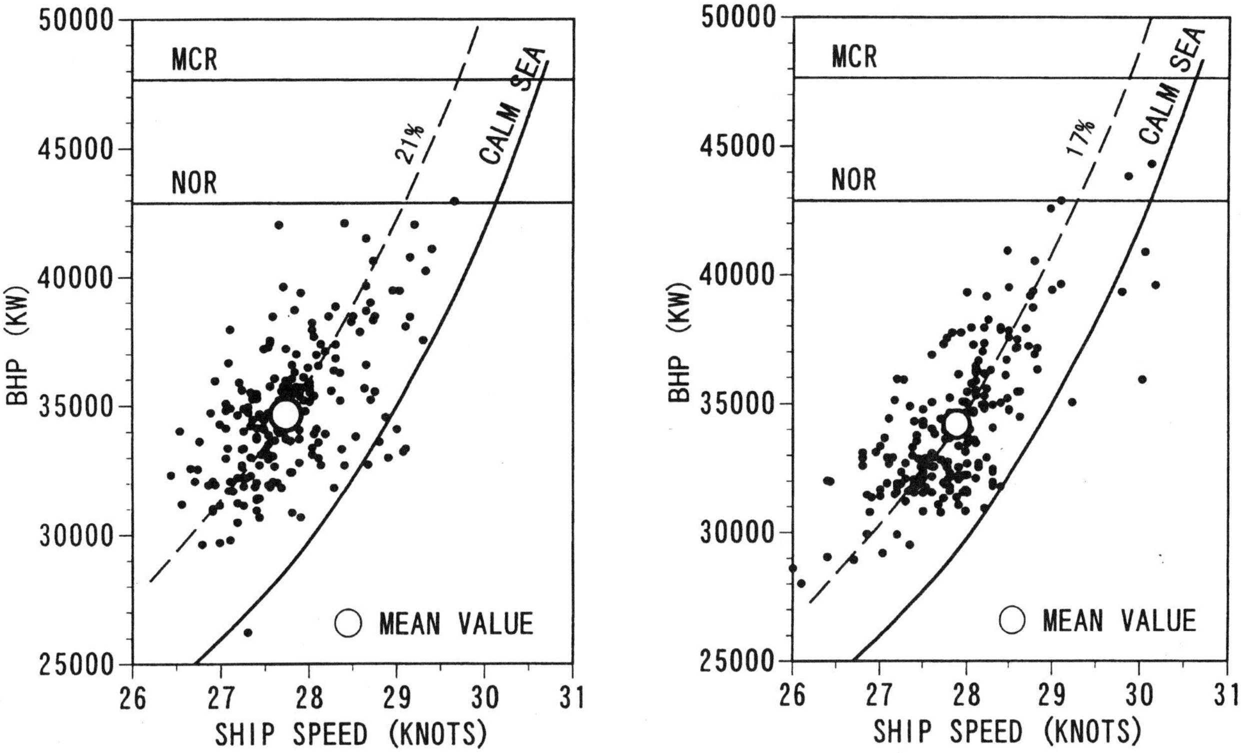

Figure 21 Speed - Power in service performance
left : "Suzuran" right : "Suisen"

was measured by a torque dynamometer attached to the propeller shaft. After calculating the average speed and average horsepower for one round trip, the speed and horsepower were corrected to the speed and horsepower at the displacement of the design draft. The speed - horsepower obtained by the analysis is shown in Fig.21. The sea margin for the "Suzuran" and "Suisen" are 21% and 17% on average, respectively, and results confirmed that the required on-schedule service was maintained. The difference of sea margin between both vessels, according to captains' remarks, seems to be attributed to the difference of frequency of operating the fin stabilizer.

CONCLUSION

The results obtained during development of the large, high-speed ferries, "Suzuran" and "Suisen" can be summarized.

- To satisfy basic requirements for the vessel, the ship dimensions were optimized by a compromise in performance to ensure sufficient propulsive performance and stability while ensuring payload and trim.
- To respond to requirements for high-speeds with a large car-ferry and RO/RO ship, a DU-S.E.M.T Pielstick 18PC4-2B model diesel engine was developed with the largest output as a single medium speed diesel engine. Meantime, various measures were taken in response to vibration and noise produced by a large output medium-speed diesel engine.
- The sectional area curve and bulbous bow form that significantly affect wave resistance were optimized using the results of theoretical calculations and towing tests, and the ratio of the wave resistance in the total resistance was reduced. Meanwhile, by improving the supporting method for the propeller shaft, appendage drag was reduced and the flow into the propeller improved.
- The propeller was designed with the intent to improve propeller efficiency and to decrease propeller cavitation. The hull efficiency increased by improvement of the flow into the propeller, and propeller cavitation was reduced to an allowable range by optimizing the expanded area ratio, blade pitch, and camber distribution of the propeller.
- On-schedule service was ensured by predicting the sea margin of the vessel through referral to weather and sea conditions on the route for a year and the sea margin of the conventional ships along with the use of results of model tests in waves.

- During hull structure design, to satisfy contradictory requirements for structural design, that is, reduction of the hull structural weight and minimization of hull vibration, a measure to avoid structure resonance was taken by defining a principal exciting force and applying FEM analysis.
- Arrangement of the car decks, main engine area, and tanks was examined so that the center of the gravity of the vessel was as low as possible and, appropriate trim maintained. The main engine room casing on the car decks and the pillars were arranged so as not to reduce the efficiency in car handling.
- Sea trials confirmed that a maximum speed of 31.4 knots could be achieved and the designed speed could be maintained at the design draft. On trials for maneuverability at high speeds of about 30 knots, allowable results were obtained for the vessel's turning ability, course keeping stability, and stopping ability. Monitoring the speed performance after commissioning confirmed that the speed performance requirements of the vessel have been fully satisfied.

"Ship of The Year 1996" was awarded to "Suzuran" and "Suisen" by The Society of Naval Architects of Japan for their contribution to improvement of domestic marine transportation and for the technology to meet challenging requirements.

REFERENCES

[1] Jinnaka, T., Tsumi, T., and Ogawara, S. "Hull form design derived from wave analysis," Proceedings of the International Seminar on Wave resistance, Feb. 1976, Tokyo

[2] Mifune, M. "The Interaction of nonlinear waves of a high-speed catamaran ship with the reflection between hulls considered," 5th Symposium on Nonlinear and Free-Surface Flows, Oct. 1997, Hiroshima University

[3] Masuko, A. "Numerical simulation of the viscous flow for complex geometries using overset method," 3rd Osaka Colloquium on Advanced CFD Application to Ship flow and Hull Form Design, May 1998, Osaka

[4] Fujino, R., Ishida S., Katagiri,T and Sato.R "Computer aided design system for marine propellers" Proceedings of Computer Applications in the Automation of Shipyard Operation and Ship Design , June 1982

Discussion

John T. Newton, Member

The authors are to be congratulated on the presentation of the paper outlining the commercial requirements for the ships and the technical development of the project to meet the commercial criteria. I offer the following comments for discussion:

1. The project background clearly recognizes the environmental issue surrounding transportation today. Clearly the Japanese Government's initiatives to encourage a modal shift from land to coastal sea transportation is welcome and could be repeated in many industrialized nations.

   Against this background could the authors indicate the fuel savings of the new ships against those that they replaced? And estimate the fuel required if the trailers had chosen to use a land based route instead of the ferry?

   The authors recognize the need for speedy transportation of perishable commodities in order to achieve the demands for high quality and longer shelf life of the products to the consumer.

2. Main Engine: It may be that the owner of the ferry wishes to be conservative in their choice of propulsion while at the same time moving towards a high speed ferry. Indeed, the authors state that they considered gas turbine propulsion but this option was rejected on grounds of high fuel consumption. What was the differential? Also, was the weight saving of the gas turbine taken into account commercially by the higher cargo deadweight that could have been achieved?

3. Clearly, the authors paid much attention to the design of the propellers especially in relation to cavitation. (It was of interest to note that Burrill's cavitation chart was still used 40+ years after its formulation.) The propellers have performed well following the ships entry into service.

   However, did the authors consider the alternatives to the CPP and if so why were they rejected? If gas turbines had been used for main propulsion would the outcome have been different? Finally, at what speed do the authors consider the conventional propeller on this type of ship would be replaced by alternatives? (35knots?)

4. Seakeeping: The authors have taken into consideration the weather conditions likely to be encountered on the ship's service route. Indeed they recognize that at high speeds the ship is liable to suffer relatively high wave impact loads. They also studied the effect using wave data for the Japan Sea. It would be of interest to have the authors' views on the maximum wave height in which the ship can maintain full speed? And also estimates of speed reduction required as the wave heights increase?

5. A monohull form was adopted for this project. Was this an owner's requirement? If not, was a twin hull form considered? And if so why was it rejected?

6. From the authors' research on this project, do they have any views on the speed at which the monohull form loses advantage against the twin hull form?

7. No mention is made of the paint protection systems used on these high speed ships. In view of the environmental issues surrounding marine anti fouls and the development of new systems to overcome these concerns this would be of interest. Additionally, the drydock interval of the ships is also relevant.

Our thanks to the authors for a most interesting submission.

[ADDENDUM] Ref: Burrils cavitation chart: Professor Burrill, Professor of Naval Architecture, Kings College, University of Durham whose work in his cavitation tunnel will be long remembered. The tunnel was adapted from a WW2 war reparation from the Meserschmitt Aircraft Company.

Robert G. Latorre, Member

The authors paper is a timely contribution to the ongoing discussion of "modal shift." There are two points on which I would like to have the authors' viewpoints:

1. In our experience, the modal shift justification can be best illustrated by a comparison of the existing truck-trailer transport cost (base) versus the ticket cost for putting the trailer on a ferry. For example, door-to-door freight rates per road trailer for a Houston to Tampa route were quoted by the trucking company at $2000. For a Fastship (33 kt.) the rate would be $1650 with a difference of –350 and a High-speed (40 kt.) vessel rate of $2400 with a difference of +400 [1]. Could the authors provide the cost comparison for Osaka to Hokkaido based on Figure 2 of the paper?

2. The authors discuss the ferry berth as 200 m. To realize the benefits of higher speed, the dock-ship truck transfer at dock time should be minimized. Could the authors provide a diagram of the marshaling yard cargo flow and indicate typical loading and unloading times. Are the trailers onboard accompanied by trucks or simply loaded?

I congratulate the authors on their successful long distance truck ferry design, which demonstrates the viability of using high speed truck-carrying vessels.

Reference

1. Latorre, R., Foley, R., "High Speed Coastal Transport Emergence in the U.S.", Proceedings of FAST'99, Seattle Washington, Aug.31-Sept.2, 1999, pp. 231–244.

Michael G. Dyer, Member

First of all, congratulations to the authors for a most thorough and interesting technical presentation. The *Suzuran* and *Suisen* are two vessels which obviously meet the requirements their owners, as put forward in the paper. The thrust of my discussion will be toward broader operational and economic aspects which were not directly addressed in the paper. I put these issues forward because the purely technical issues associated with fast ferries, while still of great interest, have receded in importance relative to several hard practicalities which the successful operator must address. I hope that the authors are willing to discuss these aspects.

This discussion therefore takes the form primarily of questions which I hope are of interest to the broad spectrum of the SNAME membership. The questions fall into two broadly defined categories, the first seeking insight into the basics of the design decision process and the second seeking information as to the performance of the craft over the last four years.

Design Requirements

It is clear from the parameters given on the paper (i.e., lift requirement, range, and reduction of round trip time from three to two days) that IHI had to go with a high speed monohull. The largest payloads on catamarans have not yet exceeded 1500 tons and certainly do not approach 5000 tons. Endurance and fuel capacity also point toward monohull design. The resulting design certainly succeeds from the point of view of "transportation efficiency" (TE) as measured by speed, payload, and power. The IHI ferry occupies a different performance niche than the smaller high speed ferries previously considered in a Volpe Center high speed market study (Ogwara et al. say as much in the brief released ahead of the paper: ". . . this new vessel was positioned as intermediate in nature between a conventional ferry and a high speed ferry."); its payload is 15 times that of the largest RO-PAX in the population considered for our study. According to the TE Index developed by Nigel Gee ([speed {knots} × payload {KG}/power {KW}), the IHI ferries rate at 3355, compared with the range of values from 106 to 477 for a population of smaller high speed craft. The TE Index here is probably reflective of what one might expect for large capacity "slow" ships, yet the IHI ferries have extraordinary speed (29.4 knots) for the job.

It is, however, the process leading to those primary owner's requirements that I hope the authors can illuminate. Specific questions are the following:

- Is the vessels' primary mission the carriage of cargo, in the form of trailers, or of passengers? Did market and demand studies indicate that the operation could be supported without subsidy? Please explain the necessity to shorten trip times, relative to the cargo and/or the passengers.
- Were alternatives considered (e.g., adding more "conventional" craft to increase service frequency, or a larger number of smaller high-speed craft)? Did the proponents conduct a comparative economic analysis for different types/sizes/numbers of vessels? If so, which factors drove the decision to go with two large high speed monohulls?
- With regard to transport of trailers by fast ferry, previous economic studies have indicated that the efficiencies may not yet exist to justify regional sea transport of cargo by fast ferry. Lindsay et al. reported that such service would not be self-sustaining in Europe, where road congestion was the impetus for the study of alternatives (15th Fast Ferry International Conference, Boston, February, 1999). Do the IHI ferries save sufficient time to justify their cargo rates to trucking companies?
- Is the passenger service designed as an alternative to regional air service, as a layover provision for truck drivers traveling with the cargo, or both? What considerations were given to sea environment, seakeeping, and passenger comfort, especially for the long sea voyages (18–20 hours) required?
- What are the safety features of the craft, and which safety code(s) do they meet: IMO HSC or domestic safety code?

Performance

- Have the vessels kept to schedule in all conditions?
- Has ridership met expectations? If not, have fares been cut to entice more custom?
- Have environmental issues arisen, with respect to air emissions, noise pollution, or wake/wave impacts?
- Has the vessels' seakeeping met expectations? Has passenger discomfort during rough weather been a problem?

These questions are put forward with the utmost respect for the authors' efforts and with the understanding that they may not be able to answer all of them. I reiterate that we must bear in mind the hard realities faced by the operator and the impact limitations imposed by society. Recent history shows us that a fast, sexy boat does not guarantee success. SNAME will be grateful if the authors can share the lessons of the *Suzuran* and *Suisen* operations.

Sathish Balasubramanian, Member

Let me begin by congratulating the authors on a very comprehensive paper describing the design evolution of their new monohull high-speed ferries *Suzuran* and *Suisen*. The paper is a very timely and appropriate addition to the growing literature on high performance vessels and highlights the unique design trade-offs essential to achieving the requisite performance parameters that are in turn driven by the ship's mission economics.

The optimization of the hull and subsystems of the two ferries to meet the performance requirements to guarantee economic success is well described. The authors have utilized a technique of subsystem optimization and compromises on selected performance criteria with a remarkable degree of success. Whole ship impact of subsystem choices is often lost, while employing subsystem optimization. The design cost and time required to evaluate competing technologies and the determination of technology forecasts to accurately predict economic feasibility over the life of a vessel are enormous. This results in the designer having to make compromises in performance, which can result in increased cost and unfavorable life cycle consequences. One approach to overcoming this hurdle is the use of whole-ship design synthesis models that allow the naval architect to rapidly arrive at answers to several design "what-ifs" in a matter of minutes. Physics-based whole ship synthesis models highlight multi-parameter optimization and have been shown to be very successful in the selection of parameters that result in a ship design that has a better economic performance index.

The authors have described a very comprehensive research effort conducted into the hydrodynamic performance of the *Suzuran* and *Suisen*. They mention that the appendage drag contribution to the overall drag as being significant and the cavitation concerns for the propellers. Did the authors consider the use of water jets, as a means of minimizing appendage drag and eliminating to a large extent, propeller cavitation issues?

William H. Falconer, Member

The authors are to be congratulated on their most interesting account of the development of a high-speed monohull ferry. I agree with the authors' choice of machinery: this must have been a difficult corporate decision as I.H.I. has extensive experience with both gas turbine and medium speed diesel propulsion.

I was also pleased to see that while the ship's speed increased by 26%, the terminal operational efficiency was improved by 53%. This is a most creditable overall performance and clearly illustrates the potential savings which can be achieved in the operation of container terminals.

I hope that the authors will build on their most valuable experience and take the next logical step by designing a high speed panamax containership for the carriage of high value cargoes, which requires a fast transit time but cannot support the cost of air freight.

A high speed panamax vessel having a genuine service speed of not less than 31 knots would be capable of crossing the Pacific in 6.25 days, the Atlantic 4.9 days and the Far East to Europe in 16 days with bunkering.

Such a ship would be much more economical to build and operate than the gas turbine propelled 'Fast Ship' and, on a door to door basis, much more competitive as the ship would be able to call at any major container port, whereas the Fast Ship has been designed to operate between dedicated terminals.

The building cost of such a ship would be about one-third of that for the gas turbine propelled Fast Ship and would have a daily fuel cost of about 26% of a Fast Ship of equivalent capacity.

On the basis that terminal operations could be improved by only half of which the authors have achieved on the Otaru/Tsuruga route, a 31 knot panamax ship could at least equal the door to door times of the Fast Ship.

Containers carrying high value cargoes have an average weight of about 12 tons per F.E.U. A well established company has indicated it is prepared to pay 0.33 cents per kg. for transport of their consolidated cargo in a 26 knot ship.

Clearly there is a niche for high speed monohulls.

I was interested to note that after considerable research the authors concluded that hulls with 'shafthosen' had a superior propulsive efficiency. This agrees with the hull forms developed by H.S.V., Hamburg for the Overseas Containers Far East Containerships built by the H.D.W. consortium in Hamburg in 1972.

These full panamax containerships had a genuine trial speed of 31.5 knots and had sufficient bunker capacity to do a round trip without bunkering.

Although the ferries have a two compartment subdivision and means of cross flooding the heeling tanks, did the authors consider a double hulled vessel and movable bulkheads to reduce the surface effect of water on the vehicle deck?

Authors' Closure

We would like to thank all of the discussers for their interest in our paper and appreciate their useful contributions.

In response to Mr. *John T. Newton*'s comments and questions:

As for fuel saving of new ships against those that they replaced, two high-speed ferries, the *Suzuran/Suisen,* were built to replace three conventional ferries having service speed of 21.8 knots. New and old fleets have almost the same cargo carrying capacity in total.

If we compare the required fuel between the new and old fleet, new fleet requires about 39% more fuel than old fleet per one ship reflecting the increase of service of new fleet up to 29.4 knots.

Propulsion performance of *Suzuran/Suisen* is improved very much by the means mentioned in the paper and the improvement in fuel consumption should be assessed by the comparison between the vessels having the same service speed.

Comparing fuel amount or cost between *Suzuran/Suisen* and transportation of the same cargo by land trucks can be made under many assumptions. Table 1 shows the results of our rough comparison. Comparison of the diesel engine and gas turbine is shown in Table 2. This comparison is based on the same engine load for both plants. Fuel costs for gas turbine is 18.1 million US$/year higher than diesel engine.

Also, if gas turbine is applied, payload may be increased about 752 ton, and this corresponds to the weight of 20 additional trailers. The cost increase for fuel and maintenance by the gas turbine cannot be recovered by the additional freight fee. Therefore, we decided to apply the diesel engine plant considering the life cycle cost.

As for the propulsor design, we considered a water jet propulsion as an alternative to CPP taking advantages such as low inboard noise and vibration and absence of propeller shaft supporting members. However, In the speed range of *Suzuran/Suisen,* the comparative study including resistance, propeller efficiency and the self-propulsion factor resulted in the total power to thrust the ship with CPP is superior than a water jet propulsion.

The maximum speed and power of a marine propeller will be restricted by the occurrence of cavitation. In order to increase speed and power of the propeller with allowable level of cavitation, thc CPP design was carried out combined with improvement of flow into the propeller. Gas turbine propulsion was not employed due to the reason mentioned above. However, if a gas turbine propulsion of higher power than the diesel engine was adopted, the authors estimate that the speed limited by the cavitation problem would be around 31 to 32 knots for the conventional propeller of this type of ship.

As for seakeeping of the vessels, I understand the discusser intention is the full speed with which the ship keeps a service schedule, so it is rather good to reply with an empirical basis of ship operation. According to the operator's remarks, the ship can maintain the service schedule up to the wave height of 6 meters measuring with eyes. We carried out the towing test in waves, and confirmed that the power due to added resistance in waves increases in proportion to the square of wave height, then the speed reduction would be approximately in proportion to the wave height.

A monohull form was adopted by the owner's requirement for its good sea worthiness at rough sea and lower cost of building compared with the twin hull form. From the authors' studies, the advantage of a monohull form will lose over the Froude number of 0.4 for the required service speed and deadweight.

As for the paint protection, tin free polishing type paint was applied for these vessels and dry dock interval is one year.

In response to Prof. *Robert G. Latorre*'s comments and questions:

Comparison of transportation hours from Sapporo City (Hokkaido) to Osaka City (Kansai) is indicated as follows:

1. *Route by sea (Suzuran/Suisen)* Highway from Sapporo to Otaru, Ferry from Otaru to Tsuruga, Highway from Tsuruga to Osaka. It takes about 24 hours between Sapporo and Osaka except waiting time to ferry. (21 hours with ferry and 3 hours on highway.)

Table 1

	By sea (*Suzuran/Suisen*)	By land (trucks)
Kind of fuel	Heavy fuel oil	Gas oil
Fuel weight	Base	47%
Fuel cost	Base	174%

Table 2

	Diesel Engine	Gas Turbine
Kind of fuel	Heavy fuel oil	Gas oil
Fuel consumption	195 ton/day	256 ton/day
Fuel cost	8.78 Mil.US$/year	26.88 Mil.US$/year
Maintenance cost (Experience basis)	0.9 Mil.US$/year	3.86 Mil.US$/year
Engine weight	800 ton/2 sets	48 ton/2 sets

2. *Route by land (Truck)* Highway from Sapporo to Tomakomai, Ferry from Tomakomai to Hachinohe, Highway from Hachinohe to Osaka. It takes about 24 hours between Sapporo and Osaka except waiting time to ferry. (9 hours with ferry and 15 hours on highway.)

Based on the above routes, comparison of transportation cost can be estimated as follows:

1. *Route by sea (Suzuran/Suisen)* It costs 1,350$ including ferry fare, fuel oil, highway fare and single driver's charge.
2. *Route by land (Truck)* It costs 3,680$ including fuel oil, highway fare, two drivers' charge and ferry fare.

Difference of cost between sea and land routes is attributed to driver's charge which counts as 50$ per person per hour.

As for loading and unloading time, 5 to 6 hours were needed for conventional ferries of this size. As the discusser points out, minimization of time for unloading/loading, refueling and cabin cleaning is also an important issue for efficient transportation. And only 3 hours for loading and unloading time was given after studying the car loading scheme and deck plan for the high speed ferries. To make loading efficient, the car loading plan is being made to meet the number of trucks and trailers before the ferry arrives to berth every day. As for the contents of cargo, almost 80% are trailer only, and others are truck and a few passenger cars. (See Fig. 1.)

Figure 1

In response to Mr. *Michael G. Dyer*'s comments and questions:

The vessels' mission is to carry both trailers and passengers and this was one of the conditions for getting the approval of Japanese Government for servicing in this route.

We are not in the position of knowing the actual contribution of trailers and passengers for the vessels' income but we suppose the carriage of trailers contribute more for the vessels' income judging from the fact that a RO-RO vessel (carrying only trailers) became more popular than ferry (carrying both trailers and passengers) for Japanese domestic service and most of them are operating successfully.

The shipowner cannot expect any operational subsidy for this kind of vessel in Japan. Operational subsidy will not be essential for these vessels.

The speed of *Suzuran/Suisen* is just matching for daily service operation in this route by two vessels. Even if the speed is increased some more, it only increases the idling time for waiting the scheduled departure time. Namely speed increase does not increase the total transport capacity of the fleet and the required number of vessels cannot be reduced. If the customers appreciate shorten arrival and accept the premium, it will be feasible but not likely.

Suzuran/Suisen were built to replace the three ferries having a service speed of 22 knots. So, the owner only studied comparative analysis between these two fleets. Other alternatives seemed not feasible by the following reasons:

Demand of cargo and passenger will not change so much by shortening the service frequency (daily service is enough) and be effected mostly by the charge. So a minimum number of larger vessels by keeping the daily service will be best.

Length overall of the ferry must be less than 200 m to meet Japanese port regulation and *Suzuran/Suisen* are designed as the largest vessels to meet this regulation.

Recent rise of fuel cost badly effected for the operational economy of fast ferries. However, many studies carried out by the Japanese Government and operation results of fast ferry and RO-RO in Japan including *Suzuran/Suisen* show the economical possibility of the introduction of a fast ferry to some Japanese domestic routes.

The passenger service was not designed as an alternative to regional air service, but for passengers who are pleased with a voyage. For seakeeping and passenger comfort, special consideration were given to keep vibration and noise level as low as possible along with sufficient control of rolling motion with fin stabilizers. As for the safety features of the craft, the ships meet domestic safety code.

The time schedule of these ships has been kept except in unusual weather such as the arrival of the typhoon. The authors consider that ridership has met expectations, since fares have not been cut from the beginning until now. Everything is as expected as to environmental issues, and environmental problems have not arisen concerning these vessels, however, analysis for greenhouse gas mitigation by marine transportation is under surveillance with national projects. Claims for the problem of wake wash also have not occurred, because the ship slows down its speed in a bay area.

As to passenger comfort, the ship motion of these vessels is relatively larger than other conventional ferries of slower speed, because *Suzuran/Suisen* have been operated with approximately 29 knots even under the severe weather condition such as the wind speed of 20 m/sec and the wave height of about 6 m. However, they also have been serving as passenger ferries for more than four years after getting into service, and the vessels are active with passengers in the summer season.

In response to Dr. *Sathish Balasubramanian*'s comments and questions:

The discusser has applied the design synthesis modeling to design optimization for a research ship and succeeded to derive a cost effective solution in term of life cycle. This methodology can be applied to the planing of ferries we feel, and the authors are interested in what solution for hull form and propulsion system will be produced by using the modeling.

As for consideration of the use of water jet propulsion, the authors give the same reply as that of Mr. J. T. Newton's discussion.

In response to Mr. *William H. Falconer*:

The authors appreciate the constructive discussion in many respects. In regard to subdivision of the vehicle deck, as the discusser is well aware, the vehicle decks are required to be as wide as possible for quick and efficient operations of trucks and particularly long trailers.

Generally in Japanese ferry terminals, only one ramp is available for each level of vehicle deck which gets even the long trailers turned around 180 degree when loading in the deck. The situation being such, provisions of the proposed double hull or movable subdivision bulkhead are not very practical in Japan; hence, other countermeasures have been applied. The authors shall consider a safer arrangement for damage stability when a terminal allows.

SNAME *Transactions*, Vol. 108, 2000, pp. 327–337

Human Factors Engineering in the Wheelhouse Design of a High Speed Catamaran Ferry

Laurel Ritmiller, Visitor, **Scott Davis** Visitor, and **Joanna Zander,** Visitor, BC Research Inc., Ergonomics and Human Factors Group, Vancouver, BC

ABSTRACT

This paper describes the integration of a Human Factors Engineering (HFE) process in the wheelhouse design of a high speed catamaran ferry. The HFE Specialists at BC Research Inc. were integrated into the design and manufacturing teams to advocate the perspective of the human operator in the design of two 300-passenger, 32-knot ferries built for the City of Vallejo, California by Art Anderson Associates and Dakota Creek Industries Inc. The HFE Specialists provided human factors expertise related to the concept, design and production of the vessels to enhance the operational ability of crew, reduce the risk of operator error, establish efficient work patterns, and allow maintenance and repair tasks to be performed in a work environment conducive to health and safety. Given the high speeds, single person operation, and heavy traffic of San Francisco Bay, emphasis was placed on visibility, well-designed layout of controls and displays, and situation awareness of the vessel operator in order to maximise collision avoidance strategies. One-tenth and full scale mock-ups, in conjunction with task simulations involving the vessel operators, were included in the process. Anthropometry, equipment usability, cognitive task requirements and environmental issues, including vessel motion were also important considerations. A number of military and marine guidelines were used to guide the HFE Specialists in the design.

HFE – Human Factors Engineering
SOR – Statement of Requirements
IMO – International Maritime Organization
ISM – International Safety Management

INTRODUCTION

Traditionally, classification societies and naval architects have concerned themselves with hardware issues. However, in light of human error issues being identified as contributing factors in major disasters, including the Exxon Valdez and space shuttle Challenger, it is now recognised that attention must also be given to the wider system, encompassing a range of human related issues [1]. Many commercial, military and government organizations are requiring Human Factors Engineering (HFE) involvement throughout the vessel design process. For example, the design of the US Coast Guard 47' Motor Lifeboats included extensive HFE in both the initial design and prototype testing to increase the usability, safety and habitability of ships' crew [2].

To further investigate the impact of human factors on maritime transportation, the International Maritime Organization (IMO) has established working groups on human factors, estimating that 80 per cent of maritime casualties are due to human error, while 20 per cent are due to technical deficiencies [3]. The International Safety Management (ISM) Code (IMO 741/18) encompasses human factors as part of an integrated safety plan, and it is believed that compliance to this code will result in reduced accidents, minimised pollution and increased protection of the environment.

Human factors is multidisciplinary to ensure a "systems approach", or that all aspects of the working environment and the impact on the human operator are considered. In transportation systems, particularly the marine environment, a number of issues are intrinsic in a human-centred design. These include:

- software usability and human computer interface design and analysis (including electronic navigational systems, electronic charts, communications, surveillance systems) in response to continuing advances in marine technological systems;
- an understanding of the working environment, in this instance, the marine environment, ship dynamics and associated human factors issues, such as the effects of motion induced sickness, motion induced interruptions, and motion induced fatigue on human operator performance;
- human reliability analysis, including functional task analysis to determine operator requirements, and human error analysis to systematically identify potential human errors and their consequences;
- function allocation and situation awareness to balance and optimise human cognitive and physical workload, for safe and efficient human-technology interaction;
- anthropometric (human body size and shape) modelling and workspace design to ensure all users can operate the system, safely and efficiently [4,5];
- environmental issues (e.g., vessel motion, temperature, noise, lighting, high frequency vibration);
- impact of the design on crew interaction, including command and control, communications, and habitability;
- if relevant, understanding of passenger requirements, including special needs passengers, involving process flow, seating, habitability, communications and safety; and
- interpretation and application of appropriate design guidelines and standards to ensure compliance with all design and safety requirements.

Background

The Ergonomics and Human Factors Group at BC Research Inc. was approached by Art Anderson Associates for HFE expertise in the wheelhouse design of a fast ferry being produced for the City of Vallejo, California. The HFE Specialists at BC Research Inc. were integrated into the design and manufacturing teams to advocate the perspective of the human operator in the design of two 300-passenger, 32-knot ferries built by Dakota Creek Industries Inc. Given the high speeds, single person operation, and heavy traffic of San Francisco Bay, the City of Vallejo required a HFE review of the wheelhouse be included in the design process. The HFE Specialists were involved throughout the process, working with the owner, vessel operators, and design and manufacturing teams, to ensure that the wheelhouse would allow safe and efficient operation.

HFE METHODS

The HFE Human – Task - Environment Model

Encompassing all of the interactions between users and their equipment, HFE employs a three-pronged approach to human-centred design. First, consideration of the human - task - environment model, with the human at the centre, ensures that all issues related to these components are addressed and that all components will function optimally under all required operational conditions. Second, a user-centred, iterative systems development approach is used during acquisition, and fielding of systems and products, helping to ensure effective, efficient, safe, and accurate performance. Third, the additional requirements of training and selection of appropriate individuals are addressed to re-enforce the safety, accuracy, and performance of the system. This HFE Model is outlined in Figure 1.

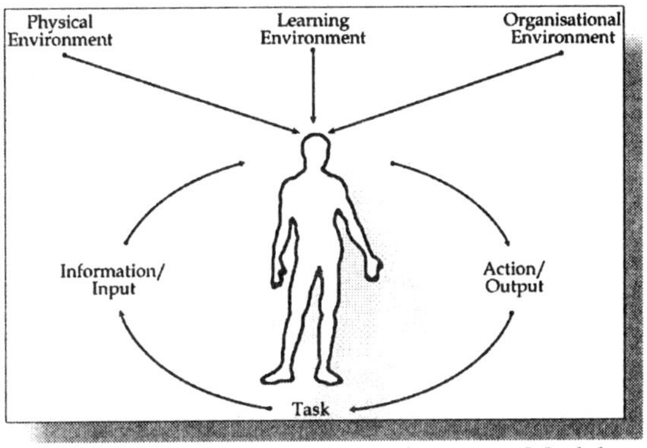

Figure 1: HFE Human – Task - Environment Model

The human - task- environment model forms the foundation for much of human factors. In this model, the human is a system, complete with subsystems interacting with, interfacing with, and influencing other systems. Performance of the human system is affected by task requirements, human activities and capabilities, the physical environment, the learning environment, and the organisational environment. The goal of human factors is to tie together each element of the human – task - environment cycle, maximising human performance in any situation.

The Task

The task may be any activity required of the human: navigating a vessel; retrieving information from and online database; or installing a child seat in a car. Each task provides information to the human and results in a response or decision applied to the task.

Activities within the Human

Within the human, the input from the sensory systems (vision, hearing, touch, taste, and smell) must be detected, recognised, and analysed. Detection requires the input to be within the capabilities of the sensory system. However, even if the input is within this range, there is no guarantee it will be recognised or analysed. The human brain continually filters information to prevent information overload. The key to optimal performance is the provision of relevant information in the proper format at the proper time. This automatic level of cognition has implications for the amount of information, and the type of information, which may be detected by the human brain (e.g., monitoring numerous types of equipment in a wheelhouse).

If the human recognises the input, it can then be analysed for its significance and relevance. In conjunction with recognised information from the physical environment, the learning environment, and the organisational environment, this "cognition" allows the person to make a decision and take action if necessary.

The Physical Environment

The physical environment affects the human's performance. But the human condition is extremely adaptable, especially when our greatest adaptation - the ability to invent and use tools and technology - is exploited. The physical environment includes not only factors such as heat, noise, light, and vibration, but also topography, the person's location, weather, time of day or night, and other factors.

People can operate in a greater range of physical environments thorough acclimatisation, physical and mental conditioning, training, and education, and especially through the use of tools and equipment (i.e., transportation methods and protective clothing). In any environment, the more relevant and useful information the person has about the physical environment, the more likely the person will be to make the best decision and choose the best action. HFE involves proper design to minimise the effects of the environment (e.g., engine noise, glare on gauges).

The Learning Environment

With proper training and education, the abilities of people are improved. Physical and mental conditioning not only make the human stronger and able to operate effectively under extreme conditions, but they also provide a knowledge of and confidence in his/her capabilities and an awareness of his/her limitations.

Human factors identifies requirements and where limitations occur and provides information on capabilities. This knowledge helps design training regimes that maximise usability, safety, and usefulness.

The Organisational Environment

Factors such as time requirements of the task, task criticality, command structure, location within the command structure, incentive systems, and leadership style affect the human's performance.

The HFE Lifecycle

The ideal design process includes HFE in the planning, evaluation, design, installation, and post-development evaluation stages of system development. Including HFE from the beginning of the design cycle ensures each part of the system is integrated with the other parts and the human and system integration is maximised. This process also optimises utility, safety, usability, and productivity. At each stage of the process, well-established techniques, relevant to the specific project, are applied to achieve an integrated design.

Five main components comprise the HFE Lifecycle:

- requirements analysis;
- research;
- market analysis;
- systems design, testing and development; and
- implementation.

Requirements Analysis

Requirements analysis identifies user needs and activities in the system being developed, including the objectives of the system and the strategy for development. There are four components of the requirements analysis:

- user profiles;
- task analysis;
- platform capabilities and constraints; and
- human factors principles.

Together, these four components are used to establish the goals of the system and acquire specific qualitative and quantitative human factors information.

The expectations, needs, and requirements of end-users can be established through the use of focus groups, mock-ups, observations, questionnaires, and interviews. The user populations are characterised through user profiles to ensure that the design team has an understanding of users' capabilities and constraints. Human factors problems with the current system or design are revealed through human factors research. Involvement and characterisation of user populations ensures support for the final design and that design specifications consider user requirements.

User Profiles

The user profile characterises the user around which the system or design will be tailored. A representative population of users must be identified prior to development of user profiles. Information about the user that is relevant to the system design is collected to be referred to throughout the design process.

Techniques used to collect this information may include market research, questionnaires distributed to a sample user population, and interviews with sample users and people knowledgeable about the users or previous systems.

Task Analysis

Task analysis is used to describe, and in some cases, to evaluate human-systems integration. Task analysis provides detailed information related to what, why, when, how, and how often task activities occur. Observation of work and interactions within the specific system provide documentation of task activities and workflow patterns and can identify the actions completed by the user and/or the cognitive processes used to achieve the system goal. Task analysis is used to obtain a user-centred model of work as it is currently performed, and to extract from this, the HFE requirements for a new system or procedure. Task analysis identifies human factors considerations (including both the strengths and weaknesses) of the existing system or model. Task analysis should be used when designing a system, evaluating a system design, or when a particular human-system performance problem has been identified.

In a design process, task analysis must be completed early in the design stages, and rarely should it be completed only once. Instead, task analysis should be performed as part of an iterative process as more detailed information about the system becomes established and as the user and system allocations become clear. As the design progresses, the task analysis will become more detailed.

Platform Capabilities and Constraints

Every system component has capabilities and constraints: things it can and cannot accomplish. In this phase of the requirements analysis, the capabilities and constraints of the system and the technologies that will limit the design alternatives of the system are established. First, a technological review is completed, followed by matching the user capabilities and requirements with the most appropriate technologies. The performance requirements for the system are outlined based on the expectations of the user, the manufacturer or designer's goal, and the operational environment in which the system will be used. The constraints and limitations are also recorded both for the user and the rest of the system. The user constraints are often based on limits in cognitive and behavioural capabilities, as well as physical abilities and characteristics. The system constraints are often based on the operational environment, technology capabilities, and/or limitations with material properties.

Determining the capabilities and constraints of the design is particularly important when considering systems to be used in extreme environments or by a selected user population. This analysis may result in either the exclusion of certain technologies because they do not meet the criteria of the user population or the exclusion of some users if they cannot use the technology.

Techniques used in this phase include a technology review, focus groups, user interviews, and documentation of the capabilities and constraints of the system and the technologies. This phase can be performed in conjunction with other aspects of the requirements analysis.

Human Factors Principles

In this phase, existing standards and guidelines are combined with both the user requirements and the functional requirements to generate design alternatives that are consistent with the user profiles and business processes. To select the design features that best meet the system requirements, design alternatives are ranked according to predetermined criteria. The highest-ranking designs are usually those that offer utility, safety, efficiency, ease of use (usability), and variability.

Statement of Requirements

The four components of the requirements analysis are combined to develop a human factors Statement of Requirements (SOR). This document summarises the system requirements with consideration of the tasks and task scenarios, user capabilities, performance requirements, user and system capabilities and limitations, and human factors principles. The statement of requirements is used to identify and clarify the human factors goals of the system. Deliverables from the SOR include:

- a description of task scenarios;
- goals of the system;
- HFE requirements; and
- criteria for the evaluation of the system.

Research

Scientific research is performed for two main reasons: to increase the level of knowledge and to provide information required to make prudent decisions. Well-performed scientific research provides valid and reliable results. Applied research provides the link between pure science and application in operational environments (e.g., the marine environment). In these cases, experience conducting field experiments provides knowledge on how experiments can be designed to maintain both validity and reliability, and how to adapt the experiment as required when demanded by the changing operational setting.

Market Analysis

Market analysis is used to identify and evaluate all systems currently available to the user. The first step in a market analysis is to complete a commercial and industry literature review to determine the different products that are available, if these products would suit the needs of the user, or if they could be easily modified to suit the needs of the user. These products are analysed to determine the characteristics of the different technology that is available.

The next step is to contact manufacturers and to interview commercial and industrial agencies. These processes also include finding users with similar needs to your own users to determine how their requirements are being met. The purpose of this step is to avoid repeating work that has already been completed by others.

The final step of the market analysis is to review and analyse the information to determine if existing technology can be purchased and integrated into the system, or if further research and development, design, or testing and evaluation are required.

System Design, Testing and Development

High Level Design

High level design provides a starting point for the development of the system and consists of function allocation, conceptual systems design and mock-ups, evaluation, and development of an updated SOR. Designs are iterated until major flaws are eliminated. An iterative process is used to eliminate any design flaws and best meet the operational requirements of both the human user and the system.

Function Allocation

The functions of a system are determined and outlined in the SOR. From this, a description of the tasks is developed to outline each individual task or function that must be realised through use of the system. Functions in the system are then allocated to either the human or the machine depending on which will perform the function best. An excessive workload will result in errors, missed information and complex decision making by the operator. However, a semi-automatic operation may result in operator boredom, inattention and a lack of situation awareness in the event that the human operator must intervene in the operation. Considerations at this stage are the complexity of the function, efficiency, usability, possibility of error, and the consequence of an error.

Conceptual System Design and Mock-ups

In this stage, a high-level design framework is defined and constructed. This framework is used to develop a high level (low-fidelity) mock-up of the design. The conceptual design represents the design plan, which is used to generate a paper and pencil or mock-up model. The design mimics the function of the actual design without necessarily working or looking like the final design. The conceptual design and the mock-up are developed as a precursor to a prototype, while considering the breadth of features and the working of the system. The mock-up is subjected to a battery of functional and usability tests to eliminate the major flaws and ensure the design meets the system requirements.

The interaction between a conceptual design and a mock-up are similar but not an exact representation with the final product design and prototype; thus, this stage is used to collect a number of iterations as feedback is obtained from users about the usability and usefulness of the system. Once obvious design flaws are eliminated, a prototype can be developed.

Evaluation

Throughout the process, products are evaluated for safety, usability, and effectiveness at several stages. This may begin with evaluation of existing systems or design concepts and proceed through evaluation of mock-ups, prototypes, and final versions. Evaluation may include formal or informal testing for a variety of metrics related to safety or functionality. Testing ideally involves a representative sample of users performing representative tasks within a typical environment. Well-documented testing can provide information useful not only for design modification, but also for cost-benefit analysis.

If the evaluation determines that the design thus far has eliminated the obvious design flaws, the SOR is modified to reflect any newly gathered information.

The revised SOR is used until the next point of evaluation.

Design Standards

This phase sees the establishment and definition of a set of standards that, along with the conceptual system design and mock-ups, set the stage for the prototype. The standards must be developed to meet the unique requirements of the system or product. Some standards may be adopted from previously existing standards or protocols. In any case, user requirements must be considered when developing standards.

Prototyping

Prototypes represent the final product and allow the analyst to test attributes of the final product before manufacture. Prototyping is used to finalise the system design prior to an expensive manufacturing and installation process. The prototype mimics the actual system or product as closely as possible. The level of detail and functionality of the prototype will increase with each iteration of the system. The initial prototype will be somewhat low level, to ensure that the main design flaws are eliminated. As each prototype is evaluated, the SOR may be modified to reflect the new detailed information and more complex prototypes may be developed. The final prototype should be a close representation of the final product, and provides a working prototype for user evaluation.

Implementation

The manufactured product is installed in the field. Implementation includes every aspect of the system, including training (if appropriate), directions, and use of the working system. The first phase of implementation may be somewhat controlled (to the extent that this is possible in a field setting) to collect user feedback.

After the system has been installed for a period of time, user feedback can be collected and iterated into the next release of the product. The next generation of design should include the evaluation information from the previous design.

HFE Methods Applied to a High-Speed Ferry Wheelhouse

The HFE design process described above was applied to the Vallejo high-speed ferry wheelhouse design. Details regarding each stage of the HFE Design Lifecycle are outlined below.

Requirements Analysis

User Profiles

In the design of the ferry wheelhouse, a user profile was developed to include the existing ferry operators, as well as a general population database of US adults (male and female) to represent future operators [6, 7]. This database provided key information including anthropometric data (i.e., human size and shape) for the physical design in the wheelhouse (i.e., reaches to equipment, visual lines, clearances for maintenance worker access). This profile included not only the physical characteristics of the user population, but also the cognitive requirements of the vessel operators.

Task Analysis

A comprehensive task analysis was conducted with vessel operators on-board the ferry scheduled for replacement. In the task analysis the HFE Specialists travelled on the vessel, interviewed operators, observed the operations conducted by the operator, and analysed the current design for HFE issues. To investigate potential merits of other designs, the HFE Specialists also conducted a task analysis on a high-speed catamaran used by Vancouver, BC law enforcement.

The information from the task analyses was invaluable during subsequent stages of the design process. This information was used to guide the design team in determining the relative importance and function of equipment required for safe navigation of the vessel, as well as developing appropriate task scenarios to evaluate the design throughout the process. These scenarios were developed to include a variety of typical vessel operations (e.g., passenger boarding, transiting operations during the day) and special case scenarios (e.g., operations under inclement weather, equipment malfunction) to thoroughly test the new wheelhouse design.

Platform Capabilities and Constraints

The equipment selected for the vessel operations was reviewed. Due to project scheduling, the HFE Specialists did not have direct input into the selection of equipment. This presented some design constraints, limiting the location and positioning of equipment.

Although the overall wheelhouse dimensions and design were developed, the HFE Specialists did have direct input into the equipment layout and presentation to the operator. This input resulted in significant modifications from the original design concept which had been a generic design for a two-person operation. From discussions with the owner, vessel operators, design and manufacturing team, and US Coast Guard, it was obvious that vessel would be operated primarily by one operator. The original design located primary vessel controls in the middle of a common console between the two operator positions, making a one person operation awkward an inefficient. Therefore, the final design was focused to optimise the wheelhouse layout for a one-person operation, as this represented the regular operation of this vessel. A secondary position was included to be used for radar, lookout and communication duties during night operations, inclement weather (e.g., fog, rain) and during high traffic periods. This shift in the design focus was important to maximise safety and efficiency with highly advanced marine technology (i.e., integrating multiple types of visual and auditory information), speed and marine traffic (i.e., closing speeds and collision avoidance). The final design is shown in Figures 2 and 3.

Figure 2: Wheelhouse Interior

Figure 3: Main Control Panel

The new design focus resulted in significant attention to the location of displays and controls to ensure an effective and safe operating environment. Information from the task analysis characterized the tasks and equipment requirements. This resulted in grouping the equipment by its use in terms of frequency, function and criticality of the operation. It also identified those tasks which may be completed by the second operator under special circumstances. This resulted in a "wrap-around" console design, placing primary controls (e.g., throttle/joystick, autopilot) within a comfortable reach of the operator (i.e., elbow to hand-grip reach), and secondary controls (e.g., vessel communication devices, radar controls) within easy arm length (i.e., shoulder to forward-grip reach). To accommodate the second operator in performing lookout and monitoring duties, a second seat was provided, with a dedicated radar display directly forward, a comfortable reach to vessel communication devices, and a clear view to the engine monitoring displays.

The console was designed primarily for operation by the pilot while in a seated position. As an example of accommodating the design for special circumstances, back-up controls were positioned on the console so as to allow a comfortable operation by the pilot while standing behind the console. The pilots requested this consideration in the design, reporting that while under such emergency operations, they will likely stand to ensure alertness and command in the wheelhouse.

Human Factors Principles

A number of military and marine guidelines were applied to the design [8, 9, 10, 11, 12]. These guidelines were selected to ensure that the design met the relevant anthropometry (human size and shape), operational requirements, design requirements, and general human factors principles.

General HFE principles were included to guide the design of the wheelhouse and layout of equipment, relative to the operator, and other equipment. These HFE principles integrated into the design process included the following.

- **Importance principle.** Important components in the workspace should be placed in convenient locations. Importance refers to the degree to which the component is vital to the achievement of the objectives of the system. The determination of importance is usually a matter of judgment by people who are experts in the system. In this design, the most important control identified by the pilots was the throttle/joystick while the most important displays were the radar and view outside the wheelhouse.
- **Frequency of use principle.** Frequently used components should be placed in convenient reach zones. The throttle/joystick, radar, and vessel communication devices were identified as high frequency controls.
- **Functional principle.** Components should be grouped according to their function. It is optimal to group displays and controls or machines that are

functionally related in the operation of a system close together.

- **Sequence of use principle.** Sequences or patterns of use frequently occur in the operation of equipment or in performing a service or task. In applying this principle, the items should be arranged so as to optimise such patterns.
- **Design for ease of manipulation and tactile feedback.** Tactile feedback includes the use of physical stop barriers, which may be accompanied by auditory feedback (i.e., audible snap). This feedback may indicate that a task-action sequence has been accomplished.
- **Design for visibility and visual feedback.** Visual information and visual feedback to the operator play a vital role in transportation systems. All features should be fully visible and provide feedback. Hidden features complicate operation.
- **Design to enhance the formation of a mental model.** Designers often have a different mental model of a product than users. Mental models play a significant role in the completion of complex operations (i.e., adjusting vessel speed and heading while monitoring radar and communication devices).
- **Design for transfer of training.** This may apply to a situation in which operators have learned to execute a similar task in a specific fashion. When a new design is developed, it is important to analyse what types of skills operators have established and to use the same set of skills for the new design. Differences in product design and workstation layout may create confusion, errors, and worker frustration. A well designed system which considers human cognitive processing and task requirements will decrease the need for an extended training period.
- **Design for job satisfaction.** Satisfied workers are more productive and comfortable workers, with a lower rate of injury and errors. Increasing job satisfaction in a working environment may include: opportunities to co-operate in the design of the workplace; opportunities to talk to others (i.e., social contact); performance feedback; control over pace of work; use of judgement and decision making; and opportunities to learn new things and develop.

Other HFE principles include:

- Design first for safety.
- Make the design "user-friendly".
- Accommodate all of the workforce (e.g., differences in body size and shape between gender and ethnic populations)
- Design jobs to be cognitive and social.
- Emphasise communication.
- Use the technology to extend, not compromise, human performance.

Statement of Requirements (SOR)

Based on the results of the task analysis, investigation of guidelines for wheelhouse design, and HFE design principles, the HFE Specialists produced a thorough description of the operator requirements and resultant design guidelines. These guidelines were provided to the design and manufacturing team, and were reviewed and modified at key stages of the development.

Research

No specific research was conducted to meet the design requirements in this project. However, the design guidelines from the marine community, as well as the HFE design principles selected for this project are based on thorough and continuing research of the interaction of the human operator with advanced technology in a complex marine environment.

Market Analysis

Because the HFE Specialists were involved in the process after the specific control and engineering systems were selected, a complete market analysis was not conducted. However, the HFE Specialists reviewed the selected equipment specifications and operation for user requirements (e.g., operator feedback, first/second/third order operation) and capabilities or constraints of the system (e.g., display options, warnings/alarms).

Systems Design, Testing and Development

High Level Design

The HFE design process was introduced in the review of design drawings and a one-tenth scale mock-up of the proposed wheelhouse design. The HFE Specialists provided feedback based on an understanding of the operational requirements, constraints and capabilities of the system, and the technological systems selected for integration into the design.

Function Allocation

As described above, a major design modification was implemented to optimise the design for a one-person operation to better meet the regular operations of the ferry. This involved a review of function allocation between the vessel operator, the secondary position operator, other crew members, and the technology of the system. This function allocation was reviewed to ensure that the requirements of the human users were not at too high, nor at too low a workload, or outside the operator's capabilities.

Conceptual System Design and Mock-ups

The design and manufacturing teams produced a full-scale mock-up of the wheelhouse using plywood, styrofoam, and cardboard. Equipment was represented by full-scale paper replicas, held in place with push-pins. The flexibility of the mock-up allowed the team to easily investigate different design options throughout the process. The vessel operators nominated one representative to be the key participant in the design process. This operator provided the design team with details of operational requirements, as well as assisting developing task simulations to evaluate the design. Other participants included the HFE Specialists, the owner and representatives of the design and manufacturing teams to provide feedback on the feasibility of certain design options.

A structured HFE design process was followed to optimise the location and positioning of the equipment following the guidelines previously described. The process used the task analysis information and link analysis methods to follow the operators visual scanning as well as his/her physical interaction with the equipment The design proceeded through several iterations which involved gross positioning of equipment based on general HFE guidelines, and repeated task simulations to test the designs and fine-tune the position and relationship of equipment. At each stage, the interim design was recorded to ensure a continuing forward path through the process, as well as to document the positive and negative features of each design option.

<u>Evaluation</u>

During the mock-up, each interim design was evaluated through task simulations and comparison to the appropriate guidelines. During sea trials, the design was re-evaluated under typical operating conditions, and final design details were completed.

As an example, during the evaluation the final position and orientation of the wing navigational controls used during docking were confirmed. The wings are outside the enclosed bridge, requiring a sealed compartment to protect the controls against meteorological conditions. User feedback under operation and HFE design principles (i.e., control positioning, anthropometry) were was used to finalize the wing control design. Small details such as cut-outs in the consoles to accommodate clearance to the ship's phone and knee clearance below the forward console, were also finalized during this stage in the design.

<u>Design Standards</u>

Design standards were documented in a final report to the design team. The standards were based on the results off the design finalized in the mock-ups and evaluation. Equipment locations were specified to ensure all individuals identified in the user profiles in the requirements analysis were accommodated in the design. As an example, This involved guidelines that the placement of critical pieces of equipment within the reach of the smallest user (i.e., 5th percentile US female adult) and clearance to controls (e.g., hand clearance between vertical joystick controls on wing control panels) for the largest user (i.e., 95th percentile US male adult). An additional design guideline was to provide space in the controls to allow new or modified equipment to be installed in the wheelhouse, as marine navigation and communication technology advances. These standards were intended for use in the development of the second vessel, as well as future refits, common in the marine industry with the continuing advances in technology.

<u>Prototyping</u>

The HFE design process, integrated into the design of the first vessel was further evaluated during operations under sea trials. The documentation of these design guidelines may be considered to be a prototyping stage which was then integrated into the design of the second vessel.

Implementation

User feedback was gathered approximately one year after the vessels were brought into services. The vessel operators reported very high rates of satisfaction with the wheelhouse design and equipment layout. They reported using the design guidelines provided by the HFE Specialists in the addition of new equipment.

CONCLUSION

The HFE Lifecycle and its intrinsic design principles are standard processes used by HFE Specialists in the any design, including transportation systems, software, tools, products, or clothing. When designing complex systems, the focus of HFE is the application of relevant principles and the involvement of expert users, in this case, the vessel operators. Combining operational expertise with knowledge about human capabilities and limitations results in a design focused on human performance.

Control and display layout and visibility were particularly important in this ferry due to its high speed and the amount of marine traffic in the San Francisco Bay area. It was imperative that the design allow the vessel operators to react in a logical and efficient way in an emergency situation.

Some constraints in the design were identified through the process. For example, the anthropometric range to satisfy the preferred range of users (5th percentile US female to 95th percentile US male adults) was not able to be accommodated under all circumstances, due to space and functional limitations. A minimum requirement of the 50th percentile US female to 50th percentile US male adults was adopted. In addition, the HFE input was limited to the console and seating design in the wheelhouse. The overall wheelhouse design included windows behind the operators, resulting in direct sunlight or glare on the forward console controls. Extended edges on the console were added to minimise the display washout. However, after the final sea trials, the engineering team investigated the use of polarizing film or other options to reduce the amount of direct sunlight entering through the rear windows, without compromising the visibility.

As an anecdote to the fact that there always exists one unanticipated factor...while visibility was emphasised as a critical factor throughout the design process, the exterior wing visibility was somewhat compromised by the positioning of the vessel running lights located directly in the operator's line of sight from the wing to the aft spring lines. The location of the lights could not be changed, and while not ideal, the operator was able to adopt a somewhat awkward position to view aft. This late realisation is an example of the need for an overall systems approach, including external issues, to ensure that all situations are accommodated.

Cost was not reported to be a limiting factor as the modifications to the original console design did not require any specialized considerations or equipment. In fact, including HFE in the design process from the beginning was reported to likely reduce the cost of making operator requested modifications to the design after the vessel was put in service.

Vessel operators have reported high satisfaction with the wheelhouse and are committed to including HFE design principles in future modifications. The design and manufacturing teams have reported that they have begun to include HFE principles into the development of other designs, and were suggesting slight modifications during the mill work stage to this wheelhouse to further apply HFE principles. The design team has also reported that they believe the new design of this wheelhouse resulted in a significant improvement of the safety and overall operability of the vessel.

The success of the design can be attributed to using an open and participatory process with vessel designers, builders and operators, involving HFE in all stages of the design process, using mock-ups and task simulations to ensure a usable design, and knowledge and attention to the unique operational requirements of a high speed marine vessel. The complete design shift from the original console layout for two operators to optimizing the design to one operator is a good example of design principles based on function and operator requirements.

The results of this project, as well as the directives of a number of maritime organizations, demonstrate the need for HFE in vessel design, not only to ensure safe operations, but to maximise vessel and crew performance.

REFERENCES

1. Schuffel, H. (1992). Designing and testing a ship's bridge layout. In: Ed. H. Kragt, *Enhancing Industrial Performance: Experiences of Integrating the Human Factors,* London - Washington, DC, Taylor & Francis.
2. Shephard, D. (1997). The Well-Tempered Wheelhouse. *Professional BoatBuilder,* August/September edition, pp. 66-77.
3. International Maritime Organization (1998). *Role of the Human Element with Regard to Pollution Presentation and Formal Safety Assessment: Report of the joint session of the Joint MSC/MEPC Working Groups on the Human Element and on Formal Safety Assessment.* IMO: Marine Environment Protection Committee, 42/16.
4. Pheasant, S. (1988). *Bodyspace - Anthropometry, Ergonomics and Design.* Taylor & Francis: New York, NY.
5. Grandjean, E. (1989). *Fitting the Task to the Man - A Textbook of Occupational Ergonomics.* (4th ed.). Taylor & Francis: New York, NY.
6. Pheasant, S. (1988). *Bodyspace - Anthropometry, Ergonomics and Design.* Taylor & Francis: New York, NY.
7. MIL-H-46855B(2)/MIL-STD-1472 (1995). *Human Engineering Requirements for Military Systems Equipment and Facilities.*
8. American Society for Testing and Materials (1995). *Standard Practice for Human Engineering Design for Marine Systems, Equipment and Facilities.* F1168-88.
9. Bittner, A.C., Wiker, S.F., Kinghorn, R.A. and Bramwell, A.T. (1993). *Crew-Based Recommendations for Improved Coast Guard Patrol Boat Design.* Battelle Research Centre, Seattle, WA.
10. National Maritime Institute (1995). *Code of Practice for Ships' Bridge Design.* EMI Ergonomics Laboratory in conjunction with the Department of Industry, UK.
11. IL-H-46855B(2)/MIL-STD-1472 (1995). *Human Engineering Requirements for Military Systems Equipment and Facilities.*
12. US Department of Transportation/US Coast Guard. (1989). *Introduction to Human Factors Engineering.* Navigation and Vessel Inspection Circular No. 4-89.

Discussion

Benjamin L. Nicholson, Member

[The views expressed here are those of the discusser, and not necessarily those of the U.S. Coast Guard or the Department of Transportation.]

The authors have provided SNAME with a well-written and timely paper regarding an emerging field of engineering. Human Factors Engineering (HFE) is a very elusive field and one that has only recently gained prominence within the design spiral process. By incorporating the user-interface within the design and construction phases of the ferry's production, a superior product is achieved in terms of both safety and efficiency. This paper first describes the intended outcome and then discusses this outcome in terms of multiple HFE methods. Thus, the authors present a very objective analysis of HFE as it is applied to a specific naval design of a high-speed ferry.

The most intriguing aspect of this paper, in my humble opinion, is its premise. Despite the application of innovative technology (pertaining to structural, mechanical, and navigational systems), the improvement in safety is only achieved once the operator is factored in as an integral component of the system. Thus, the ferry and operator can be simplistically modeled as a series circuit; one that possesses a reliability equivalent to the product of the reliability for the operator and for the machine.

As we attempt to answer the question, "what causes a human to err?" we can tailor our designs accordingly. However, each and every operator is different. How are such unique qualities taken into account? How does HFE fully assess the vast uniqueness of every operator that will pilot the ferry? The authors answer these questions and address the multiple aspects of human error. The term "user profile" is applied to convey a probabilistic approach to human reliability. This type of approach by the authors acknowledges that HFE is only as accurate as the probabilistic data that are entered into the given model.

While the authors present a very dynamic and robust model, they are humble and objective in their approach. This is surely one of the paper's strengths as the only certainty involved with manual operation and design is that humans are fallible and errors are unavoidable. The essence of the paper therefore becomes how do we account for such errors and avoid the unrecoverable failures? By employing a concurrent approach to engineering—one that combines the design, construction, *and* operation of a given vessel—we increase the odds of achieving such an end.

Peter G. Noble, Member

The authors are to be congratulated in bringing to the fore an increasingly recognized key area of naval architecture and marine design; human factors engineering.

As correctly noted by the authors, the importance of addressing the human element in maritime safety is being recognized in a more formal way through activities such as the IMO International Safety Management Code, the 1995 amendments to the Standards for Training, Certification and Watchkeeping Convention, and the USCG "Prevention Through People Program." These approaches focus on establishing safety management plans and procedures, safety training and on auditing the implementation and application of such systems.

However, human element research has found that a key factor is to design the components and systems such that they are safe for humans to construct, maintain and operate. Many engineering systems, where human factors may not have been appropriately considered, have been shown to contain features that can lead to human errors during construction, maintenance and/or operation. Management safety systems and safety training alone are not sufficient to completely eradicate these errors. Proper application of HFE principles at the design stage, with follow-up through construction and commissioning are essential to overcome these problems.

IMO has also devoted some attention to human factors during design. The sub-committee on Design and Equipment has been working on developing *the Guidelines for Engine-Room Layout, Design and Arrangement*, while the sub-committee on Safety of Navigation has a corresponding effort on *Ergonomic Criteria for Bridge Equipment and Layout*.

As one of the leading classification societies in the world, the American Bureau of Shipping has also been active in the area of human factors engineering. In 1998 ABS published *Guidance Notes* on *the Application of Ergonomics to Marine Systems*, a document that draws heavily in the American Society for Testing and Materials, ASTM, Standard Practice for Human Engineering Design for Marine Systems, Equipment and Facilities (ASTM F 1166-95a). These guidance notes emphasize fundamental ergonomic factors by taking into account physical differences in populations worldwide through the application of anthropometry to shipboard human interfaces.

ABS is continuing to work in this area with specific activity in the application of ergonomic principles to engine room arrangements, to control room arrangements, to inspectability and maintainability issues and to offshore systems.

In conclusion, the discusser fully supports the authors' conclusions that the application of Human Factors Engineering can be used to not only improve the safety of operations, but can also maximize both vessel and crew performance, leading to less downtime, lower maintenance costs and better bottom line performance.

SNAME *Transactions*, Vol. 108, 2000, pp. 339–348

Development of a 47-foot Modern Trawler Yacht

Patrick J. Bray, Member, Bray Yacht Design and Research, **George Roddan,** Member, Roddan Engineering, and **Joe Gruzling,** Member, Nautican Research and Development

ABSTRACT

Demand for safe, efficient and affordable trawler yachts has been increasing over the past decade. In response to this demand, the authors have developed a new type of trawler-type passagemaker incorporating a number of technological advances that have been made over the years in the commercial marine industry, but not widely accepted in the yacht industry.

A study was initially undertaken to define the optimal dimensions and performance of the ideal trawler yacht. An ambitious target of achieving 30 percent better running efficiency over conventional designs was also set as a goal. Safety and habitability were also major considerations to the design input and mission profile. The result was a unique 47-foot trawler-

type yacht, utilizing a new tunnel stern and semi-knozzle as well as seakeeping bulb to achieve running efficiency estimates. Fixed-fin stabilizers were also added to provide passive stabilization of roll, without the complexity of conventional active stabilizers.

This paper will present and discuss the design, performance predictions, model test results as well as construction and sea-trial results of this novel craft.

INTRODUCTION

The trawler market is very large and quite new. It is only in the last 8 years that companies are starting to take notice of the growing demand. Baby boomers are the driving force behind this new market. Demographically they have more money than ever before, are healthier and love outdoors and travel. In addition aging sailors are taking to displacement motor yachts for ease of handling and the additional creature comforts that power typically offers over sail. Both these groups want the freedom to travel world wide, at their own pace, on their own time schedule.

Like so many "new" things in the marine industry nothing is really completely new. In 1902 and 1912 the first trans-Atlantic crossings were undertaken in vessels 35 feet in length. Various other vessels made crossings over the next 50 years but the trend towards the 'passagemaker' all really started in 1974 with the publication of a book called "Voyaging Under Power" by Robert P. Beebe. This book has become a bible for offshore power cruisers. Text is frequently quoted by the interested boater, and any new vessel is compared to the technical section for "Beebe approval".

Robert Beebe had been a naval aviator during the Second World War. He spent the majority of his career in San Diego and Hawaii, with his final years in the Pacific as ship's officer and navigator on the aircraft carrier USS Saratoga, one of the world's largest ships. He spent his off duty time as an amateur designer and author, writing many articles for yachting magazines. By 1959 Beebe-designed oceangoing motoryachts were appearing. In 1962 Beebe designed his own 50 ft. motoryacht, the *"Passagemaker"*. For the next 7 years he literally circled the globe under power. From this vessel the term 'passagemaker' has now come to mean those long range motoryachts designed specifically for extended ocean cruising.

An American company building yachts in the Orient made the next major contribution when they brought out a new type of vessel to compliment their line of cruising sailboats. The Nordhavn 46, designed in 1987, was a true ocean cruising production motoryacht. Nordhavns have become the benchmark for this type of yacht, with vessels from 40' to 62' cruising every portion of the globe.

The trend towards this type of yacht has continued and grown in popularity. A new magazine, "PassageMaker", devoted to this type of vessel has begun publishing quarterly. They conducted a market study for the ideal Passagemaker yacht, drawing responses from a surprising number of interested boaters. Many boatbuilders have started production of a line of offshore trawler-style yachts in response to the growing market interest in such vessels. Boat shows aimed at this specific market, featuring trawler yachts and sponsoring seminars on offshore passagemaking are springing up across the United States and Canada.

MARKET STUDY AND MISSION ANALYSIS

Enter the new kid on the block! With the growing market interest in passagemakers and few major competitors Bray Yacht Design And Research decided to start an in-house design project to compete with the market leader. Size and features are key to the success of any new vessel. Based on market knowledge we decided to use the Nordhavn 46 as an initial benchmark. Drawing from experience gained in designing ocean cruising sailboats and live-aboard vessels, plus fishboats, small oceanographic and research vessels and with an eye to ABS rules for unrestricted offshore use, basic parameters were laid down. Shortcomings of the competition were noted and an initial design conceived, modified, and restyled. Our initial vessel fit 90% of the criteria in PassageMaker Magazine's boater survey, and we made changes to accommodate the other 10%.

Our target market includes successful business people and corporate presidents. They are well educated, well informed, and generally experienced boaters either from pleasure boating, commercial marine or Naval backgrounds. They intend to live aboard and cruise long range both, offshore and coastal. Essentially they are looking for a "two bedroom condo" on the water, but one with specific performance features. Good access to reliable mechanical equipment is a must including full standing headroom in the engine room. They prefer naturally-aspirated engines, dry exhaust, and keel cooling. Fuel efficiency and extensive range at reasonable speed are key factors. Seakeeping and range of stability are important issues; these are virtually all the characteristics that would be required for any proper offshore vessel.

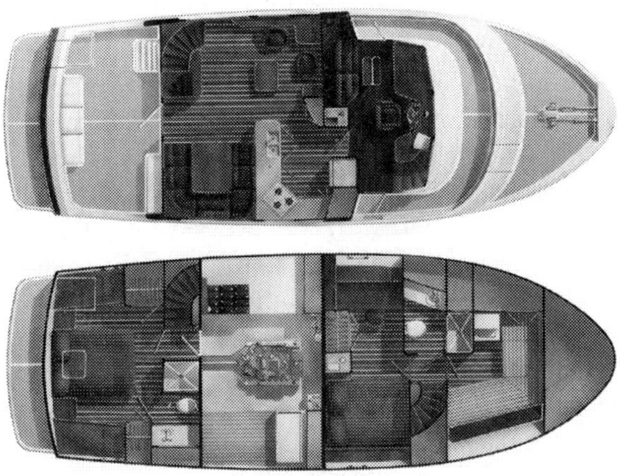

We set our goals at the following:

- Super fuel efficiency
- Range 3500 miles at 8 knots cruise with reserve
- Top speed light loaded of 12 knots for local fast cruising
- Privacy between staterooms
- Aft full width stateroom with walk around standard queen size berth
- Midship guest room/den with standard queen size berth or two standard single berths
- Each cabin to have its own wc, sink and shower
- Forepeak fitted out for storage and/or workshop as it is unsuitable for sleeping at sea.
- Stand up headroom in the engine room
- Large main salon
- Practical, good sized galley with extensive storage
- Separate eating and lounging areas
- Limited number of vertical steps on each deck
- Good initial stability combined with a good sea motion
- Long range of ultimate stability (through 180 degrees if possible) without ballast
- Collision bulkhead, watertight bulkheads and engine room door
- Effective, economical, simple roll stabilizing system
- Good styling and market appeal
- High quality yacht finish
- Modest draft

Pricing was not a major consideration in the design goals, although it was taken into account when planning the construction of the vessel to utilize efficient methods of building. The vessel is priced to meet the current market for similar boats of this type. It was felt that to price too low would infer that the vessel was of inferior quality, and to price too high would result in reduced sales.

DESIGN

Hull Design

The hull design is based on the essential characteristics used for a sailboat (or motorsailer), combined with the effective hull form of a lobster boat. The operating speed of a trawler yacht at long range cruising speed is not too much different from that of a fast sailboat. The round displacement hull form moves through the water with the least amount of fuss to show it's passing. Combine this with the occasional need to approach semi-displacement speeds and you have a hull which must operate efficiently over a wide range of speed/length ratios. By mixing in the finer features of a lobster hull with the rounded sailboat shape, this successful blend is achieved. In addition the vessel has to have a smooth rolling motion with good seakeeping abilities, and again the sailboat and lobster hull forms are a good blend for this. Add to this a large spray knocker running all the way aft and you have a vessel which is very dry on deck, has large interior volume, and excellent stability characteristics. This form also has a natural roll dampening ability.

PRINCIPAL PARTICULARS

DESIGN	Karvi 47
DATE	July 1998
CLIENT	Karvi Maritime LLC
LAUNCHED	March 2000
LOA	46' - 9"
LWL	40' - 10"
BEAM O.A.	16 '- 0"
BEAM W.L.	13' - 0"
DRAFT	4' - 6"
DRAFT (HULL)	2' - 9"
DISPLACEMENT	47,000 lbs. (S.W.) half loaded

Speed Nozzle

Also called a ducted propeller, the speed nozzle is built around the propeller to increase the efficiency thereby increasing the speed of the boat. By directing water in the tunnel to the propeller blades you are forcing more water against the blade thereby producing greater thrust through the water. This nozzle is a further development from the "Kort nozzle", which was designed for maximum thrust at zero knots. The high efficiency speed nozzle is an airfoil ring around the prop. It is like the "Kort" but thinner in section, designed to increase thrust at higher speeds with much less induced drag. This arrangement combined with the tunnel and a large diameter highly skewed propeller allows the propeller to run at a very slow rpm thereby increasing propeller efficiency.

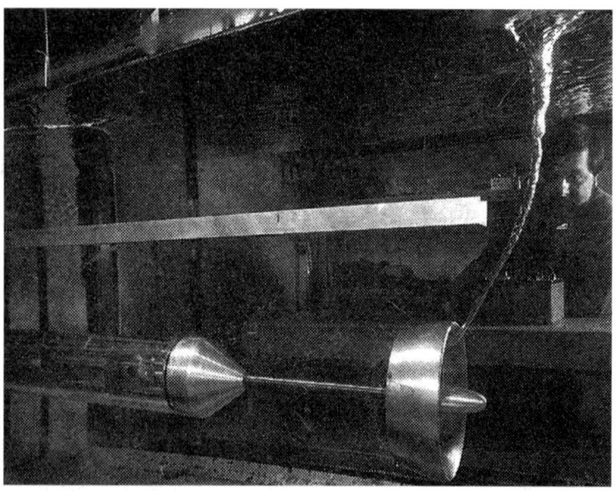

Besides increasing speed there are three other very significant benefits of the speed nozzle. These are reduced cavitation, reduction in propeller noise and provision of added protection to the propeller blades.

Cavitation occurs when low pressure develops on the surface areas of the propeller causing a breakdown in the boundary layer of water flowing over the blade. A vacuum develops along the leading edge turning the water to vapor. When these tiny pockets of air collapse back against the propeller a high screeching noise results as well as pitting of the propeller surface. This usually occurs at higher propeller loads and/or higher propeller rpm. The nozzle decreases this cavitation by containing the water mass, increasing the pressure, and controlling the outflow of the stream.

A propeller nozzle can also decrease noise within the boat. The tips of an open screw generate a vortex interacting with the hull causing noise and vibration within the boat. A nozzle directs the water out the stern away from the hull, reducing the rumble against the hull and making for quiet running.

An additional advantage of the ducted prop is the protection it gives to the propeller blade. Nobody plans on hitting underwater hazards with their propeller, but accidents happen. The protective ducting around the propeller serves as a protective casing guarding the blades against damage.

Bulb Design

The primary purpose of the bulb is to allow the craft to slip more easily through the water by reducing hydrodynamic drag. Vessels fitted with a bulb benefit by what is known as wave cancellation or destructive interference of the combination of the bulb and hull generated wave patterns. The wave form created by the bulb is generated in such a way as to partially cancel the wave train created by the hull. In effect, the bulb is helping to relieve the dynamic pressure pattern around the hull, so that the integrated effect is to reduce drag. With very full-bowed vessels, such as tankers, the main effect of the bulb is to reduce wave breaking and flow separation. Both of these phenomena represent turbulent flow effects and if these can be reduced, the result will be an overall reduction in resistance (drag) of the vessel. A bulb also has the additional benefit of reducing bow accelerations, this results in significantly improved habitability in a seaway.

With trawler type hull-forms, such as the Karvi 47, a bulb reduces drag by wave cancellation. There is also a reduction of wave-breaking and flow separation on the more full-bowed vessels. In addition, there is the important effect of squat reduction at the higher speed/length ratios. As a vessel increases speed above a speed length ratio of about 1.3, it begins to squat, or take on significant stern trim. As speed is further increased, the trim increases to over 3 or 4 degrees, leaving a large stern wave, and high resultant drag.

The parameter of major importance with respect to powering is the speed/length ratio, $V/L^{1/2}$, V = speed in knots, L = waterline length. Most trawlers operate in the speed-length ratio range 0.7 to 1.7, which is ideal for bulb applications. As the so-called hull-speed is approached ($V/L^{1/2} = 1.3$) these hull-forms tend to "squat" or take on significant trim angles, with attendant large increase in resistance. A well-designed bulb will help to delay squatting as the hull-speed is approached and result in significant resistance reduction (of the order of 8 to 10%). This translates into significant fuel savings (and also reduces the greenhouse effect as a side benefit).

What are the design parameters that influence effective bulb design? For a given vessel, it must operate in the speed/length ratio envelope from approximately 0.7 to 1.7 (this is not a fixed range but is appropriate to trawler type hulls). It should be noted that there is some evidence that the powering benefits of a bulb can extend up to speed/length ratios of up to about 2.5 but design data is lacking and may be applied only to special circumstances.

Bulb Geometry

Next, the proportions of the bulb must be defined. Bulb geometry and dimensions are influenced by the expected speed range, operating conditions, hull geometry, block coefficient, prismatic coefficient, entrance angle, depth of forefoot, bow overhang and anchor location. Each of these variables puts limits on the form that the bulb can take.

Many trawler hull-forms are amenable to a successful bulb design. The main bulb geometric parameters are: cross-sectional area, and length forward, beyond the intersection of the stem and top of bulb. Generally, bulbs are referred to by the ratio of their cross-sectional area to the area of the ship mid-section. These values can vary from about 12% to 22%. The bulb length (distance the forward-most end extends beyond the upper stem/bulb intersection) can be expressed as a percentage of the waterline length and can vary from 3% to about 8%. In most cases, a longer bulb works better for trawler hull-forms, but important considerations of anchor handling, docking and maneuvering usually limit this variable.

Bulb cross-section influences the size of the wave that is generated, while bulb length determines the phase of the bulb-generated wave. Optimal bulb design is achieved when these two factors are tuned to act in unison to minimize the net wave system of the hull.

Another important parameter is depth below the surface. Considerations of draft variability usually determine this value. As a rule of thumb, the upper surface of the bulb should remain 15 to 25 % of the bulb diameter below the still waterline. On a trawler hull if the bulb is too deep it will not be of much benefit in reducing wave drag, and if it is too shallow it may broach the surface at higher speeds. Again, a balance must be attained.

The forward end of the bulb can have a spherical or an elliptical shape. The cross-section can be cylindrical, heart-shaped or of varying cross-section along its length. Sea-keeping considerations usual determine the shape factor. Careful fairing of the bulb into the existing hull as well as around the thruster ports must be performed. It is our practice to scallop all thruster ports to prevent turbulent flow developing on the after side of the port, (possibly disrupting the beneficial effects of the bulb).

Preliminary Powering Estimates

Applying the above principles to the bulb design of the Karvi 47 resulted in a bulb which was estimated to give a drag reduction of approximately 10 percent at top speed. The preliminary powering calculation was performed using an algorithm developed by Calisal et al, and published in the SNAME publication, Marine Technology. The algorithm is based on the "UBC Series" of trawler hull forms and has been found to provide reasonable accuracy for preliminary estimates of powering performance of trawler hull forms operating at displacement speeds. Estimates of the effect of the bulb on performance are done using a proprietary algorithm based on empirical data from bulb retrofits. Figure 1 plots the results of these preliminary estimates.

OUTDOOR MODEL TESTS

Outdoor model tests were conducted to verify initial performance estimates, as well as to refine various aspects of the design in a timely and cost-effective manner. A 1:8 scale model was constructed using a strip-plank method. A removable bulb was fashioned together with removable keels. This provides for the ability to evaluate the effects of these various appendages on the performance of the design relative to the bare hull. It should be noted that at this time, no representation of the nozzle was tested.

For the outdoor tests, the model was carefully ballasted to the scale 47,000 lb. displacement and towed using a boom apparatus attached to a tender. Resistance was measured using a digital force scale attached to the towline. Running trim was monitored using a digital trim indicator, with speed through the water measured using a digital rotary speed log. Video was also recorded for all test runs.

Results of Outdoor Testing

The powering results of the outdoor tests are presented in Figure 1. As was to be expected with any outdoor model test program, there is some scatter in the data, but the general results compare well with the predictions, and confirm the general trend in the powering curve.

The benefits of outdoor testing become evident through careful observation of the model at speed. Details of bow and stern waves, midship hollow, and effectiveness of the spray knockers can be observed and recorded at length on video. The tests confirmed the effectiveness of the bulb in reducing squat, and thus resistance at speed. The tests also revealed a more optimal location for the stabilizing keels. Other changes made to the initial design were to fine up the entrance angle slightly and refine the tunnel geometry.

MODEL TESTS AT THE OCEAN ENGINEERING CENTRE

The outdoor tests provided important information and guidance for the final design. As a result, a much better design was achieved economically, without spending large sums in a model test facility. Tests in the model towing tank would be limited to the final optimized design and would largely be used as input to the design of the novel propulsion system.

The prescribed modifications were made to the initial design and the resulting model was used as a plug for a female mold. A fiberglass and foam model was then made to be tested at the Ocean Engineering Centre of B.C. Research Inc., in Vancouver, Canada.

The model was tested at a scale 47,0000 lb. displacement, in fresh water.

The following configurations were tested:

Table of Model Test Configurations – Ocean Engineering Centre

Configuration (at 47,000 lbs)	Static Trim (Degrees)	Description
A	0.00	Bulb and keel at design location
B	0.00	Bulb only
C	0.00	Bare hull
D	0.00	Bulb, keels swapped, moved aft 8 inches full scale

The results of the tests are presented in Figures 2 to 5. These results confirm the preliminary powering predictions while providing the necessary accuracy for selection of installed power and detailed design of the speed-nozzle propulsion system.

It is interesting to note that the effect of the bulb on running trim is marked at speeds over 9 knots, and the result is largely responsible for this trawler hull form being able to achieve semi-displacement speeds given sufficient power.

Listed below are some of the full-scale performance figures for the design at a nominal 10-knot speed.

Configuration (at 47,000 lb.)	Speed (knots)	*EHP	**Running Trim (deg)	Description
A	10	89.80	1.27	Bulb and keel at design location
B	10	83.31	1.20	Bulb only
C	10	93.06	1.96	Bare Hull
D	10	90.68	1.23	Bulb, keels swapped moved aft 8 inches full scale

*To determine shaft horsepower (SHP) divide EHP by the appropriate overall propulsive coefficient (OPC). For open propellers, a conservative value would be 0.55, for the speed nozzle, the theoretical OPC is 0.70 at 10 kt.
**For running trim bow-up is positive.

From the above figures one can see that with the addition of the bulb the power is reduced by approximately 10 % at 10 knots. With the addition of the Bray keels this powering advantage over the bare hull becomes about 4 %, giving an EHP of 89.8 hp at 10 knots, which still very low for a hull of this length and displacement. It should be noted that while the Bray keels add drag, it is considerably less than the drag of conventional "flopper-stopper" type or active fin type stabilizers.

DISCUSSION ON CONSTRUCTION

Fiberglass/foam core construction was chosen for manufacturing the hull and superstructure. Although steel and aluminium were also considered, fiberglass had the majority of the benefits for this application. PassageMaker Magazine's market study had shown about a 20% interest in steel, a small percentage interest in wood and aluminium, and by far the greatest interest still in fiberglass. The boating public have really been sold fiberglass as a construction medium for yachts, over and above everything else. In many ways this is because yachts typically are relatively poorly maintained (fiberglass is virtually maintenance free) and because fiberglass lends itself so well to mass production techniques. In addition, it is easier to get compound curves and detailed "yachty" styling lines into a component without relying on highly skilled labour. Once the original part is made by a skilled craftsman and a mold is taken off, the process of duplicating that part becomes a simple procedure.

Foam cored fiberglass was chosen for greater strength and safety. Although there may be considerable debate about that statement, our experience has supported this conclusion. The foam core spreads the shock loads between the outer skin and the inner skin. Damage to the outer skin is dissipated in the core and the inner skin is almost never ruptured. Also the foam core adds positive buoyancy should the vessel be completely flooded, adds a certain amount of sound deadening, and heat insulation. It is important to use a closed cell cross linked PVC foam to prevent moisture absorption. Extra care is required to make certain that the foam is in complete contact with both skins in the molding stage for without proper contact all the advantages of this material are lost. The down side is that foam core is expensive, requires more labour than solid glass, and requires a somewhat more skilled work force.

With the surprising interest in steel for this type of vessel a study was undertaken to determine what were the features people expecting from this material. The primary interest was abrasion resistance should the boat go aground and ramming strength in a collision situation. We felt that the foam core and collision bulkheads handled the latter but needed an additional feature to combat the former, so a steel shoe was laminated into the keel. This 1" thick flat bar is 6" wide and runs the full 30 feet of the keel. It is laminated in the first few layers of fiberglass so there is no need for fastenings to penetrate the hull. This was then backed up with 1 ½" of solid fiberglass, giving a substantially stronger keel section than the norm.

The bow bulb is laminated of solid fiberglass for strength and ease of construction and bonded to the hull as a separate watertight appendage. It has great strength due to it's shape and heavy laminate and essentially acts as a collision bumper, as well as contributing to reduced resistance and improved seakeeping.

The high efficiency nozzle (speed nozzle) is fabricated from stainless steel for strength and ease of construction. Stainless steel is a must for nozzles for resistance to deterioration from propeller tip erosion. The nozzle has a precise section shape which gives unique performance features. This is achieved by making up sections of the circle and welding them together to make the complete ring. The hollow foil-section rudder is also made of stainless steel over a stainless shaft.

The stabilizing bilge keels (Bray keels) are a foil section fabricated from aluminium plate. The depth of the foil and narrow cross section make it impossible to construct out of fiberglass without assembling two halves, a system which raises concerns about achieving a satisfactory joint. An accurate section shape is required to get the hydrodynamic effect from these foils so the fins were designed in the same way as sailboat keels, fabricated, and bolted onto the hull on a reinforced solid fiberglass area. The structure is strong enough for the vessel to sit upright on it's own bottom.

SEA TRIALS OF FIRST PROTOTYPE

At this time the authors are awaiting launch of the first boat. Report of sea trials will be delayed until mid June. Powering performance, seakeeping, noise and vibration as well as maneuvering will be investigated.

CONCLUSIONS AND DISCUSSIONS

This paper reports the process of development of a production trawler yacht. Presented are the necessary steps to achieve a successful design. These include market analysis, preliminary design, testing and development, use of advanced construction techniques, and final commissioning of the first prototype.

From the beginning, the goal was to produce a trawler yacht that incorporates the latest technological advances that would result in a safe and fuel-efficient design. Technical innovations borrowed from the commercial marine industry such as the bulbous bow and speed-nozzle propulsion system were applied to the design of the Karvi 47. The latest in construction methods and materials technology were used in the building process. The result is a trawler yacht that can be economically built and which exhibits performance that is significantly better than existing comparable craft.

Figure 1
Preliminary Predictions and Outdoor Model Test Results
Karvi 47

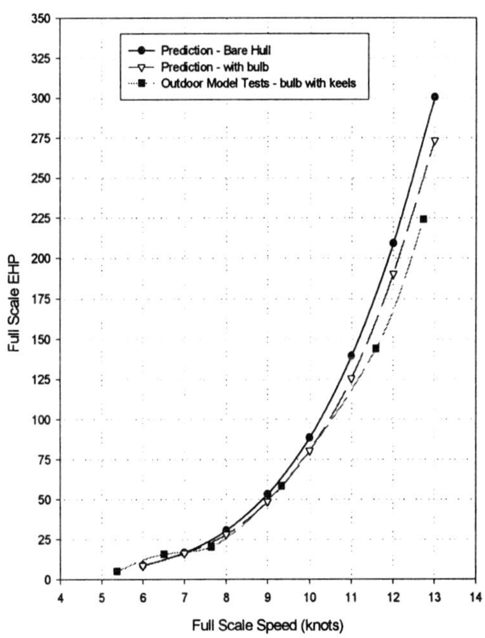

Figure 2

Resistance vs Speed
Karvi 47

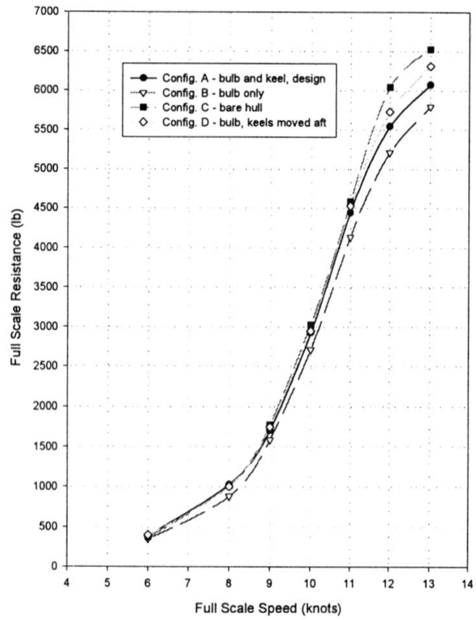

Figure 4

Trim Change vs Speed
Karvi 47

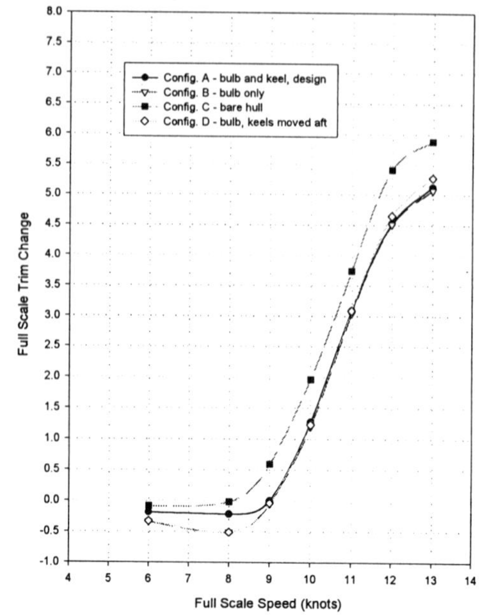

Figure 3

EHP vs Speed
Karvi 47

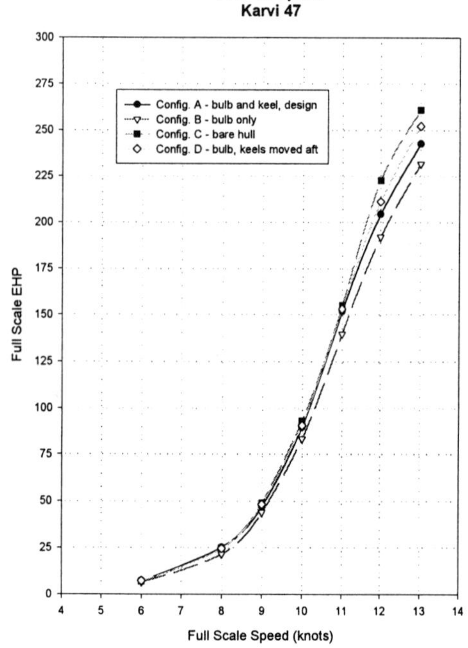

Figure 5

Heave Change vs Speed
Karvi 47

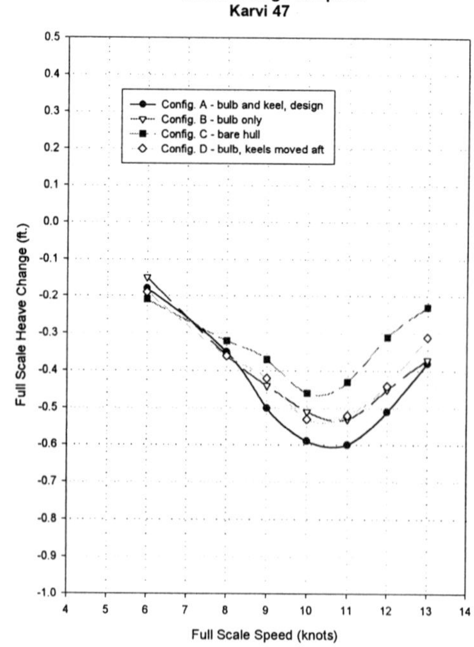

SNAME *Transactions*, Vol. 108, 2000, pp. 349–367

Design Development of a Survey Yacht

Douglas M. Wolff, Member, Chief Naval Architect, Elliott Bay Design Group, Ltd., Seattle

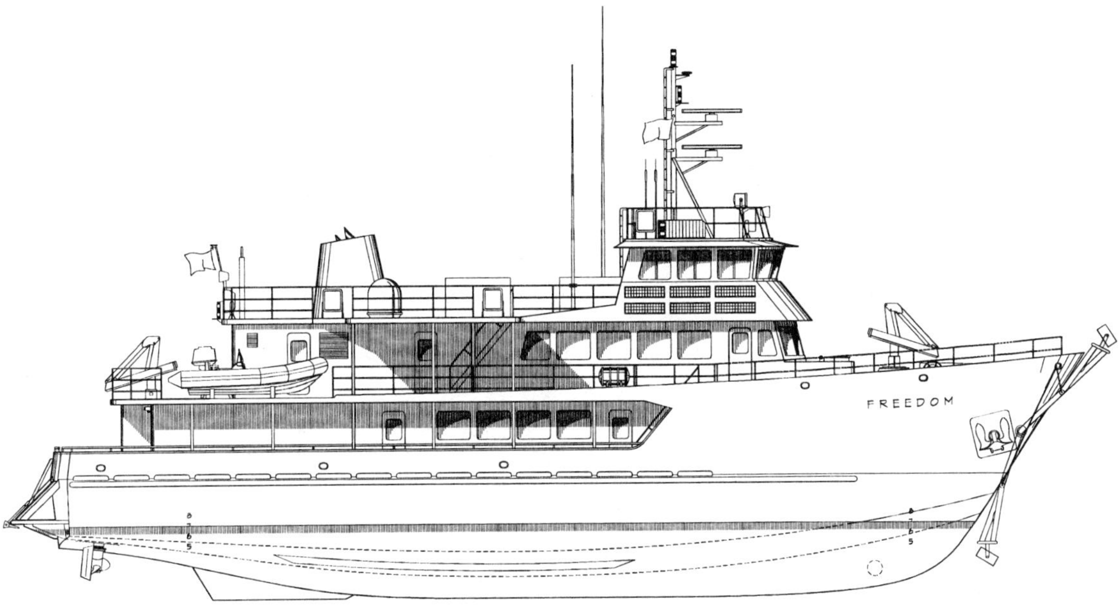

OUTBOARD PROFILE

ABSTRACT

This paper follows the contract design development of a private yacht whose owner is interested in neither the comfort nor style customary for a yacht. Priorities established by the owner at the outset of the project were: functionality, low operating cost, minimization of maintenance, ease of repair and modification, high reliability and survivability, and simplicity of construction. The resulting vessel is 32.7 meters (107'-2") long and is of all-aluminum construction, unpainted except for anti-fouling coatings. Propulsion is diesel-electric with azimuthing propulsion thrusters and a bow thruster for excellent maneuverability. Double bottoms are incorporated for protection against grounding, the engine room is divided for damage control, and the azimuthing units are in separate watertight compartments for flooding isolation. Freedom *is an unusual combination of sophistication and simplicity, a private yacht dedicated to the use of high technology for exploring and documenting some of the world's least spoiled ecosystems.*

INTRODUCTION

The booming technology sector of the Pacific Northwest economy has created the opportunity for earning wealth for hard-working, visionary and lucky entrepreneurs. Those few who have achieved great financial success find themselves in a position to retire at an early age, able to pursue their dreams. One such dream is the subject of this paper.

The resulting vessel, named *Freedom*, will serve the following missions:

- The owner plans to sever his property ties to the shore and will live aboard the vessel permanently. The vessel will accommodate four crew and guests in relatively austere accommodations. Extensive security features are incorporated into the design for protection while cruising and visiting third world regions.
- *Freedom* is a platform for the study of unsoiled shallow-water ecosystems such as reefs and estuaries. To accomplish this, the vessel is equipped with an underwater camera boom, a complete scuba support center, booms for deployment of small ROV's, and a complete suite of image processing electronics to support the underwater mission.
- *Freedom* will serve as the owner's business office and technology development center. To accomplish this end, the vessel is furnished with an office suite, a large library, and a sophisticated electronic network and communications system to facilitate new product development and efficient world-wide communications.

A most unusual aspect of this design project was the owner's personal involvement in selecting nearly all of the manufactured equipment items. To ensure that he received the highest quality at a competitive price, he performed countless hours of research to educate himself, then identify and select equipment offering good value and excellent technical support.

The owner had no background in the commercial marine industry, necessitating an in-depth education program so that important marine requirements were considered during equipment selection. This education process provided an unusual opportunity for a first-principles approach to the design process as the owner regularly called upon project engineers to explain the rationale behind even the most basic design decisions.

Owner Background and Philosophy

The owner is a partner in a successful Puget Sound area firm which designs and provides information technology systems to the software industry. Because he rose from a very modest beginning through dint of long hours and hard work, the owner places great value on initiative, efficiency and focus. Conversely, he has an intolerance of inefficiency and waste, whether it be poor project management, incompetent vendors, or unproductive attitudes exhibited by the shipyard work force.

Delivery of the vessel will fulfill a long held dream, and the owner has specific and non-negotiable requirements. In order to insure that the delivered vessel will meet those requirements, the owner first and foremost wanted an engineering firm that could share his vision.

Naval Architect Selection Process

In keeping with his experience and instincts, the owner began his search for a naval architect on the internet with the following priorities: willingness to listen and adapt the design process, technical competence, relevant experience, and location. For convenience, he selected four prominent Seattle-area firms for the first round of interviews; although a nearby location would be a plus, the owner was prepared to expand his search nationwide if required to locate the firm which would best meet his expectations.

The interview process was straightforward, and the owner selected the design firm without difficulty. His observations as a successful small businessman and an intelligent outsider of the marine industry, offer a useful perspective of client perceptions. One firm was eliminated because the owner did not sense adequate depth of staff and technical competency. Two otherwise well-qualified firms were eliminated from consideration because they asked few questions and seemed to demonstrate only a polite interest in the owner's concerns. It should be noted that the owner approached the interview process with a minimal marine background [limited to previous ownership of a 25' (7.6 m) sailboat and a 32' (9.8 m) motor yacht] and some unconventional requirements, such as a desire for only vertical ladder access between decks.

Vessel Mission

It was and is the owner's intention to, once construction is complete, sell his possessions and live aboard the vessel, which will serve as his home, office, transportation, and dive platform.

The owner is single and has minimal requirements for creature comforts. Because he intends to operate only in warm climates, he viewed heaters as unnecessary, and because he enjoys tropical weather, he viewed air conditioning as frivolous. In order to minimize consumption of fuel oil, which can be very expensive in remote ports, a variety of means to reduce fuel consumption have been incorporated into the design. These ideas will be explored in detail later in this paper.

GENERAL ARRANGEMENTS

Prior to his interviews of engineering firms, the owner had invested considerable time and energy in developing his ideas for vessel arrangements. He provided the sketches shown in Figure 2 at the project kick-off meeting. Although these sketches may be primitive by design office standards, they presented a clear view of the priorities, and served as a solid reminder of the owner's commitment to this project. These sketches were used as the road map during the early stages of design, and also provided a good tool with which to begin educating the client regarding actual space requirements and the art of design compromise. Many hours of discussion and sketching led to the contract design arrangements, Figure 3.

Office and Library

Office and business space clearly dominate the concept arrangements, accounting for more than half of the 01 level, nearly all of the main deck level, and the prime real estate below deck amidships. Included in the laboratory are specific work stations for electronic equipment repair, video editing, ship's office, and medical station. It was noted at the outset that vertical ladders were shown for all access between decks; the owner felt that this could save significant space while permitting rapid passage. This subject provided a topic for lively exchange during development of the design.

In the end, inclined ladders and a variety of other alterations were incorporated. The library was relocated to forward end of the main deck, more convenient to the office yet situated in an area of greater vessel motions and thus less desirable for working space. Lost office space was recovered on the new upper deck level, with the added benefit of providing a functional control console at the house front for development and demonstrations of ship control systems.

Dive Locker

Because underwater exploration is a key mission for this vessel, the concept arrangements allocated space for stowage of diving equipment on the main deck, contiguous with the electronics laboratory space. Although not indicated, it was intended that access to the water would be provided at the transom. In the concept arrangements, the dive air compressor was located in the equipment room aft of the library. Spare scuba tanks and the dive sleds were stored aft of the engine room, accessed via vertical ladder from the weather at the main deck aft. Consolidation of diving support functions onto the main deck was viewed as a high priority, given the expected frequency of use of the dive equipment.

This priority was accomplished in the Underwater Equipment Storeroom, which includes the scuba fill station as well as expanded metal lockers for tanks, dive gear and dive sleds. The entire space is now separated from the office and electronic areas by a watertight bulkhead in order to allow hose washdown without risk to electrical and electronic systems. Direct access is provided to the fantail through a 36" (0.91 m) weathertight door. Inside access to the starboard engine room and port propulsion room is provided via watertight trunks with vertical ladders.

Hinged doors are installed in the transom bulwarks port and starboard for access to the dive platform. The surface of the dive platform is fiberglass grating for drainage and traction, and is fitted with two hinged ladders which store against the bulwarks while underway and hang vertically into the water for diving operations.

Accommodations

Luxury is clearly not an important issue to the owner, who located the minimalist galley and mess areas below the main deck forward. As seen in Figure 2, two sleeping areas were located amidships

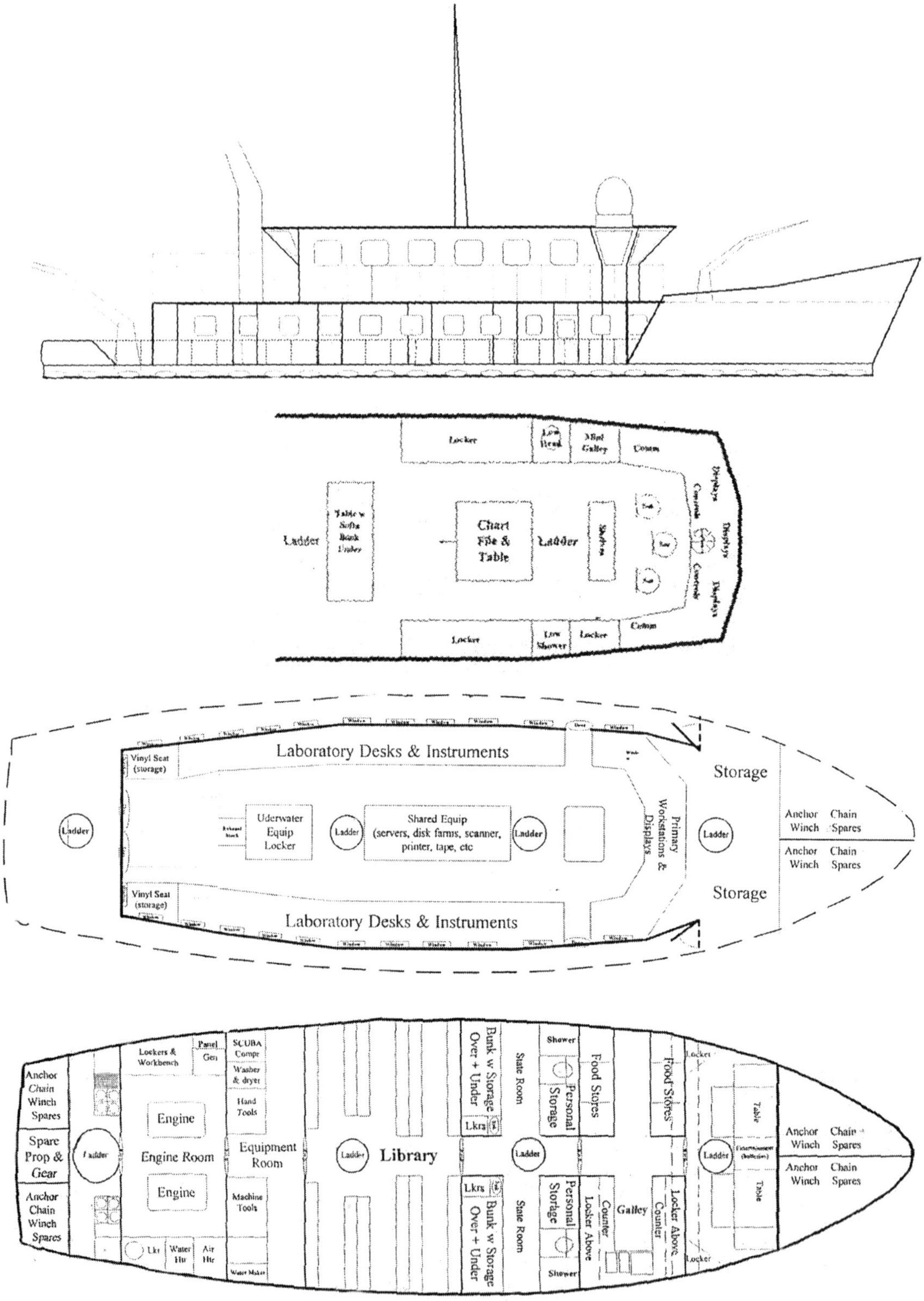

Figure 2 – Owner's Arrangement Drawings

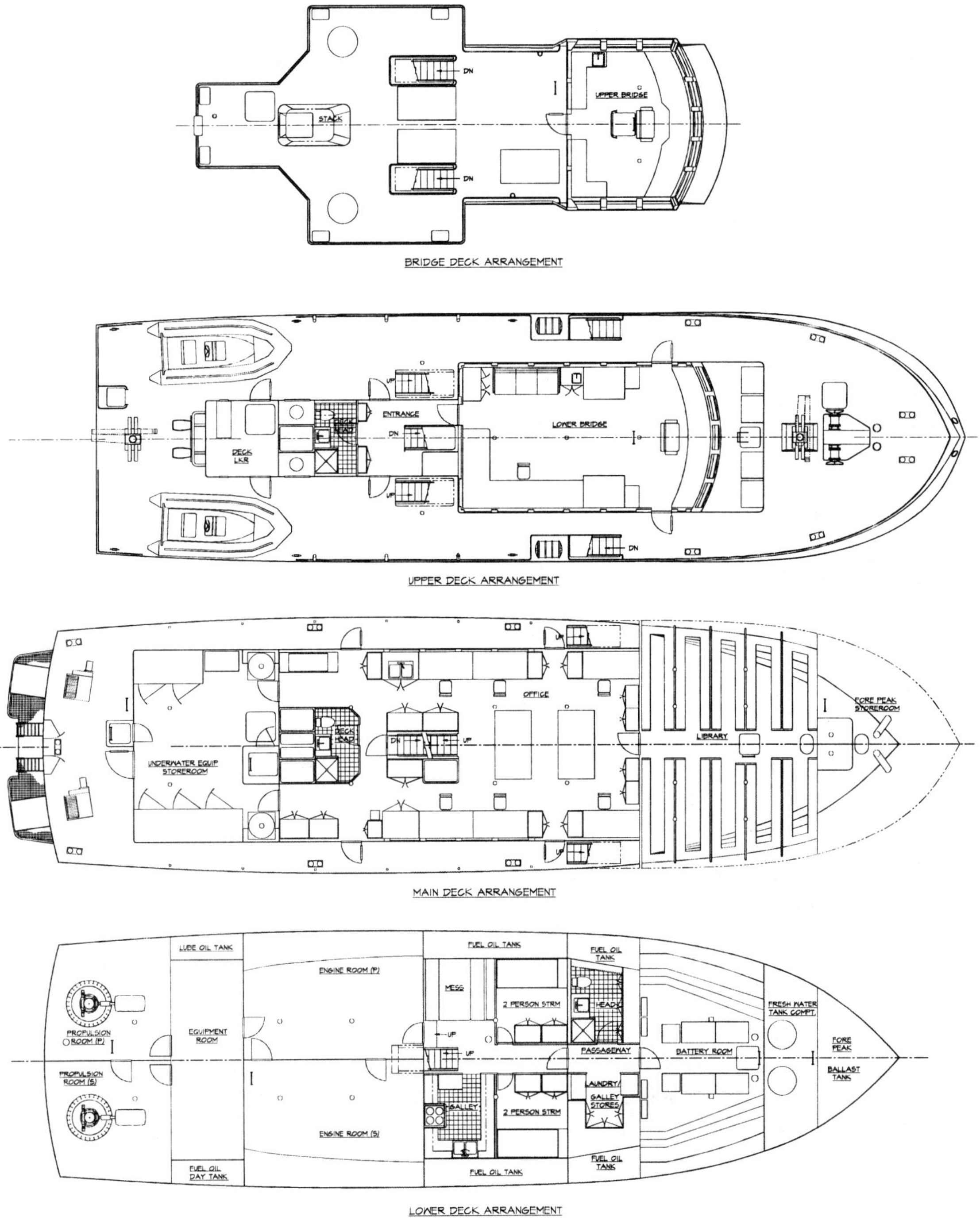

Figure 3 – Contract Design Arrangements

with direct and open access to the office above. Basic amenities included bunk beds for the four crew members with showers and toilets that opened directly into the berthing areas.

The owner's berthing area is separate from the crew area in the form of a sofa-bed aft of the chart table on the lower bridge. A full head is located aft on the upper deck for use by the owner and crew.

In keeping with the owner's priorities, creature comforts are kept to a minimum in the contract arrangements, see Figure 3. The two staterooms below decks are small at 36 square feet (3.34 m^2) of deck area per berth. Berths have been rotated for fore-and-aft orientation, and doors have been added for privacy and to reduce the airborne noise levels in the staterooms.

The mess area has been relocated from the forefoot to amidships adjacent to the galley, this location was selected to make dining more convenient and for reduced accelerations. The small galley is adequate for preparation of the simple foods preferred by the owner. An inclined ladder now provides access from the mess area to the office space on the main deck.

Galley dry storage is available under the mess settees, in cabinets forward of the starboard stateroom and on shelves in the battery room. Domestic-quality refrigerator/freezer units are located in the galley and in the galley stores area. Should it prove desirable, a chest-type freezer can be added on the upper deck; however, the size of the refrigerated spaces has been kept small deliberately in order to minimize electrical power requirements during periods of operation on the battery banks.

Water closet facilities have been expanded from those on the concept arrangements; now there are three heads, one each on the upper deck, main deck and lower deck, each containing a shower, lavatory and toilet.

Pilothouse

The concept arrangements, Figure 2, show the controls and navigation station predictably located. A second control console is located on the main deck; this console is intended for use in navigation systems integration product development; use of this console will be addressed fully later in this paper.

In the contract arrangements, Figure 3, the pilothouse is raised one level for improved visibility and protection from boarding seas. In keeping with the concept arrangements, interior access to the pilothouse is by vertical ladder only. Exterior inclined ladders from the upper deck are located port and starboard amidships.

Accommodations Insulation and Furniture

With the exception of the underwater equipment storeroom, accommodation spaces surfaces which face the weather will be thermally insulated with phenolic sandwich panels with a low maintenance decorator plastic surface. No joiner sheathing will be installed to cover insulation, piping or cableways; it is the owner's desire that all systems be in the open for ease of inspection, maintenance, repair, and modification. In addition, the owner intends to install, after vessel delivery, hard wire and fiber optic networks for central vessel control and systems testing. Totally accessible cableways will facilitate system installation and modification.

With the exception of the control consoles and the galley cabinetry, all furniture will be prefabricated to the owner's specifications and furnished as a package by the owner. In general, furniture will be all aluminum or stainless steel; exact dimensions will be provided to the manufacturer after the detail design arrangements are finalized.

Machinery Spaces

As is often the case with arrangements by a neophyte, volume allocated to machinery in the concept arrangements is woefully inadequate. In fact, the owner expressed disbelief bordering on irritation as the machinery arrangements, Figure 4, were developed. Coming from a small boat background, in which machinery is crammed into tiny spaces so that accommodation space can be maximized, the owner initially had great difficulty reconciling himself to the space required for installation and maintenance of the much larger equipment required to operate a small ship in remote locations.

The decision by the owner to subdivide the machinery spaces increased total volume requirements; the rationale for this decision will be discussed shortly. As can be seen in Figures 3 and 4, machinery is located in seven separate compartments below the main deck, which will be addressed in turn from forward to aft.

Immediately aft of the forepeak ballast tank, the fresh water tank compartment contains fresh water tanks, the pressure tank, pump, purifier, filter, and water heaters. Access is by vertical ladder from the library above.

The battery room houses the ship's supply battery banks as well as electrical equipment racks and power panels. Two transducer wells, one each port and starboard, are located at the forward end of the compartment. Access is by vertical ladder from the library above or through a watertight door from the accommodations aft. The vertical ladder is located in way of an oversized hatch aligned

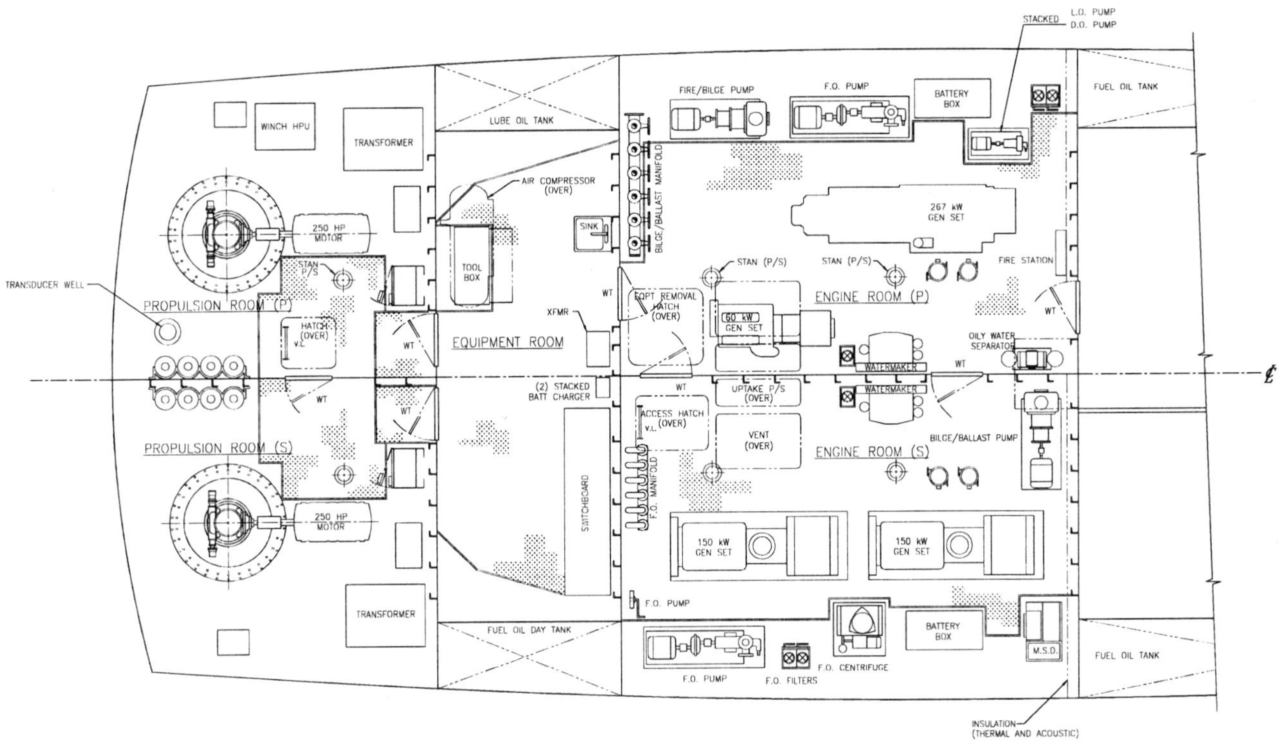

Figure 4– Machinery Arrangement

with a hatch at the weather deck so that the forward crane can be used for removal and replacement of batteries and stores.

Port and starboard engine rooms are situated immediately aft of the accommodations. A watertight door provides access from the mess area to the port engine room. Two watertight doors in the centerline bulkhead allow access to the starboard engine room, which is also furnished with a vertical ladder emergency escape up to the underwater equipment storeroom.

The equipment room, aft of the engine rooms, is accessed by a watertight door from the port engine room and by watertight doors to the port and starboard propulsion rooms. A vertical ladder from the underwater equipment storeroom also provides access and an emergency escape from the port propulsion room. A watertight door in the centerline bulkhead allows access between propulsion rooms.

Machinery Space Acoustic Treatment

Standard lead-foil acoustic insulation will be applied to the overhead and forward bulkhead of the engine rooms above the bilges. As a cost-savings measure, it is not intended to apply isolation treatment to piping or cables penetrating watertight bulkheads. Acoustic insulation will also be installed on the inside surfaces of the uptake trunk and the air intake trunks which abut accommodation spaces or the underwater equipment storeroom.

The dive air compressor and engine room ventilation supply fans will be installed on acoustic isolation mounts. Generator engines will be resiliently mounted on the skids, which will in turn be acoustically isolated from the foundation structure. Exhaust piping for the four larger generator engines will be supported by acoustical and thermal isolation hangers throughout the length of the exhaust runs.

HULL REQUIREMENTS AND OPTIONS

Maneuverability and Anchoring

It is intended that the vessel will be used for exploration and documentation of shallow water features. In addition, cramped anchorages may require excellent control both for access and to position and set anchors. As such, maneuverability was established as a paramount prerequisite for the vessel. With regard to hull form, this is achieved by use of a cutaway forefoot to allow the bow to swing more easily, combined with a small centerline skeg, giving full play for the azimuthing units; see Figure 5.

In consideration of the large wind profile, it was decided to install a 50 HP (37 kW) bow thruster to augment the steering system for positioning dockside or in confined anchorages.

A four point anchoring system is available for security and to prevent weathervaning. Two forward anchors, each weighing 1430 lb. (649 kg.), are fitted with 165 fathoms (301 m) of 5/8" (159 mm) stud link chain and are hoisted using the electric motor-powered double windlass. Two aft anchor winches are hydraulically powered and share a common hydraulic power unit which is located in the deck locker on the upper deck aft. Aft anchors weigh 750 lb. (340 kg.) each, stow on billboards, and are fitted with 117 fathoms (213 meters) of wire rope each.

Seakeeping

Although most anticipated operating areas for this vessel are in tropical waters, the ability to operate worldwide is necessary; safe passagemaking is therefore an essential feature for this vessel. The vessel's speed of 11.5 knots precludes outrunning bad weather, so good seakeeping is required for comfort, to reduce fatigue, and for the safety of the vessel and crew.

To accomplish this, a waterline entrance half-angle of 37° was combined with moderate flare forward to provide good lift in a seaway. Although the bulwark height is reduced at the bow for improved visibility during docking and close quarter maneuvering, the forward deck lockers will act as a breakwater for the house front, and the pilothouse is raised a full level above the upper deck for protection from boarding seas. Full bulwarks are provided at the main deck level at the sides and transom for crew protection at sea. Each forward and lower side window will be provided with a fitted aluminum cover for installation prior to, and for protection from, violent weather.

As can be seen in Figure 5, the hull has a double chine with a fine run aft for good flow to the propulsion units. The hull form is fully developable for ease of construction; bilge keels are provided on the chine plate to reduce rolling amplitude.

Hull Subdivision

This vessel will not be classed and no regulations mandate subdivision for a private yacht of this type. The owner, however, was acutely aware of the risk of grounding or other hull damage to a small vessel operating in remote regions, particularly around reefs. It was therefore decided to install an inner bottom for the full length of the flat of keel. A high degree of subdivision was also specified by the owner, resulting in seven transverse watertight bulkheads below the main deck.

Because the azimuthing units are electric motor-driven (to be discussed later) and are therefore particularly vulnerable to damage by seawater, a watertight centerline bulkhead isolates the units so that one unit will remain operational even if the other is flooded by shell plating damage or hull gasket failure.

Similarly, a watertight centerline bulkhead separates the port and starboard engine rooms to isolate flooding, smoke or fire. Although the owner desired complete redundancy in the engine rooms, the small size of the vessel precluded total independence. In spite of this compromise, it was felt that even in a worst case damage scenario, damage and consequent repairs would be minimized by the installation of the centerline bulkhead.

Floodable length will be addressed later in this paper.

Hull Material

In keeping with his philosophy to minimize maintenance cost and effort, the owner desired a hull requiring minimal coatings for corrosion resistance combined with ease of repair. At project inception, it was the owner's intention to have a steel hull, metalized inside and out for corrosion resistance, entirely eliminating the need for protective coatings. Metalization is the process by which an electric arc is used to vaporize a coating material (in this case, zinc) which is then deposited on the base material (steel) using 100 psi compressed air. The application thickness can be controlled by the discharge rate of compressed air; the metal-to-metal bond is reportedly excellent, lasting up to 20 years on fixed marine structures.

Recognizing that the metalizing process might be both expensive and unfamiliar to the Puget Sound area shipyards expected to compete for the construction contract, it was decided to fabricate the deckhouses from aluminum, leaving the hull material steel. While the hull scantlings were being developed, the owner visited several local shipyards to discuss the project and ascertain the prospects for application of the metalizing process. Predictably, the shipyards expressed great enthusiasm for the project but strong skepticism regarding the likelihood of their successful application of metalizing. The shipyards also were unwilling to

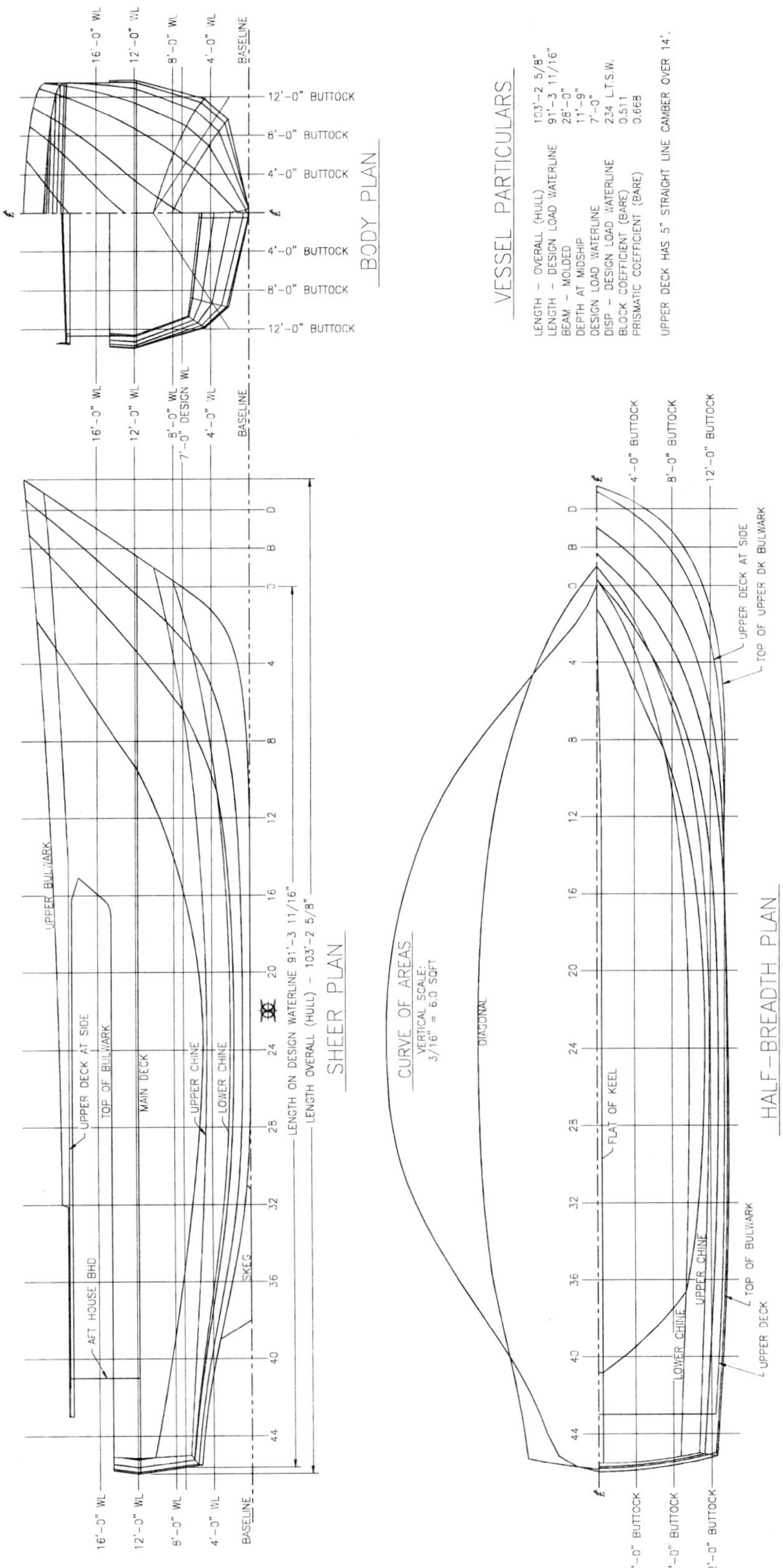

Figure 5 – Lines Plan

provide any sort of cost estimate regarding metalizing.

A new direction being required, it was decided to redesign the entire vessel in aluminum using the American Bureau of Shipping (ABS) rules [1]. Work boat rules rather than the yacht rules [2] were selected as a basis for design scantlings because a rugged hull suitable for operations in remote areas was desired. In addition to the use of work boat scantlings, it was decided to selectively increase the size of some plates and stiffeners in order to provide an extra measure of durability and impact resistance. The end result was a heavy hull with an estimated aluminum weight of 139,143 pounds (63,114 kg) which was roundly denounced as excessively heavy by all the shipyards (8 Australian, 2 Canadian and 2 in the U.S.) which eventually saw the structural design.

With his faith in the structural design badly shaken, the owner was uncertain as to how to proceed next. The design office resolved the dilemma by redesigning the structure using Det Norske Veritas rules [3], which were universally recommended for this vessel by the shipyards. Scantlings were recalculated, resulting in a total weight of aluminum of 109,914 lb. (49,856 kg), a 21% reduction. Recognizing that these calculations are commonly used for lightweight, high speed craft and did not necessarily represent the optimum design for a cruising, displacement hull, scantlings were selectively increased to provide a reasonable level of confidence.

Coatings

The change to an all-aluminum hull provided the opportunity to eliminate nearly all coatings. As discussed previously, potable water tanks are stainless steel. Integral ballast tanks and double bottom voids, now being aluminum, also do not now require protective coatings.

Preferring a utilitarian workboat finish, the owner chose not to apply any cosmetic coatings to the hull and superstructure. As a result, the only coatings applied to the aluminum structure will be antifouling paint on surfaces below the design waterline, and non-skid paint on the decks.

PROPULSION AND MANEUVERING

Propulsor Selection

Having previous experience with his twin screw yacht, the owner anticipated a similar arrangement for this vessel. Given the need, however, for minimal draft combined with precise control during navigation through shallow waters, including coral reefs, the best possible maneuverability was required, leading to the selection of azimuthing units for propulsion. An additional and significant benefit of azimuthing units was the ability to keep the propellers well above the baseline, minimizing the likelihood of damage from grounding.

Although open propellers were considered for their higher efficiency at higher speeds, the owner elected to have the azimuthing units furnished with propulsion nozzles for improved propeller protection. The selected units, Holland Roerpropeller model 300, are fitted with steel type 19A nozzles with all stainless steel inside diameter plating. The steel upper units are galvanically isolated from the aluminum hull. Propellers are nickel-aluminum-bronze.

Speed and Powering Calculations

The owner made it clear at the outset that fuel economy was to be considered a fundamental design criterion, and that vessel speed was not particularly important. With no need to push the design envelope vis-a-vis hull speed, powering calculations were straightforward. The design waterline length of 91.3 feet (27.8 m) anticipates a hull speed of about 12.8 knots. Design software indicated a required prime mover horsepower of 900 HP (670 kW) to obtain this speed in calm water, with a 10% margin added for wind and minor waves.

Considerable discussion and research was performed to resolve the dilemma of whether the propulsion drive motors should be AC or DC. Primary arguments in favor of DC motors are commonality and low cost; arguments favoring AC motors include efficiency, reliability, reduced maintenance and increased versatility. Initial discussions centered around DC and SCR in order to obtain the best propulsion motor speed control. As a result of the development of AC adjustable speed drives in the appropriate power range, however, AC motors were selected with a control system that matched drive units and motors to be designed and provided by Rockwell Automation.

Electric motors having been selected for driving the azimuthing units, the owner began research into the availability and cost of motors which deliver the appropriate horsepower. Powering requirements indicated that two motors at 300 horsepower (224 kW) each would achieve 12 knots; however, the owner's research revealed a disproportionate cost increase between a 250 horsepower (186 kW) and a 300 horsepower motor. Reiterating his willingness to sacrifice speed for economy, the owner made the decision to specify 250 horsepower (186 kW) motors as prime movers. This decision was not made lightly, but came instead after considerable discussion of the merits of different size motors, and after much research into control systems. Once the decision was finalized, it became possible to perform the electrical load analysis and select generator sizes.

ELECTRICAL SYSTEM

Generator Selection

From the early stages of the design development, the owner stressed a strong desire to minimize fuel consumption. This desire was based in large measure upon the high cost [reportedly over $4 per gallon ($1.06 per liter)] and low quality of diesel fuel available in outlying ports. This is particularly true in the South Pacific, a region the ship is expected to frequent. As the mandate to minimize the consumption of fuel oil became more established and in fact became a design requirement, various means to ensure full utilization of fossil fuel energy were discussed.

In the end, it was decided to power all machinery, including the propulsion units, with electric motors and to provide a variety of generator sizes so that generators will operate in the most efficient power ranges, minimizing fuel consumption and engine maintenance.

A sophisticated control system utilizing fuzzy logic will monitor power requirements and select the proper size generator for operation. As power needs change, the control system will automatically start, switch over and stop generator units so that the most fuel efficient unit or combination of units will be operating to provide the required electrical power. To accommodate this control system, it was necessary to select prime movers with electronic control and monitoring capability. Worldwide support infrastructure for engines and generators was also considered essential.

A number of load analyses were performed to anticipate power requirements under a variety of operating scenarios. The results of the various analyses were then reviewed for occurrence probability and then grouped and sorted until a pattern emerged, leading to the selection of five separate generator sets as follows: one 267 kW, two at 150 kW, one 60 kW, and one 20 kW. The four larger generator sets are located in the engine rooms (see Figure 4); the smallest unit is located in the deck locker on the upper deck aft. The 20 kW unit is arranged as an emergency generator serving a fire/bilge pump, navigation systems and the emergency battery charger, as well as being connected to the main bus. Individually or in combination, the generator sets provide at least a dozen power options ranging from 20 kW to 417 kW. Figure 6 illustrates the switchboard arrangement.

DC Power System

The driving consideration for the design of the DC system was the owner's desire to operate the vessel for several days without running the generator sets. The result of this requirement is a DC and AC inverter system which is capable of independently powering basic habitability functions and electronic systems, with capacity for future expansion.

Each of the two main battery banks, located in the battery room, consists of flooded marine grade deep cycle batteries, which are charged by one of two battery chargers, one of which is supplied from the bridge deck panel and the other from the lower deck panel. Refer to Figure 7 for a diagram of the DC system.

The main battery banks supply a 24 VDC bus for distribution to DC equipment, emergency lights, navigation equipment, inverters and DC outlets located throughout the vessel. A crossover is also provided so that the main battery banks can supply power to the emergency circuits. Solar panels installed on the house side and top provide a float charge to either the emergency batteries or the main battery banks, as selected.

The inverter bank is rated at 9 kVA and feeds the clean AC power bus, which provides clean electrical power to the electronics throughout the vessel.

SHIP SYSTEMS

Potable Water

Two 500 gallon (1900 l) stainless steel potable water tanks are located in the compartment forward of the Battery Room. Two 15 gallon (57 l) water heaters, each of which can use either engine cooling water or electricity as an energy source, are installed to provide hot water. The small size of the water heaters was selected to both save space and reduce water consumption by discouraging long showers.

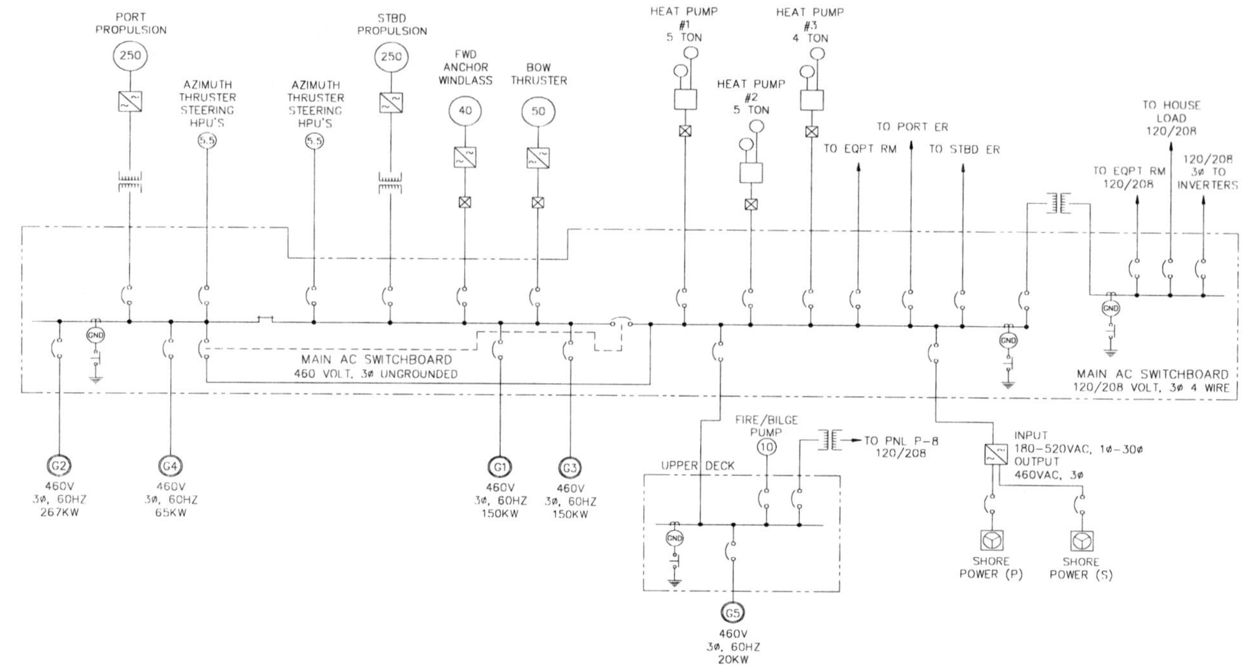

Figure 6 – AC One Line Diagram

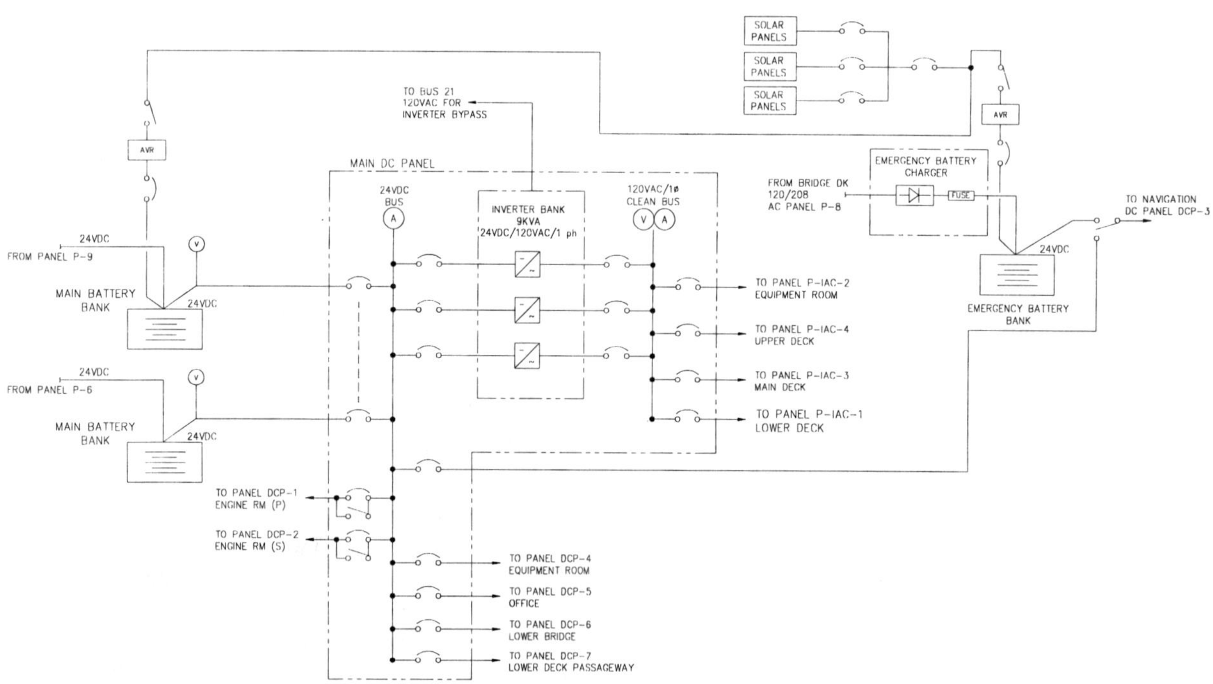

Figure 7 – DC One Line Diagram

In keeping with the project philosophy for redundancy of essential systems, the potable water system is supplied by two reverse osmosis watermakers, each with a capacity of 200 gallons (760 liters) per day.

Guaranteed water purity being of paramount importance, all potable water passes through a simplex strainer, a filter, an ultraviolet purifier and finally a bromide purifier.

Fuel Oil

Fuel oil storage capacity is 115,200 gallons (435,900 l), providing a range of 5000 NM at 8 knots or 3600 NM at 11 knots.

The fuel oil transfer system is largely conventional with several features to suit the owner's concerns regarding the quality of fuel oil generally available in remote ports. Fill stations are provided port and starboard on the main deck; the fill pipes are fitted with butterfly valves and camlock fittings. Fill pipes combine into a common header which drains through a 50 gallon (190 l) sight tank with a glass for visually inspecting fuel quality.

Two electric motor-driven fuel oil pumps are provided, one in each engine room, and a backup manual pump is also installed for emergency use. Motor driven pumps are 3 HP (2.2 kW) with a capacity of 81 gallons (306 l) per minute, and are protected by a simplex strainer.

To replenish the two fuel oil day tanks (one serving the four generators in the engine rooms and one serving the emergency generator), fuel oil is transferred from one or more of the six storage tanks by a 200 gallon (760 l) per hour centrifuge, which is also protected by a simplex strainer.

To complete the protection from poor quality fuel, each of the five generator engine suction lines is fitted with a fuel oil catalyst, a fuel oil purifier and duplex coalescing filters. In order to accurately measure and record fuel consumption, both the supply and return lines to each engine are fitted with a flow meter which is constantly monitored by the vessel control system.

Security

Experienced in both travel through and living in remote and third world locations, the owner approached the issue of security with strong concern and a variety of ideas. As well as the usual fears with regard to breaking and entering, the owner has specific concerns regarding piracy, as well as pilferage by both visitors and officials. Vessel security is addressed through both passive and active means.

The exterior profile and finish of the ship are deliberately designed to resemble a working vessel, which is expected to be a less tempting target than a yacht because of the larger crew size and fewer valuables. Quartz floodlights are installed to provide 360° illumination around the vessel to discourage uninvited approaches at night. Opening windows and screen doors are fitted with welded-in-place grilles to preclude entry even when open for ventilation.

An obvious security measure is to provide all exterior entrances with locks so as to exclude uninvited visitors. Less obvious is the risk of theft by invited guests such as immigration and customs officials. In the experience of the owner, remote ports may send a large contingent of officials aboard to inspect; these officials may disperse throughout the vessel, leaving with more than just paperwork. This concern has been addressed by providing locks for each interior door, drawer, locker and cabinet. Prior to arrival at a new port, all spaces and lockers can be locked, allowing the owner to maintain complete control over access to any items which might be stolen.

Electronic security measures include optical and acoustic sensors to detect and track the movement of people throughout the ship. Additional security measures to be installed by the owner remain confidential.

HVAC

Heating

It was the owner's original desire not to provide heating for the accommodations because he had no intention of straying into the cooler climes, and because he enjoys cool night air. It was realized, however, that some heating might be required when working through the night, and that even in the home Puget Sound region, temperatures can become uncomfortably cool on a summer evening.

Accordingly, three domestic-type heat pumps are arranged to heat the accommodations (5 tons), working spaces (5 tons) and pilothouse (4 tons). Heat pumps were selected for their simplicity, flexibility, and low capital cost, as well as the abundance of electrical power.

Dampers are installed in the ducting for each space so that the control center can regulate compartment temperatures individually, and so that heat can be turned off to spaces which are not occupied. Additionally, the control center can be programmed to reduce heating loads so as to maintain optimum loading on the generator in use.

Air Conditioning

In a similar vein, there was no initial intention to air condition to the vessel interior, the owner being satisfied with circulating outside air. Upon consideration of the effects of high humidity upon electronics and electrical systems, in addition to the potential for serious mildew problems, it was decided to use the heat pumps to cool and dehumidify accommodations spaces. As with the heating system, the control center may be used to reduce electrical consumption by regulating cooling loads when the on-line generator is approaching full capacity.

MISSION SUPPORT

Forward Camera Boom

A key vessel mission is the documentation of sensitive littoral areas by high-resolution video and still photography. The owner performed extensive research regarding the merits of a transparent camera dome located at the forefoot. Despite the obvious advantage of being able to dry-mount cameras inside the hull, the dome concept was ruled out because of the following disadvantages:

- The dome surface would be susceptible to rapid biological fouling in tropical waters, creating the requirement for regular cleaning by divers.
- The dome's forward location makes it particularly vulnerable to damage by impact with floating or semi-submerged objects. Setting aside concerns of massive damage which might flood the forepeak, the significant issue vis-à-vis photography is surface scratching caused by abrasion and/or cleaning. These inevitable scratches would reduce the quality of images captured through the transparent material, degrading the results of the photographic survey.
- Reflected and refracted light through the transparent dome, a particular concern in clear tropical waters, would increase the potential for reduced image quality.

These disadvantages, coupled with the recent significant improvement in capability of underwater cameras, led to the decision to perform photographic surveys using cameras mounted on a hinged boom, mounted on the vessel's stem. Inspiration for this boom comes from a similar but larger and more sophisticated arrangement used on Cousteau's research vessel *Calypso*. In its current configuration, the camera boom is capable of supporting up to three cameras with selectable mounting angles which control the fields of vision.

When not in use, the camera boom will be secured in a raised position against the stem. For photographic surveys, the boom will be lowered using the forward crane. The boom will be held in the operating position securely against the stem in an integral boom crotch by hydrodynamic force. By design, it is intended that the camera boom will only be operated at slow speeds and in protected or calm waters. Power and camera controls will be portable cables from each camera to sockets located at the upper deck bulwarks near the stem.

ROV Booms

Under certain conditions it may be desirable to tow one or more small arrays, or to operate remote operated vehicles (ROV's) from the vessel side. To accommodate this eventuality, transverse sliding aluminum booms are provided port and starboard below the upper deck at the aft end of the diving support space. Booms will be extended manually from fixed sockets, with wire rope stays extending forward to provide support during towing operations.

Rigid Inflatable Boats (RIB's)

Two RIB's are provided at the aft end of the upper deck, one port and one starboard. The aft deck crane is located so as to launch and retrieve both boats, which can be conveniently boarded from the transom swim platform. Outboard motors have diesel engines so as to eliminate the need for gasoline storage; the fuel oil day tank serving the emergency generator engine in the aft deck locker is provided with a connection for filling the portable fuel tanks. A rack for storing outboard motors is located on the aft side of the deck locker; it is anticipated that at least one spare motor will be carried on board, as well as the usual spare parts.

Diving Support Space

The diving support space is located at the aft end of the main deck and has access to the aft deck through a 36" (0.91 m) wide weathertight door. The space is provided with direct overboard deck drains and can be hosed down to wash away sand and mud which come aboard with the diving equipment; watertight doors isolate the space from the adjacent interior of the vessel. Large lockers are provided for storage of dive equipment, including sea sleds. Lockers are fitted with expanded metal shelves for cleaning and drainage, and lockable expanded metal doors for easy visibility of locker contents.

Electrical charging stations are provided for sea sleds and an air charging octopus is located in the space for refilling scuba tanks.

Electronic Systems Development

Office space is arranged on the upper and main decks. Upper deck space is devoted to system integration with a single long counter for flexibility of component arrangement. An experimental control console is located at the forward end of the space with house front windows above, similar to a working control station. The console is fabricated from aluminum angles in an open framework designed to fit Electronics Institute Alliance (EIA) standard 310-C racks and enclosures underneath. The console top is marine plywood for ease of component installation and relocation to suit revisd configurations. It is intended that the console top will be replaced regularly as the number of cutouts becomes excessive.

The main deck office space is fitted with six desks, two tabletop work stations, and numerous cabinets for storage of components and materials. As in the concept arrangement, the office layout includes specific work stations for electronic equipment repair, video editing, ship's office, and medical services.

115 VAC clean power receptacles are located at each electronics work station. Outlets are also provided for 230 VAC and 24 VDC; small transformers are located adjacent to 12 VDC equipment as required.

The library, located immediately forward of the office for convenient access, contains ten storage bays with a total of 650 lineal feet (198 meters) of non-magnetic stainless steel shelving. Shelves are fitted with sea rails and netting as required to contain and secure the variety of video tapes, CD-ROMs, and books.

In addition to the usual ship to ship communications, this vessel carries a full suite of satellite equipment, including Inmarsat transceiver, satellite weather receiver, and satellite television receiver.

STABILITY

Because the vessel will not be classed and a flag state has not been selected, the choice of stability criteria was left to the naval architect's discretion. Accordingly, vessel stability was calculated for compliance with the following criteria:

- 46 CFR 170.173 Vessels of Unusual Proportion and Form, subpart (c)
- 46 CFR 170.170 Weather
- 46 CFR 171.050 Passenger Heel

Stability was calculated for full load (100% tankage and provisions), transit (50% tankage and provisions) and burned out (10% tankage and provisions) conditions. Icing was not considered because the owner intends to operate only in tropical and temperate climates. In all cases the actual free surface moment of slack tanks was calculated and applied.

Based on these conditions and the estimated lightship weight of 131 long tons (133 MT), the vessel complies with the selected criteria.

Given the owner's concern for vessel safety during shallow water operations, it was decided to calculate the floodable length curve based on a very conservative 7 foot (2.14 m) draft. As can be seen in Figure 8, the vessel compartmentation falls well below the floodable length curves.

SHIPYARD SELECTION

As the contract design approached completion, the owner began, with his usual diligence, to research capable shipyards. Extensive experience in aluminum construction was viewed as an essential prerequisite, as were receptivity to new ideas and willingness to work closely with the owner during the development of the design details. Compatibility was to prove one of the most important criteria because the owner intends to act as his own on-site agent during the detail design and construction phases, and he intends to relocate to the shipyard during the period required to complete the design and build his ship.

Conversely, shipyards which displayed lax management or projected an attitude of intolerance of new ideas and methods would undoubtedly receive no further consideration. The owner's observations during the interview process are particularly illuminating because he had never visited a shipyard prior to this project and had few preconceptions.

One further factor in the evaluation of shipyard capabilities was the design department within each company. It was the intention of the owner to select the shipyard after completion of the bidding process, but with a major difference from the usual practice.

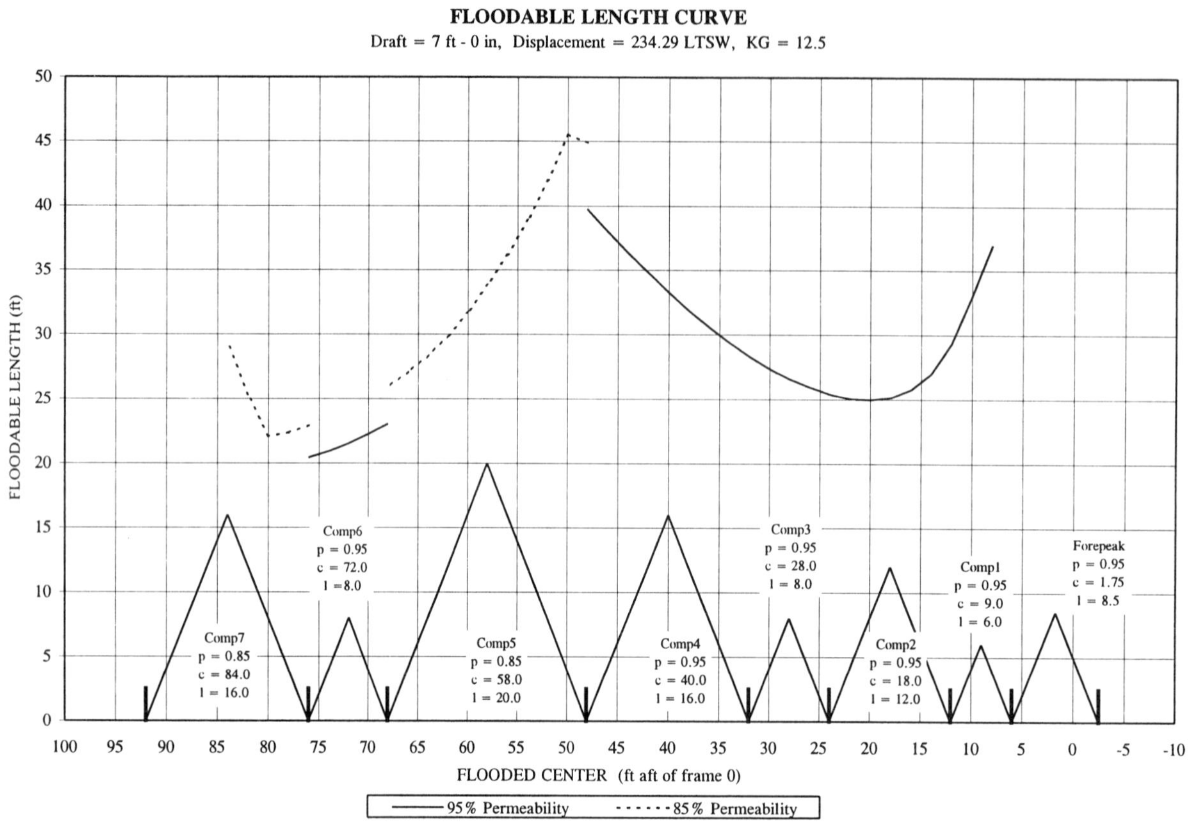

Figure 8 – Floodable Length Curve

The selected company would first be awarded a contract to produce the detail design of his vessel. This design would include all calculations and plans necessary to build the vessel (lofting excepted); these plans would then be released for best and final offers by two or three finalists from the previous bidding. In this way, the owner hoped to iron out all details before the start of construction so that final bids would reflect the delivered price; completion of the design would eliminate change orders due to unforeseen issues.

Location

Interestingly, the owner approached the selection process with the one unshakable conviction that Gulf Coast production facilities would not be considered. The reasons for this opinion were not pursued; however, elimination of the numerous otherwise-qualified shipbuilders certainly had the effect of simplifying the selection process.

In a similar manner to his process for interviewing naval architects, the owner elected to first interview shipyards close to home in the Pacific Northwest. Appointments were arranged at facilities in Washington and British Columbia; each interview consisted of a design review and discussion of the project followed by a tour of the shipyard.

The purpose of the design review and discussion was not only to establish the interest, relevant experience, and capabilities of the yard, but also to ascertain the organization's flexibility to incorporate and fabricate unconventional ideas. Needless to say, any shipyard which projected the attitude "this is the way we have always done it" faired poorly in the owner's scoring system. Favorable reviews resulted when open dialogue revealed an aggressive management style that encouraged innovation and efficiency.

Touring each facility provided the owner the opportunity to learn about construction methods. More significant, in the owner's mind, was the apparent efficiency of production as revealed by yard layout, general tidiness and organization, and – most importantly – worker ethic. Most impressive were the shipyards where workers were visibly active and productive. Several shipyards were eliminated from consideration because of observed worker idleness and low productivity in aluminum.

Completing his tour of facilities in the Northwest, the owner decided to visit the highly regarded Australian shipyards, which were seriously considered because of their reputation for innovative technology and efficiency. Previous time spent in Australia had provided fond memories for the owner, so relocation "down under" during design and construction would not be a hardship.

Construction in Australia appeared to present certain significant advantages in that government subsidies for ship construction and in support of high-technology projects would reduce vessel cost. In addition, the hourly labor rate was reported to be well below that paid in Canadian and American shipyards.

Unfortunately, the highly competitive relationship between the Australian shipyards precluded a willingness to allow engineering to be shared with competitors. Shipyards refused to participate in the scheme to put the detail design out to bid for construction, forcing a plan modification.

CONTRACT AWARD

During the detail design process, the owner will purchase virtually all equipment and outfit items for the vessel, thereby providing the shipyard at project start with all of the long lead items. Because the detail engineering is expected to generate precise specifications for all of these items, misunderstandings regarding owner-furnished equipment are expected to be minimized.

Following completion of the detail design engineering and successful construction contract negotiations, construction is projected to commence in the last quarter of 2000, with delivery of *Freedom* in the summer or fall of 2001.

SUMMARY AND LESSONS LEARNED

The owner's intense enthusiasm and energy overcame some initial skepticism regarding his ability to see this project through. The design firm was forced to reexamine the rationale underlying many of the routine design decisions and in attempting to justify the design process to a questioning client, we were able to revise the process to accommodate unique requirements.

Probably the most difficult adaptation demanded was the owner's insistence on involvement in every design decision, including his personal selection of each and every major piece (and many minor pieces) of equipment on the vessel. The interactions and research required to verify mission suitability consumed much time and energy, but rewarded the designers with exposure to non-traditional ideas.

Small and medium-sized consultant firms regularly receive "walk-in" potential clients, a large percentage of whom have highly unconventional ideas. In spite of the temptation to dismiss these people as dreamers, the prudent course involves investing the time to listen carefully and fully explore the ideas before passing judgement. Our industry is not noted for its innovation; it is important to remain open and objective to the unusual and creative.

ACKNOWLEDGEMENTS

The author wished to express sincere gratitude, first and foremost, to the vessel owner Joe Quagliani, for entrusting his dream and vision of *Freedom* to Elliott Bay Design Group, thus providing (forcing) the author with the opportunity to view the design process with zero-sum balance perspective.

Appreciation is also due to Mr. Pete Randall, Senior Electrical Engineer, for working closely with the owner to design and resolve the myriad details of the electrical systems.

REFERENCES

1. American Bureau of Shipping "Rules for Building and Classing Aluminum Vessels," 1975
2. American Bureau of Shipping "Guide for Building and Classing Motor Pleasure Yachts," 1990
3. Det Norske Veritas "Rules for Classification of High Speed and Light Craft," 1996

Discussion

Dirk H. Kristensen, Member

The author is to be congratulated for an interesting paper describing a challenging vessel design process. The paper describes a vessel design that is certainly unique in how it fulfills a demanding set of functional owner's requirements.

The author describes a number of design features that address vessel safety concerns. For example, the possibility of operating in shallow waters, e.g., "reefs" and "sensitive littoral areas" has prompted a high degree of subdivision consisting of a number of transverse watertight bulkheads and a full double bottom. An issue related to ship safety in shallow areas is environmental safety. Can the author describe what consideration was given in the design for preventing pollution from the vessel in these sensitive areas? For example: Are fuel, oil or other polluting liquids carried in the double bottom? Are containments provided at fuel fill stations? How are sewage and gray water handled? How is garbage handled? Were exhaust emissions (NO_x) an issue in engine selection?

The author states that substantial structural weight savings were realized by using DNV High Speed and Light Craft rules as opposed to ABS aluminum rules but that some scantlings were selectively increased to "provide a reasonable level of confidence." Can the author elaborate on the scantling system chosen, i.e., transverse or longitudinal framing, and which scantlings were increased above rule requirements? Although the vessel is not to be classed, it does not appear to meet the DNV criteria of a "Light Craft" or a "High Speed Craft." Did the investigate the suitability of the rule scantlings by any first principles analyses?

Lastly, the shipyard selection scheme describing the selection of one yard to develop the detail design package then distributing this package to other yards for final bids, is an intriguing idea. However, I was not surprised to read that Australian yards were unwilling to participate in this scheme due to a highly competitive environment. I would be interested to know the reaction of the U.S. yards that bid on the construction.

Peter L. Randall, Visitor, Elliot Bay Design Group

Many of the features on *Freedom* are unique—for now. The AC electric "power plant" arrangement, diesel electric propulsion with pulse width modulated (PWM) AC drives and induction motors, clean power busses, networking of programmable logic controllers (PLCs), high emphasis on fuel efficiency and survivability; they are not common on vessels this size, let alone a yacht. Since the time of the completion of the contract design for *Freedom* less than a year ago, however, mega yachts with similar features have been built in Europe. Features that are frequently associated with military and cruise ships are now entering the small vessel marketplace.

Perhaps without even knowing it, the client's foresight and personal involvement may have put *Freedom* at the forefront of what will be commonplace a few years from now.

One feature that was considered but could not be incorporated were azimuthing pod electric propulsors instead of azimuthing propellers with fixed AC induction motors. Units small enough for *Freedom* simply are not available yet. Commercially available units are rapidly becoming smaller, and may soon be an option for small vessels having the prerequisite generating capacity.

A few comments on the AC One-Line Diagram (Figure 6) are appropriate. The distribution concept is not far removed from those found in IEEE 45-1998 for large vessels. The robust nature and ability to satisfy single failure criterion made it an obvious choice for an owner insistent on reliability.

The client expressed a strong dislike for the central hydraulic systems that are common on vessels of this size, yachts and workboats alike. The combination of the "power plant" electrical system and PWM AC drives allows the forward anchor windlass and bow thrusters to be sufficiently (or efficiently powered) without long runs of pipe, and without potential leakage and maintenance.

The use of PWM AC drives offers tremendous flexibility and high efficiency but not without certain tradeoffs. Induced motor bearing circulating currents and associated bearing failure, insulation stress, EMI, harmonic voltages generated in the AC distribution and motor regenerative energy should all be carefully considered. The reader is referred to numerous recent articles on these topics in *IEEE Industry Applications Magazine,* as well as other sources. It is strongly recommended that the motors, drives, and either line reactors or isolation transformers be matched equipment engineered, manufactured, and applied by a single manufacturer.

Author's Closure

Gratitude is extended to Mr. Peter Randall for his elaboration on the philosophy and choices made during the development of the propulsion and control systems. It should be noted that, in the time since this design was completed, smaller azimuthing pod units have been developed so that this type of propulsion is now available for vessels of this size and power. The author expects that the owner will seriously consider azimuthing pod propulsion for his next vessel.

Mr. Kristensen's insightful comments and questions are appreciated. Regarding pollution prevention, the owner, as might be expected, has both practical and unconventional solutions. The small size and large range of the vessel mandated the use of double bottom tanks for fuel oil; however, the owner believes that the combination of sophisticated electronic charting, extraordinary maneuverability, and prudent seamanship will provide reasonable protection against grounding. Should a damaging grounding occur, double bottom compartmentalization will minimize loss of fuel oil.

The issue of NO_x pollution was not raised during the design process; however, the owner's desires to minimize fuel consumption and control the generator sets with the most current electronic controls would likely result in low emissions.

Standard containment coamings are provided at fueling stations. Sewage is treated with a USCG-approved type III marine sanitation device. Gray water discharges directly overboard; however, only biodegradable soaps will be used on board. Garbage disposal will be minimized by elimination of all plastics and unnecessary packaging at the dock during reprovisioning. At sea, the owner intends to dispose of biodegradable garbage overboard after packaging it in large weave burlap bags, tied with cotton string, and weighted with rocks.

With regard to scantlings, the vessel is framed longitudinally on 10 inch centers, with transverse webs spaced 48 inches. The use of DNV High Speed and Light Craft Rules was demanded by the owner after becoming convinced by Australian shipyards of the excessive material requirements of ABS aluminum rules. Recognizing that these two sets of rules are intended for different services, scantlings were increased at the shell forward, house front, and in way of the engine room and propulsion units. Unfortunately, the use of first principles analyses were not affordable within the scope of the contract.

Regarding the issue of distribution of the shipyard-produced detail design package, the author is unaware of any comments from the U.S. yards. It should be noted, however, that the prices from U.S. yards were substantially higher than those provided by Australian yards, so that the U.S. yards were not finalists in the selection process and discussions were undoubtedly brief.

SNAME *Transactions*, Vol. 108, 2000, pp. 369–395

Mast/Snorkel Performance of the *DOLPHIN* AUV

Christopher D. Williams, Visitor, Institute for Marine Dynamics, **Neil Bose,** Member, Memorial University of Newfoundland, **Mark-Andrew Butt,** Visitor, Instrumar Limited, and **Mae L. Seto,** Visitor, International Submarine Engineering Research

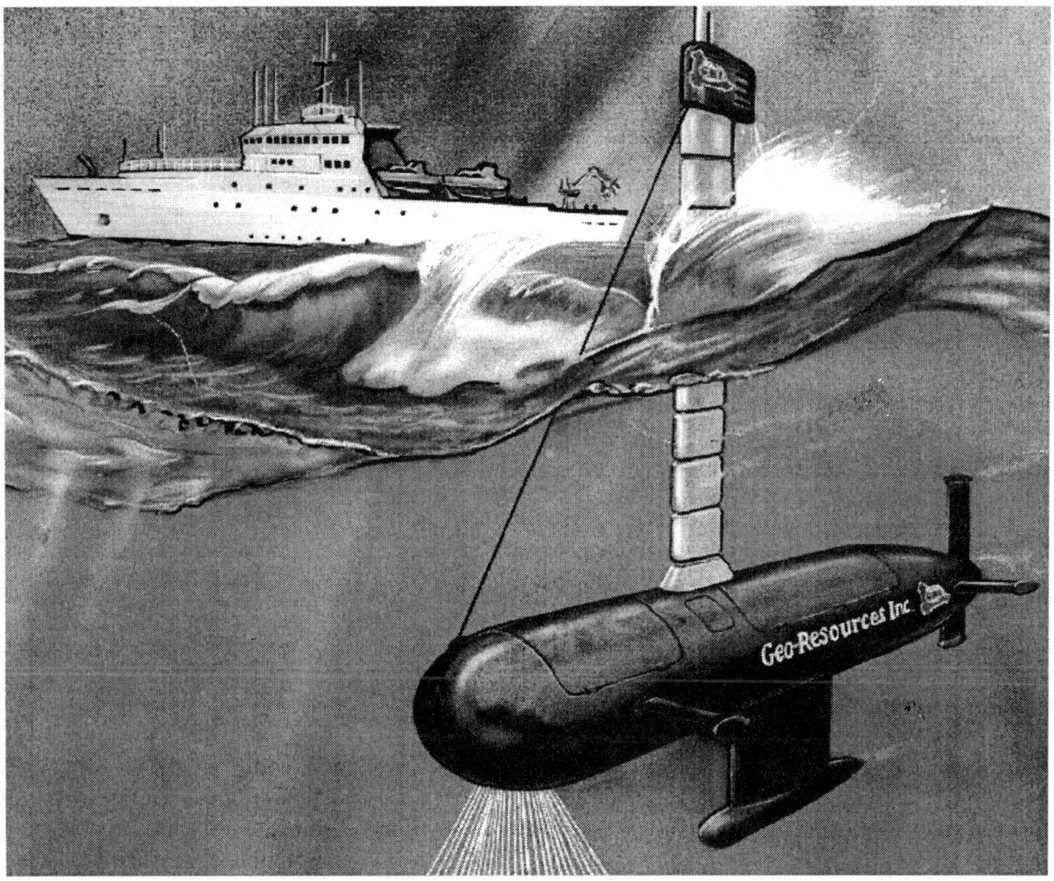

ABSTRACT

The DOLPHIN is a self-propelled semi-submersible vehicle that operates under but near to the sea surface; typically one-third of the mast projects above the sea surface. As such, the hydrodynamic performance of the complete vehicle is affected by the hydrodynamic performance of the mast. The purpose of the mast is two-fold: to act as a streamlined fairing for the snorkel for the air-breathing Diesel engine, and to support the antenna that is used for communications with the vehicle.

This paper examines the effect of the mast on the hydrodynamic performance of the vehicle and shows what improvements might potentially be possible through the use of an alternative design of mast. The results presented are from captive-model towing tank experiments with a half-

scale model of the DOLPHIN. Two mast configurations were examined: the original mast; and an alternative mast that includes near-leading-edge air ejection from the upper half of the mast, and a 25-percent-chord trailing-edge flap on the lower half of the mast.

The scale model included the main hull, keel and bulb, and two different mast assemblies; the forward horizontal fins, the aft vertical and horizontal fins and the propeller were not included. The free-flooding model was supported on a sting-mounted internal six-component balance. This balance measured the overall hydrodynamic loads that were exerted on the whole vehicle; no measurements of loads on the mast alone were attempted. All experiments were performed in calm water so the effect of waves (other than the bow wave that the mast produced) was not examined.

The measured vehicle forces and moments (axial force, lateral force, normal force, pitch moment, yaw moment and roll moment) are presented as functions of the vehicle yaw angle, vehicle draft and vehicle speed. In addition, for measurements with the alternative mast, vehicle forces and moments are presented in terms of the air-ejection flow rate and flap angle.

INTRODUCTION

Through continuing research and development by ISE (International Submarine Engineering), the *DOLPHIN* vehicle has been constantly evolving since its first deployment. In December 1996, ISER (International Submarine Engineering Research) requested that an alternative surface-piercing mast design be investigated. A collaborative research, design and testing plan was set up between ISER, IMD (The NRC Institute for Marine Dynamics) and MUN (Memorial University of Newfoundland) to design and test an alternative surface-piercing mast for the *DOLPHIN*.

The *DOLPHIN* (Deep Ocean Logging Platform with Hydrographic Instrumentation for Navigation) is a self-propelled semi-submersible vehicle that was designed and built by International Submarine Engineering Research Limited (ISER) of Port Coquitlam, BC, Canada for the Canadian Hydrographic Service (CHS). Since its initial deployment in the early 1980s, the *DOLPHIN* has seen both civilian and military applications. It has been adapted to perform varied tasks such as hydrographic surveys, oceanographic surveys, remote mine-hunting and electronic warfare. The *DOLPHIN* is a unique vehicle. Unfortunately, development funding from the CHS did not provide for extensive model testing or examination of design alternatives. In the 1990s an ongoing effort has been made to understand its performance. Figure 1 shows a more recent version of the *DOLPHIN* vehicle when it is used with a towed body. In this configuration the towing vehicle is named *DORADO* and the towed body has the name *AURORA*. Further details (and colour photographs) of this remote-mine-hunting configuration can be found on the ISER web site at 'www.ise.bc.ca'.

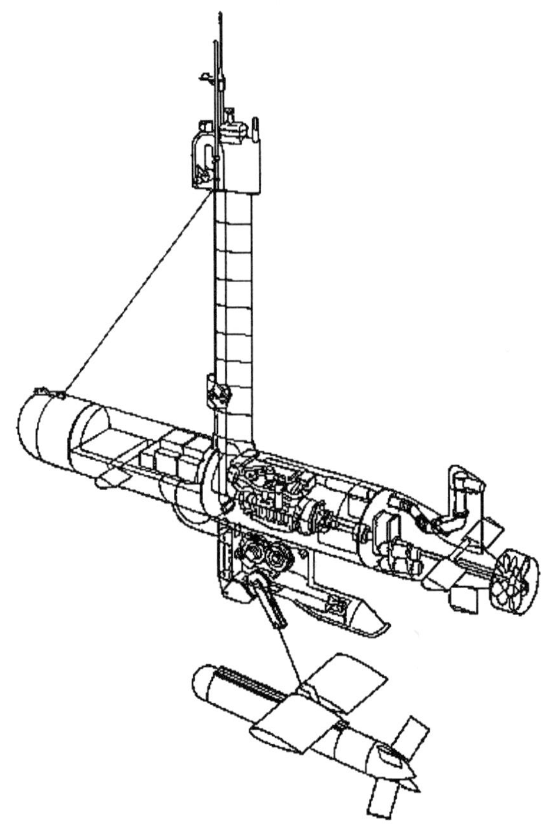

Figure 1. The ISER Remote Mine-Hunting System: *DORADO* tow vehicle with the *AURORA* towed body

The *Mark 1 DOLPHIN* vehicle (as shown in the frontispiece and Figure 2) is 7.3 m long, has a main hull diameter of 1.0 m and a mass of 3350 kg; this vehicle uses a snorkel to supply air to its Diesel engine. The vehicle was designed so that at most operating speeds the hull centreline would be submerged to a depth of from 3 to 4.5 m, that is, from 3 to 4.5 hull diameters.

The original mast used nine fairing sections that swivel freely around a central pipe; the fairing sections were based on the IfS-61-TR-25 rudder section, see (Watt et al, 1997). A cable forestay was used to help support the drag load on the mast.

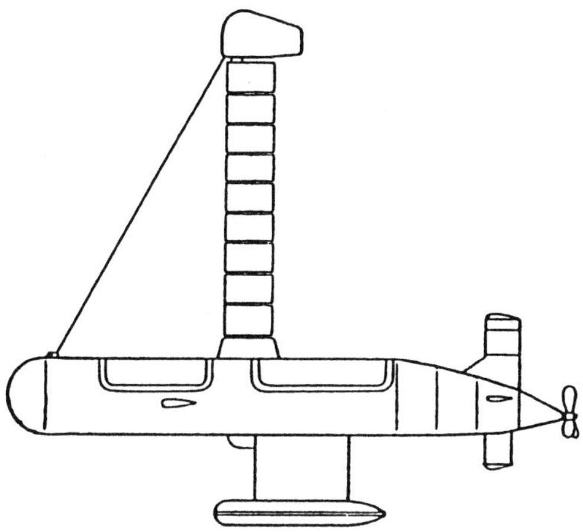

Figure 2. *Mark 1 DOLPHIN*

While turning, the keel on the *DOLPHIN* produces a large side-force that acts well below the longitudinal axis of the vehicle; this side-force induces a large roll moment about that axis and undesirable vehicle roll angles result. The primary focus of this investigation was to design a mast that would mitigate the keel-induced roll moment.

When the *DOLPHIN* operates in a choppy sea, the effective draft of the vehicle varies. As a result, the submerged length of the mast also changes, and this, in turn, causes the hydrodynamic side force and roll moment produced by the mast to vary. Moreover, the net side force will vary at the uppermost portion of the mast (i.e. where the mast pierces the water surface) where the rolling arm is the greatest. The net effect on the stability of the vehicle will depend on: the height of the oncoming waves; the encounter frequency of the waves; the natural roll period of the *DOLPHIN*; the amount of roll damping. Any waves with periods near the natural roll period of the *DOLPHIN* will therefore require that the vehicle have an additional means to maintain stability.

SUMMARY OF THE DESIGN FEATURES OF THE ALTERNATIVE MAST

The primary purpose of the alternative mast was to reduce the overall vehicle drag force on the *DOLPHIN* by reducing the drag of its surface-piercing mast. The secondary focus of this investigation was to design a mast that could counteract the keel-induced roll moment, and to alleviate (or eliminate) the mast-induced side-force and pitch moment due to variable submerged mast length in waves.

The final design of the alternative mast, as tested, included the following components:

a. The cross-sectional shape of the upper section of this mast was optimized for reduced spray drag and reduced section drag, in that order.
b. The cross-sectional shape of the lower section of this mast was optimized for structural strength (no guy wire was used) and reduced section drag, in that order, with an appropriate sectional shape for lift-generating purposes.
c. The vertical span of the lower mast section was selected such that it alone would provide sufficient roll moment to counteract the roll moment that is generated by the keel.
d. An air ejection system was incorporated into the port and starboard surfaces of the upper mast section, near the leading edge; the purpose was to provide a device that generated near-zero side-force over the range of turn-induced yaw angles. On the full-scale *DOLPHIN*, the source of air for the air ejection would be the exhaust gases from the Diesel engine aboard the *DOLPHIN*.
e. A 25-percent-of-chord, partial-span, trailing-edge plain flap was included in the lower mast section; the role of such a flap was to compensate for the varying mast-induced lift force and roll moment during variations in the submerged length of the mast when the vehicle operates in waves.

DETAILS OF THE HALF-SCALE MODEL

A 0.516-scale model of the *DOLPHIN* body, complete with keel and bulb, was designed by the third author, and was fabricated at the Memorial University of Newfoundland (MUN), in St. John's, Newfoundland. The fore-planes, aft-planes, rudder, vertical stabilizer and propeller were not included as part of this model. The diameter of the main hull was 511 mm and the overall length was truncated to 4040 mm in order to accommodate the 'sting' mounting.

Experiments were performed with this model in the Towing Tank at the Institute for Marine Dynamics (IMD) of the National Research Council Canada (NRCC) in St. John's, Newfoundland. The purpose of these experiments was to test and compare the relative performance of the existing mast with an alternative mast. Since it was the relative performances of the two masts that was the focus of these experiments, it was decided to omit all minor appendages from the model.

The existing mast (M1), as fabricated for this model, is shown in Figure 2. The nine fairing sections were based on the same IfS-61-TR-25 sections that are used on the *Mark 1 DOLPHIN* (Figure 3). Each fairing had a chord length of 324 mm, a height (span) of 203 mm and a maximum thickness that is 25 percent of the chord length; the fairings were spaced 3 mm apart vertically, and were free to swivel around a 38 mm diameter shaft that was located 56 mm aft of the leading edge. A small support stub (Figure 3) was affixed to the hull at the base of the mast; this support had a chord length of 324 mm, a height of 102 mm, a leading-edge radius of 44 mm and a trailing-edge radius of 13 mm. The cable forestay was made from 4.7 mm steel braided wire. The cable fore-stay was affixed to the fairing shaft above the fairings and was attached at the bottom end to just aft of the nose of the hull.

The conceptual design for an alternative mast on the scale-model *DOLPHIN* is shown in Figure 4. The alternative mast (M2) is depicted in Figure 5 alongside the existing mast. The alternative mast was fixed (it was not free to swivel) and had two distinct sections: a surface-piercing upper section, and a lower fully-immersed section, separated by a solid horizontal sheet-metal 'fence'. The choice of shape for each of the mast sections was a compromise between the drag performance of the fully-submerged portion and the drag performance of the surface-piercing portion.

The shape of the upper mast section (Figures 5 and 6) was based upon a symmetric double-arc profile with a thickness-to-chord ratio, t/c, of 16 percent and a chord length of 279 mm; the height of the upper mast section was 1067 mm.

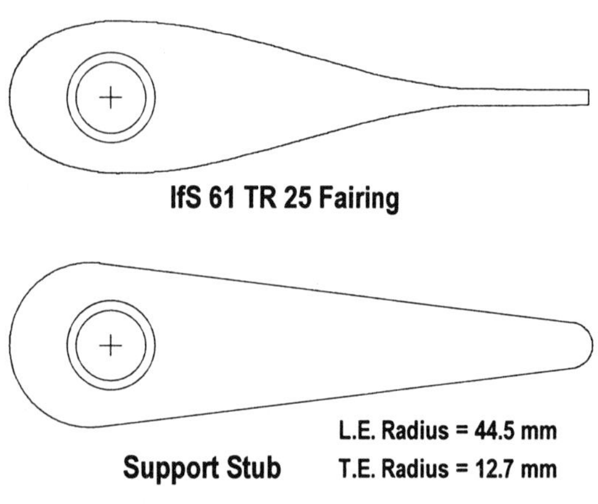

Figure 3. Profiles of the sections for the existing mast for the half-scale model.

The shape of the lower mast section (Figures 5 and 6) was based upon a modified symmetric double-arc profile with a thickness-to-chord ratio of 17 percent and a chord length of 330 mm. A leading-edge radius of 8 mm and a trailing edge angle of 22° were incorporated into the lower mast profile. The lower mast section was 826 mm in height. A solid horizontal brass plate (ventilation fence) with a thickness of 4.7 mm was located at the juncture of the lower and upper masts (Figures 5 and 6); the purpose of this fence was to ensure that the depth of the 'ventilation pocket' (that formed behind the mast) did not extend beyond the base of the upper mast section.

Figure 4. Conceptual *DOLPHIN* half-scale model.

As shown in Figure 5, a plain flap was integrated into the lower mast section. The flap length was 25 percent of chord and was 673 mm in height. As can be seen in Figure 6, the shape of the lower mast section was a straight-sided wedge for the last 25 percent of chord; such a shape also simplifies the fabrication of the mast-flap combination.

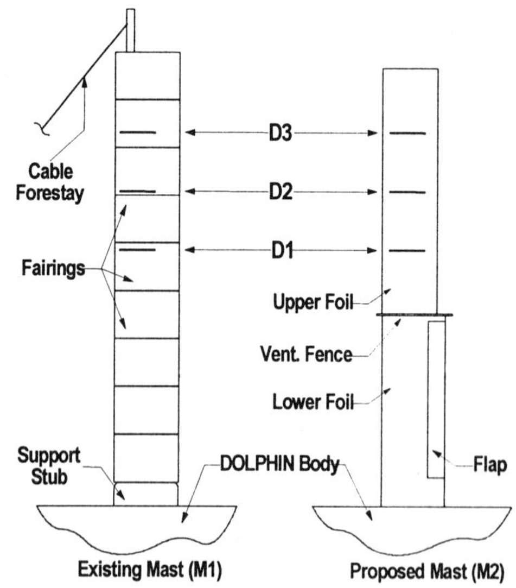

Figure 5. Existing and alternative masts for the half-scale *DOLPHIN* model.

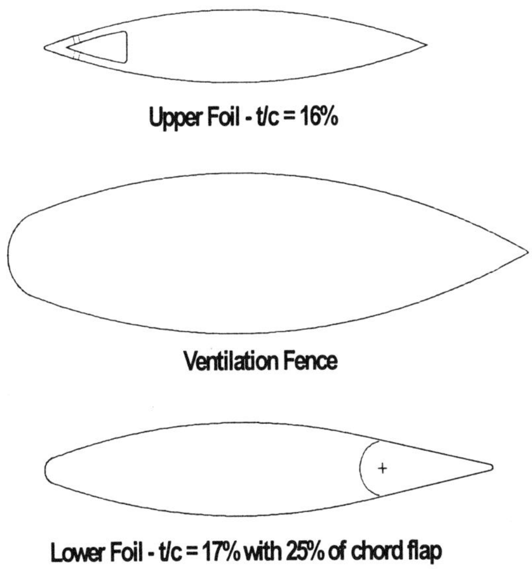

Figure 6. Profiles for the sections for the alternative mast for the half-scale *DOLPHIN* model.

Air was ejected through a row of equi-spaced holes that were located 13 mm aft of the leading edge (4.5 percent of chord) of the upper mast section; there were 300 holes on each of the port and starboard sides. The holes were 1.0 mm in diameter and were spaced 3.2 mm apart. Air flow rates of 9.4 and 14.2 litre/second were fed to a chamber within the upper mast during testing by a 3.75 kW compressor with a 300 litre accumulator. The accumulator was fully recharged with air after each test run.

EXPERIMENTAL METHOD

A series of captive-model experiments was carried out in the IMD Towing Tank (200 m length, 12 m width, 7 m depth). The purpose of these experiments was to measure the loads and moments experienced by the model. The model was suspended from the towing carriage via two streamlined vertical struts and a 'sting'; the model was attached to the 'sting' via an internal six-component balance, see Figure 7. The 'live' portion of the balance was attached to the inner framework of the *DOLPHIN* model. The 'ground' portion of the balance was in turn flange-mounted to the 152 mm diameter 'sting' that extended out through the aft end of the *DOLPHIN* model. Note that the diameter of the 'sting' was less than 30 percent of the diameter of the model hull so that there was minimum interference from the 'sting' on the model. The overall length of the 'sting' was 4.5 m so that the aft end of the model was a distance 825 mm forward of the forward strut so that there was negligible interference from the support struts (Mackay, 1993). The depth to the horizontal centreline of the model was varied to provide three different drafts for each mast; the hull centreline was immersed to depths of 2.6, 3.1 and 3.6 times the hull diameter. Yaw runs were performed at each of these depths.

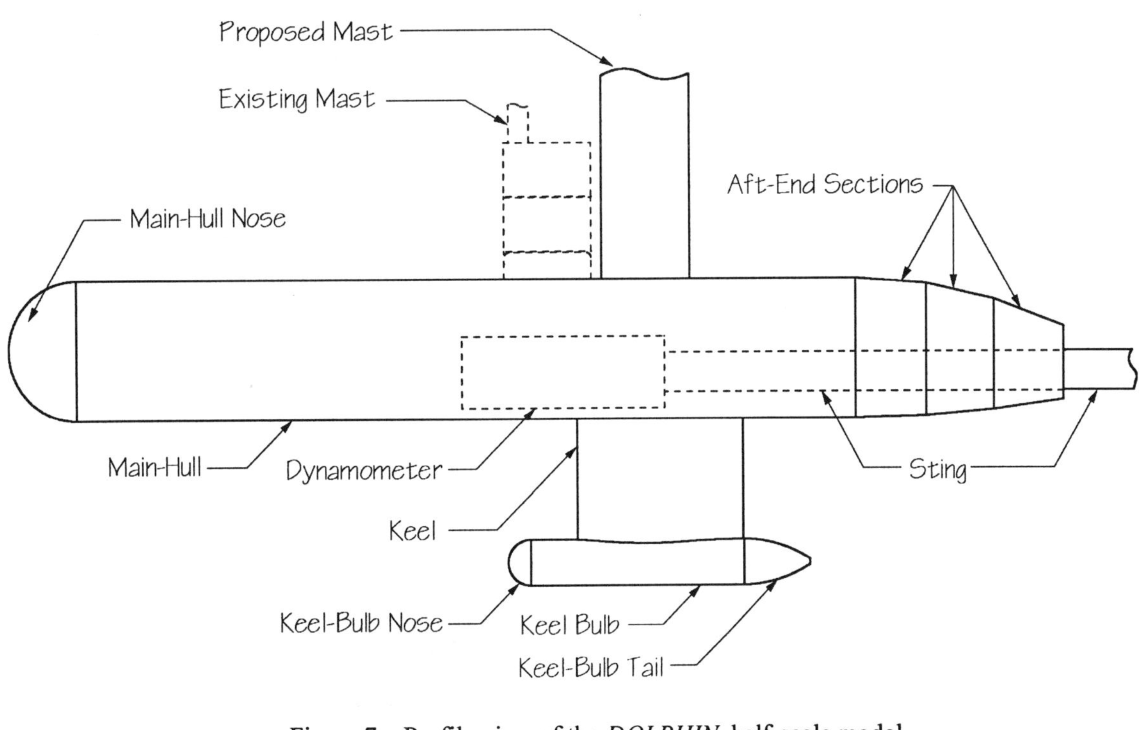

Figure 7. Profile view of the *DOLPHIN* half-scale model.

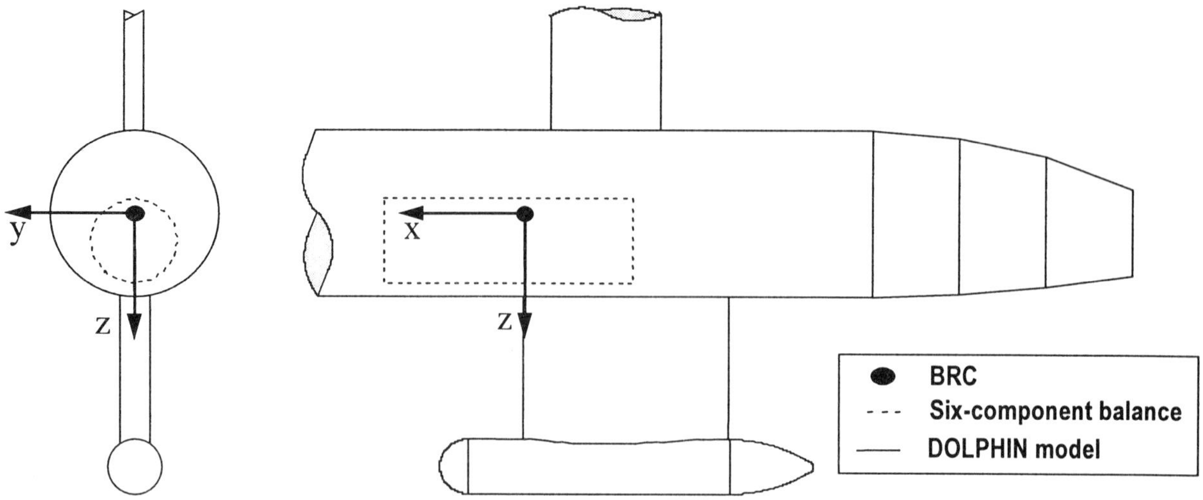

Figure 8. Front and profile view of the *DOLPHIN* model with location of the Balance Resolving Centre.

An array of three orthogonal accelerometers was attached to the forward end of the 'live' portion of the balance; an identical array was attached to the aft end of the 'live' portion of the balance. The purpose of the accelerometers was to detect and measure any vibrations due to flow separation from the hull, keel or mast when at large yaw angles, or from the flap at large flap angles. The signals from the six waterproof load cells in the balance and from the six waterproof accelerometers were fed to the data acquisition system onboard the towing carriage.

Digital photographs were taken of the water surface disturbances for the different mast configurations. In addition, S-VHS video recordings were used to observe the disturbance of the free surface as well as to monitor the sub-surface effects due to the air ejection.

TEST PROGRAM

For the test program the vehicle yaw angle was varied in increments of 2° from -2° (port) to +10° (starboard). The model was level (zero pitch angle) and upright (zero roll angle) during all the experiments. The vehicle was towed at six speeds of 2.00, 2.83, 3.46, 4.00, 4.47 and 4.90 m/s. Based on a model length of 4.040 m, these speeds corresponded to Froude Numbers of 0.32, 0.45, 0.55, 0.64, 0.72 and 0.78 respectively; based on the mast chord length the Froude Numbers were 1.1, 1.6, 1.9, 2.2, 2.5 and 2.7.

Three vehicle drafts were used during the test program. Figure 5 depicts these drafts as 'D1', 'D2' and 'D3' (the dark horizontal lines on each mast correspond to 6 mm wide pin-striping tape applied to each mast, for visualization purposes; the stripes were spaced 127 mm apart. These drafts were respectively 1105 mm (D1), 1359 mm (D2) and 1613 mm (D3) as measured from the upper surface of the main hull of the *DOLPHIN* model.

The model was tested with no mast at draft 'D1'; this test configuration was referred to as 'M0'. The existing mast (configuration 'M1') and the alternative mast (configuration 'M2') were tested at all three drafts. The alternative mast was tested at all drafts and with two rates of air ejection of 9.4 and 14.2 litre/s; these were referred to as flow rates 'Q1' and 'Q2' respectively. Note that at draft 'D1' about one quarter of the air holes were immersed, about one half at 'D2' and about three quarters at 'D3'.

Tests with the trailing-edge flap deflected used eight flap angles ranging from -2° to +12°, with three vehicle yaw angles of -2°, 0 and +2°, but only at draft 'D2'. In plan view, clockwise rotation of the flap resulted in a positive flap deflection (trailing edge deflected to port) and clockwise rotation of the vehicle corresponded to a positive yaw angle. When tests were performed with the flap deflected, the test configuration was referred to as 'M2F'.

The remaining details of all the experiments can be found in the thesis by the third author (Butt, 1999).

RESULTS AND OBSERVATIONS

The internal six-component balance was designed to measure, in a body-fixed coordinate system, the axial force (FX), lateral force (FY), normal force (FZ), roll moment (MX), pitch moment (MY) and yaw moment (MZ); see Figure 8. From the measured body-axes loads we computed the hydrodynamic lift and drag forces on the vehicle; in this paper the drag force is parallel to the direction of travel of the towing carriage (parallel to the longitudinal centreline of the towing tank) and the lift force is perpendicular to that direction. In this paper these forces and moments are given with respect to a 'load centre' which is on the centreline of the hull at the leading edge of the keel; this 'load centre' is 2145 mm aft of the nose of the model, and is 1895 mm forward of the aft end of the model. We adhere to the SNAME sign convention for underwater vehicles when reporting these loads: the axial force FX is positive when directed forward, the lateral force FY is positive when directed to starboard, the normal force FZ is positive when directed downward through the keel, the roll moment MX is positive when the vehicle lists to starboard, the pitch moment MY is positive when the nose of the vehicle rises, and the yaw moment is positive when the nose of the vehicle yaws to starboard; see Figure 8. Note that all the dimensions and loads are reported at model-scale, not full-scale, unless otherwise indicated.

In the following sections, the results from the experiments with the existing and alternative masts are discussed. The focus of this paper is the presentation of the measured hydrodynamic forces and moments that are exerted on the vehicle. These results are presented as functions of the vehicle yaw angle, vehicle draft and vehicle speed. In addition, for measurements with the alternative mast, vehicle forces and moments are presented in terms of the air-ejection flow rate and flap deflection angle.

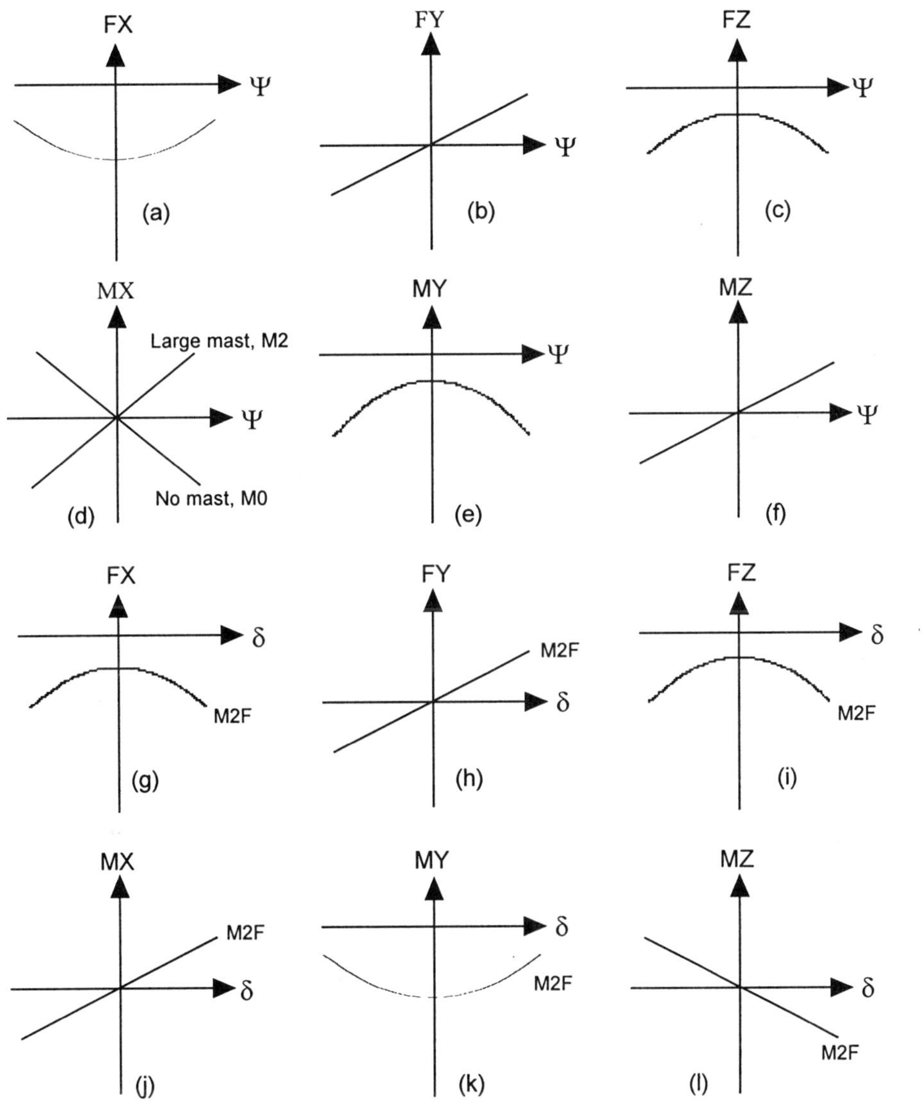

Figure 9. Generic load graphs for the *DOLPHIN* half-scale model.

Figure 9 provides a set of 'generic' force and moment plots that show how each load varies with vehicle yaw angle 'Ψ' and flap deflection angle 'δ'. It is assumed that the *DOLPHIN* vehicle is port-starboard symmetric. As a function of vehicle yaw angle, the lateral force, roll moment and yaw moment graphs are straight lines through the origin; FY and MZ have positive slope; MX has negative slope with no mast (M0) and positive slope with a large 'lifting' mast. Each of the other graphs (versus vehicle yaw angle) is a symmetric parabola; only the graph of axial force is concave upward; the graphs of normal force and pitch moment are concave downward with negative intercepts of the ordinate axis with no mast and positive intercepts with the existing mast (M1) and alternative mast (M2). As a function of the flap deflection angle 'δ', again the lateral force, roll moment and yaw moment graphs are straight lines through the origin; FY and MX have positive slope; MZ has negative slope. Each of the other graphs (FX, FZ, MY versus flap deflection angle) is a symmetric parabola; some are concave upward and some are concave downward.

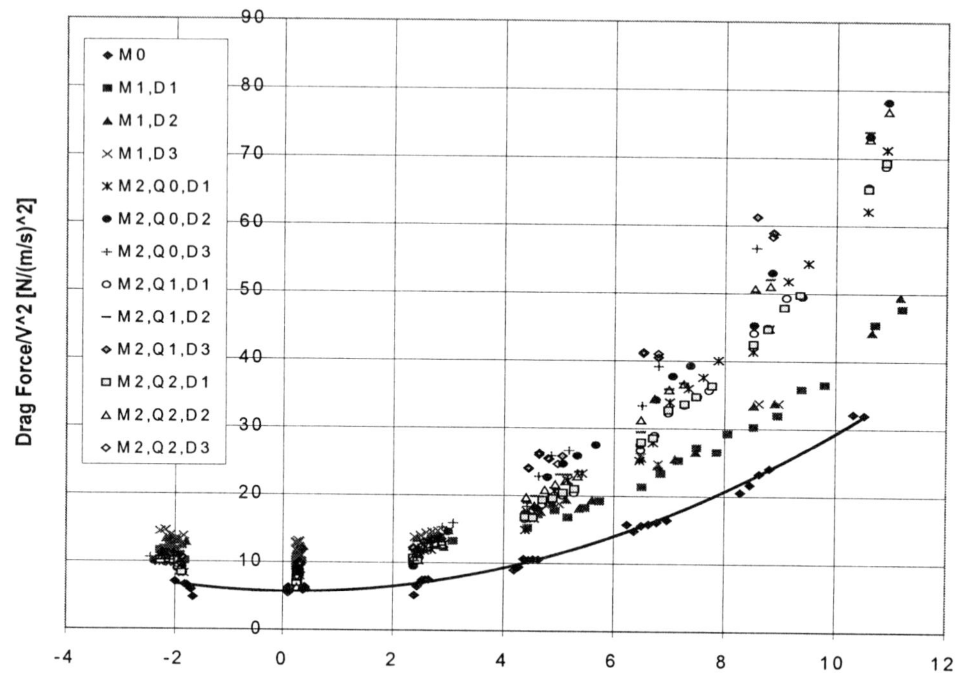

Figure 10. Normalized drag force vs vehicle yaw angle: masts M0,M1,M2; drafts D1,D2,D3; airflows Q0,Q1, Q2.

Mast Configuration	Draft	Drag Force Intercept 'A' Equation 1	Drag Force Slope 'B' Equation 1	Percent of Overall Drag at Zero Vehicle Yaw Angle	Percent Change Relative to Mast M1 at Zero Yaw	Equi-Drag Angle Relative to Mast M1
No Mast (M0)	D1	0.224	5.950	n/a	n/a	n/a
Old Mast (M1)	D1	0.274	10.330	42	n/a	n/a
	D2	0.268	11.792	50	n/a	n/a
	D3	0.252	12.828	54	n/a	n/a
Alt. Mast (M2,Q0)	D1	0.502	7.944	25	-23	3.2
	D2	0.577	8.066	26	-32	3.5
	D3	0.657	8.087	26	-37	3.4
Alt. Mast (M2,Q1)	D1	0.466	8.110	27	-21	3.4
	D2	0.554	8.109	27	-31	3.6
	D3	0.653	8.537	30	-33	3.3
Alt. Mast (M2,Q2)	D1	0.480	8.268	28	-20	3.2
	D2	0.540	8.290	28	-30	3.6
	D3	0.650	8.682	31	-32	3.2

Table 1. Summary of relative drag results for various masts and vehicle drafts.

Summary of Drag Performance of the Basic Mast Configurations

The details of the drag performance of the various mast configurations were presented at the 25th ATTC (Butt et al, 1998); only a summary of the drag results will be presented here, see Figure 10 and Table 1. Figure 10 shows that the 'normalized' drag force on the vehicle (in all configurations) increases quadratically with increasing vehicle yaw angle. The solid line through the data points for the vehicle with no mast (M0) indicates the lowest drag of all configurations tested and this drag is therefore due to the effect of the flow around the hull and the keel. Since the keel is designed to be an efficient "lifting surface", it is expected that most of the lateral force on the vehicle will be produced by the keel and, for yaw angles below "stall" the lateral force that is generated by the keel will increase linearly with the yaw angle. In anticipation that the drag force on the keel will increase with the square of the lift force generated by the keel (this is true for most three-dimensional "lifting surfaces"), it is reasonable to expect that the drag force due to the keel will increase with the square of the yaw angle. We also expect that the major portion of the total drag force on the vehicle will be produced by the keel.

The fact that the vehicle overall drag force is expected to increase with the square of the vehicle yaw angle 'Ψ', can be expressed as follows:

$$\frac{Drag.Force}{V^2} = A + B \cdot \Psi^2 \left[\frac{\mathrm{N}}{\mathrm{m}^2/\mathrm{s}^2}\right] \tag{1}$$

This relation was fitted to the 'normalized' drag forces for all 13 mast configurations and the results are summarized in Table 1 which contains the fitted values of the coefficients 'A' and 'B'. Figure 10 and Table 1 show that the existing mast, M1, contributes from 42 to 54 percent of the overall drag (at zero vehicle yaw angle) depending on the draft. Since the segments of the existing mast are self-aligning with the direction of the local flow, they exhibit little variation of 'normalized' drag force with vehicle yaw angle. Thus the data points for the vehicle with the existing mast lie above those for the vehicle with no mast; the data points are essentially "shifted upward" by the additional drag force that is generated at zero vehicle yaw angle. For example, at a vehicle yaw angle of 10°, the existing mast contributes from 25 to 27 percent of the overall drag, depending on the draft.

For the alternative mast (M2), the 'normalized' drag force increases more quickly with increasing yaw angle than does the drag for the vehicle with the existing mast. However, at zero yaw angle Table 1 indicates that the basic alternative mast, M2, contributes from 25 to 26 percent of the overall drag, depending on the draft; these values indicate that the overall vehicle drag is less with the alternative mast. However at a vehicle yaw angle of 10°, the alternative mast contributes from 51 to 62 percent of the overall drag, depending on the draft. At some angle between zero and 10° the drag curves (for each draft) for the existing and alternative masts, cross each other. Here the drag forces for the two mast configurations are equal so we refer to the vehicle yaw angle at which this occurs as the 'equi-drag' angle; the values of this angle are given in Table 1 for the various mast configurations. In fact, above yaw angles of from 3.2° to 3.6° (depending on the draft) the drag of the alternative mast was higher than that of the existing mast; see Table 1. It should be noted that with the existing mast (M1), cross-flow angles greater than 2° are not maintainable during steady turns, therefore the increased drag of the alternative mast over the existing mast at higher yaw angles is not considered to be a performance handicap.

Lift and Roll Performance of the Basic Mast Configurations

This section concentrates on the measured lift force and roll moment that are produced by: the basic mast configurations (no mast, M0; existing mast, M1; alternative mast, M2), the air ejection configuration (alternative mast M2 with airflow rates Q1 and Q2) and the flap configuration (M2F). The following relation expresses how the 'normalized' lift force varies with vehicle yaw angle: here 'V' is the forward speed of the vehicle and 'Ψ' is the vehicle yaw angle, measured in degrees:

$$\frac{Lift.Force}{V^2} = C_0 \cdot \Psi \left[\frac{\mathrm{N}}{\mathrm{m}^2/\mathrm{s}^2}\right] \tag{2}$$

In Figure 11 the measured lift force (as 'normalized' by the square of the vehicle forward speed 'V') on the model is plotted versus the model yaw angle (at drafts D1,D2,D3) for the case with no mast (M0), the existing mast (M1), and the alternative mast (M2) respectively. A straight line (with zero intercept) was fitted through the data, and the slope was computed. The slopes, for each mast and draft, are given in Table 2 as the coefficient C0.

Similarly, the following relation expresses how the 'normalized' roll moment varies with vehicle yaw angle:

$$\frac{Roll.Moment}{V^2} = C_1 \cdot \Psi \left[\frac{\mathrm{N \cdot m}}{\mathrm{m}^2/\mathrm{s}^2}\right] \tag{3}$$

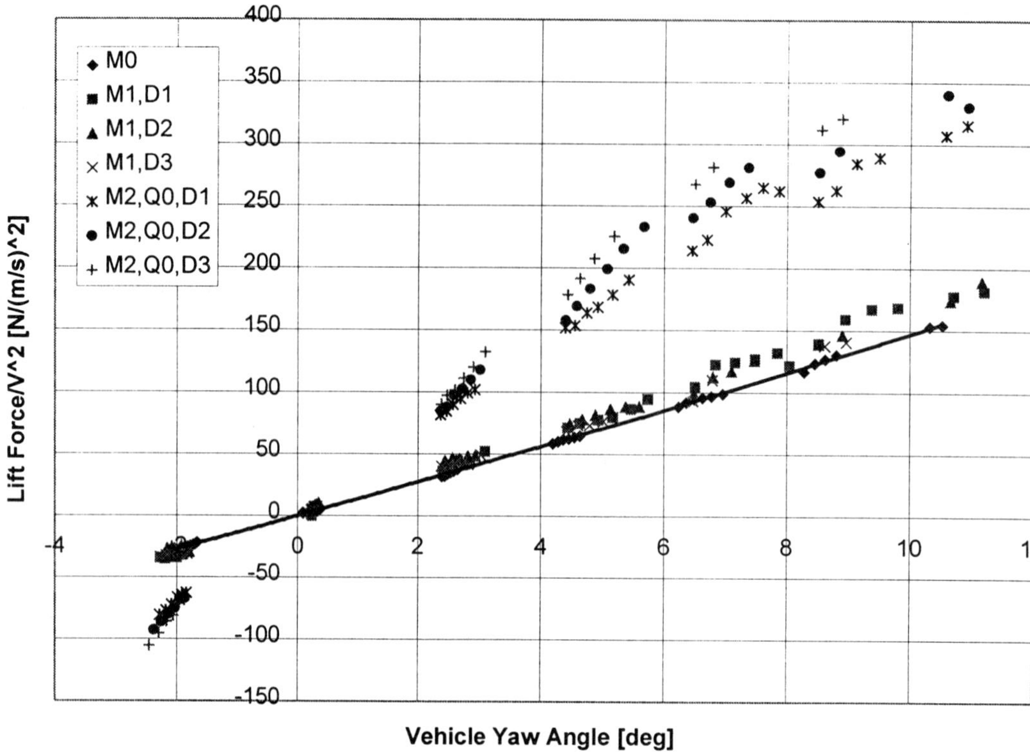

Figure 11. Normalized lift force vs vehicle yaw angle: masts M0, M1, M2; drafts D1, D2, D3.

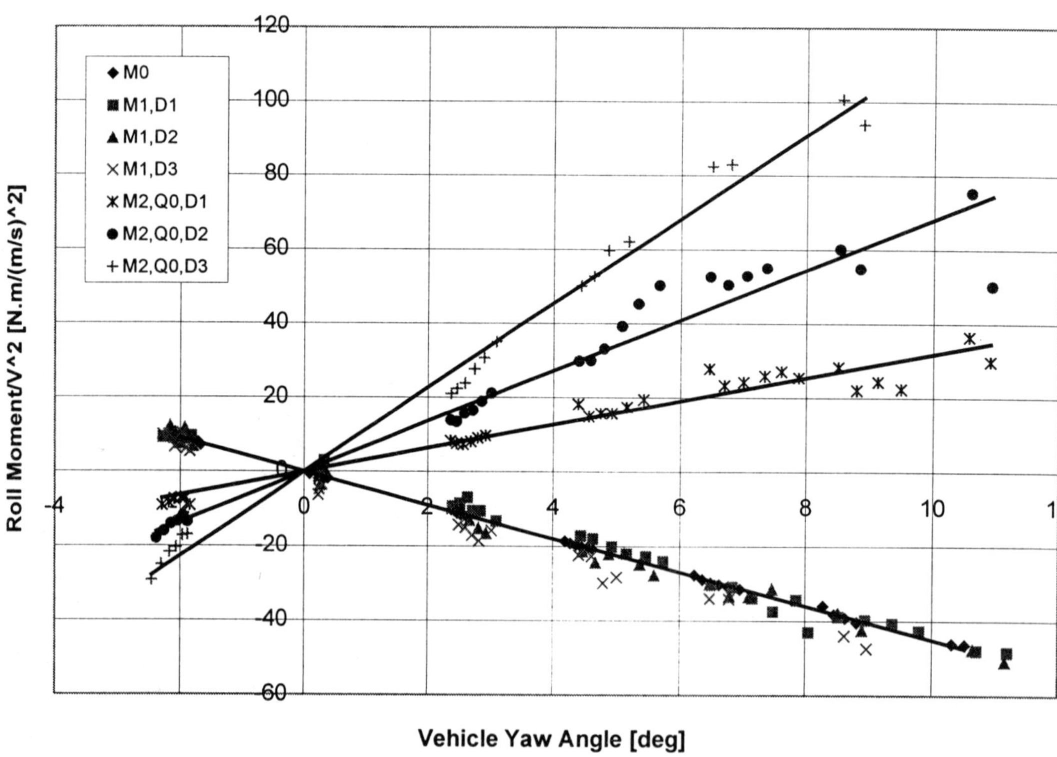

Figure 12. Normalized roll moment vs vehicle yaw angle: masts M0, M1, M2 at drafts D1, D2, D3.

In Figure 12 the measured roll moment (as 'normalized' by the square of the vehicle forward speed 'V') is plotted versus the yaw angle of the model (at drafts D1,D2,D3) for the configuration with no mast (M0), the existing mast (M1), and the alternative mast (M2) respectively. Again straight lines (with zero intercept) were fitted through the data, and the slopes were computed.

From the values of coefficient C0 in Table 2 it is seen that the slope of the curve of the normalized lift force versus vehicle yaw angle for the alternative mast configuration (M2) is about double the slope of the lift curve for either the no-mast (M0) or the existing-mast configuration (M1), at all three drafts D1, D2 and D3. Also note that the magnitude of the lift force produced by the alternative mast (M2) increases with increasing draft, as expected, while the magnitude of the lift force that is produced by the existing mast (M1) decreases with increasing draft. The latter result was unexpected and may be related to the 'splaying' of the fairings, that is, the 'sticking' (or 'jamming' of one segment against the other) and resulting misalignment of the free-swivelling, self-aligning fairing segments. These segments were observed to straighten out only at the higher towing speeds; this can be seen in some of the photos in (Butt et al, 1998) and (Butt, 1999), one of which is included here as Figure 13.

From the values of coefficient C1 in Table 2 it can be seen that the slope of the curve of the normalized roll moment versus vehicle yaw angle is negative, for the vehicle with no mast, M0, (due to the action of the keel) and for the vehicle with the existing mast, M1, (due to the action of the keel and the self-aligning fairings), at all three drafts. For the vehicle with the alternative mast (M2), this slope is positive for all three drafts; a plot of the slope versus the draft shows that there is a draft, call it D0, at which the slope is zero which means that the alternative mast can be used to cancel the roll moment that is generated by the keel and the resulting vehicle should remain upright for all vehicle yaw angles and all vehicle speeds. The value of the draft D0 is about 85 cm.

Therefore, a condition of zero roll moment exists (all else being equal) at a draft of about 85 cm, with the alternative mast M2. The waterline for a draft of 85 cm on this mast is located roughly 10 mm above the ventilation fence. Recall that the lower section of mast M2 was designed so that it alone would generate sufficient roll moment to counteract the roll moment that is produced by the keel. Therefore, the design of the lower section of the alternative mast configuration is considered to meet this criterion closely.

Lift and Roll Performance of the Alternative Mast (M2) with Air Ejection (Q1, Q2)

The goal of air ejection was to reduce the effective lift force to zero over the upper section of the alternative mast. Ideally, in order for this to occur, a thin, evenly-distributed air sheet is required on both sides of the upper mast section, as described in (Rothblum et al, 1976). However, during our experiments with air ejection it was noticed that this "ideal" condition was not satisfactorily achieved.

Mast Configuration	Draft	Airflow Rate [litre/sec]	Lift/(V^2) Slope C0 [per degree]	MX/(V^2) Slope C1 [per degree]
No Mast (M0)	D1	0	14.1	-4.48
Old Mast (M1)	D1	0	16.2	-4.28
	D2	0	15.8	-4.63
	D3	0	15.1	-5.01
Alt. Mast (M2,Q0)	D0	0	30.5	0.00
	D1	0	33.6	3.53
	D2	0	37.7	7.53
	D3	0	40.4	10.96
Alt. Mast (M2,Q1)	D1	9.4	31.4	1.18
	D2	9.4	30.4	0.11
	D3	9.4	28.3	-2.47
Alt. Mast (M2,Q2)	D1	14.2	30.5	2.03
	D2	14.2	32.8	4.93
	D3	14.2	37.6	9.29

Table 2. Coefficients C0 and C1 for masts M0, M1, M2 and air flow rates Q0, Q1, Q2.

Figure 13. Existing Mast (M1), Draft D1,Yaw = 0°, V = 4.47m/s.

Firstly, it was observed that the distribution of the ejected air had a variation in sheet thickness along the vertical span of the upper mast section. The thickness of the air sheet was greatest on the lower third (roughly from D0 to D1) of the upper mast section, while the thickness was minimal over the middle third (roughly from D1 to D2). The upper third of the upper mast section (roughly from D2 to D3) had an intermediate air sheet thickness. Secondly, the air sheet on the high-pressure side of the upper mast became thinner as the vehicle was yawed away from zero degrees. At large yaw angles the air sheet on the high-pressure side collapsed, with all the ejected air exiting from the low-pressure side; however, for yaw angles of less than 4°, the low/high-pressure symmetry of the air sheet was satisfactory. Thirdly, it was observed that the air sheet was excessively thick during some of the lower vehicle speed experiments.

Figure 14 shows the ('normalized') measured lift force plotted versus the vehicle yaw angle for the alternative mast with air ejection rates Q1 and Q2, of 9.4 and 14.2 litre/s, respectively, for vehicle drafts D1, D2 and D3. A straight line (with zero intercept) was fitted through each set of data, for yaw angles of less than 4°, and the slope was computed. Note that as the draft is increased, the data points above 4° diverge from the straight-line relation.

Figure 15 shows the ('normalized') measured roll moment plotted versus the vehicle yaw angle for the alternative mast air ejection rates Q1 and Q2, for vehicle drafts D1, D2 and D3.

The variation of measured roll moment (using air ejection) with vehicle yaw angle is not as "linear" as in the previous datasets. As a result, a cubic spline was used to indicate the "trend" of each dataset as the yaw angle was increased. For vehicle yaw angles above 4° the thickness of the air sheet on each side of the mast is not well behaved and this can be seen in the plots.

The slopes of these straight-line sections are given in Table 2 as the coefficients C0 and C1, for the normalized lift force and normalized roll moment, respectively. In determining these slopes, only yaw angles of less than 4° were considered. In addition to this, the data from the two lowest vehicle velocities, 2.00 and 2.83 m/s, were excluded from the analysis.

From the results for the alternative mast, M2, for flow rates Q0 (no air ejection), Q1 (9.4 litre/s) and Q2 (14.2 litre/s) in Table 2, one sees that for a given draft the air ejection reduces the overall vehicle lift force (C0). Note that for the air ejected at rate Q1, the overall vehicle lift force (C0) decreases as the draft is increased. Note also that for the air ejected at rate Q1, at a given draft, that the magnitude of the overall vehicle roll moment (C1) is lower than with Q0 (no air ejection) and with the existing mast M1. In fact, the magnitude of the roll moment at draft D3, for air ejected at rate Q1, is less than the magnitude of the roll moment at draft D1 for either Q0 (no air ejection) or for the existing mast M1. These results show that air ejection has the potential to control both the overall vehicle lift force and roll moment, at any draft.

Some additional conclusions can be extracted from Table 2 for the case with air ejected at rate Q1. Note that the overall vehicle roll moment (C1) is nearly zero for draft D2. If the overall vehicle lift force (slope C0(Q1,D2)) with air ejected at rate Q1 at draft D2 is compared to the overall vehicle lift force (slope C0(Q0,D0)) with no air ejected (Q0) at draft D0, it can be seen that there is not much difference in the amount of lift that is produced. Similarly, if the overall vehicle roll moment (slope C1(Q1,D2)) with air ejected at rate Q1 at draft D2 is compared to the overall vehicle roll moment (slope C1(Q0,D0)) with no air ejected (Q0) at draft D0, it can be seen that there is not much difference in the amount of roll moment that is produced.

In fact, in these two cases, (Q1, D2) and (Q0, D0), the difference in lift force is less than 0.5 percent. Therefore it can be concluded that at a draft D2 for the alternative mast with air ejected at rate Q1, that the net lift force generated by the upper mast section (i.e. D0 to D2) is essentially zero. Due to the positive slope of the roll moment curve (slope C1) at draft D1 for air ejected at rate Q1, and the fact that the overall lift force (slope C0) with air ejected at rate Q1 at draft D1 is greater than with no air ejected (Q0) at draft D0, it can be concluded that the there is still some "leftover lift force" generated by a portion of the upper mast section (i.e. D0 to D1). However, because of the negative slope of the roll moment curve (slope C1) at draft D3 for air ejected at rate Q1, and because the overall lift force (slope C0) with air ejected at rate Q1 at draft D3 is less than with no air ejected (Q0) at draft D0, it can be concluded that there is a "reverse lift force" generated by the upper mast section (i.e. D0 to D3). In other words the ejected air produces a counter thrust which acts in the opposite direction than would normally be expected, at a given angle of attack.

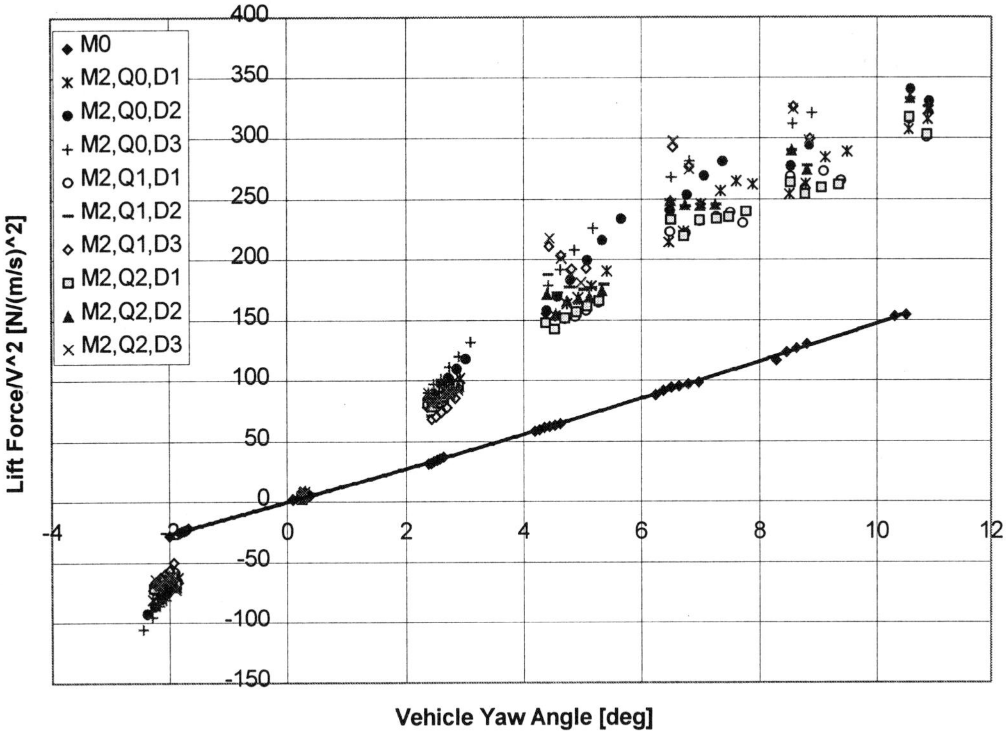

Figure 14. Normalized lift force vs vehicle yaw angle: masts M0, M2; drafts D1, D2, D3; airflows Q0, Q1, Q2.

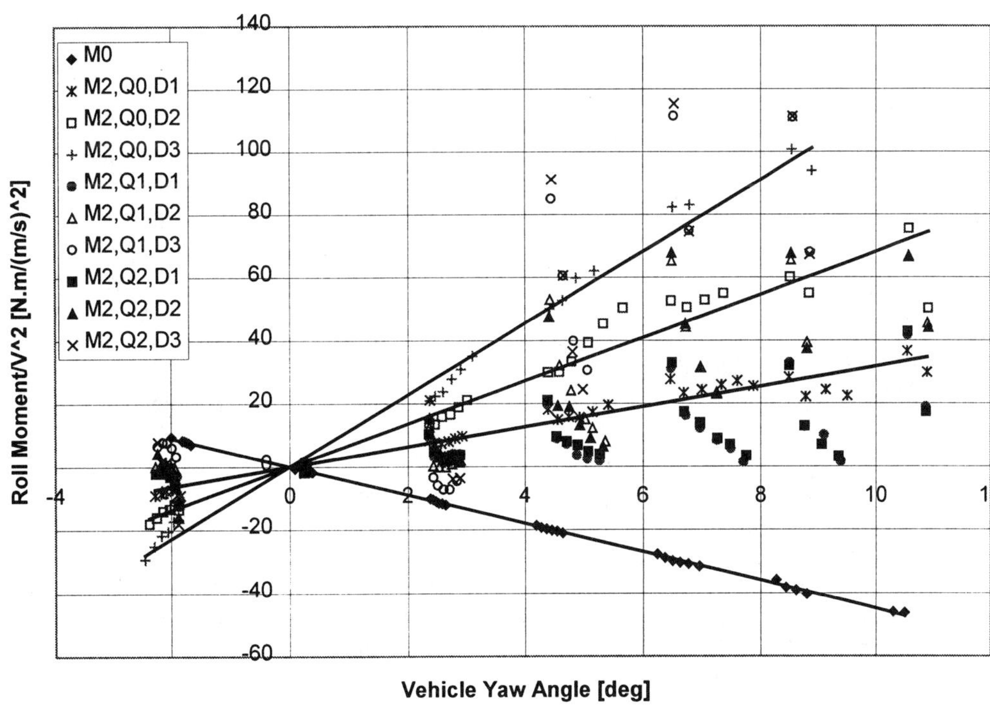

Figure 15. Normalized roll moment vs vehicle yaw angle: masts M0, M2; drafts D1, D2, D3; airflows Q0, Q1, Q2.

From Figures 14 and 15, and Table 2, it can be concluded that:

a. The slope C0, for the 'normalized' lift force, for each of three drafts, is only a weak function of flow rate (Q0,Q1,Q2) so whether you use air ejection or not you are not going to get any major reduction in the amount of lift generated by the alternative mast (M2).

b. The greatest change in reducing the roll moment to zero (slope C1) was achieved by going from zero air flow rate (Q0) to flow rate Q1; any subsequent change that was achieved by increasing the flow rate by 50 percent from Q1 (9.4 litre/s) to Q2 (14.2 litre/s) was counterproductive. In fact, for flow rate Q1, Table 2 shows that the sign of the slope changes with draft from positive to almost zero to negative as we go from draft D1, to D2 to D3. The latter result is why we can find a draft D0 at flow rate Q1 for which there is complete cancellation of the keel-generated roll moment such that the net hydrodynamic roll moment that is exerted on the vehicle will be zero, at least for any yaw angle in the range +/-2° (and probably for a much larger range of vehicle yaw angles). We see that as a function of air flow rate, the slope C1 is always minimal for Q1, for all three drafts.

c. In Figure 15, by examining the data points in the range of +/-2°, it can be seen that the slopes for flow rates Q1 and Q2 (for all three drafts) are close to zero for curves of roll moment vs vehicle yaw angle. Also, over all the vehicle yaw angles, the data points for flow rates Q1 and Q2 are essentially coincident. The conclusions are: (i) that zero roll moment was achieved over a wide range of vehicle speeds and vehicle yaw angles, and, (ii) that the air flow rate Q1 functioned as a 'threshold' flow rate in that increasing the flow rate above the value of Q1 provided no further change in roll moment and served only to increase the drag due to the mast.

Lift, Roll and Yaw Performance of the Alternative Mast with Flap (M2F)

The effect of the flap on the lower section of the alternative mast can be thought of as an additional appendage. Therefore, the performance of the vehicle with the flap can be added to the performance of the overall vehicle without the flap.

One can rewrite equations (2) and (3) to include the effect of the flap, in terms of 'V', the forward speed of the vehicle, 'Ψ' the vehicle yaw angle and 'δ' the flap deflection angle, both measured in degrees. The expression for the 'normalized' vehicle lift force is given by:

$$\frac{Lift.Force}{V^2} = C_0 \cdot \Psi + C_2 \cdot \delta \left[\frac{\mathrm{N}}{\mathrm{m}^2/\mathrm{s}^2}\right] \quad (4)$$

while the expression for the roll moment is given by:

$$\frac{Roll.Moment}{V^2} = C_1 \cdot \Psi + C_3 \cdot \delta \left[\frac{\mathrm{N \cdot m}}{\mathrm{m}^2/\mathrm{s}^2}\right] \quad (5)$$

In the above equations, C2 and C3 represent the slopes of the 'normalized' vehicle lift force and roll moment, respectively, when plotted versus flap deflection in degrees.

The 'normalized' lift force and roll moment are plotted versus vehicle yaw angle in Figures 16 and 17 respectively, for flap angles in the range from -2° to +12°, for three vehicle yaw angles, -2, 0 and +2° and for six vehicle speeds in the range from 2.0 to 4.9 m/s.

In Figures 16 and 17, straight lines were fitted to the data, as indicated by equations (4) and (5). The values of the slopes C0 and C1 are found in Table 3.

In Figures 18 and 19, the 'normalized' lift force and roll moment, respectively, are plotted versus flap angle, for flap angles in the range from -2 to +12°, for three vehicle yaw angles, -2, 0 and +2°, and for six vehicle speeds in the range from 2.0 to 4.9 m/s. It is seen that the data are clustered into three 'bands', one for each vehicle yaw angle. Straight lines were fitted to the data, and the positive slope of each line in these two Figures is represented by a value of the coefficient C2 or C3 in equations (4) and (5); the values of the slopes C2 and C3 are to be found in Table 3.

In Figure 20, the 'normalized' yaw moment is plotted versus flap angle, for flap angles in the range from -2 to +12°, for three vehicle yaw angles, -2, 0 and +2°, and for six vehicle speeds in the range from 2.0 to 4.9 m/s. Again the data are clustered into three 'bands', one for each vehicle yaw angle. Note that a positive vehicle yaw angle produces a positive yaw moment while a positive flap deflection (trailing edge to port) produces a negative yaw moment. Thus the slopes of the yaw moment curves (vs flap deflection) are negative for all three 'bands'.

The results of the experiments with the partial-span 25-percent-chord plain flap indicate that this flap, when mounted on the lower section of the alternative mast, is adequate to provide vehicle roll control. This is demonstrated in Figure 17 by the range of roll moments for the mast with flap, in comparison with the roll moment measured for the vehicle with the existing mast, M1, as shown in Figure 12.

Using the roll moment data presented in Figure 19, for each vehicle yaw angle and vehicle speed, a straight line was fitted to the data. Where each straight line intersected the axis of zero roll moment, the flap angle was determined.

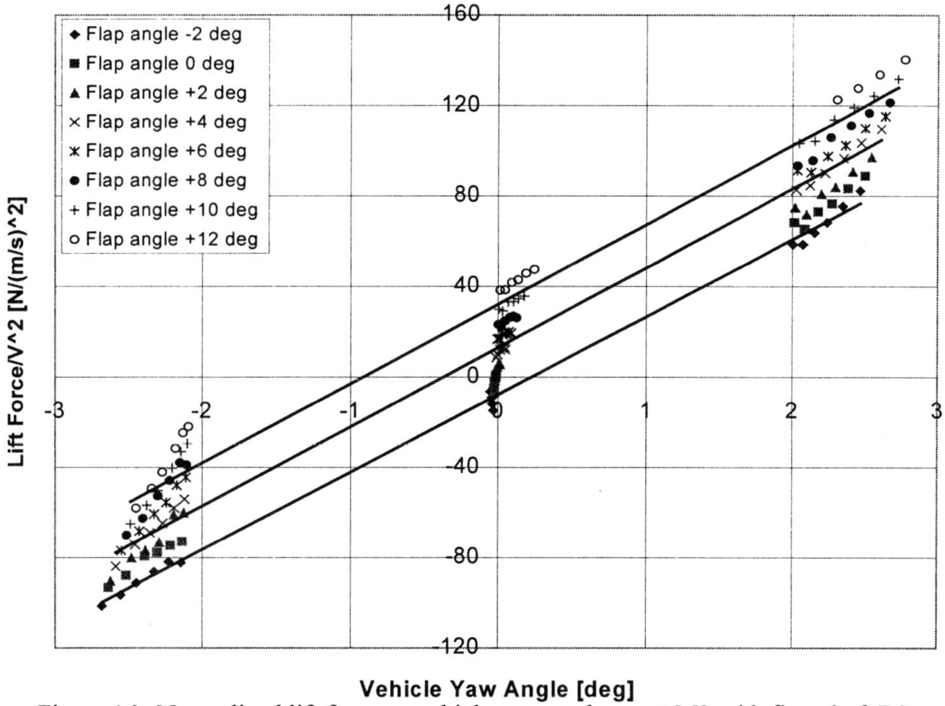

Figure 16. Normalized lift force vs vehicle yaw angle: mast M2 with flap, draft D2.

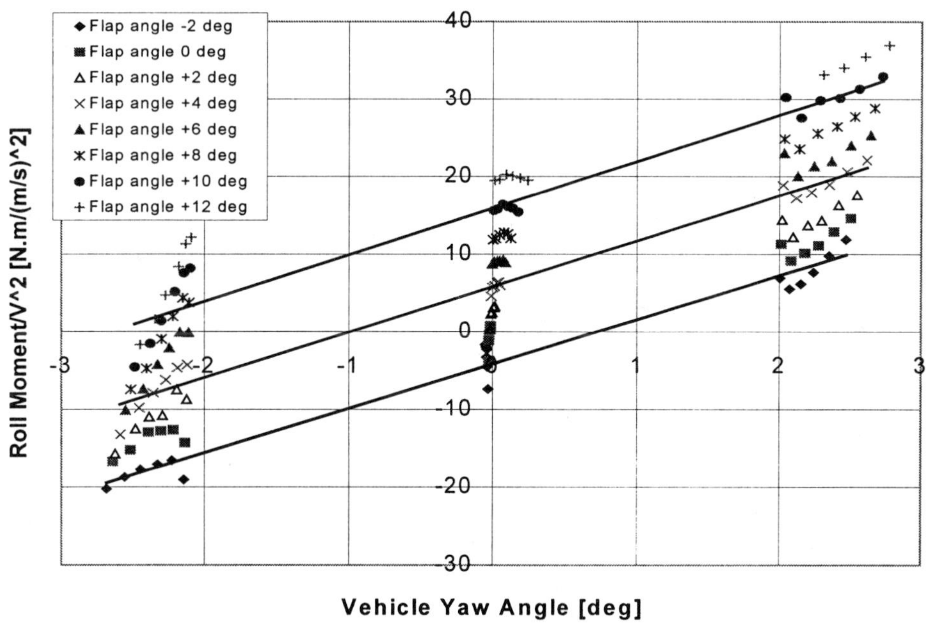

Figure 17. Normalized roll moment vs vehicle yaw angle: mast M2 with flap, draft D2.

Mast Configuration	Draft	Lift/(V^2) Slope C2 [per degree]	MX/(V^2) Slope C3 [per degree]
Alternate Mast with Flap (M2F)	D2	3.600	1.706

Table 3. Flap effectiveness coefficients for the alternative mast (M2F).

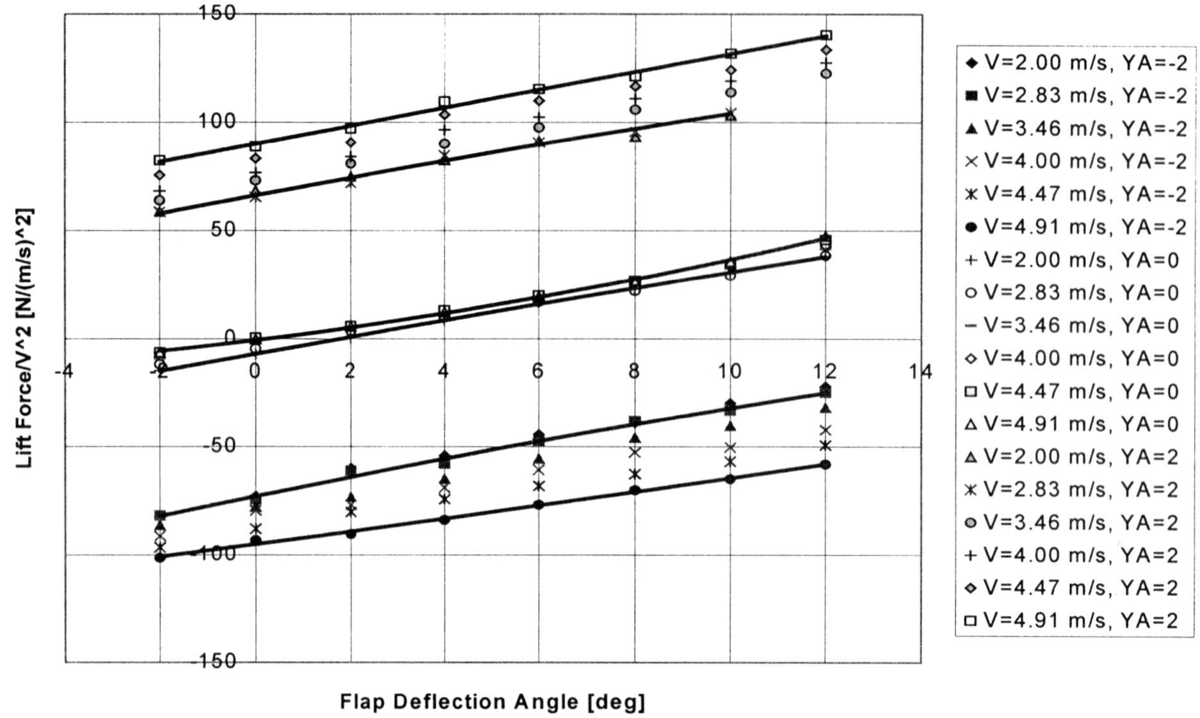

Figure 18. Normalized lift force vs flap angle: alternative mast M2; draft D2.

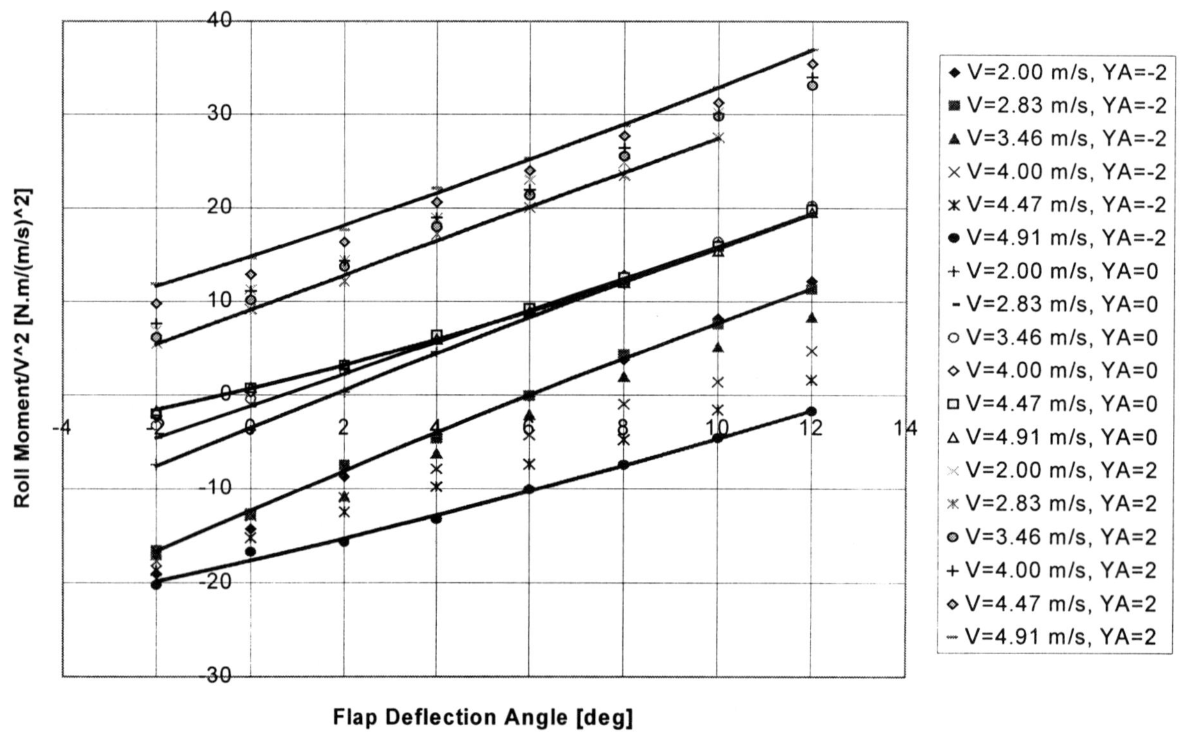

Figure 19. Normalized roll moment vs flap angle: alternative mast M2; draft D2.

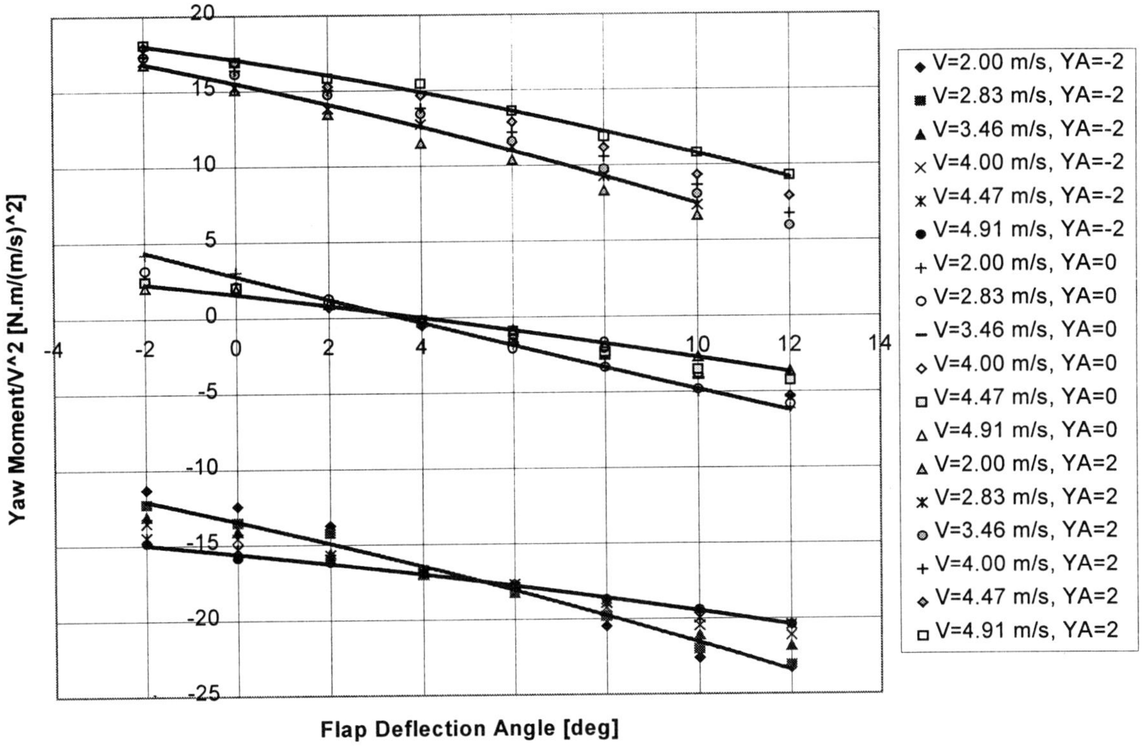

Figure 20. Normalized yaw moment vs flap angle: alternative mast; draft D2.

Mast Configuration	Draft	Speed [m/s]	Vehicle Yaw Angle [deg]		
			-2	0	+2
Alternate Mast with Flap (M2F)	D2	2.00	6.1	1.9	-5.6
	D2	2.83	6.0	0.7	-5.0
	D2	3.46	7.1	0.3	-5.2
	D2	4.00	8.7	-0.2	-6.6
	D2	4.47	10.9	-0.5	-7.3
	D2	4.91	13.2	-0.6	-9.0

Table 4. Flap angles required for zero vehicle roll moment for various vehicle speeds and vehicle yaw angles

As a guide to the range of flap angles required to counter-act the overall vehicle roll moment (about the hull longitudinal centreline), Figure 21 and Table 4 show the flap angle that is required to produce zero vehicle roll moment for various vehicle speeds and three vehicle yaw angles. Although these results do not exhibit perfect port-starboard symmetry, the results are representative of the magnitude of the flap angles required. From Figure 21 it can be seen that:

a. The flap angle that was required, for a given vehicle yaw angle, to produce zero vehicle roll moment, depends on the vehicle forward speed.
b. At low vehicle forward speeds, between two and three m/s, about 5 or 6° of flap angle was required to produce zero vehicle roll moment, when the vehicle yaw angle was 2°.
c. The variation of flap angle (required for zero vehicle roll moment) with forward speed is weak, when the vehicle yaw angle is in the range of +/-2°.
d. For the largest speed tested, 4.9 m/s, about 11° of flap angle was required to produce zero vehicle roll moment, when the vehicle yaw angle was 2°.

It should be noted that the gap between the mast and the circular leading edge of the flap was about one millimetre at all flap deflections since no sealing strips were used; thus there may have been some 'leakage flow' through the gap. This gap may account for the unexpected loss of efficiency of the flap that was observed at the higher forward speeds and at the larger flap deflections.

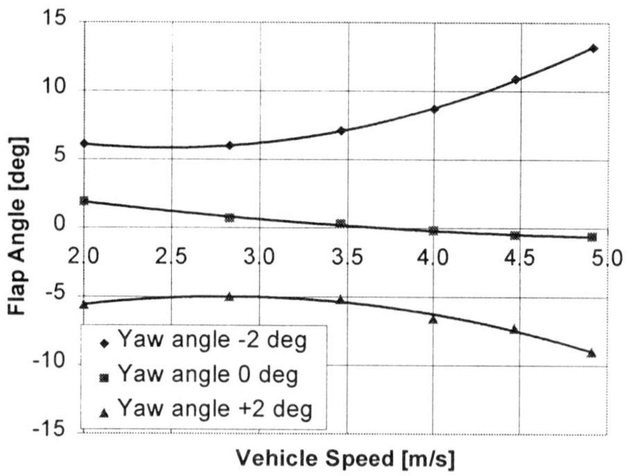

Figure 21. Flap angle required for zero roll moment for various vehicle speeds and yaw angles.

Conclusions Concerning the Measured Lift Force and Roll Moment

One of the goals of this investigation was to use the alternative mast to control the vehicle roll moment. It was found that the basic alternative mast configuration, M2, (i.e. with no air ejection and no flap) provided too much counter-roll for drafts D1, D2 and D3. This was expected, since the mast was designed such that only the lower portion would provide sufficient counter-roll to eliminate the roll moment.

With air ejection, it was found that the magnitude of the overall vehicle roll moment was lower than with either the basic alternative mast, M2, (with no air ejection and no flap) or with the existing mast, M1. Although one would expect that the roll moment (which is due to the 'lift' force generated by the airfoil-like mast) would increase in proportion to the length (span) of mast that was submerged, in fact, the magnitude of the roll moment at the deepest draft (D3) with air ejection was less than the magnitude of the roll moment at the shallowest draft (D1), for both the basic alternative mast and the existing mast. However, because it was not possible to maintain a thin, evenly-distributed air sheet (along both the port and starboard surfaces of the mast) during testing, the experimental conditions were not ideal for determining conclusively the effect (benefits) of the air ejection. In particular, it was not possible to achieve the zero-lift benefits from the air ejection over as wide a range of yaw angles as were reported in (Rothblum et al, 1976).

In these experiments, the airflow was supplied to a single plenum chamber inside the mast, so the effect of the flow past the mast on the 'pressure side' was to make the sheet (of air) thinner on that side and to increase the airflow rate out of the holes on the 'suction side' of the mast. Had it been possible to equip the interior of the upper section of the mast with two independently-supplied chambers, then a similar volume flow rate of air would have exited the two surfaces of the mast at all vehicle yaw angles. So instead of having a balanced airflow and hence zero lift from each side of the mast (Rothblum et al, 1976), there was a counter-lift due to the increased proportion of air ejected from the suction side.

The ratio of the length of the air holes to their diameter was approximately 1:1. For values of this ratio greater than about 6:1 the air flow rate should be insensitive to the outside pressure. A better performance from the air ejection system could have been achieved by increasing the length of the air holes, or by reducing their diameter and by increasing the size of the plenum chamber.

Note that in practical terms it will probably be simpler to provide active roll control via the trailing-edge flap than to try to vary the volume flow rate of the exhaust gas to the air ejection system. An active control device on the supply of exhaust gas might be over-complicated since it would have to be designed to deliver (and limit) the flow rate to the 'threshold' value Q1 over a wide range of engine RPM.

Normal Force, Pitch Moment and Yaw Moment for the Basic Mast Configurations

Figures 22 and 23 show, for all 13 cases, how the 'normalized' normal force (FZ) and pitch moment (MY), respectively, vary with vehicle yaw angle. In reality, since the model, as tested, has port-starboard symmetry, the graph of normal force (or pitch moment) vs vehicle yaw angle should be a parabola that is concave downward, symmetrical about Ψ=0; in the plots only the measured data for (-2°,+10°) are shown and the curve that has been fitted through the measured data was not forced to have port-starboard symmetry. In order to show the effect of the air ejection, curves were fitted through five sets of data: (i) no mast (M0); (ii) existing mast (M1) at draft D3; (iii) alternative mast (M2) at draft D3 with no air ejection (Q0); (iv) alternative mast (M2) at draft D3 with air ejection at 9.4 litre/s (Q1); (v) alternative mast (M2) at draft D3 with air ejection at 14.2 litre/s (Q2). Note that the common draft for the fitted curves is 'D3', the maximum draft, which is the draft that has the maximum number of air holes submerged and should therefore show the largest effect of the air ejection.

In Figures 22 and 23 the solid line indicates the trend of the data for the vehicle with no mast (M0). The uppermost curve (the dotted line), shows the trend of the data for the vehicle with the existing mast (M1) and draft D3, but the trend is similar for the other two drafts. The other three curves, the dashed lines, show the trend of the data for the vehicle with the alternative mast (M2) with (or without) air ejection at draft D3.

Figure 22 shows that the effect of the ejected air is minimal on the 'normalized' normal force; the three curves for flow rates Q0, Q1 and Q2 are essentially coincident. The effect of draft is stronger and the trend (not shown with fitted curves) is as expected. The only configuration that produced a negative (upward) normal force over all vehicle yaw angles was with no mast (M0, solid line). The existing mast (M1, dotted line) produced a positive (downward) normal force over a wide range of vehicle yaw angles, at all three drafts. The configurations with the alternative mast (M2, dashed lines) produced a negative (upward) normal force at vehicle yaw angles greater than about 3°.

Figure 23 shows that the only configuration that produced a negative (nose-down) pitch moment over all vehicle yaw angles was with no mast (M0, solid line); this is due directly to the drag force on the keel which is acting at some distance below the vehicle centreline axis. The existing mast (M1, dotted line) produced a net positive (nose-up) pitch moment over a wide range of vehicle yaw angles, at all three drafts. The configurations with the alternative mast (M2, dashed lines) produced a negative (nose-down) pitch moment at vehicle yaw angles greater than about 2.5 to 4°, depending on the amount of air ejected.

Figure 23 shows that the effect of the ejected air is stronger on the 'normalized' pitch moment than it is on the normal force. For the pitch moment, the curve for the alternative mast (M2) with no air ejection (Q0) is the lowermost curve, and the two curves with air ejected at rates Q1 and Q2 are seen to be displaced upward relative to the Q0 curve. Thus the effect of the alternative mast with air ejection is to create a nose-up pitch moment (which is due directly to the drag force on that mast) in the range from zero to 4° of vehicle yaw angle, which is well within the normal operating range of the *DOLPHIN* during turns; with no air ejection, the pitch moment is nose-up only up to about 2.5°.

Figure 24 shows the 'normalized' yaw moment plotted versus the vehicle yaw angle, for all 13 mast configurations. In this figure, the solid curve indicates the situation when the vehicle has no mast (M0), where the yaw moment is due to the hydrodynamic forces on the keel and hull. This curve effectively provides an upper bound for the yaw moment, for all other mast configurations, regardless of draft and the use of air ejection. Again the largest effect on yaw moment (with the alternative mast, M2) is that of draft; the effect of air ejection (or not) is relatively minor.

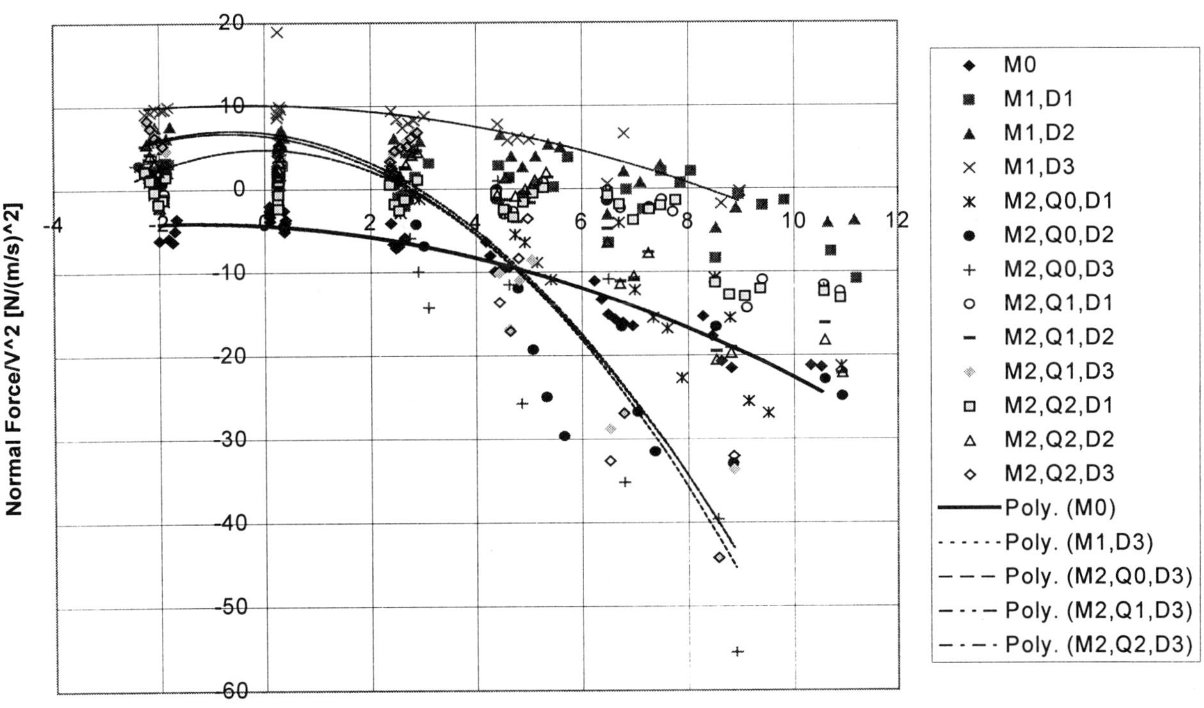

Figure 22. Normalized normal force vs vehicle yaw angle: masts M0, M1, M2; drafts D1, D2, D3; airflows Q1, Q2, Q3.

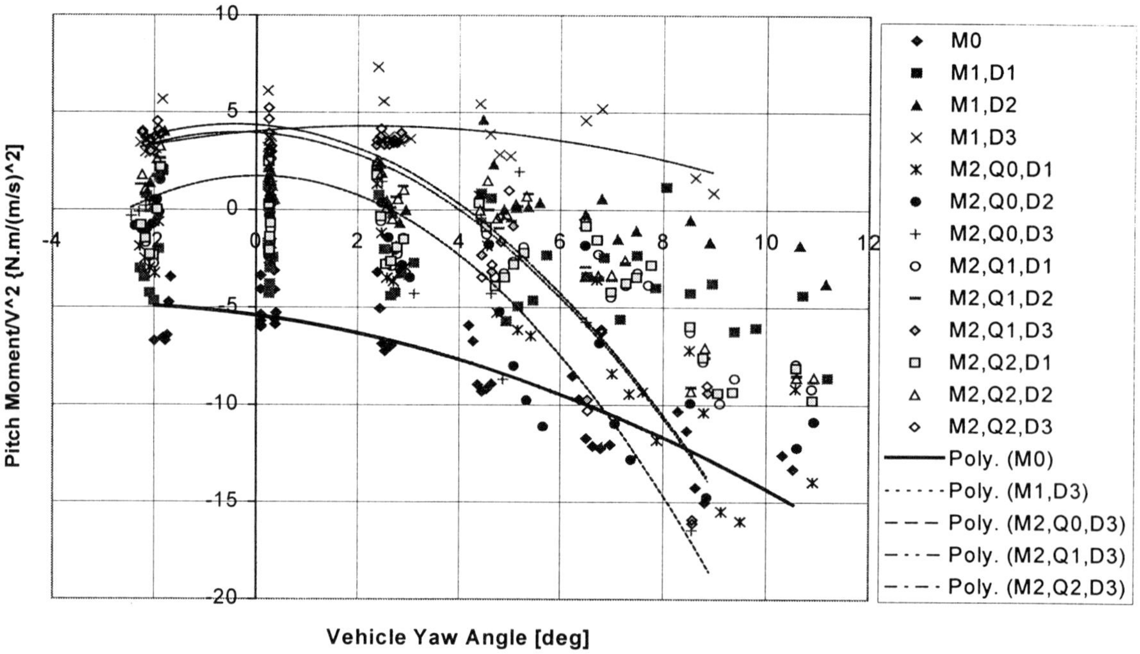

Figure 23. Normalized pitch moment vs vehicle yaw angle: masts M0, M1, M2; drafts D1, D2, D3; airflows Q1, Q2, Q3.

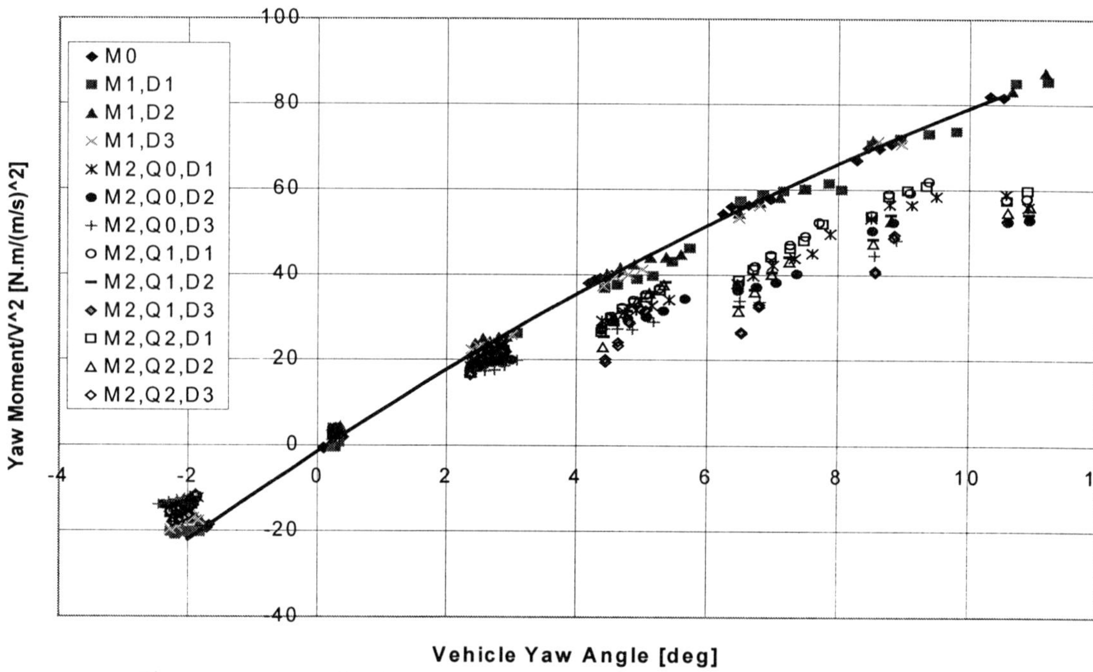

Figure 24. Normalized yaw moment vs vehicle yaw angle: masts M0, M1, M2; drafts D1, D2, D3; airflows Q1, Q2, Q3.

Location of the Pivot Axes and the Centre of Effort

From a combination of the yaw moment and side force data (the latter are similar in form to the lift force data, hence they are not shown here), we can compute the longitudinal location of the "pivot axis", which is effectively the axis about which the vehicle "pivots" when in a turn.

The yaw moment was measured about a vertical axis through the leading edge of the keel. The yaw moment about any vertical axis that is forward of that origin can be computed; that calculation includes the measured values of both the yaw moment and the lateral force. Since for positive vehicle yaw angles the lateral force and yaw moment are both positive, there exists a location forward of the measurement origin about which the counterclockwise moment due to the lateral force just balances the hydrodynamic yaw moment; at small yaw angles this is known as a "neutral point" and it defines the location of the "pivot axis" and this longitudinal distance is denoted by 'XCP'

The computed values of 'XCP' are given in Table 5, for 13 different mast conditions. Note that 'XCP' is measured forward from the leading edge of the keel.

From a combination of the measured hydrodynamic roll moment and side force data, using a procedure identical to that for 'XCP', the vertical location of the "centre of effort" was computed for the 13 different mast conditions. The values of 'ZCP' are given in Table 5. Note that 'ZCP' is measured positive downward from the longitudinal centreline of the hull.

Mast Configuration	XCP from Keel Leading Edge [m]	ZCP from Hull Centreline [m]	XNF from Keel Leading Edge [m]
No Mast (M0)	+0.6	+0.3	-1.2
Old Mast (M1;D1,D2,D3)	+0.5	+0.25	-0.5,-0.2,-0.4
New Mast (M2;D1;Q0)	+0.2	-0.10	-0.5
New Mast (M2;D1;Q1,Q2)	+0.2	-0.05	-0.6,-0.6
New Mast (M2;D2;Q0)	+0.2	-0.20	-0.1
New Mast (M2;D2;Q1,Q2)	+0.2	-0.10	-0.5,-0.6
New Mast (M2;D3;Q0)	+0.2	-0.30	-0.3
New Mast (M2;D3;Q1,Q2)	+0.2	-0.30	-0.7,-0.7

Table 5. Location of pivot axes and centre of effort for masts M0, M1, M2 for drafts D1, D2, D3 and airflows Q0, Q1, Q2 (see Figure 25).

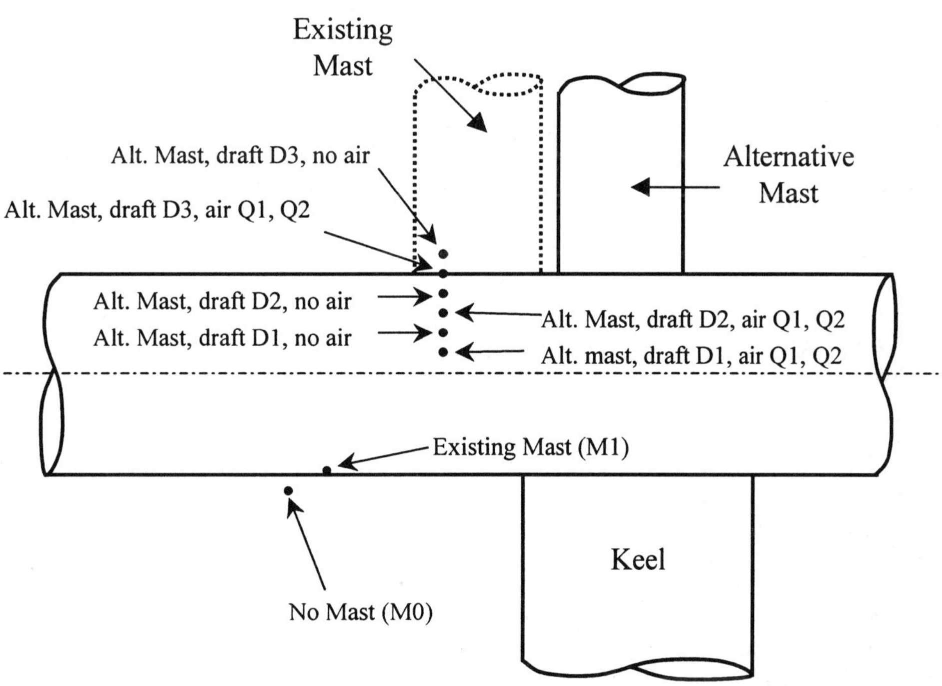

Figure 25. Profile view of the *DOLPHIN* half-scale model with centres of effort indicated: masts M0, M1, M2; drafts D1, D2, D3; airflow rates Q0, Q1, Q2.

Figure 25 and Table 5 show the locations of the "pivot axis" and the "centre of effort"; the values of 'XCP' and 'ZCP' in Table 5 are the intercept values at zero vehicle yaw angle. With no mast (M0), the pivot axis was found to be 0.6 m (about 1.2 hull diameters) ahead of the leading edge of the keel. With the existing mast (M1), the pivot axis was found to be 0.5 m (about 0.98 hull diameters) ahead of the leading edge of the keel. With the alternative mast (M2), the pivot axis was found to be 0.2 m (about 0.4 hull diameters) ahead of the leading edge of the keel. So for all cases the quarter-chord axis (the expected location of the lift force that was generated by the keel) of the keel was aft of the "pivot axis" of the vehicle.

With no mast (M0), the vertical location of the "centre of effort" (COE) was found to be 0.3 m below the longitudinal centreline of the hull, so with no mast the COE is below the hull but ahead of the keel. With the existing mast (M1), 'ZCP' was found to be 0.25 m, which means that the COE is close to the lower surface of the hull. For the alternative mast (M2), with no air ejection (Q0), 'ZCP' was found to range from -0.10 to -0.30 m (depending on the draft), which means that the COE is inside the hull and above the centreline. Similar results were found with air ejection rates Q1 and Q2; the details are given in Table 5. From these results it can be concluded that the effect of the air ejection is to move the COE closer to the hull centreline, that is, to reduce the magnitude of both the side-force and roll moment that is generated by the alternative mast.

In the same way that 'XCP' locates the axis about which the vehicle pivots during turns to port or starboard ("yaw axis"), and that 'ZCP' locates the "roll axis", 'XNF' locates the axis about which the vehicle pivots during pitching motions.

The value 'XNF' gives the longitudinal location of the line of action of the normal force, FZ. This location was determined by computing the ratio of the measured pitch moment to the measured normal force, and finding the intercept at zero vehicle yaw angle. The purpose of computing 'XNF' is to show how (a) the draft, and, (b) the amount of air ejected, affect the pitch attitude of the vehicle when running at zero yaw angle. The estimated values of 'XNF' are given in Table 5. From this Table one sees that with either the existing mast or the alternative mast, the farthest forward location of the "pitch axis" is about 0.1 m (about 0.2 hull diameter) aft of the leading edge of the keel; the farthest aft location of the 'pitch axis' is about 0.7 m (about 1.2 hull diameter) aft of the leading edge of the keel. With air ejection, the location of the "pitch axis" moves aftward. For each air flow rate, the minimum shift of the "pitch axis" appears to occur at the middle draft 'D2'.

These results (the measured values of the pitch moment, normal force and 'XNF') can be used to determine the size and location of the horizontal control surfaces (foreplanes, sternplanes) for maximum effectiveness in counteracting the force exerted by the tow cable (Figure 1), with either the existing mast (M1) or the alternative mast (M2), with or without air ejection. The foreplanes and sternplanes can be used together (inclined in the same direction) to offset the downward component of the tow force (at a given speed), and can be used in opposition (inclined in opposite directions) to offset the nose-down pitch moment due to the tow force. Knowing the value of 'XNF' assists the designer to select the relative sizes of the horizontal control surfaces, and the fore-and-aft location of these control surfaces relative to the axis of pitch, for maximum control effectiveness.

DISCUSSION AND CONCLUSIONS

A collaborative project between ISER, IMD and MUN was initiated to investigate an alternative mast design for the *DOLPHIN* vehicle.

A symmetric double-arc shape was selected as the cross-sectional shape for the upper portion of a two-part mast. Active air-ejection was implemented into the upper mast as a secondary means of regulating vehicle side force. The profile for the lower half of the mast was based on a modified double-arc shape. In addition to this, a 25-percent-of-chord plain, partial-span flap was integrated into the lower half of the mast to control vehicle side force and roll moment.

The primary goal of this project was to investigate the overall vehicle drag reduction on the *DOLPHIN* by reducing the drag of its surface-piercing mast. The drag results are summarized in Table 1. It was found that the drag from the existing mast represented from 42 to 54 percent of overall vehicle drag at zero vehicle yaw angle, depending on the vehicle draft. In comparison, the alternative mast drag represented only 25 to 26 percent of overall vehicle drag at zero vehicle yaw angle, depending on the vehicle draft. With air ejection, the drag that was contributed by the alternative mast was slightly higher, being between 27 and 31 percent of overall vehicle drag at zero vehicle yaw angle, depending on the vehicle draft. This indicated that although there was a small penalty of increased drag by using air ejection, there was still a large reduction in overall vehicle drag relative to the existing mast.

The overall drag of the vehicle with the alternative mast increased at nearly twice the rate of the existing mast as the vehicle yaw angle increased. In fact, above yaw angles of from 3.2° to 3.6° (depending on the draft) the drag of the alternative mast was higher than that of the existing mast; see Table 1. It should be noted however, that with the existing mast (M1) cross-flow angles greater than 2° are not maintainable during steady turns, therefore the increased drag of the alternative mast over the existing mast at higher yaw angles is not a performance handicap.

The secondary goal of this project was to use the alternative mast to control vehicle roll during turns. It was found that the basic alternative mast configuration (with no air ejection or flap) provided too much counter-roll for all three drafts tested. This was expected, since the mast was designed such that the lower portion alone was to provide sufficient counter-roll to eliminate the roll moment. The draft at which zero roll moment occurred, D0, was found to be about 85 cm, which is within 10 mm of its design location, that is, slightly above the ventilation fence that separated the lower portion of the mast from the upper portion.

With air ejected from the upper portion of the alternative mast, it was found that the magnitude of the overall vehicle roll moment was lower than with either the basic alternative mast or the existing mast. Although one would expect that the roll moment would increase in proportion to the length (span) of mast that was submerged, in fact, the magnitude of the roll moment at the deepest draft (D3) for air ejection was found to be less than the magnitude of the roll moment at the shallowest draft (D1), for both the basic alternative mast and the existing mast. Because the experimental set-up could not provide separate control of the air exiting from the two sets of holes (on the port and starboard sides of the mast), a thin, evenly-distributed air sheet (of the same thickness) was not always observed during testing, so the observed behavior of the air ejected from each set of holes was neither what was desired nor considered ideal. However, these results show that air ejection can be used to control both the overall vehicle lift force and roll moment, at any draft.

The results of the flap configuration tests indicate that the 25-percent-of-chord partial-span plain flap included on the lower section of the alternative mast was adequate to provide vehicle roll control. However the flap deflection required (to counter-act the keel-induced roll moment) for a given vehicle yaw angle was not unique, but depended on vehicle speed. As a result, in order to eliminate the roll moment, different flap angles would be required for each vehicle speed. Therefore, it would be necessary to incorporate the roll control via mast flap deflection into a portion of the active roll-control system on the *DOLPHIN*.

Table 5 and Figure 25 summarize how the location of the "pivot axis" and the centre of effort (COE) vary with mast configuration. The combination of the hull and keel alone (no mast, M0) has a pivot axis that is 0.6 m ahead of the leading edge of the keel. The addition of the existing mast (M1) moves the pivot axis rearward by 0.1 m, and by 0.4 m for the alternative mast. The draft, as expected, has negligible influence on the longitudinal location of the pivot axis, nor does the ejection of air.

The combination of the hull and keel alone (no mast, M0) has a COE which is below the bottom of the hull. The presence of the existing mast (M1) raises the COE by 0.4 m for draft D1, by 0.5 m at draft D2 and by 0.6 m at draft D3. The effect of the ejected air is to lower the COE at each draft.

Table 5 contains a summary, for each configuration, of the longitudinal location 'XNF' of the "pitch axis". With either the existing mast or the alternative mast, the farthest forward location of the "pitch axis" is about 0.1 m (about 0.2 hull diameter) aft of the leading edge of the keel; the farthest aft location of the 'pitch axis' is about 0.7 m (about 1.2 hull diameter) aft of the leading edge of the keel. With air ejection, the location of the "pitch axis" moves aftward.

Since the *DOLPHIN* vehicle is under continuous development by ISER, it is to be understood that the reported loads were measured for one particular configuration of the *DOLPHIN* vehicle, and not for the other versions of *DOLPHIN* that are available from ISER. Thus the conclusions and recommendations apply only to the configurations that were experimented and not to any version of the *DOLPHIN* that has been developed subsequent to these experiments.

ACKNOWLEDGEMENTS

The towing tank experiments were performed in late 1997 and early 1998 and were the subject of a masters thesis by the third author (Butt, 1999), under co-supervision of the primary authors. The fabrication of the model, and use of the experimental facilities, were primarily co-funded by IMD, MUN and ISER. A small amount of funding was provided by the Natural Sciences and Engineering Research Council, Canada, NSERC. The frontispiece image was graciously provided by Geo-Resources Inc. of St. John's, Newfoundland.

REFERENCES

1. Butt, M-A. (1999), 'Hydrodynamic Improvements for the *DOLPHIN*: A Surface-Piercing Semi-Submersible AUV', Thesis for Master of Engineering, Faculty of Engineering and Applied Science, Memorial University of Newfoundland.
2. Butt, M-A., Bose, N., Datta, I., Williams, C.D. (1998), 'Mast/Snorkel Tests for the *DOLPHIN* Semi-Submersible AUV', 25th American Towing Tank Conference, Iowa City.

3. Watt, G.D., Seto, M.L., Brockett, T. (1997), 'Hydrodynamic Considerations for Semi-Submersible Minehunting Vehicles', 4th Canadian Marine and Structures Conference, Ottawa.
4. Mackay, M. (1993), 'A Review of Sting Support Interference and Some Related Issues for the Marine Dynamic Test Facility (MDTF)', DREA report 93-107.
5. Rothblum, R.S., Jeffers, M.F. and Smith, R.P. (1976), 'Methods for Controlling the Sideforce on Surface-Piercing Hydrofoil Struts', DTMB report SPD-700-01.

AFFILIATIONS

1. National Research Council Canada, Institute for Marine Dynamics, St. John's, Newfoundland, Canada
2. Memorial University of Newfoundland, Faculty of Engineering and Applied Science, Ocean Engineering Research Centre, St. John's, Newfoundland, Canada
3. Instrumar Limited, St. John's, Newfoundland, Canada
4. International Submarine Engineering Research, Port Coquitlam, BC, Canada.

Discussion

Edward M. Lewandowski, Member

Congratulations to the authors on a very complete presentation of an ambitious experimental program. The unique data concerning the effects of air ejection constitute a valuable contribution to our knowledge of submersible performance.

My comments concern the discussion of the "pivot axis" beginning on Page 20 of the paper. The authors define the pivot axis, or the "axis about which the vehicle pivots when in a turn," as the axis about which the yaw moment vanishes when the vehicle moves at a drift angle. This is incorrect. The "pivot point" is defined in maneuvering as the point (i.e., vertical axis) at which the local *lateral velocity component* vanishes (see sketch below):

$$v(x) = v + xr = 0$$

where v and r are the lateral and yaw angular velocity components, respectively. Solving for x, we obtain:

$$x_P = -v / r$$

Using the linearized yaw-sway equations, we can find x_P as a function of the linear stability derivatives and rudder force and moment rates in a steady turn (this expression is good for wide turns in which nonlinear terms are negligible):

$$\frac{x_P}{L} = -\frac{N'_\delta (Y'_r - m') - Y'_\delta (N'_r - m'x_G')}{N'_v Y'_\delta - Y'_v N'_\delta}$$

using the standard maneuvering nomenclature. Thus knowledge of the yaw-induced side force and yaw moment rates is necessary to compute the location of the pivot point. Note that the quantity

$$\frac{N'_v}{Y'_v}$$

which could be described as a "longitudinal center of effort," does not appear in this expression, and in fact has no physical significance, since the yaw moment contains a pure couple (the destabilizing "Munk moment") mitigated by real-fluid effects, which generally act near the stern, as well as by fin forces.

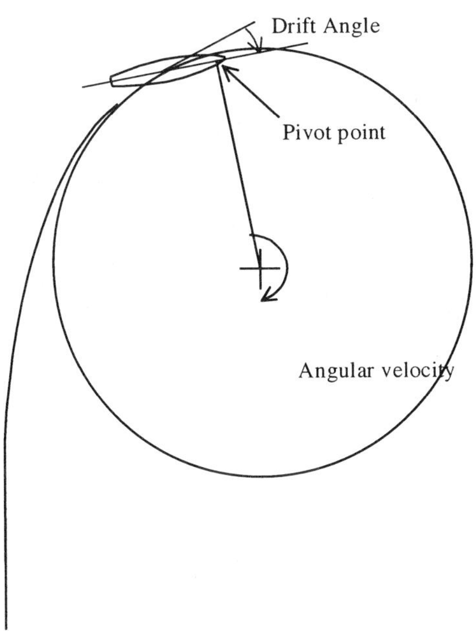

Sketch of pivot point location in a turn

Meyer Nahon, Visitor

The authors are to be congratulated on this insightful paper relating to the mast design of the *Dolphin* semi-submersible.

The *Dolphin*'s performance is itself impressive primarily because of its mast, which allows it to use a powerful diesel engine, and the extremely high energy density of a liquid fuel. However, much like the tether of a remotely-operated underwater vehicle, the mast is both a great advantage, as well as a dominant determinant of the vehicle's motion. This paper addresses the very important issue of the performance tradeoffs inherent in various approaches to mast design. It also gives enough systematic data to help a vehicle designer determine mast placement and geometry. In their discussion, the authors have presented a very balanced perspective of the advantages and disadvantages of each of the mast modifications considered.

As with any good piece of work, it inspires questions based on the results shown. For example, the vertical force and pitch moment variation with yaw angle are considered at some length, but it is unclear how the authors perceive this point. Would a vehicle with zero normal force and pitch moment be optimal? Or would, perhaps an upward force and nose up pitch moment be more desirable for fail-safety?

Much of the paper confines itself to the operational regime where the magnitude of the cross-flow angle is less than 2 degrees because the existing vehicle is confined to this regime. It would be interesting for the authors to comment on whether this is due primarily to mast effects, or whether it is due to other reasons. This raises the question of whether an improved mast might allow a wider operating range.

By the end of the paper, one is left with the impression that, perhaps, an ideal mast design might be a combination of the existing mast (M1) and the alternative mast (M2). In particular, using a fixed lower mast of the same size and shape as in M2, combined with an upper mast composed of swiveling sections as in M1, but using the cross section of M2. This would result in minimal rolling moment in all operation; would avoid the complexity of air injection or an active flap; and would remain safe in operation at higher cross-flow angles.

Finally, it should be noted that, in Table 1, the data of the third and fourth columns ('A' and 'B') should likely be interchanged so as to correspond to Figure 10.

George D. Watt, Visitor, Defence Research Establishment Atlantic

This paper describes a novel method to reduce lift on a streamlined fairing piercing the free surface in a practical engineering application. Much innovative work has been done and an important contribution has been made. I concentrate my comments on the results associated with the primary focus of the work.

1. I find your principal conclusion, that the rolling moment on the top section of the new mast is effectively negated by the air ejection, to be compromised by the lack of any error analysis. Your conclusion is based on the data shown in Figures 11 through 15, and on the slope of a straight line fitted through only the yaw = ±2.5 degree data in Figure 15. Because force magnitudes are small at low yaw angles, this is the least accurate of your data from which to extract

lift curve slopes. The higher yaw angle data are excluded from the fit because they do not follow the linear trend set by the low yaw angle data. You suggest the reason for this is that above 4 degrees of yaw the air cavity formed by the ejected air has partially collapsed. This seems reasonable, but another explanation for the inconsistency is the variability in the rolling moment data in these figures and, in particular, the inherent error in the low yaw angle data.

What is the inherent error in your data, and how does this error translate through to error estimates for the fit parameters listed in Table 2?

You state that less accurate, low velocity data are excluded from the fit; are these discarded data shown in these figures?

2. In item (c) on page 14 you state: "...zero roll moment was achieved over a wide range of...vehicle yaw angles." I suspect this is a mistake as you have previously restricted this conclusion to the ±4 degree yaw angle range.

3. Your data do show that air ejection has some lift reduction effect, and this is worth pursuing. Before implementing it as you have in a vehicle application, it seems worthwhile to first optimize the air ejection technique on a simple two-dimensional section, perhaps at a free surface, measuring section forces only. I do not think that air ejection is viable if its beneficial effects are restricted to very small yaw angles as is currently the case. The main purpose of a semi-submersible is to provide a stable operating platform in rough weather. This means the mast hydrodynamics must be robust, predictable, and preferably linear at the free surface in the presence of wave action that will subject the mast to high angles of attack.

C. D. Williams, Visitor

This paper reports the results of a series of careful and thorough experiments on a scale model of the *Dolphin* AUV. This pioneering autonomous vehicle has a great advantage over battery-powered AUVs in both its power and its energy capacity. But this advantage comes at a price—the need for a snorkel. Here, the authors show how incremental design changes to the mast can result in improved performance through lower drag and improved stability during turns.

As few AUV developers have access to the superb facilities at IMD used for this series of experiments, I had hoped that the authors could have drawn out generic observations and results from their work. Such generic results might well have benefited groups in the UK and Europe working on surface piercing AUVs. Unfortunately, as they conclude in their last sentence, the work can only be applied to the particular configuration of *Dolphin*, as tested. Is there any hope that through combining tank and computational simulations on a 'typical' body, the need for the extensive tank testing described here can be reduced?

Two issues came to mind on reading the introductory paragraphs: How was the design of the alternative mast optimized? And were numerical simulation experiments performed before embarking on the tank trials to assist with experiment design?

As a seagoing oceanographer, I'm interested in the authors' views on the robustness of the paper's results and conclusions to the physical processes in the upper ocean; processes that were not simulated in the experiments. For example: the wave climate; the combination of wind and waves, related to the course of the vehicle; the current shear over the depth of the mast. In such circumstances, the vehicle could encounter cross-currents resulting in cross-flow angles of over 3°. What effect might this have on the 'optimum' mast design?

To me a key conclusion is that active control of the flap angle might eliminate the keel-induced roll moment. Given the dynamic characteristics of the vehicle and typical wave and current fields, might we expect the vehicle to cope with the short (wave) period perturbations, as well as the longer (turn) periods?

Finally, how might changes in the bulb design (for example, to accommodate different payloads) affect the results? Are the results at all applicable when towing the Aurora body?

Authors' Closure

First of all we would like to thank the discussers for their thoughtful comments and input on our paper.

Mr. Griffiths raises the important point as to whether the results from these tests can be generalized to other vehicles. The simple answer is that these tests were designed to provide a specific answer concerning the mast design of the *Dolphin* vehicle and it is not easy to draw general conclusions for other vehicles. A way in which more generic results might be approached would be to test hull geometries in isolation of the mast and keel and to compare results with CFD (RANS) predictions for the same bodies. Bodies with differing levels of parallel middle body might be tested and comparisons made to gain confidence with numerical predictions. Results for further geometries might then be calculated using CFD simulations alone. However, this was not the aim of the present tests.

In this work no full numerical simulations were done prior to testing. In addition, there was insufficient experimental time available to optimize the mast design. However, estimates were made of the overall mast performance prior to design and construction.

Concerning the wave climate encountered, the aim of the air bleed system design was to reduce the sensitivity of the surface piercing portion of the mast (in terms of side force or lift production) to changes in immersed area in order that the lift from the mast would be insensitive to waves. The tests show that this design aim was only partially successful due to difficulties in control over the air bleed system. No tests were done in sheared cross currents and no predictions were made for this condition. However, again a design of mast which is insensitive to angle of attack changes (such as an effective air bleed system) would perform well in these conditions.

One of the conclusions of the work was that effective control of the vehicle would be more directly achieved through active control of the mast flap. However, this control is at the cost of a more complex control system and a further drain on the on board energy sources.

Dr. Watt's comments are interesting and support several of the conclusions of the paper and the views of the authors. The air bleed system does have merit, but it is clear that further design work is necessary to achieve full effectiveness. Even at the design stage, it was known that some of the limitations were the result of geometric and functional restrictions on the design which it was not possible, or difficult, to circumvent. Some of the data will be analyzed with a view to assessing the levels of error in the tests.

A recent re-analysis of the data of Figure 10 now includes error bars for the measured loads; see Figure 26 below. From each time-series of a measured load we computed not only the mean value but also the standard deviation. These new plots, with one-sigma error bars plotted above and below the mean-value data point, show that there is no substantial overlap of the error bars with the result that the curves in Figure 10 are indeed distinct curves. For example, at 10 degrees yaw the standard deviation of the normalized drag force is about 4 units; in the range of ±3 degrees the standard deviation is typically 2.8 units. These new results reinforce our conclusion that at vehicle yaw angles of less than 3 degrees the drag of the alternative (fixed) mast is lower than the drag of the existing (swiveling) mast.

Dr. Nahon suggests that an improved mast might result from a combination of the new mast base section with an existing style freely rotating upper section. However, one of the features of the new mast design was to reduce the spray drag of the surface piercing section and this was successfully achieved with the new design. In addition, the freely rotating mast sections in the existing design are often misaligned with the actual flow in practice. On reflection, our own conclusions are tending towards the idea that more design effort should be spent to perfect the air bleed system as it is this system that when working effectively can be tuned to supply minimal lift (side force) over a range of conditions.

Dr. Nahon poses the question "Would a vehicle which operates in a condition of zero normal force and pitch moment be 'optimal'?". The results concerning vehicle normal force and vehicle pitching moment are provided so that the designer (or operator) can see how *Dolphin* behaves by itself and with a tow-cable. There is no obvious need for the vehicle to be designed to operate in a 'zero normal force and pitch moment' condition since the pitch controller will need to be tuned differently for these two operating conditions and for the corresponding range of forward speeds.

Dr. Lewandowski's comments on pivot axis are noted. This may be a problem of terminology and we believe that the paper clearly states what is meant by the center described. In this paper we have identified three pivot centers (axes):

a. "yaw moment null axis" as computed from the ratio of the measured vehicle yaw moment to the measured vehicle side force; this is the quantity "XCP" in the paper;
b. "roll moment null axis" as computed from the ratio of the measured vehicle roll moment to the measured vehicle side force; this is the quantity "ZCP" in the paper;
c. "pitch moment null axis" as computed from the ratio of the measured vehicle pitch moment to the measured vehicle normal force; this is the quantity "XNF" in the paper.

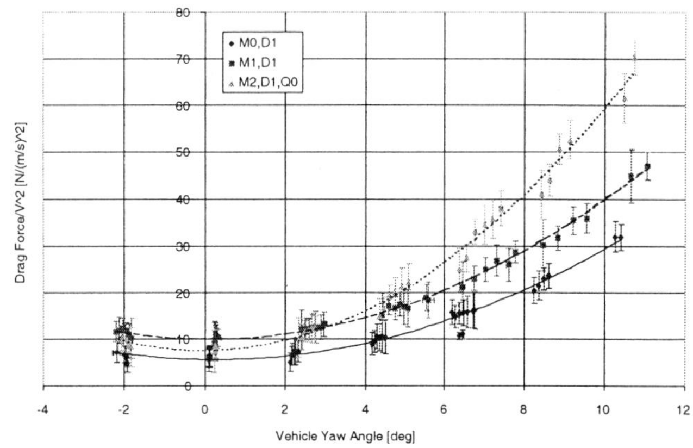

Figure 26 Error bars for normalized drag force vs. vehicle yaw angle

Dr. Calisal observed that the results presented were for the vehicle in calm water, and for large-radius turns in calm water the maximum vehicle yaw angle may indeed be limited to 2 or 3 degrees. In practice, in higher sea states, the vehicle can experience quite large effective yaw angles due to the water particle velocities and Dr. Calisal and his students have published wind tunnel results for the *Dolphin* vehicle for yaw angles up to 20 degrees. Thus he would be interested in obtaining additional results at larger vehicle yaw angles; unfortunately in these experiments the results were limited to maximum vehicle yaw angles of about 11 degrees and to maximum flap angles of about 12 degrees.

SNAME *Transactions*, Vol. 108, 2000, pp. 397–418

Responsibility Sharing Management (RSM) of Ship Safety

Chengi Kuo, Life Member, University of Strathclyde, Glasgow, UK, and **H. Paul Cojeen,** Member, US Coast Guard Headquarters, Washington, DC

ABSTRACT

The paper presents a new approach for treating ship safety entitled Responsibility Sharing Management (RSM). The text begins by outlining the importance of safety in maritime activities and in the context of meeting a specific goal. The prescriptive regulatory approach is described by detailing its emphasis, merits and drawbacks. Special attention is given to the main weaknesses associated with the devolvement of responsibility and the absence of a management system. Next, a brief outline of the goal-setting approach, based on application of the safety case concept, is given together with its use, merits and drawbacks. On the basis of this review and lessons learnt, a set of desired safety features for a ship is presented. The principle of the RSM approach is then explained and its key features are examined. To demonstrate how the RSM approach can be used effectively to improve safety standards in marine activities two examples are used. The main conclusion is that this balanced and flexible RSM approach incorporates a robust management system and provides scope for collaboration by all stakeholders while blending novel features with the strengths of the existing approaches.

INTRODUCTION

Everyone associated with maritime activities such as shipbuilding, shipping, ship repair and offshore work recognises the importance of safety and it has received even more attention following a number of major accidents and tragic disasters during the past two decades, see, for example Refs. [1] to [5]. In recent times it has been recognised that safety is not simply a technical matter but one that is extensively affected by what are called human factors, involving human behaviour, actions, decisions and violations, [6]. It would be helpful to consider safety within the context of an activity that must meet a specific goal, such as the transport of people and cargoes, or doing work offshore. More generally, the goal of a provider, such as a ship operator or shipbuilder or insurer of shipping, is as follows, see [6]:

> "To be competitive in meeting the client's specification with solutions that are cost-effective at an acceptable level of safety."

Competitive relates to profit and quality. Competitiveness is defined by the client and will be affected by various forms of competition. There are many types of *client*. For example the clients of a shipbuilder are likely to be ship owners/operators; and their clients in turn are freight transporters and passengers. *Specification* describes the work to be done, or services to be provided and deadlines to be met. *Cost-effectiveness* implies value for money. The determination of *an acceptable level of safety* will constitute the main emphasis of this paper. It should be noted that everyone is responsible for safety and this responsibility must be shared and managed if success is to be achieved.

The paper begins with a definition of the term *safety*. The prescriptive regulatory approach to safety and its practical application are then considered.. This is followed by an examination of the goal-setting approach using the safety case concept. The merits and drawbacks of both approaches are compared. Finally, a new approach is presented that incorporates the strengths of both. This new approach is called Responsibility Sharing Management (RSM) of Ship Safety. RSM is based on the idea that the best results will be achieved when the stakeholders share responsibility for safety. Application of RSM for dealing with ship safety is demonstrated with the aid of two examples.

SAFETY AND ITS UNDERSTANDING

Safety is a very broad concept and, while everyone appreciates its importance, understandings of the meaning behind the actual term tend to vary widely. Those with engineering training tend to believe safety is a technical matter and their publications make little mention of features such as management or human factors. Data for marine insurance claims collated over successive five-year periods since 1987 show that human error is directly accountable for about 60% of te accidents and is indirectly responsible for an additional 30%, see [7] for further details. Other studies on the importance of human factors in maritime safety can be found in [8] to [11]. The safety definition given in [12] states that:

> "*Safety* is a perceived quality that determines to what extent the management, engineering and operation of a system is free of danger to life, property and the environment."

There are three points worth noting here. Firstly, safety is not something absolute, but rather a quality which has to be specified and can be continuously enhanced over given periods as a result of increased experience. Secondly, perception of safety depends on the actual circumstances and the judgement, competence and experience of those involved in the situation. Lastly, because the concept embraces so many factors, the subject has to be dealt with in a systematic way.

THE PRESCRIPTIVE APPROACH TO SAFETY

The principle of this approach, also called the prescriptive regulatory approach, is very straightforward and well understood. It can be stated as follows:

> "One party prescribes the safety requirements in the form of rules and regulations to be obeyed, and the other parties comply with them often unthinkingly."

The prescribing party is usually the government of a country or its representative, or an international organisation in which a number of governments participate. The obeying parties are those who wish to provide the general public and other clients with products or services

The development and implementation of a set of prescriptive regulations involve the following main steps:

Step 1: RESPOND to a need such as a major disaster.
Step 2: SEARCH for suitable existing regulations.
Step 3: DISCUSS with parties interested in the specific circumstances.
Step 4: DEVISE a set of regulations.
Step 5: IMPLEMENT new regulations and publicise to potential users.
Step 6: OBTAIN three main responses:

6.1 – compliance
6.2 – misinterpretation
6.3 – complaints

Step 7: MODIFY changes may be made in the light of feedback.

The relationship between these steps is illustrated graphically in Figure 1.

The main emphasis of the prescriptive regulatory approach is to ensure that regulations are obeyed. Modifications are usually done with reluctance because in many cases these regulations have to go through a statutory process e.g., obtaining parliamentary approval in the United Kingdom or other parliamentary government or, for some countries, prescribed regulatory procedures, before implementation.

MERITS AND DRAWBACKS OF THE PRESCRIPTIVE APPROACH

The prescriptive regulatory approach has both merits and. The principal advantages of this approach are as follows:

- Reference standards for everyone.
- Incorporation of experience from accidents.
- A straightforward concept that is readily understood.
- Helpful to those who are new to a safety-critical activity.

The drawbacks can be grouped under two headings, *Major Weaknesses* and *Other Drawbacks*.

Major Weaknesses

- Devolved Responsibility: Responsibility for safety is vested in the prescriber, and the user complies with the devised regulations. When an accident or failure occurs, the "fault" lies with the prescribing authority who will, as a result implement more stringent regulations. Because safety is both underpinned and undermined by human factors, however, it is impossible to anticipate all the likely outcomes of applying the regulations. This, therefore, is not a satisfactory way to deal with safety and further discussion of the matter will be found in the next Section.

- Lack of a Management System: As the prescriptive regulation diagram shows, see Figure 1, the emphasis is on ensuring compliance, but there is no management system to measure performance and review experience before it is fed back into policy formulation. Furthermore, the prescriptive regulatory approach tends to be "reactive" to ɔecific events. Further discussion of the management system will be found in a later Section..

- Assumed "Infallibility" of Regulations: It is usually assumed that once a regulation is published and implemented it must be correct, and those who question of its correctness are looked on unfavourably and, indeed, considered as trouble-making. Bearing in mind the impact of human factors on safety, this is not a healthy situation. Robust and continuing debate is necessary in order to ensure that the best possible regulations can be derived.

- Cost to the User: For the reasons already discussed, prescribers tend to take no account of the cost of implementing a set of regulations. It is not considered to be part of their job, as their primary role is to "safeguard the well-being of society". This, however, is a very unhelpful attitude, since users' competitiveness needs to be taken into consideration if a high standard of safety is to be achieved.

Other Drawbacks

- Dealing with New Developments: The approach tends to provide answers before all the questions have been posed. The ship operator has to satisfy many rules and regulations before it has been ascertained whether some of them are applicable.

- Problems with Keeping Up-to-Date: The formulation of rules and regulations tends to lag far behind advances in technology. It can also be ineffective to apply the existing versions of regulations in new projects involving novel features.

- Scope for Innovative Treatment: Prescriptive regimes tend to inhibit innovative solutions. Indeed the existence of regulations can deter designers from proposing novel solutions for clients' requirements.

- Possibility of Imbalance: Regulations are strongly influenced by practical events, and after a major disaster, changes and increased stringency can be expected in areas that do not appear to offer a problem.

For further discussion, see [6].

THE QUESTION OF RESPONSIBILTY

In simplest terms responsibility means liability. Since, in the case of the prescriptive regulatory approach, the prescriber carries full liability for safety, operators or users of a system only have to meet the prescriptive requirements detailed in the regulations. This is often done without any real thought as to whether the regulations are correct or if they are relevant to a specific situation. Users know that even if an accident occurs they can still argue that they met the relevant regulations, and for that reason should not be liable. In other words they cannot be blamed for the accident. However, even though the prescriber devises regulations using the best information and experience available, these regulations cannot cover every eventuality, because human factors are always involved. Regulations, therefore, are not adequate for all situations.

It is not unusual for a designer to meet a given regulatory requirement while having serious doubts about the scientific correctness of the regulations in question. For example, a fishing vessel may fully meet the stability requirements and yet capsize while another vessel may operate satisfactorily, even though it does not "meet the regulations".

THE ROLE OF A MANAGEMENT SYSTEM

Most people adopt a form of management system for doing a task, but very few would describe this process as "applying a management system". An explanation of what is meant by the term *management system* would be helpful. Although many definitions of *management* and *system* are available, few can be found for *management system*. A concise one can be stated as follows:

> "A process that co-ordinates various activities and resources in order to meet an objective, aided by the presence of feedback."

It is important to recognise that there has to be "feedback" in order to ensure that continuous improvement takes place.

An example on making travel arrangements will be used to illustrate the basic concept. This could be applied by someone who wishes to explore alternatives to his or her normal mode of travel or is making a particular journey for the first time. In this example the objective is to travel to a location in another city for a project meeting involving all the contractors, at a given time in the morning of the day specified. To ensure a successful arrival involves examining the various transport options, such as bus, plane or car. The schedule, cost and convenience would all be taken into account before a decision is ma . The preferred option, e.g., air travel, is then implemented. Before and during the journey checks are made about departure and arrival times, and interface with other modes of transport at the destination. Once the trip is completed, and certainly before the next one is undertaken, a review of the procedure would be done and notes compared with colleagues. This would determine the success of the selected option and assist in selecting the best one for getting to the next meeting. *Review* in this case might include considering the location and timing of the meeting itself.

The key elements of a typical management system are as follows:

Element 1: DEFINE the objectives of an assignment, project or policy.
Element 2: ORGANISE the resources, human and financial, to ensure that the objective can be achieved.
Element 3: IMPLEMENT the various decisions made in practice.
Element 4: MEASURE performance and how fully the objective has been achieved
Element 5: REVIEW the results obtained and experience gained before feeding them back into Element 1 for future refinement.

Since this is an iterative process, the relationships of these elements can be illustrated by means of a revolving circuit, see Figure 2.

It would be useful to note two points. Firstly, in management activities, unlike most scientific ones, there are no absolute correct or incorrect answers. The principal method of measuring "success" is to assess how well the objectives have been met or the policies have been fulfilled. Secondly, the process is an iterative one because it is possible to define either undemanding or very ambitious targets, and it will take time before the correct balance can be achieved.

Management systems are needed in an organisation so that everyone can focus their effort towards the same objectives. As a result, these systems have become formalised. For example, application of Total Quality Management Systems [13].

In the context of safety it is essential for an effective Safety Management System (SMS) to be also on a revolving circuit and to incorporate *Measure* and *Review* elements. Thus, *measurement of performance* and *review of results* and *experience* become a positive and "natural" process with a strong emphasis on improving safety standards. This in turn avoids the negative effects of auditing solely for compliance.

THE GOAL-SETTING APPROACH

The aim of the goal-setting approach is to seek an appropriate level of safety. This approach is based on the application of the safety case concept which, in turn, is derived from application of the principles of systems engineering. The emphasis is on dealing with the safety of systems for which little or no previous operational experience exists. In such a situation it would be logical to seek answers to the following questions:

Q1: What aspects of this system could go wrong?
Q2: What are their chances/effects of going wrong?
Q3: How can these chances and effects be reduced?
Q4: What should be done if an accident does occur?
Q5: How can safety be managed?

In scientific terms these five questions respectively become the following elements:

Element 1: Hazard identification
Element 2: Risk assessment
Element 3: Risk reduction
Element 4: Emergency preparedness
Element 5: Safety management system

The relationship of these elements is illustrated in Figure 3, and further details an be found in [6].

The safety case concept was first used in the nuclear power industry when there was little or no past experience of the operation of nuclear installations. The responsibility for conducting operations in a way that was safe for the public and employees fell on the shoulders of operators and/or owners. Any approach selected required the operator to deal with safety from first principles and hence the five basic questions. The second industry to select this approach was the UK chemical industry in which it was introduced in 1984 via the Control of Industrial Major Accident Hazards (CIMAH) Regulations, see [14]. Following the explosion on the offshore installation *Piper Alpha* in the North Sea in 1987 a public inquiry under Lord Cullen recommended that the offshore industry should embrace the safety case concept, see [2], and this was approved by the UK Government. The shipping industry was attracted to the concept but had difficulty in adopting it. The main reason for this was that shipping is an international activity unlike nuclear, chemical and offshore activities which can fall under the jurisdiction of a single government.

Merits

The goal-setting approach has three merits:

- The Presence of a Safety Management System (SMS): This approach includes a clearly-structured SMS which ensures that all hazards are controlled within the tolerable or negligible regions.
- Focus on Significant Hazards: Safety effort is focused on the most significant hazards.
- Suitability for New Projects and Activities: Because the approach starts from fundamental principles it can also be applied in activities such as ship design and ship production.

Main Drawbacks

The following are the main drawbacks of the goal-setting approach:

- Responsibility for Safety is Vested in the Provider: With this approach the responsibility for safety is vested in the provider, with the prescriber having a passive role. This is equally as undesirable as its opposite, the prescriptive regulatory approach, because all the responsibility for safety is in the hands of a single party.

- Possibility of Inconsistent Application: It would be very difficult for all the parties concerned from many different countries to derive the same level of safety, and this lack of consistency is unhelpful in an international activity such as shipping.
- Difficulties Linked with Using Risk Methodology: The term *risk* is one that is poorly understood, and for this reason the safety case concept can be incorrectly applied. For example, many people use *risk* and *hazard* interchangeably, as if they had the same meaning.

DESIRED FEATURES OF A NEW APPROACH

There is a need to devise an approach that will overcome the basic difficulties and incorporate most of the desired features from both approaches. This approach should be balanced and minimise drawbacks. The desirable features for such an approach can be summarised as follows:

- Clearly-defined goals: The first requirement is to have clearly defined safety goals, policies to be implemented, a standard to be met and a positive safety culture to be developed. Each of these needs to be effectively communicated to every member of the organisation.
- Shared responsibility: All stakeholders should share responsibility for safety. Where possible, they should work as a team to derive the most appropriate solution for a given situation.
- Focusing on significant hazards: The effort required to manage the risk levels should be proportional to the significance of key hazards identified, i.e., more attention should be given to the more serious hazards.
- Greater understanding of practical issues: When safety solutions are adopted, due weight should be given to their practicability in terms of cost-effectiveness and likely benefit to all parties including the users, rather than offering an undefined "benefit to the quality of life".
- Robust safety management system: A robust safety management system should be set up to define safety objectives, organise resources, assign responsibilities, ensure stakeholder collaboration, control operations, actively measure performance and review results and experience so as to feed them back for continuous improvement.
- Transparency and clarity: It is important that every step in the safety management methodology should be both transparent and readily understood by all stakeholders.

PROPOSING THE RSM APPROACH

To meet the criteria outlined above, there was clearly a need for fresh thinking. This led to the development of the Responsibility Sharing Management (RSM) approach which has been applied to ship safety. The basic concept behind the approach is the recognition that effective management, i.e., the co-ordination of activities and resources to meet an objective, requires the participation of stakeholders, and RSM is one way of achieving this. In the RSM approach, once the safety goals have been defined and resources organised, collaboration with stakeholders then takes place to identify hazards; assess their significance; to reduce risk levels as appropriate from the intolerable to the tolerable risk level. Since a management system is an integral part of RSM, continuous improvement enables stakeholders to share responsibility for every aspect of safety.

The principal elements of the RSM approach will now be outlined and their relationships with each other are illustrated in Figure 4.

Element 1: DEFINE the safety goals and policies.

Element 2: ORGANISE the resources - human and financial - so as to ensure the safety goals are met and policies are fulfilled. Tasks coming under this heading include assigning responsibilities to different individuals and developing communication mechanisms.

Element 3: COLLABORATE with all stakeholders to address safety issues requiring attention. The key steps are:

Step 3.1: *Identify* hazards in the system, activity or project by appropriate methods such as "What If?", "Brainstorming" or "HAZOP".

Step 3.2: *Assess* the risk level of the identified hazards by qualitative and quantitative methods with the aim of establishing their significance and placing each hazard in one of the following regions: intolerable, tolerable, or negligible.

Step 3.3: *Reduce* intolerable risk levels of hazards through a combination of management, engineering and operational methods, addressing both consequence and probability of occurrence. If cost-effective, reduce tolerable risk levels in a similar way.

Step 3.4: *Prepare* for emergency situations by giving thought to how to move people to safer areas and evacuate them, and devise ways of rescuing them.

Element 4: IMPLEMENT the output from the previous element. Implement this in practice so as to ensure that safety goal and policies are satisfied. It is possible to opt for prescriptive regulations for routine tasks, guidelines for situations that require more flexible treatment, and other methods if they are more appropriate.

Element 5 MEASURE all aspects of safety performance objectively and actively. Devise a reporting mechanism to assist data collection and use.

Element 6 REVIEW the results obtained, experience gained, and lessons learnt. This information is then fed back into Element 1 DEFINE so that goals and policies can be refined.

KEY FEATURES OF THE RSM APPROACH

The Responsibility Sharing Management approach has a number of merits. Key ones will now be examined under separate headings.

Developing a Positive Attitude

The prescriptive approach has been familiar to all of us since childhood. In the early days parents and school teachers prescribed codes of behaviour. The prescriptive regulatory approach is based on the DAD principle (Devise-Apply-Defend), i.e., the prescriber devises regulations, publishes them and defends the decision regardless of whether they are right or relevant. There is little provider involvement and the process tends to generate negative attitudes. In practice, companies that implement safety measures have only a cursory involvement in regulation formulation. This means that the regulatory bodies have to exercise "heavy" control in order to ensure compliance. The RSM approach, on the other hand, encourages the collaboration of all the stakeholders. This is a more effective way forward. Thus, in place of the DAD principle, RSM encourages the joint application of the MUM principle (Monitor-Understand-Manage), which means that all stakeholders can jointly monitor a situation and understand the issues involved before devising ways of effectively managing safety. This approach will greatly assist in the development of positive attitudes.

Responsibility for Setting Safety Standards

It is important to understand that in the RSM approach the emphasis of *responsibility sharing* is on jointly arriving at decisions on safety issues. This feature will be illustrated later in two case studies. It is different from taking responsibility for complying with safety standards and regulations prescribed by others, e.g., with the International Safety Management (ISM) Code, a shipping company has to devise its own safety management methods to handle the responsibility for complying with prescriptive requirements, see [15]. In that case it is responsibility for compliance with prescriptive regulations (with all the drawbacks indicated earlier) that is being implemented.

A Robust Management System Adopted

A fundamental weakness of the prescriptive regulatory approach is its lack of a management system and the placing of emphasis on compliance with regulations once they have been established - regardless of whether or not they are appropriate for a given situation. RSM is different as it has a robust management system which allows safety performance to be measured objectively. Results and experience are also regularly reviewed and benchmarked before they are fed back into the DEFINE element. In this way continuous improvement becomes the norm, which is essential for safety issues because of the predominance of human factors in safety.

Flexibility in Implementation

The RSM approach does not specify how a safety standard should be implemented in practice. Instead it encourages adoption of the most appropriate one for a given situation. For example, in emergency situations the most appropriate method of implementation could be the rigid following of safety-critical procedures. However, for a new design or a new operational arrangement implementing the best method could involve a mixture of basic requirements and decisions based on collaboratively-agreed solutions or wisdom appropriate to the situation. With experience comes confidence and the likelihood that all parties will become more imaginative in future applications and be less afraid of the unknown. In this way it would be possible to set the level of safety that is appropriate to the company and its particular operations. This could lead to a reduction in the number of regulations and avoid the tendency to create additional prescriptive regulations for special situations.

Improving Design

In the RSM approach the REVIEW element considers the results obtained and the experience gained before feeding them back into the DEFINE element. In practice there is no reason why the same information cannot be given to those involved in various phases of a ship's life cycle. This would be advantageous for two reasons. Firstly, many of the activities in the shipbuilding process, e.g., design and construction, influence the safety of a vessel. Secondly, safety information is required for each phase of the life cycle, and RSM is the most appropriate way of providing this.

Assisting Effective Risk Management

For many marine activities, success is critically dependent on the ability of the organisation or individual to manage effectively the risk levels of hazards. In the RSM approach, Step 3.3 of Element 3 *risk reduction* can readily enable the provider to correlate "money spent" with "return on investment" because the methods used address the root causes of accidents. In Step 3.4 of the same Element, *emergency preparedness* can be regarded as comprising the two components of *Emergency Planning* and *Emergency Response*. *Emergency planning* would involve generating some "unthinkable" scenarios with which risk reduction cannot deal completely because of the dominance of human factors in safety, and the contingency plans could include solutions based on management, engineering or operational methods or on a combination of these. The selection of the method or methods to be used would be made after considering cost-effectiveness and efficiency.

Emergency response would address the situation when an "unthinkable" becomes a reality, i.e., the need to evacuate personnel and rescue them or to clear up an oil spill. In practice, it is extremely difficult to correlate "money spent" with "return on investment", but it is vital to allocate some resources to address prevention of the "unthinkable" situation.

By applying the RSM approach, which involves all stakeholders, it is likely that the derived solutions will not only be better than those suggested by an individual company or prescribing authority, but they will be more acceptable because of the team effort. In other words, all parties will feel they have had input in the selected solution. An additional advantage is that all stakeholders will acquire skill in risk management and become familiar with the way results can be put into in practice.

ILLUSTRATIVE EXAMPLES OF RSM USE

Two examples have been selected to show how RSM could be helpful in achieving a higher standard of safety by means of collaboration between the all the stakeholders. The cases themselves were implemented without using the notion of RSM but these examples show how this approach could be used in future projects. In each case, key features are highlighted here while further details are provided in the Appendices.

Example 1: Replacing Lifeboats by an Evacuation System

This example is concerned with the use of the RSM approach in the process of replacing traditional lifeboats on inshore passenger ferries with some other type of evacuation unit where people can disembark via various types of slide. The key features will be summarised here and further details are given in Appendix A.

Background

Traditionally, lifeboats have been installed on ships for evacuation purposes under mandatory prescribed requirements. Practical experience over many years has shown that this method may not always be the most effective solution for passenger ferries, and in particular for those operating in inshore waters. For example, lifeboats are difficult to launch and are very sensitive to adverse weather conditions. For inshore passenger ferries, where rescue services are readily at hand, it is important to explore alternative methods of evacuation. This is particularly relevant since the International Maritime Organization's High Speed Safety Code does not prescribe a specific evacuation system, see [16].

A shipping company that operated passenger ferries in the inshore waters of a European country was under pressure to be more competitive. It developed a plan to replace lifeboats with a more effective evacuation system that occupied less space and was easier to launch. Initial discussion with the country's marine administration agency was negative because the regulations prescribed lifeboats and there was no guidance on how to deviate from that requirement. Determined action by the shipping company and the prescribing authority led to discovery of a regulation in the EC Council Directive entitled, *Safety Rules and Standards for Passenger Ships* [17] concerning Class B for passenger ferries operating within 32 kilometres of a point at which survivors can be taken ashore. Under this classification, passenger vessels can replace lifeboats with dryshod evacuation methods such as a marine evacuation system (MES) with inflatable liferafts and rescue craft to marshal them. After the effectiveness and flexibility of the MES had been demonstrated, approval was given to the shipping company to use an evacuation system in place of lifeboats on their new vessel.

The RSM approach is used to show that a process can be enhanced with input from stakeholders.

Assumptions

The example assumes that the European Class B Classification is not yet in place, and an operator of passenger ferries who fulfils all the requirements of the Class is wishing to introduce the new evacuation units. It also assumes that in the prescriptive regulatory regime the maritime authorities would treat any suggestion about replacing lifeboats suspiciously. They would probably pose a series of questions such as: "What is the ship operator's motive - saving money or improving safety?" and "How is the system going to replace lifeboats?"

Practical Implementation of RSM

The six elements of RSM: DEFINE, ORGANISE, COLLABORATE, IMPLEMENT MEASURE and REVIEW, will be considered and highlighted in Appendix 1. In the COLLABORATE element the hazards associated with the use of a marine escape system (MES) and helicopters will be identified and their risk levels assessed by qualitative methods. For hazards with intolerable risk levels, management, engineering and operational methods will be used to reduce their risks. Emergency preparedness will address the "unthinkable" scenario with an emphasis on emergency planning and emergency response.

As far as the IMPLEMENT element is concerned, there are the following options:

- Mandatory Regulations to permit the use of alternative solutions in place of lifeboats, but specifying the use of approved designs only.
- Letting ship operators decide on their preferred system in the light of the COLLABORATIVE element's findings.
- Prescribers and ship operators would jointly judge the effectiveness of the evacuation method using both scientifically-based and commercial criteria. A set of guidelines would be devised for use.

Comments on the Results

Application of the RSM approach has shown that better solutions can be derived via collaboration between prescribing authorities and ship operators, and these solutions will also improve safety standards. Hopefully, efforts which were previously devoted to ensuring compliance with regulations will now transferred to seeking improved ways of rescuing survivors from the sea and protecting them from exposure to adverse weather conditions.

Example 2: Innovative Tanker Design

This example shows how an innovative tanker design could be developed using the RSM approach. The key features will be summarised here and further details are given in Appendix B.

Background

The grounding of the ***Exxon Valdez*** in Prince William Sound, Alaska, in March 1989 [3], led to the US House and Senate passing the Oil Pollution Act in 1990 [18], along with the birth of a mandatory requirement for double hull tanker designs and the setting of liability fines for oil spillage. The adverse effects and outright costs of an oil spill are such that many oil companies have opted out of the oil transportation business altogether. Others have risen to the challenge by building and operating tankers with features that far exceed prescribed regulatory requirements. One US shipping company decided to build its domestic fleet with double tankers with highly redundant propulsion, steering and electrical power generation and other critical systems.

Assumptions

This example assumes that the shipping company developed the new design in close collaboration with the prescribers (government representatives), who did not discourage them from introducing safety features above existing requirements. The envisaged collaboration in the RSM approach would also involve the ship operator, specialist equipment suppliers and the shipbuilders.

Practical Implementation of RSM

Implementation of the six elements of RSM is described in Appendix 2. In the COLLABORATE element the hazards associated with operating double hull tankers are identified and their risk levels are estimated. The focus of attention is on how risk reduction is achieved primarily by engineering methods. As far as the IMPLEMENT element is concerned, the options for application are: modified mandatory regulations, the operators being given special dispensation, or prescribers and operators together judging the effective of the system against scientific and commercial criteria - this leading to guidelines with less onerous requirements than for other tankers.

Comments on the Results

The results show that the RSM approach can be applied as a basis for collaboration between all the stakeholders, and the solution would be more attractive in both operational and safety terms.

DISCUSSION

A number of issues deserve brief discussion under separate headings.

The Need for Continuous Improvement

The effectiveness of RSM requires that the six elements be located on a "revolving circuit". This means that the first steps in the process are to define the goal, and then organise resources. This will ensure that the defined goal can be achieved. Collaboration of all stakeholders then follows. Thereafter, the most appropriate way to implement the agreed safety standard is selected. The performance of the safety standard is measured objectively, preferably by independent parties. The results are reviewed and benchmarked by the company against other standards. The findings are then fed back to the DEFINE elements. This ensures that the process of improvement is continuous, and becomes the "norm". This factor is important because safety is dominated by human factors and human beings require continual stimulation or competition in order to maintain high standards.

Introducing RSM of Ship Safety in Practice

Initial informal discussions with colleagues who work on the safety of maritime and other transportation systems regarding the concept of responsibility sharing management of safety have resulted in two main groups offering different sets of views. Those in first group tended to be very positive, and wanted to see the RSM approach actively promoted. Those in the other group did not object to the concept, but felt it was too idealistic for the real maritime world. Aspects of the RSM approach have already been implemented in practice. One example is the FishSafe project in which the offshore oil and fishing industries are working together to improve fishing efficiency near oilfields in the North Sea while minimising the risk of fishing nets being snagged by pipelines and subsea equipment.

It can confidently be stated that the RSM approach can greatly assist the implementation of such collaboration and ensure its continuous enhancement.

The Role of Education

A number of the ideas outlined here will be "new" or "unusual" to many people, and no doubt there are implementational difficulties. For example, a senior manager could block any safety approach that is not prescriptively based. There is, however, no doubt about the soundness of the principle of the RSM approach for use in the 21st century. However, a continuous programme of education will be required to enable all stakeholders to understand the thinking behind the approach and persuade them to apply it in practice.

Education is a "long-term" task that must start early and remain a constant part of the whole process. Parents and teachers achieve the best results when they develop an attitude of responsibility sharing with children and pupils. To maintain success, this should be followed up throughout the period when young people are at college or university. It is important that the principle of responsibility-sharing is embedded in their minds and applied in all their activities. Later on, the techniques for implementing the RSM approach in general can be acquired via safety workshops that combine of knowledge-delivery, discussion, interaction, group exercises and presentation of findings. See [19] for an example of practical implementation.

CONCLUSIONS

The main conclusions of this paper are:

- The prescriptive regulatory approach to safety is suitable for routine tasks, but the responsibility is vested solely in the prescriber and there is a lack of a management system.
- The goal-setting safety case concept is the most effective choice for new projects, systems, activities and designs where little operational experience exists but safety responsibility is vested solely in the provider.
- The balanced and flexible Responsibility Sharing Management (RSM) approach incorporates a robust safety management system and provides scope for the collaboration of all stakeholders, while blending novel features with the strengths of the existing approaches.

ACKNOWLEDGEMENT

The authors would like to thank Christine Hutcheon, Simon Craufurd and Molly Moore for their assistance in the preparation of this paper. Thanks are also due to Randy Gilbert for his helpful suggestions and incisive critique, and Robert Levine and Jim Read of ARCO Marine for sharing the concepts described in the tanker example.

REFERENCES

1. MV Herald of Free Enterprise: *Fatal Accident Investigation (Sheen Report).* Report of Court No. 8074, UK Department of Transport, 1987. HMSO 1987.
2. *The Public Inquiry into the Piper Alpha Disaster (The Cullen Report).* HMSO Cm 1310, November. 1990.
3. *Marine Accident Report - Grounding of US Tankership* Exxon Valdez *on Bligh Reef, Prince William Sound, Near Valdez, Alaska. March 24, 1989.* National Transportation Safety Board Report NTSB-90-04, Washington, DC.
4. MV Estonia: *Final Report on the Capsizing on 28th September 1994 in the Baltic Sea of the Ro-Ro Passenger Vessel,* Edita Ltd, Helsinki. ISBN 9

51531 6111. (See also website: http://www.edita.fi/aib_finland/estoneng/

5. *The Loss of the Semisubmersible Drill Rig* Ocean Ranger *and its Crew*. The Government of Canada:: Report One. *Royal Commission Report 1984*, Coun. For Publ. Cit. Ottawa.

6. KUO, C. *Managing Ship Safety*. LLP Ltd, 1998. ISBN 1 85978 8416.0.

7. *Analysis of Major Claims 1992*. UK Mutual Steam Ship Ass. (Bermuda) Ltd. P&I Club 1992.

8. BONIFACE, D. and BEA, R. G. Assessing the Risks of and Countermeasures for Human and Organisational Error. *Trans. SNAME*, Jersey City, 1996.

9. MOORE, W. H. and BEA, R. G. Management of Human and Organisational Error Throughout a Ship's Life Cycle. *IMAS 95*, Inst. of Mar. Eng., London. May 1995.

10. KUO, C. Introducing Human Factors into the Maritime Safety Framework. Ship Structure Symposium - *Human and Organizational Error in Marine Structures*. Virginia, USA, November 1996.

11. The Human Element, *Surveyor*, Journal of the American Bureau of Shipping, Vol. 27 No. 3. Sept 1996.

12. KUO, C. *A Preventive Framework for Achieving Effective Safety*. Invited Address, 4th Int. Conf. on Stability of Ships and Marine Vehicles, Naples, Italy, September 1990.

13. OAKLAND, J. S. *Total Quality Management*. Heinemann, 1989.

14. *The Control of Industrial Major Accident Hazards (CIMAH) Regulations*, UK Government Statutory Instrument No. 1902, 1984.

15. *International Safety Management (ISM) Code*. International Maritime Organization Publication, 186E, 1994.

16. *International Code of Safety for High Speed Craft - HSC Code*. International Maritime Organization, 1995. ISBN 92-801-1326-7.

17. Safety Rules and Standards for Passenger Ships. *Official Journal Of the European Communities*. Council Directive 98/19/EC. March 1998.

18. Oil Pollution Act 1990 (OPA 90). US Coast Guard. Washington DC. 1990.

19. KUO, C. *Ship Safety Fundamentals*. Interactive Workshops for the UK Ministry of Defence, 1998-2000.

20. *OPA 90 Alive and Kicking - Exxon Valdez Ten Years On*. Fairplay.Vol. 335. Issue 6013. March 25, 1999. Pages 22-24.

21. National Research Council (NRC), *Tanker Spills: Prevention by Design*. Marine Board, Washington, DC: 1991.

22. NRC. *Double-hull Tanker Legislation: An Assessment of OPA 90 (Section 4115)*, Marine Board, Washington, DC: 1998.

23. SIRKAR, J et al: A Framework for assessing the Environmental Performance of Tankers in Accidental Groundings and Collisions, *Trans SNAME*, 1998.

24. CHELMINSKI, R. *Super Wreck* Amoco Cadiz*: the Shipwreck that Had to Happen*. William Morrow. New York. 1987.

25. *The Oil Spill Intelligence Report, Arlington, MA, USA*. http://www.usg.mil/hq/g%2dm/nmc/response/stats/aa.htm

26. SULLIVAN, J LEVINE, *The Millennium Class Tanker Design, The Motor Ship Conference, Athens, Spring 1999.*

27. CARLSON, L. *Shallow Draught, Deep Impact - Sunoco Refines Future Tanker Design*. Fairplay pgs. 22-23, December 24/31 1998.

APPENDIX A: REPLACING LIFEBOATS ON INSHORE PASSENGER FERRIES

This example was selected to show the use of the Responsibility Sharing Management approach in the process of replacing traditional lifeboats on inshore passenger ferries with marine evacuation units. The different features will be examined under the following headings:

- Background
- Highlights of Alternative Systems
- Application of the RSM Approach

BACKGROUND

A shipping company operating inshore Ro-Ro passenger ferries in a European country was under commercial pressure to be competitive in its operations. The company's main routes attract tourists in the summer months, but in the winter months its operations are essentially a service to the permanent inhabitants of the islands. During the design stage of a new ship for an existing route it was decided that one way of increasing competitiveness would be to reduce costs while maintaining, or even improving, safety standards, specifically by replacing lifeboats with other evacuation systems. The thinking behind this idea was as follows:

- The ferries operate in inshore waters, and never go further than 30 km. from mainland or island, i.e., rescue services are readily at hand and in the event of an accident the distance to be covered by rescue personnel would normally be less than 30 km.
- Lifeboats are heavy, difficult to launch and sensitive to adverse weather conditions.
- The IMO High Speed Craft Safety Code, see [16], does not specify lifeboats as the evacuation system, but gives ship operators the flexibility to propose viable alternatives.

The technical director and his team went to discuss the plan with the country's maritime administration but it was not received favourably by the official surveyors. Instead of seeing this initiative as an opportunity to develop their expertise and long-term maritime viability, the majority of the staff did not like the idea of a conventional passenger ship not having a lifeboat. They also claimed that none of the existing regulations showed that the suggested alternative was open to this shipping company or any other.

Undeterred by the initial negative response, the company pressed the matter further and found that a set of guidelines from the European Classification of Ferries does allow lifeboats to be replaced by other evacuation systems. The company was therefore given permission to interpret these guidelines in the way it understood and to provide the new Ro-Ro passenger ferry with a marine evacuation system (MES) which would operate along with fast rescue craft.

In this case the company managed to achieve its objective However, if the RSM approach had been understood and applied by all the parties concerned, it can be envisaged that a different way of arriving at an agreed solution would have been possible.

HIGHLIGHTS OF THE ALTERNATIVE SYSTEM

A specialised evacuation system would be adopted in place of a set of lifeboats. Various types of such systems exist and the following are typical examples:

- A Marine Escape System (MES): The MES consists of an inflatable slide connected with a boarding platform at the seaward end which is also inflatable. Once people reach the boarding platform they are transferred to liferafts which are then towed away.
- Liferafts: Typical designs include "throw-overboard" rafts taking between 6 and 100 people, and there are both conventional and automatically self-righting versions of this design.
- Evacuation Chute: This is a variant of the MES where the chute takes the place of the slide and a liferaft replaces the boarding platform. The chute is an almost vertical flexible and "expandable" hose within a hollow column to protect those descending from the weather conditions.

The main advantages of these systems over lifeboats are that they are lighter, more flexible, can carry a large number of people in a short time, and occupy less deck space.

APPLICATION OF THE RSM APPROACH

Element 1: DEFINE
Objective: that the ship operator and prescribing agency agree to explore the possibility bo installing an evacuation system such as the Marine Escape System (MES) on inshore passenger ferries in place of traditional lifeboats. The desired outcome is to establish that such a system - coupled with back-up from on-shore rescue services - would be better than, or at least as good as, the existing evacuation system.

Element 2: ORGANISE
Objective: to use resources to ensure that the goal can be realised.

Many tasks are involved, including:

- Allocating resources and management effort to taking the project forward.
- Considering in detail both the technological and cost implications of such a decision.
- Examining in detail the range of alternative systems and comparing their performance with each other and with other similar systems.
- Identifying training needs for the crew and other staff.
- Linking with on-shore rescue services and inviting views on practical issues in relation to potential emergencies.

Element 3: COLLABORATE
Objective: to ensure that the ship operator and the prescribing agency of the government concerned will work together through the four steps of this element. If appropriate, other stakeholders may also be invited to take part or provide input.

The four steps are as follows:

Step 3.1: *IDENTIFY*
Objective: to identify hazards associated with every aspect of using MES.

Hazard Identification: made of "brainstorming" and "What if?" techniques. The following are typical examples of hazards in relation to MES evacuation:

a) System release lever jams

b) Shutter/door will not open and drop to the sea surface
c) Insufficient pressure to inflate chute
d) Passenger descent halted
e) Leading passengers injured by those following
f) Passengers falling into the water
g) Passengers washed off the boarding platform
h) Shuttle-raft capsizes
i) Panic at crowded assembly station
j) Insufficient life-jackets

Some typical examples are given here *in relation to helicopter evacuation*:

k) Helicopter ditching because of adverse weather
l) Pilot falling ill
m) Engine failure
n) Malfunction of hoist
o) Passenger injured by down-draught from helicopter
p) Rescued passenger hits helicopter.

***Step 3.2:** ASSESS*

Objective: to assess the risk level of each hazard identified in Step 3.1 so that they can be categorised into the intolerable, tolerable and negligible risk regions.

Risk Assessment: This would involve the use of both qualitative and quantitative techniques. In the former, the risk matrix is used, see [6]. The latter involves analysis of the consequence and probability of occurrence, see [6]. Since the number of failures providing meaningful statistics can be limited, it may be necessary to make a first approximation using reliability probabilities.

The popular risk relation used gives (R) as a product of the consequence (C) of an event being realised and (P) as the probability of occurrence, i.e., R = C x P. It is possible to carry out risk assessment using both qualitative and quantitative method. In this example the qualitative approach would be recommended. Scales for the consequences of an accident and the probability of its occurrence would be adopted as follows:

Consequence Scale:
Z: Catastrophic: many deaths
Y: Hazardous: one death
X: Major: e.g., limbs lost
W: Minor: cuts, bruises

Probability Scale:
5: Frequent - 10^{-2}
4: Probable - 10^{-3}
3: Remote - 10^{-4}
2: Extremely Remote - 10^{-5}
1: Extremely Improbable - 10^{-6}

Risk Levels:
I : Intolerable
T : Tolerable
N : Negligible.

The risk matrix is as follows:

Z	T	T	T	I	I
Y	N	T	T	I	I
X	N	T	T	T	T
W	N	N	N	T	T
	1	**2**	**3**	**4**	**5**

It is possible to derive the following risk levels for the identified hazards (a) to (p):

Intolerable level: (f), (i), (k).
Tolerable level: (a), (b), (c), (e), (g), (h), (m), (n).
Negligible level: (d), (j), (l), (o), (p)

The three hazards with intolerable risk levels can be checked by doing a simplified quantitative assessment:

Available and related data show that the consequence for each of these three scenarios is serious, and that attention should be focused on the probability of their occurrence. The probability of occurrence of each hazard can be established firstly by constructing a fault tree diagram using a combination of AND and OR logic gates. As an example, the principal components for the fault tree for the case of passengers falling into the water are:

- Weather conditions
- The motion of the boarding platform and the position of the rescue raft
- The type of inflatable evacuation chute
- The experience and qualifications of the crew members supervising the evacuation.

The next stage is to estimate numerically the probability values and evaluate the overall probability. It should be noted that the absolute values of the derived P should be used very carefully as the input data are at best generally only accurate to within ±10%. It is more beneficial to use the fault tree as a means of doing sensitivity tests so as to find which parameters have the greatest influence on probability of occurrence.

Step 3.3: REDUCE

Objective: to reduce the risk levels of hazards with intolerable risk levels and, if appropriate, to reduce those with tolerable risk levels cost-effectively, i.e., if the ALARP (As Low As Reasonably Practicable) principle applies.

Risk Reduction: It is possible to reduce risk levels via consequence and probability of occurrence by management, engineering or operational methods or a combination of all three.

Possible risk reduction methods for the identified hazards with intolerable risk levels would be as follows:

For (f) Passengers falling into the water In high seas, passengers may fall into the water while transferring from the MES platform at sea level to the rescue raft. The risk level of this hazard can be reduced by improved crew training in transfer (management), using the MES only on the lee-side of the vessel (operational) and improving the design of the boarding platform at water level so as to ensure greater protection for the passengers (engineering).

For *(i) Panic at a crowded assembly station*: This situation can occur in an emergency when a number of passengers are overcome by stress. Its risk level can be reduced by using experienced crew members at assembly points and giving them special training in crowd management (management), avoiding overcrowding by careful planning (operation) and providing more space for the assembling passengers (engineering).

For *(k) Helicopter ditching*: This is something that can occur in adverse weather conditions, or because of a technical failure in the helicopter. Its risk level can be reduced by setting operational limits for helicopters (management), a more frequent maintenance schedule (engineering) and regular practice of the helicopter rescue exercise (operation).

***Step 3.4**: PREPARE*

Objective: to be prepared for unexpected or "unthinkable" situations, and have contingency plans available.

Emergency Preparedness: "Unthinkable" situations would include the crew not being able to launch the MES because of the nature of the accident, e.g., fire destroying some of the system or collision occurring in way of the MES.

Emergency planning would include: outlining the procedures to be followed, assigning crew responsibility for specific tasks and improving the communication mechanism.

Emergency response would include: mobilisation of a helicopter on location some distance away from the immediate inshore area and the use of another ferry or ship to act as a rescue vessel.

Element 4: IMPLEMENT

Objective: to consider the most appropriate method of implementing the derived solution.

The options available include:

Adopting a Prescriptive Regulatory Solution: In this case the stakeholders have the opportunity to provide substantial input and the resulting solution takes effectiveness into account as well as cost implications. The solution selected also reduces the effort needed to monitor its practical implementation.

Ship Operators Decide on the System to Use: In this case it is the ship operators and their insurers who derive the "best" way forward, e.g., in the form of guidelines. The prescribing agency, having had a significant role in the development of the alternative approach, measures performance on a random basis so as to ensure that safety standards are met.

A Combined Approach: The prescribing agency draft a set of basic regulations allowing operators flexibility in various special circumstances to adopt the system best suited to the operating area. The ship operators will have to support any suggested system with scientifically-based evidence.

Element 5: MEASURE

Objective: to measure the performance of the new evacuation system using independent parties to do the evaluation.

The measurement needs to be done during training sessions in harbour and also at specified sea-states. Typical measuring parameters would include:

- samples of "typical" passengers
- the qualifications and experience of the crew involved
- the time taken for each stage of the evacuation process.

Element 6: REVIEW

Objective: to analyse the results obtained and the experience gained for all the previous elements.

The results obtained from the previous five elements will be critically reviewed and comparisons will be made with other evacuation systems in order to establish benchmarks. The advantages of the alternative system over lifeboats would be critically assessed. The lessons learnt will be fed back to Element 1 to modify the goal or policies as deemed appropriate.

APPENDIX B: INNOVATIVE TANKER DESIGN

This marine engineering and ship design example was selected to illustrate how the initiative of a ship owner was able to reduce the risk level of hazards associated with the operation of a tanker from a significantly high or "intolerable" region to a "tolerable" region. The emphasis was heavily tilted towards the environmentally friendly operation of crude oil tankers in the pristine waters of Alaska, Puget Sound in Washington State and the boundary waters of the Canadian province of British Columbia. This example depicts a low structure with a high degree of powering and propulsion redundancy designed quite differently from the traditional Very Large Crude Carrier or VLCC with a single engine room. Each feature will be examined under the following headings:

- Background

- Highlights of the new design
- Application of the RSM approach.

BACKGROUND

As most readers are aware, following the grounding of the ***Exxon Valdez*** in Prince William Sound Alaska, in March 1989 with the release of nearly 42,000 m³, 11 million gallons, of North Slope crude oil, major changes in the tanker industry have taken place. Granted, the U.S. is known for its tendency towards litigious reactions, but the public is simply not patient with this degree of spill in heretofore unblemished waters and wildlife habitats. The U.S. House and Senate reacted, and in August 1990 the *Oil Pollution Act of 1990* (referred to simply as OPA 90) [18] was signed into law by President George Bush. When the statistics of oil spills are compared [20], this is not the largest oil spill in the history of tankers but only the 40th worst spill. However, it set the scene for new attitudes and approaches for tanker operators. Hundreds of papers and reports have since been published. Three are particularly worth referencing: two are by Committees of the Marine Board of the National Academy of Sciences, *Prevention by Design* [21] and *Assessment of the Double Hull Legislation* (OPA 90 Section 4115) [22] and the third is a SNAME paper by Sirkar et al [23]. Alert tanker operators recalled the fouling of the beaches of Normandy in 1978 by the ***Amoco Cadiz*** which drift-grounded [24] with a cargo of 278,000 m³ of crude. The Cadiz ranked second in the forty incidents noted above.

Some tanker operators buy and operate VLCCs and Suezmaxs, as one would with any other commodity:- judging its present value based on charter rates and its value after a ten or fifteen year charter. In the U.S., due to cabotage laws, a tanker carrying North Slope crude from Alaska to the "lower 48s" must be U.S. built, flagged and crewed. This changes the formula for designing and building a new tanker for exclusive U.S. service. In general what is now needed is a tanker with close to thirty or more years of service life. The Alaska-west coast service does not operate in a benign environment but in one far more than dangerous than that of the Persian Gulf-Japan service. Transiting the Gulf of Alaska, whether in winter or summer, entails wave climatologies of a severity similar to winter North Atlantic conditions.

The adverse effects and outright costs of an oil spill [25] cannot be sustained by most oil companies. Many oil companies have therefore opted out of the oil-carriage business altogether. Others have risen to the challenge to build and operate tankers with the least risk to the corporation. This example on the use of innovative design illustrates the analyses and measures one such company has taken to re-build its domestic fleet with double-hulled tankers which exceed the international requirements, and further have ensured their large capital investments with highly redundant propulsion, steering and electrical power generation systems. By all accounts this is far from the industry norm.

There is a unique twist where the Administration is concerned. A U.S. program gives ship operators "one-stop shopping" and permits the classification societies to take on the duties of the prescriber. So, for this example, the term "prescriber" is to represent both the Administration and the classification society.

HIGHLIGHTS OF THE NEW DESIGN

The Millennium Class tanker [26] is a first in the world with completely separate engine rooms. Their separation is complete: keel to the top of the stack, watertight and fireproof bulkhead. The two engine rooms are operated as separate entities in all phases of the voyage. There are two sets of "rules" available for redundant design, DnV and ABS. Twin screw, slow speed MAN B&W engines driving Kamewa controllable pitch propellers, each with a shaft PTO. In addition to the PTO, each engine room has a power converter unit (PCU) and diesel generators, along with the corresponding switchboards.

From the control aspect each engine room is designed as an independent space, with dual control rooms configures such that: (a) both port and starboard engine rooms can be controlled from the Starboard Machinery Control Room, and (b) port engine room can be controlled from the Port Machinery Control Room.

The maneuverability of the Millennium Class is equally designed for redundancy, with the tanker capable of operation with a single propeller at zero pitch and rudder hard over.

The usual voyage "times" include 24 hours in coastal water, 28 hours loading/unloading and about 200 hours at sea for each 10.5 day round trip, making coastal transits the most risk for collision and grounds - nearly 10% of each voyage time. A Swedish independent tanker operator has developed the Stena V-Max [27] VLCC with many of the redundancy features discussed in this example.

APPLICATION OF THE RSM APPROACH

Element 1: DEFINE

Objective: To achieve a high standard of safety for this vessel.

This tanker frequently operates in confined waters where the potential for a major oil spill has a low probability but does have extremely high consequences. The goal is therefore to ensure that the risk of spillage is reduced via redundant main propulsion systems and controls on board the vessel. It is also to ensure as much inherent design as possible through double hull configuration in excess of international requirements. The nature of a 10.5 day round trip is very similar to that of shuttle tanker operations in the North Sea and other places, and to capitalise on their pertinent, proven features.

Element 2: ORGANISE
Objective: To ensure the defined goal can be realised. This involves a number of tasks such as:

- Making a management decision to develop the new tanker design
- Giving attention to the technological details and cost implications
- Examining various technological options before selecting a preferred one
- Working out the various steps to be taken
- Interacting with the prescribers in order to meet existing design requirements.

Element 3: COLLABORATE
Objective: To ensure that the ship operator and the prescriber as represented by the government and classification society will work together through the four steps of this element. If appropriate, other stakeholders may also be invited to take part or provide input. The four steps are summarised as follows:

***Step 3.1**: IDENTIFY*
Objective: To identify hazards associated with the operational aspects of the tanker.

Hazard Identification:, The following typical examples of hazards were identified by means of brainstorming techniques:

a) Grounding in the harbour area
b) Being hit by another ship
c) Collision with a another ship
d) Collision with a stationary vessel or reef/isle mount
e) Fire in engine room
f) Engine breakdown
g) Fire in cargo tank
h) Explosion in cargo tank
i) Propeller damaged by floating debris
j) Navigation system giving incorrect information
k) Steering gear malfunction
l) Rudder jamming
m) Excessive structural deformation
n) Structural failure due to sea loadings
o) Protective paint coating deteriorates
p) Static electricity causes sparks
q) Inert gas system ineffective
r) Inappropriate loading/unloading sequence
s) Oil spillage during loading/unloading of cargo
t) Crew fatigue
u) Person overboard
v) Injury to crew member caused by slippery deck
w) Poor communication between crew and shore.

***Step 3.2**: ASSESS*
Objective: To assess the risk level of each hazard identified in Step 3.1 and place it in the appropriate risk region, i.e., intolerable, tolerable or negligible.

Risk Assessment: Both qualitative and quantitative techniques would be employed for this purpose. Qualitative assessment involves the combination of experience and a risk matrix of consequence and probability of occurrence. For quantitative assessment, fault tree techniques are used. A typical result would see the listed hazards grouped as follows:

Intolerable level: (a), (d), (g), (k), (l).
Tolerable level: (b), (c), (e), (f), (h), (i), (j), (m), (n), (p), (q), (r), (s), (t), (u).
Negligible level: (o), (v), (w).

The qualitative assessment of hazards with an intolerable risk level, e.g., (a) and (g), can be checked using quantitative techniques, such as fault tree analysis.

***Step 3.3**: REDUCE*
Objective: To reduce the risk levels of hazards in the Intolerable risk region, principally using technological and design features.

Risk Reduction: A number of features are introduced to reduce these risk levels.

- *Redundant Propulsion, Power and Control*: These systems are the real essence of reducing the risk both in coastal transit and off-shore where a drift grounding could cause terrible pollution of beach and waterfowl habitat. See Figure B.1, control and propulsion arrangements, for an appreciation of twin engine, twin screw, PTO and PCU, diesel generators, as well as the high and low voltage switchboards. The two modes are "running free" and maneuvering". Engines use combinator controls to maintain 10r RPMs during 12 to 15 knots maneuvering speeds, and run at 127 RPMs for open sea (running free) above 15 knots. The plants are split and completely on-line for both these maneuvering modes, Figure B.2. This is common control system setup for PTOs where the engine RPM is set so the PTO can produce either 50Hz or 60Hz. At sea, if the ship is not cleaning tanks or exchanging ballast, there is sufficient capacity on the single engine/PTO for the PTO in the other engine to be declutched. Figure B.3, cargo operations mode, has been included to illustrate the use of a single engine to power all cargo and ship hotel functions, with the other engine but with the propeller engaged.
- *Machinery Control Rooms*: Although there is Wheelhouse control, the company sought to reduce the risk by designing independent and dual control rooms; should the Starboard Machinery Control

Room become disabled, port plant can be independently controlled from the Port Machinery Control Room.

- *Power Management and Automation Systems*: ABB automation and power management systems.
- *Fuel and "Steaming" Capability*: Each engine room has fuel storage - combined capacities, MDO and HFO - for 72 hours. The designers note that all fuel and lubricating oils are separated from the hull by cofferdams, i.e., no pollutant against the skin of the ship.
- *Human Factors*: In consideration of winter operations in the North Pacific/Gulf of Alaska environment the company chose a completely enclosed wheelhouse arrangement; both wings extend a meter out from the molded sides. Additional human factor consideration lies in the one-man operation arrangements, as well as the multi-mode maneuvering - joystick, thrusters depending on location - docking, slow-speed maneuvering or coastal transit.
- *Enhanced maneuverability*: Rudder areas based on performance requirements from IMO were substantially increased. High angle rudders with 45-degree capability were specified for maneuvering speed operation, as compared to the usual 35 degrees, these clearly exceed the norm.
- *Separate steering gears, rudders and controls*: Completely separate hydraulic pumps, controllers and steering gear rooms are provided.

Step 3.4: *PREPARE*

Objective: To be prepared for totally unexpected or "unthinkable" situations and have contingency plans available.

Emergency Preparedness: "Unthinkable" situations would include those where one or more hazards with "tolerable" risk levels interact with each other and so result in an emergency, e.g., fire destroying some of the systems. Thus the following should be given attention:

Emergency Planning: Having a clear, carefully prepared emergency procedure to be followed, which has been communicated to everyone so it can be readily put into practice.

Emergency Response: This would involve working out how the various organisations would respond in the event of an accident so as to reduce a potential major incident to a minor one.

Element 4: IMPLEMENT

Objective: To consider the most appropriate method of implementing the derived solution. The available options include:

- *Adopting a Prescriptive Regulatory Solution*: Instead of the traditionally prescribed regulations, a modified requirement would be given to the new design, taking into account its novel safety features.
- *Ship Operators Play a Much Fuller Role*: In this case the ship operator is recognised for the additional features introduced and is compensated by lower insurance premiums (though not for the actual case) or given greater flexibility. The ship is considered such an integral part of the oil delivery system that this "insurance" ensures delivery of feed-stock to the refineries.
- *A Combined Approach*: The prescribers and the ship operator work together to monitor operations without having to meet additional prescriptive regulations.

Element 5: MEASURE

Objective: to measure the performance of the Millennium Class designs by independent parties, or by people not directly involved in the development of the system.

Aspects to be measured would include the following:

- Technical specifications
- Human-factors related activities such as crew performance using the new system
- Operational experience from safety considerations.

Element 6: REVIEW

Objective: To analyse the results obtained and experience gained from all the previous elements.

The results obtained from the previous five elements will be critically reviewed and the lessons learnt will be fed back to Element 1 to modify the goal or policies as deemed appropriate for future designs.

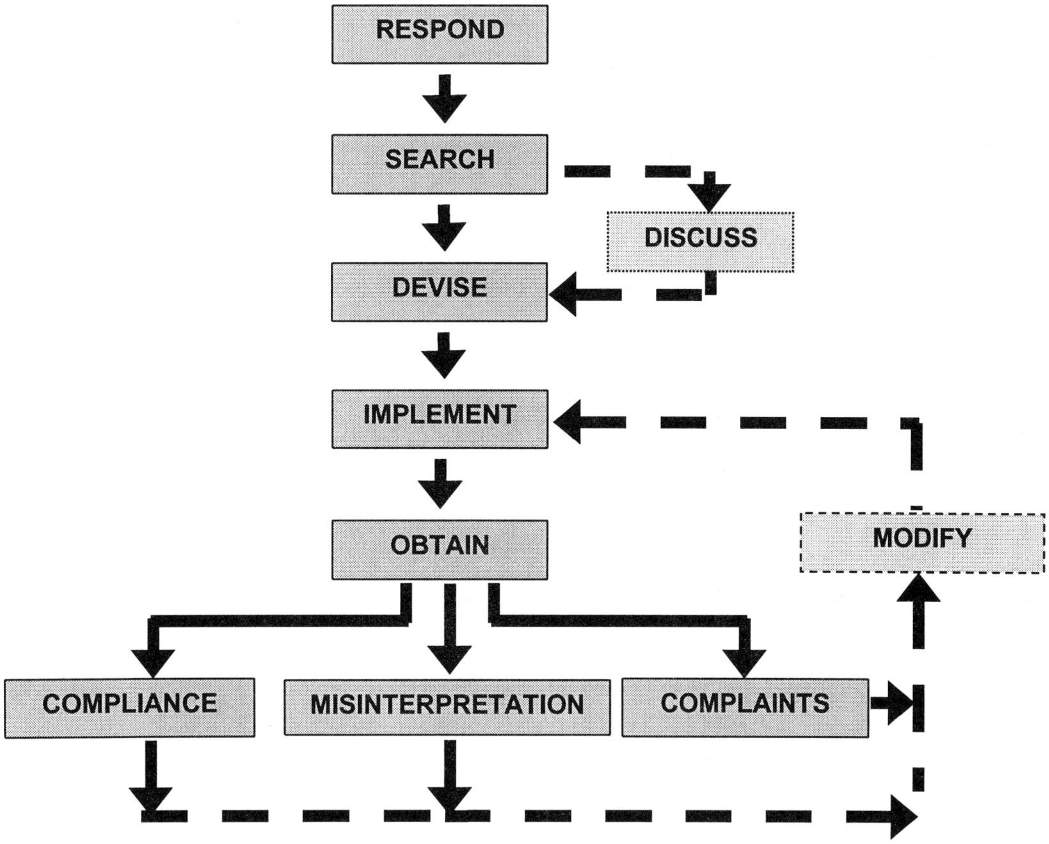

Figure 1

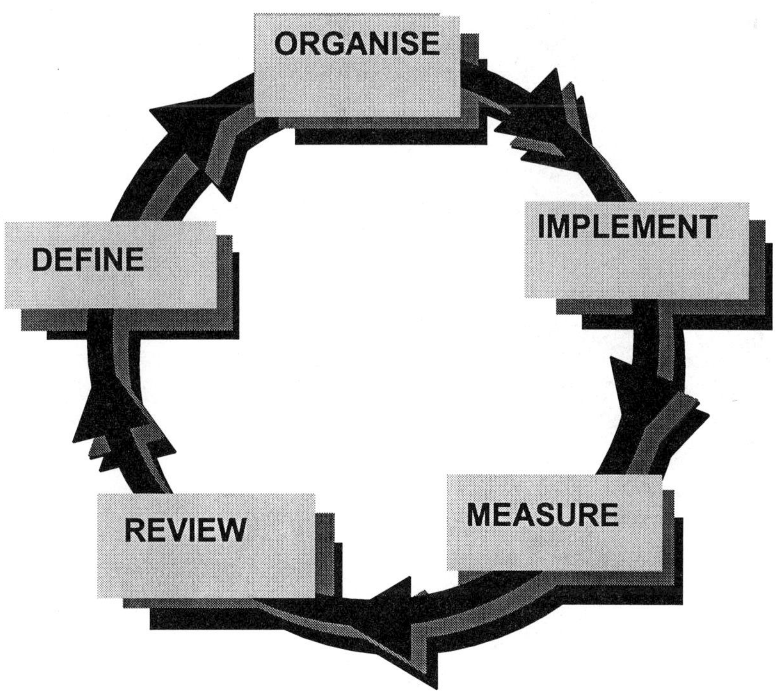

Figure 2

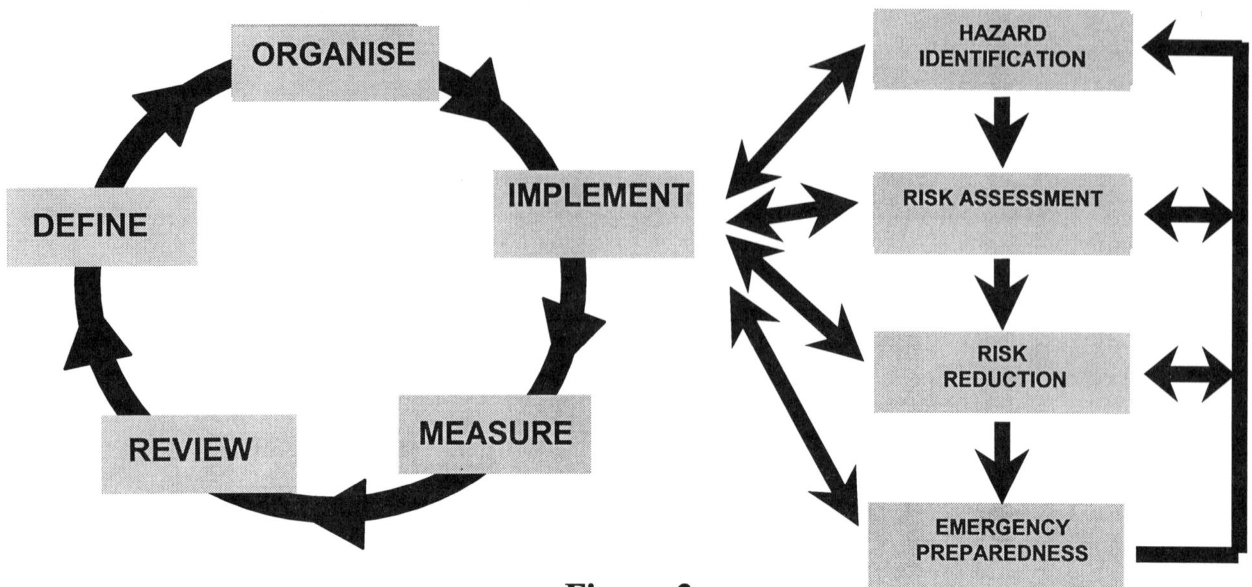

Figure 3

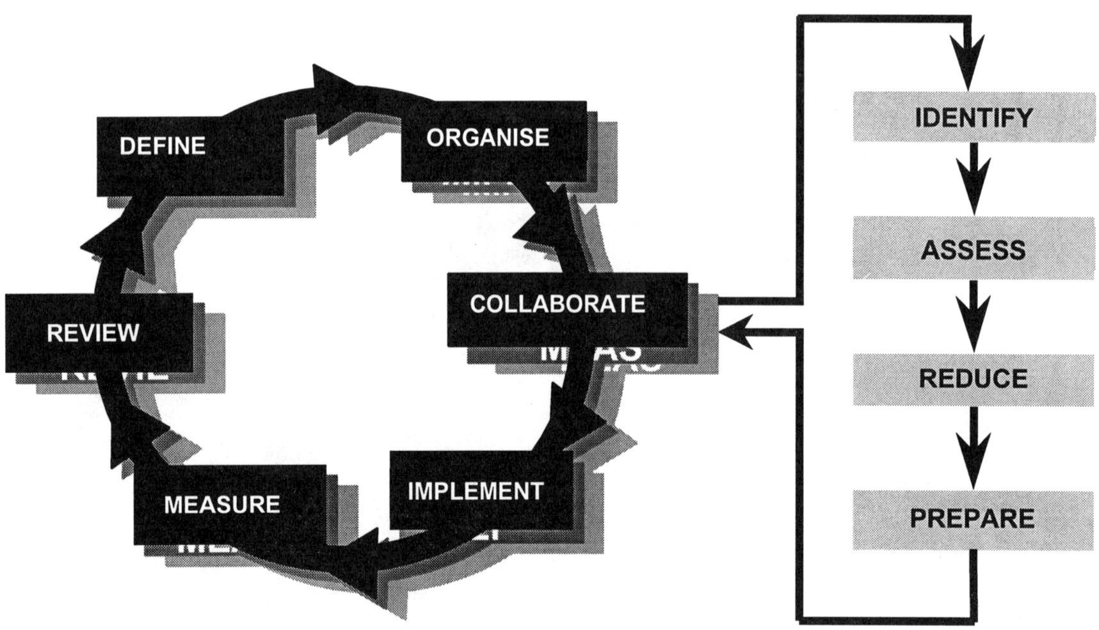

Figure 4

Discussion

Andrew Kendrick, Member

I should like to thank the authors for providing us with a paper that explores an important and often poorly understood subject. Individually and jointly, their contributions have always been thought-provoking, and the current paper is no exception. However, although I can accept many of the points that are made, on this occasion I have considerable difficulty in buying into either their overall diagnosis of the problem or their proposals for solutions.

The description of the Prescriptive Approach on page 2 includes in the 'principle' that: ". . . other parties comply with [the requirements] often unthinkingly."

While this may be true in practice, it is certainly not part of the underlying principle. Any approach to safety, including those favored by the authors, can be undermined by incomprehension, negligence, deliberate evasion, or other influences.

The discussion of weaknesses and drawbacks of the prescriptive approach also contains some surprising assertions and some which are (or should be) caricatures. For example, the statement that responsibility under a prescriptive system is vested in the prescriber would startle many owners and operators who have found that responsibility in this sense does not appear to correlate with liability. Also, in my own experience it is quite untrue that prescribers take no account of the cost of implementing regulations. In many jurisdictions, there are formalized procedures for conducting impact assessments of regulatory proposals, and these are generally taken very seriously. Of course, they can be subverted by political pressure in the aftermath of highly public disasters, but it is naïve to imply that this would not be the case in any conceivable safety management system.

The discussion of the Goal-Setting (performance-based) approach in the paper is relatively brief, but notes three main drawbacks. I agree with two of these—the possibility of inconsistent application and the difficulty of using risk methodology—but I will come back to these later. The third—the vesting of active safety responsibility in the provider—I consider to be entirely appropriate. If the regulator is setting the goals, this cannot be considered as avoiding an underlying responsibility, but rather as accepting a fundamental (and demanding) role. As in the prescriptive approach, it can and should involve complex trade-offs considering societially acceptable levels of various sorts of risk.

The 'new' concept of Responsibility Sharing Management (RSM) promoted by the paper does not appear to me to be significantly different from current best practices in either prescriptive or performance-based regulation. Indeed, the authors themselves note that it is intended to incorporate their desirable features. Equally, I doubt that most regulators would disagree with the list of these desirable features, or deny that their own initiatives were trying to follow these to the greatest degree possible. What is extremely unclear to me is how RSM can avoid many of the drawbacks of the other two approaches in all but the simplest of applications.

To illustrate this, I will discuss a few aspects of the authors' two examples, the evacuation system and the innovative tanker. As an initial observation, it can be noted that both of these are atypical of much of the marine industry, in that they each fall within the jurisdiction of a single, well-resourced national administration and have fairly high public profiles. Collaboration will always be vastly easier under such circumstances than in many other applications.

In both examples, a risk assessment methodology is proposed. It is difficult for me to see how this avoids the drawbacks for risk assessment noted for Goal-Setting. Two key elements in the process are always hazard identification and probability assessment. Hazard identification is extremely difficult, and probability assessment is even more so, even where a project is a direct descendent of a long tradition. Most accident databases are incomplete and of poor quality. Compounding this, many of them are not generally available. I cannot see how, for example, the risk associated with helicopter ditching due to adverse weather could have been assessed as intolerable on the basis of anything other than individual perceptions (unless these were current Canadian SAR helicopters). It is equally difficult for me to see how a responsibility sharing approach could develop solutions to mitigate this risk, since the required resource will almost never be under the (full or partial) control of the parties involved in the risk assessment.

In the second example, the nature of the innovative tanker design does not convince me that RSM is more cost-conscious than prescriptive regulation—rather than reverse. This project was only feasible within the unique US domestic shipping regime, which, coupled with the legal system, allows operators to pass all regulatory and super-regulatory costs on to the consumers. Meanwhile, the existing regulatory context obviously put many potential approaches out of bounds, such as adopting design options other than double-hulling.

Although my comments to this point may seem fairly negative, I should emphasize again that I do feel that the paper makes many valid points and outlines some useful principles. I am mainly concerned that it appears to promote a 'new' approach rather than proposing ways in which existing systems could be applied more efficiently and effectively to address the points made in the authors' own discussion.

Safety management, and its subset of regulatory development do need to mobilize the expertise of the community as a whole. SNAME's recent successes with ad-hoc T&R panels provide one good model for this. It must be recognized that the process of developing standards and methodologies for handling new developments is always likely to be protracted, unless a major accident leads to knee-jerk reaction. Government, industry, and professional organizations all share in the responsibility of identifying areas in which new approaches are likely to be required, and in dealing with these proactively (on a precautionary basis) rather than reactively, when the collective expertise of our community many be undervalued by the political actors, due in part to our inability to foresee and prevent the problem. Unfortunately, for this to happen all parties have to truly put into the precautionary principle, rather than to give issues low priority until the accident has occurred.

Risk assessment is a powerful tool, and one that is increasingly important to marine projects. Its weaknesses include a lack of expertise in application, and lack of data. The former may be mitigated by education and experience. The latter may actually be getting worse. When operators are required to pay for the collection of data and the development of methods to validate their safety proposals, they will wish to keep this information confidential. Similar projects can end up with quite different solutions (and, probably, safety levels) resulting from different approaches to the problem, but very little of this finds its way into the public domain for review or to be used in future work. In my opinion, where regulators allow (or insist on) risk assessment, they need to enforce full disclosure of the supporting information. This may require new attitudes on the part of all players, but is necessary to the objective of continuous improvement.

I am somewhat apologetic for the length of this discussion,

but trust that the authors recognize that this results from my own belief in the importance of the subject.

H. Hormann, and **S. Gumpel,** Members

It was with great interest and indeed pleasure that I read this paper and I congratulate both authors, Prof. Kuo and Mr. Cojeen for the way they have initiated fresh thinking.

For me as a pragmatist in my professional task to enhance safety it is a sort of relief that here an amalgamation of the traditional prescriptive and the "modern" goal setting approach is tried.

I have seen with some concern over the last years that frequently the merits of the new tools for managing safety and quality were promoted at the expense of the prescriptive structure for safeguarding a sufficient level of safety. Sooner or later—no doubt, there are many examples to prove this—the industry will find an optimum in combining the methods available. This paper marks a promising step.

There is one part of the paper to which I have a few critical remarks, (this is certainly no contradiction to my general praise!).

In the section "The Prescriptive Approach to Safety" the development of the prescriptive regulations is characterized by seven steps, with the first one being: Respond to a need such as a major disaster. This suggests that here is, at least normally, no other way of progress. This may historically be correct, but as someone who has spent his professional career more or less on the rulesetting side and overseeing the implementation of prescriptive regulations, I say with some pride that long since the normal course of development is not just in leaps after a disaster, but the routine is adaptation of regulations in step with technological progress and advances in knowledge.

In this context, when the authors say, the "fault" leading to an accident lies with the prescribing authority, I would add: yes, that is the welcome and common excuse—though mostly not true.

Finally I wish say: with all the justified enthusiasm for a goal based approach which I share, I would underline as the prime merit of the prescriptive side what also the authors mention first: it gives a reference standard for everyone. This is a pro which cannot be overestimated in a sector which is highly competitive and by its global nature requires an equal treatment no matter where in the world a relevant decision has to be taken.

Rod K. Pudduck, Member

My congratulations to the author on his very well argued paper.

Until recently I was Head of Ship Safety for the United Kingdom Ministry of Defence and many of the principles he expresses closely match those that my team has been promoting for managing the safety of ships of the Royal Navy and its auxiliary fleets.

There is general recognition that companies (providers) have an obligation to exercise a duty of care over their employees, the public and increasingly the environment. For the marine environment this principle is upheld by the International Maritime Organization. Thus the main onus of responsibility is presently laid upon the company that owns or operates the vessel. They are seen as owning the risk and must therefore deal with it. The author proposes that safety responsibility be shared by all the stakeholders.

Many will question the practicability of arriving at a consensus in an industry that involves so many interests. No doubt everyone could add to this diagram.

Stakeholders in the Shipping Industry

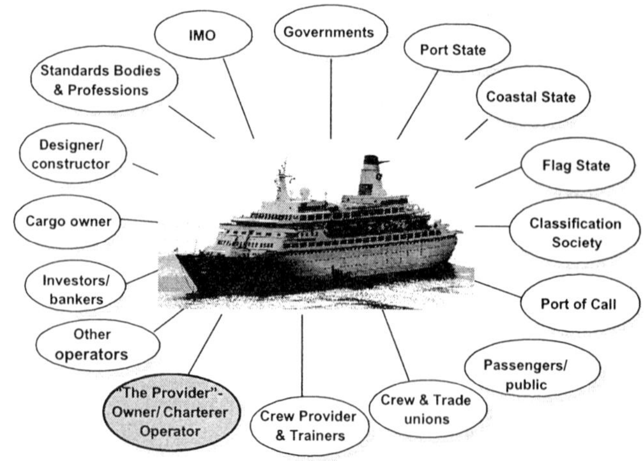

While strongly supporting the team approach to assessing risks, it is unlikely that this will always achieve unanimity and the decisions of even the most representative group are likely to be challenged by outsiders (usually fuelled by the media). In most cases a strong executive decision will be needed to settle things. There is a danger in sharing responsibility that the unscrupulous or weak provider may attempt to "pass the buck," particularly if tempted to avoid heavy costs of safety measures.

It seems right to retain the main burden of responsibility for safety on the owner's "controlling mind"—that is the chief executive rather than the master. Despite virtually all accidents being due to human error, the lookout or helmsman is usually exonerated from blame and there is an increasing desire to punish companies, i.e., the management board, rather than individual employees. Identifying others to share safety responsibility would exercise the courts in deciding the apportionment of blame and it is unlikely that regulators, for example, would be willing to be become involved.

The shipping industry is moving forward from relying solely on prescriptive rules and regulations. The IMO has recently adopted a risk-based approach to formulate rules using a process known as Formal Safety Assessment. The process uses multi-stakeholder input and takes account of the cost to implement safety measures in compliance with the new regulation. A few companies have recognized the benefits of safety management in avoiding the collateral costs of accidents and have gone further by carrying out risk assessments—some have produced whole ship safety cases. Even the classification societies now speak of goal-setting standards.

In an international business with a spectrum of cultures and local issues, prescriptive regulations serve to ensure at least a base-line minimum safety standard—imperfect perhaps, but something to progress from as the desire and means allows. Compliance with statutory requirements does not shift overall safety responsibility from the provider, neither should the input and agreement of other team players in the risk debate.

Robin C. Gehling, Member

I would like to congratulate my friends and colleagues, the authors, on their excellent paper and express my appreciation to the Society for inviting me to participate in this discussion.

I would like to bring the authors' attention to the regulatory systems applying to Floating (Oil) Production, Storage and

Offloading facilities, known as FPSOs. These are facilities moored over the sub-sea completion of one or more oil wells and fitted with processing equipment on deck to separate off gas and water before storing the produced oil in the ship's tanks on preparation for offload to a trading tanker. Often they are converted tankers. They are not arranged to undertake voyages but are subject to offshore oil development legislation, which in Australia and many other countries is based on a "safety case" or goal-setting approach. I understand that one of the shortcomings of this approach is that the safety case is generally *not* subject to on-going review through a Management System as the authors propose.

A number of FPSOs are employed in waters off northern Australia that are seasonally susceptible to tropical revolving storms. These facilities are capable of detaching from the riser and steaming away from such extreme weather patterns, returning and re-connecting to the riser as soon as permitted by the prevailing wind and sea conditions. Accordingly, they also need to comply with appropriate international ship safety requirements during those voyages, namely the prescriptive requirements for oil tankers from the SOLAS, MARPOL and STCW Conventions.

It seems to me that many of the prescriptive tanker safety requirements are inappropriate to a vessel that spends on average about 98 percent of each year on station attached to the riser. Examples of inappropriate regulatory requirements in this case include carrying a full marine crew in anticipation of disconnection, mandatory 5-yearly drydockings and the double-hull/bottom requirements of MARPOL.

In contrast, those FPSOs that are *permanently* moored are *not* subject to these prescriptive requirements and could remain on station indefinitely despite being subject to tanker-like cargo operational hazards.

In my view, the hazards involved with both permanently moored and disconnectable FPSOs would be more appropriately and more economically addressed by a system which draws on both the prescriptive and goal-setting approaches, namely the *Responsibility Sharing Management* approach.

I also support application of RSM to the safety requirements of Wing-In-Ground Effect (WIG) craft, which would involve combining the best features of the prescriptive approach of traditional maritime regulation and the aviation industry's goal-setting approach.

Would the authors please comment on these two possible applications of RSM?

Authors' Closure

We thank the contribution of *Mr. Pudduck* and would like to say that the work of UK Ministry of Defence's Ship Safety Management Office under his leadership is well appreciated. The task of moving from Prescriptive Regulatory Approach to adopting the Goal Setting Approach is a very challenging one since the majority of us are "ingrained" with this approach all our lives, i.e., as a child it is the parents who prescribe, at school it is the teacher and in society there are prescribed codes of behavior to be followed. The progress made in the Royal Navy and its contractors is very encouraging. One of the authors has the privilege to have worked with Mr. Pudduck and his team in offering a ship safety training course that addresses the fundamental issues and is very aware of the achievement and, in particular, to persuade many experienced technically trained personnel that safety involves more than engineering and that management and human factors have crucial roles. The main theme of the discussion can be summarized as follows:

"Is it really practical to manage safety by reaching a consensus in an industry that involves so many different interested parties?"

Our response is that we believe that stakeholders should share responsibility for the following main reasons:

Firstly, in the practical implementation of any decisions, policies or issues, the Pareto (or 80-20) principle applies. In the present case 20% of the interested stakeholders dominate 80% of business activities, and it is they to whom major efforts should be directed. In this way, the task is not as overwhelming as it seems.

Secondly, the collaboration is mainly on identifying hazards, assessing the risk levels, reducing those with intolerable risk levels and seeking solutions for dealing with emergencies. We have developed software to ensure this process can be put into practice efficiently from many groups of interested parties before an executive organization could use the findings.

Thirdly, at the Mare Forum held in Athens on 25–26 September 2000, that addressed safety and sub-standard ships with the breakup of the tanker *Erika* being the central issue, Mrs. Netelenbos, the Minister of Transport for The Netherlands, ended her keynote address as follows:

"In ten years time, we visualize two possible scenarios in place for transport safety. In one scenario the provider is surrounded by a large number of regulations, formulated as results of high profile accidents and other needs. In the other scenario, the solutions to many safety issues have been derived by the collaboration between the transport authority and providers plus selected stakeholders so that responsibility would be shared. It is our vision that the latter scenario will be the one we shall strive to achieve...."

It is gratifying to note there are others believing in the RSM concept and are in a position to do something about it. We hope to persuade Mr. Pudduck and UK MOD (Navy) to adopt this approach as well.

We would like to thank *Mr. Gehling* for his very kind remarks. He raises a very relevant point relating to FPSO (Floating Production Storage Offloading) and the application of the safety case concept. As we stated in our paper, one drawback of the safety case concept is that responsibility is predominantly vested with the operator and there is a need for more active collaboration with the authority so that issues raised by Mr Gehling concerning permanently moored nature of FPSO can be addressed properly. We agree that the existing tanker regulations are inappropriate. This is an example where application of RSM will be appropriate and yield the desired outcome.

It is our belief that RSM approach would also be ideal for examining the safety of a Wing-In-Ground (WIG) craft as a prescribing authority, operators and other key stakeholders would be actively collaborating in order to achieve optimum result.

As Mr. Gehling has noted, RSM approach aims to incorporate the merits of both prescriptive regulatory and goal-setting approaches. Element 4—Implementation is where the actual method of applications can be determined after the hazards are identified, their risk assessed and reduced, and emergency considered.

We would also like to thank *Mr. Hormann* for his very kind remarks. We agree that the first step of prescriptive regulatory approach, "respond" could be explained as follows: "Respond to a need such as a major disaster, advances in knowledge and technology." The point we wish to make is that it is a response to a need and major disasters tend to dominate attention. We, of course, are very aware of Mr. Hormann's contributions in advancing ship safety and his work in IMO (International Maritime Organization).

Mr. Kendrick has given his perceived, in places self contradictory, views on safety and our response will address the main issues raised under the following headings:

(a) Insufficient information on the goal-setting approach

We have not given a full description and discussion of this approach as the safety case concept has been outlined in detail in the book entitled *Managing Ship Safety* (Published in 1998 by LLP Ltd) and referred to in the paper as Reference [6]. The main elements of the safety case concept as presented in that book consist of hazard identification, risk assessment, risk reduction, emergency preparedness and a safety management system in which the five components are a location revolving circuit. Mr. Kendrick would find it useful to study the details from that reference.

(b) Where responsibility is vested for safety

In the prescriptive regulatory approach, responsibility is vested with the prescribing authority and, in general, operators meet the minimum standard. In the goal-setting approach, which is used in the nuclear, chemical and offshore industries but *not* as yet by the shipping industry, the responsibility is vested in the operator. This is a good feature because the "ownership" is now with the operator. In the UK offshore industry, the operator proposes how he will meet a safety standard and the authority, in UK's case the Health and Safety Executive (HSE), notes the solution and makes random checks offshore and if the inspectors are not satisfied that the operator has done what he says he is doing, the installation can be shut down. In our paper, we believe a better arrangement would be for both parties to collaborate, i.e., HSE plays a more active role. Please refer to the discussion of Mr. Gehling which highlights an example of the problem.

(c) RSM is not new

We are somewhat amused by the sentence "The 'New' Concept of RSM does not appear to me to be significantly different from current best practice in either prescriptive or performance-based regulation." We are sorry that he cannot see the new features and suggest that perhaps the many regulation trees are blocking his view. What is definitely new is his "performance-based regulation"! The statement reminds us of an observation made by the Dean of Management in a famous business school. He noted that there is a tendency in some technically trained people to downgrade anything relating to management studies as no more than "common sense," and in mid-life quietly enroll for a part-time MBA course (Master of Business Administration). On completion, these persons do not use the knowledge gained to advance their profession—instead, it is used to show to the world what they know!

Can we draw Mr. Kendrick's attention that "newness" is often mistaken to be something which is totally unfamiliar to the person. In other professions, e.g., pharmaceutical, patent/new drugs frequently differ from existing drugs by 5%–15%. The maritime industry needs innovations, and it is important for everyone in the maritime industry to appreciate this fact and not self-deprecate its own achievements.

(d) Extreme difficulties on identifying hazards and assessing risk

It can be a challenging task to do hazard identification, but there are established techniques such as Brainstorming, What If?, and HAZOP. In fact, technically trained people quickly grasp the mechanics of applying these techniques and our experience with around 800 experienced personnel confirm this fact. Risk assessment can be difficult but not in the way Mr. Kendrick is thinking, i.e., the importance of numerical value alone. It has to be realized that both consequence and probability are functions of management, engineering, operation, human factors and time, and numerical values must be treated with great caution. For further details, we refer Mr. Kendrick to Reference [6].

(e) Mr. Kendrick's belief in the importance of safety

To progress from this belief it is helpful to have a good understanding of the subject, its perceived nature and role of human factors. Our advice would be that one should be positive, open-minded and have an attitude to recognize the need to advance from prescriptive regulatory approach and explore how one can add value to existing progress made in ship safety.

SNAME *Transactions*, Vol. 108, 2000, pp. 419–445

Recent Developments and Application of a Formalized Design for Safety Methodology in an Integrated Environment

Dracos Vassalos, Member, **Ivan Oestvik,** Student Member, and **Dimitris Konovessis,** Student Member, The Ship Stability Research Centre, Department of Ship and Marine Technology, University of Strathclyde, Glasgow, UK

The paper describes a formalisation of a Design for Safety methodology in an integrated environment, outlines early developments of a software tool, and presents the results of an application of the methodology to a case study. The approach adopted attempts to link safety performance prediction through the utilisation of appropriate technical tools, safety assessment deriving from risk-based methodologies and disparate design activities and issues. Blackboard systems have been utilised as the platform in the development of the integrated design environment, allowing safety assessment to become an integral part of the design process. Finally, the case study addresses the application of the developed methodology to three different arrangements of a conventional passenger Ro-Ro vessel, with the aim to demonstrate the validity of the process and methodology adopted. The findings are presented and discussed, and recommendations given for the way forward.

INTRODUCTION

Safety is a broad concept and understanding of what it actually means tends to vary widely depending on the context and the individual. From a philosophical point of view the need of human beings for safety and security is fundamental. As a result humans become deeply involved with events and phenomena which can have a long or short-term impact on life, property and the environment. Deep feelings are aroused by the controversy over safety, especially when technology fails us and human lives are lost or when ships disappear in the knowledge that simple precautions could have greatly reduced loss of life such as in the recent well-published Ro-Ro accidents.

Such emotions, however, can weigh heavily on how rationally and systematically maritime safety is approached, from any of technological, operational and managerial points of view. After investing for decades in ships' hardware for the purposes of increased returns, emphasis must now be shifted towards the human element (humanware) and the organisation and management (software) before a marked improvement of safety can be achieved. With human regard for the environment at an all time high, maritime safety has to be further extended to account for environmental issues, as shown in Figure 1. In this respect, the following definition is now deemed appropriate:

"**Ship safety is the quality that reflects a state of acceptable[1] risk concerning humanware, hardware, software and the environment**".

This broadening of safety necessitates changes in attitude and the adoption of a new approach capable of striking a balance between all the elements involved cost-effectively and throughout the life cycle of the vessel. The implications deriving from this are many, bringing to surface many challenges:

- A change in attitude must come first and, hence, the role of education must be pivotal in this process.
- An approach to safety must be based on a comprehensive assessment of the risks involved and risk mitigation measures and must utilise routinely cost-benefit analyses to aid decision

[1] In terms of societal perception and economic considerations

making, for safety costs money, particularly poor safety!

- Safety improvement necessitates investment in people.

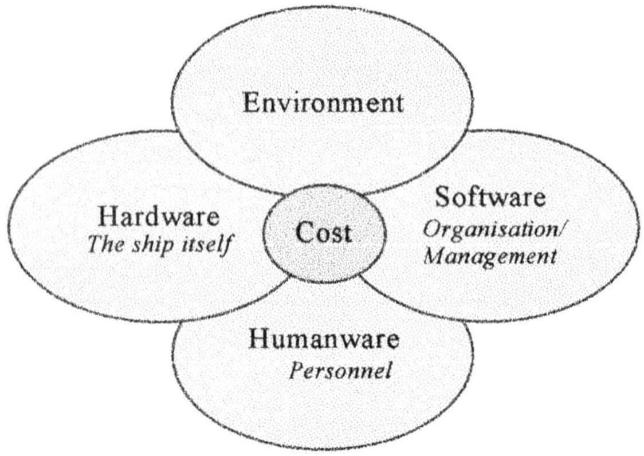

Figure 1: Elements of Maritime Safety

Approaching safety this way must derive from a logical framework and offer the means to take into consideration both the operating environment and the hazards specific to the vessel in question [1], [2]. With Ro-Ro vessels, for example, one of the tasks should be to quantify the probability of damage with water ingress in a given service area and, another, to quantify the consequences of damage by identifying and analysing all the important factors using probabilistic methods. In this case, however, even though it is self-evident that the risks involved can be minimised by reducing either the probability of damage or the consequences of damage, or both, there is a level beyond which consequences cannot be tolerated. Reducing the probability of damage alone will not suffice. It will be necessary, therefore, to address key questions, seeking answers concerning definition of acceptable risks, definition and management of maximum tolerable consequences and procedures for dealing with residual risks. The thought process adopted by the Joint North West European Project is illustrated in Figure 2, [17].

The challenge concerning the need for investment in people is not straightforward either. For whilst investment on hardware is easily understood as it can be directly linked to business goals and a return envisaged and quantified, investment on humanware and software is not easy to justify or quantify nor is it easy to fully explain or understand. Moreover, the human element does not lend itself easily to quantification. The axiom "if something cannot be measured it cannot be improved" lies at the route of many of the problems faced in the past and no doubt will exercise our minds in the future in the strife to improve safety.

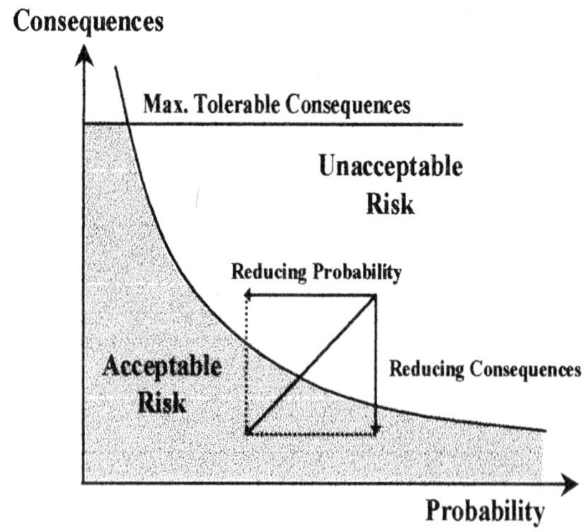

Figure 2: Risk Cost Boundaries

SAFETY IN SHIP DESIGN

Approaches to ship safety are in a transitional state, reflecting the evident shift on the treatment of ship safety from design periphery to a core design factor and from a post-design consideration to a through-life design imperative. A review of the relevant fundamental concepts and methodologies is given in [3]. Conventional approaches are under scrutiny and potential new approaches come under the microscope as the shipping industry is forced to respond positively. Table 1 summarises the various "conventional" and "new" approaches to shipping safety.

However, safety in design is still treated haphazardly by designers in the way that safety conformance is rule-based and not an integrated part of the design process. In this respect, safety is excluded from the creative design phase and measures are taken as an afterthought or as a post-design compliance with regulations. To deal with this problem a new philosophy has been adopted through the theme *Design for Safety*, which aims to integrate safety cost-effectively in the ship design process.

In general, ship design is characterised by the fact that some of the most important decisions regarding the ship are taken at the early stages of the process where later design actions are inevitably restricted by the set frame of earlier decisions allowing for little possibility to positively affect cost and performance. Figure 3 illustrates this situation. As the design process proceeds, the knowledge on the design object increases, whilst the freedom to make changes decreases due to the large costs associated with these changes. There is evidently a need for knowledge feedback to the early stages of the design process where the assigned costs are lower and the freedom to make changes are greater.

The philosophy adopted is illustrated in Figure 4, which aims to facilitate the development of a formalised *Design for Safety* methodology by linking:

- safety performance prediction through the utilisation of appropriate technical tools
- safety assessment deriving from risk-based methodologies
- disparate design activities and issues

The underlying theme promotes safety assessment as the backbone for safe-ship-designing to be formalised as a process within an iterative procedure that provides a two-way dynamic link between technical "tools" and design. The procedure, on the one hand, gathers and assimilates technical information, prioritises safety issues, identifies practical and cost-effective safeguards and sets requirements and constraints for the design process. On the other hand it provides feedback from the design process to stimulate validation and refinement of the tools, in the light of the experience gained from simulation, implementation, and/or practical applications.

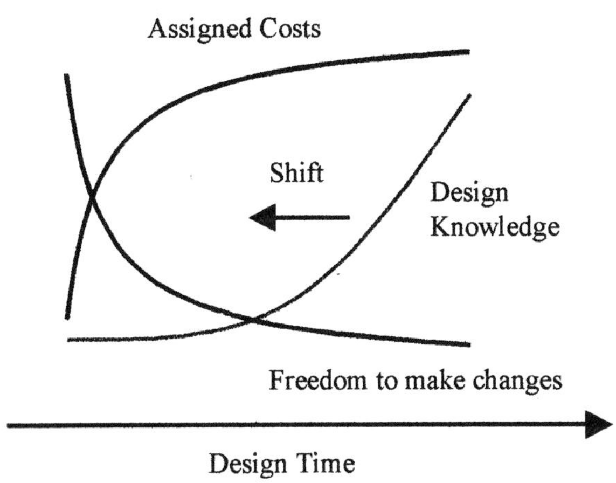

Figure 3: Elements of the Design Process

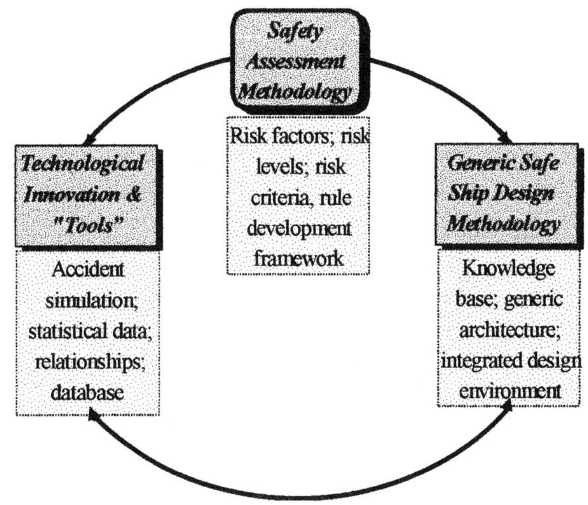

Figure 4: Design for Safety Philosophy

Table 1: Approaches to Ship Safety

"Conventional"	"New"
Reactive	Pro-active
Prescriptive	Goal-setting
Regulation	Self-regulation
Deterministic	Probabilistic (risk-based)
Conformance-based	Performance-based
Compulsory	Safety Culture
Discipline-oriented (sectorial)	Total (integrated)
Experiential	First principles (calculation/simulation)
Hardware focus	Balance of safety elements
Short-term	Life-cycle
Irrational (subjective/emotional/political)	Rational (scientific/cost-benefit analysis)

A METHODOLOGY IN AN INTEGRATED DESIGN ENVIRONMENT

A top-down approach is advocated, governed by high-level events and their frequencies and consequences, in order to design for safety. There is literature on design for safety describing it as a bottom-up approach focusing on component failure and system reliability. However, the latter approach cannot address safety in the early design phases as an integral part of the design process in line with traditional naval architecture disciplines, such as resistance and general arrangement, but is rather targeted in the detail design phase. The bottom-up approach focuses on design for reliability rather than the *Design for Safety* approach described here.

The relationships between risk reduction measures and ship performance must be established in the early design phases, as keeping this relationship outside the design process will only result in local optimisation of safety. The effects of risk reducing design features on resistance, seakeeping, loading/unloading, stability, etc. should be determined by utilising relevant tools in the design process. This aspect is fundamental in the *Design for Safety* philosophy. A ship is a compromise between many conflicting parameters and past research has managed to integrate most naval architectural issues very well, e.g. stability and hydrodynamics. It is argued that this is not the case for safety and research is needed to both identify relationships with other technical issues in naval architecture and implement this knowledge in the design process. Safety may then develop from being an afterthought to becoming an integral part in the design process.

A chief concern in integrating safety in the design process, particularly when claiming that this must be done in a way that safety "drives" design relates to the presumption that any investment on safety does not compromise returns. This concept is ill-based. Figure 5 illustrates the relationship between economic and technical issues in a safe ship design process. The outer boundary corresponds to a design solution that achieves a perfect balance among all safety and cost criteria and constraints, which is presently unattainable. Today's practice is represented by the inner boundary, whilst it is argued that a safety-effective and cost-effective solution could be achieved by adopting first-principles-based design. The enhanced awareness on safety-related issues and the improved appreciation of how safety and cost interrelate and interact is slowly beginning to drive home the simple fact that scientific approaches to dealing with safety is the key to increasing competitiveness.

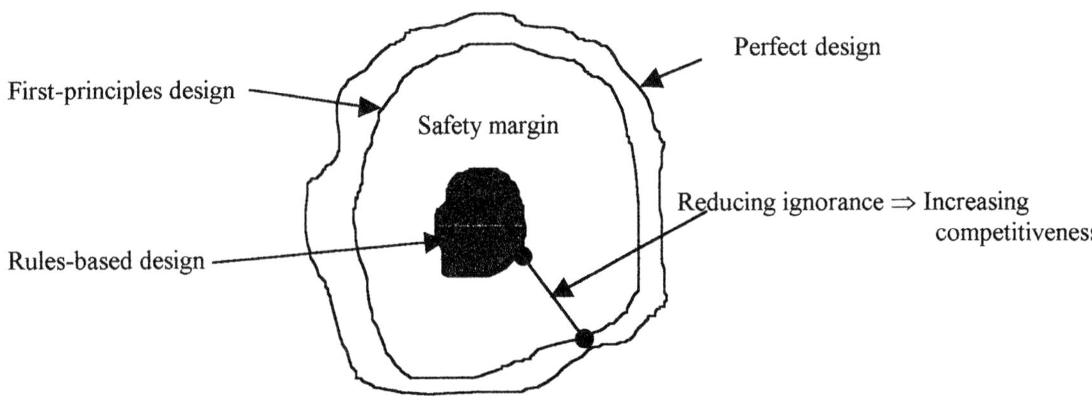

Figure 5: Relationship between Safety, Costs, and Design Constraints

The safety assessment process

An accident is the result of a chain of several undesired events, whilst the seriousness of the accident is a compound set of technical failures, operating errors, fundamental design errors, and management errors. The removal of any contributing links, or causes, may be sufficient to prevent accidents. The chain of events leading to a catastrophic accident for Ro-Ro ships is illustrated in Figure 6.

There are two types of technical safeguards. The one prevents accidents from happening, whilst the other mitigates the effects of accidents. They should be treated separately as their contributions to risk reduction are distinct and not comparable. The difference between these two concepts is that by preventing risks the probability, or frequency, of accident occurrence is reduced, whilst by mitigating risks the consequence of the accident is reduced.

On this background, a safety assessment process is established as illustrated in Figure 7. This process is an integral part of a larger *Design for Safety* methodology and targets optimised design of safe and cost-efficient Ro-Ro ships. The safety assessment framework accommodates the various definitions of hazard and risk assessment and their numerical approach, known as Quantitative Risk Assessment (QRA). The merit function of the framework is based on a cost-benefit analysis (CBA) comparing the costs of manufacturing and operating the proposed safety measures with the benefits of enhanced operational safety. Combined, these components comprise the safety assessment process. The main difficulty in this approach is to quantify the benefits of the safety features.

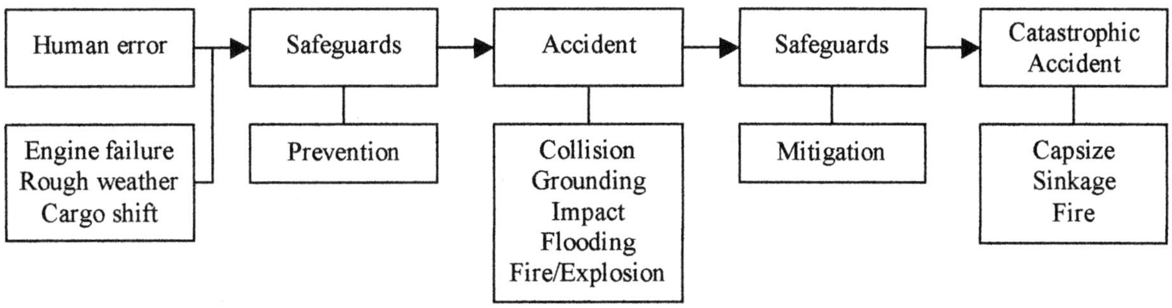

Figure 6: Chain of Events

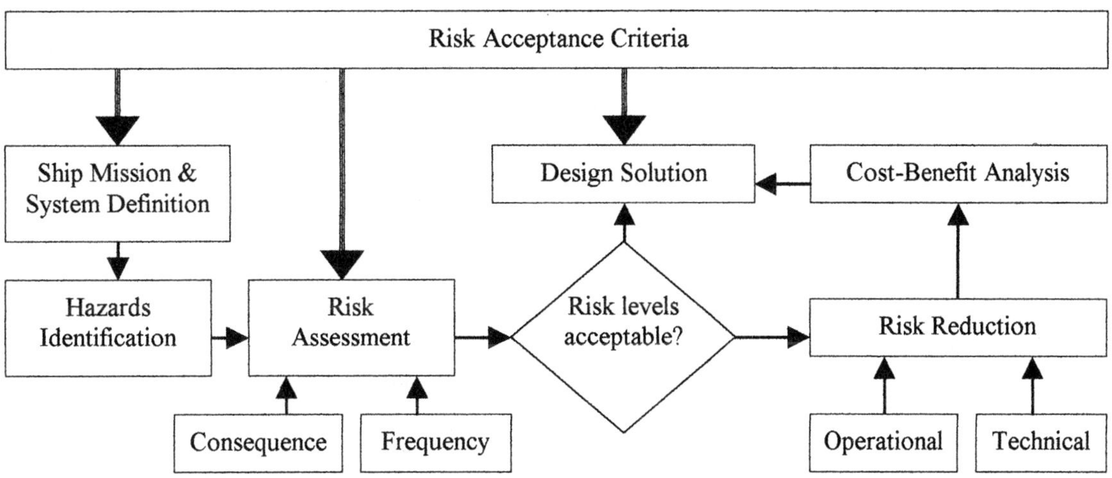

Figure 7: Safety Assessment Process

A Design for Safety methodology

The *Design for Safety* methodology, illustrated in Figure 8 [4], is an iterative process whereby an optimal solution for a ship design is sought that is safe-, performance- and cost-effective using a top-down approach. The input required is a ship design, which is developed using information modelling techniques.

Risk analysis is performed for the design concept and the resulting quantified risk level is controlled against established risk acceptance criteria. Risk reduction measures, or design features, are considered when a ship fails to meet these criteria. There is a general distinction between risk reduction and mitigation means and both must be considered in order to develop an optimal design. On the basis of applying risk reduction measures "new ship designs" are developed and the effects of the changes are again evaluated against risk acceptance criteria.

Designs that are considered to be safe are put forward in the procedure and cost-benefit analysis of the risk reduction measures are performed. Using

ICAF (Inverted Cost of Averting a Fatality) criteria the new design solutions are evaluated based on their cost-benefit performance and economic viable design solutions are put forward in the process. The safe and cost-effective design solutions are thereafter assessed for their effect on other performance factors, such as seakeeping, cargo capacity, operational efficiency, turnaround time, etc. The resulting solutions of this process are weighted and the best design is put forward in the process for further development.

The methodology has potential to accommodate multiple accident events, where the effects from the various event-driven design configurations are assessed. In such a scenario, event-driven design features may be conflicting necessitating the use of decision support models in order to derive the best overall design configuration.

A Design for Safety Information Model

This section targets the development of a design for safety (DFS) information model on the background of the design for safety methodology outlined in the previous sections. The model is an object-oriented representation of the information associated with design and safety aspects. The model provides a framework for integrating relevant safety issues in the early phases of the ship design process and it provides a means for identifying relationships and enabling decision support.

The DFS information model pertains to the survivability of a passenger Ro-Ro vessel following a collision or grounding event, however, the model is generic and events and consequences utilised in this section may be replaced.

The development of the DFS model is innovative and the objective here is to define the philosophy and method for establishing information models.

The DFS information models

The DFS information models comprise the generic knowledge block shown in Figure 8. The model illustrates that an event has taken place, e.g. collision or grounding, and that the consequence is flooding of the ship. The design for safety procedure thus targets the safeguards, or design features, needed to avoid sinking or capsize of the ferry, the cost-benefits of these design features and the effects on other factors affecting ship performance. Thereafter the resulting solution is put forward in the design process for further consideration.

The DFS information model describes a framework for assessing survivability following collision and grounding events. Similar models may be developed for other safety issues and these models combine to form a complete DFS information model for passenger Ro-Ro vessels.

Following the development of the sub-models, their respective attributes are mapped in order to establish suitable relationships, which in turn function as a guide in the design process.

Design sub-model

The design sub-model of a passenger Ro-Ro vessel is explained in Figure 9. The model illustrates the main survivability issues, which must be targeted when designing a ferry. The described attributes are the most relevant when designing passenger Ro-Ro vessels:

- Subdivision: A passenger Ro-Ro vessel has to comply with a number of rules for subdivision, on and/or below the freeboard deck, as laid out in SOLAS and classification society's rules. The rules are generally deterministic in nature and must be accommodated when designing. Compartmentation on the freeboard deck may be side or central casings, transverse bulkheads, freeing ports, etc. Compartmentation below the freeboard deck may be transverse and longitudinal subdivisions, cross-flooding arrangements, stepped transverse bulkheads, watertight doors, etc.
- GM and GZ curve: The metacentric height GM holds information on the stability of a passenger Ro-Ro vessel. The GZ curve holds information on the ferry's ability to maintain a stable ship in a seaway. The GM value is a function of the hull shape and the vertical centre of gravity and is used to calculate the GZ curve for various degrees of heel.
- Length and beam: The length and beam of a passenger Ro-Ro vessel are set by the owner's demands for payload and speed in addition to hydrodynamic laws for the L/B relationship.
- Depth and draught: The depth and draught are set by the owner's demands for payload and speed, in addition to hydrodynamic laws for the L/T and B/T relationships, stability rules for the depth height and draught constraints in ports.
- Speed: The speed is set by the owner's demands for time at sea between ports, in addition to a compromise between fuel usage and the transport operation in question.
- Lane meters: The lane meters is set by the owner's market analysis aiming to accommodate the market demand for trucks, busses and cars.

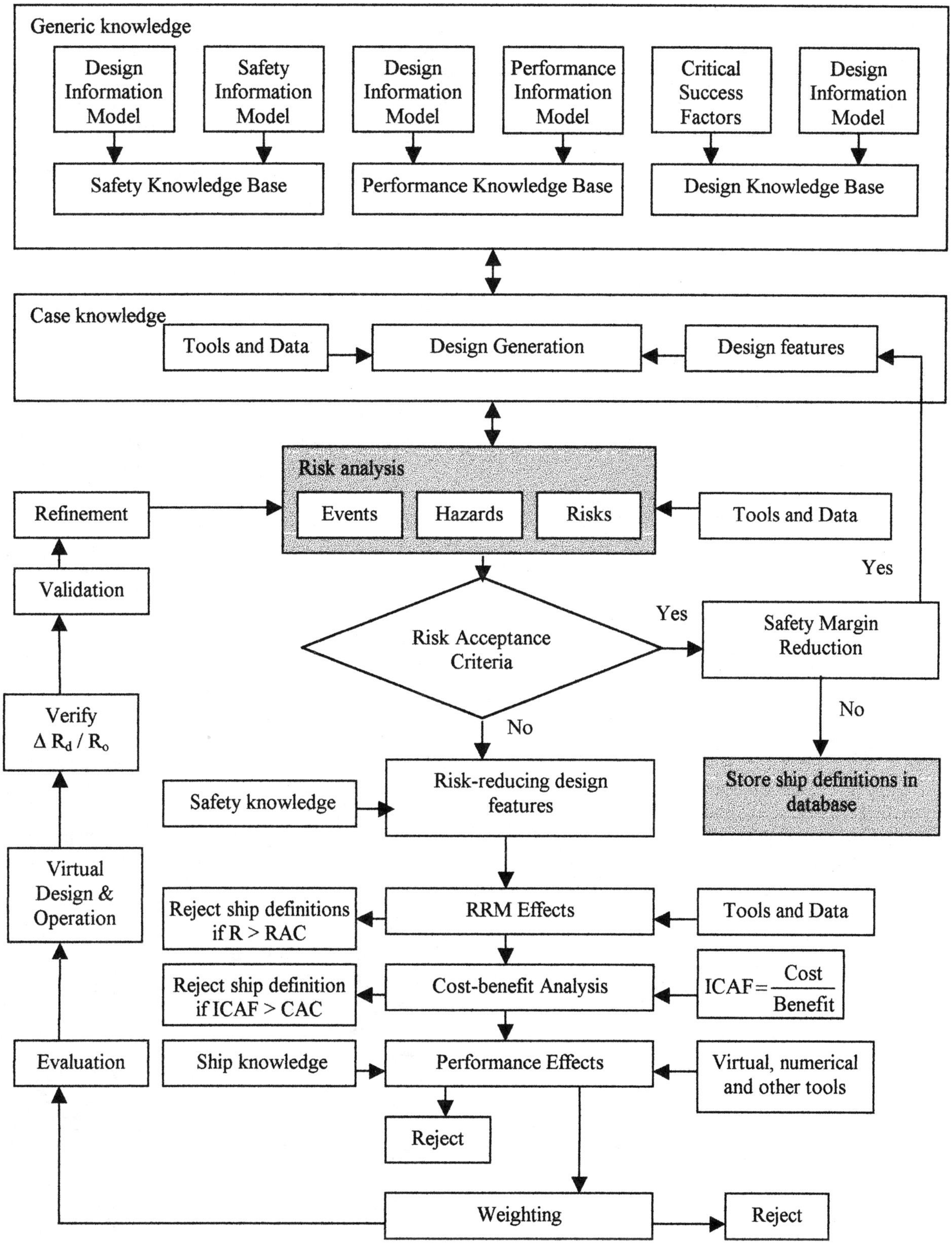

Figure 8: A Generic Design for Safety Methodology

Safety sub-model

The safety information model, explained in Figure 10, holds information on the most important issues regarding passenger Ro-Ro vessels survivability. The model aims to illustrate the main issues, which are used in ship survivability assessment in current practice:

- Evacuation equipment: There are guidelines and rules for the evacuation procedure from a passenger Ro-Ro vessel, which results in certain evacuation equipment to be carried, such as rafts, lifeboats, slides, etc.
- Fire equipment: There are rules for fire monitoring and fighting onboard a passenger Ro-Ro vessel, which results in certain fire equipment to be carried, such as sprinkler systems, sensors, foam systems, signals, etc.
- Crew: The number and training of the crew is a large factor when assessing the safety of a passenger Ro-Ro vessel. The crew must be able to handle all situations that may occur.
- Flag: The safety standard of the ship is often related to which flag she carries and there are known to be differences in safety attitude among the various flag nations.
- Number of passengers: The number of passengers onboard a passenger Ro-Ro vessel is important for the handling of critical situations. The number of passengers determines which class the ship is given and which rules she must adhere to.
- Weather: The effects of the operational environment on a passenger Ro-Ro vessel must be evaluated during the design process in order to maintain the required safety level consistent with the weather conditions experienced in the operating area.
- Rule compliance: There are a number of rules and regulations a passenger Ro-Ro vessel must comply with. The general issues are covered in the rules of the classification societies, which builds upon IMO and SOLAS regulations. Most rule compliance is today based upon a deterministic approach where requirements are fixed. The alternative is to make use of probabilistic rules where the designer has to document that the ship is safe for the intended operation.

Performance sub-model

The performance information model, explained in Figure 11, shows the most important attributes, which a ferry must accommodate in a satisfactorily manner. The attributes are assessed utilising standard techniques and software based on the resulting design. The performance attributes are general ship design issues, which are commonly assessed in design offices:

- Seakeeping: The seakeeping of a passenger Ro-Ro vessel is related to her ability to maintain a stable, comfortable and safe operation during the intended service. Guidelines to adhere to sustainable/tolerable motion amplitudes and accelerations by ship/cargo and crew/passengers are set by IMO.
- Manoeuvring: The manoeuvring of a passenger Ro-Ro vessel is related to her ability to steer at sea and manoeuvre in port. The manoeuvring capability is directly related to the hull shape, windage area and type and size of propulsion unit and/or rudder.
- Intact/damage stability and survivability: The intact and damage stability characteristics of a passenger Ro-Ro vessel determine her ability to undergo large motions safely, to resist capsize under external and/or internal excitation and to be evacuable in the worst scenarios.
- Powering: The power requirements on a passenger Ro-Ro vessel relate to the propulsion unit, support systems and accommodation. The objective of the designer is to achieve maximum speed at minimum power requirements.
- Structural strength: The structural strength requirements relate to maintaining the structural integrity of the hull during the operational life and to providing "habitable" environment free of excessive noise and vibration. The dimensions of the structure are given by calculations according to rules and regulations.

Critical success factor sub-model

The information model of the critical success factors, explained in Figure 12, targets the attributes, which determine the competitiveness of the transport system within which the ferry will operate. The attributes of the model are standard issues in logistic studies, as performed by transport researchers. The model consists of the most general attributes and is not comprehensive:

- Costs: A passenger Ro-Ro vessel traditionally covers two markets. One market is the cargo market where the ferry is a part of a transportation chain where the shipper uses the ferry in this transport chain. The other market is the passenger market where service and sales onboard are the main drivers, in addition to car freight for a further destination beyond the harbour. To be competitive the ferry must compete on costs for both cargo and passenger markets. The income for the ship operator must cover capital costs, operation costs and maintenance costs and allow for earnings.

- Time: The shipper considers the door-to-door transport time and not the ship transit speed. The ship operator should thus pay due respect to minimising the turnaround time, as savings in transit speed may be achieved without influencing door-to-door transport time. The passengers are concerned with the departure and arrival times, which must be convenient, whilst the transit time between two ports is of less importance.
- Frequency: The shipper demands a sailing frequency satisfying logistic needs. The optimum route frequency-speed relationships must be decided based on a market study identifying the actual/potential flow and volume of cargo on the route. A frequent service is bound to attract more cargo and passengers than less frequent services, as planning is made easier for the users.
- Quality of Service: Cargo must be delivered to the consumer without any damage. This relates to the handling of the cargo, the vessel's motion characteristics and securing systems. Customer service in terms of ease of booking cargo freight and procedures for ordering and payment is important for having a successful route. Passenger service must be efficient and satisfy demands. On longer sailings more shops and attractions must be put onboard. Booking and payment of tickets should be made easy.
- Reliability: It is important that the ferry keeps the schedule time of arrival/departures without delays due to weather conditions, maintenance, bunkering, supplies or breakdowns. In order to keep a high reliability of the service the vessel must be designed to have good seakeeping qualities, good arrangements for supply of stores and optimum maintenance schedules, which keep the probability of any breakdowns to a minimum.
- Environmental Friendliness: The environmental issues are becoming progressively more important and a "green" ship is seen as an advantage. The environmental factors are divided into emissions, wash and noise where the objective is to reduce their adverse effects.

Relationships between attributes

The described attributes in the four sub-models of the larger DFS information model could be realised in a large number of ways and the task at hand is to define and structure these relationships.

The way to perform this, is by factorising relations in such a way that factors equalling 100 show highest importance, whilst factors equalling 0 lowest importance. Therefore, relationship factors describe the importance of the respective relationships and hence illustrate which attributes the designer should be more aware of in the design process.

The relationship factors can be determined at least qualitatively by interviewing ship designers, operators and owners. Factor analysis is a statistical technique applied to a single set of attributes in order to discover their relationships. This work is on-going at the University of Strathclyde.

Application of information models in ship design

The described process in this section may be adopted when targeting the designing of a safe passenger Ro-Ro vessel. The DFS information model and its sub-models are developed and the attribute relationships are factorised, which provides the designer with knowledge about the importance of the various attributes in the early design phases.

An Integrated Design Environment Architecture (IDEA) for Safety

An Integrated Design Environment Architecture (IDEA) provides the designer with a means to assess the technical and analytical characteristics of the design using relevant tools. The IDEA must be formalised indicating that entities, attributes and relationships for the relevant issues are generic. This allows information to dynamically change, for altering design input and innovations can be readily implemented. The design information is stored in object-oriented knowledge bases, which are updated independently as required. A control and management function is needed to accommodate these issues. The *Design for Safety* procedure outlined in the foregoing has been accommodated in an IDEA using blackboard systems as the platform. An IDEA is illustrated in Figure 13, having a central blackboard to control and manage the overall ship design process applying the appropriate knowledge bases, tools and methodologies as and when required in building a design solution.

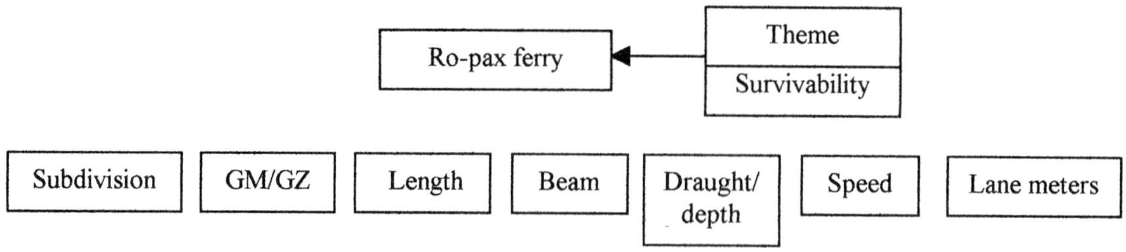

Figure 9: The design sub-model

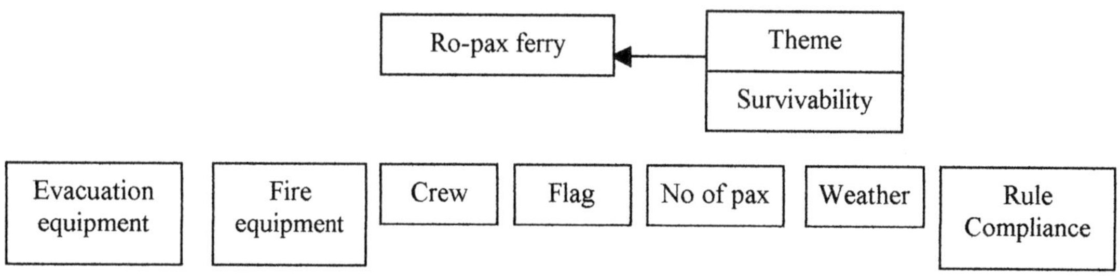

Figure 10: The safety sub-model

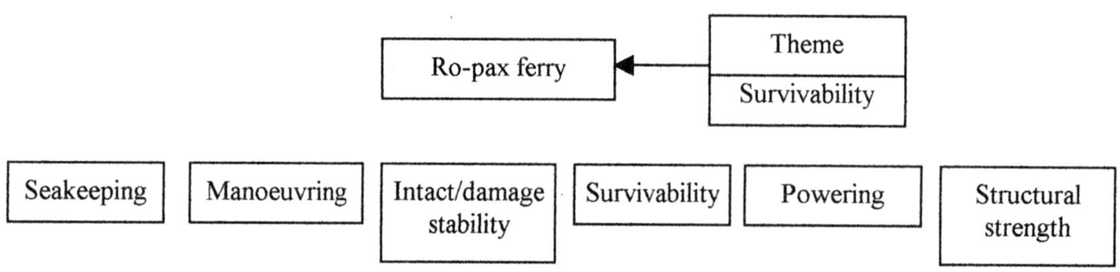

Figure 11: The performance sub-model

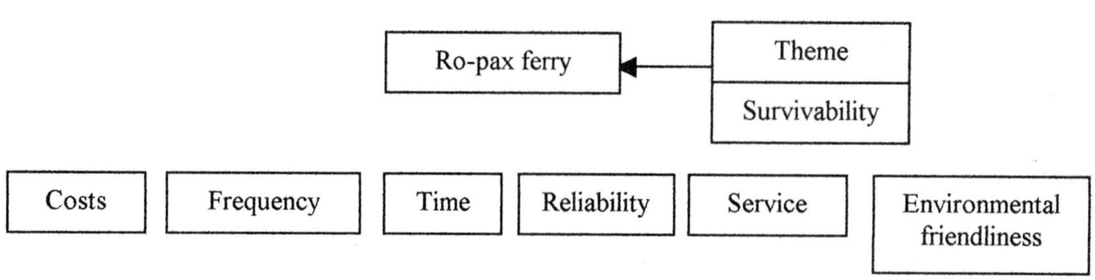

Figure 12: The critical success factor sub-model

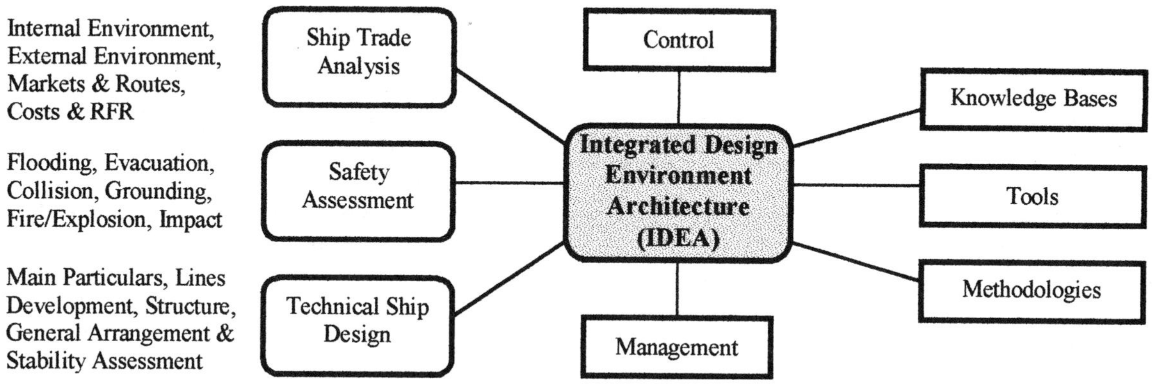

Figure 13: An IDEA for Safety

Blackboard Systems

Ship design is a complex engineering decision-making process involving the integration of many subsystems into a final design solution. There are many ship design methods, which have been developed based upon the ingenuity and experience of separate design capabilities and experience around the world. The basic constituent and of central importance to all methods is the *ship design spiral*.

This was introduced decades ago by Evans [5], and in more recent years attempts have been made for enhancement by applying various means of computer technology and modelling techniques [6, 7, 8, 9]. The ship design spiral advocates an iterative, step-wise solution procedure that is time-consuming and produces results, which may be acceptable but not necessarily optimal. It certainly does not take into account the whole life cycle of the ship and is not capable of accommodating effectively and efficiently contemporary and future ship design tools and concepts.

There are two major trends evident in ship design research and development today: (i) the pursuit of a definite theory of ship design, (ii) the development and application of computer-based tools in design. Regarding the latter, a great deal of work has been undertaken by industry and academia world-wide resulting in quality software packages, advances in technology and scientific understanding, which combined have shortened considerably the time required during the ship design process. Regarding the former, ship design is still in need of a formalised framework to control and manage the design process in an efficient way with the potential to take advantage of new emerging computer technologies. To this end, blackboard systems have been identified to have such a potential and are being utilised as the platform to overcome the stated deficiencies.

Fundamentals of Blackboard Systems

Blackboard systems (BBS) are regarded as a part of the Artificial Intelligence family and originated with the Hearsay project in the USA twenty years ago [10, 11]. BBS constitute the fundamental assumption of design co-ordination, a formalisation within concurrent engineering. Design co-ordination emphasises that tasks must not necessarily be carried out concurrently, but rather in such fashion as to achieve optimum performance.

Design co-ordination is defined as a high level concept of the planning, scheduling, representation, decision-making and control of product development with respect to time, tasks, resource utilisation and design aspects [12].

The philosophy of blackboard systems is to opportunistically piece together a solution on the blackboard by using external knowledge sources, which are working co-operatively and are activated by a control mechanism, be it human or software programs, applying the right knowledge at the right time [13].

In this respect, various sources of knowledge participate in forming and modifying the emerging solution by knowledge sources contributing opportunistically when called upon. Furthermore, as steps are taken towards the solution, the processing commitments are minimised since the solution is built incrementally and steps of forward chaining can be arbitrarily interleaved with steps of backward chaining. Blackboard systems are particularly effective for incremental solution generation, which is typical for the ship design process, where the knowledge sources contribute to the solution as appropriate outperforming a problem solver that uses the traditional ship design approach to generate a solution.

A blackboard system consists of three components:

Knowledge sources: Independent modules that contain the knowledge needed to solve the problem, having a formal structure to receive and send information in order to communicate with the blackboard. Knowledge sources are static repositories of knowledge activated in a predetermined way by trigger functions, or instances, which are the active functions looking for a triggering command or condition on the blackboard. Knowledge sources can be referred to as agents in engineering applications where the technology supports the integration of fully autonomous software and human expertise. Knowledge sources can be added or deleted from the blackboard system as required without influencing the overall performance of the blackboard, they can have a wide diversity in ways of representing information and problem solving techniques, but must operate based upon a common interaction language. With regard to ship design, lines development could be such a knowledge source, engine selection another, both performed in the ship design process based upon separate methods, tools, and information.

The blackboard: A global structure that is available to all knowledge sources and serves as a storage medium of raw input data, partial and final solutions, and control information, as a communication medium and buffer, and as a knowledge source trigger mechanism.

A control component: It directs the problem-solving process by allowing knowledge sources to respond opportunistically to changes on the blackboard database. The control triggers any knowledge source based upon a defined ranking, and it can change the focus of attention based on the state of the solution.

Control Shells

To execute the code of a Knowledge Source (KS), a control shell is required to activate it. This knowledge-source activation (KSA) represents the use of the KS in response to particular conditions in the application. A given KS can be activated as often as is appropriate in an application. Events represent the occurrence of actions within the application or external occurrences. The occurrence of an event can cause the control shell to trigger a KS or perform other activities, such as revalidating a KSA immediately before execution, obviating a KSA (removing it from the queue of KSAs waiting execution), or re-triggering a KSA. The control shell is a collection of control components that supports defining and activating KSs, and scheduling and executing KSAs. Events serve as the interface between the control shell and the blackboard.

Importantly, the control component must be able to make its selection among pending KS executions without sharing the expertise of the individual KSs. Without such a separation, the modularity and independence of KSs would be lost. Therefore, the control component must be able to ask for estimates from triggered KSs in decision-making.

Application of BBS to Ship Design

Blackboard systems offer a powerful problem-solving framework particularly in view of the arguments put forward in Table 2. Blackboard applications are not widely used despite their long history and documented advantages. The reason for this is "that blackboards are only worth pursuing for complex applications" and "while useful for prototyping an application, once developed and understood, the application can be re-implemented without the blackboard structure" [14]. The former statement indicates that for a ship designer a blackboard tool would be invaluable. The latter statement is valid for any type of development since human experience is hard to beat in any case. The best example on this is ship designers who produce ship designs based on intuition and experience. Three situations are defined in Table 3 when a blackboard system is not the best solution. However, these arguments are used to justify the introduction of blackboard systems in the ship design process.

An Integrated Design for Safety Environment

As already explained in the foregoing, current ship design practice is a design-check (trial-and-error) procedure, which is insufficient in itself as it keeps safety considerations outside the creative design process. It is here proposed that the vehicle for shifting safety from design periphery to the core of design is an integrated design environment utilising the outlined blackboard system philosophy. To this end, a prototype ship design blackboard system has already been developed, as reported in [15], accommodating the *Design for Safety* procedure described in the foregoing. The *Design for Safety* process has been embedded in an integrated environment, using blackboard systems (BBS) as the platform in order to function as a decision support tool for the designer. The blackboard system program specification is outlined in Table 4, which formed the basis for development.

The integrated *Design for Safety* environment, illustrated in Figure 14, accommodates a methodological assessment of relationships between safety, cost, and design features adopting a top-down approach by assessing hazards and risks at the event level, i.e. for collision, grounding, impact, flooding, and fire/explosion. Furthermore, the IDEA accommodates the identification of risk prevention and mitigation measures, cost-benefit quantification of design features/safeguards, assessment of ship performance effects and provides decision support.

Table 2: Applicability of Blackboard Systems to Ship Design

Blackboard Systems	Ship Design
Many diverse, specialist knowledge representations are needed	There are few engineering activities that are more diversified and specialised than ship design
An integration framework is needed that allows for heterogeneous problem-solving representation and expertise	A ship design evolves today by performing heterogeneous tasks in a sequential order. A framework for control and management is needed
The development of an application involves numerous developers	Ship design activities involve numerous organisations, suppliers, consultants, etc.
Uncertain knowledge or limited data inhibit absolute determination of a solution	Alternative solutions can be designed with little extra effort absorbing uncertainties
Multilevel reasoning or flexible, dynamic control of problem-solving activities is required in an application	Ship design will profit immensely by enabling dynamic control between two or more applications such as stability and resistance

Table 3: Inapplicability of Blackboard Systems to Ship Design

Blackboard Systems	Ship Design
All knowledge could be easily represented in a framework with which a user is already familiar	This has yet to be achieved within ship design - the knowledge exists in various places and forms
The application does not need to make dynamic control decisions	When implemented this function leads to savings on time, effort, and fewer errors. Moreover, with the implementation of simulation in design this function is needed
The complete application will not be combined with other systems	Ship design comprises numerous systems - a framework controlling and managing these systems is called for

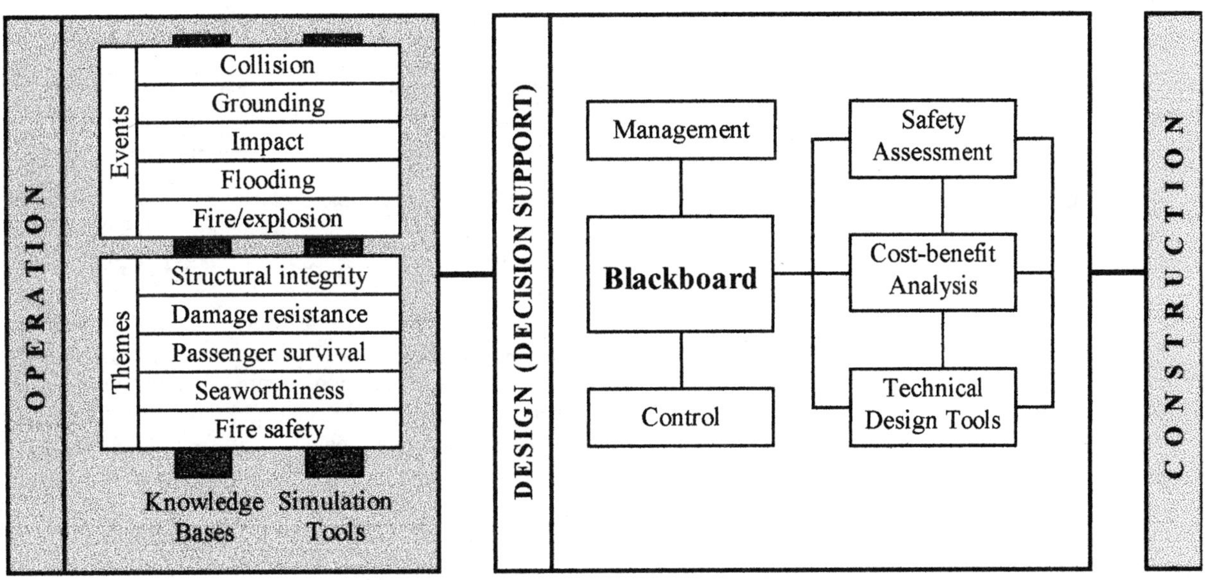

Figure 14: An Integrated Design for Safety Environment

Table 4: Blackboard System Specification

	KS	Task	Work comments
10	Start	Start program	Introduce user to program
20	Ship type and configuration	Identify ship type and import ship configurations	Identify ship type on the blackboard
30	Events	Select events	Select events for assessment for the ship type in question
40	Hazard	Identify event-specific hazards	Identify event-specific hazards using databases or hazard identification techniques (if data do not exist)
50	Frequency	Identify event frequency	Identify overall event frequency using databases or fault tree techniques.
60	Consequence	Identify consequences of events	Identify consequences of events using databases or event tree techniques.
70	Design Risk	Identify probability of fatal events	Identify probability of fatal incidents for designs using databases or event tree techniques
80	RAC	Check risk acceptance criteria	Use databases to check risk for ship design configuration vs. fatal event criteria - when r < rac then store ship definition in database and go to 160, when r > rac then proceed
90	Risk Reduction	Reduce risks	Identify preventive and mitigating design features (rrm) to reduce risks for specific hazards and events using a safety knowledge base
100	Design Generation	Generate new ship designs	Implement rrm on unsafe design configuration resulting in a number of new ship design configurations for the considered events
110	RRM Effects	Identify effects	Determine the effects of rrm on risk using programs or databases and assess effects on other events
120	RAC	Check risk acceptance criteria	Use databases to check risk for rrm ship design configuration vs. fatal event criteria - when r < rac then proceed, when r > rac then reject rrm design
130	Cost-Benefit	Determine ICAF ratio	Determine ICAF ratio for safe rrm design configuration - when ICAF<CAC then store in database, when ICAF > CAC then reject
140	Ship performance effects	Safety and performance relationship	Identify relationship between rrm designs and design performance using virtual and numerical tools or data – when performance is OK proceed, otherwise reject design
160	Weighting	Weight event-based design configurations	Weight event-based design configurations and identify the best configuration based on route characteristics
170	Virtual design and final selection	Design virtual ships and select final ship design configuration	Build up virtual representations of the remaining ship design configurations and select final design configuration to be put forward to virtual operation
180	Virtual operation	Identify operational hazards and risk	Identify operational hazards and risks for ship configuration in virtual ship system
200	Refinement	Refine tools	Compare risk used in design process with operational risk (go to 70 to refine design risk); when no refinement needed then proceed
210	Finish	End program	Terminate program or go to 10

A DESIGN FOR SAFETY PROCEDURE

A case study on a passenger Ro-Ro vessel collision and grounding has been utilised to illustrate the design for safety procedure. Data generation has not been targeted and assumptions have been made where data was lacking, as focus was on the process and not the validity of the outcome.

Passenger Ro-Ro vessels are used as example ship and collision and grounding as example events. Consequences in the form of fatal/serious accidents are focused upon. The effects from other events may be implemented using the same approach as for collision and grounding. Risk reduction measures, or design features, are considered when a ship fails to meet risk acceptance criteria. In this case study, collision avoidance system (CAS), routing systems (RS) and internal arrangement below the freeboard deck (IA) are used as example design features. CAS influences collision frequency, RS influences grounding frequency, whilst IA influences collision and grounding consequences. Input to the process are four different design configurations, which are assumed to be generated by mapping information models on safety attributes, design features and market requirements, as illustrated in Figure 8.

The procedure is illustrated herein with a safety assessment case study implemented in a blackboard system application, which provides decision support. This application makes it possible to accommodate multiple accident events by adopting the same approach, whereby the effects from the various event-driven design configurations are assessed relative to each other. In such a scenario, it may be conflicting event-driven design features where the blackboard controls the process in order to derive the best overall design configuration.

Input

Four design configurations are used as input to the design for safety procedure. The type of ship guides the identification of operational events, hazards and risks.

Hazard identification

A passenger Ro-Ro vessel may experience accidents (events) when operating on a route with certain event frequency characteristics. Hazard identification is the normal process applied in order to determine potential hazards for the ship and operational area in question. Generic collision and grounding sub-events and their related hazards are listed in Table 5 and Table 6, respectively.

Risk Assessment

Hazards are the cause of events, which are assigned with a probability of occurrence, i.e. frequency, and an outcome, i.e. consequence, which combined make up a risk value for a serious/fatal accident to occur. The collision and grounding risks for the ship are thus compared with risk acceptance criteria and if the risk level is satisfactory, and it is not overly safe, the design configuration is accepted and stored in a database for later evaluations. If it is overly safe, design modifications are made. On the other hand, if risk acceptance criteria are violated risk reduction measures are considered.

Frequencies and consequences should be determined using risk assessment tools. However, this case study utilise existing data from such assessments [16, 17], based on statistical estimates, where the average frequency for a collision incident is 1.4 x 10^{-2} per ship year and for grounding is 1.0 x 10^{-2} per ship year. There is work underway to develop collision models estimating collision frequencies on any route replacing the use of statistics [1].

The probability of having serious/fatal casualties following a collision, i.e. the consequence, is estimated to be 3.7 x 10^{-3} per ship year and for grounding to be 2.0 x 10^{-3} per ship year. Examples of the consequences for passenger Ro-Ro vessel collisions and grounding are given in Table 7 and Table 8, respectively. In the future when virtual ship systems are developed collision and grounding consequences can be simulated and knowledge/information will replace/ complement the use of statistics/experience.

Table 5: Generic Collision Sub-events and Related Hazards

Generic collision events	Hazards
Collision under way causing minor damage to ferry	Poor visibility
Collision under way causing impact damage to ferry	Rough weather
Collision under way causing flooding on ferry	Navigational confusion
Collision under way causing fire on ferry	Unauthorised route
Ferry strikes other berthed ship	Failure to use radar
Passing ship strikes ferry at berth	Excessive speed
	Steering failure
	Engine failure
	Power failure

Table 6: Generic Grounding Sub-events and Related Hazards

Generic grounding events	Hazards
Grounding under way causing minor damage to ferry	Poor visibility
Grounding under way causing impact damage to ferry	Rough weather
Grounding under way causing flooding on ferry	Navigational confusion
Grounding under way causing fire on ferry	Unauthorised route
	Failure to use radar
	Excessive speed
	Steering failure
	Engine failure
	Power failure

Table 7: Collision Consequences

Collision consequences
Extensive damage and holes to starboard side aft
Capsized and grounded – total loss
Severely damaged on side and holed in car deck above waterline
Holed on side aft above and below the waterline
Hole in side, flooded, caught fire, towed, and beached

Acceptance criteria

Risk acceptance criteria may vary between the probability of having an accident and the consequences in form of fatal accidents. The criteria for probability may also vary between the different types of accidents; e.g. criteria for the probability of fire in the accommodation area may be different to the criteria for the probability of collision with the quay. Criteria for consequences when having an accident may vary between regions in the world and also between different industries and companies. In this respect, it is important that the relevant risk criteria (probability and consequences) for the ship and its functions are determined prior to the safety assessment process.

This case study focuses on the consequences when having an accident and ways of reducing the risk of having fatal accidents. In such a scenario the maximum tolerable risk for passengers is estimated to be 1×10^{-4} whilst the risk is considered negligible when it is lower than 1×10^{-6}. When the risk is in the ALARP region, i.e. between these values, risk reduction measures must be applied to reduce it to a level as low as reasonable practical utilising design (technical) measures and cost-benefit analysis.

Results

A fictitious risk assessment results in that three ships are assessed to be collision safe, whilst one ship is considered unsafe and is put forward in the process where risk reduction measures pertaining to collision must be evaluated. Similarly, three ships are assessed to be grounding safe, whilst one ship is considered unsafe and put forward in the process where risk reduction measures pertaining to grounding must be evaluated. This is illustrated in Table 9 and Table 10, respectively, where the fatality risks listed are assumptions and based upon subjective judgement. The safe design configurations are stored in a database for later evaluations.

Table 8: Grounding Consequences

Grounding consequences
Extensive damage on bottom plating
Grounded leading to total loss
Grounded and stuck for three days
Grounded and heeled causing harm to passengers
Grounded and damaged onshore facilities

Risk reduction measures

Utilising the defined risk acceptance criteria and the risk assessment results, safety critical aspects can be identified using event trees to identify possible outcomes of the initiating event (ref. Appendix) and thus create information on issues to be targeted in the design process.

The ship designs assessed to be unsafe for the considered events will be treated separately in the risk reduction process and the resulting solutions be weighted and further evaluated in a virtual design process. There is a general distinction between operational and technical risk reduction measures, where the latter is targeted in this work and the former is subject to ship operation procedures. It should be noted that there is a fine line when choosing between operational or technical risk reduction measures in real-life scenarios.

Design features

Technical risk reduction measures, or risk-reducing design features, must be considered for design configurations that do not to comply with risk acceptance criteria for the considered events. The effects of these design features are based upon experience or by utilising tools to simulate their effects.

In the collision case study, two different collision avoidance systems (CAS) and three different internal arrangements below the freeboard deck (IA) are utilised to illustrate the design for safety process considering the "unsafe" Ship B in Table 9. The grounding case study utilises two different routing systems (RS) and three different internal arrangements below the freeboard deck (IA) to illustrate the design for safety process considering the "unsafe" Ship C in Table 10.

New design configurations

The CAS and IA provide eleven different ship design configurations for the collision case study, listed in Table 11, which are assumed as generated.

The RS and IA provide the same set of assumed design configurations, as listed in Table 12.

The generation of ship design configurations by varying the design features may be automated in order to make this process more efficient and improve the result. This is not considered in this paper.

Effects of risk reducing measures

Accident frequencies and consequences are affected by implementing risk-reducing design features. The task at hand is to estimate/determine the new frequencies and consequences. Simulation/ technical tools can be used to estimate the effect of the design features, if data do not exist. Collision and grounding simulator programs can obtain accident frequencies for the design configurations, whilst a safety/survivability program can determine their consequences, in terms of serious/fatal casualties. The risk levels of the various design configurations are compared with risk acceptance criteria. The configurations violating the criteria are rejected, whilst the configurations complying with the criteria are used for trade-off analyses in further evaluations.

The event and consequence effects of CAS and IA design features on Ship B+rrm are presented in Table 11, and similarly the event and consequence effects of RS and IA design features on Ship C+rrm are presented in Table 12. The numbers presented are estimates, as no tools have been utilised in this case study, and must not be read as anything else than examples.

The effects of the applied design feature configurations on other events, i.e. event effects, are in this case study assumed to be acceptable and do not negatively influence the safety level of other events. This is illustrated in Tables 11 and 12. In further work emphasis must be paid to evaluate the effects of design features, which reduces risk for one event, on other safety critical events.

Cost-benefit analysis

The cost-benefits of the accepted collision and grounding design features are illustrated in Tables 13 and 14, respectively. The benefits are derived from the effects of the design features in terms of reducing the consequences of an accident and presented in terms of averted fatalities per ship year. Note that the relationship between risk of fatalities and fatalities averted per ship year is based upon assumptions in this case study and further work must be assigned to find true and valid relationships. The costs of implementing the design features on the ship can be estimated using information from manufacturers. The ICAF is calculated based upon the cost/benefit relationship, which is compared with a cost criterion for averting a fatality, standing currently at approximately £2m in Northwest Europe.

The illustrated cost-benefit analysis for the collision study in Table 13 results in two acceptable configurations and the grounding study in Table 14 results in three acceptable configurations. The accepted configurations are further evaluated for their effects on ship performance.

Effects on ship performance

The effects of the utilised risk reduction measures, or design features, on ship performance must also be assessed in order to determine their practical viability. The design configurations deriving from the above process for the collision and grounding studies must be assessed for a number of design tasks and this case study fictively assesses the influence on seakeeping, resistance, cargo space and economy, as illustrated in Tables 15 and 16, respectively.

An overall decision is made on the applicability of the design features based on the influence of the individual performance criteria. A fictitious performance criterion of 0.5 is applied and utilised where all effects must satisfy the criterion. In this case study no real assessments are carried out and it only illustrates the process of how the effects on ship performance should be carried out. For a real-life case study thorough performance tests must be carried out using the appropriate tools.

Weighting

Five design configurations derive from the above case study, one configuration that results from the assessment pertinent to collision, two configurations that result from the assessment pertinent to grounding and two configurations assessed to be safe at an earlier stage.

The resulting five configurations are then evaluated using a weighting function, e.g. decision support tools, in order to select suitable candidates, which will be put forward in the design process. The weighting process ensures that the effort into building virtual models is limited.

The weighting function has not been developed in this case study and the process only assumes that such function, e.g. Saaty's Analytical Hierarchical Process, can be utilised in order to select the suitable design configurations to put forward for virtual design. Fictitious weighting results, with an assumed selection criterion of 0.75, are illustrated in Table 17 where two design configurations are put forward to virtual design and three rejected.

The virtual design and virtual operation processes are assumed processes in this case study and not further elaborated upon. More work is needed to include these two areas in the overall design for safety process.

Table 9: Collision Risk Assessment Results

Design configuration	Risk of fatalities per ship year	Risk Acceptance Criteria	Risk assessment	Judgement
Ship A	1.2×10^{-5}	1.0×10^{-4}	Safe	Store
Ship B	1.0×10^{-3}	1.0×10^{-4}	Unsafe	Consider rrm
Ship C	4.5×10^{-5}	1.0×10^{-4}	Safe	Store
Ship D	7.1×10^{-5}	1.0×10^{-4}	Safe	Store

Table 10: Grounding Risk Assessment Results

Design configuration	Risk of fatalities per ship year	Risk Acceptance Criteria	Risk assessment	Judgement
Ship A	1.2×10^{-5}	1.0×10^{-4}	Safe	Store
Ship B	4.1×10^{-5}	1.0×10^{-4}	Safe	Store
Ship C	1.0×10^{-3}	1.0×10^{-4}	Unsafe	Consider rrm
Ship D	7.1×10^{-5}	1.0×10^{-4}	Safe	Store

Table 11: Effects of Collision Risk-reducing Design Features on Ship B+rrm

Collision Design Feature Configurations	Event effects	New Consequence (Risk of fatalities per ship year)	Risk Acceptance Criteria	Decision
CAS 1	OK	1.2×10^{-4}	1.0×10^{-4}	Reject
CAS 2	OK	1.0×10^{-4}	1.0×10^{-4}	OK
IA 1	OK	4.4×10^{-4}	1.0×10^{-4}	Reject
IA 2	OK	2.1×10^{-5}	1.0×10^{-4}	OK
IA 3	OK	4.5×10^{-4}	1.0×10^{-4}	Reject
CAS 1 & IA 1	OK	7.7×10^{-4}	1.0×10^{-4}	Reject
CAS 1 & IA 2	OK	8.0×10^{-6}	1.0×10^{-4}	OK
CAS 1 & IA 3	OK	9.1×10^{-4}	1.0×10^{-4}	Reject
CAS 2 & IA 1	OK	7.2×10^{-5}	1.0×10^{-4}	OK
CAS 2 & IA 2	OK	3.4×10^{-6}	1.0×10^{-4}	OK
CAS 2 & IA 3	OK	8.1×10^{-5}	1.0×10^{-4}	OK

Table 12: Effects of Grounding Risk-reducing Design features on Ship C+rrm

Grounding Design Feature Configurations	Event effects	New Consequence (Risk of fatalities per ship year)	Risk Acceptance Criteria	Decision
RS 1	OK	1.2×10^{-4}	1.0×10^{-4}	Reject
RS 2	OK	1.0×10^{-4}	1.0×10^{-4}	OK
IA 1	OK	4.4×10^{-4}	1.0×10^{-4}	Reject
IA 2	OK	2.1×10^{-5}	1.0×10^{-4}	OK
IA 3	OK	4.5×10^{-4}	1.0×10^{-4}	Reject
RS 1 & IA 1	OK	7.7×10^{-4}	1.0×10^{-4}	Reject
RS 1 & IA 2	OK	8.0×10^{-6}	1.0×10^{-4}	OK
RS 1 & IA 3	OK	9.1×10^{-4}	1.0×10^{-4}	Reject
RS 2 & IA 1	OK	7.2×10^{-5}	1.0×10^{-4}	OK
RS 2 & IA 2	OK	3.4×10^{-6}	1.0×10^{-4}	OK
RS 2 & IA 3	OK	8.1×10^{-5}	1.0×10^{-4}	OK

Table 13: Cost-benefits of Collision Design Feature Configurations on Ship B+rrm

Collision Design Feature Configurations	Cost (£m per ship year)	Benefits (Fatalities averted per ship year)	ICAF (£m)	Cost-criteria (£m)	Decision
CAS 2	0.05	0.02	2.5	2.0	Reject
IA 2	0.20	0.05	4.0	2.0	Reject
CAS 1 & IA 2	0.23	0.11	2.1	2.0	OK
CAS 2 & IA 1	0.25	0.13	1.9	2.0	OK
CAS 2 & IA 2	0.17	0.04	4.2	2.0	Reject
CAS 2 & IA 3	0.16	0.06	2.7	2.0	Reject

Table 14: Cost-benefits of Grounding Design Feature Configurations on Ship C+rrm

Grounding Design Feature Configurations	Cost (£m per ship year)	Benefits (Fatalities averted per ship year)	ICAF (£m)	Cost-criteria (£m)	Decision
RS 2	0.10	0.02	5.0	2.0	Reject
IA 2	0.20	0.11	1.8	2.0	OK
RS 1 & IA 2	0.17	0.04	4.3	2.0	Reject
RS 2 & IA 1	0.23	0.12	1.9	2.0	OK
RS 2 & IA 2	0.16	0.06	2.7	2.0	Reject
RS 2 & IA 3	0.25	0.13	1.9	2.0	OK

Table 15: Effects of Collision Design Features on Ship Performance

Collision Design Feature Configurations	Resistance	Seakeeping	Cargo space	Economy	Overall Decision
CAS 1 & IA 2	0.9	0.71	0.58	0.76	OK
CAS 2 & IA 1	0.8	0.65	0.45	0.78	Reject

Table 16: Effects of Grounding Design Features on Ship Performance

Grounding Design Feature Configurations	Resistance	Seakeeping	Cargo space	Economy	Overall Decision
IA 2	0.4	0.7	0.6	1.0	Reject
RS 2 & IA 1	0.8	1.0	0.7	0.8	OK
RS 2 & IA 3	0.6	0.9	0.9	0.7	OK

Table 17: Weighting of Resulting Design Configurations

Ship design	Risk reducing design features	Risk of fatalities per ship year	Weight	Decision
Ship A		1.2×10^{-5}	0.88	OK
Ship B	CAS 1 & IA 2	8.0×10^{-6}	0.87	OK
Ship C	RS 2 & IA 1	8.0×10^{-6}	0.67	Reject
	RS 2 & IA 3	3.4×10^{-6}	0.64	Reject
Ship D		7.1×10^{-5}	0.72	Reject

APPLICATION

The case study addresses the application of the developed methodology, with the aim to demonstrate the validity of the process adopted. It mainly draws from the area of *Design for Survivability*, a field of particular importance within the broad spectrum of *Design for Safety*. The application focuses on collision incidents and possible outcomes, by carrying out a study of safety-critical design features using damage stability and survivability calculations and criteria as the means to obtain the benefits on risk reductions and concludes with the selection of the more appropriate alternative through design trade-offs between different safety and cost-effectiveness or performance measures.

An existing passenger Ro-Ro vessel, of typical size and layout for the Northwest European routes[2], has been used as the example ship. The vessel has been built at the early 1980s and complies with the SOLAS 74 two-compartment damage stability standard. The three upgrading alternatives considered are providing marginal compliance to the SOLAS 90 two-compartment standard. The particulars of the alternatives are as follows:

- **Sponson alternative (S):** Sponson fitted at the midship area of the ship with length of 51.75 m. width of 0.7 m and of conventional configuration (vertical extent from below waterline to a depth above the car deck level).
- **Transverse bulkheads alternative (B):** Four transverse bulkheads of 4 m height have been considered installed on the car deck.
- **Combination alternative (C):** Two transverse bulkheads of 4 m height have been considered installed on the car deck (at the same locations with the aft and fore bulkheads of the previous alternative), together with the fitting of a sponson at the midship area of the ship with length of 31.5 m, width of 0.8 m and of conventional configuration.

Hazard identification and analysis of the associated risks have not been performed for the case study. Instead, generic figures for collision incidents obtained from the safety assessment in the Joint North West European Project have been used. These figures are obtained from relevant accidental data, theoretical models and expert judgements, and are therefore applicable in the main for this study. More specifically, the event tree for collision outcomes contained in [16] is used. For ease of reference, the event tree is reproduced in the Appendix. The considered final outcomes for a collision incident have been associated with their potential to cause a number of fatalities or no fatalities at all. In this respect, a categorisation of the final outcomes that can potentially cause minor, considerable, or major number of fatalities is possible. In Table 18 the aggregation of the frequencies per ship year relevant to the final outcomes of a collision incident is shown.

Damage stability calculations have been carried out for the existing vessel and the three upgrading alternatives for one-, two- and three-compartment damages cases, assuming B/5 inboard penetration for all the cases (standard SOLAS notation). Elements of the probabilistic framework developed during the Joint North West European Project have been utilised [17] for the aggregation of results to form comparable measures of safety levels achieved by the different alternatives considered.

The Attained Index is considered as follows:

$$A_i^* = \sum_1^i p_i^* \times s_i^* \qquad (1)$$

with i = 1, 2, 3 and

$$p_i^* = \int_{l/L-\varepsilon}^{l/L+\varepsilon} f(x)dx \qquad (2)$$

the probability that a compartment or combination of compartments are being flooded, with l the flooded compartment(s) length, f(x) the distribution for damage length used in the Joint North West European Project, and ε a small positive number. In the current context, this probability is used as a weight that a certain damage will occur. Finally,

$$s_i^* = \sqrt[4]{GZ_{max} \times Range \times Area} \qquad (3)$$

with GZ_{max}, positive range of and area under the GZ curve corresponding to the actual attained values of the damage case in question. In the current context, this factor is not expressing a probability, but a figure which when aggregated will express the full benefit offered by an alternative.

Table 19 contains the calculated increments on the attained indices of the three upgrading alternatives (**S**, **B** and **C**) against the existing ship's corresponding indices. The results clearly suggest that the installation of transverse bulkheads on the Ro-Ro deck is the best solution from safety enhancement point of view.

These increments can be directly applied to calculate the benefits and residual risks for the three alternatives, based on the original risk levels contained in Table 18. It is assumed therefore, that the risk levels in Table 18 correspond to the existing ship configuration. Furthermore, a correspondence mapping is possible if we consider that one-compartment damage cases relate to incidents involving minor number of fatalities, and one- and two-compartment damage cases, and one-, two- and three-compartment damage cases relate to considerable and major number of fatalities respectively. Table 20 contains the benefits and the residual risk levels for the three alternatives.

---

[2] Main Particulars: L_{OA}=167.75 m, L_{BP}=154.5 m, B=24 m, D=8.1 m (main car deck), T=5.7 m, centre casing on the main car deck, 16 watertight compartments below the main car deck.

The risk levels contained in this table can be compared against societal risk criteria, applicable to passenger Ro-Ro vessels. Such criteria have been proposed as part of the work of the safety assessment study of the Joint NW European Project, in the form of the F-N curve. As it can be seen from Figure 15 the original level of risks for the existing ship is generally high in the ALARP region, even in the intolerable risk region for minor number of fatalities, which coincides with the general observation for the level or risks for existing passenger Ro-Ro ferries.

In order to calculate the ICAF, a cost model has been developed appropriate for the problem. It takes into account the cost associated with procurement and installation or fitting of the devices (a period of 15 years of remaining operational life with a 5% discount rate have been assumed), and the effects on payload, manning, maintenance and fuel consumption. Table 21 contains the comparisons of the different alternatives with respect to the calculated ICAF. These results now suggest that accounting for costs incurred following the installation of safety enhancing devices, the sponson solution is the clear favourite.

Measures costing less than £2m per fatality averted are considered cost-effective and should be adopted. Measures costing more than £50m are not considered cost-effective and will not normally be adopted. Measures having a cost in the range between £2m and £50m must be carefully evaluated according to their benefits, which is the case for all three alternatives considered.

In order to decide on the alternative to consider for the upgrading of the vessel, trade-offs between safety measures and cost-effectiveness or performance indicators are necessary. Safety measures can be expressed with the levels of residual risks or the benefits obtained (Table 20). Cost-effectiveness can be expressed with the annual cost or the calculated ICAF as contained in Table 21. A performance indicator can be obtained through the considerations of Table 22, where it is now shown that the combination alternative is the solution that must be preferred. The ranking of the alternatives is achieved when considering the five areas (resistance, manoeuvring, loading time, survivability in waves and capacity) and the effect expressed on the scale 1 to 3, with the notation that the higher the figure the more effective the alternative is. The ranking is based on judgements and on the actual performance of the vessel. No weighting among the areas has been considered, in order to maintain the generic nature of the application. In this respect, the highest possible score for a design is 15.

The trade-offs contained in Table 24 prove that the alternative that utilises a combination of bulkheads and sponson can be considered as the best solution to the problem. This result coincides with the decision taken by most of the owners/operators that have already undertaken upgrading of their vessels.

CONCLUSION

A formalised *Design for Safety* methodology and procedure have been described with particular reference paid to the essential elements necessary for the required endeavours to be met. Safety assessment should become an integral part of the ship design process. An appropriate integrated environment has been defined and the vehicle to act as the platform for the development has been identified, through a detailed account of the benefits that blackboard systems technology can offer to ship design. Based on the above, initial software tools development has been briefly described. Finally, a case study has proven the validity of the methodology adopted, in a process that technical elements of safety (i.e. damage survivability in waves) have been considered in parallel with the classical ship design tasks.

The approach adopted developing the *Design for Safety* methodology can be made generic to accommodate other issues. The approach describes the current situation for the topic in question (reality), the development and utilisation of information models and integrated design environments resulting in an overall methodology and procedure for future application in industry/academia.

Table 18: Final Outcomes of Collision Incidents (frequencies per 1000 ship year)

Incidents involving	ID Codes for Final Outcomes	Frequencies
No Fatalities	C2.1, C3.1.1, C2.3, C3.2.1, C5/C6	3.9916
Minor Number of Fatalities	C1, C2.2, C4.1, C4.3	26.8212
Considerable Number of Fatalities	C3.1.2, C4.2, C3.2.2, C4.4	0.1398
Major Number of Fatalities	C3.1.3, C4.5	0.08991
Total		**31.04251**

Table 19: Attained Indices of Upgrading Alternatives

Alternative	A_1*	A_2*	A_3*
Sponson (S)	32.7%	36.5%	38.9%
Bulkheads (B)	45.5%	51.3%	53.4%
Combination (C)	44.7%	49.1%	51.9%

Table 20: Benefits and Residual Risks for the Three Alternatives (frequencies per 1000 ship year)

Incidents involving	Existing Ship		Sponson Alternative (S)		Bulkheads Alternative (B)		Combination Alternative (C)	
	Benefit	*Risk*	*Benefit*	*Risk*	*Benefit*	*Risk*	*Benefit*	*Risk*
No Fatalities		3.992		3.992		3.992		3.992
Minor Number		26.821	8.744	18.077	12.177	14.644	11.989	14.832
Considerable		0.139	0.051	0.088	0.071	0.068	0.068	0.071
Major Number		0.089	0.035	0.054	0.048	0.041	0.046	0.043
Total		**31.041**	**8.83**	**22.211**	**12.296**	**18.745**	**12.103**	**18.938**

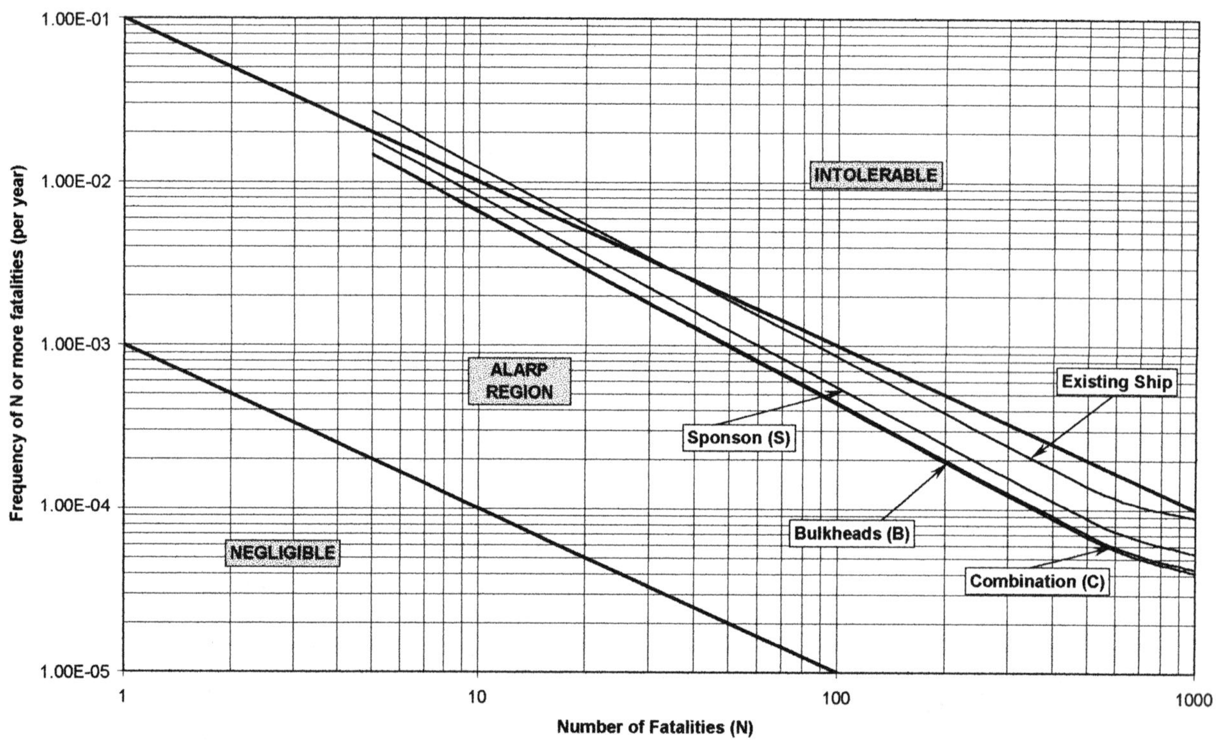

Figure 15: F-N Curve for the Existing Ship and the Three Upgrading Alternatives

Table 21: Comparison of Alternatives with respect to ICAF

Alternative	Benefit	Cost (£ per Year)	ICAF (£m per Fatality Averted)
Sponson (S)	8.83×10^{-3}	240,000	27
Bulkheads (B)	12.296×10^{-3}	605,000	49
Combination (C)	12.103×10^{-3}	408,000	34

Table 22: Performance Indicators

Design Aspect	Sponson Alternative (S)	Bulkheads Alternative (B)	Combination Alternative (C)
Resistance	1	3	2
Manoeuvring	1	3	2
Loading Time	3	1	2
Survivability	1	2	3
Capacity	3	1	2
TOTAL	**9**	**10**	**11**

Table 23: Design Trade-offs

Trade-offs	Sponson Alternative (S)	Bulkheads Alternative (B)	Combination Alternative (C)
Safety (Risk Benefit) Vs. Performance	(8.83×10^{-3}, 9) (0.718, 0.6) 0.489	(12.296×10^{-3}, 10) (1, 0.667) 0.333	(12.103×10^{-3}, 11) (0.984,0.733) **0.267**
Safety (Residual Risk) Vs. Performance	(22.211×10^{-3}, 9) (0.844, 0.6) 0.429	(18.745×10^{-3}, 10) (1, 0.667) 0.333	(18.938×10^{-3}, 11) (0.989, 0.733) **0.267**
Safety (Risk Benefit) Vs. Cost-Effectiveness (Cost per Year)	(8.83×10^{-3}, 240000) (0.718, 1) **0.282**	(12.296×10^{-3}, 605000) (1, 0.397) 0.603	(12.103×10^{-3}, 408000) (0.984, 0.588) 0.412
Safety (Residual Risk) Vs. Cost-Effectiveness (ICAF)	(22.211×10^{-3}, 27) (0.844, 1) **0.156**	(18.745×10^{-3}, 49) (1, 0.551) 0.449	(18.938×10^{-3}, 34) (0.989, 0.794) 0.206
Performance Vs. Cost-Effectiveness (Cost per Year)	(9, 240000) (0.6, 1) **0.4**	(10, 605000) (0.667, 0.397) 0.689	(11, 408000) (0.733, 0.588) 0.491
Performance Vs. Cost-Effectiveness (ICAF)	(9, 27) (0.6, 1) 0.4	(10, 49) (0.667, 0.551) 0.559	(11, 34) (0.733, 0.794) **0.337**

Explanation: The values in the first line for each trade-off are the actual values attained by the alternative. In the second line the corresponding relative values are contained where (1, 1) is the ultimate best design. Finally, the distance from the point (1, 1) is calculated. The alternative that is closer to point (1, 1) for each trade-off is highlighted in bold.

REFERENCES

[1] Vassalos, D. (1997), "A Thematic Network on 'Design for Safety' – An Integrated Approach to Safe Ship Design", Proceedings of the Sixth International Marine Design Conference (IMDC 1997), University of Newcastle upon Tyne, 23-25 June.

[2] Vassalos, D. (1999), "Design for Safety", Keynote Address, Proceedings of the WEGEMT Design for Safety Conference, University of Strathclyde, 25-28 October.

[3] Papanikolaou, A. and Konovessis, D. (1999), "Safety in Ship Design: Review of Fundamental Concepts and Methodologies", Proceedings of the WEGEMT Design for Safety Conference, University of Strathclyde, 25-28 October.

[4] Oestvik, I. (1999), "Integrating Safety Assessment in the Ship Design Process", Proceedings of the WEGEMT Design for Safety Conference, University of Strathclyde, 25-28 October.

[5] Evans, J.H. (1959), "Basic Design Concepts", Journal of the American Society of Naval Engineers (ASNE), November, pp. 671-678.

[6] Buxton, I.L. (1987), "Engineering Economics and Ship Design", BMT Publication, 3rd Edition.

[7] Andrews, D. (1981), "Creative Ship Design", Transactions of the Royal Institution of Naval Architects (RINA), Volume 123, pp. 447-471.

[8] Mistree, F., Smith, W.F., Bras, B.A., Allen, J.K. and Muster, D. (1990), "Decision-Based Design: A Contemporary Paradigm for Ship Design", Transactions of the Society of Naval Architects and Marine Engineers (SNAME), Volume 98, pp. 565-597.

[9] State of the Art Report on "Design Methods", the Sixth International Marine Design Conference (IMDC 97), Volume 2: State of the Art Reports, 23-25 June 1997, University of Newcastle upon Tyne.

[10] Erman, L.D., Hayes-Roth, D.F., Lesser, V.R. and Reddy, D.R. (1980), "The Hearsay II Speech – Understanding System: Integrating Knowledge to Resolve Uncertainty", ACM Computing Surveys, Vol. 12, No. 2.

[11] Nii, H.P. (1986), "Blackboard Systems: The Blackboard Model of Problem Solving and the Evolution of Blackboard Architectures" (Part 1), "Blackboard Systems: Blackboard Application Systems, Blackboard Systems for a Knowledge Engineering Perspective" (Part 2), The AI Magazine, Summer.

[12] Duffy, A.H.B., Duffy, S.M., Andreasen, M.M. and Rimmer, D. (1995), "Team Engineering within the Context of Product Design Development", Strathclyde University CAD Centre Paper, CADC\P\95-13, 3 August.

[13] Corkill, D.D. (1991), "Blackboard Systems", AI Expert, Vol. 6, No. 9.

[14] Fiegenbaum, E.A. (1988), "Foreword to Blackboard Systems", edited by R. Engelmore and T. Morgan, Addison Wesley Publishing Company.

[15] Oestvik, I., Konovessis, D. and Vassalos, D. (1999), "Application of Blackboard Systems to Ship Design", International Conference on Computers and Ships, The Institute of Marine Engineers, London, 11-12 May.

[16] Spouge, J. (1996), "Safety Assessment of Passenger RoRo Vessels", International Seminar on The Safety of Passenger Ro-Ro Vessels (presenting the results of the NorthWest European R&D Project), The Royal Institution of Naval Architects, London, 7 June.

[17] Rusaas, S., Jost, A.E.E. and Francois, C. (1996), "A New Damage Stability Framework Based Upon Probabilistic Methods", International Seminar on The Safety of Passenger Ro-Ro Vessels (presenting the results of the NorthWest European R&D Project), The Royal Institution of Naval Architects, London, 7 June.

APPENDIX: Final Outcomes for Collision Incidents [16]

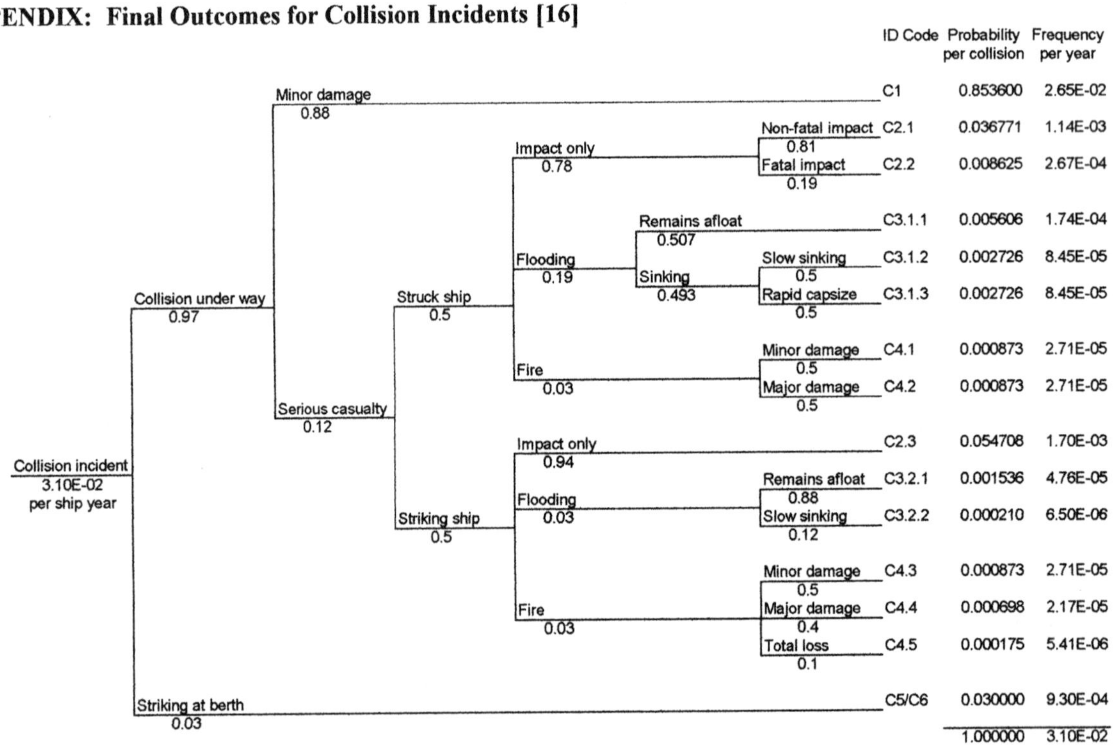

Discussion

Rolf Skjong, Member

The authors have described a Design For Safety (DFS) methodology, implemented in a virtual environment called Blackboard systems.

The DFS methodology is presented as a standard Risk Based Design (RBD) methodology, including thoughts about risk perception and human factors. The paper is hard to grasp by risk specialists, as the terminology deviates from most standards, e.g., IMO (1997), ISO (1999).

We agree that RBD methodologies represent a paradigm where all important issues may be integrated. Within RBD, and relating to structures, the classical Probabilistic Design (PD) method is applicable. We have a lot of experience with these methods in the offshore sector, where the Safety Case Regimes allows for the application of RBD and PD. For DNV classed marine structures PD may be approved according to DNV (1992).

Within the Flag State Regime this is currently not a legal option, except as methods for documentation of safety equivalency. SOLAS currently only allows for technical equivalency, still safety equivalency is sometimes accepted. To demonstrate safety equivalency is currently a high-risk option for the owner, because of the risk of presenting a design that will not be approved, or where changes must be implemented late in the project.

The authors' description of the application of the methods in a virtual environment, is virtual in more than one way. For example, it is assumed that risk acceptance criteria exist and may readily be used. This is not the case. The status is that such criteria will be required in Formal Safety Assessment (FSA). A suggestion on risk acceptance criteria was prepared by Norway, MSC 72/16. When IMO has agreed on such criteria, their main use is within IMO as a basis for their own decisions based on FSA, resulting in Risk Based Rules/Regulations (RBR). As a side effect, the use of a safety equivalency principle will be made easier.

Sometimes, the authors could benefit from presenting more evidence. For example, one of the main reasons for going back to more use of rules in the offshore sector, is the fact that risk assessment tended to deliver results far too late, resulting in costly modifications. Therefore, development of RBR by FSA, and subsequent design by RBR may be a better option than RBD. PD methods as a basis for RBR development are also a mature technology, see Skjong et al. (1996). Some documentation of cost differences using the various options may be found in Hauge et al. (1992).

References

DNV (1992) Classification Note 30.6 "Structural Reliability Analysis of Marine Structures."

Hauge, L., Løseth, R. and Skjong, R. "Optimum Code Calibration and Probabilistic Design", OMAE-92.

IMO (1997) "Interim Guidelines for the Application of Formal Safety Assessment (FSA) to the IMO Rule-Making Process." MSC Circ.829/ MEPC/Circ.335.

ISO (1999) "The Third Working Draft if Risk Management Terminology," ISO/TMB WG in Risk Management Terminology.

MSC 72/16 "Decision Parameters Including Acceptance Criteria."

Skjong, R. E., Bitner-gregersen, E., Cramer, A., Croker, Ø., Hagen, G., Korneliussen, S., Lacasse, I., Lotsberg, F., Nadim, K. O., Ronold (1996) "Guideline of Offshore Structural Reliability-General," DNV Report 95-2018.

Benjamin L. Nicholson, Member

[The views expressed here are those of those of the discusser, and not necessarily those of the U.S. Coast Guard or the Department of Transportation.]

The authors have provided SNAME with a thought-provoking paper about a very broad and tenuous area of engineering. Safety is no longer viewed as an afterthought—one that is taken into account well after a vessel is launched. Furthermore, safety is no longer a luxury—one that can be incorporated by those with the resources and wherewithal to employ safety management *after* financial stability. Simply put, safety is an attitude . . . one that must be cultivated within the highest levels of management and one that must be carefully fostered throughout a given company all the way to the deck plates of a given vessel.

The authors provide a robust methodology for improving the safety climate of today's integrated environments. Because companies have become increasingly multi-dimensional, fostering a culture that truly embodies safety becomes more and more difficult. The authors' approach is one that accounts for safety concurrently throughout the design spiral process. While this may sound broad and elusive, it is well grounded in its philosophy. How can a holistic level of safety be improved if each and every aspect of a vessel is not designed with safety in mind? Thus, after reading this paper, it is manifest that safety must be integrated into *all* aspects of a given design in order to achieve the most robust and reliable outcome. Simply put, a safe design is one that is a sum of safe parts.

However, these "parts" are much more that steel, pipes, and casings. Rather, these "parts" include the human element with respect to labor, supervision, and executive management. Ultimately, all of these aspects will continually have influence over a given design throughout its lifecycle. Such an approach reveals the true premise contained within this paper: that safety is dynamic and the culture that surrounds a ship must embrace the attributes of safety and be utterly proactive in their operations.

The long-term effects are multi-fold. Not only is the maritime community and environment better protected by a company that embraces safety in such a way, but the company will ultimately realize greater financial success (fewer accidents, fewer potential violations, fewer malfunctions and breakdowns). The initial investment is minor when compared with the ultimate return. This paradigm shift has not been fully realized throughout the maritime community. Therefore, the authors' presentation of such a rational and robust methodology will most certainly improve the recognition of safety as an instrumental element of maritime management and operation.

Pierre C. Sames, Visitor, Germanischer Lloyd AG, and **Anneliese Jost,** Member

We congratulate the authors for their fine work. They presented a new approach to integrate safety in the design from the start. The following discussion concentrates on the example application of the method proposed by the authors. Germanischer Lloyd, as partner of the running thematic network SAFER EURORO (funded by the European Commission), monitor ongoing research related to structural safety of passenger RO-RO vessels and integrate the findings in the overall Design-for-Safety theme of the network.

Classification societies are usually not involved in the early design phase. Normally, they check—at a later design phase—

a given design for rule conformity and with regard to statutory requirements. This approach is comparable to adding safety features in the later design stage. If designers include safety features in the beginning of the design phase, classification societies will probably follow suit. Then rules should become suitable for application in a Design-for-Safety method, implying that compliance may have to be probabilistically approved.

The application presented by the authors focuses on events that are initiated either by collision or grounding. Existing data for probability of collision and grounding are used. Current research [1] funded by the European Commission resulted in a simulation tool that can be employed within the presented framework to predict probabilistic collision damage distributions [2]. Appropriate input parameters account for traffic conditions and route layout. Predefined causation factors account for environmental conditions and bridge layout. Results comprise probability of collision (for a given route with a fixed traffic pattern) and probabilistic distributions of damage size. Considering grounding, simulation tools are not available for the prediction of probabilistic damage distributions that depend on conditions along the route, i.e., distribution of obstacle size. To fully exploit modern probabilistic prediction tools, an effort should be made to gather up-to-date traffic data. In addition, accurate information on seabed topology is needed for grounding analyses.

The probability of occurrence and the probabilistic damage distribution are prerequisites for the determination of consequences. For each event (described by, e.g., damage size and location) an analysis of damage stability is required to assess whether the vessel remains afloat. Table 7 and 8 list five classes of collision consequences, ranging from minor to substantial damage including the total loss of the vessel. To assess the consequence of a given damage, a fast and reliable method to determine damage stability for each damage scenario is required. In particular, the prediction of total loss must be extremely accurate if risk analysis should yield sensible results. This link between collision analysis and damage stability has not been established in a way suitable for quantitative risk analysis. Some of the necessary research was conducted within the Joint North West Research Project on RO-RO Ferry Safety and is currently underway within the European research project HARDER [3].

Effort is underway to establish a Design-For-Safety method. If this new approach were employed in the future, we ask the following question: how should the ship operator respond to a change in risk level when traffic density changes?

Additional references

[1] Work program of DEXTREMEL (1997), "Design for structural safety under extreme loads", Brite Euram III project BRPR-CT97-0513, http://research.germanlloyd.org/Projects/DEXTREMEL/index.html.

[2] P. T. Pedersen and S. Zhang (1999), "Collision Analysis for MS Dextra," open workshop of thematic network SAFER EURORO, April 1999, Nantes, France.

[3] Work program of HARDER (1999), "Harmonisation of rules and design rationale," GROWTH project G3RD-1999-00028.

Authors' Closure

The authors are most grateful to the contributors to the discussion and for all the encouraging views expressed. The discussers have raised many valid points, which we will try to address in turn as reference to each individual contributor.

The authors would like to thank *Dr. Skjong* for his contribution and for the unintended compliment of referring to us as non-risk specialists. However, his comment on Blackboard Systems and his prompting for consultation of structural reliability papers for risk-based design demonstrate the need for designers and risk specialists to start being engaged into closer exchanges. In this respect, and in his position, Dr Skjong, a self-confessed risk specialist, will be in a prime position to help.

As indicated in the paper and described in detail in the quoted references [10–14], blackboard systems provides a platform for managing and controlling information flow in design, in supporting the development of a Design for Safety methodology based on concurrent engineering principles. It is certainly not a virtual environment! More specifically, it follows a design co-ordination paradigm, which can accommodate classical techno-economic aspects of ship design and operation while accommodating safety aspects assessed on the basis of first-principles and performance-based methods through suitable trade-offs. Our overall goal is to achieve maximum safety at minimum cost. In this respect, safety itself is treated as a goal, as it should, rather than a constraint or a penalty that the owner has to pay for. As a side-development, blackboard systems have the capability of supporting virtual prototyping, through their extensive connectivity capabilities.

We are certain Dr. Skjong needs no reminding that Probabilistic Design methods for structural assessment is only one area of concern, among other areas that include ship survivability, passenger and crew survival, seaworthiness, fire safety and environmental protection. Of course knowledge from application to the offshore industry must be taken into account, but in a way that distinguishes the fundamental differences between shipping and offshore industries.

On the question of the safety criteria used in the paper, we are well aware of the problems towards establishing risk acceptance criteria and that a great deal of work is under way in this area. The safety criteria used in the paper follow closely the work of the Risk Assessment Team of the EU-funded funded Thematic Network SAFER EURORO, led by Dr. Skjong. Improving upon this will have to follow further developments, which of course are being eagerly awaited. Our present concern and that of the paper focus on the development of a design methodology.

Finally, it is not our research goal to propose a Design for Safety methodology that accounts only for modifications (or risk control options) on an already developed design, which seems to be the norm in the Formal Safety Assessment (FSA) jargon. Furthermore, rules have always been risk-based, whether formally or informally, but have all the same failed to promote cost-effectiveness in safety or a safety culture in shipping. The authors would disagree with Dr. Skjong on this point. Clearly, rules no matter how they derive, do not (cannot) take onboard the full benefit offered by the application of a methodology ***to*** design.

The contribution by *Dr. Sames* and *Mrs. Jost* is most welcome, and the authors would like to take this opportunity to record here their appreciation for the important contribution and impact of Germanischer Lloyd to rendering SAFER EURORO a successful EU Thematic Network, and thus indirectly, greatly influencing the development of the new Design for Safety methodology presented in our paper.

Responding to the first point of the discussers, it is our opinion that classification societies, through the accumulated knowledge and expertise they possess, should play an active role in the early stages of the ship design process. As it is widely known, most design decisions are taken during the very early stages of the design process, which in later stages is extremely expensive to alter. Feedback from the construction and operation phases is important in order to derive a safer, and in general more finely tuned ship design. Classification societies' role is paramount in this process, and we anticipate that the Design for Safety methodology proposed is capable of supporting such feedback.

We are well aware of the theoretical models for the prediction of collision and grounding frequencies, as well as for their input requirements. The example presented in the paper only focuses on the possible outcomes of a collision incident, using a generic overall frequency derived during the Joint Nordic Project [16, 17]. Due to the fact that the relevant research is still on going, theoretical models are now being implemented in the overall framework of Design for Safety, with particular reference to the area of Damage Survivability. A method to predict the damage survivability of a vessel in the design process, using the Static Equivalent Method [27], statistical distributions for damage extent and operational patterns, which is based on Monte Carlo simulations, is currently under development. The method has the potential of being extremely fast and reliable. We agree that the example presented in the paper is very general, but it presents most of the essential elements of the Design for Safety methodology at an acceptable level, as well as it allows for the important conclusion that trade-offs among various, contradicting design criteria should be present in order to reach a solution that is acceptable from any technical, economical or societal point of view.

Finally, only a partial answer is possible on the last question of Dr. Sames and Mrs. Jost, as a complete answer must await many years of focused research. This involves accommodating in the whole design process market analyses for the growth of the traffic in the area of operation, and apply the projections during the life cycle of the vessel.

The authors would like to thank *LTjg. Nicholson* for his very encouraging words for their work. It will undoubtedly be a major achievement if safety could be included within the ship design process. Furthermore, the potential of realizing a formal framework where safety considerations pertinent to both managerial and technical aspects of the whole life-cycle of a ship are taken into account in a systematic and all-embracing manner during design will do much to change today's haphazard practice, as this is skillfully described by the contributor.

In concluding, the authors hope that their presentation has raised safety awareness among the shipping community. Design for Safety is a concept that will take a great deal of effort to be fully developed, and this should include the realization that increased safety is by no means associated to increased incurred costs, which is considered to be the norm. It is our belief that as these ideas mature and materialize the full potential of applying the described concept will be more fully appreciated.

Additional reference

[27] Vassalos, D., Pawlowski, M. and Turan, O. (1996), "Criteria for Survival in Damaged Condition," International Seminar on The Safety of Passenger Ro-Ro Vessels (presenting the results of the NorthWest European R&D Project), The Royal Institution of Naval Architects, London, 7 June.

SNAME *Transactions*, Vol. 108, 2000, pp. 447-473

Structural Design and Response in Collision and Grounding

Alan Brown, Member, Virginia Tech, **Kirsi Tikka,** Member, Webb Institute, **John C. Daidola,** Fellow, AMSEC LLC, **Marie Lützen,** Student Member, Technical University of Denmark, and **Ick-Hung Choe,** Student, Pusan National University, Korea

Abstract

The results summarized in this paper represent the work of SNAME Ad Hoc Panel #6 convened under the SNAME Technical and Research Program. This is a summary and overview paper. Topics discussed will be addressed individually and in more detail in later publications. The 2nd International Conference on Collision and Grounding of Ships, to be held in Copenhagen, July 1-3, 2001, will also present and discuss many of the results of this panel and other related research. The paper discusses four primary areas of panel work: collision and grounding models, data, accident scenarios and design applications. A probabilistic framework for assessing the crashworthiness of ships is presented. Results obtained from various grounding and collision models are compared to validating cases and to each other. Data necessary for proper model validation and probabilistic accident scenario development are identified. Deformable striking-ship bow models and their application are described. Potential design applications and alternatives for improving crashworthiness are discussed.

1 INTRODUCTION

On May 14, 1998, SNAME Ad Hoc Panel #6, Structural Design and Response in Collision and Grounding, was formed under the SNAME T&R Steering Committee with support from the Hull Structure Committee, Ship Design Committee and Marine Safety and Environmental Protection Panel (O-44). The objectives of the panel are:

- Investigate, compare and assess tools for predicting grounding and collision damage (including collision with ice). Emphasize simplified methods, which can be readily applied to the design and rapid analysis of new structural concepts, including applications using probabilistic scenarios. Finite element and other more rigorous methods may be considered and used to validate simplified methods. Include comparisons to available data.
- Propose a format for a collision and grounding database. The database may include actual, experimental and simulated data as long as it is properly identified. Gather initial data.
- Define standard scenarios and conditions for grounding and collision analyses. This may include probabilistic and/or deterministic descriptions of ship speed, trim, collision angle, striking ship displacement, striking ship bow characteristics, impact point, bottom, rock characteristics, rock height above baseline, sea state, ice mass, etc.
- Define standard criteria and methodologies for applying structural performance in grounding and collision including oil outflow, carriage of nuclear waste, damaged stability, ultimate strength and serviceability. Provide a basis for performance specifications in these areas.
- As tools, standard scenarios, methodologies and performance criteria are evaluated and defined, apply them to a matrix of typical and innovative structural designs and concepts. Identify promising concepts for future study.

This paper is a summary report of work performed by the panel. It is the first of the deliverables specified in the panel's charter. Additional papers and reports will follow. The panel has three working groups studying: 1) tools for predicting damage in grounding and collision; 2) collision and grounding scenarios; and 3) innovative design concepts. Funded research is centered at Virginia Tech and Webb Institute.

Other Members of the Ad Hoc Panel contributing to this paper are:

Donghui Chen	Jeom Paik	Bo Simonsen	Surya Vakkalanka	Ge Wang	Rong Huang	Rick van Hemmen
John Sajdak	Arne Stenseng	Brett Fox	Ryszard Kaczmarek	Peter Gooding		

The work of the panel, as presented in this paper, consists of four essential elements:

- Benchmarking of grounding models
- Benchmarking and development of collision models
- Definition of grounding and collision scenarios including striking bow geometry and bow stiffness.
- Innovative design concepts

2 MOTIVATION

The serious consequences of ship grounding and collision necessitate the development of regulations and requirements for the subdivision and structural design of ships to reduce damage and environmental pollution, and improve safety.

The International Maritime Organization (IMO) is responsible for regulating the design of oil tankers and other ships to provide for ship safety and environmental protection. Their ongoing transition to probabilistic performance-based standards requires the ability to predict the environmental performance and safety of specific ship designs. This is a difficult problem requiring the application of fundamental engineering principles and risk analysis.

IMO first introduced probabilistic standards in damage stability regulations for passenger ships [1] and later for cargo ships [2]. IMO's first attempt to apply a probabilistic methodology to tankers was in response to the US Oil Pollution Act of 1990 (OPA 90). In OPA 90 the US required that all oil tankers entering US waters must have double hulls. IMO responded to this unilateral action by requiring double hulls or their equivalent. Equivalency is determined based on probabilistic oil outflow calculations specified in the "Interim Guidelines for the Approval of Alternative Methods of Design and Construction of Oil Tankers Under Regulation 13F(5) of Annex I of MARPOL 73/78" [3], hereunder referred to as the Interim Guidelines.

All of these regulations use probability density functions (pdfs) to describe the location, extent and penetration of side and bottom damage. These pdfs are derived from limited historical damage statistics [4], and applied identically to all ships without consideration of their structural design. Figure 1 is the IMO probability density function used to define the longitudinal extent of bottom damage from grounding in oil outflow calculations. The histograms represent the statistical data collected by the classification societies and the linear plot represents IMO's piece-wise linear fit of the data. Other IMO pdfs are constructed in a similar manner.

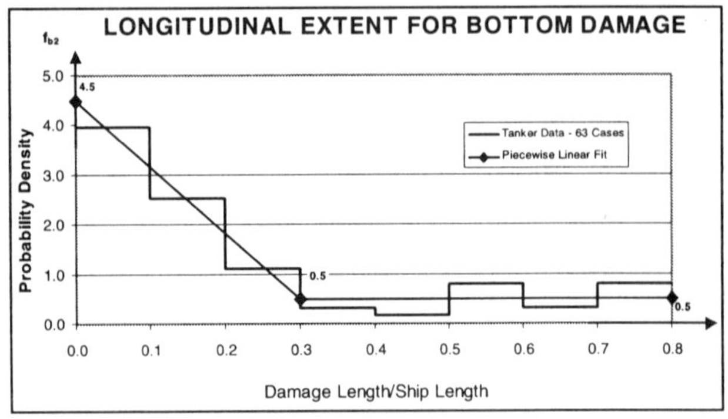

Figure 1 - Damage Probability Density Function [4]

A major shortcoming in IMO's current oil outflow and damage stability calculation methodologies is that they do not consider the effect of structural design or crash-worthiness on damage extent [5,6]. The primary reason for this exclusion is that no definitive theory or data exists to define this relationship.

Other deficiencies include:

- Damage pdfs consider only damage significant enough to breach the outer hull. This penalizes structures able to resist rupture.
- Damage extents are treated as independent random variables when they are actually dependent variables, and ideally should be described using joint pdfs.
- Damage pdfs are normalized with respect to ship length, breadth and depth when damage may depend to a large extent on local structural features and scantlings vice global ship dimensions.

It is logical and essential that crashworthiness is considered in oil outflow and damage stability calculations. **An analytical method is required to define the relationship between structural design and damage extent in collision and grounding. This is a primary objective of the work of this panel.**

A method that considers crashworthiness must be sensitive to at least the basic parameters defining unique structural designs while maintaining sufficient generality and simplicity to be applied by working engineers in a regulatory context for a ship in worldwide operation. It should not require detailed finite element analysis or be limited to a single accident scenario.

An additional advantage of performance-based models is the potential for their application to different and innovative designs. This is a second primary objective of this panel.

3 PROBABILISTIC FRAMEWORK

Figure 2 illustrates the process proposed to predict the probabilistic extent of damage in collision as a function of ship structural design [7]. A similar process is proposed for grounding.

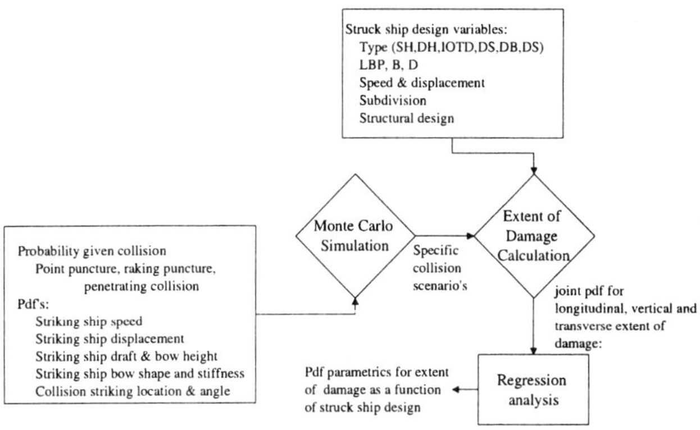

Figure 2 - Process to Predict Probabilistic Damage [7]

The process begins with a set of probability density functions (pdfs) defining possible grounding or collision scenarios. Using these pdfs, a specific scenario is selected in a Monte Carlo simulation, and combined with a specific ship structural design to predict damage. This process is repeated for thousands of scenarios and a range of structural designs until sufficient data is generated to build a set of parametric equations relating probabilistic damage extent to structural design. These parametric equations can then be used in oil outflow or damage stability calculations.

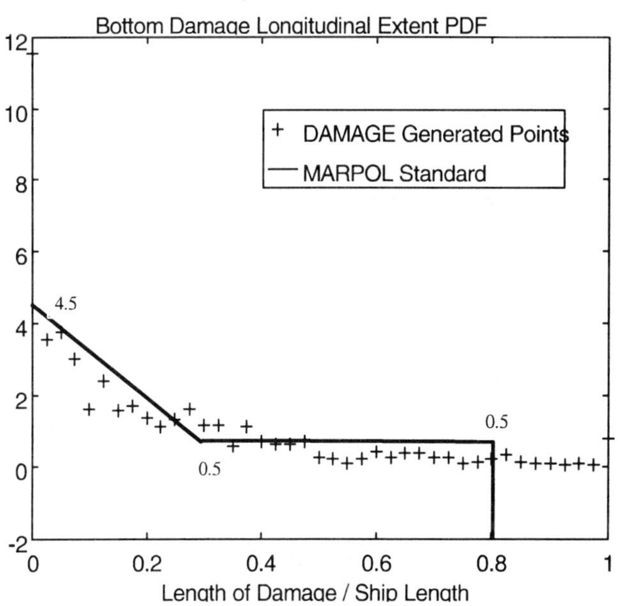

Figure 3 – Bottom Damage pdf [6]

A damage pdf generated in an early application of this approach is shown in Figure 3 [6]. This pdf was generated for a MARPOL single hull tanker typical of the struck ships represented in the data used to develop the current MARPOL damage pdfs.

4 GROUNDING MODELS

4.1 Background and Plan

The Exxon Valdez accident in 1989 and the regulatory events following it lead to active research on structural behavior in grounding. The earlier studies had been mostly empirical, the best-known being a study by Card [8] who surveyed 30 grounding incidents to determine the effectiveness of a double bottom in reducing pollution. More recent studies have included large numerical simulations, small and large scale experiments, and the development of simplified methods. Many of these studies were supported by the project "Protection of Oil Spills from Crude Oil Tankers" carried out by the Japanese Association for the Structural Improvement of Shipbuilding Industry (ASIS). In the United States, the Carderock Division of the Naval Surface Warfare Center (NSWCCD) conducted grounding experiments, and the MIT-Joint Industry Project on Tanker Safety developed software to analyze structural damage in grounding. Development of analytical models for the software was supported by experimental studies. Active research effort on grounding analysis has been carried out also in Europe, mainly in Denmark and the Netherlands.

The Specialist Panel V.4 of the 1997 International Ship and Offshore Structures Congress (ISSC 1997) [9] reviewed the state-of-the-art of the research and concluded that nonlinear finite element analysis coupled with calculation of ship motions had reached a level at which a fairly accurate prediction of the structural response was possible. However, the report noted that since the method is time consuming and requires a high-level of expertise, it is not suitable for the design or regulatory environment. The report also concluded that the simplified methods that existed at the time required further validation.

This study concentrates on evaluation of the existing simplified methods, which have been further developed since the ISSC 1997 report. The objective is to assess their suitability for design and regulatory work.

4.2 Models Included in the Current Study

The methods evaluated in this study include *DAMAGE*, a software tool developed by the MIT-Joint Industry Project on Tanker Safety, and a simple analytical method developed by Dr. Wang at the University of Tokyo (method by Wang). Both methods calculate grounding force and extent using closed-form solutions, but they differ in the detail in which the structure is defined and its behavior is analyzed. The methods also differ in their treatment of ship motions. The program *DAMAGE* was selected for further testing and analysis, because it is applicable for a wider range of grounding scenarios.

4.2.1 DAMAGE [10]

The computer program DAMAGE was developed at MIT under the Joint MIT-Industry Program on Tanker Safety. This project, lead by Professor Tomasz Wierzbicki, was initiated in 1991, and in addition to the program DAMAGE the project has produced more than 70 technical reports about prediction of grounding and collision damage. The program DAMAGE Version 5.0 can be used to predict structural damage in the following accident scenarios:

- Ship grounding on a conical rock with a rounded tip (rigid rock, deformable bottom)
- Ship-ship collision (deformable side, deformable bow)

Compared to previous models for prediction of grounding and collision damage, a major advantage of DAMAGE is that the theoretical models are hidden behind a modern graphical user interface (GUI) [10]. The program has been developed with the objective of making crash analysis of ship structures feasible for engineers that do not have any particular experience in the field of crashworthiness. Figure 4 shows the input screen for defining the "ship-ground interaction", the relative position and velocity of ship and ground, coefficient of friction, etc.

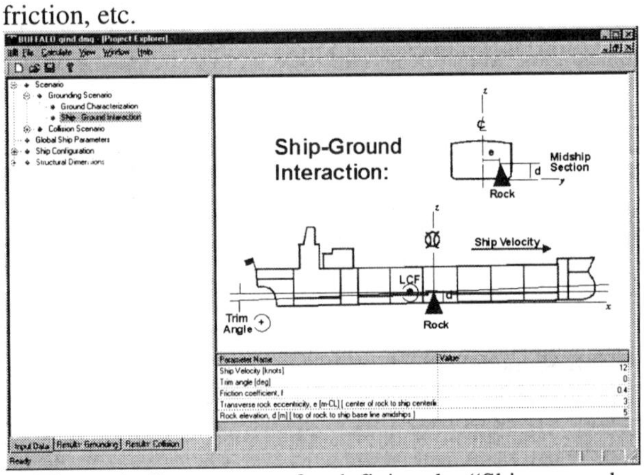

Figure 4 - Input screen for defining the "Ship-ground Interaction" in the program DAMAGE [10]

Initially, the ship is assumed to be on a straight-line course with a known velocity and trim. The rock is at some distance to port or starboard of the ship's centerline, and at some height above the ship's baseline. During the event, the contact force between ship and ground induces heave, roll and pitch motion on the ship, and eventually causes the ship to stop.

The primary mechanisms of energy dissipation are:

1. A change in potential energy of the ship and the surrounding water as the ship is lifted by the ground reaction.
2. Friction between the ground and hull.
3. Deformation and fracture of the hull.

Since the lifting of the ship off the rock generally causes a reduction in crushing and tearing forces it is important to include the coupling between the global ship motions (external dynamics) and the local damage process (internal mechanics). Detailed descriptions of the model for the external dynamics can be found in References [11] and [12].

For the model to be applicable to optimization for structural crashworthiness it is important that it captures the effect of :

- Material strength and ductility
- Dimensions and arrangement of major structural components such as bulkheads, decks, girders and large stiffeners.

The theoretical model is based on a set of super-element solutions, i.e. closed form, analytical solutions for the energy absorption of rather large structural components undergoing large deformations. All structural components are assumed to follow the same overall mode of deformation around the rock. This way it is possible to take into account the structural resistance of all typical members in a ship bottom. The fracture criterion is based on analytical solutions for plastic response of a membrane. Figure 5 shows a window from an animation of the ship motion over the rock.

A description of the procedure including detailed derivations of the super-element solutions can be found in References [13] and [14].

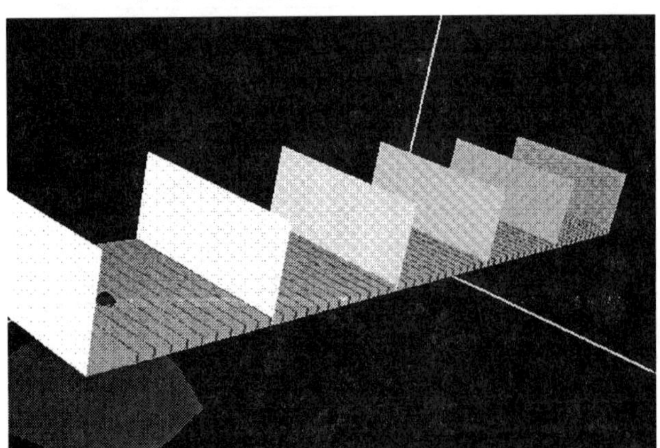

Figure 5. DAMAGE simulation of ship motion over a rock (only selected structural elements are shown).

4.2.2 Method by Wang [15]

Dr. Wang developed a method to predict structural resistance in raking type grounding in his doctoral work under Professor Hideomi Ohtsubo's supervision. The method was further developed and published in 1997 [15].

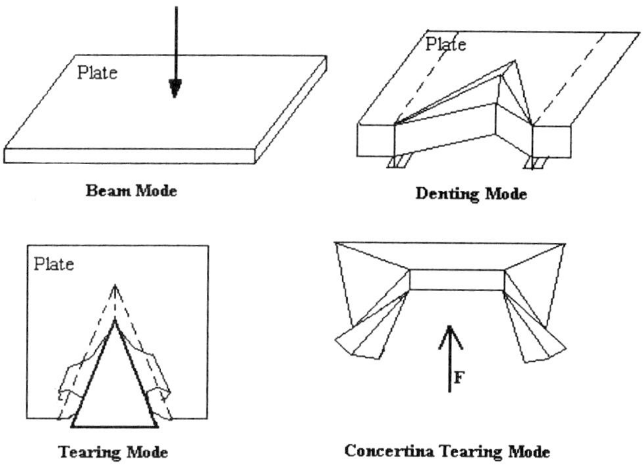

Figure 6 - Four Failure Modes in the Method by Wang

The method includes closed-form solutions for four failure modes: stretching (beam mode), denting, tearing and concertina tearing illustrated in Figure 6.

The rock is modeled as a wedge, and the ship's bottom structure is modeled with periodic structural members where the period corresponds to transverse frame spacing. Only horizontal ship motions are considered in the calculation.

The grounding damage is calculated by combining the failure modes to model periodic resistance of the structure. The resistance of the transverse structure is predicted with a beam model. The denting failure mode is assumed for the plate immediately behind the transverse structure. As the wedge advances in the plating and a crack develops at the tip of the wedge, the rupture is modeled by tearing failure mode. Concertina tearing is used to model the accordion like behavior of the plating, which can lead to cracks at other locations than the tip of the wedge. The entire calculation can be carried out by hand or with a simple spreadsheet program, but the analyst must decide how the failure modes are combined to achieve the final damage.

4.3 Application

DAMAGE was selected for further testing and analysis, because of its applicability to a wider range of grounding scenarios. Although the method by Wang is elegant in its simplicity, its application in the current formulation is limited to raking type damage only. The major limitations of the current version of *DAMAGE* are the type of obstruction (pinnacle only), and the structural model (the structure is modeled for the cargo block only).

4.3.1 Validation of *DAMAGE*

Simonsen presented verification of the theory behind *DAMAGE* models by comparing calculated results with US NAVY 1/5-scale grounding experiments, with large-scale grounding experiments carried out in the Netherlands, and with an actual grounding of a VLCC [14,16].

Based on the limited number of validation cases *DAMAGE* is found to predict the damage extent well. In the case of a VLCC grounding, the difference between the calculated damage length of 177 meters and the observed damage length of approximately 180 meters is only 1.7%. The result is sensitive to the transverse location of the rock relative to the ship's centerline as is illustrated in Figure 7. As the rock moves away from the centerline the effect of ship motions increase and the predicted damage length increases.

DAMAGE predictions for average rock penetration and grounding force are also excellent. Predictions for minimum and maximum values are not as good. Figures 8-10 illustrate a comparison of *DAMAGE* results with a large-scale grounding test carried out in the Netherlands by ASIS. Large differences in the penetration at the initial stages of grounding are probably due to simplifications in the global motion calculations.

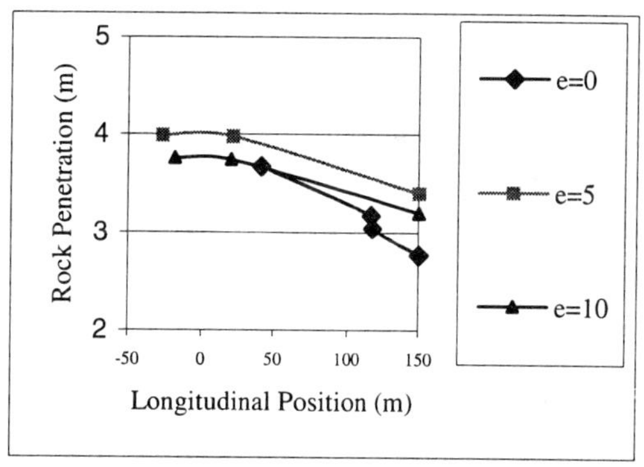

Figure 7 - Rock Penetration vs. Rock Eccentricity

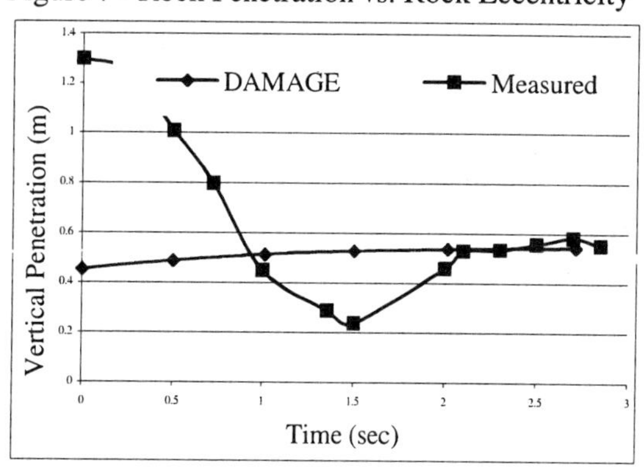

Figure 8 - Vertical Penetration Comparison (ASIS Test 2)

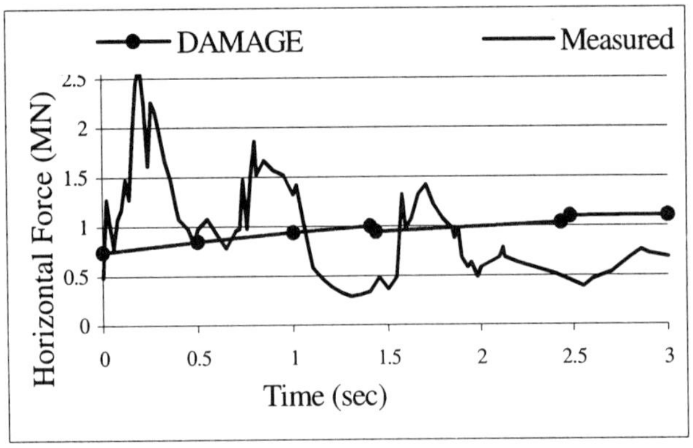

Figure 9 - Comparison of Horizontal Forces (ASIS Test 2)

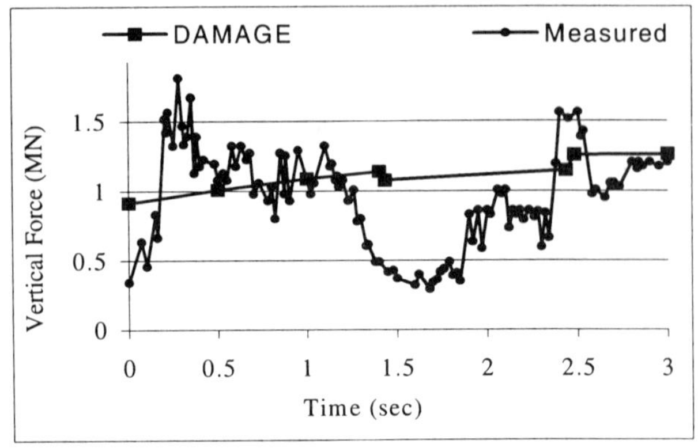

Figure 10 - Comparison of Vertical Forces (ASIS Test 2)

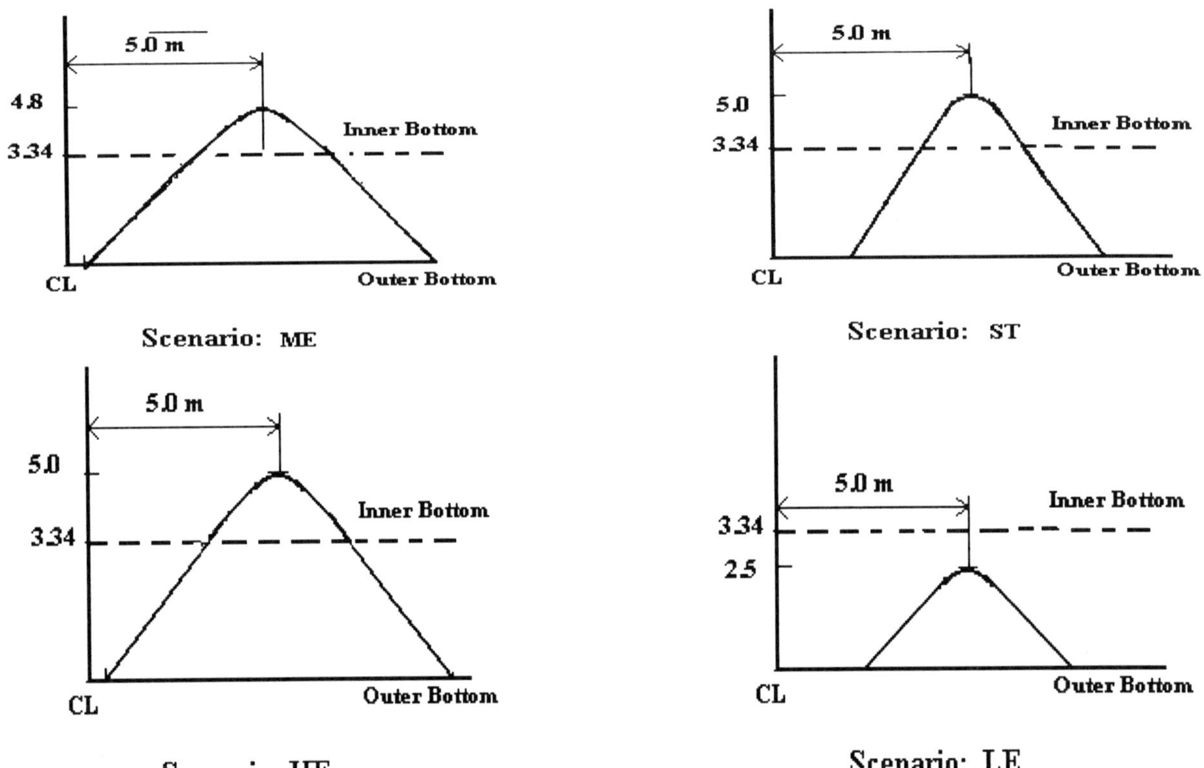

Figure 11 - Sketch of Four Grounding Scenarios

4.3.2 Sensitivity Analysis

The sensitivity of DAMAGE results to changes in grounding parameters was studied and presented in [17]. The base ship used in the sensitivity analysis was a 237,000 DWT single-hull VLCC and the grounding case studied corresponded to the VLCC validation case.

In the sensitivity analysis only one parameter was changed at the time and the other parameters were kept at their initial values. The objective of this analysis was not to study the effect of the structural design on the damage extent, rather the objective was to investigate the sensitivity of the DAMAGE model to the parameters defining the grounding event.

As can be expected the results are very sensitive to the ground characteristics and to the parameters defining the ship-ground interaction (rock elevation, rock shape and transverse location, ship velocity and displacement, friction coefficient and trim angle).

The parameters affecting the global ship motions (longitudinal center of floatation and the longitudinal and the transverse metacentric height) had little effect on damage results. Surprisingly, the tank spacing as well as the characteristics of both the transverse bulkhead and the longitudinal bulkhead had very little effect on damage results. The results were also not very sensitive to material characteristics.

Damage results were sensitive to the thickness of the outer bottom. Reducing the spacing and increasing the scantlings of longitudinal girders and transverse floors also affected damage results, but not as effectively. The same applied to longitudinal stiffeners.

4.3.3 Structural Modifications

DAMAGE was found to provide a good tool for comparative studies based on the validation cases and the sensitivity analysis. The next step in the study was to investigate the effect of structural modifications on the damage extent in selected grounding scenarios. The study was limited to eight grounding scenarios, and the structural changes were limited to changes in scantlings of a conventional double-bottom structure. The analysis was intended to provide insight into the effects of the modifications and scenarios to help design a probabilistic analysis, which will be the next step in the study.

The base ship used in the analysis was a 150,000 DWT double-hull tanker. The grounding scenarios were selected to represent high, medium and low rock elevations relative to the ship's baseline. One of the scenarios had a sharp rock tip (30 degree semi-apex angle), whereas the shape of the rock was kept constant in the other scenarios (45 degree semi-apex angle). Two velocities were analyzed: 7 knots to represent port speed and 14 knots to represent service speed of the tanker. The rock shapes and elevations are illustrated in Figure 11. More details on the base ship and the grounding scenarios can be found in [17].

Eight structural modifications, all designed according to the requirements of ABS SafeHull (97/98), where investigated. The scantlings met the minimum requirements except for the dimension under investigation. The modifications included:

1. Increase in the outer bottom plate thickness.
2. Increase in the inner bottom plate thickness.
3. Increase in both outer bottom and inner bottom plate thickness.
4. Additional longitudinal girder.
5. Two additional longitudinal girders.
6. Reduced spacing of transverse floors.
7. Decrease in double bottom height.
8. Increase in double bottom height.

The steel weight (longitudinal structure only) was calculated for each design, and the effectiveness of a design modification was measured in terms of the reduction in inner bottom rupture as a function of the steel weight. Changes in the transverse structure caused by the modifications were not taken into account.

Some conclusions based on the selected scenarios include:

- In the low rock elevation case, which represents raking type grounding, none of the designs had inner bottom rupture. At service speed the outer bottom ruptured throughout the entire length of the ship. In this type of scenario, the structural modifications have little impact on the damage extent and no effect on the oil outflow assuming that the vessel survives the raking damage.
- In the sharp rock tip, high-rock elevation case the rock penetrated through the entire cargo block in all designs at service speed. Even at port speed all designs had serious damage. However, increased outer bottom or inner bottom thickness, two additional girders, or reduced spacing of transverse floors did reduce outer bottom and inner bottom rupture. This was the only grounding scenario where increasing double bottom height had very little effect on the inner bottom rupture.
- In high and medium rock elevation cases most of the design modifications improved structural performance. Reducing the double bottom height resulted in the worst performance. The design with two additional longitudinal girders performed best in reducing inner bottom rupture, but it also had the heaviest steel weight among the studied designs. If the steel weight was taken into account, increasing the thickness of the outer bottom was found to be most effective in reducing the inner bottom rupture. Increasing both the inner bottom and outer bottom thickness reduced the inner bottom rupture more but at the cost of a higher steel weight.

4.4 Ongoing and Future Work

The work discussed above was carried out to provide a basis for a probabilistic analysis, which will take into account a range of possible grounding scenarios. However, to add credibility to the results further validation is needed. Data needs for validation work will be discussed later in the paper. Further review of the results from the presented validation cases is ongoing.

The analysis on the effect of structural modifications discussed above measured the performance in terms of the damage extent without taking into account the subdivision of the vessel. Therefore the observed changes in the damage extent may have no effect on the resulting oil outflow. To determine the effect of the structural modifications on oil outflow the *DAMAGE* calculations should be coupled with outflow analysis.

The next step in the study is to carry out a probabilistic analysis of damage in grounding scenarios defined by probability density functions of the input parameters. This analysis will be coupled with outflow calculations. The objective of the study will be to provide a tool for comparative studies between vessel designs.

5 COLLISION MODELS

5.1 Background and Plan

In 1979, the Ship Structure Committee (SSC) conducted a review of collision research and design methodologies [18]. They concluded that the most promising simplified collision analysis alternative was to extend Minorsky's original analysis of high-energy collisions by including consideration of shell membrane energy absorption. A simple and fast model is important in probabilistic analysis because thousands of different scenarios must be run to develop statistically significant results.

A more recent review of the literature and of the applicability of available methods for predicting structural performance in collision and grounding was made in the 1997 International Ship and Offshore Structures Congress (ISSC 97) by Specialist Panel V.4 [9]. Their report states: "Knowledge of behavior on a global level only (i.e., total energy characteristics like the pioneering Minorsky formula) is not sufficient. The designer needs detailed knowledge on the component behavior (bulkheads, girders, plating, etc.) in order to optimize the design for accident loads."

The approach taken in this study is to progressively increase the complexity of SIMCOL (Simplified Collision Model), the Ad Hoc Panel's baseline collision model, until results with sufficient accuracy and sensitivity to design characteristics are obtained. In order to assess the model's consistency and sensitivity, the model is applied in a series of collision scenarios with a range of tanker sizes and designs. SIMCOL results are compared to results obtained using three other collision models: DAMAGE [10,19], ALPS/SCOL [20,21], and a Technical University of Denmark (DTU) model. The DTU model and ALPS/SCOL were also further developed during this study. Determining the ideal trade-off between simplicity and sufficiency is a primary concern of this process.

5.2 Models Included in the Current Study

Three types of collision were identified for investigation by the panel:

- Low energy puncture at a point
- Low energy raking puncture
- Penetrating Collision – right angle and oblique sufficient to penetrate outer hull with significant damage extending in at least 2 directions (penetration, horizontal, vertical)

Work thus far is focused on models for penetrating collision.

Each collision model must consider two very different aspects of the collision event:

- External problem – includes the global characteristics of the striking and struck ships and their motion before, during and after collision;
- Internal problem - includes the internal deformation mechanics and structural response of the struck ship and the striking ship bow during collision.

The four models studied by the panel take different approaches to solving and coupling these two problems. Although not a formal validation, comparison and agreement between these models provides useful insight into their performance and increases confidence in the validity of their results.

5.2.1 Simplified Collision Model (SIMCOL) [22]

SIMCOL uses a time-domain simultaneous solution of external dynamics and internal deformation mechanics similar to that originally proposed by Hutchison [23]. Figure 12 shows the SIMCOL simulation process. The external sub-model uses a three degree-of-freedom system for ship dynamics shown in Figure 13. The internal sub-model determines reacting forces from side and bulkhead structures using detailed mechanisms adapted from a 1981 Rosenblatt study [24,25]. It determines the energy absorbed by the crushing and tearing of decks, bottoms and stringers using the Minorsky correlation [26] as modified by Reardon and Sprung [27].

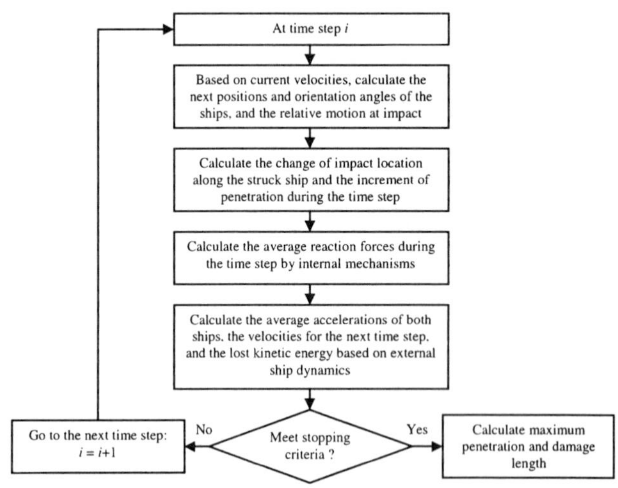

Figure 12 - SIMCOL Simulation Process

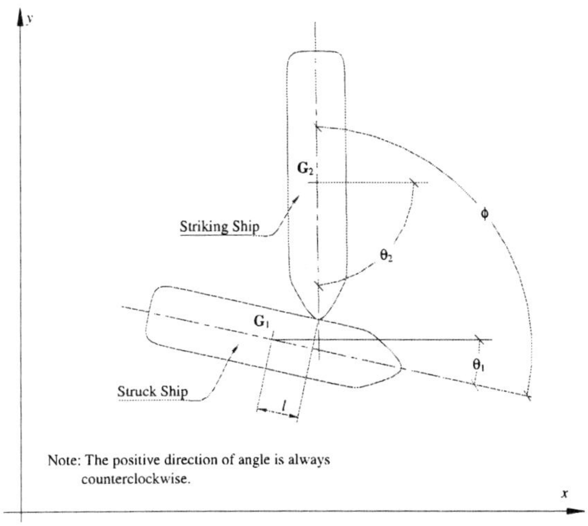

Figure 13. SIMCOL External Ship Dynamics

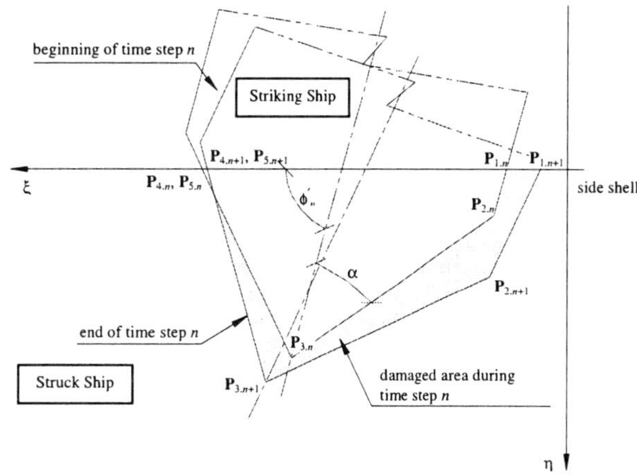

Figure 14. Sweeping Segment Method

V.U. Minorsky conducted the first and best known of the empirical collision studies based on actual data [26]. His method relates the energy dissipated in a collision event to the volume of damaged structure. Actual collisions in which ship speeds, collision angle, and extents of damage are known were used to empirically determine a linear constant. This constant relates damage volume to energy dissipation. In the original analysis the collision is assumed to be totally inelastic, and motion is limited to a single degree of freedom. Under these assumptions, a closed form solution for damaged volume can be obtained. With additional degrees of freedom, a time-stepped solution must be used.

Crake and Brown developed SIMCOL Version 0.0 as part of the work of SNAME Ad Hoc Panel #3 [6,28]. Based on further research, test runs and the need to make the model sensitive to a broader range of design and scenario variables, improvements were progressively made by Chen and Brown at Virginia Tech [22]. A sweeping segment method, shown in Figure 14, is added to the model in SIMCOL Version 1.0 to improve the calculation of damage volume and the direction of damage forces. Models from the Rosenblatt study [24,25] are applied in Version 1.1 assuming rigid web frames. In Version 2.0, the lateral deformation of web frames is included as shown in Figure 15. In Version 2.1, the vertical extent of the striking ship bow is considered. Table 1 describes the evolution of SIMCOL over the course of the study thus far.

Design data required for the striking ship includes bow half-entrance angle, bow height, length, beam, draft and displacement.

Scenario data required includes striking and struck ship velocity, collision angle, and longitudinal location of impact in the struck ship.

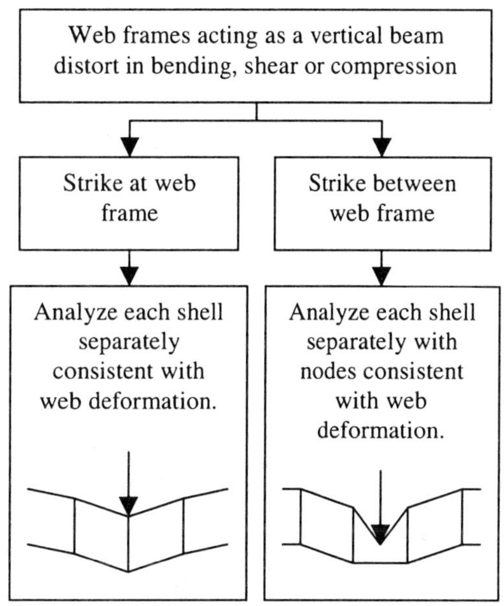

Figure 15. Web Deformation in SIMCOL 2.0 [24,25]

5.2.2 DAMAGE 4.0 [19]

The DAMAGE 4.0 collision module solves the external problem uncoupled from the internal problem, and applies the calculated absorbed energy to plastic deformation of the struck ship. Structural components, motions, masses etc. are described in ship coordinate systems local to each ship and in one global coordinate system. Degrees of freedom in DAMAGE include striking ship surge and struck ship sway and yaw.

The following assumptions are applied in DAMAGE 4.0:

- Both ships are perpendicular before and during impact, i.e. only right angle collisions are considered.
- The forward motion of the struck ship is assumed to be zero.

Table 1. SIMCOL Evolution

	Version	0.1	1.0	1.1	2.0	2.1
Simulation		Simulation in time domain				
External Model		Three degrees of freedom (Hutchison and Crake)				
Internal Model	Horizontal Members	Minorsky mechanism as re-validated by Reardon and Sprung				
		Crake's model	Sweeping segment method to calculate damaged area and resulting forces and moments			
	Vertical Members w/o rupture of plate	Jones and Van Mater		McDermott / Rosenblatt Study methods		
		Crake's model (Jones)	Van Mater's extension of Jones	Does not consider deformation of webs, friction force and the force to propagate yielding zone	Considers deformation of webs, friction force and the force to propagate yielding zone	Striking bow with limited depth
	Vertical Members w/ ruptured plate	Neglected			Minorsky method for calculating absorbed energy due to longitudinal motion	

In DAMAGE 4.0, the striking ship bow is assumed to be rigid. DAMAGE Version 5.0 (Released June 2000) includes crushing of the bow.

Based on conservation of linear momentum, angular momentum and energy, the velocities after the impact are calculated as well as the loss of kinetic energy, which is available for the structural deformations.

In order to determine the deformation of the bow and the side, the striking ship is moved into the struck ship in small increments. In each increment, the total resistance forces from crushing of the bow and penetration into the side are compared. The actual crushing/penetration increment takes place in the ship with lowest resistance.

DAMAGE 4.0 considers the material and structural scantlings of all major structural components for the side structure.

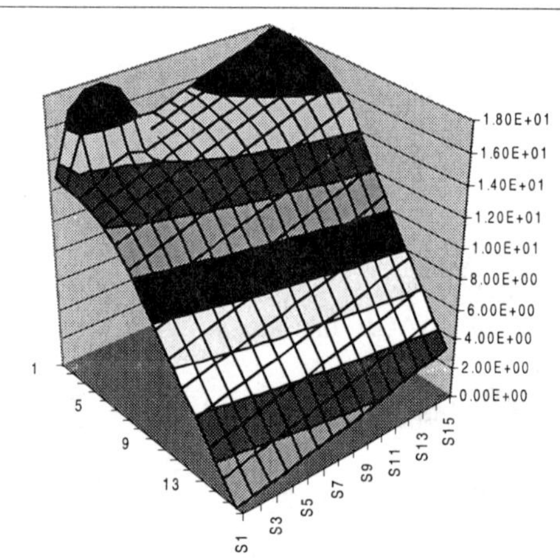

Figure 16. DAMAGE Bow Geometry

The model for the internal mechanics is based on the direct contact deformation of super-elements. The super-elements used to model the side in DAMAGE are:

- Shell and inner side plating (laterally loaded plastic membrane)
- Deck and girder crushing
- Beam loaded by a concentrated load
- X-, L- and T-form intersections crushed in the axial direction

The bow geometry is defined by eight parameters. Figure 16 shows an example the bow geometry.

5.2.3 ALPS/SCOL

ALPS/SCOL is a coarse-mesh 3-D non-linear finite element code using super-elements based on the Idealized Structural Unit Method (ISUM) [20,21]. The geometry of the striking and the struck ships is described in a global (three-dimensional) rectangular coordinate system. The stress in an ISUM unit is described in a local element coordinate system. ALPS/SCOL considers sway and yaw of the struck ship with the following assumptions:

- The added masses of the striking and the struck ships are calculated based on ships of similar type and size using a linear strip theory-based computer program.
- The striking ship is assumed to be rigid.
- The analysis of the external and the internal dynamics is undertaken separately.
- The longitudinal velocity of the struck ship is not considered.
- Since ALPS/SCOL is based on a simplified 3-D nonlinear finite element approach, damage in the three directions (i.e., including penetration, vertical and horizontal damage) are considered.
- The geometry of the striking ship bow shape is described by gap/contact elements. The bow is assumed to be rigid.
- One cargo hold of the struck ship is taken as the extent of the analysis.
- ISUM stiffened panel units are used to model the struck vessel structure.

The geometry of the struck ship is described using about 600 rectangular or triangular ISUM units. If the deformation of the struck ship is symmetric, the total degrees of freedom in the numerical model are reduced by half. Each node has 3 degrees of freedom.

Figure 17 shows damage calculated in an ALPS/SCOL simulation.

Design data required for the striking ship includes a detailed bow geometry description, length, beam, depth, draft and displacement.

Design data for the struck ship includes, length, beam, depth, draft and displacement, transverse bulkhead location, COG, and detailed structural design and scantlings.

Scenario data required includes striking ship velocity and longitudinal location of impact in the struck ship.

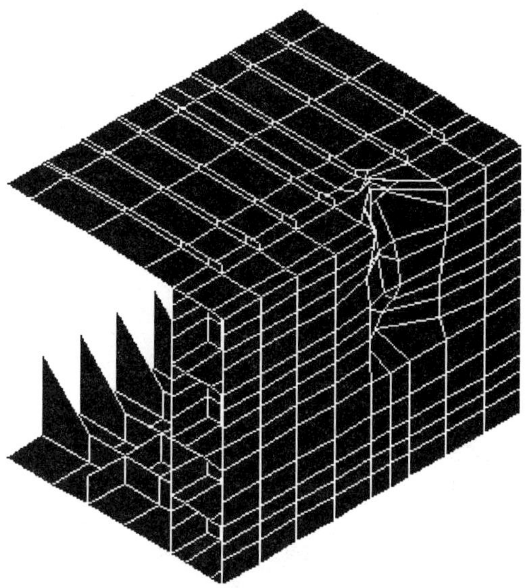

Figure 17. Damage from ALPS/SCOL Simulation

5.2.4 DTU Model

The Technical University of Denmark (DTU) model also solves the external problem uncoupled from the internal problem, and applies the calculated absorbed energy to plastic deformation of the struck ship.

Solution of the external dynamics is accomplished based on an analytical method developed by Pedersen and Zhang [29]. This method estimates the fraction of the kinetic energy that is available for deformation of the ship structure. The energy loss for dissipation by structural deformation is expressed in closed-form expressions. The procedure is based on a rigid body mechanism, where it is assumed that there is negligible strain energy for deformation outside the contact region, and that the contact region is local and small. This implies that the collision can be considered instantaneous as each body is assumed to exert an impulsive force on the other at the point of contact. The model includes friction between the impacting surfaces so those situations with glancing blows can be identified. Both ships have three degrees of freedom: surge, sway and yaw. The interaction between the ships and the surrounding water is approximated by simple added mass coefficients, which are assumed to remain constant during the collision.

The loss in kinetic energy by the method is determined in two directions, perpendicular and parallel to the side of the struck ship. Both right and oblique angle collisions are considered and both vessels may have velocity before the collision.

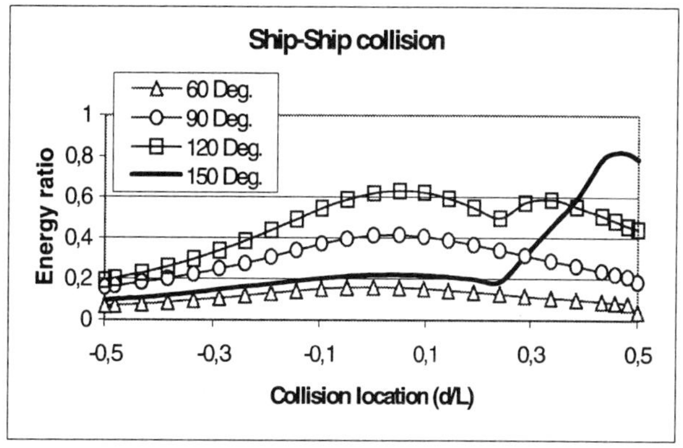

Figure 18 - Energy ratio defined as the ratio between energy released for crushing and the total kinetic energy of the two ships before the collision as function of the collision location, d, for collision angles equal 60, 90, 120 and 150 degrees. The two tankers are identical and have the same speed prior to the collision [29].

The model for the internal mechanics is based on a set of super-elements, where each element represents a structural component. The calculation method is based on the principle that the area of the struck vessel affected by the collision is restricted to the area touched by the striking vessel. The super-elements are:

- Lateral plate deflection and rupture. Large deflections are assumed; this implies that the bending resistance can be neglected
- Crushing of structure intersection elements (X- or T-elements)
- In-plane crushing and tearing of plates
- Beam deflection and rupture

The design data for the struck vessel includes length, beam depth, draft, displacement, COG and detailed structural design and scantlings.

The bow of the striking vessel is assumed to be rigid. The basic data for describing the striking ship bow are stem angle, breadth and bow height. The horizontal shape of the deck and the bottom are assumed to be parabolic. If the striking vessel is equipped with a bulb, this is assumed to have the form of an elliptic parabola.

Scenario data required includes striking and struck ship velocity, collision angle and longitudinal location of impact at the struck vessel.

5.3 Application

In order to assess the models' consistency and sensitivity, they are tested in a series of collision scenarios with a range of struck tanker sizes and designs. The Baseline

Tanker design is a 150000 dwt double-hull tanker. It was developed to be consistent with the dimensions of the 150000 dwt reference tanker in the IMO Interim Guidelines. HECSALV and SafeHull were used to develop the details of the design, and to insure that the arrangement satisfies IMO regulations and the structural design satisfies ABS classification requirements. A 60000 dwt double hull tanker and a 283000 dwt double tanker were also developed in accordance with the Interim Guidelines [3]. Results for the baseline 150000 dwt double-hull tanker are presented in this paper.

5.3.1 Struck Ship

The 150000 dwt Baseline Tanker design shown in Figure 19 is the primary struck ship used for the initial model testing. Tables 2 and 3 list principal characteristics and structural data for this design.

Table 4 lists the input data for each test matrix. The first test matrix considers damage for a series of strike locations on the web at the center of each cargo tank. This represents a large global variation in strike location. The second test matrix considers damage for a series of strike locations on either side of the web at the center of the midship cargo tank. This represents a relatively small local variation in location on and between webs. The third test matrix considers damage for a series of collision angles with a strike location on the web at the center of the midship cargo tank.

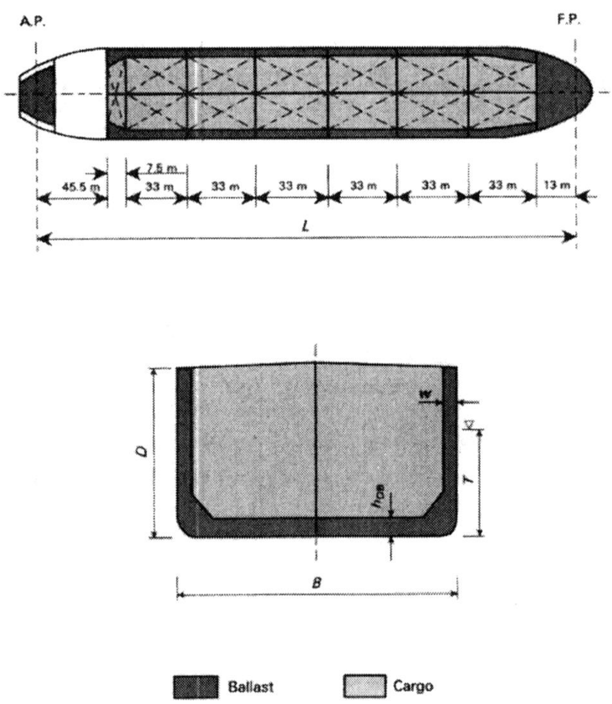

Figure 19 - Baseline Tanker Design [3]

Table 2. Baseline Tanker Principal Characteristics

Deadweight, tonnes	150,000
Length L, m	264.00
Breadth B, m	48.00
Depth D, m	24.00
Draft T, m	16.80
Double Bottom Ht h_{DB}, m	2.32
Double Hull Width W, m	2.00
Displacement, tonnes	178,867

5.3.2 Model Test Matrices

Three scenario test matrices were used in the initial study. All matrices use the Baseline Tanker as the struck ship and a 150000 dwt bulk carrier as the striking ship. Principal characteristics for the striking ship are listed in Table 4 and the resulting vertical alignment of the two ships is shown in Figure 20. Since the focus of the initial tests is on penetration damage, zero struck ship speed is used in all cases. Scenarios for the test matrices are described in Table 5.

Table 3. Baseline Tanker

Ship		150,000 dwt double hull tanker
Web Frame Spacing L_s, m		3.30
Smeared Thickness t_h, mm	Deck	47.32
	Inner Bottom	26.92
	Bottom	28.29
	Stringers	3 × 15.34
Smeared Thickness t_v, mm	Side Shell	21.92
	Inner Skin	22.94
	Bulkhead	22.28
Web Thickness t_w, mm	Upper	12.00
	Lower	18.00

Table 4. Striking Ship Principal Characteristics

Ship Type	150,000 dwt bulk carrier
Length *L*, m	274.00
Breadth *B*, m	47.00
Depth *D*, m	21.60
Bow Height *H*, m	26.00
Draft *T*, m	15.96
Displacement, tonnes	174,850
Half Entrance Angle, α	38°

5.3.3 Model Results

Representative model results for struck ship penetration are shown in Figures 22-27. The figures show transverse penetration into the struck ship as a function of the particular variables in each matrix. The results show good agreement between the models. DAMAGE generally predicts the lowest penetration, and ALPS/SCOL generally predicts the highest.

Figures 22 and 23 (Matrix 1) show the effect of the external dynamics. More energy is absorbed in strikes around midship. SIMCOL shows a larger variation in penetration as a function of global strike location than the other models, particularly at low energy. SIMCOL's coupled dynamics predict larger changes in the relative motions at the strike point of the two ships as the strike location moves away from midship. This results in more longitudinal damage and less penetration. ALPS/SCOL shows a similar, but lesser trend. This has the greatest effect on shell and web deformation, and is most evident in the lower energy case, Figure 22.

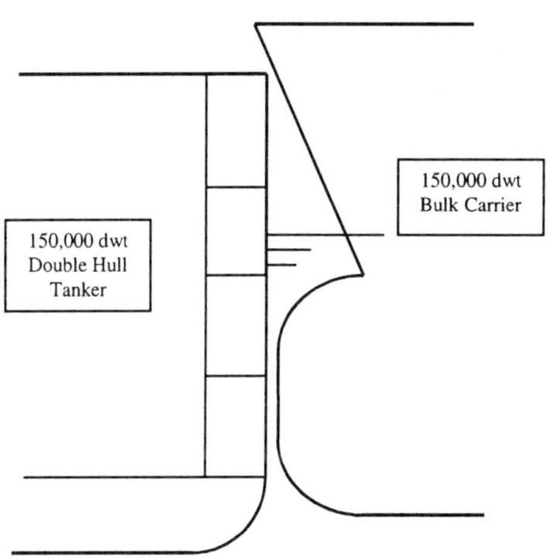

Figure 21 - Collision Strike Vertical Alignment

Table 5 - Test Matrices

	Struck Ship Speed (knt)	Striking Ship Speed (knt)	Collision Angle (deg)	Strike Location (m fwd MS)
Matrix 1	0	3,4,5,6,7	90	-62.5,29.5, 3.5,36.5, 69.5,102.5
Matrix 2	0	3,4,5,6,7	90	1.85,2.675, 3.5,4.325, 5.15
Matrix 3	0	3,4,5,6,7	45,60,75, 105,120,135	3.5

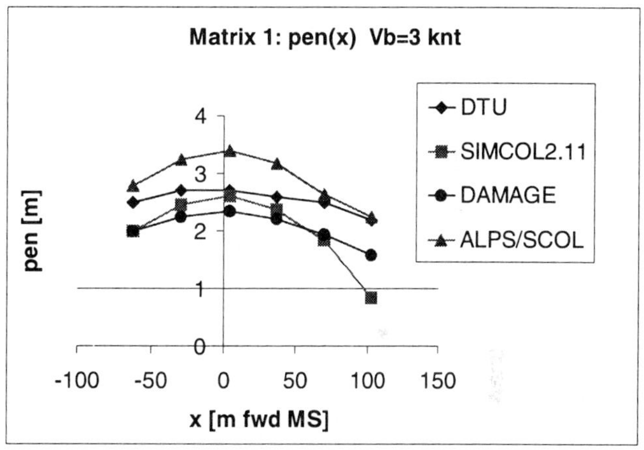

Figure 22 – Matrix 1 Low Energy Collision

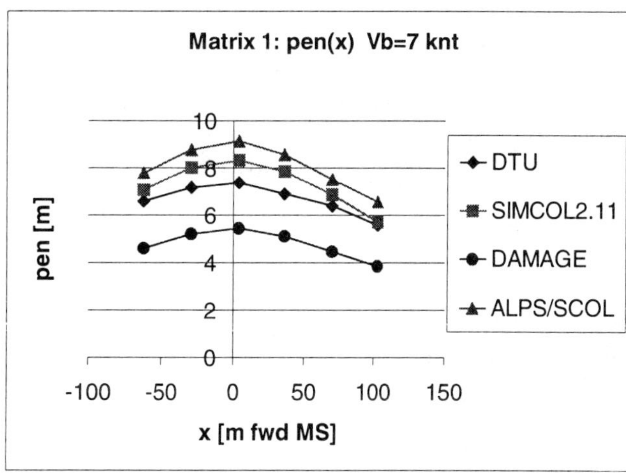

Figure 23 – Matrix 1 High Energy Collision

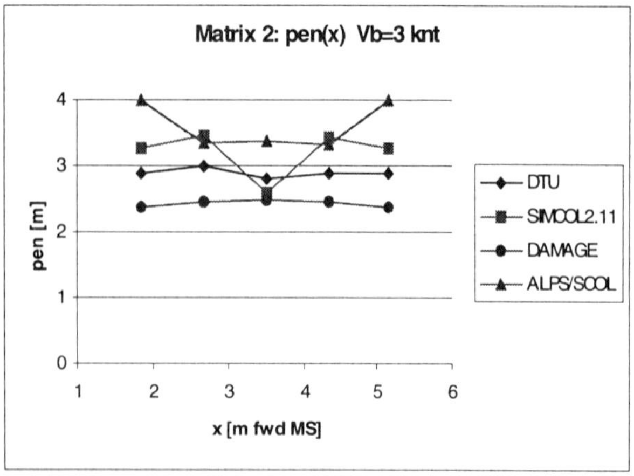

Figure 24 – Matrix 2 Low Energy Collision

In Figures 24 and 25 (Matrix 2) the results for SIMCOL, DTU and ALPS/SCOL are remarkably consistent. In the lower energy case, Figure 24, SIMCOL shows a significant reduction for a perfect right angle strike, on the web, at midship. Membrane tension is balanced on both sides of the strike in this case, and in combination with a direct strike to the web, this substantially reduces penetration in lower energy cases where membrane tension dominates. ALPS/SCOL and DTU show a similar, but less pronounced decrease.

Figures 26 and 27 (Matrix 3) also show the effect of the external dynamics, i.e. more energy must be absorbed in a right angle collision. SIMCOL, DTU and ALPS/SCOL show excellent agreement. SIMCOL again shows a marked reduction at 90 degrees for the lower energy case, Figure 26. The current version of DAMAGE is only able to consider right angle collisions, so DAMAGE is not used in Matrix 3.

5.3.4 Collision Model Discussion and Conclusions

Although not a validation, the results from four very different models are remarkably similar, and this increases confidence in their results. SIMCOL results are less homogeneous for different scenarios because of coupling to external mechanics. The SIMCOL bow geometry may be oversimplified, but at least for these test cases, it provides consistent and sufficient results. The advantage of this simplified geometry is its single parameter description (half-entrance angle) which facilitates its application for probabilistic analysis. SIMCOL's simple geometry and coupling with external dynamics should also facilitate calculation of longitudinal extent of damage that is very important in damage stability and oil outflow analysis.

5.3.5 Ongoing and Future Work

The single most critical next step is model validation using actual or large-scale model test data. This will be discussed in the next section of this paper. Validation using FEA can be used to fill in the gaps in actual data, but only if data and FEA are consistent. FEA using LSDYNA has just begun. Collaboration and comparison of SIMCOL results with DAMAGE, ALPS/SCOL and DTU will also continue.

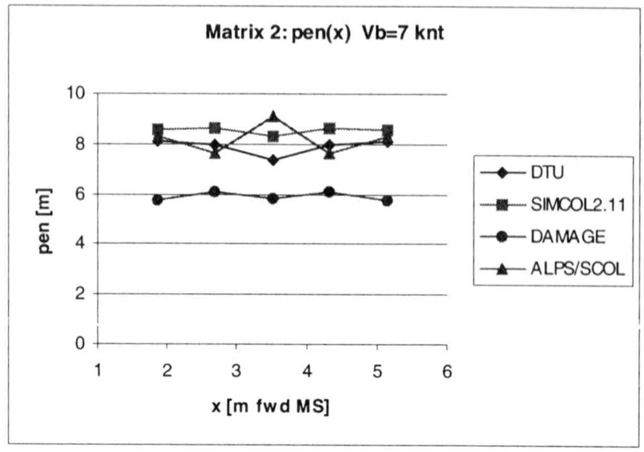

Figure 25 – Matrix 2 High Energy Collision

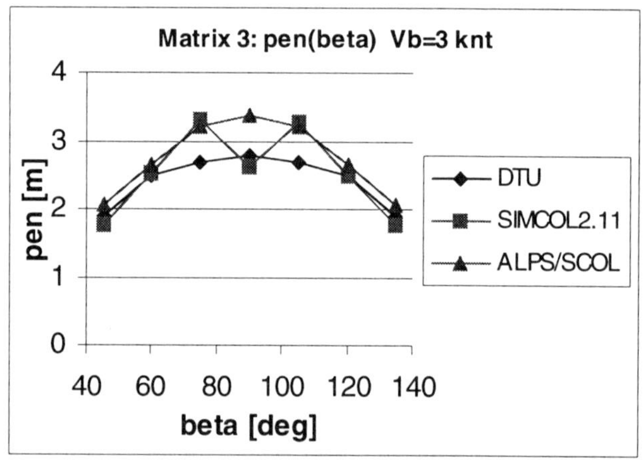

Figure 26 – Matrix 3 Low Energy Collision

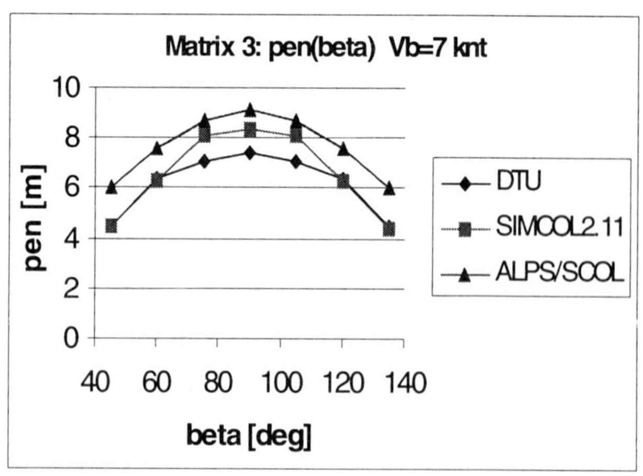

Figure 27 – Matrix 3 High Energy Collision

Other ongoing and future work includes:

- Improved modeling of longitudinal extent of damage including the effect of transverse webs and transverse bulkheads.
- Application and comparison of models in cases where the struck ship has forward speed.
- Application of models to other designs and assessment of damage length, beam and depth scalability with struck ship principal characteristics.
- Struck ship design-parameter sensitivity analysis.
- Probabilistic analysis using scenario descriptions discussed in the next section of this paper.
- Modeling and application of a deformable striking ship bow.

6 COLLISION AND GROUNDING DATA

Two basic types of data are required to support the goals of this panel:

1. Data for developing pdfs for grounding and collision scenarios.
2. Data for grounding and collision model validation, particularly for double hull damage.

6.1 COLLISION AND GROUNDING SCENARIOS

As illustrated in Figure 2, probabilistic descriptions of grounding and collision scenarios are required to develop probabilistic descriptions of grounding and collision damage.

The following data are required to define grounding scenarios:

- Bottom or obstacle description
- Depth of water
- Grounding ship displacement, trim, draft and speed
- Location of the obstacle relative to the ship's centerline

A longer list is required for collision:

- Striking ship
  - Speed
  - Displacement
  - Draft
  - Bow height
  - Bow shape
- Collision angle
- Strike location
- Struck ship
  - Speed
  - Displacement

The pdfs for these variables are not independent. Possible relationships for the collision variables are illustrated in Figure 28. Given the struck ship design and requirement:

1. Since specific struck ships trade in specific ports on specific routes, it is expected that they will encounter a related subset or distribution of other ships (striking ships) that may <u>not</u> be a distribution representing all ships in worldwide trade.
2. A specific struck ship with known design characteristics in a specific trade will also have related distributions for draft, trim and speed. Note: this speed is the speed at the moment of the collision, and not necessarily operating speed.
3. Given a specific type and tonnage of striking ship, its other characteristics will also be related including displacement/mass, bow half entrance angle, bow height, draft, bow stiffness or structural design and speed. Again, this is speed at the moment of collision, not operating speed.
4. When two ships are maneuvering to avoid a collision (in-extremis), the resulting collision angle and strike location are expected to be related.

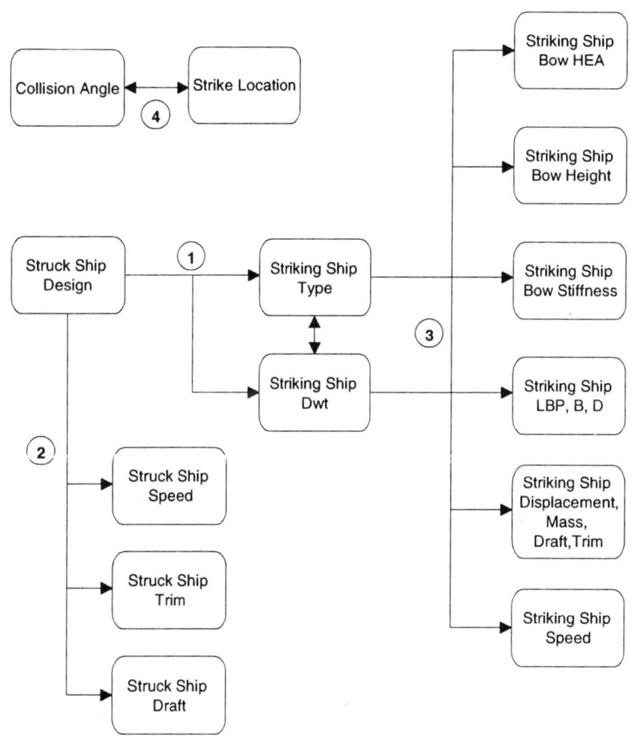

Figure 28. Collision Scenario Variable Relationships

The data necessary to establish these relationships is very limited. Collection and analysis of this data is ongoing.

6.2 DATA FOR MODEL VALIDATION

The second type of data required by the panel is for model validation. In collision the following data is required:

- Collision Scenario
  - Striking ship
    - Speed
    - Displacement
    - Draft
    - Bow height
    - Type
  - Collision angle
  - Strike location
- Struck ship design
  - L,B,D,T,Δ
  - Speed
  - Structural design
    - h_{DB}, w_{DS}, y_{LBHD}, x_{TBHD}
    - web spacing, z_{string}, z_{strut}
    - scantlings
- Damage description

In grounding, the following model validation data is required:

- Grounding Scenario
  - Geometry and elevation of the rock
  - Location of the rock relative to the ship's centerline
  - Ship speed, draft, displacement and trim
- Grounded ship design
  - Hull type (single hull, double hull, mid-deck, ...)
  - L, B, D, LCG, LCF, GM_T, GM_L
  - Waterplane area
  - Transverse and longitudinal bulkhead locations
  - Frame spacing, spacing of longitudinal girders
  - Scantlings
- Damage description

The availability of this data is extremely limited. Accident data from USCG databases for the period of 1980-1998 was screened for sufficiency and applicability to this problem. Useful validation cases must include all of the data specified above (scenario, structural design, damage description). In the case of collision, the following additional criteria were applied:

- Striking location in struck ship away from bow or stern.
- Collision angles of 45-135 degrees.
- Penetration greater than one meter.

Collection and evaluation of this data is also ongoing.

7 STRIKING SHIP BOW

7.1 SIMCOL Bow Model Study

All of the collision models used thus far in this study, and most models used by other researchers, assume that the striking ship bow is rigid. This is changing. Panel work in this area is focused on the following questions and problems:

1. Is the rigid bow assumption rational? Is significant energy absorbed by the striking ship bow? Is this energy variable in different collision scenarios?
2. If bow deformation is important, can this problem be solved uncoupled from the side damage problem?
3. Identify a simple, but sufficient bow model for future use in SIMCOL.
4. Develop a course-mesh bow model to speed up LSDYNA modeling for SIMCOL validation.

Most collision accidents that result in significant struck ship damage also result in significant damage to the striking ship bow. Minorsky's original analysis [26] considered absorbed energy in the striking bow. A reanalysis of his data indicates that absorbed bow energy in these cases accounted for between 8 and 53 percent of the total absorbed energy [30]. More recent studies by Reckling [31], Akita and Kitamura [32], and Valsgard and Pettersen [33] indicate absorbed bow energies in the range of 42 to 55 percent of the total absorbed energy. LSDYNA analyses conducted at Virginia Tech with deformable bows and sides show absorbed bow energies in the range of 22 to 44 percent [30]. We must conclude that absorbed bow energy can be significant and variable, and therefore should be considered in collision analyses.

STRUCK OBJECT CONVENTIONAL = FINE MESH FEA SIMPLIFIED = COURSE MESH SUPERELEMENT FEA		BOW MODEL							
		CONVENTIONAL FINE MESH FEA		INTERSECTION ELEMENT COURSE MESH FEA				RIGID	
		23K DWT	150K DWT	500 DWT	23K DWT	40K DWT	150K DWT	23K DWT	150K DWT
RIGID WALL	Closed Form			I-1	I-2	I-3	I-4		
	LS-DYNA	C-1	C-2	I-5	I-6	I-7	I-8		
100K SH TANKER SIMPLIFIED	LS-DYNA Fixed	C-7			I-13				
	LS-DYNA Sway	C-8			I-14				
100K SH TANKER CONVENTIONAL	LS-DYNA Fixed	C-3			I-9			R-1	
	LS-DYNA Sway	C-4			I-10			R-2	
150K DH TANKER SIMPLIFIED	LS-DYNA Fixed		C-9				I-15		
	LS-DYNA Sway		C-10				I-16		
150K DH TANKER CONVENTIONAL	LS-DYNA Fixed		C-5				I-11		R-3
	LS-DYNA Sway		C-6				I-12		R-4

Table 6. Bow Model Test Matrix

Table 6 lists the bow and struck ship model cases considered in the SIMCOL bow model development. Four rigid bow tests were run to validate SIMCOL calculations. These are labeled R-1 through R-4 in Table 6. Ten conventional FEA bow analyses were run to validate the simpler intersection bow model results. These are labeled C-1 through C-10 in Table 6. Two types of intersection model analyses were accomplished. Tests I-1 through I-4 use closed-form equations from Pedersen [34], Amdahl [35], and Yang and Caldwell [36]. Tests I-5 through I-16 use intersection elements applied in LSDYNA simulations.

Simplified LSDYNA intersection-element bow models were developed to increase the speed of LSDYNA finite element solutions used in SIMCOL collision model validation. In these models, only intersections of sides, decks, longitudinal bulkheads and girders were included in the model with longitudinal stiffener area smeared into plate thickness based on plastic bending moment. Intersection elements are modeled as truss elements in LSDYNA with material properties replaced by properties derived for crushed L, T and cruciform sections as illustrated in Figure 29 [35]. Transverse frames were modeled as normal truss elements. Nodes were only allowed a single degree of freedom in the striking ship longitudinal direction. The 150kdwt bulk carrier bow is shown in Figure 30 [34]. Figure 31 shows the intersection model for this bow (I-8). Figure 32 shows the conventional fine-mesh FEM for this bow with beam and panel elements (C-5). Figure 33 shows a comparison of the force-indentation plots for the intersection and conventional models (I-8 and C-2). The results compare very well. Similar results were obtained for the other test cases. It is concluded that the bow intersection model is sufficient for LSDYNA collision simulations. Work is continuing to apply these results to the development of a SIMCOL bow model.

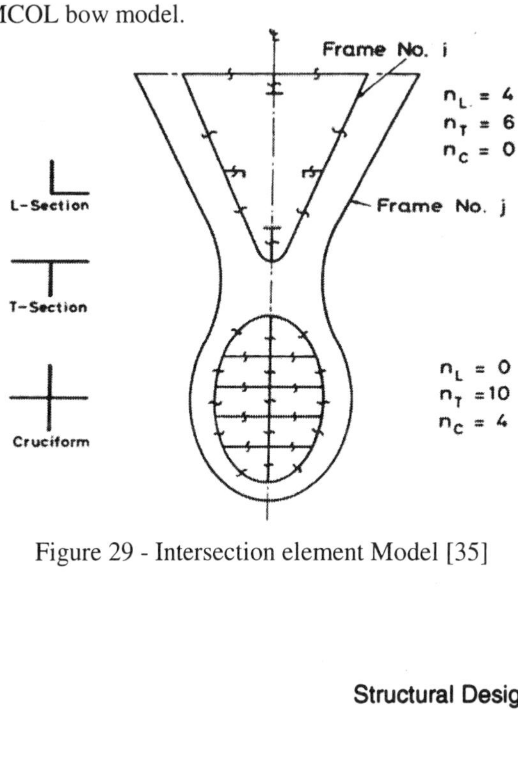

Figure 29 - Intersection element Model [35]

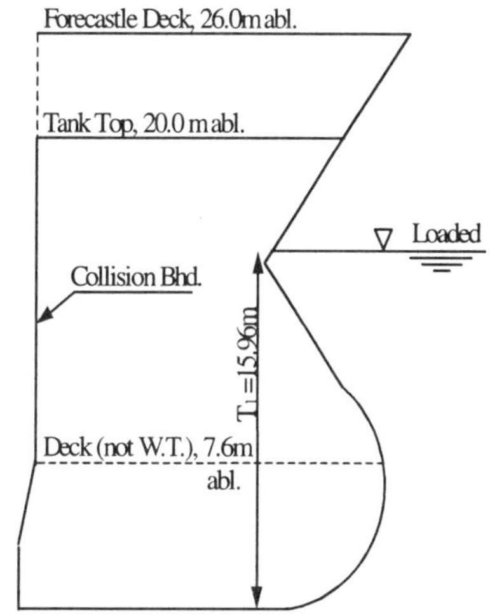

Figure 30 - 150kdwt Bulk Carrier Bow [34]

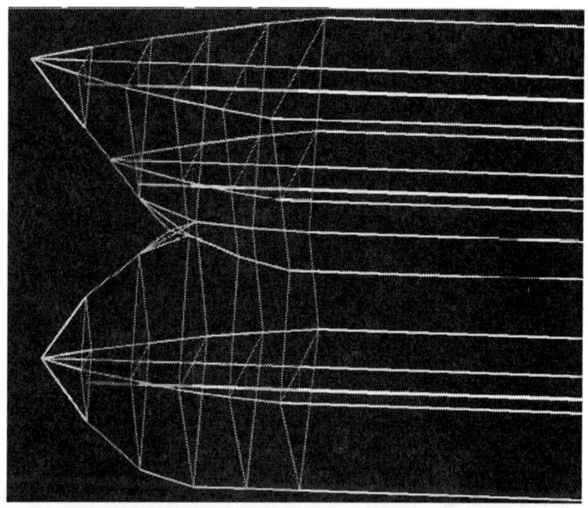

Figure 31 - Bulk Carrier Intersection Element Model [30]

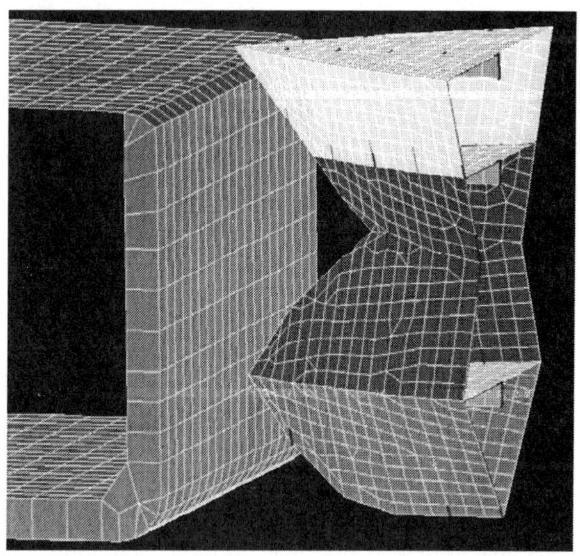

Figure 32 - Bulk Carrier Conventional FE Bow

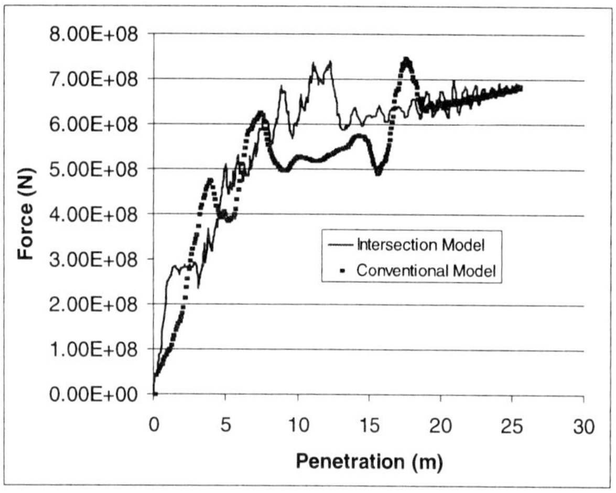

Figure 33 - 150kdwt Bulk Carrier Striking Rigid Wall [30]

7.2 DAMAGE Bow Model

DAMAGE 5.0 includes a deformable bow. The initial bow geometry is the same as in Damage 4.0, illustrated in Figure 16. Bow crushing is modeled using L-, T- and X-form super-elements in an 'intersecting unit method' [37]. Bow and side force-deformation calculations are completed separately (uncoupled). These calculations are performed assuming that the bow strikes a rigid wall and that the side is struck by a rigid bow. The results are then compared incrementally with deformation applied to the weakest component (bow or side) at each increment. This results in deformation and energy absorption in both components.

7.3 DTU Bow Model [38]

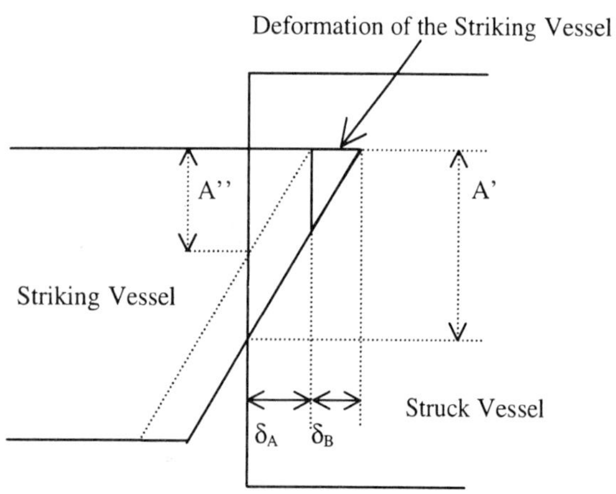

Figure 34 - Deformation of vessels during collision. The A's relate to areas not lengths

A more recent DTU model also predicts damage to struck and striking vessels in a collision event [38]. The analysis is carried out in penetration steps. Only one of the involved vessels can be deformed in each step. By a comparison of the crushing forces for respectively the bow and the side, it can be determined which vessel deforms during the considered step.

Before calculation of the deformation of the two vessels the following calculations are carried out:

1. The Force-Penetration curve $F_{struck}(\delta_A)$ for the struck vessel is calculated, where the striking vessel is rigid.
2. The Force-Penetration curve $F_{striking}(\delta_B)$ for the striking vessel is calculated, where the struck vessel is assumed rigid.

If the striking vessel has a bulbous bow, the analysis of the crushing forces is separated into a bulb analysis and an analysis of the top of bow above the bulb.

The force-deformation curve for the struck vessel is determined by the procedure described in Section 5.2.4 and for the striking vessel by the procedure described in Pedersen et. al [34].

A commonly used procedure for taking into account the deformation of the bow is to compare the two force-penetration curves, $F_{struck}(\delta_A)$ and $F_{striking}(\delta_B)$, at each step. This approach, however, only includes a very limited level of interaction. In reality, the force-penetration curve for the side of the struck vessel is a function of the deformation of the bow, and vice versa. This stronger interaction is taken into account by comparing the forces F_A and F_B, which is determined as:

Struck vessel: $$F_A = F_{Struck}(\delta_A)\frac{A'}{A''} \quad (1)$$

Striking vessel: $$F_B = F_{Striking}(\delta_A + \delta_B) \quad (2)$$

where

F_A force to crush the struck vessel;
F_B force to crush the striking vessel;
F_{struck} force from the force-penetration curve for struck vessel, where the striking vessel is rigid;
$F_{striking}$ force from the force-penetration curve for striking vessel, where the struck vessel is rigid;
δ_A penetration into the struck vessel;
δ_B deformation of the striking vessel;
A' cross-sectional area of the striking vessel taken at a distance of $\delta_A+\delta_B$ from bow or bulb tip;
A'' cross-sectional area of the striking vessel taken at a distance of δ_A from bow or bulb tip;

See also Figure 34.

The forces at the struck and the striking vessel F_A and F_B are compared

- If $F_A > F_B$
  Deformation of striking vessel, δ_B is increased
- If $F_B > F_A$
  Deformation of struck vessel, δ_A is increased

The reason for correcting the resistance of the struck vessel is that if the bow is deformed, the resistance is approximately equal to the force at the side times the ratio between the areas. For a single hull vessel the correction will have nearly no influence, but for a double hull vessel, there will be some corrections when the bow penetrates the inner side. See Figure 34.

When the deformation patterns of the struck and the striking vessel are known, the total absorbed energy can be calculated and compared with the energy calculated by the external dynamics. See Pedersen and Zhang [29].

8 STRUCTURAL DESIGN FOR COLLISION AND GROUNDING

The function of a tank vessel's structural system may be viewed from the standpoints of normal operation and casualty operation. In providing adequate resistance for normal operations, the objective in structural design is to maintain structural integrity of the hull girder, of bulkheads, decks, etc., and of plating, stiffeners and details. Other considerations relate to vessel size, complexity and heaviness of structure, producibility and maintainability. In terms of casualty operations, the objective is to maintain vessel integrity and to protect cargo, or conversely to protect the environment from oil pollution in case of a casualty. In this case the primary considerations should include:

1. Resistance to fire and explosion damage and its containment,
2. Resistance to collision and grounding damage,
3. Containment of petroleum outflow if damage does occur, and
4. Maintenance of sufficient residual strength after damage in order to permit salvage and rescue operation and to minimize further damage and spilling of oil.

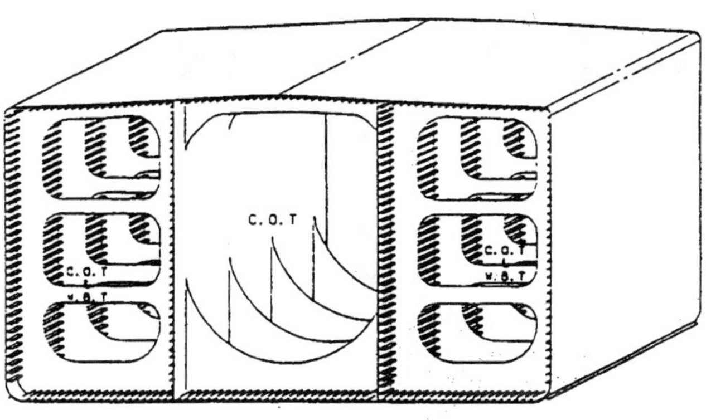

Figure 35 - Single-Hull Tanker [39]

Tank vessels have traditionally been designed as single-skinned hulls, Figure 35 [39]. Depending on the size of the vessel, longitudinal bulkheads are often present, and the overwhelming majority of single-skinned designs are longitudinally stiffened. OPA '90 mandates the use of double-skinned tanker designs, Figure 36 [39].

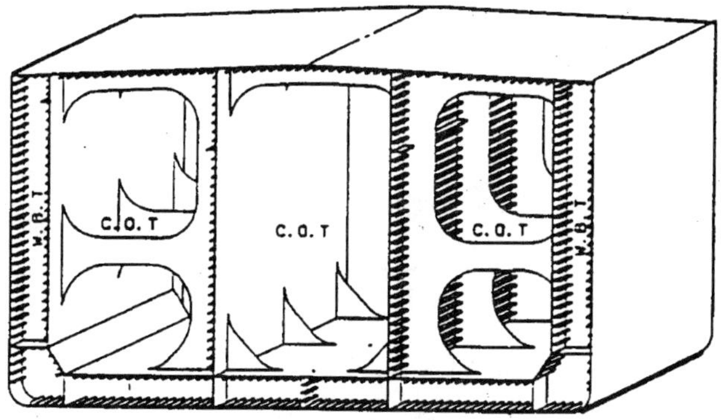

Figure 36 - Double-Hull Tanker [39]

The Marine Board of the National Research Council has recently convened a "Committee on Evaluating Alternative Tanker Designs". The committee is to establish, for the first time, a rational methodology to evaluate alternatives to the double hull. The US Coast Guard is sponsoring this Committee, and its report is expected in the first quarter of 2001.

The emphasis of new and proposed requirements for reducing the probability of oil cargo outflow has been on the subdivision of cargo spaces. The impact of hull structure on the reduction of outflow has had more limited attention. The complexity of determining the contribution of structure to cargo protection and the unpredictability of structural response under the variety of potential damage scenarios have no doubt contributed to this set of circumstances.

The following subsections address the subject of vessel structural design characteristics for collision, grounding and bow impact. The intent is to consider what has been learned from analytical studies regarding structural design features that may suggest future modifications and alternatives to enhance resistance to cargo outflow.

8.1 Collisions

8.1.1 Analytical Indications – Struck Ship

The sequence of structural events during a collision varies according to the nature of the collision and vessel design. Figures 37 and 38 [24] provide general flow diagrams representing the phenomena for single and double hull tankers. This information is useful in identifying where greater resistance can be addressed.

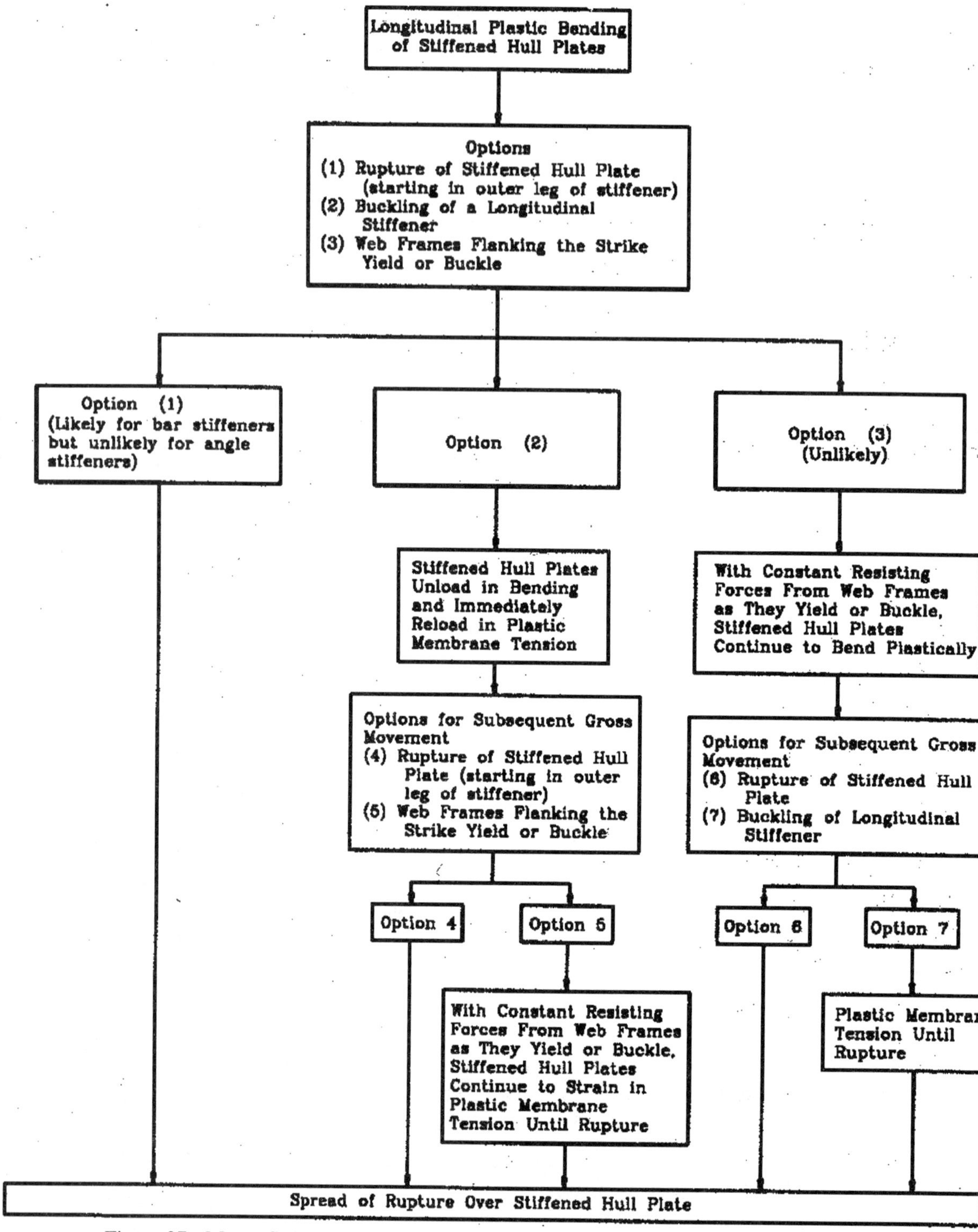

Figure 37 - Macro flow diagram for side collision plastic-energy analysis of a single hull [24]

Initially, the stiffened hull plating distorts in a plastic bending phase, with plastic "hinges' forming in the vicinities of the strike and the web frames flanking the strike. During this phase, insignificant membrane tension develops. For a typical tanker with longitudinal angles stiffening the hull plating, the longitudinal angle-shaped stiffeners then buckle in the vicinity of the flanking web frames, and possibly "trip" in the vicinity of the strike. Subsequently, the stiffened hull unloads momentarily as the strike continues, but reloads in a membrane-tension phase. The hull ruptures at the end of this phase, with possibly the flanking web frames yielding or buckling before the hull ruptures. In such cases, the membrane-tension phase includes two sub-phases: (1) there is no transverse movement of the web frames flanking the strike; and (2) the web frames flanking the strike move inward toward the ship's centerline, and the damage extends into the adjacent web frame spaces. During these phases, the deck is also distorting in membrane tension. However, the deck behavior is assumed not to affect the sequences of the options.

Other sequences of phenomena are possible. A hull with longitudinal stiffeners, such as rectangular bars that are not apt to buckle or trip, tend to rupture before significant membrane tension has a chance to develop.

Alternatively, with any type of hull stiffeners, very weak web frames can yield or buckle before rupture or buckling of the longitudinal stiffeners, in which case the damaged length increases during the bending phase. These phenomena are unlikely, however, for typical ships.

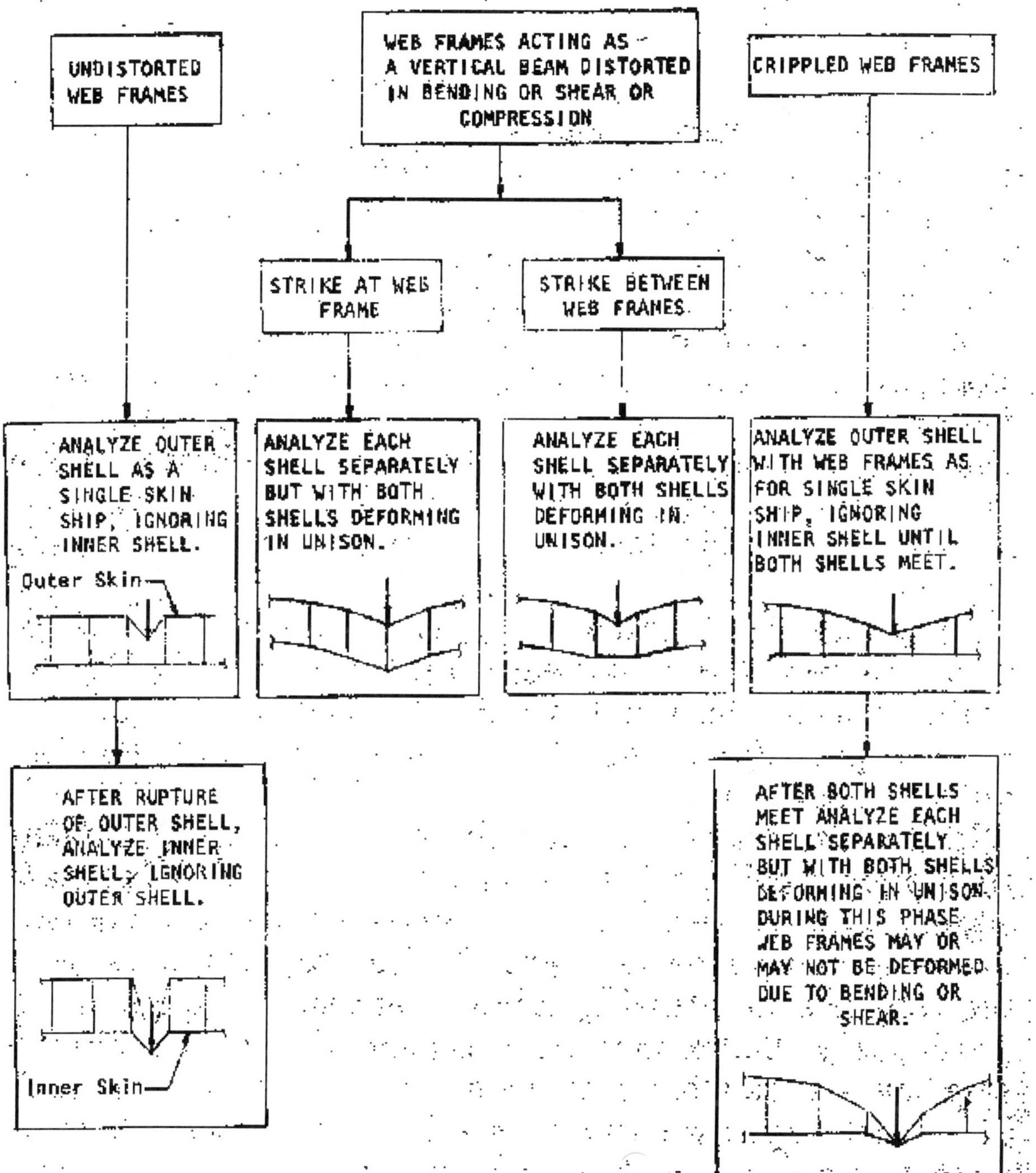

Figure 38 - Macro flow diagram for side collision plastic-energy analysis of a double hull [24]

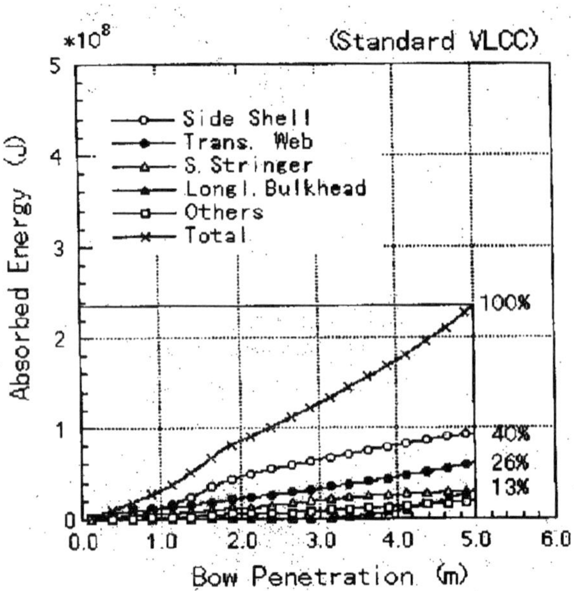

Figure 39 - Energy Absorption of Structural Components (Standard VLCC) [40]

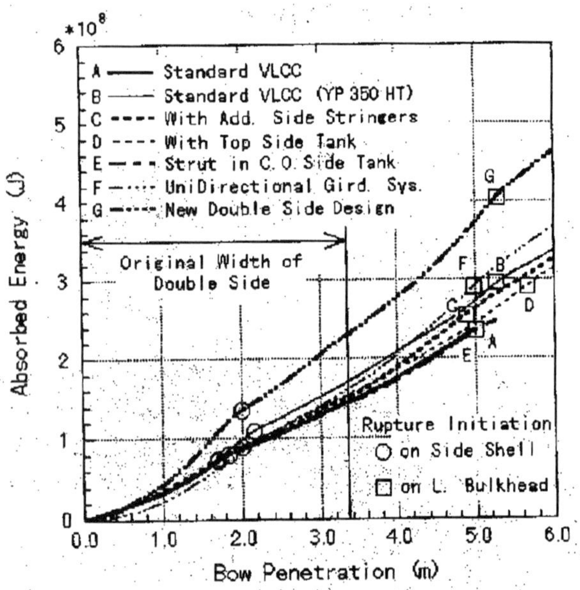

Figure 40 - Energy Absorption Capacity [40]

The contribution of various structural components for a ULCC with a more conventional double side structure is shown in Figure 39 [40].

8.1.2 Application

With the knowledge acquired from understanding the process identified by the analytical evaluations, it is possible to postulate hull structural features that may enhance resistance to collision. These fall into two categories:

- Conventional hull structural arrangements that are modified to enhance energy absorption.
- Unconventional hull structural arrangements that include structurally actuated mechanisms for energy absorption.

As part of the research project for improved tanker safety against collision and grounding, the Association for Structural Improvement of the Shipbuilding Industry (ASIS) of Japan conducted a study of varying structural characteristics of a conventional VLCC design and applying a more unconventional structural feature as well [40]. The more conventional alterations included use of high strength steel, additional shell stringers, a longitudinal strut in the outboard cargo tank supporting the double side, and a double side unidirectional stiffening system. As the side-shell has shown to be a significant contributor to energy absorption, an unconventional new skin composed of double plate panels internally stiffened by web frames, essentially a cored panel skin providing additional net shell thickness, was substituted for the conventional stiffened plate. The results are shown in Figure 40. Other ideas of the more conventional nature included:

- Bulkhead stools on the wing tank side to reduce hard spots and increase damage length.
- Reduced stiffness of web frames to increase damage length.
- Increased side-shell plating thickness, especially from the ballast waterline up, to increase membrane tension in the side-shell.
- Decreased spacing of side-shell longitudinals and added intermediate vertical web frames on the side shell to increase the integrity of the side-shell.
- Increased number of horizontal stringers in the wing tank to increase side skin integrity and provide additional material for membrane tension.

Although the practicability and acceptance from a regulatory point of view of very unconventional approaches may be seriously in doubt, it is nevertheless possible to postulate alternatives, and in fact the greatest step in the level of possible energy absorption is likely with an unconventional approaches:

- Controlled pressure fluid chamber in the wing tanks wherein the venting of fluid under pressure to other spaces when compressed during collision results in energy dissipation.
- Introduction of material in wing ballast-tanks that does not reduce ballast capacity significantly, but increases energy absorption, like ultra-large open-cell foam. Permeability and inspection are possible problems with this approach.
- Truss structure separating wing tank skins. Such an arrangement can enhance the integrity of the inner

hull longer as the truss deforms while supporting the outer skin and more slowly loading the inner skin.

- Pre-planned failure or "crumple zones" as have been developed by the automobile industry for the more critical and predictable areas of damage.
- A foam release system actuated in the event of an impending collision within the expected area of contact to increase stiffness and energy absorption.

8.2 Bow Impact

8.2.1 Analytical Indications

The significance of the energy absorption in bow damage and the effects of the bow interacting with the struck ship's side structure make it a critical element in considering collision results. Calculations assuming rigid bows with vertical stems and no bulbs have simplified analyses and fostered understanding, but do not depict the current and likely future nature of bows. Ship resistance, powering and seakeeping requirements are likely to continue to foster the existence of bows with raked stems and bulbs into the future.

Analytical evaluations do point a way to considering bows. If they are deformable, they will account for a sizeable portion of the total energy absorption in a collision. If they can be made to deform in a manner that does not cause premature failure of important struck ship energy absorbing structure, such as the side shell, then the same attributes of a rigid vertical stem bow will be achieved at the struck vessel.

8.3 Application

The bows of ships incorporate a significant amount of stiff horizontal fore and aft reinforcing structural elements that enhance the rigidity of the structure. Alternatively, arrangements with transverse stiffening coupled with appropriate spacing can lead to significant folding deformation, which can effectively reduce or eliminate the protrusions of the upper bow and bulb as well as their structural hard-spots. Recent studies [44] have confirmed that lower crushing pressure of a bulbous bow is preferable to avoid the early rupture of the struck sideshell and the crushing strength should not exceed that of the side structure.

8.4 Grounding

8.4.1 Analytical Indications – Grounding

The potential phases of grounding are depicted in Figure 41. In grounding and raking, the vessel has forward speed. In stranding it does not.

Model testing and computer finite element simulation of a stranding for a 280000 dwt tanker double bottom resulted in the lateral load versus indentation of up to 2m shown in Figure 42. The peak point, point 1, denotes the buckling of the center longitudinal girder, and point 2, the fracture of the outer bottom plating. Point 3 indicates fracture of the center girder-floor connection. It is interesting to note that the response shown in Figure 42 is similar to that predicted for collision, with the principal difference being the nature of the obtrusive loading, which may be a line load for a collision and a point load for a grounding.

Raking bottom damage is of significantly different character than the grounding damage heretofore discussed. In raking damage, the obstruction ruptures the bottom structure and then continues destructive shearing, tearing and buckling as the vessel continues to move in its direction of motion, Figure 43. The length of bottom raking is based on an energy balance concept where the kinetic energy of the vessel is entirely absorbed by the destruction of bottom structure.

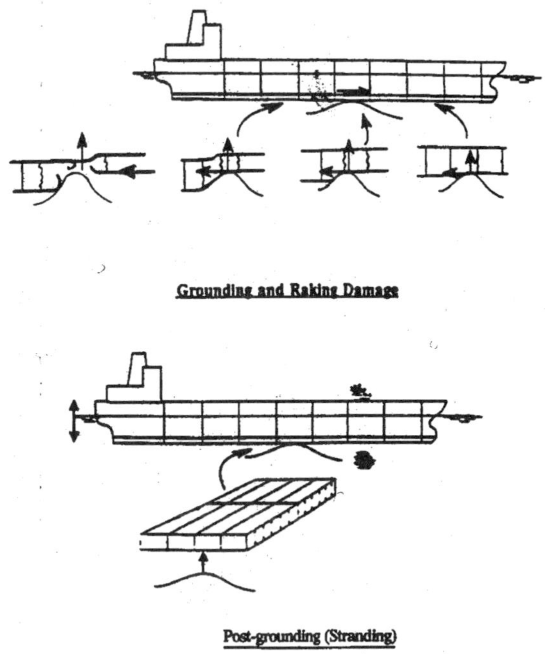

Figure 41 - Model of the grounding process [41,42]

An energy analysis for grounding which takes into account energy due to the tearing that occurs during tracking-type collisions presumes that in such cases the damage will be long and shallow, involving plate tearing, but relatively little volume distortion of the inner structure. However, Figure 42 indicates that if there is a vertical component to the striking force, after fracture of the outer plating at point 2, there is an increase in stiffness leading up to point 3 and beyond as more girders and floors come into contact with the obstruction. Figure 44 also indicates a retention of stiffness.

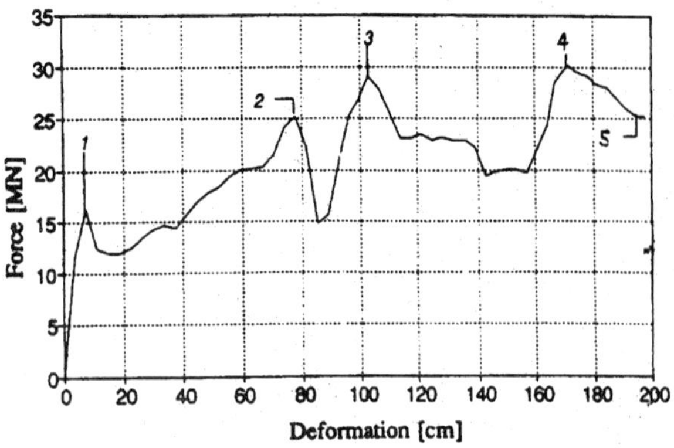

Figure 42 - Load versus lateral indentation for 280000 dwt tanker double bottom [42]

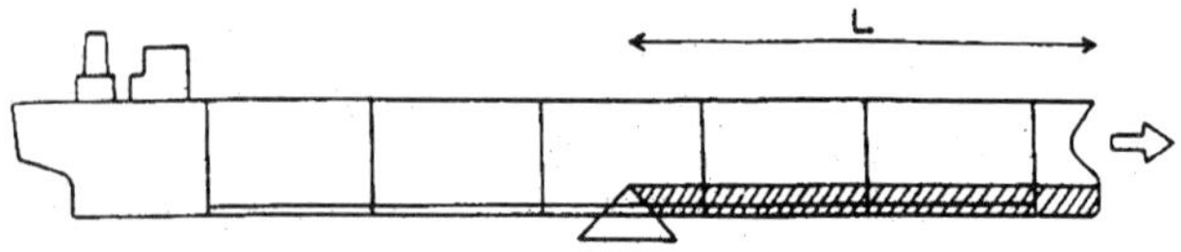

Figure 43 - Bottom raking due to grounding

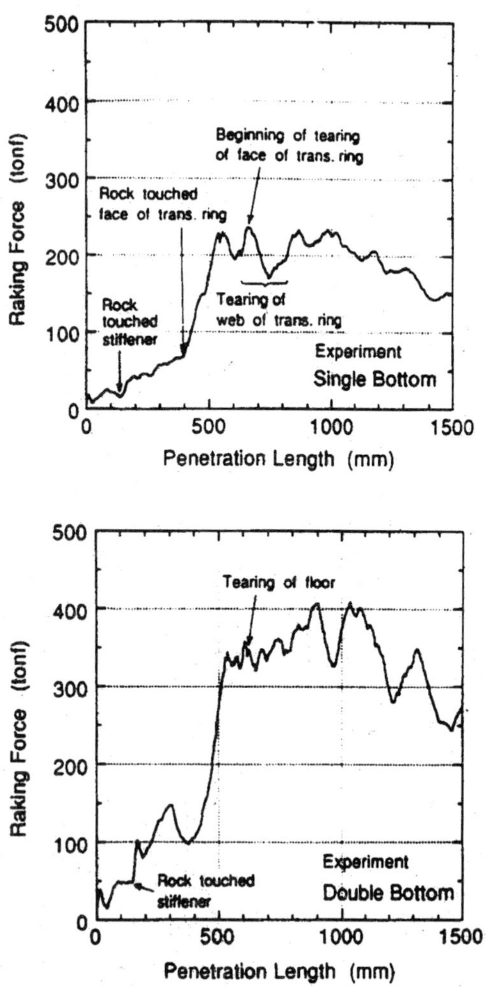

Figure 44 - Raking force versus penetration length by experiments [41]

8.4.2 Application

In resisting both grounding and raking damage, it is clear that the bottom shell and the internal grid of the transverse floors, longitudinal girders and bulkheads is critical.

In the case of raking damage, the longitudinal stiffness of floors and their ability to absorb energy in distortion are more critical than their support of the shell. As in this case the distorting force is applied in the weakest direction of the floor, the situation is further complicated. Clearly, the more floors and the smaller the spans between their longitudinally resistive structure, the better. On the other hand, this arrangement will serve to limit the lateral deflection and therefore, the energy absorption in a stranding. Other ideas have included:

- Increase the scantlings of the lower one-third part of the hull.
- Slope the inner bottom, higher at the collision bulkhead and lowering to the regulatory requirement at the end of the first cargo tank.
- Preclude web frame and bulkhead damage propagation through the inner bottom by introducing a discontinuity in the rigidity between the two. A bulkhead stool would be an example.

9 CONCLUSIONS

The almost two-year effort of the panel has achieved or partially achieved many of the original objectives stated in the introduction to this paper:

- Simplified grounding models have been investigated, compared and assessed. DAMAGE 4.0 is shown to be an excellent tool for comparative analysis of grounding damage, and based on limited validation, provides good results for cases with similar idealized rock geometry. More validation is required. Limitations include the ability to consider only conventional ship geometries and an idealized pinnacle geometry. Grounding on other rock geometries, reefs, etc., cannot be analyzed using DAMAGE, but should be considered in the future as an important part of a complete probabilistic analysis. Only a preliminary attempt at probabilistic analysis using DAMAGE was completed.
- Four simplified collision models have been investigated, compared and assessed. The models provide similar penetration results for the cases considered. Although some limited validation has been accomplished for specific mechanisms used in the models, more validation is required. The prediction of longitudinal extent of damage, particularly at transverse bulkheads, remains largely unexplored. This is very important for oil outflow and damage stability calculations. This prediction

requires consideration of various collision angles and struck ship speed. Consideration of striking bow deformation is a relatively new addition to the collision models. Further application and validation of bow models and their coupling to struck ship models is required. Most of the researchers are using finite element analysis as part of simplified model validation. Only a preliminary attempt at probabilistic analysis using SIMCOL was completed. Additional analysis is in progress.

- Data for model validation and scenario definition is very limited. A proposed data requirement is specified in this paper. The search for data to fill the requirement goes on with the support of the USCG, SSC and SNAME. Funding and significant IMO and class society support may ultimately be necessary to fully satisfy this data requirement.
- Some progress has been made in the definition of scenarios and conditions for grounding and collision analyses, but limited data is also hampering this effort. Data describing the worldwide population of ships that might be involved in collisions is available, but since specific struck ships trade in specific ports on specific routes, it is expected that they will encounter a related subset or distribution of other ships (striking ships). Sufficient data has not been obtained to quantify or assess this hypothesis. More data is also required for collision angle, strike location, grounding and collision ship speeds and grounding bottom description.
- A number of possibilities for design improvement are presented in the paper. As tools, standard scenarios, methodologies and performance criteria are evaluated and defined they must be used to assess innovative designs, and to optimize and evaluate conventional designs. Damage models should be applied in a total ship design framework to trade-off variations in structural design, subdivision and system redundancy.
- The application of this work in a regulatory framework requires that a very simplified, preferably parametric relationship be established between structural design variables and probabilistic damage. This work remains to be done.

10 SIGNIFICANCE

This work is not done. Grounding and collision models have been improved, data requirements have been defined, and design possibilities for improved crashworthiness have been identified, but this work must be finished and validated before regulations and designs can be improved. This is our ultimate goal. Based on the progress thus far, the following important areas of work can proceed in parallel with the remaining model development and validation:

- Probabilistic analysis
- Sensitivity analysis and the development of parametric equations relating damage extent to structural design variables in conventional designs
- Modification of the grounding and collision models to consider unconventional designs

11 CONTINUING WORK OF THE PANEL

Individual sections of this paper have discussed possibilities for future work. This panel has been very active and very productive, but in many ways has just scratched the surface. Individual papers will be written in most all areas covered by this summary paper. Subsequent work will include: completion of sufficient grounding and collision models, validation of the models, specification and acceptance of standard accident scenarios, and ultimately application of models and scenarios to improved structural design in collision and grounding. A two-year continuation of the panel will be proposed to the T&R Committee. This continuation will include a new charter with new objectives and deliverables. The Ad Hoc framework has proven very effective for meaningful collaboration and effort. The panel chairman thanks all those who have contributed.

12 REFERENCES

[1] IMCO, *Regulations on Subdivision and Stability of Passenger Ships as Equivalent to Part B of Chapter II of the International Convention for the Safety Of Life At Sea*, IMCO Resolution A.265 (VIII), adopted on 20 Nov 1973.

[2] Gilbert, R. and Card, J.C., "The New International Standard for Subdivision and Damage Stability of Dry Cargo Ships", *Marine Technology*, Vol. 27, No. 2, March 1990, pp. 117-127.

[3] "Interim Guidelines for Approval of Alternative Methods of Design and Construction of Oil Tankers under Regulation 13F(5) of Annex I of MARPOL 73/78", Resolution MEPC.66 (37), Adopted September 14, 1995.

[4] "IMO Comparative Study on Oil Tanker Design," IMO paper MEPC 32/7/15, Annex 5, Distribution of Actual Penetrations and Damage Locations Along Ship's Length for Collisions and Groundings.

[5] Sirkar, J., et al., "A Framework for Assessing the Environmental Performance of Tankers in Accidental Groundings and Collisions", *1997 SNAME Annual Meeting*, October 1997.

[6] Rawson, C., Crake, K. and Brown, A.J., "Assessing the Environmental Performance of Tankers in Accidental Grounding and Collision", *1998 SNAME Annual Meeting*, November 1998.

[7] Brown, A. and Amrozowicz, M., "Tanker Environmental Risk - Putting the Pieces Together", *SNAME/SNAJ International Conference on Designs and Methodologies for Collision and Grounding Protection of Ships,* August 1996.

[8] Card, J. C., "Effectiveness of Double Bottoms in Preventing Oil Outflow from Tanker Double Bottoms", *Marine Technology*, January 1975.

[9] ISSC, Report by Specialist Panel V.4 – Structural Design Against Collision and Grounding, Proceedings of the 13th International Ship and Offshore Structures Congress 1997, Trondheim, Norway, 1997.

[10] W. Abramowicz and M. Sinmao, "User's Manual and Modeling Guide for the Program DAMAGE, Joint MIT-Industry Program on Tanker Safety, Report No. 66, June 1999.

[11] B. Cerup Simonsen and T. Wierzbicki. "Grounding Bottom Damage and Ship Motion over a Rock". Int. J. of Offshore and Polar Engineering, Vol. 6, no. 3, Sep. 1996.

[12] B. Cerup Simonsen, *"Mechanics of Ship Grounding"*, PhD thesis, Department of Naval Architecture and Offshore Engineering, the Technical University of Denmark, February 1997.

[13] B. Cerup Simonsen, "Ship Grounding on Rock: I Theory", *Marine Structures*, Vol. 10, pp. 519-562, 1997

[14] B. Cerup Simonsen, "Ship Grounding on Rock: II Validation and Application", Marine Structures, Vol. 10, pp. 563-584, 1997.

[15] Wang, G., Ohtsubo, H., and Liu D., "Simple Method of Predicting the Grounding Strength of Ships", *Journal of Ship Research*, Vol. 41, No. 3, Sept. 1997.

[16] B. Cerup Simonsen. "DAMAGE Theory Validation", Joint MIT-Industry Program on Tanker Safety, Report no 63, May 1998.

[17] Tikka, K.K., and Chen Y.J., "Prediction of Structural Response in Grounding – Application to Structural Design", SSC/SNAME Ship Structural Symposium June 13-14, 2000.

[18] Giannotti, J.G., Johns, N., Genalis, P. and Van Mater, P.R., "Critical Evaluations of Low-Energy ship Collision Vol. I - Damage theories and Design Methodologies", Ship Structure Committee Report No. SSC-284, 1979

[19] Simonsen, B.C., "Theory and Validation for the DAMAGE Collision Module", Joint MIT-Industry Program on Tanker Safety, Report No. 67, June 1999.

[20] Paik, J.K., et al, "On Rational Design of Double Hull Tanker Structures against Collision", *1999 SNAME Annual Meeting*, 1999.

[21] Paik, J.K. and Pedersen, P.T., "Modeling of the Internal Mechanics in Ship Collisions", *Ocean Engineering*, Vol. 23, No. 2, pp. 107-142, 1996.

[22] Chen, D., "Simplified Collision Model (SIMCOL)", Dept. of Ocean Engineering, Virginia Tech, Master of Science Thesis, May 2000.

[23] Hutchison, B.L., "Barge Collisions, Rammings and Groundings - an Engineering Assessment of the Potential for Damage to Radioactive Material Transport Casks", Report No. SAND85-7165 TTC-05212, 1986.

[24] Rosenblatt & Son, Inc, "Tanker Structural Analysis for Minor Collision", USCG Report, CG-D-72-76, 1975.

[25] McDermott, J.F., et al, "Tanker Structural Analysis for Minor Collisions", *SNAME Transactions*, Vol. 82, pp. 382-414, 1974.

[26] Minorsky, V. V. (1959) "An Analysis of Ship Collisions with Reference to Proteciton of Nuclear Power Plants," Journal of Ship Research.

[27] Reardon, P. and Sprung, J.L., "Validation of Minorsky's Ship Collision Model and Use of the Model to Estimate the Probability of Damaging a Radioactive Material Transportation Cask During a Ship Collision", *Proceedings of the International Conference on Design and Methodologies for Collision and Grounding Protection of Ships*, San Francisco, August 1996.

[28] Crake, K., "Probabilistic Evaluations of Tanker Ship Damage in Grounding Events", Naval Engineer Thesis, MIT, 1995.

[29] Pedersen P.T. and Zhang S. "On Impact Mechanics in Ship Collisions", Marine Structures, Vol. 11, pp. 429-449, 1998.

[30] Vakkalanka, Suryanarayana, "Simplified Bow Model for a Striking Ship in Collision", Dept. of Ocean Engineering, Virginia Tech, Master of Science Thesis, July 2000.

[31] Reckling, K.A. (1983), "Mechanics of Minor Ship Collisions", International Journal of Impact Engineering, Vol. 1, No. 3, pp. 281-299.

[32] Akita, M. and Kitamura, K. (1972), "A Study on Collision by an Elastic Stem to the Side Structure of Ships", Trans. SNAJ, 131, 307-317.

[33] Valsgard, S. and Pettersen, E. (1982), "Simplified Non-Linear Analysis of Ship/Ship Collisions", Norwegian Maritime Research, No. 3, pp. 2-17.

[34] Pedersen, P.T., et al, "Ship Impacts: Bow Collisions", *International Journal of Impact Engineering*, Vol. 13, No. 2, pp. 163-187, 1993.

[35] Amdahl, J., "Energy Absorption in Ship-Platform Impacts", Dr. Ing. Thesis, Report No. UR-83-84, The Norwegian Institute of Technology, Trondheim, 1983.

[36] Yang, P. D. C., and Caldwell, J. B., "Collision Energy Absorption of Ships' Bow Structures," International Journal of Impact Engineering, Vol. 7, No. 2, pp. 181-196, 1988.

[37] Abramowicz, W., "Crush Resistance of "T", "Y", "X" Sections," Joint MIT-Industry Program on Tanker Safety, Report #24, 1994.

[38] Lützen M., Simonsen B.C, Pedersen P.T., "Rapid Prediction of Damage to Struck and Striking Vessels in a Collision Event", Proceedings of Ship Structures for the New Millennium: Supporting Quality in Shipbuilding, Arlington, June 13-14 2000.

[39] Daidola, J.C., "Tanker Structure Behavior During Collision and Grounding," SNAME Marine Technology, January 1995.

[40] Kitamuru, O., "Comparative Study on Collision Resistance of Side Structure," SNAME Marine Technology, October, 1997.

[41] Kuroiwa., T., Kawamoto, Y., and Yuhara, T., "Study on Damage Ship Bottom Structure Due to Grounding," Conference on Marine Safety and Environment—Ship Production, The Netherlands, June 1992.

[42] Amdahl, J. and Kavlie, D., "Experimental and Numerical Simulation of Double Hull Stranding," DNV-MIT, Det norske Veritas, Oslo, Sept.1992.

[43] Okamoto, T. et al, "Strength Evaluation of Naval Unidirectional-Girder-System Product Oil Carrier by Reliability Analysis," Trans. SNAME, Vol. 93, 1985.

[44] Kitamuru, O., "Buffer Bow Design for the Improved Safety of Ships," Ship Structure Symposium, June, 2000.

Obituary

Ioannis C. Agouras

Life Member

Ioannis C. Agouras of Kifissia, Greece died July 2000.

Born May 5, 1931 in Greece, Mr. Agouras served his apprenticeship at S.G. Embiricos, London, England where he learned the skills of marine engineering and electricity.

After completing his apprenticeship in 1953, Mr. Agouras joined the Greek Navy as Petty Officer Engineer where he remained for three years. From 1955 through 1966 Western Shipping Corporation employed Mr. Agouras, initially as 3rd assistant engineer, rising to chief engineer. From 1968 through 1974 Konkar Maritime N.Y. Agencies Ltd. employed Mr. Agouras as vice president, Marine Superintendent Engineering.

Mr. Agouras joined the Society in 1974 and became a Life Member in 1987.

Roy G. Anderson

Life Member

Captain Roy G. Anderson, USN (Ret.) of Riverside, California died November 4, 2000.

Born February 3, 1919, CAPT Anderson obtained a bachelor of science degree in naval architecture and marine engineering from the Webb Institute in 1940. He was awarded the professional degree of naval engineer from Massachusetts Institute of Technology in 1952. CAPT Anderson earned a master's degree in engineering from Penn State University in 1969.

His 27-year career in the U.S. Navy began with the rank of Ensign in 1941, progressing ultimately, in 1961, to the rank of Captain. His principal areas of expertise covered ship design, ship structure, ship maintenance and power plant operation and education and research environment. In 1968 CAPT Anderson retired from the Navy and pursued a teaching career at City College of San Francisco, California specializing in mechanical engineering.

CAPT Anderson joined the Society in 1941, was upgraded to Member in 1971, becoming a Life Member in 1985 and finally in 1991 was awarded a 50-Year Golden Award Membership Certificate.

William J. Burns, Jr.

Member

William J. Burns, Jr. of Mount Pleasant, South Carolina died December 30, 1998.

Born September 16, 1916 in Detroit, Michigan, Mr. Burns received a bachelor of science degree in mechanical engineering from the University of Michigan and a master's degree in business administration from Columbia University. He also completed the executive training program in nuclear physics and engineering at the University of Pittsburgh.

Mr. Burns joined Newport News Shipbuilding & Dry Dock Company in 1939. During World War II he worked at a subsidiary, North Carolina Shipbuilding, in Wilmington, N.C. In the post-war years he served as a chief engineer with United States Lines until 1951, when he returned to Newport News Shipbuilding.

He entered the Company's nuclear program in 1953 when he became a Junior Design Supervisor in Engineering Technical, a Senior Design Supervisor in the Atomic Power division in 1955, and was named Assistant Production Engineer in 1960, Production Engineer in 1965, and manager of the program division in 1967. In 1972 Mr. Burns was promoted to Construction Manager responsible for steel hull outfitting and machinery in the welding engineering, non-destructive testing and ship repair departments.

After retiring from Newport News he was employed by Todd Shipyard Corporation first at Galveston, Texas and then at Los Angeles, California. He then became associated with Pacific Gulf Marine following which he became a marine consultant.

Mr. Burns joined the Society in 1939 and was awarded a 50-Year Golden Award Membership Certificate in 1989.

Delma L. Crook

Life Member

Delma L. Crook of Bella Vista Village, Arkansas died in November 1999.

Born in Asheville, North Carolina, on January 12, 1924, Del Crook was a graduate of the United States Merchant Marine Academy in 1947, and after graduation entered the United States Navy. By the early 1960s he was assigned to the Supervisor of Shipbuilding at Newport News, Virginia, as head engineer of the Design Division.

In 1963, he left naval service and joined the Maritime Administration Office of Research and Development, where, as Chief of the Division of Nuclear Projects, he participated in the operation of the N. S. *Savannah* and other projects related to nuclear propulsion.

His later career included employment with Amoco Marine Transportation Company, National Bulk Carrier, Inc., at the company's Kure Shipyard Division, and Beker Industries Corporation.

Mr. Crook joined the Society in 1947 while a Cadet Midshipman at Kings Point. He became a Life Member in 1990.

Roger H. Derusha

Life Member

Roger H. Derusha of Menominee, Michigan died May 17, 1999.

Born September 14, 1930 in Menominee, Wisconsin, Mr. Derusha graduated from the University of Wisconsin in 1957 with a degree in marine engineering. He had served in the Army during the Korean Conflict.

He commenced his career in 1947 with Marinette Marine Corporation as a draftsman and project engineer. From that modest beginning he rose to the position of President and CEO in 1968, and served in that capacity for many years.

Mr. Derusha was renowned for being an avid and respected

yachtsman on the Great Lakes. He was a founder and board member of the M& M Ice Yacht Club and the Menekaunee Ice Boat Club. He also held the position of president of the technical advisory committee of the Renegade Ice Yacht Association. He was a major promoter and main organizer of the Ensign Class National Championships. He had sailed 32 Chicago to Mackinaw races and won two times overall. Mr. Derusha was the proud owner of the 83-year old ice yacht Debutante III which had set a world land speed record in 1938 at 143 miles per hour. He served on numerous community boards and was widely known for his interest in helping people start new businesses, an avocation which gave him much enjoyment.

Mr. Derusha joined the Society in 1976 and became a Life Member in 1991. He served on the Ship Production Committee from 1978 to 1981.

Albert J. Dieudonne

Life Associate Member

Albert J. Dieudonne of New Orleans, Louisiana died January 2, 1998.

Born September 20, 1910 in New Orleans, Louisiana, Mr. Dieudonne was a graduate of Holy Cross College, New Orleans, Louisiana. Mr. Dieudonne commenced his career with Todd-Johnson Dry Docks, Inc. At the time of his death, Mr. Dieudonne was employed as Vice President of Violet Dockport, Violet, Louisiana.

Mr. Dieudonne joined the Society in 1948 and became a Life Associate Member in 1981.

Joseph J. Dimock IV

Member

Joseph J. Dimock IV of Savannah, Georgia died October 18, 2000.

Born October 15, 1911 in New York City, Mr. Dimock graduated from the Massachusetts Maritime Academy with a masters license. He received an honorary diploma from United States Merchant Marine Academy, where he served as a department head. He served also as a lieutenant in the U.S. Navy during World War II, and later was a lieutenant commander in the U.S. Naval Reserve.

At the time of his death, Mr. Dimock was a retired self-employed marine surveyor.

Mr. Dimock joined the Society in 1976.

Jesse E. Eastman

Life Member

Captain Jesse E. Eastman, USCG (Ret.) of Shoreline, Washington, died November 12, 2000.

Born July 21, 1910 in Duluth, Minnesota, CAPT Eastman worked aboard Dolomite Marine Corporation and Ford Motor Company vessels before joining the United States Coast Guard in 1943. He served for over twenty-five years, rising to the position of Officer in Charge, Marine Inspection.

Between 1975 and 1983, Captain Eastman was an active member of the Sections and Membership Committees of the Society, serving as chairman and member of the Pacific Northwest Section and Membership Committees.

CAPT Eastman joined the Society in 1956 and became a Life Member in 1970.

D. Wayne Eberhardt

Member

D. Wayne Eberhardt of Cuyahoga Falls, Ohio died February 17, 2000.

Born April 15, 1927 in Massillon, Ohio, Mr. Eberhardt graduated from the U.S. Merchant Marine Academy, Kings Point, New York in 1948 where he earned a bachelor of science degree in engineering.

Mr. Eberhardt worked for over twenty years with Fairbanks Morse in a variety of engineering positions, ranging from Engineering Sales Correspondent to Acting General Sales Manager.

Mr. Eberhardt joined the Society in 1969.

Alfred E. Gallant, Jr.

Life Member

Captain Alfred E. Gallant, Jr., USN (Ret.) of Kaneohe, Hawaii died February 5, 1999.

Born February 27 in Los Angeles, CAPT Gallant received a bachelor of science degree in marine engineering from the California Maritime Academy at Vallejo in 1940. He also completed various graduate level Industrial Management Courses sponsored by the United States Navy.

After graduation, and for the duration of World War II, CAPT Gallant served in the Engineering Department of the USS *Pennsylvania* (BB-38) as chief engineer. He also served on the USS *Yorktown* (CVS-10), and on the USS *Tarawa* (CS-40). His tours ashore included Machinery Repair and Ship Superintendent for Long Beach Naval Shipyard. CAPT Gallant also served as Naval Liaison Officer for the Office of Assistant Secretary Defense at the Pentagon.

CAPT Gallant's civilian activities included Port Engineer and Senior Marine Surveyor for MSC Honolulu, Hawaii, and design manager for Litton Ships at Culver City, California. For many years he served as a senior marine engineer and marine surveyor for his own consulting firm in Kaneohe, Hawaii, in addition to duties for Cooper and Company of Santa Monica, California. CAPT Gallant retained his USCG chief engineer's license throughout most of his professional career. As a tribute to his generosity the California Maritime Academy is planning a permanent memorial in his honor.

CAPT Gallant served on the SNAME Sections Committee – Hawaii for the period September 1, 1980 through August 31, 1981.

CAPT Gallant joined the Society in 1964 and became a Life Member in 1969.

Kenneth McRae Garland

Associate Member

Kenneth McRae Garland of Rockland, Massachusetts died December 31, 2000.

Born August 12, 1916 in Worcester, Massachusetts, Mr. Garland worked as a marine surveyor at the American Bureau of Shipping.

Mr. Garland joined the Society in 1944 and was awarded a 50-Year Golden Associate Membership Certificate in 1994.

Adolph Gemmet

Member

Adolph Gemmet of Bremerton, Washington died in December 1999.

Born July 31, 1918 in San Francisco, Mr. Gemmet graduated from the University of California at Berkeley in 1947. From 1958 through 1966 he took additional maritime courses at the University of California, San Mateo. Mr. Gemmet served in the U.S. Army during World War II.

Mr. Gemmet commenced his career as a naval architect in 1964 with San Francisco Bay Naval Shipyard. He worked for over 36 years as a naval architect for multiple shipyards in the states of California and Washington, winning many awards for his accuracy and professionalism. Toward the end of his career, Mr. Gemmet was employed as a staff engineer at M. Rosenblatt & Son, Bremerton, Washington. Upon retirement, Mr. Gemmet continued his studies in the field of engineering. He was known as a mentor in his local area, encouraging and assisting the young to enhance their goals and careers. His widow comments that Mr. Gemmet left a legacy of honor, integrity and service to his community.

Mr. Gemmet joined the Society in 1967, and was upgraded to Member in 1983.

Warren A. Goodwin

Member

Warren A. Goodwin of Hampton, Virginia died December 19, 1997.

Born in Cambridge, Massachusetts December 24, 1927, Mr. Goodwin received his education at the Massachusetts Maritime Academy, from which he graduated in 1949.

Mr. Goodwin spent twelve years at sea as an engineer with Moore-McCormack Lines, American President Lines and Peninsular and Occidental Steamship Company. Following his seagoing experience, he came ashore in August 1961 to join the American Bureau of Shipping as a surveyor attached to the Newport News office.

His career with the Bureau resulted in assignments at a number of ports, including Boston, Piraeus and San Juan, where he was principal surveyor for the Caribbean area. Mr. Goodwin later returned to Newport News as principal surveyor for the Atlantic Coast and the construction of the "Seabee" vessels for Lykes Bros. Steamship Company—possibly his most challenging assignment. Mr. Goodwin retired January 1, 1988.

Warren Goodwin was the possessor of a wry New England sense of humor—but conducted his business activities in a thoroughly professional manner.

Mr. Goodwin joined the Society in 1978.

Otho H. Haunschild

Member

Otho H. Haunschild of Orange, Texas died September 20, 2000.

Born March 7, 1916 in Lockwood, Missouri, Mr. Haunschild was employed by the Levingston Shipbuilding Company for twenty-five years. His starting position at Levingston was Plate Shop Superintendent, and he rose to president within his tenure there.

Mr. Haunschild joined the Society in 1969.

E.W. Hazard

Life Member

E. W. Hazard of Boca Raton, Florida died in 1998.

Born April 23, 1920 in Baltimore, Maryland, Mr. Hazard obtained his degree in mechanical engineering from Johns Hopkins University, School of Engineering in 1937.

From 1946 to 1950 Mr. Hazard was employed as assistant president, Arnot & Company. He next moved to Leary Manufacturing Co. as general manager–Fabrications of Ships Furniture, then in 1951 reverted back to Arnot & Company as vice president. In 1954 he moved again to become president, E. W. Hazard Associates.

Mr. Hazard joined the Society in 1960 and became a Life Member in 1988.

Leroy V. Honsinger

Past President of the Society,
Life Honorary Member, Fellow

Rear Admiral Leroy V. Honsinger, USN (Ret.), a Past President, Life Honorary Member and Fellow of the Society died Sunday, October 1, 2000 at Wallingford, Connecticut.

Born September 5, 1905, Admiral Honsinger graduated from the United States Naval Academy in 1927 and received a degree in naval architecture from Massachusetts Institute of Technology in 1932. As a young naval officer, he served a tour of duty in Panama among other assignments.

During World War II, Admiral Honsinger was stationed in the Hampton Roads area, and was deeply involved in battleship and LST (landing ship tank) construction at the Norfolk Navy Yard, in repairing British warships damaged in the Battle of the Atlantic, and in aircraft carrier construction at Newport News Shipbuilding and Dry Dock Company.

Assigned to Europe during the last year of the war, he served as acting chief of the United States Technical Mission responsible for examining German shipyards and their wartime shipbuilding programs, and comparing their production methods with our own. Admiral Honsinger then served several tours of duty in the Bureau of Ships in Washington. Among his tasks during this period was detailed responsibility for the Navy's shipbuilding program during the Korean War.

After a tour as commanding officer of the Navy shipbuilding at Long Beach, California, he was promoted to Rear Admiral and returned to the Bureau of Ships for four years, first as division head of ship design, construction, and repair, and then for two years as deputy chief of the Bureau. Admiral Honsinger was associated early and for many years with the nuclear submarine program. He completed his thirty-nine-year naval career as commanding officer of the Mare Island Naval Shipyard, whose mission during that period was the construction of nuclear submarines. Admiral Honsinger retired from the Navy in 1962 while at Mare Island, and was awarded the Legion of Merit at that time.

After retirement at the end of his distinguished Navy career, Admiral Honsinger entered the commercial world and spent ten years with Todd Shipyards Corporation at its New York headquarters. For eight years he was the company's vice president for shipbuilding.

In later years he was a trustee and vice chairman of World Education, an organization set up to support the literacy work of his aunt, Dr. Welthy Honsinger Fisher. Always civic-minded, Admiral Honsinger was active in church work wherever he lived and for many years was an active Rotarian, including a year as president of the Tri-Bury Rotary Club in Southbury, Connecticut. While at Southbury, he developed a great interest in environmental issues, and in the 1980s became an expert on solid waste management and recycling programs.

Admiral Honsinger joined the Society in 1943, served as President of the Society in 1975–1976. He was elected an Honorary Member for Life in 1977 and a Fellow in 1984, and received his Fifty-Year Golden Membership Certificate in 1993. He also served as President of the American Society of Naval Engineers.

Universally known in the Society as "Mike," Admiral Honsinger will be remembered by the Society for his leadership, his commanding presence and assertive voice. Endowed with a keenly analytical mind, he noted that during his naval career he transited the Panama Canal fifty-four times.

Leroy V. Honsinger

Adm. Honsinger married Norma Chapin, in 1930, and the couple had just celebrated their 70th wedding anniversary on September 6, 2000. In addition to his wife, he is survived by a son, retired Capt. Vernon Chapin (Chip) Honsinger of Lexington, Virginia; a daughter, Valerie Kirkpatrick Joiner of Atlanta, Georgia.; a sister, Helen Halstead of Peru, Vermont; five grandchildren; and six great-grandchildren.

Ivar T. Jorstad

Member

Ivar T. Jorstad of Walnut Cove, North Carolina died March 2000.

Born November 3, 1928 in Oslo, Norway, Mr. Jorstad graduated in 1963 from the Technische Hochschule Friedericiana University, at Karlsruhe, Germany, where he was awarded a master's degree in mechanical engineering.

In 1964 Mr. Jorstad emigrated to the U.S. where J. J. McMullen Associates, Inc. employed him as senior project engineer responsible for the preparation of machinery and piping plans on numerous ship conversions.

Mr. Jorstad joined the Society in 1970.

William H. Jory

Affiliate

Mr. William H. Jory, of Eugene, Oregon died in July 1998.

Mr. Jory was born in Glen Burnie, Maryland, on September 19, 1910. His early education was at the Baltimore Polytechnic Institute, and in 1939 he completed a four-year course in naval architecture at the Maryland Institute.

His long association with Maryland Shipbuilding and Drydock Company began in late 1928. His first position was as a clerk in the Specifications Department. He quickly advanced to junior estimator, then estimator, eventually working his way upwards to senior vice-president and a member of the board of directors. He resigned in 1962 to become president of the American Shipbuilding Company, Lorain, Ohio, a position he held until he retired in 1968. He was later elected a member of the board of directors of Norfolk Shipbuilding and Drydock Corporation, and served as that company's President and General Manager in 1974–1975. He was also associated with J. J. Henry Company.

Mr. Jory served on the Pension Plan Committee of the Society from 1958 to 1971. During the latter portion of this period he was the committee's chairman. He was also a Member of Council representing Affiliates.

Bill Jory joined the Society in 1937 as a Junior Member, later advancing to the grades of Associate Member and Affiliate.

Karl W. Kahlenberg

Member

Karl W. Kahlenberg of Two Rivers, Wisconsin died February 4, 2000.

Born October 29, 1922 in Two Rivers, Mr. Kahlenberg was

a graduate of Denison University, Granville, Ohio in 1952. He served with the United States Army from 1956 to 1958. Mr. Kahlenberg's career with the family-owned Kahlenberg Brothers Company, since 1895 a well-known manufacturer of marine air horns and other marine-related products, was for more than forty years. At the time of his retirement, Mr. Kahlenberg held the position of president.

Mr. Kahlenberg was very much a community-minded person. A few of his many community contributions include member of the Board of Directors of the American Cancer Society, president of the Two Rivers Chamber of Commerce, and a member of and an advisor to the Wisconsin Maritime Museum at Manitowoc. The Chamber of Commerce awarded Mr. Kahlenberg the 1999 Joseph P. Schmitt "Speak Up" Award in 1999. He was an avid golfer, fisherman, hunter, bridge enthusiast and bird watcher.

Mr. Kahlenberg joined the Society in 1976.

William F. Kirkpatrick

Member

William F. Kirkpatrick of Livingston Manor, New York died March 10, 2000.

Mr. Kirkpatrick was born in Ann Arbor, Michigan, on May 24, 1927. He was educated at St. Joseph's College, Collegeville, Indiana, and later at the University of Michigan, where he received a bachelor of science degree in naval architecture and marine engineering in June 1952.

His first employment was as a draftsman at New York Shipbuilding Corporation. After a short time, he joined J. J. Henry Company, moving in 1955 to Caltex, where he specialized in the design and construction of large tankers and LPG carriers. He resided for a number of years in Japan as the company's assistant manager and later manager of new construction. During this period he was also a director of Tokyo Tanker K. K., a Caltex affiliate.

In 1971, he returned to the United States to become vice president for Ogden Marine, Inc., a position he held until his retirement.

Mr. Kirkpatrick joined the Society in 1952 as an Associate Member, upgrading to Member in 1976.

Nari G. Kirpalani

Member

Nari G. Kirpalani of Surrey, England died November 2000.

Born December 10, 1930 in Bombay India, Mr. Kirpalani attended the University of Bombay, graduating in 1953 with a degree in marine engineering.

Mr. Kirpalani spent most of his working life employed by the Great Eastern Shipping Co. Ltd. in Bombay, India. He began his career there as 5th engineer, steamship and was elevated to technical manager, Development and Projects. Later on Mr. Kirpalani moved to England where he was employed as chairman, Seascan International Ltd.

Mr. Kirpalani originally joined the Society in 1974 as Affiliate Associate Member and became a Member in 1998.

Luis Kazuo Kumasaka

Associate Member

Luis Kazuo Kumasaka, of Fortaleza, Brazil died October 28, 1999.

Born April 15, 1956 in Sao Paulo, Mr. Kumasaka graduated in 1978 from the Politechnic School of Sao Paulo University with a degree in naval architecture.

Mr. Kumasaka began his career in the Production Department of Ishimkawajima do Brazil Shipyard. He later worked for the Brazilian Navy Engineering Division, Hanseatica Shipyard, and lastly was partner of Piccolo, Kumasaka Ltd., a marine and cargo survey company at Fortaleza.

Mr. Kumasaka became an Associate Member of the Society in 1982.

Earl Emmet Lagergren

Member

Earl Emmet Lagergren of Port Orchard, Washington, died October 6, 2000.

Born July 27, 1912 in Treadwell, Alaska, Mr. Lagergren graduated from the University of Washington in June 1940 with a degree in mechanical engineering and was elected to Tau Beta Pi, engineering honorary society.

Mr. Lagergren joined the Puget Sound Naval Shipyard at Bremerton, Washington in 1940 where he worked for thirty-three years, retiring in 1973. During his career in the shipyard he worked on many major projects, including replacement of CO_2 refrigeration plants on naval vessels with Freon systems, and the modernization of main control systems, distilling plants and main circulating water cooling systems.

Mr. Lagergren was the lead engineer in the installation in 1954 of the first steam catapult on an American aircraft carrier USS *Hancock*. This was accomplished after consultation with Lieutenant Commander Colin Mitchell, RN, recognized as the inventor of the steam catapult.

At the time of his retirement Mr. Lagergren held the position of department head in charge of main propulsion machinery at Puget Sound Naval Shipyard. He was a member of the Port Orchard Yacht Club.

Mr. Lagergren joined the Society in 1970.

Leonard J. Lucas

Member

Leonard J. Lucas of Key West, Florida died February 2000.

Born in Oakland, California, Mr. Lucas graduated from the University of California at Berkeley in 1950 with a bachelor of science degree in mechanical engineering.

Mr. Lucas worked at San Francisco Naval Shipyard for six years doing design work on various types of machinery for naval vessels. In 1959, Mr. Lucas went into private practice, specializing in the design of unusually large or unusually complex marine machinery. Most of Mr. Lucas's work entailed the redesign of new machinery built by others but which was not operating satisfactorily.

In the mid-60s Mr. Lucas was employed as technical consultant to the Crawler-Transporters which were being built by Marion Power Shovel Company of Marion, Ohio to carry the

Apollo moon rockets from the assembly building to the launch pad. These transporters are still carrying the space shuttle to the launch pad today.

Employed as a technical consultant to Schuller and Allan, Houston, Texas, Mr. Lucas designed a 2500 horsepower paddlewheel propulsion system for a Mississippi riverboat. Mr. Lucas was a registered mechanical engineer in the state of California.

Mr. Lucas joined the Society in 1958.

Reuven Leopold

Life Fellow

Dr. Reuven Leopold, of Potomac, Maryland, passed away on December 24,2000. Reuven Leopold was born in Hungary on May 5, 1938. During World War II, he was in Romania and in 1950 moved to Israel where he attended the Israeli Naval Academy, graduating in 1959.

After his emigration to the United States in 1959, he entered Massachusetts Institute of Technology, where he received the degree of Bachelor of Science in Naval Architecture and Marine Engineering in 1961, followed by a Master of Science degree in Naval Architecture and Marine Engineering from MIT in 1963. During this period he was engaged in research at the same institution, where in 1965 he received a Marine Mechanical Engineering degree.

From 1965 to 1966, he was on the staff of Arthur D. Little, Inc., where he was employed in the development of early computer-aided ship design (CASDOS) systems.

His first foray into the world of ship design and construction came in 1966, when he joined Litton Ship Systems Division of Litton Industries, then engaged in the DD963 and LHA programs. From his first position, Assistant to the Technical Director, he advanced rapidly to Director of Engineering of the company's Advanced Marine Technology Division.

In 1971, he was appointed technical director of the United States Navy's Naval Ship Engineering Center (NAVSEC), in charge of the preliminary design of advanced concepts for both surface vessels and submarines, as well as research and development programs related to future naval vessels. During his tenure at NAVSEC, many of today's advanced fighting ships were conceived and built.

Further academic recognition came in 1977 with an MBA degree from George Washington University and a PhD degree from Massachusetts Institute of Technology.

In 1978, Dr. Leopold's expertise was directed to aircraft engines when he joined the Pratt and Whitney Division of United Technologies as Vice President of Advanced Systems. Three years later he moved to the General Electric Company as Manager of Plans and Business Development, in charge of business development and planning for all military and small commercial aircraft engines.

He returned to engineering technology and research in 1984 when he became Chairman and CEO of NKF Engineering Inc., a prominent firm in the field. This was followed by a period commencing in 1986 as Chairman and CEO of JJH, Inc. (later the JJH/SMA Group) and as Chairman and CEO of ODSI Defense Systems, Inc.

Later, he founded Syntek Technologies, Inc. As the company's President and CEO, he continued his service to government and industry in the fields of transportation, oceanography, information technology, energy and the environment.

He served on many advisory groups and task forces on naval issues throughout his career. These included the Defense Science Board, the CNO Executive Panel, the Naval Studies Board of the National Academy of Sciences and many others. His dedicated work was recognized by the presentation of numerous awards, typified by the Navy's Superior Civilian Service Award in 1978.

A prolific author of technical papers relating to naval vessel design and construction, he presented papers at the Society's annual meetings in 1964, 1966.1971 and 1974, as well as the 1969 Spring Meeting. Among his other contributions to the Society, he was a member of the Ship Production Committee from 1971 to 1979, the Fellows Committee from 1979 to 1994, the Publications Committee from 1990 to 1994 and a member of the Council in 1989.

Dr. Leopold joined the Society in 1963 and was elected a Fellow in 1977 at the age of thirty-nine.

John Rymer Mallahan

Member

Mr. John Rymer Mallahan of Sarasota, Florida died November 26, 1999.

Born October 18, 1913 in Chicago, Mr. Mallahan graduated from the University of Illinois in 1940 with a degree of bachelor of science in mechanical engineering. He was employed by the American Bureau of Shipping for thirty-seven years before retiring.

Mr. Mallahan joined the Society in 1941. In 1991 he received his 50-year Membership Certificate.

Daniel Turney Mallet

Daniel Turney Mallet of Mamaroneck, New York died April 19, 2000.

Born December 10, 1912 in Hackensack, New Jersey, Mr. Turney graduated from Stevens Institute of Technology with a bachelor of science degree in mechanical engineering. In 1934 Mr. Turney worked as a naval architect for George G. Sharp Inc. in New York City, remaining with the company until his retirement in 1979.

Mr. Turney was an avid sailor and boat builder. As a young man he built the 23-ft sloop *Cachalot,* an experience he wrote about for *Yachting Magazine*. Later he built *Lena,* a 35-ft sloop that he and his family enjoying sailing for twenty summers.

Mr. Turney joined the Society in 1938 and was enrolled as a Life Member in 1969.

Feleciano Manzutti

Member

Captain Feleciano Manzutti of Sault Ste. Marie, Ontario, Canada died in January 2000.

Born July 27, 1910, Captain Manzutti began his seafaring career in various capacities. He graduated from deck hand to earning a Master's Certificate of Competency–Great Lakes and Gulf of St. Lawrence, and later earned a Master Home Trade Certificate. From there Captain Manzutti went on to ownership and operation of four commercial crane-equipped ships, running this business for twenty-five years.

Later Captain Manzutti continued in the marine profession as Marine Salvage Master and Marine Surveyor for Ships and/or Cargo. Captain Manzutti served both commercial and pleasure craft in the geographical areas of U.S. and Canadian Great Lakes and Rivers, Gulf of St. Lawrence and East Coast Canadian Sub-Arctic Water, and had occasional calls to the West Coast areas

Captain Manzutti joined the Society in 1998.

John McCormick

Associate Member

John McCormick of Shelter Island Heights, New York died in April 2000.

Born April 10, 1949, Mr. McCormick held the professional license of Chief Marine Engineer, U.S. Coast Guard. Sadly, Mr. McCormick died aboard the MV *Islander* while bringing her home from a spring refitting.

Mr. McCormick joined the Society in 1996.

Edgar Carlyle Outten, Jr.

Life Member

Edgar Carlyle Outten, Jr. of Long Beach, California died November 20, 2000.

Born July 25, 1921 in Virginia, Mr. Outten was a graduate of the University of Southern California. He was employed for thirty years by North American Aviation and Rockwell International working on the Apollo and space shuttle programs. A colonel in the California State Military Reserve, he also served in the armed forces during World War II.

Mr. Outten joined the Society as a Member in 1971 and was enrolled as a Life Member in 1984.

W. H. E. Peters

Life Member

W. H. E. Peters of Vancouver, Canada died June 21, 2000.

Born December 10, 1912 in Karachi, India, Mr. Peters graduated as a marine engineer's apprentice from the R.I.M. Dockyard, Bombay, India. He later studied at the College of Marine Engineering, Liverpool, U.K., obtaining his 2nd class engineer's certificate of competency. Subsequently, Mr. Peters earned his 1st class engineer's certificate of competency at the Marine Engineering Academy, Belfast, Northern Ireland.

In 1943, Mr. Peters emigrated to Canada where he worked as chief engineer in Ship Delivery Service, and later as Supervisor, Ship Repairs, Department of Transport, Victoria, B.C.

From September 1, 1993 through August 31, 1994 Mr. Peters served on the Society's National Membership Committee under the chairmanship of Mr. Bruce S. Rosenblatt. Mr. Peters was a contributor to the paper "The Motorship" published in the *Transactions* of ImarE.

Mr. Peters joined the Society in 1956 and was enrolled as a Life Member in 1982.

Alberto A. Rivas

Member

Alberto Quintin Rivas of Madrid, Spain died in October 2000.

Born May 17, 1934 in Algorta, Vizcaya, Spain, Mr. Rivas graduated in 1962 from the University of Michigan with a bachelor of science degree in naval architecture and marine engineering.

Mr. Rivas began his career as a junior engineer with Maryland Shipbuilding Company, Baltimore. From there he was employed as naval architect, manager of engineering, and senior design engineer for a variety of shipbuilding companies throughout the United States. In 1969 Mr. Rivas returned to Spain where he was employed as Manager, Vessel Construction, for Amoco International Oil Company.

Mr. Rivas joined the Society in 1961.

Johnston F. Robb

Associate Member

Dr. Johnston F. Robb of Greenock, Scotland died September 10, 2000.

Born May 17, 1921 in Greenock, Scotland, Dr. Robb graduated with honors from the Paisley Technical College, London, U.K., with a bachelor of science degree in engineering.

From 1953 to 1958 he was employed as chief draughtsman at Scotts Shipbuilding and Engineering Company in Greenock, Scotland where he was in charge of sixty staff members. He was engaged in the design and drawing of main and auxiliary propelling machinery for both steam turbine and diesel-propelled cargo vessels, oil tankers and destroyers.

Subsequently he rose to the position of director at Scotts Shipbuilding. At the time of his retirement, Dr. Robb held the position of general manager and naval director of the company.

Dr. Robb joined the Society in 1958.

James B. Robertson, Jr.

Life Fellow

James B. Robertson, Jr. of St. Michaels, Maryland died June 14, 2000.

Mr. Robertson was born in Chicago, Illinois, on January 15, 1909. He was a graduate of the University of Michigan in 1932, earning a bachelor of science in naval architecture and marine engineering. His early career included a period devoted to yacht design at Henry C. Grebe and Company, as well as employment with the National Advisory Committee for Aeronautics at the experimental tank at Langley Field, Virginia.

In 1938 he commenced a long career with the Bureau of Marine Inspection and Navigation (succeeded by the United States Coast Guard.) His first position was as an assistant naval architect. Steady advancement followed until he was Chief Naval Architect of the Merchant Marine Technical Division.

He was a technical advisor to the U. S. delegation to the 1948 and 1960 International Safety of Life at Sea Conferences, and to the 1966 International Load Line Conference. He participated in numerous working groups in IMCO relating to subdivision and stability

He retired from his position at the Coast Guard in 1967, culminating a career which was, in his own words, "an extended period of intensely satisfying work and many pleasant associations, as well as many frustrations." After retirement, he devoted much energy to pursuing "even more intensely...the unfinished business of trying to bring the International Passenger Ship Subdivision Rules out of the dark ages."

There is no better way to summarize Jim Robertson's career than by quoting him once again. "Professionally and otherwise, life has been rewarding."

Mr. Robertson's contributions to the Society's literature consist of three important papers. At the 1950 Annual Meeting he was co-author, with Vito L. Russo, of "Standards for Stability of Ships in Damaged Condition." He received the Captain Joseph H. Linnard Prize in 1962 for a 1961 Annual Meeting paper entitled "Survival of Collision Damage Versus the 1960 Convention on Safety of Life at Sea," co-written with John P. Comstock. At the 1974 Annual Meeting he was the lead author of the paper, "The New Equivalent International Regulations on Subdivision and Stability of Passenger Ships."

In recognition of his indefatigable devotion—professional and personal—to the improvement of passenger ship subdivision and stability and many other aspects of maritime safety, he was the recipient of the Society's prestigious David W. Taylor Medal in 1975.

Mr. Robertson joined the Society in 1936, became a Life Member in 1967 and was elected a Fellow in 1976.

David Saveker

Life Member

David Saveker of Pismo Beach, California died November 4, 1999.

Born January 10, 1920 in San Jose, California, Mr. Saveker attended Cal State, San Jose and graduated from Stanford University with a bachelor of science degree in mechanical engineering in 1941. He was commissioned in the U.S. Navy and attended the Naval Academy Post Graduate School to study naval architecture. During World War II he was assigned to the New York Naval Shipyard. After the war he was promoted to lieutenant commander and received a master of science degree in naval architecture and marine engineering from Massachusetts Institute of Technology.

Mr. Saveker was engaged in research on underwater explosions, was assigned to the Norfolk Naval Shipyard and did research for the *Nautilus* prototype nuclear reactor. Promoted to Captain, he served in Puget Sound Naval Shipyard as construction superintendent, later doing a tour of duty for COMSERPAC in Hawaii. His final assignment included a tour in charge of the mechanical engineering curriculum at the Naval Post-Graduate School in Monterey, and planning officer at the San Francisco Naval Shipyard. After retiring from the Navy, Mr. Saveker completed graduate studies in architecture at Cal Poly and taught architectural engineering there for 10 years. He was a research fellow for NASA in system design at Stanford.

Mr. Saveker joined the Society in 1948 and became a Life Member in 1988. In 1998 he was awarded his Fifty-Year Golden Membership Certificate.

Jack L. Schmidt

Member

Jack L. Schmidt of Beaumont, Texas died in July 2000.

Mr. Schmidt graduated from Columbia University, School of Business, New York City.

At the time of his death Mr. Schmidt was employed as project manager with Oceaneering International, Inc., Houston, Texas. His assignment at that time was construction of the Ocean Intervention II. Mr. Schmidt was well liked at Oceaneering International, Inc. and his presence is sadly missed.

Mr. Schmidt's career covered many challenging assignments, a few of which were: corporate president, Gulf Copper Manufacturing Corporation; director of shipyard operations, INTERMARINE USA; assistant general manager/director of production at Lockheed's Gulfport Marine Facility.

Mr. Schmidt originally joined the Society in 1976 and was re-instated later in 1995.

Frank Sellars

Member

Frank Sellars of Montclair, New Jersey died March 25, 2000.

Born in Hartford, Connecticut on June 19, 1935, Mr. Sellars received a bachelor's degree from MIT in 1957, a master degree from the University of California, Berkeley in 1960, and an engineering degree from MIT in 1967. He was a licensed professional engineer in the Commonwealth of Virginia.

After receiving his final degree from MIT, Mr. Sellars worked with MPR Associates in Arlington, Virginia until 1989. Shortly thereafter he joined Flume Stabilization Systems as a senior project manager, a position he maintained until his death.

Mr. Sellars had a continuing and active interest in wave theory and seakeeping, and served on SNAME's various Technical & Research Committees from 1983 to the time of his death. Mr. Sellars was the recipient of the SNAME Distinguished Service Award in 1988 for publication of numerous technical papers and contribution to texts and reference works. Mr. Sellars also served on the Metropolitan New York Sections Committee as Section Member. Mr. Sellars was honored by the Society of Professional Engineers and awarded a Certificate of Technical Recognition for his published technical papers.

Mr. Sellars joined the Society in 1958.

Theodore L. Soo-Hoo

Life Member

Theodore L. Soo-Hoo of Gary, North Carolina died December 18, 2000.

Born March 18, 1904 in San Francisco, Mr. Soo-Hoo graduated from the University of California with a bachelor of science degree and later earned his master's degree in marine engineering at Massachusetts Institute of Technology.

Mr. Soo-Hoo was a contributor to what subsequently became the specific guidelines for construction of naval vessels. For his accomplishment in this area, Mr. Soo-Hoo won an award from the Bethlehem Shipbuilding Corp. in the 1930s.

Mr. Soo-Hoo joined the Society in 1930 as Student Member, and was awarded a 50-Year Golden Award Membership Certificate in 1980.

Pasquale Spinosa

Associate Member

Pasquale Spinosa of Gaeta, Italy died in 1999.

Born in Gaeta, Italy on April 14, 1954, Mr. Spinosa graduated from the University degli Studi di Napoli in 1981, earning a degree in marine engineering and naval architecture.

In the early 80s, Mr. Spinosa was employed as an officer with the Italian Navy. While in that position, Mr. Spinosa participated in various maritime activities such as design of GRP motoryachts ranging in size from 9 to 22 meters, design of GRP naval planning crafts, and also performed teaching assignments.

From 1983 to 1995 Mr. Spinosa worked at Cantieri Navali Italcraft in Gaeta as a naval architect. The Cantieri Navali Italcraft is a famous Italian shipyard that has been producing highly sophisticated motoryachts since the fifties. These are high-speed luxury craft with careful attention given to their fine woodwork and luxurious interiors.

From 1997 Mr. Spinosa was self-employed as an independent naval architect, working as a consultant, designer and ship surveyor for shipyards and marinas in the Gaeta area.

Mr. Spinosa joined the Society in 1997.

John D. Wilson

Member

John D. Wilson of Summerland Key, Florida died in 1997.

Born September 6, 1904 in Antwerp, Belgium, Mr. Wilson served his apprenticeship at the Staten Island Shipbuilding Company where he learned the skills of steam and diesel engineering.

After completing his apprenticeship in 1935, Mr. Wilson was employed by U.S. Lines, and later by American Bureau of Shipping.

Mr. Wilson joined the Society in 1941 and was awarded a 50-Year Golden Award Membership Certificate in 1991.

John A. Zalite

Member

John A. Zalite, of St. Petersburg, Florida died November 29, 2000.

Born August 9, 1907 in Moscow, Russia, Mr. Zalite graduated in 1934 from the University of Latvia, Faculty of Mechanics, Shipbuilding Branch.

After graduation, Mr. Zalite was employed by A/B Gotaverken, Gothenburg, Sweden, as a marine designer. Subsequent to that he joined shipyards in Libau and Riga. During the Second World War Mr. Zalite returned to A/B Gotaverken. In 1945 Mr. Zalite emigrated to Canada and was employed at Canadian Vickers as chief marine designer. In 1966 he joined the tanker department of Esso International in New York. His last overseas assignment was to Kobe, Japan. After his retirement in 1973, Mr. Zalite was active for several years as a consulting naval architect.

Mr. Zalite joined the Society in 1952.

Memorial Resolutions

Leroy V. Honsinger

The Society of Naval Architects and Marine Engineers unanimously adopted, at a meeting of its Executive Committee, the following memorial resolution:

Whereas, The Society of Naval Architects and Marine Engineers, the maritime industry, academia, and our Government have suffered a great loss in the death of Leroy V. Honsinger on October 1, 2000, a greatly respected, highly honored, and deeply beloved Past President, Honorary Member and Fellow of the Society, it desires to record these memorial resolutions.

Admiral Honsinger first joined the Society in 1943, and was a dedicated and long serving member. He served as President of the Society, 1975-1976, and elected an Honorary Member for Life in 1977. He was elected Fellow in 1984 and received a 50-Year Golden Membership Certificate in 1993.

Admiral Honsinger was born in 1905, in Syracuse, New York. He graduated from the U.S. Naval Academy in 1927, and received a Master of Science degree in naval architecture from Massachusetts Institute of Technology in 1932. As a young officer, he served in Panama and, in his long life, he transited the Panama Canal 54 times. During World War II, Admiral Honsinger was deeply involved in battleship and LST (landing ship tank) construction in the Norfolk, VA, Navy Yard, in repairing damaged British warships, and in aircraft carrier construction at Newport News Shipbuilding, Newport News, VA. He was sent to Europe in 1945, and served as acting chief of the U.S. Technical Mission responsible for investigating German shipyards and shipbuilding programs. Admiral Honsinger served several tours of duty in the Bureau of Ships in Washington, D.C., including responsibility for the Navy's shipbuilding program during the Korean War. After a tour as Commanding Officer of the Navy shipbuilding at Long Beach, CA, he was promoted to Rear Admiral and returned to the Bureau of Ships for four years, first as division head of ship design, construction, and repair, and then for two years as deputy chief. Admiral Honsinger was associated early, and for many years, with the nuclear submarine program. He completed his 39-year naval career as Commanding Officer of the Mare Island Naval Shipyard, whose mission then was building nuclear submarines. Admiral Honsinger retired from Mare Island in 1962 with the Legion of Merit. He then spent ten years with Todd Shipyards Corporation in New York City, eight of those years as vice president for shipbuilding. In his later years he was a trustee and vice chairman of World Education, an organization set up to support the literacy work of his aunt, Dr. Welthy Honsinger Fisher.

In addition to his SNAME activities, he served as President of the American Society of Naval Engineers in 1958. He was active in church work wherever he lived and for many years was an active Rotarian, including one year as president of Tri-bury Club in Southbury, CT.

In tribute to his long, distinguished, and invaluable service to the Society, the maritime industry, the Government, and his fellowman, hereby be it

Resolved, That The Society of Naval Architects and Marine Engineers, the maritime industry, and the nation have lost, in the death of **Leroy V. Honsinger,** a beloved member of the Society, a distinguished, capable, and selfless servant of our country, a fine gentlemen and a true friend.

Further, be it

Resolved, That these resolutions be recorded in the TRANSACTIONS of the Society and that a copy suitably engrossed, be sent to his family as a token of the Society's deep regard for **Leroy V. Honsinger.**

Richard Lowery

The Society of Naval Architects and Marine Engineers unanimously adopted, at a meeting of its Executive Committee, the following memorial resolution:

Whereas, The Society of Naval Architects and Marine Engineers, the maritime industry, academia, and our Government have suffered a great loss in the death of Richard Lowery on March 29, 2000, a greatly respected, highly honored, and deeply beloved Honorary Vice President and Fellow of the Society, it desires to record these memorial resolutions.

Mr. Lowery first joined the Society in 1936 and was a dedicated and long-serving member. He was the first Chairman of the Eastern Canadian Section serving from 1952–1953, and was a member of the Sections Committee from 1954–1957. He was a member of Council and served on the Steering Committee for the Society's 1968 Diamond Jubilee International Meeting.

In recognition of his "outstanding accomplishment in the marine field" the Society awarded Mr. Lowery the Vice Admiral "Jerry" Land Medal in 1964. He was elected to the grade of Fellow in 1974.

Mr. Lowery was born and educated in Newcastle-upon-Tyne, Great Britain. He attended Heaton Technical School, Rutherford Technical College, Armstrong College and Durham University. He was a pupil Naval Architect from 1927–1931, with Hawthorn Leslie Shipbuilders, Tyneside, Assistant Naval Architect, 1931–1936, Rutherford Technical College; Naval Architect, 1936–1938, Alfred Holt and Co., Liverpool, England; Naval Architect and Assistant Dockyards Manager, Singapore Harbour Board, Singapore; Naval Architect and General Manager, Melbourne Harbor Trust Shipyards, Australia; and from 1946–1951, he was Naval Architect and Vice President, Canadian Vickers, LTD, Shipbuilders and Marine Engineers, Montreal, Canada; In 1951 he joined Canada Steamship Lines as Vice President and Chief Technical Officer, and President of its subsidiary shipbuilding operations until he retired in 1972. Mr. Lowery gained an international reputation as a world class naval architect, marine engineer and business leader. As head of Davie Shipbuilding and a number of other related companies, he was responsible for much of Canadian Industries' success during the construction of the St. Lawrence Seaway, a critical time in the development of the Canadian Merchant Fleet. His invaluable technical assistance, given to the Allied Fleets in the war in the Pacific, has been officially recognized by the United States. He was awarded the "Officer of the Order of Canada" by the Chancellery of Canadian Orders and Decorations.

In addition to his SNAME activities, Mr. Lowery was a Fellow of The Royal Institution of Naval Architects; past Chairman, Canadian Shipbuilding and Ship Repairing Association; past Chairman, Canadian Committee, Lloyd's Register of Shipping; and a Member of the North East Coast Institution of Engineers and Shipbuilders.

In tribute to his long, distinguished, and invaluable service to the Society, the maritime industry, the Government, and his fellowman, hereby be it

Resolved, That The Society of Naval Architects and Marine Engineers, the maritime industry, the nation and Canada have lost, in the death of **Richard Lowery,** a beloved member of the Society, a distinguished, capable, and selfless servant of our country, a fine gentlemen and a true friend.

Further, be it

Resolved, That these resolutions be recorded in the TRANSACTIONS of the Society and that a copy suitably engrossed, be sent to his family as a token of the Society's deep regard for **Richard Lowery.**